Guidebook for Power Line Technicians

Guidebook for Power Line Technicians

Third Edition

Wayne Van Soelen

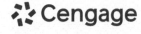

Australia • Brazil • Canada • Mexico • Singapore • United Kingdom • United States

Guidebook for Power Line Technicians, **3e**
Wayne Van Soelen

SVP, Product: Cheryl Costantini

VP, Product: Thais Alencar

Portfolio Product Director: Jason Fremder

Portfolio Product Manager: Emily Olsen

Product Assistant: Janell Whitted

Content Project Manager: Kugan Vasudevan

VP, Product Marketing: Jason Sakos

Director, Product Marketing: Neena Bali

Content Acquisition Analyst: Erin McCullough

Vendor Content Manager: Sangeetha Vijay

Production Service: Lumina Datamatics Ltd.

Designer: Tim Biddick

Cover Image Source: Andriy Popov / Alamy Stock Photo

For product information and technology assistance, contact us at
**Cengage Customer & Sales Support, 1-800-354-9706
or support.cengage.com.**

For permission to use material from this text or product, submit all requests online at **www.copyright.com.**

Library of Congress Control Number: 2023915399

ISBN: 978-0-357-93458-6

Cengage
5191 Natorp Boulevard
Mason, OH 45040
USA

Cengage is a leading provider of customized learning solutions. Our employees reside in nearly 40 different countries and serve digital learners in 165 countries around the world. Find your local representative at: **www.cengage.com.**

To learn more about Cengage platforms and services, register or access your online learning solution, or purchase materials for your course, visit **www.cengage.com.**

Printed Number: 1   Print Year: 2024
Printed in Mexico

Preface

The Guidebook for Power Line Technicians 3rd edition is an updated comprehensive resource dealing with managing risk in power line work. Managing risk is incorporated in description of work methods and the details of working in an electrical environment. Electrical utilities have an abundance of manuals and reference material for the people who work on power lines. Most of the information supplied by the utility or employer involves work procedures, rules, and regulations and widely scattered electrical reference material. This book concentrates on essential immediate "must know" topics that a power line technician is required to understand to work effectively and minimize risk in an electrical. It is expected that a power line technician will review specific work procedures provided by the utility or contractor.

The Guidebook for Power Line Technicians can be used as both a training manual and a reference guide. Used as a training manual, it is important that the content is studied chronologically so that the concepts presented in later chapters is fully understood. When training, the information in the book needs to be backed up with "in the field" training. Reference is made to equipment and material with the assumption that the reader knows or will be shown what they look like. The early chapters are intended for a person entering electrical-utility work. Later chapters assume some experience as a power line worker. Much of the book content is valuable for arborists, ground workers, and line truck drivers.

The Guidebook for Power Line Technicians fills the need for a convenient single-volume reference source on the operation of an electrical-utility system. Existing sources of information tend to be very basic electricity and magnetism theory or complex theory at an engineering level. This book deals with equipment and situations that power line workers are exposed to in their daily work. This book also provides a foundation that will aid in understanding procedures, rules and regulations. This foundation includes the mechanical aspects of the job, such as working with rigging, trucks, stringing wire, and tree felling.

The Guidebook for Power Line Technicians is intended to help power line workers meet the expectation that he or she will have the knowledge and skills to construct, operate, and maintain the lines and cables in an electrical utility system.

Acknowledgments

The publisher would like to thank the author, Wayne Van Soelen, for his perseverance in the improvement of this book and for doing the necessary hard work to bring this edition to the next level. Also, special thanks to John Bellows for the many hours spent reviewing the manuscript, and for his dedication to seeing the project through; to Steve Anderson for his encouragement, guidance, and commitment to the revision; and to Jim Simpson for his help in keeping the project running smoothly and on schedule.

A special acknowledgment is due to the following people for helpful suggestions made during the development process of this edition.

Albert Baca
NJATC, 3211 Regal Drive, Suite A, Alcoa, TN 37701

Bob Bass
California/Nevada Line Constructors, 9846 Limonite, Riverside, CA 92504

Rocky Clark
IBEW 8th District, 2225 West Broadway, Suite H, Idaho Falls, ID 83402

Daniel Dade
American Line Builders Joint Apprenticeship and Training Program, P.O. Box 370, 1900 Lake Road, Medway, OH 45341

Jim Fawcett
Utility Risk Management Inc., 171 Charles Street, Arnprior, ON K7S 3V5, Canada

Huel Gunter
Salisbury (by Honeywell), 7520 North Long Avenue, Skokie, IL 60077-3226

Jason Iannelli
NEAT, 1513 Ben Franklin Highway, Douglassville, PA 19518

Dave McAllen
Alaska Line Constructors, 5800 B Street, Anchorage, AK 99518

Virgil Melton
Southeastern Line Constructors, P.O. Box 2004, Newnan, GA 30264

John Schroeder
Missouri Valley Line Constructors, 600 South Jefferson Way, P.O. Box 271, Indianola, IA 50125

Bill Stone
Northwest Line Constructors, 9817 NE 54th Street, Suite 101, Vancouver, WA 98662

Charley Young
Southwestern Line Constructors, 825-C Vermont Street, Lawrence, KS 66044-2665

About the Author

Wayne Van Soelen is the president of Utility Innovations Inc., whose main products have been the preparation of electrical utility work procedures, safety standards, training material, work method audits, and expert witness reports. He also works in association with Utility Risk Management Ltd., as a consultant in safety management training, accident investigations, safety management audits, job planning training, risk analysis, and a safety management approach to work observation training for supervisors.

Mr. Van Soelen started his working career as a lineman with Hydro One in Ontario, Canada, in 1964, where he worked on projects such as restringing and resagging energized transmission lines, tension stringing, hot sticking, rubber glove, and barehand work. He progressed within the company to supervisor, safety professional, distribution engineering, and then to a senior administrative position. He formed Utility Innovations Inc. in 1994. Wayne is also the author of Electrical Essentials for Linemen and the Field Guide for Linemen.

About the Technical Reviewer

Jason Iannelli started his career as a Line Clearance Tree Trimmer in the early 1990's out of IBEW Local Union 126. Transitioning from tree trimmer to apprentice lineman through the NEAT AJATC program and graduated to Journeyman Lineman in 2001.

His career has progressed from Journeyman lineman to General foreman, to Assistant Training Director, on to Executive Director of the NEAT program for nearly 10 years.

The past 8 years, Jason's role at the Electrical Training Alliance has been Director of Curriculum. Jason lives in Pennsylvania with his wife, five children and two grandkids and is still a proud member of IBEW Local Union 126.

Brad Dozier graduated from Bell County High School in Pineville, Kentucky. Directly after finishing High school, he completed Somerset Community College's Lineman Program where he would begin his career in the Outside Electrical Industry. From there he completed The American Line Builders Apprenticeship Program, where he would later return to instruct and help manage the largest Outside Apprenticeship in the Nation. Brad currently has over 12 years of experience in the industry and holds an extensive list of certifications. He now works with the Electrical Training Alliance and resides in Kentucky with his wife Shayla. Brad is proud to be a Journeyman Lineman through IBEW Local 369 and is a strong advocate for the Outside Electrical industry.

Contents

Chapter 1
Controlling Risk for Power Line and Cable Technicians 1

Orientation to Power Line Work 1
Risk in Line Work 6
Managing Personal Safety Risk 7
Manage Your Risk in the Working Environment 13
Reduce Risk When Climbing and Working at
 Heights 30
Managing Emergencies 40

Chapter 2
Electrical Units 50

Introduction 50
Electrical Potential 51
Electrical Current 54
Electrical Resistance 57
Electrical Power 63

Chapter 3
Electrical Power System Overview 66

Introduction 66
Electrical Energy 67
Generation of Electrical Energy 67
Solar Energy Converted to Electrical Energy 74
Transmission of Electrical Energy 78
Electrical Distribution 81

Chapter 4
Working in Substations 87

What Is a Substation? 87
Maintaining and Operating a Substation 100
Constructing a Substation 104

Chapter 5
Alternating or Direct Current 108

Introduction 108
Characteristics of AC 109
Reactance in AC Circuits 113
AC Power 119
Why Direct Current (DC)? 122

Chapter 6
Three-Phase Circuits 125

Introduction 125
Characteristics of Three-Phase Circuits 125
Delta-Connected Systems 130
Wye-Connected Systems 131
Three-Phase Power 134

Chapter 7
Awareness in an Electrical Environment 137

Introduction 138
Connecting a Load in Series with a Circuit 138
Connecting a Load in Parallel with a Circuit 141
Electrical and Magnetic Induction 145
Voltage Gradients 149
Working with Neutrals 152
Vehicle Grounding and Bonding 154
Electromagnetic Fields 158
Minimum Approach Distance 163

Chapter 8
Constructing Overhead Power Lines 167

Before a Work Order Is Issued 167
Constructing a Pole Line 175
Constructing an Overhead Transmission Line 217

Chapter 9
Working on Underground Power Lines 233

Underground Distribution Lines 233
Underground Transmission Lines 244

Chapter 10
Working with Conductors and Cable 251

Electrical Properties of a Conductor 251
Overhead Conductors 257
Working with Overhead Conductors 265
Underground Cable 273
Working with Underground Cable 282
Working with Fiber-Optic Cable 290

Chapter 11
Operating Switchgear 298

Switching Characteristics and Switching Hazards 299
Using Maps to Locate Switchgear 307
Operating Isolating Switchgear 310
Operating Protective Switchgear 313
Underground Distribution Switchgear 328

Chapter 12
Circuit Protection 337

Introduction 337
Transmission System Protection 338
Distribution Protection 340
Specifying Protection for a Distribution Feeder 345
Overvoltage Protection 355
System Grounding for Protection 366
Protection from Corrosion 371

Chapter 13
Installing Personal Protective Grounds 374

Reasons to Install Personal Protective Grounds 374
Applying the Grounding Principle to Control Current 375

Applying the Bonding Principle to Control Voltage 381
Controlling Induced Voltage and Current from Electromagnetic Induction 386
Procedures for Applying Protective Grounds 388
Specific Grounding Hazards 393
Protective Grounding of Underground Cable 395

Chapter 14
Connecting Transformers 400

Introduction 400
Transformer Basics 401
Transformation Effect on Current 407
Transformer Losses and Impedance 409
Transformer Protection 410
Single-Phase Transformer Connections 415
Three-Phase Transformer Connections 419
Three-Phase, Secondary-Voltage Arrangements 431
Troubleshooting Transformers 436
Working on a Voltage Conversion 438
Other Transformer Applications 440
European Secondary Systems 441
Urban Secondary Network Grid 444
Secondary Network (Banked Secondary) 445
Specific Hazards Working with Transformers 446

Chapter 15
Supplying Quality Power 449

Introduction 449
What Is Power Quality? 450
Factors Affecting Voltage in a Circuit 451
Voltage on the Transmission Lines System 454
Distribution Substation Voltage 454
Distribution Feeder Voltage 455
Feeder Voltage Regulators 456
Capacitors 463
Troubleshooting No Power, High Voltage, or Low Voltage 467
Harmonic Interference 472
Voltage Flicker 473
Ferroresonance 475
Tingle Voltage 478
Investigating a Radio and Television Interference (TVI) Complaint 482

Chapter 16
Working with Aerial Devices and Digger Derricks 485

Checking Out the Truck 485
Monitoring a Hydraulic System 488
Stabilizing a Boom-Equipped Vehicle 491
Electrical Protection for Working with Noninsulated Booms 493
Electrical Protection for Working with Insulated Booms 497
Operating a Digger Derrick 500
Operating an Aerial Device 505

Chapter 17
Rigging in Power Line Work 510

Introduction 510
Using Ropes and Rigging Hardware 511
Lifting a Load 527
Working with Tensioned Conductors 534

Chapter 18
Working It Hot 543

Safety Strategy for Hot-Line Work 543
Working on a Hot Secondary 550
Rubber-Glove Work 551
Hot-Line Tool Work 558
Barehand Work 563

Chapter 19
Outdoor Lighting Systems 570

Types of Outdoor Lighting 570
Outdoor Lighting Infrastructure Systems 577
Maintenance and Troubleshooting Outdoor Lighting Systems 583
Safety and Environmental Hazards Working with Outdoor Lights 583

Chapter 20
Revenue Metering 585

Introduction 585
Determining Cost to the Customer 586
Types of Revenue Metering 594
Single-Phase Metering 597
Polyphase Metering 601
Transformer-Rated Metering 604

Chapter 21
Managing Vegetation in an Electrical Environment 611

Vegetation Management in Electrical Utilities 611
The Hazards of Tree Work 617
Tree Work Near Electrical Circuits 618
Essential Skills for Tree Work 622

Appendix 638
Glossary 659
Index 664

Guidebook for Power Line Technicians

Chapter 1

Controlling Risk for Power Line and Cable Technicians

Objectives

After completing this chapter, you should be able to:

1. Take steps needed to control the high risks involved in working on or near power lines and cable.
2. Participate in safety meetings and job planning sessions that identify high-risk hazards and controls.
3. Identify personal protective equipment (PPE) needed for the planned work.
4. Put in place controls to prevent electrical contact.
5. Set up for traffic control per regulatory requirements in the work zone.
6. Establish the controls needed to work safely in utility vaults, trenches, and confined spaces, such as in a large transformer.
7. Identify the safety controls needed when working with helicopters.
8. Identify equipment and controls to reduce risk when climbing and working at heights.
9. Participate in training sessions for managing on-the-job emergencies.

Orientation to Power Line Work

The power line work environment is one of a kind. Even when first starting out, you will be exposed to high-risk hazards on a daily basis. You will be introduced to material and tools not found in a typical hardware store. Many of the tools and equipment are unique to the "line" work environment, and an orientation to each new setting is essential.

The contents of this chapter, along with explanations from a mentor, can serve as part of an initial orientation. This text supports training but does not take the place of classroom or field training. Training and confirmation of training (testing) are mandatory before doing the tasks described in this book.

The Nature of Line Work

A huge variety of work is available for power line technicians.

Outdoor Environment

People coming into line work recognize that the work is outdoors. On larger construction-type jobs, the work requires being outdoors all day. Some power line tech will find themselves standing on top of a tower, at –20°, waiting for a helicopter to pick them off the top. Power line technicians will be found working in high heat and humidity while wearing flame-retardant clothing, rubber gloves, and rubber sleeves. Others will work on trouble and service work and will spend much of their days—or nights—driving in all kinds of weather and traffic. **See Figure 1–1.**

Figure 1–1 Power line technician work is required under all weather conditions. *(iStock.com/Bryngelzon)*

Rugged Work

This work requires upper-body strength for climbing, as well as for lifting heavy crossarms, for lifting a conductor on an insulator, and for closing a mechanical press. While back-saving devices are available, many lifting tasks straddle the boundary between lifting manually or using a mechanical aid—and you probably can guess which course of action many power line tech take.

Independent Work

The nature of this work is such that power line technicians are frequently part of a small crew or are working alone and without supervision. Each job is different and requires planning and decision making that can have major consequences on safety or on a community's power supply.

From Shovel to Computer

A power line tech can spend an hour or two trying to get boulders out of the bottom of a wet hole, then an hour or two retrieving complex switching information or importing data on a mobile field computer. Line work ranges from very ordinary labor to complex, high-tech work.

Safety Responsibilities and Expectations

Occupational Safety & Health regulations decree both the employer and the employee have safety responsibilities and rights. The regulations may be worded differently in each jurisdiction, but the basics are the same everywhere.

Employer's Responsibilities

An employer has a general duty to provide work and a workplace free from recognized hazards and provide standards, rules, and regulations that apply to the work. An employer must also ensure that employees have, maintain, and use safe tools and equipment. An employer that sets up a safety management system will find it effective for managing the many safety responsibilities required.

An employer must inform employees of their rights and duties under the employer's safety and health program and, of course, must not discriminate against employees who exercise their rights under Occupational Safety & Health regulations.

Employers have other duties, such as informing employees of the existence of their medical and exposure records. Employers also must provide these records to employees upon request.

Employee Safety Responsibilities and Rights

Occupational Safety & Health regulations are written to protect employees. An employee should become very familiar with the regulations, so that there is little doubt as to their responsibilities and rights. In summary, employees are responsible for performing their jobs safely, complying with all Occupational Safety & Health regulations, following their employer's safety and health program, following their supervisors' instructions, participating in required training classes, reporting any hazardous conditions to their supervisors, and wearing personal protective equipment where instructed at all times. Employees also must report any job-related injury or illness to the employer and seek treatment promptly.

An employee has legal rights and is entitled to protection for safety on the job. One such right is an employee is not allowed to be punished or discriminated against an employee for such acts as complaining to the employer, union, or any government agency about job safety or health hazards.

The Safety Committee

A safety committee is a great vehicle for promoting workplace safety. It gets employees, supervisors, and management together to solve safety problems. Typically, a safety committee encourages employee involvement in improving workplace safety and gives employees input into making recommendations to the top of an organization.

An ideal safety committee would have its function and role within the organization well defined. Committee members should be trained on safety and health legislation, as well as receive safety management training on topics such as, how to prepare a safety strategy, conduct a risk assessment, and investigate an accident.

The committee should meet at least every 3 months, and a prepared agenda and the minutes should be recorded, distributed, and posted on bulletin boards. The committee's role should include assisting management with developing and monitoring safety programs by:

- Having committee members involved in accident investigations
- Reviewing accident investigation reports
- Monitoring the status of any action plans from safety meetings and accident investigations
- Initiating problem-solving teams to address priority safety issues
- Participating in regular physical condition inspections of the workplace
- Participating in regular internal assessments of the effectiveness of safety programs

Managing Safety

An employer will have a complete safety management system in place. A safe work model can illustrate the types of safety functions or elements that must be managed, as well as show the relative influence of each element on supporting safe work. **See Figure 1–2.**

Safety Elements That Affect the Worker Culture

The elements shown within the inner band of the model have a direct impact on worker safety. Because people in electrical utility work are without direct supervision, having the elements of the inner circle in place supports the worker when decisions are being made that impact safety. The elements and actions that most impact the worker culture are as follows:

Rules and regulations are developed, introduced, easily retrievable, and enforced.

Lockout/Tagout procedure is developed, workers are trained, responsibilities assigned, operating diagrams available, all switchgear assigned nomenclature, and application of the procedure monitored.

Employee training is well coordinated, new employees are given a thorough orientation, and employees receive skills proficiency training and safety-related training.

Figure 1–2 The worker culture is the focus of the safe work model. *(Courtesy of Utility Risk Management Inc.)*

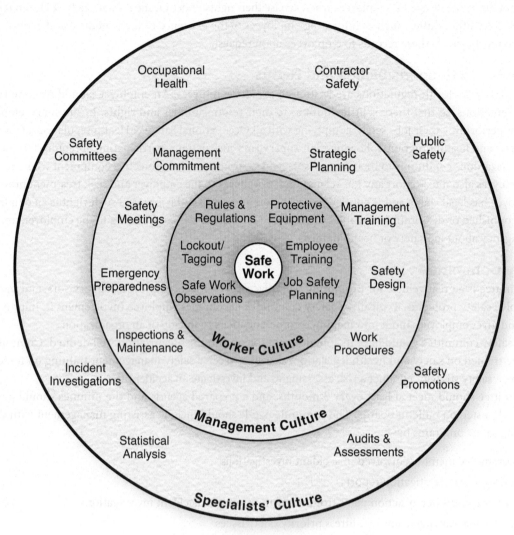

A JSA (job site analysis) policy is in place, and people are trained to do project safety planning, JSA planning, and process monitoring.

Work observations policy is in place, and supervisors and managers are trained in conducting effective work observations. Protective equipment is provided, and people are trained in their use and inspection, and are monitored.

Safety Elements That Establish the Management Culture

The elements within the middle band of the model have a direct impact on the management culture but are one step removed from influencing the worker culture. Consequently, deficiencies identified within these elements should be given second priority.

Management commitment is demonstrated by having a health and safety policy and by showing visible participation, responsibility, and accountability.

Strategic planning is demonstrated by setting performance objectives and a strategy for achieving them. A safety program manual, available to all, outlines coordination of the complete safety program.

Management and supervisors receive safety management training along with supervisory skills training. A programmed orientation for new managers and supervisors is in place.

Safety meetings are scheduled, an agenda set, and minutes kept, and a company document with guidelines for safety meetings is in place. Written work procedures for critical technical and maintenance tasks are developed, documented in a procedure format, and introduced.

Safety design analyses—such as job safety analysis, hazard analysis, ergonomic analysis, and a hazard registry (risk assessment)—are tools used regularly to reduce risk in the workplace.

Inspections and maintenance of critical tools and equipment, mobile equipment, and system equipment are scheduled and carried out.

Emergency preparedness plans are prepared for fire prevention, first aid response, spills response, and evacuation.

Safety Elements That Affect the Specialists' Culture

The elements within the outer band are generally administered and carried out by specialists such as safety professionals. Deficiencies identified within these elements should be given third priority.

Occupational health activities are administered to ensure air and water quality, noise control, proper management of hazardous materials, and that biological and physical agents are identified and controlled.

Health and safety promotion is administered for programs such as the employee assistance program (EAP); health, safety, and wellness communication; and off-the-job safety and safety performance recognition. These programs should be introduced to all new employees during orientation.

Joint health and safety committees' role and function are defined, committee members receive training, meetings are scheduled, and committee activities are monitored.

Contractor safety policy is in place with a requirement for contract administrators to be trained in contract administration, contract monitoring, and conducting a contractor orientation.

Public safety is administered so that public access is controlled at worksites, public hazards involving utility plants are identified and controlled, any accidents involving the public with a utility plant are investigated, and public electrical safety awareness is promoted.

Incident investigations are carried out based on an established policy requiring an initial reporting and investigation, a detailed investigation for defined serious incidents or incidents with a high potential for harm, follow-up activities, and injury management.

Statistical analysis is carried out and incident summary reports, incident statistical reports, and proactive measures statistical reports are prepared.

Audits and assessments such as employee surveys, internal assessments, and external audits are carried out.

Orientation for the New Employee

An employer should have a programmed orientation for new employees, temporary employees, and those going into a new position within the organization.

The orientation should include an overview of applicable safety policies, health and safety programs, and rules. Certain training cannot wait for a time when there are enough employees to make a class. Depending on the work given, there may be an immediate need for safety training such as first aid, CPR, hazardous materials management, defensive driving, fire response, confined space entry, electrical safety awareness, forklift operation, traffic flagging, trenching/shoring, and chainsaw operation.

Participate in Safety Meetings and Job Briefings

Power line technicians can and should provide feedback, suggestions, and questions at safety meetings and at job briefings.

Characteristics of a Good Safety Meeting

A good safety meeting must be planned ahead and scheduled. An agenda should be set, minutes kept, attendance recorded, and follow-up actions recorded and posted on bulletin boards. A safety meeting schedule should include critical safety topics. Safety meetings are good vehicles for updating staff on current safety issues, technical issues, new policies, new tools, and equipment and for jointly solving technical and safety problems.

Characteristics of a Good Job Briefing

Occupational Safety & Health regulations require that, at minimum, a job briefing must cover the hazards associated with the job, the work procedures involved, special precautions, energy source controls, and personal protective equipment.

The foundation of a good job briefing is a discussion, on the job site, structured around a daily job briefing checklist or written job plan.

Conduct

Power line technicians are highly visible because they work in the public eye. They drive in vehicles that resemble traveling billboards. Driving recklessly or acting out, such as engaging in road rage, in a company truck would reflect on the company.

The work can include sensitive communications with customers, such as work that affects private property, notification of a power outage, or a disconnect for nonpayment.

To the public, you are the power company. Behave responsibly.

Reporting Incidents and Accidents

Employers require that all accidents and close calls be reported. If a minor injury is not reported and later results in blood poisoning or a need for knee surgery, for example, you may have to prove that it resulted from an on-the-job accident. Even if the injury seems so minor that it does not need to be reported to the insurance company or for compensation, having it recorded in a supervisor's log or diary offers you financial protection from a future flare-up.

Any incident with a high potential for harm, even without injuries, should be reported and investigated as though it were a severe accident. The lessons learned may prevent a fatality in the future.

The Line Work Culture

A power line technician who has braved the elements, worked at heights in the wind and snow, worked on live circuits, stepped out of a helicopter onto a structure, operated large equipment, and perhaps experienced a potentially fatal near-miss accident tends to become overly self-confident. The line work culture lends itself to a bold image, where the risk of the job becomes an attraction. This culture has traditionally resisted change, especially new safety equipment. Over time, power line techs have less willingness to accept risk or to take the extra steps needed to use additional safety equipment.

When Charlie White, a first baseman for the Boston Red Sox, came onto the field wearing a thin protective glove after baseball had been a barehanded game for 25 years, with broken hands and split thumbs, he was laughed off the field.

Similarly, in line work, there has been major resistance to wearing hard hats or traffic vests, the use of potential testers, consistently applying grounds and using fall-arrest equipment. Such elements of this culture will probably always be present, but an old slogan from aviation should be part of the power line tech work code: "There are old pilots and there are bold pilots, but there are no old bold pilots."—Unknown author.

Risk in Line Work

Line work is a potentially high-risk job. Power line technicians suffer a high rate of fatal or permanently disabling injuries. This working environment is very unforgiving. High exposure leads to potentially fatal hazards such as burns from an electrical contact, a fall, asphyxiation in a vault, or wayward vehicular traffic bursting into a work zone.

The definition of risk shows why power line technician work is considered a high-risk occupation:

$$Risk = Consequence \times Exposure \times Probability$$

Where:

Consequence = the damage or injuries caused when an accident occurs

Exposure = the amount of time a person is within a hazardous area

Probability = the likelihood of making contact with a hazard

Power line technician work is considered high risk because the consequences of accidents are often severe. Risk is more than merely the odds of having an accident. Not much can be done about consequences or exposure. Risk is reduced by using controls (barriers), such as safety equipment, that will lower the probability of contact with a hazard.

Managing Personal Safety Risk

Reduce Risk by Using Only Approved Equipment

An employer is responsible for providing approved personal protective equipment, tools, and hardware that meet regulatory requirements. An employee is responsible for inspecting and using the equipment provided as required by regulatory jurisdictions and the employer. An employee is responsible for reporting equipment defects and for taking defective equipment out of service.

This text assumes that only trained, qualified people carry out power line technician work, using only approved equipment, and that the equipment is inspected as required. Therefore, these requirements will not be repeated in this text for each situation.

Rubber Gloves as Personal Protective Equipment

Rubber gloves and sleeves can be used as tools for carrying out hot line work or as personal protective equipment. Wearing rubber gloves as a backup protection when working in the vicinity of energized circuits has the same function as a hard hat, in that the equipment provides backup protection if something goes wrong. In some jurisdictions, rubber gloves used as personal protective equipment are referred to as "truck gloves"; and because they are used to set poles and other things, they are not used as tools for hot line work.

The following are typical rules for the use of rubber gloves as personal protective equipment in distribution work:

1. *Ground to ground rule:* Rubber gloves shall be worn continuously while working on any structure carrying energized conductors.
2. *Cradle to cradle rule:* Rubber gloves shall be worn from the time a bucket truck boom leaves its cradle until it is back in the cradle.
3. *Extended reach rule:* Rubber gloves shall be worn any time a worker and the length of any noninsulated tool being used are able to contact an energized conductor. The extended reach is considered to be 5 ft (1.5 m) and is called a five-foot rule in some jurisdictions.
4. *Lock to lock rule:* Rubber gloves shall be worn when a vault or cabinet with an energized cable is opened and shall be worn continuously until the vault or cabinet is locked.

Examples of Rubber Gloves Used as Personal Protective Equipment

When opening and closing air-break switches, circuit breakers, or electronic reclosers at an equipment site. (Note that there are no rubber gloves rated high enough for the line voltage of many switches that are operated. The rubber gloves are used as backup protection only.)

While using hot-line tools, though this varies among utilities, from requiring them for all hot-line tool work, to prohibiting them for all hot-line tool work, to requiring them on distribution voltages only, to requiring them when using switch sticks in wet weather. They are used:

1. When moving or handling energized underground primary cables
2. When removing sheaths and sleeves from cables and splices and when opening or cutting cables unless proven de-energized
3. When working on energized secondaries and services
4. When removing vines and weeds that have grown into energized equipment
5. When handling ropes and conductors while stringing in the vicinity of energized circuits

Types of Rubber Gloves

Rubber gloves (and in many places rubber sleeves) are used every day in distribution line work. Rubber gloves come in different classes or voltage ratings. **See Figure 1–3a.** The voltage ratings of the gloves are expressed as nominal voltages from one energized wire to another, expressed as phase-to-phase (Ø to Ø) voltage. Rubber gloves are more likely to be exposed to phase-to-ground voltages, which can be calculated by dividing the phase-to-phase voltage by √3, or 1.73.

Figure 1–3a Rubber glove class and voltage ratings. *(Courtesy of Salisbury by Honeywell)*

Class Color	Proof Test Voltage AC / DC	Max. Use Voltage* AC / DC	Rubber Molded Products Label	Insulating Rubber Glove Label	Insulating Rubber Dipped Sleeve Label
00 Beige	2,500 / 10,000	500 / 750		10 SALISBURY ANSI / ASTM MADE IN D120 CLASS 00 U.S.A. TYPE I MAX USE VOLT 500V AC	
0 Red	5,000 / 20,000	1,000 / 1,500	SALISBURY MAX USE VOLTAGE 1,000 V AC CLASS 0 TYPE I	10 SALISBURY ANSI / ASTM MADE IN D120 CLASS 0 U.S.A. TYPE I MAX USE VOLT 1000V AC	SALISBURY ANSI / ASTM MADE IN D1051 CLASS 0 U.S.A. TYPE I MAX USE VOLT 1000V AC
1 White	10,000 / 40,000	7,500 / 11,250	SALISBURY MAX USE VOLTAGE 7,500 V AC CLASS 1 TYPE I	10 SALISBURY ANSI / ASTM MADE IN D120 CLASS 1 U.S.A. TYPE I MAX USE VOLT 7500V AC	SALISBURY ANSI / ASTM MADE IN D1051 CLASS 1 U.S.A. TYPE I MAX USE VOLT 7500V AC
2 Yellow	20,000 / 50,000	17,000 / 25,500	SALISBURY MAX USE VOLTAGE 17,000 V AC CLASS 2 TYPE I	10 SALISBURY ANSI / ASTM MADE IN D120 CLASS 2 U.S.A. TYPE I MAX USE VOLT 17000V AC	SALISBURY ANSI / ASTM MADE IN D1051 CLASS 2 U.S.A. TYPE I MAX USE VOLT 17000V AC
3 Green	30,000 / 60,000	26,500 / 39,750	SALISBURY MAX USE VOLTAGE 26,500 V AC CLASS 3 TYPE I	10 SALISBURY ANSI / ASTM MADE IN D120 CLASS 3 U.S.A. TYPE I MAX USE VOLT 26500V AC	SALISBURY ANSI / ASTM MADE IN D1051 CLASS 3 U.S.A. TYPE I MAX USE VOLT 26500V AC
4 Orange	40,000 / 70,000	36,000 / 54,000	SALISBURY MAX USE VOLTAGE 36,000 V AC CLASS 4 TYPE II	10 SALISBURY ANSI / ASTM MADE IN D120 CLASS 4 U.S.A. TYPE I MAX USE VOLT 36000V AC	SALISBURY ANSI / ASTM MADE IN D1051 CLASS 4 U.S.A. TYPE I MAX USE VOLT 36000V AC

Rubber gloves come in different lengths, can be straight or have a bell cuff, can have five fingers or one finger (mitten style), can have two contrasting colors or one color, and come in various sizes. They are always used with a protective leather glove. Because the cover has no insulating value and can easily provide a path for electricity to flow, check that the rubber glove extends past the leather protector glove. Check the minimum distance between the top of the protector glove and the rolled top of the rubber glove. **See Figure 1–3b.**

Figure 1–3b Minimum distance between the top of the protectors and rubber gloves. *(Courtesy of Salisbury by Honeywell)*

CLEARANCE TABLE FOR LEATHER PROTECTORS PER ASTM F496 - Table 4		
Glove Class	Min. Distance Between Protectors and Rubber Gloves	
	in.	mm
00, 0	1/2	13
1	1	25
2	2	51
3	3	76
4	4	102

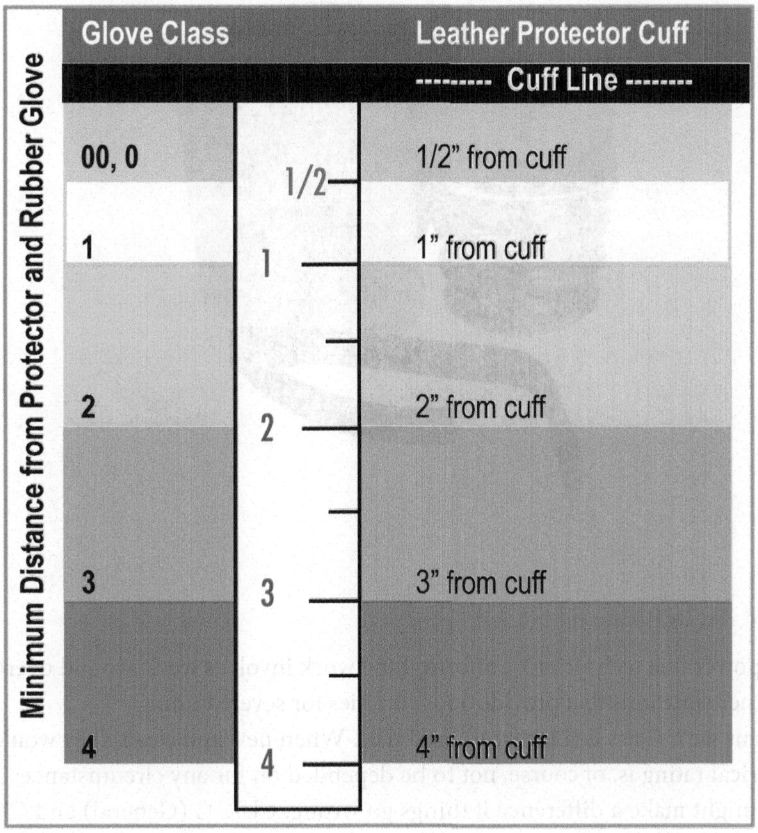

Care and Inspection of Rubber Gloves

Rubber gloves and sleeves are not very rugged, and special care and inspection are needed when using them. Rubber gloves can be punctured easily, and, therefore, rings and jewelry should never be worn with them. The rubber can be damaged by contact with chemicals such as hand creams and cleaning and petroleum products; therefore, only mild soap and warm water should be used to clean rubber gloves. The rubber must be thoroughly rinsed after cleaning and left to air dry. The leather covers are rubber glove protectors and must be inspected for embedded objects, such as wire particles. The cuff of the leather glove must be kept clean of substances such as conductive joint compound. Even after the leather covers are removed from service as a rubber glove protector, they are not to be used as work gloves on the job. It is otherwise difficult to warn workers who fail to wear the rubber gloves. Rubber gloves should be stored in their rubber glove bag, away from ultraviolet light, and not in a truck tool bin. Rubber gloves must be electrically tested at least once every 6 months and rubber sleeves at least once every 12 months.

Rubber gloves must be inspected before use. The most effective way to inflate the glove is with a portable mechanical inflator. **See Figure 1–4.** This inflator allows a much closer inspection of a glove. Listening for air leaks is another method of inspection but may not be effective around noisy streets and highways. Inflating a rubber glove by rolling the cuff toward the palm to trap air (roll-up method) is a common but hurried method, because preventing the air from escaping while inspecting the glove and listening for air leaks requires skill and perseverance. Water may be substituted for air if required. The roll-up method must be taught in the field by demonstration.

Figure 1–4 The use of a rubber glove inflator is a preferred method to inspect rubber gloves. *(© Salisbury by Honeywell)*

Wear Head Protection

Hard hats are part of a power line technician's uniform. Line work involves work around overhead structures, hoisting operations, and many other situations that provide opportunities for severe bumps.

Power line technicians wear Class E (Electrical) hard hats. When new and clean, they would have passed a 20,000 V electrical test. The electrical rating is, of course, not to be depended on for any circumstance; it is only an indication of backup protection that might make a difference if things go wrong. Class G (General) and Class C (Conductive) hard hats are not to be worn by power line tech.

Both Type 1 and Type 2 hard hats meet the same top-impact standards, but Type 2 also has protection in the sides, front, and rear. Some utilities are now requiring Type 2 hard hats.

Wear Eye Protection

Eye injuries are relatively easy to avoid by wearing appropriate eye protection. Most employers require eye protection to be worn 100% of the time on the job.

Fortunately, safety glasses are relatively comfortable and stylish. Prescription safety glasses, or eye protection that is worn over prescription lenses, are available. Filter lenses that have a shade number protect the eyes from light radiation, such as electric arcs.

Always do the following:

- Wear eye protection during switching, installing, and removing protective grounds, and when working on an energized primary or secondary circuit for protection from the ultraviolet light of an electrical arc or flash and from flying copper or aluminum particles. (The latter are nonmagnetic and are difficult for a doctor to remove.)
- Wear eye protection when there is a risk of flying particles, such as striking steel with steel or working with toughened glass insulators.
- Wear eye protection when exposed to strong alkalis or acids during battery boosting, when cleaning rubber hose, when refurbishing hot-line tools, or when cleaning an aluminum conductor with lye in hot water.
- Wear eye protection when firing on wedge connectors.
- Wear eye protection when working with high-pressure hydraulic hoses and other hydraulic tools.
- Wear eye protection when working with power tools, such as a chainsaw or drill.

Wear Safety Footwear

A variety of safety footwear is used in utility work, such as footwear that provides mechanical protection, electrical resistance, electrical conductivity, and chainsaw resistance.

Mechanical Protective Footwear

Reduce the risk of foot injuries by wearing footwear with a steel toecap and puncture-resistant sole for mechanical protection. "PR" is the designation for "puncture resistant." Carbon fiber safety toe boots may also be used.

Electrical Hazard (EH)–Rated Footwear

Like a hard hat with an electric rating, electrical hazard (EH)–rated boots may provide backup protection from step potentials. The initial electric test of up to 18 kV for a new clean boot would be much less or nonexistent in long, wet grass or in mud, but there have been incidents where the EH boots have made a difference.

Some employers require EH overshoe footwear. Usually, overshoe boots are specified for use on a particular job, such as working around a truck with an uninsulated boom. **See Figure 1–5.**

Work boots with nailed soles or stitching that goes all the way through the sole are highly conductive and should not be used for work around power lines.

Figure 1–5 Dielectric overshoe footwear is personal protective equipment that reduces exposure to voltage gradients. (© Salisbury by Honeywell)

Electrical Conductive Footwear

Electrical conductive footwear is used for two purposes:

1. For barehanded work from a bucket (The boots help to bond a worker to the metal grid in the bucket and avoid painful electric shocks.)
2. To prevent the buildup of static electricity on a person's body and painful electric shocks each time a grounded object is touched when working with hot-line tools from steel transmission line structures and in high-voltage substations

Chainsaw-Resistant Footwear

For regular chainsaw users, wear boots that are designed specifically for minimizing foot injuries caused by accidental contact with a running chainsaw.

Slip-Resistant Footwear

While there is slip-resistant footwear available for certain conditions, for utility workers the most extreme situation is walking on ice. The most common aids to traction on ice are devices that can be attached to regular footwear.

Wear Arc-Rated (AR) Clothing

Electrical arcs and flashes can occur when switching, energizing equipment (transformers), accidental short circuits, working on energized lines, installing meters or racking out a breaker. Wearing materials such as nylon, polyester, or rayon during an electric arc or flash can increase the severity of the injuries because the clothing keeps burning and the melted "plastic" material burns into the skin. For severe burns, preventing infection is a large part of medical care and melted clothing in the skin is a major complicating factor.

Occupational Safety & Health regulations require that workers exposed to electrical arcs and flashes wear Arc Rated (AR) clothing. A flash hazard analysis would show that there are very high explosive fault currents available for example in or near a substation, at the secondary side of a large transformer and on heavily loaded circuits and at some metering locations. Racking out a breaker in a substation (one without a remote operating feature) is likely a worst-case scenario for a power line technician. The employer would have an exacting documented work procedure and specify the maximum AR clothing, probably more than one layer, a balaclava, full face shield and more.

Wear Hearing Protection

Typical situations that require hearing protection are working with or near a portable two-cycle engine, small four-cycle engine, chipper, compressor, rock drill, jack hammer, or helicopter.

The most common hearing protection for power line technician work is disposable earplugs. **See Figure 1–6.** They must be inserted properly with clean hands. Hold the earplug between the thumb and forefinger, then roll and compress the ear plug into a small cylinder. To make a snug fit, straighten the ear canal by using one hand to reach over your head and pull up and back on your ear and use the other hand to insert the earplug. The earplug will now expand and fill the ear canal.

Remove an earplug slowly by twisting it to break the seal, thereby reducing the risk of injury to your ear drum.

Identify and Protect from Chemical Hazards

Programs to manage chemicals are in place everywhere because of required regulations. Employers are required to establish a written hazard communications program that includes ensuring that containers are labeled, safety data sheets (SDS) are available, and employees are trained. Employers are required to maintain a list of the hazardous chemicals in the workplace. Any chemicals used therein require a hazard evaluation, SDS, labeling, and training.

As a power line technician, identify any high-risk chemicals used in your work and ensure that you have the training, know the relevant work procedures, and use the required personal protective equipment to work with these chemicals.

- Polychlorinated biphenyls (PCBs) are not an immediate hazard to a person but breathing in the by-products of a fire involving PCBs is a major health hazard. Stay upwind from any fire involving equipment with insulating oil, which

Figure 1–6 This is the proper way to insert earplugs for hearing protection. *(Delmar, Cengage Learning)*

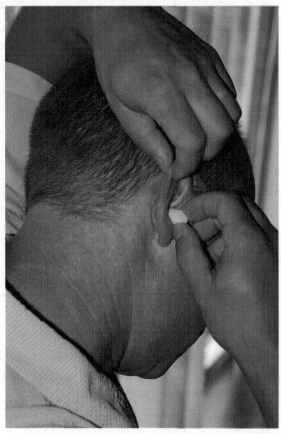

may be contaminated with PCBs (e.g., pyranol, askarel, inerteen). Report all incidents involving PCBs. Utilities have stringent procedures and training involving spills, cleanup, transportation, and disposal.

- Methyl hydrate (methanol) is used as a drying agent for insulated booms and energized sticks. It is a Class I flammable liquid, so it must be handled as though it is gasoline. Drinking it is fatal.
- Working with lead cables or cables wrapped with asbestos requires training and under certain conditions, respiratory protection.

Manage Your Risk in the Working Environment

Participate in Job Planning and Job Briefing

A good job planning format allows documentation of a job sequence, identification of high-risk hazards and effective barriers for each step, and a minimum amount of writing. To be effective, a job plan must be written down. **See Figure 1–7.**

A good job site analysis (JSA) includes the following:

- *Upstream documentation:* Job planning must include a study of existing conditions, especially in relation to identifying hazards. A line crew should have been issued an applicable instruction order, engineering drawings and layout, and a materials list.
- *Written job sequence:* Everyone shall know how a job is to proceed. A plan laying out the job steps should be kept in a simple format and scaled to the complexity of the job.
- *Identification of major hazards:* Each job step should be checked for potential high-risk hazards. Effective barriers should be chosen to control each identified hazard.

Figure 1–7 This form is a sample format for planning smaller jobs. *(Delmar, Cengage Learning)*

JOB: Tree fallen on circuit, cutouts open on two phases, one conductor broken.		
JOB STEPS	**MAJOR HAZARDS**	**REQUIRED BARRIERS**
1) Take an isolation certification on the circuit and place grounds.		
2) Remove tree.	Tree leaning heavily on line, uncertain of how it will drop if cut.	Rope the tree and pull it off the line with the truck. Keep every one clear.
3) Repair broken conductor.	Potential between ends of broken conductor.	Install grounds on each side of break and use the intact conductors as a jumper.
4) Remove grounds and surrender isolation certification		

- *Pre-job briefing:* A pre-job briefing is much more focused and effective when there is a written job sequence with high-risk hazards and appropriate barriers for each step identified and discussed, such as in a tailboard conference plan. **See Figure 1–8**. In summation, "Plan the work, work the plan."

Eliminate the Probability of an Electrical Contact in Line Work

To protect themselves from electrical contact injuries, power line techs must either isolate themselves from an energized electrical circuit or insulate themselves from the circuit. Isolating oneself from an energized electrical circuit means either keeping beyond the minimum approach distance or isolating and grounding the circuit. The most common method of insulating an electrical circuit is with the use of cover-up on the conductors. Cover-up can be rubber, typically put on with rubber gloves, or rigid, put on with a shotgun stick. Cover-up is available only up to 69 kV.

Power line technicians can insulate themselves somewhat by using rubber gloves and sleeves. These protect only the hands and arms and are available only for distribution-level voltage. Live-line tools can be considered isolating or insulating, depending on one's perspective.

A third method of protection is to bond oneself to the electrical energy. Barehand work and working from ground gradient matting will bond a person to the electrical energy or fault current.

Apply the Minimum Approach Distance as an Electrical Energy Control

Maintaining a minimum approach distance to exposed energized conductors is the most common barrier a power line technician uses to avoid electrical contact. Typical Occupational Safety & Health regulations do not allow an unqualified person to come within 10 ft (3 m) of a distribution electrical circuit and an additional 4 in. (10 cm) for every 10 kV over that, which works out to 14 ft (4.3 m) for 169 kV, 16 ft (4.9 m) for 230 kV, and 25 ft (7.6 m) for 500 kV. To work closer, a person either must be trained and declared qualified by an employer or must be undergoing on-the-job training under the direct supervision of a qualified person. A power line technician apprentice becomes qualified by electrical awareness orientation training and over the course of time by demonstrating the ability to perform work safely in the vicinity of energized circuits.

Orientation to electrical awareness should include learning to identify power conductors, neutral, open-wire bare secondary, wrapped secondary, tree wire, telephone cable, streetlight circuit, and the voltage on each. Some structures will have three or four circuits, each with a different voltage. Although the thickness or diameter of insulation or the voltage

Figure 1–8 A JSA conference plan form is essential for a structured job briefing. *(Courtesy of Utility Risk Management Inc.)*

TAILBOARD CONFERENCE PLAN

Prepare, discuss, and review the job plan with the crew. Use this form daily and whenever a change is introduced to the job.

Job Being Performed:

Date	Crew Members Present

Hazard Identification List

GRAVITY	ELECTRICITY	MECHANICAL	KINETIC/VEHICULAR
Falling from a height Falling objects Falling structures Climbing obstructions	Electrical contact Induction/backfeed Static charge Ground gradients Flash potential Boom contact	Equipment failure Lifting with the boom Max. working loads on rigging Conductor/guy tensions Vehicle stability	Traffic control Driving conditions Moving loads

Have We Considered?

PEOPLE	PROCEDURES	HARDWARE/EQUIPMENT	ENVIRONMENT

Person in charge
Qualification of personnel
Job coordination with other
  work groups
Communication
Worker fatigue
Pedestrian control
General public
	Isolation of apparatus
	Adequate grounding
	Work protection/Hold off
	Vehicle grounds
	Distribution standards
	Confined space entry
	Emergency rescue
	  procedures
		Work equipment
		Tools and Personal
		  Protective Equipment
		  (PPE)
		Vehicles
		Structures
		Safe loads for rigging
		Warning devices
		Physical barriers
			Other utilities
			Weather conditions
			Soil conditions
			Lighting conditions
			Work schedules

MAJOR HAZARDS	BARRIERS TO ELIMINATE OR CONTROL

How will we execute a rescue? _____

Exact location for emergency aid: _____

reading on a transformer may determine the voltage of a circuit, it is best to ask the owner of the lines. It is the voltage level that determines the allowable minimum working distance. A power line technician also must be able to recognize hazards such as a broken crossarm, a broken insulator, or a fallen conductor that could cause contact even when working outside of the qualified minimum approach limit. Government regulators and utilities have developed tables listing the minimum approach distance from various voltages for different levels of qualified people and equipment. Minimum approach distances apply to work in the vicinity of an energized circuit and live-line work. **See Figure 1–9.** The flashover voltage for an energized-line tool is the same as it is for air. A fiberglass energized-line tool may be better insulation than air, but the distance needed on a tool is an air gap between the hands on a live-line tool and the energized conductor.

Figure 1–9 Typical Minimum Approach Distances

Maximum Phase-to-Phase Voltage (Maximum Phase-to-Ground Voltage)	Minimum Approach Distance Phase-to-Ground Exposure	Minimum Approach Distance Phase-to-Phase Exposure
0.05 to 1 kV	Avoid contact	Avoid contact
Up to 15 kV (8.7 kV)	2′ 1″ (64 cm)	2′ 2″ (66 cm)
Up to 36 kV (20.8 kV)	2′ 4″ (72 cm)	2′ 7″ (77 cm)
Up to 46 kV (26.6 kV)	2′ 7″ (77 cm)	2′ 10″ (85 cm)
Up to 72.5 kV	3′ (90 cm)	3′ 6″ (1.05 cm)
Up to 121 kV	3′ 2″ (95 cm)	4′ 3″ (1.29 m)
Up to 145 kV	3′ 7″ (1.09 m)	4′ 11″ (1.5 m)
Up to 169 kV	4′ (1.22 m)	5′ 8″ (1.71 m)
Up to 242 kV	5′ 3″ (1.59 m)	7′ 6″ (2.27 m)
Up to 362 kV	8′ 6″ (2.59 m)	12′ 6″ (3.8 m)
Up to 550 kV	11′ 3″ (3.42 m)	18′ 1″ (5.5 m)
Up to 800 kV	14′ 11″ (4.53 m)	26′ (7.91 m)

Plan Your Work When Working Alone

Typical Occupational Safety & Health regulations require more than one person when installing or removing, lines, equipment, where a worker is exposed to contact with an energized circuit and/or apparatus at more than 600 V (in some jurisdictions 300 V). This regulation should not be a license for all work under 600 V be carried out alone. Job planning will identify hazards where the nature of the work should require more than one.

Regulations may list exceptions and allow one person to perform switching for example. Another exception is allowing a worker to safe guard the public in an emergency. An example is using a live-line tool to lift or cut a fallen conductor from a hazardous situation. A power outage should not be considered an emergency that would allow someone to, for example, change out a transformer alone. A person working alone should plan and preferably document the work to be performed as though a job briefing was held.

When a second person is required, they must have the skills to perform a pole top rescue, lower a bucket or rescue someone from a vault. The second person must know emergency communications and procedures and be qualified to perform cardiopulmonary resuscitation (CPR.)

Apply Lockout/Tagout to Work on a Deenergized Circuit

One of the most sacred procedures for line work is the procedure that isolates an electrical circuit or electrical equipment from all electrical sources and ensures that the isolation will stay in effect while working on the line or equipment. A lockout/tagout procedure ensures that all potential electrical sources are locked and/or tagged.

When a circuit or equipment has been isolated from all sources of reenergization a certificate/credential/document signed by an authorized person is issued by a controlling authority. A document signed by an authorized person certifying isolation is complete must be obtained before it is safe for a crew to apply grounds and go to work.

Be aware that when you cross utility boundaries on storm trouble, the lockout/tagout procedure in other utilities may not have the same terms and meaning. For example, a "hold off" tag can be used to tag an isolating device for an isolation certification, whereas it may be used in another utility to block the automatic reclose on a breaker. "De-energize" can mean isolating a circuit in one utility but can mean applying protective grounds on a circuit in another. An isolation certificate can be named as a clearance, work permit, markup, LTO or other document. This book will use the term "isolation certification."

Common Elements of a Lockout/Tagout Procedure

Common Elements	Details
1. There is a controlling authority (system control or dispatch) responsible for granting an isolation certification on a circuit or equipment.	1. The controlling authority is normally the system control (dispatch) for the system. 2. In some utilities, power line tech are given control of the distribution system, especially for a switchgear that is not remotely operated. They will act as their own dispatchers and will prepare their own switching orders and will apply tags on applicable switchgear. To open or close some switchgear, such as a tie point between two substations, may require information only available to system control personnel.
2. An application is made out by an authorized person/ work authority to isolate a circuit or apparatus when it is needed to provide safe working conditions.	1. An isolation certificate for work is given to an authorized person who is deemed responsible for obtaining and maintaining a circuit in an isolated and grounded state. 2. The application (usually on a prepared form) for an isolation certification on a circuit is made out in advance so that the controlling authority can study it in relation to other applications and can determine the impact on the system (especially for transmission lines). 3. The application can be done verbally when an outage is needed on short notice. Both parties should write down the verbal application on an application form. 4. There is a joint responsibility for the controlling authority and the authorized person/work authority to ensure that the proper element is identified for an isolation certification.
3. Switching is carried out and locks and/or tags (cards) are placed on the switchgear providing isolation.	1. A switching instruction or switching order form is prepared as a plan for the switching sequence needed to provide the isolation certification for the work. 2. If verbal switching instructions are used (radio, etc.), the message should be repeated back for confirmation. 3. After operating any switchgear, it must be visibly checked to ensure that it is in the correct position. 4. Switchgear that serves as an isolation point for an isolation certification for work on a circuit over 750 V must have a visible open point and be tagged. 5. In addition to tagging, there may be situations where an authorized person will reduce risk of an inadvertent operation of a switchgear by using an additional safety measure, where applicable, such as a lock, removing a switch element, removing a loop, disabling a motor-operated disconnect, or opening an extra disconnecting device.
4. The controlling authority certifies and signs a formal document that the circuit is isolated from all sources and will remain isolated until the isolation certification is formally surrendered by the authorized person/ work authority.	1. A formal communication, confirming that the isolation certification is in effect, is made, often issuing an isolation certification number to keep the isolation certification distinct from others issued by the controlling authority. 2. Authority to install protective line grounds and go to work is given to the authorized person/work authority by the controlling authority.
5. The work authority verifies that isolation is complete.	1. Potential checks are carried out to ensure the absence of a normal line voltage before protective line grounds are installed. 2. In some jurisdictions, the protective grounds are controlled and tagged by the controlling authority, while in others the work authority controls the placing of protective line grounds.
6. When work is completed, the work authority surrenders the isolation certification for the work.	1. The person holding the isolation certification ensures that all workers are clear and that protective line grounds are removed before surrendering the isolation certification.

Use a Switching Order

To reduce the risk of an operating error, use a written switching order form whenever a sequence of switching is to be carried out. For example, the form can be used to isolate a set of voltage regulators. **See Figure 1–10.**

Figure 1–10 Switching Order Form

Switching Order				
Purpose	Isolate Voltage Regulator 7R16			
Ordered by	District Control	**Time**		
Completed by	John Lineman	**Time** 10:30		
Sequence Number	**Device Being Operated**	**Operation**	**Tag #**	**Initials**
1	Auto/Manual Switch	Neutral Tap		JL
2	Input & Output Terminals	Voltage Test		JL
3	Auto/Manual Switch	Turn Off		JL
4	Bypass Switches – 3 phases	Close		JL
5	Load Switches – 3 phases	Open		JL
6	Source Switches – 3 phases	Open		JL
7	Source Risers	Remove		JL
8	Load Risers	Remove	NA	JL

When acting as a switching agent for a controlling operator (dispatch), the operator should have a written switching order; however, the power line technician should also get a copy or write down the sequence when received by radio.

Avoid These Typical Failures When Implementing a Lockout/Tagout Procedure

- Incorrect identification of the required circuit
- Incorrect identification of switchgear being operated
- Failure to test for potential before installing protective grounds
- Failure to install protective grounds to prove isolation
- Working outside the scope of the isolated and grounded circuit or equipment
- Neighboring energized circuits or equipment encroaching into the isolated and grounded work zone
- Taking a circuit off-potential, as an extra safety barrier, with no intention to place protective grounds (even if the intent is to treat the ungrounded circuit as energized, the hot-line work methods are compromised. If a circuit is de-energized to provide a safe (or safer) working condition, you must apply the full lockout/tagout procedure, including the installation of protective grounds.)

Protection from Traffic

Exposure is high risk when working along a roadway because the consequences of a wayward vehicle entering a worksite can be severe. Local jurisdictions dictate the type of traffic control needed when utility work will affect the free flow of traffic. Many jurisdictions provide pocket-sized field manuals showing minimum requirements for various types of work. In the United States, the Federal Highway Administration issues the *National Manual on Uniform Traffic Control Devices* (MUTCD). This document is used as a reference for most of the utility rules governing traffic control.

Being highly visible with signs, flags, cones, warning lights, reflective clothing, flaggers, and police presence is usually effective for channeling traffic around a work area. However, motorists are often distracted by rerouted lanes, signs, workers, and equipment and will not see any reason to stop for work operations such as low-hanging wires being strung across a roadway.

The selection, placement, and spacing of traffic control devices will depend on the proximity of work to traffic, the roadway type (number of lanes and traffic speed), and the length of time workers will be exposed to traffic. A physical barrier such as a large truck (crash truck) parked in a closed traffic lane should be considered for highly vulnerable locations.

Manuals and standards for temporary traffic control cover rules for the training of flaggers, high-visibility clothing, and placement of devices that regulate, warn, and guide road users.

Utility-Specific Traffic Hazards and Barriers

Hazards	Barriers
Traffic makes contact with a conductor or rope being strung across a roadway during a slack stringing operation.	Stop traffic *before* running wire or rope across a roadway. Waiting for a break in traffic and then stopping vehicles entering the work zone increases the risk. Drivers often do not see any obstruction when only a wire is suspended across the road.
Traffic makes contact with a conductor or rope being strung across a roadway during a tension stringing operation.	*Caution:* If a road crossing is a much longer span than others in a pull, it is vulnerable to extra conductor sagging into that span. Rider poles and/or truck booms can be set up to reduce the risk of a conductor coming down into traffic.
The elbow of an aerial bucket stretches over an open lane of traffic, and a contact is avoided only if an approaching truck is alerted to the hazard.	Block off enough lanes to cover the potential of the boom elbow over the roadway or use a dedicated observer. The truck boom has to be clear of traffic, and the operator must be alert when changing work position. **See Figure 1–11**.

Figure 1–11 Proper traffic control is in place because the traffic cones are placed, and the truck boom elbow is kept out of the travelled portion of the street. *(al clark/Shutterstock.com)*

Working in a Confined/Enclosed Space

Depending on the jurisdiction, underground utility vault stubs, and open top spaces more than 4 ft deep such as trenches (one interpretation) are considered to be confined/enclosed spaces. **See Figure 1–12**. Because most electrical utility

Figure 1–12 A qualified attendant is placed at the entrance of a confined space to provide rescue capability. *(iStock.com/Mccaig)*

confined spaces are designed for entry under normal operating conditions, Occupational Safety & Health regulations declares them to be enclosed spaces rather than permit-required confined spaces. An enclosed space still needs to be tested for oxygen deficiency, flammable gases, and vapors. It is not necessary to prepare a confined space entry permit for an enclosed space as compared to confined spaces that are entered infrequently, such as tanks, vessels, silos, or hoppers.

For workers who enter an enclosed space infrequently, preparing a permit before entering will reduce the risk of overlooking an important step in the procedure. **See Figure 1–13.**

Testing is done by a qualified person using equipment that has been calibrated. Testing is done for oxygen deficiency, flammable gases, and vapors before removing a cover so that a spark created by the opening of a cover or hatch will not cause an explosion.

Remote probes are used to check for explosive gases. *Caution:* Flammable gases or vapors may not show up on a test until oxygen is introduced into the space. After opening, test again for oxygen deficiency, flammable gases, and vapors, as well as for common hazardous chemicals.

Ventilate to Reduce Risk

The risk of a contaminated atmosphere is greatly reduced through continuous ventilation in all areas of the enclosed space, so that the ventilation provides clean air at a rate of at least 200 cubic feet per minute (cfm) per occupant. Or in an enclosed space larger than 2,000 cubic feet, six air changes of the confined space volume per hour.

Typically, use a blower with a capacity of 1,000 cfm or greater. The blower should be located to ensure that engine exhaust gases are not blown into the confined space. **See Figure 1–14.** Ventilate the confined space by blowing air into the space for at least 5 minutes before entering.

Historically, about 60% of fatalities involving confined spaces occur when attendants perform a rescue. Ideally, all entrants wear a full-body harness, which is kept attached to a hoist to allow a rescue without the need to enter the confined space. **See Figure 1–15.**

Reduce Your Risks Working in Vaults

- Do not climb into or out of a vault by stepping on cables or hangers.
- Before lowering tools or material into the opening of a vault, ensure that the area directly under the opening is clear.
- Maintain reliable communications, through two-way radios or equivalent means, among all employees involved in the job.

Figure 1–13 A filled out form ensures an entry into an enclosed place is planned. *(Delmar, Cengage Learning)*

ELECTRICAL UTILITY ENCLOSED SPACE ENTRY PERMIT

Site Location and Description: _____
Supervisor: _____
Date and Time Issued: _____
Date and Time of Expiration: _____
Authorized Entrants: _____ _____
                   _____ _____
Attendants: _____ _____

Identify Potential Hazards	CONTROL MEASURES	ADDITIONAL CONTROL MEASURES
Inadequate Electrical Clearances	Lockout/Tagout/Ground	_____
Limited Exits	Set Up Rescue Capability	_____
Traffic and Pedestrian Control	Set Up Barriers	_____
Lack of Natural Ventilation	Ventilate	_____
Airborne Combustible Dust	Ventilate	_____
CO from Vehicle Exhaust	Ventilate	_____
CO_2 from Rotting Vegetation	Ventilate	_____
Nitrogen from Pressurized Cable	Ventilate	_____
H_2S	Ventilate	_____
The vault is over 15 ft. (4.5 m) deep	Ventilate	_____
Hot Work (i.e., soldering, pouring compound, cutting, or heating)	Ventilate	_____
Other: _____	Purge—Flush and Vent	_____

Test and Monitor Atmosphere (record monitoring every 2 hours)

TEST TO BE TAKEN	LIMIT	COMPLETED	TIME	TIME	TIME
% oxygen	19.5–23.5%	_____	_____	_____	_____
% LEL (LFL)	10% Max.	_____	_____	_____	_____
Carbon monoxide	35 ppm	_____	_____	_____	_____
Hydrogen sulfide	10–15 ppm	_____	_____	_____	_____
Other toxics	PEL	_____	_____	_____	_____

RESCUE READINESS

Communications Setup _____   Lifelines, Hoisting Equipment _____
Full-body Harness _____   Explosion-proof Lighting _____
Protective Clothing _____   Respiratory Protection _____
Fire Extinguishers _____   Special Tools _____
Other Information/Specific Requirements_____

This information and the work covered by this permit have been reviewed with all entrants and attendants. Safety procedures have been received and are understood. All appropriate items have been completed.

Entry Authorized _____
Title _____
Date/Time _____ / _____

THIS PERMIT SHOULD BE KEPT POSTED AT THE JOB SITE.

- If duct rods are used, install them in the direction presenting the least hazard to employees. A worker should be stationed at the far end of the duct line being rodded to ensure that the required minimum approach distances are maintained. When multiple cables are present, identify the cable to be worked on by testing or spiking.

- An impending fault is indicated when a cable in a vault has one or more abnormalities, such as oil or compound leaking from cables or splices, broken cable sheaths, hot localized surface temperatures of cables or splices, or splices that are swollen beyond normal tolerance. The defective cable should be de-energized before anyone works in the vault.

- When work is performed on cables in vaults, ensure that the metallic sheath continuity is maintained. A lethal voltage may be present between the ends of a break in the sheath.

Figure 1–14 Vault ventilation prevents the accumulation of explosive gases, oxygen displacement gases and noxious gases. *(Delmar, Cengage Learning)*

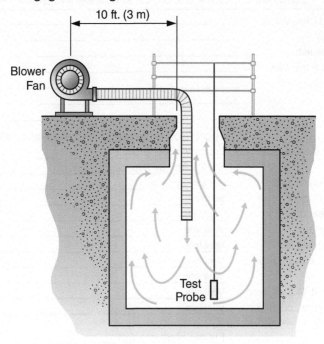

Figure 1–15 A rescue rig is needed on site in case a casualty needs to be rescued from a utility hole without a rescuer having to enter it. *(Delmar, Cengage Learning)*

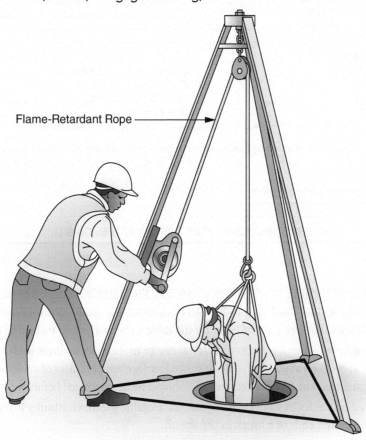

Fire in a Vault

There is at least one product, Stat-X First Responder, that will put down a fire in an enclosed space instantaneously. To activate, simply pull the actuator of this handheld aerosol-based device and throw it into the vault. **See Figure 1–16.** It may provide time for workers to escape from the vault. There is no oxygen depletion, and it can prevent a reflash for a short period. It suppresses the fire at very low concentrations by interference with free radicals.

Figure 1–16 A fire suppressant for a confined/enclosed space is a good first responder device. *(© Fireaway LLC)*

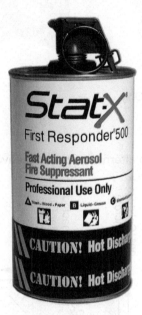

Protection for Work in Excavations and Trenches

At an excavation site, protect yourself from hazards:

- Get locations for gas, water, sewers, and telephone, television, and power lines.
- Before working on or moving energized cables, inspect them for damage.
- Carry out air sampling in trenches more than 4 ft deep where oxygen deficiency or other hazardous atmospheres could reasonably be expected to exist.
- Reduce the risk of cave-ins from material that could fall or roll from an excavation face or from material piled next to an excavation.
- Place a protective system, such as sloping, benching, shoring, or a trench box, when you are exposed to the hazard of falling or sliding material from an excavated bank or side more than 4 ft (1.2 m) above a worker's head. **See Figure 1–17.**
- There are different slope angles needed for certain types of soils. **See Figure 1–18.** Refer to applicable regulations regarding the minimum slope, bracing, and piling needed for the various types of soil in your work area. Keep excavated material and work equipment at least 3 ft (1 m) from the trench to prevent overloading and stress cracks. If caught in a cave-in, run or jump up the bank, not down. Do not work within reach of an excavating machine. Leave a trench immediately when a sudden downpour occurs, because water can fill a trench and cause rain-soaked soil to give way.
- Provide a stairway, ladder, or ramp for a hurried escape route at a distance of no more than 25 ft (7.5 m) in any direction. When working in a trench, it is not unusual to be on the receiving end of a stone "shot" from between a truck tire and pavement. Have loose stones swept from your work area.

Figure 1–17 Protective barriers are needed working in a trench to protect workers from a cave-in. *(Delmar, Cengage Learning)*

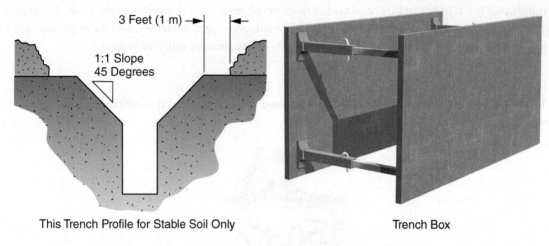

This Trench Profile for Stable Soil Only

Trench Box

Figure 1–18 Slope angles of a trench is an alternative to protective barriers. *(Delmar, Cengage Learning)*

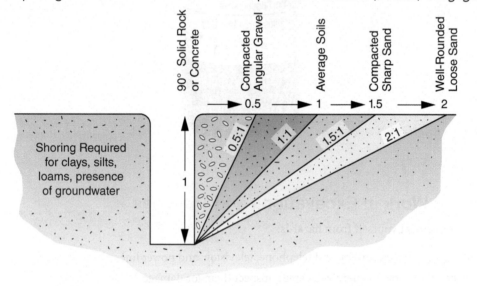

Working in Heat and Cold

Power line technicians work outdoors, in all kinds of weather. A human body is most efficient within a temperature range of $+/-3°F$ ($+/-2°C$) of the body's normal temperature of 98.6°F (37°C). The internal human thermostat works very well but will be stressed or overwhelmed when working outdoors on very hot days or very cold nights.

Working in the Heat

Symptoms of heat stress are headaches, dizziness, lightheadedness, irritability, confusion, upset stomach, fainting, and pale, clammy skin. Heat stress left untreated can lead to a fatal heat stroke. To avoid heat stress, do the following:

- Before you get thirsty, drink a lot of water.
- Wear light, loose-fitting, breathable clothing, such as cotton or FR clothing designed for high temperatures.
- In extreme heat, take short breaks in air-conditioned vehicles or buildings or in the shade.

- Eat smaller meals and avoid caffeine and sugar.
- Flame-retardant clothing may increase the risk of heat stress; therefore, these precautions become more critical.

Working in the Cold

Symptoms of cold stress that could lead to dangerous hypothermia are severe shivering, slurred speech, clumsiness, poor judgment, confusion, apathy, slow pulse or slow breathing, excessive fatigue, drowsiness, reduced sense of touch, less grip strength, and less ability to sense heat, cold, and pain.

To avoid cold stress, do the following:

- Have regular warm-up breaks in a heated truck cab or indoors with an opportunity to remove clothing to prevent sweating. A person who has become damp or sweaty will chill quickly and be susceptible to hypothermia.
- Drink hot beverages, which will provide energy and warmth and will prevent dehydration.
- Use a buddy to check each other's face for frostbite.
- Dress for the cold by wearing layers, including an outer shell to protect from the wind and inner insulated layers. Layers can be added and removed to stay comfortable and to avoid getting damp.
- Wear a good winter hard hat liner covered with a parka hood; half a person's heat loss can occur through the head.

First Aid for Frostbite and Hypothermia

Superficial frostbite will show up as white skin areas, typically on the face. The buddy system is needed because victims will not feel or know that they have frostbite. Frostbite is treated by placing a warm hand over the white spots. Do not rub the skin because it will break the small frozen capillaries and make a face quite ugly.

More severe frostbite can occur, first in the toes, fingers, or face. When an area becomes cold, white, and numb, it must be heated. When heated, first-degree frostbite will cause the area to become red and is not unlike a first-degree burn; second-degree frostbite will form blisters, not unlike a second-degree burn. Keep the body warm and supply hot drinks free of alcohol and caffeine to treat any accompanying hypothermia.

If frostbite shows up as dark skin, or cold and lifeless extremities, keep the body warm to treat hypothermia and seek medical aid. Do not warm the body part yourself. The outer surface will thaw first, and the artery that normally supplies the needed oxygenated blood to keep the thawed area alive will still be frozen.

Protection from Drowning

Reduce the risk of drowning while working on or near water by taking the following precautions:

- Wear a personal flotation device (survival suit). If the work involves very cold water or traveling across ice, this device protects against hypothermia.
- Any boat used for line work should be of a size that meets or exceeds regulations and is equipped with safety gear such as paddles, a fire extinguisher, an anchor, and a first aid kit.
- Tie up when working from boats, barges, or on dams.

Reduce the Risk of Drowning When Traveling on Ice

No ice crossing is without some risk! Ice conditions vary and can change from day to day, hour to hour. Watch for changing water levels on the headwater for a power dam. Avoid slushy ice, ice near rivers or currents, ice that has thawed and refrozen, layered or rotten ice caused by sudden temperature change, ice covered by a heavy blanket of snow, and pressure ridges from wind or current pressure.

Talk to local people with knowledge about ice conditions and routes to take, and test for the thickness of clear ice. **See Figure 1-19.** Wear a survival suit to protect against hypothermia.

When walking on ice, walk tied together in pairs 10 ft (3 m) apart. Carry a long pole and ice picks. In an emergency with poor ice conditions, the lead person should wear a body harness with a rescue rope tied to it.

Figure 1–19 The minimum clear hard ice thickness shown is considered safe when used in conjunction with all other ice travel rules. *(Delmar, Cengage Learning)*

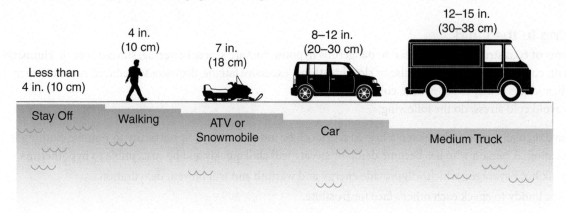

Rule of Thumb: Safe Ice Thickness in Inches = Gross Load in Tons + 6

Snowmobile or all-terrain-vehicle (ATV) travel on ice should be done with at least two machines spaced 100 ft (30 m) apart. Ice travel with large trucks or large off-road equipment requires preparation and testing to reduce risk.

- Prepare an ice road by scraping away snow and flooding it to build up the ice thickness.
- The rule of thumb "Safe ice conditions in inches = Gross load in tons + 6 in." (15 cm) applies to clear ice. A crack in the ice reduces the load-bearing capacity by 50%; intersecting cracks reduce the load-bearing capacity by another 25%.
- Travel slowly. The wave created under the ice by a moving vehicle can break the ice. The heavier the vehicle, the slower it should travel.

Working around Helicopters

Helicopters are used to carry out line patrol, string conductor (pilot line), set line workers on a structure, hover to support a line worker carrying out barehand work on a conductor, set poles, hang crossarms, build towers, and carry loads into difficult terrain. The long line method carries power line techs from the ground to the working position on a transmission structure.

If flying as an observer, a power line technician should help identify flying hazards in a wire environment, such as a crossing circuit, nearby shield wire, and lateral taps. A warning sign placed on the top of a structure one span before the known hazard would alert a distracted pilot and observer.

Before a job involving a helicopter begins, a briefing supported by a written job plan is essential. The ground crew, signal person, flying power line techs, and pilot need to know the specific hazards and safety controls to the job, work methods, signals, and communication methods to be used. It could be argued that flying is a hazard, but flying is seen as risk that is as low as reasonably achievable. The hazard of securing material and tools on the external platform of the helicopter cannot be overlooked. Three power line techs and a pilot lost their lives when an unsecured empty canvas tool bag blew off the platform into the tail rotor. **See Figure 1–20.**

Pilot Training

Ideally the pilot has had training flying in an electrical utility wire environment. When flying by a structure, it is easier to see the position of each wire in a span. When mid-span it can be difficult to see some wires, especially a shield wire at the top of a structure. It can also be difficult to judge the distance to a wire when there is only open sky behind it.

Personal Protective Equipment Specific for Helicopter Work

- Wear eye protection to prevent material caught in the downwash from contacting the eyes.
- Wear a hard hat, secured by a chin strap or other device.

Figure 1–20 Hand Signals Used for Helicopter Work. *(Delmar, Cengage Learning)*

Move Right

Left arm extended horizontally; right arm sweeps upward to position over head.

Hold Hover

The signal "Hold" is executed by placing arms over head with clenched fists.

Move Left

Right arm extended horizontally; left arm sweeps upward to position over head.

Tackoff

Right hand behind back; left hand pointing up.

Move Forward

Combination of arm and hand movement in a collecting motion pulling toward body.

Land

Arms crossed in front of body and pointing downward.

Move Rearward

Hands above arm, palms out using a noticeable shoving motion.

Move Upward

Arms extended palms up; arms sweeping up.

Release Sling Load

Left arm held down away from body. Right arm cuts across left arm in a clashing movement from above.

Move Downward

Arms extended, palms down; arms sweeping down.

- Wear tight-fitting clothing to prevent clothes flapping in the downwash from being caught in a hoist line or from interfering with other work.
- Wear hearing protection.

Approaching a Helicopter

- When approaching a helicopter, stay in a position visible to the pilot or be in radio communication with the pilot.
- Keep away from the tail rotor. **See Figure 1–21**.
- Stay in a crouched position when in range of the main rotor.

Figure 1–21 The proper approach to a helicopter is from a position where a person can be seen by the pilot and away from the hazard of the tail rotor. *(Delmar, Cengage Learning)*

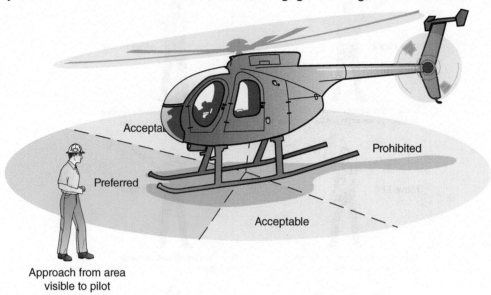

- Carry all tools, slings and equipment below waist level.
- Good housekeeping is essential, secure or move loose gear outside of the downwash area, at least 100 ft (30 m) from the helicopter to reduce the likelihood of flying materials.
- Secure any load placed on the outside platform on either side of a helicopter. A loose tool bag drawn into the tail rotor of a helicopter while coming into land resulted in the death of all four occupants.

Attaching or Receiving a Load on the Cargo Hook

- The pilot/operator must be told the weight of the load to be lifted. Ensure the load cannot snag onto an object on the ground or be frozen to the ground.
- Only trained workers may hook or unhook loads under a helicopter. All other workers on the ground must stay at least 50 ft (15 m) from the helicopter. Do not work under a hovering helicopter, unless necessary to hook or unhook loads. Prepare a preplanned route to and from the helicopter.
- Only a designated signal person trained in radio and/or hand signals may communicate with the pilot when lifting or lowering a load.
- Cargo hooks are tested daily to ensure that pilot can release the load when it is delivered or in a flight emergency. The cargo hook release is designed to reduce the risk of the pilot releasing the load inadvertently. The cargo hooks can also release the load with a mechanical control in an emergency.
- Drain the static charge that builds up on a helicopter before touching any suspended load that has a direct metallic connection to the helicopter.
- A spinning load can unravel a hand splice or an eye finished with cable clamps. Wire and fiber ropes used for lifting must have a pressed, swaged or plug lock eye, preferably with a swivel hook.
- The length of the tag lines attached to the load must be short enough to prevent them from being drawn up into rotors.

Fueling a Helicopter

If involved in refueling a helicopter in the field, ensure that all metal parts are bonded together to prevent a static discharge (spark). Follow the sequence described in **Figure 1–22**.

Figure 1–22 A proper grounding/bonding sequence will prevent a static discharge when fueling a helicopter. *(Delmar, Cengage Learning)*

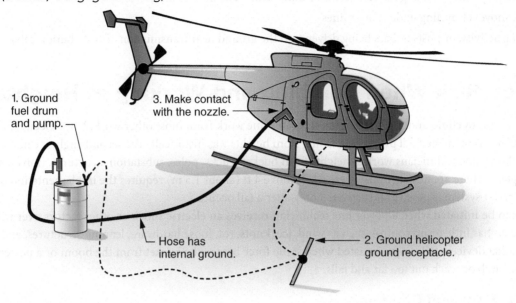

1. Ground fuel drum and pump.

3. Make contact with the nozzle.

Hose has internal ground.

2. Ground helicopter ground receptacle.

Working Near Gas Pipelines and Communications Cables

While underground electric utilities are hazardous when contact is made with an auger or backhoe, hitting a gas line can be very explosive, and hitting communications cables, sewers, water lines, or underground tanks can be very disruptive and expensive. Dial 811 from anywhere in the United States to be connected to your local One Call Center and arrangements will be made to mark any underground hazards.

Below-ground utilities are usually located when planning a job, but contact with an underground pipe or cable utilities is not unusual. When in doubt, dig by hand or use hydro-vac truck, which pneumatically sucks up solids, sludges, and slurries.

Identify and Protect from Biological Hazards

Almost every location has biological hazards that must be controlled. Most people are familiar with biological hazards in their own locations, but biological hazards may be unique to other regions of the country. Although contact with these hazards is quite rare, it is prudent to inquire.

Poisonous snakes, killer bees, swarming wasps, ticks that cause Lyme disease, mouse/rat droppings that cause the hantavirus, poison ivy and poison oak, mosquitoes that carry West Nile virus, and various diseases from bird/bat droppings may be present. These hazards cause symptoms that can vary from mild irritation to death.

Reduce the Risk of Public Electrical Contact

Overhead and underground power lines are constructed to be out of reach of the general public. While utility work is carried out, the public is further protected by barricades around stringing equipment, covers over dug holes, traffic and pedestrian controls, grounded conductors that are tied down temporarily, and covers placed over hot meter bases.

Power line technicians are the most qualified people to identify potential public electrical contact hazards with utility circuits and should report any of the following:

- Flagpoles, antennas, and ladders within striking distance of an electrical circuit
- Cranes, boom trucks, and ladders within 10 ft (3 m) of an energized distribution circuit
- Construction and buildings being erected within established safe limits
- Trees that people may climb near power lines, such as those in suburban streets, school yards, and parks

- Farm equipment, such as grain augers and high loads, being moved under lines
- Irrigation systems with large solid streams of water near transmission lines
- House movers traveling under power lines
- Digging activity, or posts or bars being driven into the ground near transmission or distribution cables

Reduce Risk When Climbing and Working at Heights

The skills needed to climb and work at heights separates line work from most other work. Until Occupational Safety & Health regulations mandated fall protection, falling from heights was historically the second-highest cause of fatalities in line work. Power line technicians work at heights from bucket trucks, poles, substation structures, transmission towers, and helicopters. There are jurisdictions where work above 4 ft (about 1.5 m) requires the implementation of fall protection. A fall arrest system is passive and activated only after a fall occurs.

A fall can be initiated when a power line technician receives an electric shock or other factors over which a power line technician has little or no control, such as wind, ice, knots, rot, loose hardware, leaning structures, and rigging failure. In an aerial device a fall can be initiated when some force breaks the bucket from the boom or a power line technician tries to climb or reach out too far and falls.

Falling Hazards and Controls

Hazard	Control	Details
Falling when climbing a pole.	Wear an approved fall arrest system. Receive training in the fall arrest system chosen.	Fall arrest equipment is available for climbing all pole types, including concrete, steel, and wood.
Falling when climbing a transmission line structure.	Wear an approved fall arrest system. Receive training in the fall arrest system chosen.	There are step bolts and ladders, especially on transmission line poles, and detachable steps. There are nut fasteners embedded into steel poles that allow step bolts to be screwed in as needed, as well as various designs of climbers.
Falling while being held in place by a work positioning system while both hands are free working.	When in a working position and supported only by a body belt and pole strap/safety rope the pole strap safety belt position must not allow more than a 2 ft (0.6 m) drop. A self-retracting lifeline (SRL) is considered a device that limits a fall to 2 ft (0.6 m).	This control is for all work aloft, pole, lattice steel tower, or bucket. To have fall protection while changing position on a structure, a fall arrest system is used along with a work positioning system.
Falling from an aerial device bucket.	When working from the bucket of an aerial device, a fall arrest system must be worn. An approved fall arrest consists of a full-body harness with a lanyard in series with a shock absorber attached to a D-ring on the back of the harness. The other end of the lanyard is attached to a specifically designed anchor point on the boom.	Falling from a bucket can occur by overreaching or being ejected from the bucket by a leveling cable failure; the bucket truck being struck by another vehicle; the bucket catching on a structure or hardware and suddenly releasing; an object, such as a pole, crossarm, or tree, striking the bucket; and lifting by using the bucket instead of a jib.
The fall arrest or work positioning system fails when the anchor to which it is attached fails.	Choose a designed anchor point or an anchor point that you are confident that it will hold in case of a fall.	Occupational Safety & Health regulations can be very specific about the strength of an anchor. One example: A work positioning anchor must be able to support at least twice the potential impact load of a fall or 3,000 lbf (about 1,400 kgf). The nature of line work requires anchor point choices to be made spur of the moment without benefit of a study. Shock absorbers ease stress on an anchor. Do not anchor to any rigging. One purpose of fall arrest is to protect a worker from a rigging failure.

Injuries falling more than 2 ft (0.6 m) into a waist belt.	Wear a full body harness when a potential fall exceeds 2 ft (0.6 m). A full-body harness distributes falling forces throughout the torso.	Safety belts that are worn around the waist are acceptable for a travel restraint system or to stop a fall less than 2 ft (0.6 m) but not as part of a fall arrest system. When wearing a waist belt properly, with the D-ring in the middle of the back, a fall greater than 2 ft (0.6 m) will cause internal injuries to the abdominal area.
Falling while working from a suspension system.	To have complete fall protection, a fall arrest system or a secondary backup must be used.	Work from a suspended system includes tree work using a rope and saddle, barehand work while suspended by a link stick/live-line rope, work on transmission line suspension insulator from a saddle and rope or from a rope ladder.

Specific Hazards Initiating a Fall from a Pole

Hazards	Controls
A hand line securely fastened to the body belt is snagged by a passing vehicle or by a work vehicle on the job site.	Carry the hand line on a breakaway hook or similar device. Although the probability of a vehicle catching on a hand line may seem rare, it has caused at least one fatality while a power line technician was belted into a tower.
An energized conductor is in contact with the pole.	Eliminate possible "pole shock" hazard. Do not climb until the conductor is lifted clear from the pole or the circuit is isolated and grounded.
The pole to be climbed has workers at the bottom of the pole driving ground rods or using a tamper.	Eliminate a possible serious consequence of falling. Do not climb a pole while this work is in progress.
One climber is interfering with another climber.	Wait until the other climber is belted into position before starting to climb.
The pole has knots, weather cracks, rot, or loose hardware.	These climbing hazards can be controlled best by using a fall arrest system.
A power line technician is climbing while not looking up, which increases the hazard of making contact with an energized conductor.	Confidence in climbing ability allows a line hand to look up while climbing. Use a dedicated observer for novice climbers.
A pole strap can slip over the top of a pole. Exposure to this hazard is increased when climbing a new pole with no hardware on it.	Install at least one bolt near the pole top when setting a pole. This can act as both a physical and a visual barrier. Always belt in below a physical barrier when working near the top of a pole.
The integrity of the hardware supporting an energized conductor is suspect and could cause an energized phase to fall.	Recognize known hazards such as aluminum-capped dead-end insulators, wood pins, potentially weak conductors such as #4 ACSR or #6 copper, damaged guy wires, and porcelain insulation on switch gear.
Weather-related conditions, such as high wind or ice-coated poles, increase the risk of climbing a pole.	Fall arrest systems are the most effective barriers for these climbing hazards. **See Figure 1–23.**

Specific Hazards Initiating a Fall with a Pole

Poles have an average service life of 30 to 40 years. One safety code requires pole replacement when one third of the load-carrying capacity of the pole is lost. Unless there is confidence in a pole testing instrument, use the following guidelines before climbing a pole. A hammer test, or bore test may not be an adequate control when the consequence of being wrong can be fatal.

Figure 1–23 Two types of wood pole fall restraint belts.

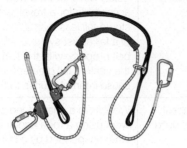 

If	Then
The pole is older than 20 years or is obviously unsafe. The pole species has a history of early failure. There are burn marks, woodpecker holes, a large knot, and/or several smaller knots at the same height on the pole. The anchor rods and down guys have lost their galvanized coating or are corroded (e.g., near pipeline). The pole is beside a trench and/or is in soft, wet, or loose soil. There are indications of a shallow setting (e.g., a former ground line above the existing ground level may be seen). The strain will be altered while working aloft. The pole will be stripped of all conductors and guys. (The pole holds up wires, but the wires also hold up the pole.) The pole leans more than 5°.	Support the pole with a line truck boom, ropes or guys, or by lashing a new pole alongside it.

Falling/Dropping Hazards Specific to Transmission Structures

Working on steel transmission line structures creates hazards distinct from working on wood poles. Take action to ensure that you are protected from these hazards. **See Figure 1–24.**

Hazards	Controls
Tower climbing hazards include slippery soles, slippery steel, long reaches to the next handhold, frost, ice, wind, or a missing step bolt. A fall from a tower is usually fatal.	Fatal falls have happened to the best climber on a crew. Fall protection is the only true control available to prevent a fall while climbing. Climb up and down towers with three points of contact on the tower at all times. Do not hold onto the step bolts, but slide the hands on either side of the main member at the corner so that you always have hands on the steel.
Objects falling from a tower and striking a worker below can be lethal.	Ensure that no one is under a tower while work is in progress. Workers involved in running a hand line or tag line should work from a position that is clear of any potential falling material.
Material being sent up strikes the tower or spins uncontrolled on the load line.	Tag lines should be used to maintain control of material being raised or positioned. During tower erection, the load line must not be detached from a member or section until the load is secured.
High winds, snowstorms, and ice- or frost-covered steel increase the risk of falling from a tower.	Unless specific training or controls have been put in place for the hazard encountered, work aloft should not be started during these conditions.

Tripping or loss of balance occurs while carrying tools or material across a tower bridge.	Fall protection is the only true control to arrest a fall.
The jolt from a static electric shock while climbing from an insulated ladder onto steel or when contacting a ground wire while climbing a wood structure can result in inadvertent movements.	Being prepared and making a fast slap with the hand to bond onto the grounded object will reduce the shock and the risk of an inadvertent movement. Fall protection is the only true control to arrest a fall.
Climbing or crawling over an insulator string is using an unstable platform for climbing, as well as a potential source for a serious gash from a broken insulator.	Use a hook ladder, platform, or scaffold to work at or beyond an insulator string. Fall protection is the only true control to arrest a fall.

Figure 1–24 A fall arrest system is required when climbing a transmission line structure. *(iStock.com/Itobi)*

Fall Protection Working on a Hook Ladder

To work on an isolated and grounded conductor of a transmission line, hook ladders and in many cases rope ladders are hung on the tower above the conductor and then used to climb down to the conductor level. To work on an isolated and grounded dead-ended conductor of a transmission line, hook ladders or suspension platforms are used to climb from the tower to the conductor. **See Figure 1–25.**

Insulated hook ladders are used for energized-line tool work on transmission lines. The ladder is hung at a proper distance from the energized conductor, and the live-line tool work is carried out by a worker belted into the ladder. Hook ladders are also used to carry out energized-line barehand work on both tangent and dead-end structures.

Figure 1–25 A hook ladder is being used as a horizontal platform to provide access to the dead-end hardware on a tower. *(Courtesy of Robert Jackson)*

Safety Controls for Working from a Hook Ladder

For work on an isolated and grounded conductor, a full-body harness is worn with a retractable lanyard tied directly to the structure. The positioning device should be woven through the rungs so that the body belt will provide a secure work positioning system. When two persons are working from the same ladder, only one person at a time should be climbing up or down the ladder.

When working on the insulated ladder near an energized circuit, an electrostatic charge will build up on a person's body. When transferring from the steel to the ladder, an electric static shock will be felt as the last hand breaks contact with the grounded steel. Similarly, when leaving the ladder, the first hand to make contact with the grounded steel will draw an electric arc. To avoid the static shock, make the first and last contacts using a bare steel wrench or other tool in your hand or slap the steel or remove the hand very quickly.

Working Aloft in a Bucket

When working from the bucket of an aerial device, a fall arrest system must be worn. An approved fall arrest consists of a full-body harness with a lanyard in series with a shock absorber attached to a D-ring on the back of the harness. The other end of the lanyard is attached to a specifically designed anchor point on the boom.

Falling from a bucket truck has not been an uncommon occurrence. Causes include over-reaching or being ejected from the bucket by a leveling cable failure; the bucket truck being struck by another vehicle; the bucket catching on a structure or hardware and suddenly releasing; an object, such as a pole, crossarm, or tree, striking the bucket; and lifting by using the bucket instead of a jib.

Specific Hazards When Working from a Ladder

Fixed Ladders

Fixed ladders are found on transmission line, microwave, and cell towers. Fixed ladders are also found in the holes to access a maintenance vault. A cage around the ladder is not an adequate fall arrest. Unless the tower has a ladder safety system such as a fixed rail system, a personal fall arrest system using a full body harness and equipment similar to that used when climbing a lattice steel tower is used.

Extension Ladders

An extension ladder does not have fall protection systems that apply very well to power line work. Climbing and working from a ladder requires concentrating on fall prevention instead of fall protection.

Hazards	Controls
Electrical contact	All ladders used by utility workers should be nonconductive, preferably fiberglass.
	A nonconductive ladder reduces the risk of injury in case of contact with an energized conductor and can provide an insulated platform when working on secondary voltage.
Falling from a ladder	Face the ladder while climbing.
	Maintain at least three points of contact—for example, two feet and one hand or, preferably, belt onto the ladder with a positioning device.
	Avoid overreaching. Keep hips within the ladder uprights.
	Never stand above the fourth rung from the top.
	Climb the ladder with both hands free. Use a tool belt or pull up material after climbing.
	Belt into the ladder with the pole strap/work-positioning strap around both rails and one wrapped around a rung.
Falling with a ladder	Stabilize the bottom of the ladder. On grass or gravel, use the ladder feet in the spike position. On hard, stable surfaces, use the feet in the flat-foot position.
	Tie the top of the ladder when practical, especially if it is on an aerial cable. Throw a rope over the aerial cable and use it to help set up and tie the ladder into position.
	The minimum length of overlap between two sections of an extension ladder should be 3 ft (1 m) for ladders up to 36 ft (11 m) high and 4 ft (1.2 m) for ladders over 36 ft (11 m) high.
	Set the ladder at a proper angle; about 75°. The ladder angle is correct when a person standing at the foot of the ladder can reach the ladder with outstretched arms. **See Figure 1–26**. Another rule of thumb is to set the horizontal distance of the foot of the ladder out one-fourth of the working length of the ladder.
Transferring from the ladder to a roof and vice versa	Extend the ladder above the roof by at least three rungs to provide a handhold.
	Transfer to the roof from the side of the ladder, not over the top rung.

Figure 1–26 Set the ladder at a proper angle for maximum stability. *(Delmar, Cengage Learning)*

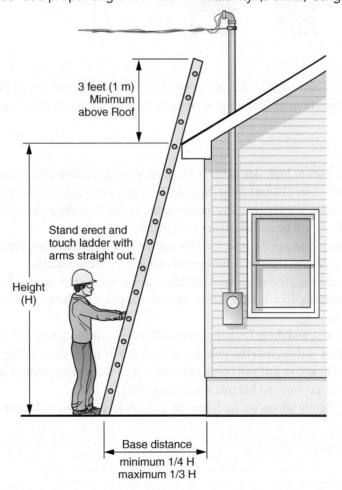

3 feet (1 m)
Minimum
above Roof

Stand erect and
touch ladder with
arms straight out.

Height
(H)

Base distance
minimum 1/4 H
maximum 1/3 H

Fall Protection Working on Substation Transformers

The top of a substation transformer is at a height requiring fall protection. Working on the top is like working on a platform where typical fall protection would call for a railing. A more popular acceptable fall protection system is a fall restraint system. Workers are tied to a post (temporary or welded) with a length of lanyard preventing encroachment to the edge of the transformer.

Initial Training and Fitting for Climbing and Working on a Wood Pole

This section is critical for people still free climbing a wood pole. It concentrates on North American–style climbers. There is a style more popular in Europe that looks like a cant hook coming out from the front of the foot. **See Figure 1–27.** Cutting out would be rare, but these are not as maneuverable as the climbers worn in North America.

Figure 1–27 European-style pole climbers eliminate the problem of a gaff cutting out of the pole. *(Delmar, Cengage Learning)*

Climbing a pole safely is an essential skill for power line technicians and the discussion that follows must be accompanied by field demonstrations and training. Wear an approved fall arrest system. Receive training in the fall arrest system chosen.

Climbing Techniques

When possible, climb up the back or high side of the pole. Climb with short steps approximately 8 to 10 in. (20 to 25 cm) apart, keeping the knees about 8 in. (20 cm) from the pole. Long steps will automatically bring the knees into the pole, increasing the risk of the climber cutting out. The hands on the backside of the pole should be at shoulder level. If the hands are high on the pole there will be undue strain on the arms.

Aim the gaffs at the heart (center) of the pole, with the toes pointed outward. Raise the left foot and left arm together, and then the right foot and right arm together, a movement that is opposite to walking.

Look up to see where you are going and to avoid climbing hazards on the pole such as knots, bolts, cracks, and climbing too high into electrical hazards. If your knees are away from the pole and the climbers are properly sharpened, there is no need to look at your feet.

When descending, lock, aim, and drop. Lock the knee so that the leg is straight, aim the gaff to the center of the pole, and then drop your full weight and cut into the pole. Long steps can be taken descending, and the steps will get longer with increased confidence. The upper gaff will break out easily when taking longer steps and rolling the knee outward. Look at your feet to avoid obstructions and hazards when descending.

Sneaking up the pole by lightly setting the gaffs into the pole looks good but will cause an increased risk of the gaffs cutting out.

Sizing and Maintaining Climbers

The top of the climber shank must be in the hollow between the knee and the calf, as high as possible without it touching the bottom of the kneecap. If it is too short, the shank will dig into the calf of the leg, and if it is too long, it will touch the bottom of the kneecap and will be cripplingly painful. Instructions will say to measure the leg from the instep to 0.5 in. (13 mm) below the kneecap for the shank length, but the length is critical for comfort. Some trial and error is best, if possible. Adjustable climbers allow trial and error for determining the final comfortable length, and then you can install rivets to eliminate the rattle and movement common with the pins or bolts.

Climber guards should be used when walking with climbers, especially through ditches and when storing them in a truck. A puncture wound from a climber gaff is very painful.

The gaffs need to be kept sharp while maintaining their proper shape. Sharpening is a precise task that initially requires an instructor, a gaff gauge, and the manufacturer's instructions. There seem to be two differing opinions about sharpening a gaff. One suggests filing the gaff on the two outer surfaces to make a point while never sharpening a gaff on the underside because it changes the angle of the gaff. The other is to file the underside only, filing it flat and straight except for the end tip. **See Figure 1–28.**

Figure 1–28 The gaff must be sharpened to its proper shape to reduce the risk of cutouts. *(Delmar, Cengage Learning)*

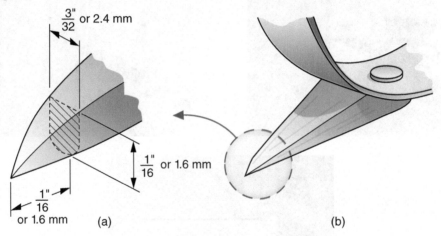

$\frac{3"}{32}$ or 2.4 mm

$\frac{1"}{16}$ or 1.6 mm

$\frac{1"}{16}$ or 1.6 mm

(a)

(b)

The gaff length is 1⅝ in. (4 cm) for new pole climbing climbers and 2¾ in. (7 cm) for new tree climbing climbers. A gaff is measured on the underside and should not be less than 1¼ in. (3 cm) for pole climbing climbers or less than 2⅜ in. (6 cm) for tree climbers. Some utilities do not allow replaceable gaffs. The instructions that come with a gaff gauge must be consulted because the procedure is not self-explanatory. **See Figure 1–29.**

Figure 1–29 Use a gaff gauge to ensure proper sharpening. *(© Buckingham Manufacturing Co Inc)*

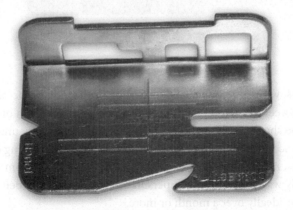

In addition to the use of a gauge, a pole cutout test will show whether the gaffs are sharpened properly. Fasten the foot strap of the climber without fastening the leg strap. **See Figure 1–30.** While standing on the ground and supporting the top of the climber with one hand, insert the gaff lightly into the pole (about 0.25 in., or 0.5 cm) at the normal climbing angle. Maintain enough pressure to keep the gaff in the pole; and with your hand, push the top of the climber against the pole and then push down. A gaff that is sharpened properly will cut into the pole and hold at a distance of 2 in. (5 cm) or less. A gaff that is shaped properly but is dull will cut into the pole and hold, but usually at a distance greater than 2 in. (5 cm). A gaff that is very dull or not shaped properly will cut out or plow through the wood for a distance greater than 2 in. (5 cm). **See Figure 1–30.**

Figure 1–30 A cutout test is another method to ensure that the gaffs are sharpened correctly. *(© Buckingham Manufacturing Co Inc)*

Sizing the Body Belt

Body belts come with gut straps, with suspenders, and as part of a full-body harness, with floating D-rings. Body belt size is based on the distance between the two D-rings around a person's back from one hip bone to the other hip bone. **See Figure 1–31.** If the belt it too short, it will pinch the hip bones and be instantly uncomfortable. If the belt is too long, it will be like a choker and feel uncomfortable after a few minutes or longer, depending on the length. The D-rings should be just forward of the hip bones, not on them. A rule of thumb says, "Measure from seam of pant, around buttocks to other seam, and add 2. The only way to ensure that a person has the proper-size belt is to try some different sizes and various work positions on a pole, ideally over a month or more.

Figure 1–31 The measurement of a body belt must be very precise to ensure comfort when working aloft. *(Delmar, Cengage Learning)*

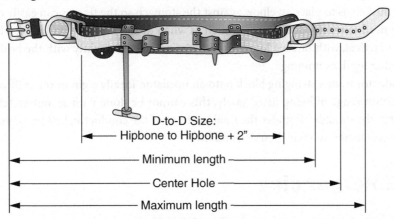

D-to-D Size:
Hipbone to Hipbone + 2"

Minimum length

Center Hole

Maximum length

The Pole Strap

Pole straps (safety straps, positioning straps) are generally made with nylon impregnated with neoprene to provide good wear resistance. Pole straps have an internal layer with a contrasting color that shows through when the belt is considered worn and needs replacement. The colored layer may become discolored because of abrasion and, therefore, requires regular and thorough visual inspection for wear. Pole straps wear out much quicker when used on lattice tower steel. A pole strap is not used with certain pole-climbing fall arrest systems. **See Figure 1–32.** It is used along with the fall arrest systems on transmission line towers. Some pole straps have slide buckles that allow for easier length adjustments.

Figure 1–32 The adjustable pole strap is a work positioning device, not a fall arrest device. *(Delmar, Cengage Learning)*

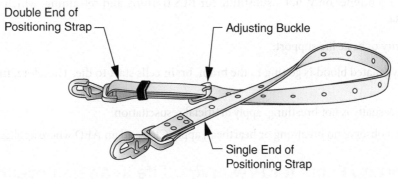

Double End of
Positioning Strap

Adjusting Buckle

Single End of
Positioning Strap

The pole strap snap hook openings should face away from the body to prevent accidental opening. This was much more important before the introduction of locking snap hooks. **See Figure 1–33.**

Figure 1–33 A locking snap hook reduces the risk of a snap roll out. *(Delmar, Cengage Learning)*

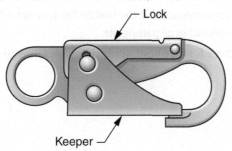

Lock

Keeper

Work Positioning

The positioning device is adjusted so that a person cannot fall more than 2 ft (0.6 m). Work close to the pole—a rule of thumb, while seated in the belt, is to place an elbow against the stomach so the fingers can easily contact the pole.

Pole workers often need to reach out a long way, such as when dead-ending a conductor. The positioning device is lengthened, as long as practical, with one leg low, locked, twisting around slightly with the body belt, one can face the work to be done. The other leg does nothing.

When lifting a conductor from a stringing block onto an insulator, ideally a gin or other lifting device is used. However, if the weight is within range of being lifted safely, this cannot be done with an outstretched arm. Lengthen the positioning device so that the shoulder is under the final position of the conductor, belt in as high as possible above the conductor, then put the conductor on your shoulder, and lift.

Managing Emergencies

Managing and Communicating Emergencies

A utility/employer shall have in place plans, procedures, and training for such emergencies as an electrical system failure, fire, injuries on the job, and the need for rescue from aloft or from a confined space.

Communication can make the difference between life and death during an emergency. Emergency code words should be designated that will open up a radio and allow a control center the authority to isolate a circuit without question.

Basic Life Support Summary

Basic life support (BLS) is the application of emergency measures such as first aid, cardiopulmonary resuscitation (CPR), and an automated external defibrillator (AED) to preserve life and stabilize the person from the effects of an injury or harm until professional help arrives.

This summary is a reminder only, not a substitute for BLS training and retraining, which are regulatory requirements. **See Figure 1–34.**

Apply the "ABC" priority of basic life support:

Airway—If no oxygenated blood is getting to the brain, brain cells start to die. Therefore, first check the airway for breathing.

Breathing—If the casualty is not breathing, apply artificial resuscitation.

Circulation—If you observe no breathing or heartbeat, apply CPR and an AED where applicable.

Critical Supporting Facts for Preserving Life in an Emergency

1. Training in the use of an AED will include the proper location for the electrode pads. **See Figure 1–35.** Automatic units will apply a shock without the rescuer's command. Semiautomatic units will tell the rescuer that a shock is needed, and the rescuer must press a button.

2. To reduce the risk of traumatic shock, keep the casualty comfortable and provide reassurance.

3. When providing first aid, avoid direct contact with another person's blood or other bodily fluids. Practice universal precautions, which means treat every exposure as if the bodily fluids contain hepatitis B virus (HBV), human immunodeficiency virus (HIV), or other bloodborne pathogens.

4. If the casualty is breathing, then check for bleeding. Apply direct pressure on any bleeding wounds to prevent excessive blood loss.

Figure 1–34 A current CPR flow chart. *(Courtesy of Coyne First Aid)*

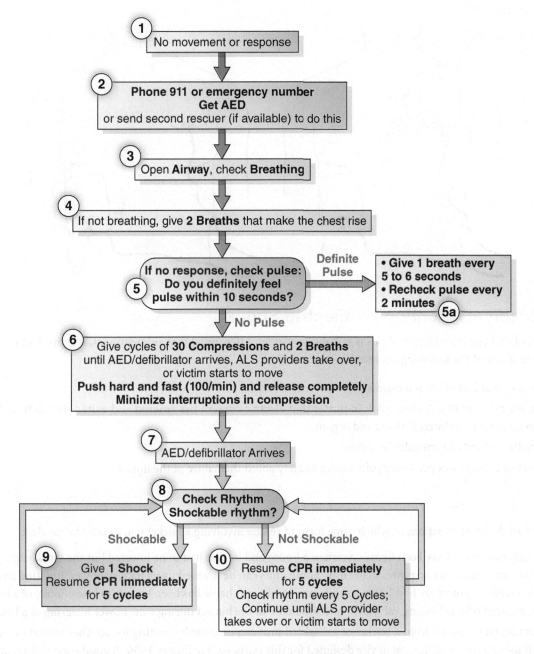

5. If the casualty is breathing and not bleeding, then check for fractures. Apply the first aid approach "do no harm" when dealing with potential fractures. Unless the broken bone has ruptured a major artery, first aid for broken bones does not require speed. If the accident scene suggests a possible back injury, do not move the casualty. If an injured person must be moved before the arrival of professional help, avoid twisting or bending of the body.

6. Treat a casualty that has fallen into the body harness or has been suspended in a body harness without moving for more than 5 minutes for "suspension trauma" (harness hang syndrome). There is a dangerous reduction of blood flow to the brain in someone left suspended. Place the casualty in a sitting position (never lying down). Anyone suspended for more than 10 minutes should be sent for medical aid even if there are no obvious injuries.

Figure 1–35 Proper placement of the pads (electrodes) of an AED to stimulate the heart. *(Courtesy of Coyne First Aid)*

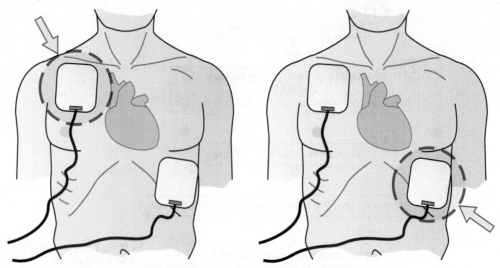

Seek Medical Attention for Electric Shock

The extent of damage from electric shock is not always visible. After initial first aid treatment, it is prudent to seek medical attention if any of the following occurred:

1. There is or was loss of consciousness.
2. There are burn marks. A burn mark indicates that electric current has entered and exited the body and has done unknown damage to internal tissue and organs.
3. The victim exhibits an irregular heartbeat.
4. The victim experiences persistent pain and/or anxiety about the nature of the injury.

Bucket Rescue

There are four different situations in which some form of rescue involving a bucket truck could be needed:

1. Although rare, there have been occasions when a boom and bucket had to be lowered but the truck engine had broken down and there was no power to the unit. Some units can be lowered using battery power. Rotating and lowering a bucket with no power are best left to a mechanic. Even a mechanic has been known to lose control of a boom while trying to bleed off the holding valves. No one should be in the bucket during a no-power lowering of a bucket.
2. Evacuating (self-rescue) from a bucket of a disabled unit can be done by waiting for another bucket or by lowering oneself using a rope and descent device designed for this purpose. **See Figure 1–36.** A good rope and a taut line hitch will also do the job, but may be slower to get prepared. Bucket escape should be practiced from a safe height or with a fall arrest backup on a scheduled basis.
3. If the person aloft becomes incapacitated, the lower controls can be used to bring down the bucket. This should be practiced so that it becomes second nature to lower the bucket to a flat surface or another bucket rest where the casualty can be pulled out.
4. History has shown that when a victim slumps in a bucket, one strong person is often not enough to pull the victim out. Various devices are used to pull a victim from a bucket. **See Figure 1–37.** Using the device chosen by the utility/employer requires practice. Some buckets tilt, however, so it is critical to be able to lower it to a position where the tilt function will work. Pulling people out of a bucket using the device chosen by the utility/employer needs a scheduled practice. Even for buckets that allow tilting to remove a casualty needs practice, especially to learn the quickest way to get a bucket to ground level. Devices to remove a casualty from a bucket include a bracket that can be mounted on the lip of the bucket, which has a small hand-cranked winch to pull out the casualty. Rope blocks (done up in a bag) can be anchored on top of the upper boom where the blocks would be above the casualty, when the lower boom is up and the upper boom slants down.

Figure 1–36 A worker in a bucket must have access to a self-rescue method to escape from a disabled bucket. *(Delmar, Cengage Learning)*

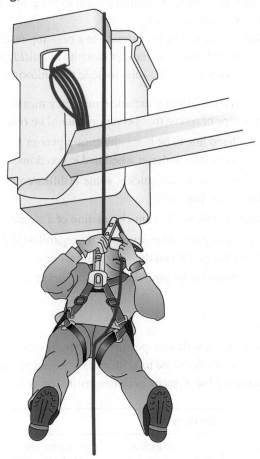

Figure 1–37 One method to rescue someone from a bucket is to hoist a person out of the bucket using rope blocks mounted on the boom above. *(Delmar, Cengage Learning)*

Rescue from Pole Top, Tower Top, or Substation Structure

The primary goal for a rescue from aloft is to lower an injured person to the ground as quickly as is safe to do so and to start first aid or CPR while waiting for medical assistance.

A tower top or substation structure rescue will be very similar to a pole-top rescue but with different rigging to lower the victim. In all cases, the rescue procedure involves lowering a casualty as quickly as possible.

Some lessons learned that have evolved into specific rules include the following:

- Some utilities/employers insist that a job briefing include emergency measures, including having a copy of emergency telephone numbers and the type of rescue that personnel should be ready to perform.
- Some utilities/employers require a hand line to be hung any-time a person is working aloft. The rope and hardware must have a 10-to-1 working load factor and undergo scheduled inspections.
- To remove a casualty from an electrical contact quickly, some utilities/employers require a live-line cutter to be readily accessible on the pole during any hot work.
- Code words are set up with system control to allow quick isolation of a circuit.
- Some utilities/employers require rescue packs on every aerial lift regardless if the baskets tilt because they assume it may not always be possible to lower the baskets all the way to the ground.
- Some utilities/employers place a priority on keeping the deck of a bucket truck free of hardware, especially near the lower hydraulic controls.

Pole-Top Rescue Procedure

The pole-top rescue procedure shown is a generic one-person climbing rescue. If an aerial basket or more people are available, the steps needed for the rescue are the same in principle but easier to manage. The number of steps and the details may appear lengthy in an emergency, but a rescue becomes instinctive with regularly scheduled practice sessions.

Step	Action	Details
1.	Call loudly to the injured employee.	At the first indication of a problem, call "Are you okay?" loudly to the worker aloft. If the casualty responds but seems stunned or dazed, a rescue is probably needed, but the timing will not be as critical because the victim is breathing. Conscious victims of an electrical contact will often say they can descend on their own. Try to convince them to at least wait for a pole-top rescue rope arrangement as a backup for fall protection. If there is no response, timing is critical.
2.	Call for help.	At the first indication of the need to perform a rescue, call for help using a prearranged call or code words on the company radio or get help from nearby observers. The utility/employer should have prearranged code words that immediately take priority over all other radio traffic.
3.	Evaluate the situation.	If the casualty is still in contact, call for system control or dispatch to drop the circuit, preferably using prearranged code words to avoid a lot of discussion. If that fails, the situation is the casualty is energized and the pole may be energized. There is a risk to any rescuer who climbs the pole. Unfortunately, it may be obvious from the ground that it is too late for a rescue. By climbing with rubber gloves on and taking short steps, a rescuer may be able to get high enough to push the casualty with a hot-line tool, cut the conductor with hot-line cutters, install a ground or toss a rope around the casualty, and pull clear. *Note:* This procedure assumes a worst-case scenario of a one-person climbing rescue. Removing a casualty from a hot line using a bucket truck would certainly be much safer for a rescuer. If the casualty is in a precarious position, well within the minimum approach limit, decisions about the need for rubber gloves, a shotgun stick, an energized-line cutter, and/or grounds will be needed to clear the victim. Sometimes sliding an existing hose to cover the contact spot will work. If the casualty is clearly not in contact, proceed with the rescue.

4.	Plan the rescue.	Prepare to use the rigging that the utility/employer has specified and used during training and practice.
		If a hand line is to be used, either get it from the truck or use the one already up the pole. (A utility/employer who specifies a hand line would ensure that the hand line design and hardware are suitable and inspected on a regularly scheduled basis.)
		If a special rescue rope is to be used, it is usually 1/2-in. nylon, double the length of the highest pole climbed. Some have a fork spliced in with a snap at each end that can be snapped into the body belt of the casualty.
		A sharp knife should (already) be a regular tool in the tool belt.
		Decide where the hand line/rope will be hung before going aloft. Putting the rope over a crossarm about 1 foot (30 cm) away from the casualty is ideal; however, many poles are without crossarms. When all else fails, a screwdriver may need to be driven into the pole and the rope wrapped around the pole.
		If the casualty is still in contact, or an energized circuit complicates the rescue, a decision needs to be made as to which tools will have to be carried or pulled up the pole. Choices will be rubber gloves, protective line grounds, a hot-line cutter, and shotgun stick. Ideally, you will have dispatch/system control isolate the circuit.
5A.	Climb up and belt in below the casualty, if the casualty is in contact or in another precarious position.	If the casualty is still in contact, there is a high probability that there are voltage gradients along the pole. Wear rubber gloves take short steps, and plan to stop below the casualty. Climb up and belt in just below the casualty so that a hot-line tool (cutter) or rubber gloves can be used to move the hot conductor or the casualty.
5B.	Climb to the rescue position.	Climb to a position slightly above and to one side of the casualty, keeping in mind that this position may be very close to the hot circuit.
6.	Assess the casualty's condition.	If the casualty is breathing and/or conscious, the speed of the rescue is less urgent. Reassure the casualty while preparing to lower him or her to the ground. If a casualty insists on climbing down, the rescuer should insist on rigging the pole-top rescue rope as a backup.
		If the casualty is not breathing, it is urgent to get oxygenated blood to the brain as soon as possible. The best place to do that is on the ground. Before lowering, some utility/employers specify four quick mouth-to-mouth breaths to fill the lungs; some specify lower only and others specify starting CPR aloft.
7.	Rig the rescue rope or hand line.	Different methods are used. It is important to use and practice the method specified by your utility/employer.
		One method: Place the rescue line over an object that will allow an obstruction-free descent (ideally a crossarm) 2 to 3 ft (about 1 m) from the pole, and wrap the short end twice around the fall line. Pull the slack from the short end and pass it around the casualty under the arms, then tie it with two half hitches (some specify three). The knot should be in front of the casualty, near one armpit and high on the chest. Take up the slack in the rope, grip firmly, and cut or unsnap the casualty's safety strap. **See Figure 1–38**.
		Some use a hand line where the short side rope is used, and they drop the hand line pulley to the ground.
		Some have a hand line that is designed with a snap below the hook, and they split the hand line there. The pulley is left in place and the hook is woven through one of the casualty's D-rings and hooked into the other D-ring. The fall line is wrapped anywhere to get friction.
		Warning: When the rope is wrapped twice around an object (crossarm) to get friction, the rope end bearing the casualty's weight must stay under the slack rope used to let the casualty down or the casualty will get hung up.

(Continued)

Step	Action	Details
8.	Lower the casualty to the ground.	Take the slack out of the line, then cut the safety strap. Cut the strap on the side opposite the desired swing. Lower the victim, and control the descent with one hand on the rope fall line and the other hand guiding the casualty through any lower obstructions.
		If a lot of obstructions are present, a rescuer may need to follow the casualty down to push and pull around obstructions.
9.	Administer first aid and/or CPR.	When the casualty reaches the ground, apply first aid and, if needed, CPR (look, listen and feel, two breaths, 30 compressions, check pulse).

Figure 1–38 The Keiling Hitch requires practice. *(Delmar, Cengage Learning)*

| 1. Pass Line through Both "D" Rings. | 2. Then Behind and Twist Lines. | 3. Return to Front and Tie Bowline. |

Fighting Fire in the Electrical Environment

The only real electrical fire would be an electric arc, but that is not the type of fire put out with firefighting equipment. An electrical fire refers to a fire in an electrical environment.

It is not unusual for a line crew to be called out to a pole or transformer fire. They are also called by fire departments that cut off the power to buildings that are on fire. *Isolating a circuit in or near a fire creates the safest environment for line crews and firefighters.* Safe methods exist for fighting a fire in an electrical environment, but they should be carried out only by trained personnel in an emergency.

Cutting down conductors will change the stress on the pole, so secure the pole, as necessary. Check for hot conductors lying on the ground. Water flowing near energized conductors will be conductive and hazardous to anyone walking on the wet ground.

There are three ratings important to a power line technician when fighting a fire around a power line. **See Figure 1–39**. Any extinguisher used in an electrical environment must have a C rating in combination with another rating. The extinguishers normally carried on electrical utility trucks are rated as ABC, which can be used for transformer

Figure 1–39 A fire extinguisher used in an electrical environment must have a "C" label. *(Delmar, Cengage Learning)*

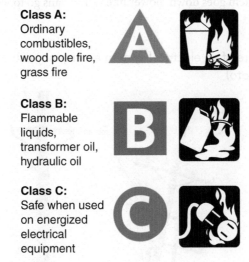

Class A:
Ordinary combustibles, wood pole fire, grass fire

Class B:
Flammable liquids, transformer oil, hydraulic oil

Class C:
Safe when used on energized electrical equipment

fires, pole fires, or vehicle fires. Stay back when using a fire extinguisher. Even a small extinguisher can have pressure equal to a large extinguisher; it only contains less powder.

A transformer fire is actually a flammable liquid fire with the added hazard of it being in an electrical environment. However, if a transformer is on fire, it is not salvageable. Stay upwind—remember, if the transformer oil is contaminated with PCBs, the smoke contains very dangerous chemicals.

A pole fire is a wood fire in an electrical environment. Be aware, however, that any spent dry powder lying on a cross arm or on other hardware can become contaminated and conductive, especially in damp conditions. Wet powder could become a path to ground and cause an explosive arc near any worker trying to put out a fire.

Water and Electricity

Most people know that water and electricity do not mix. However, exposed electrical apparatus and water do mix every time it rains. When insulators are clean, the rain does not cause the circuits to trip out. It is the combination of water and some kind of contaminant that causes electrical current to flow. Water becomes a conductor when it dissolves salts from contaminants.

When a line crew washes insulators on energized circuits, the water does not normally short out the circuit. A high-pressure pump (600 to 850 psi) is used to blast the dirt off the insulators. Instead of a solid stream of water, a special nozzle is used to break up the stream into fine particles. This allows a closer limit of approach for the power line technician. The nozzle must be grounded for the following reasons:

1. A grounded nozzle drains the static charge that can be generated by the water flowing through the nozzle.

2. A grounded nozzle drains any current flowing from the energized circuit through the stream of water to ground.

Firefighters can be trained to use water for fighting fires near power lines. Some fire crews have been trained to use water with a spray or fog stream on 230 kV from 15 ft away or up to 50 kV from 10 ft away. A solid stream using a 5/8-inch nozzle can be used on circuits up to 230 kV from 30 ft. Before firefighters are trained, the water in their jurisdiction is tested for unusually high salt or mineral content. Live conductors lying on the ground are very dangerous when water is applied to them. Water flowing on the ground is conductive and is a hazard to anyone touching the wet soil. There will, however, be less voltage between the ground gradients when spread over a longer distance.

Electrical System Emergencies

When storms hit and an electrical system goes down, power line technicians go to work. **See Figure 1–40.**

Figure 1–40 When arriving on site of an electrical system emergency caused by a storm, good job planning is required. *(Courtesy of iStock Photo)*

Notes for Power Line Technicians

- A major storm with many utility customers out of power is a crisis but not an emergency. Any institution or person that has a life-and-death need for uninterruptible power has the responsibility to have backup generation. Line work must be planned, job briefings carried out, and complete lockout/tagout procedures used for isolation certification.

- Many utilities insist that any outside crews work to the host utility standards. Some utilities provide an orientation and a booklet for outside crews with a brief description of the voltages, the type of system control, the hours of work, fusing requirements, and much more.

- Ideally, the host utility will give an outside crew work that is confined to a circuit or specific geographic area where the work is independent of other crews. Line work requires everyone on a job site to know the same job plan, take part in the same SA, use procedures that will not conflict with others, and stay within the confines of isolation certification covered by a lockout/tagout.

- Fatigue, namely lack of sleep, will be an issue. Rest during the delays that frequently occur in the administration of outside crews. Ideally, work will be scheduled to take advantage of the daylight hours available.

Review Questions

1. Based on the safe work model shown in Figure 1–2, name three elements that have the most direct influence on safe work.

2. Why is it important to report all accidents or near misses?

3. Based on the risk formula, if the probability of tipping over while working from a bucket truck is low, but the consequence could be very high, why is the work considered high risk?

4. What is the rubber glove rule for your employer?

5. What is the purpose for inflating a rubber glove when doing a pre-use inspection?

6. What is the difference between a job briefing and a job plan?

7. Name three high-risk eye hazards in line work.

8. Name two ways to positively identify a neutral wire.

9. How can an isolated line be proven to be de-energized?

10. What tests are performed before entering a confined space?

11. Name four hazards when working in excavations or trenches.

12. What are two signs that a coworker may be suffering from heat stress?

13. How can a power line technician help protect the public from making contact with power line?

14. How can a power line technician reduce the risk of a pole falling to the ground while working aloft?

15. Name four safety courses that are a necessity for line work.

Chapter 2

Electrical Units

Objectives

After completing this chapter, you should be able to:

1. Measure the voltage on a circuit.
2. Evaluate the reading compared to a standard.
3. Use work procedures to prevent an uncontrolled discharge of the pressure (voltage) in a circuit.
4. Measure the current in a circuit in amperes.
5. Determine the ground resistance network at a transformer meets standard, by using a Megger.
6. Calculate the expected current in amperes to supply any piece of equipment using Ohm's Law.
7. Explain "Keeping away from a second point of contact is a fundamental rule for working on or near energized circuits."
8. Explain the difference between electrical energy and electrical power.

Introduction

What Is Electricity?

All materials are made up of atoms. Each atom has a nucleus, and each nucleus has electrons in orbit around it in the same way as the planets orbit the sun. The positive charge of the nucleus and the negative charge of the electrons keep the electrons in orbit and keep the electrical charge of the atom neutral.

The electrons of good conducting material, such as copper or aluminum, are dislodged fairly easily. With an external force, the electrons can be bumped from their own orbits into the orbits of adjacent atoms. An atom that loses an electron will then have a net positive charge and will be susceptible to gaining another electron. The atom that gains an electron will then have a net negative charge and will be susceptible to losing an electron. A charged atom has potential and is called an *ion*.

The transferring of electrons from one atom to the next is **electrical current**. In other words, electrical current is the flow of electrons.

Summary of Electrical Units

Electricity is a current or flow of electrons. Electrical current will only flow in a circle (circuit) and must always return to its source. The **ampere** is the unit used to measure the rate of flow.

Electrical current must have a pressure pushing it. Pressure, an electron-moving force, which is also called an electromotive force (emf), is measured in **volts** (V). Pressure is needed to overcome any resistance, which impedes the current flow in a circuit. Resistance or impedance is measured in **ohms**.

A combination of electromotive force (volts) and current (amperes) is a measure of the rate of work being done. The unit of work is a watt (W) (1 volt × 1 ampere), which is more commonly measured in blocks of 1,000 watts or kilowatts.

When the rate of work is at 1 kilowatt and it lasts 1 hour, 1 kilowatt-hour (kWh) of work is completed. The quantity of electricity used is measured in kilowatt-hours.

Electrical Potential

Voltage Basics

To introduce electrical concepts, reference is often made to the properties of a water system. For example, water flows in a garden hose when some force or pressure is pushing the water from a high-pressure area to a low-pressure area. Electricity also needs a pressure or a potential difference to have a current flow. Water pressure comes from a water pump, and electrical pressure comes from an electrical generator.

Electrical potential is measured in volts. The symbol for electrical potential is E (from electromotive force) or V (from volts).

Measuring Voltage

To measure the amount of potential or voltage that is available in a circuit, a voltmeter is used. In the line trade, voltage checks are frequently made because an improper voltage is the first indication that something is wrong. To measure voltage, the voltmeter leads must be connected across (parallel) two different potentials. Most voltmeters are rated up to 750 V. A multimeter that also has current jacks should not be used for power line technician work. It is too easy to mistakenly plug the test leads into the current jacks when testing voltage. This would result in a short circuit through the meter, causing a flash and destroying the meter. Measurement of higher voltages is done at substations using voltage transformers (VT) to bring a representative voltage into the control room. A potential tester is used to determine if the circuit is energized or isolated. A potential test is an essential step before placing protective working grounds on a circuit. A potential tester is used with an appropriate length of live line to check for potential on a higher voltage circuit. **See Figure 2–1.**

Figure 2–1 A potential tester does not measure voltage but will confirm if a circuit is energized or deenergized. *(Courtesy of Salisbury by Honeywell)*

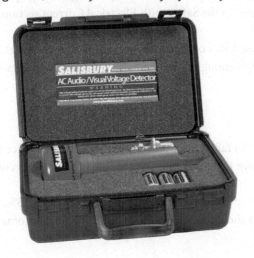

A power line technician can also measure higher voltages using phasing tools. Phasing tools are used for tasks such as checking the voltage across an open point to ensure that connections are made to the proper phase. A wireless phasing tool shows zero volts when both tools are on the same conductor. **See Figure 2–2.**

Figure 2–2 Phasing tools can be used to measure high voltage and can be used as potential testers. *(Courtesy of HASTINGS Fiber Glass Products, Inc.)*

A power line tech has no real reason to measure higher voltages. For example, there is no requirement to check if a 46 kV circuit is operating at a proper voltage. There is a need to check for the existence of voltage on a circuit before installing personal protective grounds. There are various voltage (potential) testers that will give an indication that a circuit is energized. Phasing testers are used on higher voltages, and they can give a voltage reading between phases and between phases and the neutral. Phasing testers (also called phasing sticks or phasing tools) are often used to confirm that there is no voltage difference across an open point before closing a switch between two circuits, making a connection or splice.

Safety with Electrical Potential

Where there is a difference in potential, receiving an electrical shock is also a possibility. Any electrical potential is always looking for a path to a different potential. For example, when putting one hand on each post of a car battery, a person normally would not feel anything because the potential is only 12 V, which is not high enough to overcome the resistance of the skin. Most people can feel 40 V from hand to hand. A common voltage, such as 120 V, is a high enough potential to drive a fatal current through a person's body, and as will be explained in the Safety with Electrical Current section, it takes very little current to cause a fatality.

Power line technicians are exposed to much higher voltages. Safe contact can only be made with a high-voltage circuit when the resistance between the circuit and a person is high enough to prevent a current flow. Such resistance is provided by live-line tools or rubber gloves.

Maintaining Good Voltage

Voltage in a circuit is susceptible to many influences that cause it to fall or rise. Good quality electrical service requires that a customer's voltage be kept within an acceptable range. Voltage that is too high or too low will damage a customer's motors, appliances, and electronic equipment.

The American National Standards Institute (ANSI) standard for proper voltage is a range of +6% to −13%. When a voltage is found to be lower or higher than extreme voltage, power should be shut off to avoid damaging customer equipment. **See Figure 2–3.**

Figure 2–3 Voltage Standard

Nominal Voltage (V)	Extreme Low Voltage (V)	Normal Low Voltage (V)	Normal High Voltage (V)	Extreme High Voltage (V)
Single Phase				
120/240	106/212	110/220	125/250	127/254
240	212	220	250	250
Three-Phase Four-Wire				
120/208 Y	110/190	112/194	125/216	126/220
277/480 Y	240/418	258/446	288/500	293/508
347/600 Y	306/530	318/550	360/625	367/635
Three-Phase Three-Wire				
240	212	220	250	254
480	418	442	495	508
600	530	550	625	635

Voltage Drop

When an electric circuit feeds a load, for example, a 120 V light, the current goes through the light and the current returns to the source on the neutral at zero volts. In other words, all the voltage is dropped across the total resistance in the circuit. In line work, when the term **voltage drop** is used, it usually refers to the voltage drop between the source and the customer. As the load increases on a circuit, there is an increase in the voltage drop as can be seen in the formula for a voltage drop.

A voltage drop is equal to $I \times R$,

Where:

I = Current

R = Resistance

Similarly, when the current in a circuit increases, there is increased energy loss, which is normally called line loss. Line loss refers to wasted energy used up in a power line before it gets to the customer. The amount of line loss varies with the amount of current squared (I^2). If the current doubles, the line loss quadruples.

$$\text{Line loss in watts} = I^2 \times R$$

Where:

I = Current

R = Resistance

It can be concluded that voltage drop and line loss are reduced by serving a customer with short, large-diameter conductors. Electrical system planning engineers calculate predicted voltage drop and line loss when choosing a cost-effective conductor size.

Typical Utility Voltages

A large variety of standard voltages are used in the electrical utility business. A person working on distribution must be alert when choosing transformers, surge arrestors, switches, and insulators.

Power line technician can differentiate between familiar voltages within a utility by referring to utility operating diagrams. When the voltage of a circuit is in doubt, reference can always be made to a nameplate on an existing transformer. Insulators and cutouts are not reliable indicators of the system voltage because of standardization of materials and prebuilding for future voltage conversions.

When a circuit voltage is given, it is normally the "nominal" phase-to-phase voltage. **See Figure 2–4.** Where applicable, a phase-to-ground voltage follows the slash. In Europe and many other countries, common transmission line voltages are 400, 300, 275, 230, and 132 kV. Common distribution lines are 6.6, 11, and 33 kV. In Europe, the distribution lines are like subtransmission lines because they feed small substations (up to 2,000 kVA) that step down to a standard utilization voltage or 400/230 V, feeding many customers.

Figure 2–4 Typical North American Voltages

Transmission Line Voltages (kV)	Subtransmission Line Voltages (kV)	Distribution Voltages (kV)	Utilization Voltages (V)
765	13.8	34.5/20	240/120
500	23	27.6/16	208/120
230	34.5	13.8/8	480/240
138	46	12.5/7.2	480/277
115	69	8.3/4.8	600/347
69		4.16/2.4	

Note that distribution voltage equipment is also commonly called *medium voltage equipment,* and utilization voltage equipment under 1,000 V is commonly called *low-voltage equipment.*

Electrical Current

Current Basics

The flow of electrical current can be compared with the flow of water in a garden hose. A garden hose conducts water, whereas a wire or conductor conducts the flow of electrical current. Just as a large pipe can conduct more water than a small pipe, a large-diameter electrical wire can conduct more electrical current. It is the flow of current that does the work. For example, when electrical current meets resistance, heat is produced.

Electrical current is actually the flow of electrons jumping from one atom to the next. While electricity is known to travel at the speed of light, which is 186,000 miles (300,000 km) per second, the electrons themselves do not actually travel at that speed. The actual speed of electron travel in a conductor is about 0.0001 in. (0.003 mm) per second. It is the electrical charge or voltage that travels at the speed of light.

The symbol for electrical current is *I,* from the French word intensité. The unit of measure is the ampere, which is represented by the symbol *A.*

Measuring Current

Current is measured with an ammeter. An ammeter is usually described as a meter that is connected in series with a circuit and measures the current flowing through it. This means that the circuit must be opened and reconnected to go through the meter so that current can be measured. To measure current with a multimeter, requires that the circuit be opened. **See Figure 2–5.** This type of meter may be suitable for electronic work but not for line work. Currents measured by power line techs are very high, and the typical type of ammeter used is a clamp-on meter. **See Figure 2–6.** This type of ammeter should be used because it is convenient and can measure the current without having to connect the meter in series with the conductor. The magnetic field around the wire induces a representative current into the clamp-on

Figure 2–5 A common multimeter can measure voltage but is not practical to measure current in power line work. *(iStock.com/StanRohrer)*

Figure 2–6 A clamp-on ammeter is considered to be safer and more useful for measuring current. *(Courtesy of iStock Photo)*

ammeter coil. A clamp-on ammeter is not voltage sensitive and can be used on all voltages as long as a line worker uses the meter while wearing rubber gloves or using an energized-line tool suitable for the voltage of the circuit.

Electrical Current Needs a Circuit

Electrical current will not flow unless it is in a circuit. The current that leaves the source must make a complete "circle" and return to the source. An electrical circuit, therefore, needs a return path to the source. Depending on the type of circuit, the return path can be a ground, a neutral, or another "hot" wire. A break anywhere in the circuit, including in the return path, opens the circuit and the current flow stops.

Opening a ground wire or a neutral on an energized circuit is dangerous because an electrical current could be interrupted. When a current is interrupted, a voltage will appear across the open point in the circuit.

Safety with Electrical Current

It takes a certain amount of voltage to break down the initial skin resistance of a human body before a current path is established. Once a current path is established, it is the amount of current, the path the current takes through the body, and the duration of exposure that do the damage.

100 milliamperes (mA) can cause damage to a human body. **See Figure 2–7.** Considering that a typical household circuit is fused at 15 or 20 A, which translates into 15,000 or 20,000 mA, the potential for a lethal electrical shock is available on all electrical circuits.

Figure 2–7 Note the typical effects of electrical current on the human body. It takes a very small amount of current through a human body to cause a fatality. *(Delmar, Cengage Learning)*

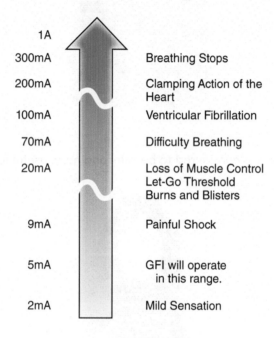

1A	
300mA	Breathing Stops
200mA	Clamping Action of the Heart
100mA	Ventricular Fibrillation
70mA	Difficulty Breathing
20mA	Loss of Muscle Control Let-Go Threshold Burns and Blisters
9mA	Painful Shock
5mA	GFI will operate in this range.
2mA	Mild Sensation

Note that a ground fault interrupter (GFI) on a utilization circuit will open a circuit before any dangerous current can flow. A device as sensitive as a GFI is not available on utility circuits.

Load Current

Load current in a circuit is the current needed to supply the load demands of the customer. When an ammeter is used to take a reading on a circuit, the reading measures the load current at that moment, along with a small amount of current due to line loss.

Voltage is *supplied* by an electrical utility, and load current is *drawn* by the customer. A utility can control the voltage, but it cannot control the load current because load current is based on customer demand.

Peak Load Current

The load current in a system fluctuates with the seasons and the time of day. Peak load current is measured by recording ammeters found at substations and portable recording ammeters installed on lines. Peak demand for power tends to occur on the coldest and hottest days of the year at the time of the evening meal.

An electrical system must be built to meet these peak demands even though they may occur infrequently. More trouble calls occur during peak load conditions because any weakness, such as an overloaded transformer, an overloaded circuit, or a low voltage, will appear then.

Utility conservation programs try to reduce peak loads by offering reduced rates during off-peak hours, by installing demand meters that penalize customers with high peak loads, or by encouraging the efficient use of electricity to bring down the total base demand.

Cold Load Pickup

When trying to restore power, especially during peak load periods, a fuse or breaker will sometimes trip out even though the cause of the original outage has been fixed. It is difficult to pick up a cold load after a circuit has been out of service due to the loss of diversity and the initial inrush of current into the circuit.

Diversity in a circuit refers to the normal situation in which everyone on the circuit is not using the furnace, air conditioner, appliance, or water pump at the same time. After a circuit has been out of service for a while, much of this equipment could be set to come on at the same time when power is restored. The load current on the circuit could be as much as two times the diversified load current. An outage of 30 minutes may be enough to lose load diversity, and it could take as much as 45 minutes to restore the circuit to normal diversity.

There is a short period of high inrush current to loads such as transformers, motors, and heaters. The initial inrush current is considerably higher than the current needed to maintain these loads.

Educating customers to switch off much of their electrical load when the power goes off can go a long way toward solving the cold load pickup problem. After power is restored, customers could switch their appliances back on one at a time. Meanwhile, without customer help in reducing load, a trouble crew has to rely on a time-consuming process of sectionalizing the circuit to pick up the load in smaller chunks.

Fault Current (Short-Circuit Current)

When a circuit is faulted or short-circuited, all the current which the electrical system is able to supply to the faulted location goes to that fault. Fault current can be explained by using an automotive electrical system as an example. The normal usage of the starter and lights would draw load current equivalent to what a battery is designed to deliver. However, if a wrench were dropped across the battery terminals, an explosive fault current would go from one terminal to the other. All the current available in the battery would feed the fault.

Similarly, in an electrical system, a fault current due to a short circuit can be extremely high. The resultant flash and heat generated by the fault current can be very dangerous to a worker in the vicinity. Eyes are the most vulnerable to a large flash, and safety glasses along with arc-rated clothing should be worn any time there is a potential for a flash.

The highest fault current is near the source of a circuit, such as close to a substation or secondary conductors close to a distribution transformer. Using our car electrical system as an example, a short circuit at the battery is much more explosive than a short circuit at a taillight. The electrical system cannot supply nearly as much power to a fault at a taillight because the small wire and the distance to the taillight add resistance, preventing a large current flow.

Electrical Resistance

Resistance Basics

Using our garden hose example again, consider that as a garden hose provides resistance to the water flowing in it, an electrical wire also provides resistance to the flow of electrical current. Resistance in a conductor causes electrical energy to be converted into heat. This will be useful or wasteful, depending on whether the heat is a desired product or a line loss.

Some of the current intended to run a motor or light a building is converted to heat because there is resistance in every part of an electrical circuit. Resistance can also be added to a circuit intentionally, such as when a heating element is used in an electric range.

The symbol for electrical resistance is R. The unit of measure is *ohms*, which is represented by the Greek letter Ω (omega). A circuit has a resistance of 1 ohm if 1 volt causes a current of 1 ampere to flow.

Measuring Resistance with an Ohmmeter

Resistance is measured with an ohmmeter, which is normally found as one function of a multimeter. Insulation testers are like ohmmeters and are more suitable and user-friendly for line work. For example, like an ohmmeter, they can be used to check the insulation in a meter base before installing a meter. When an insulation tester or ohmmeter is hooked to two different wires and the meter reads zero, there is no resistance between the two wires being measured, and they must be connected together or shorted somewhere. When an ohmmeter reads "∞" (infinity), the resistance is high, and the two wires being measured can be considered insulated from one another.

If a multimeter is used to check for continuity or short circuits, there is some probability that sooner or later the leads will be put into an energized meter base or other location while the dial on the multimeter is set on the ohmmeter function. This will expose the worker to a possible serious flash and will destroy the ohmmeter/multimeter.

Measuring Earth or Insulation Resistance

Earth resistance testers and insulation resistance testers are like ohmmeters except that they put out a higher voltage than a regular ohmmeter. A higher output voltage is needed to test the relatively higher resistance of earth or the very high resistance of an insulator or cable insulation. These testers are commonly called meggers (which is easier to say than "megohmmeters"). Earth resistance testers can have various scales, such as 0–1 ohm, 0–300 ohms, or 1–1,000 ohms. It is important to have a low-resistance grounding network at station sites, transformer locations, and customer premises. Ground rods are driven until the earth resistance tester verifies that the grounding network meets specifications. An ideal grounding network at a transformer should read 25 ohms or less before connecting it to the neutral. More ground rods need to be driven and connected together until a minimum of 25 ohms is reached. A clamp-on earth resistance meter is much easier to use than previous models. **See Figure 2–8.** The meter induces a current into the ground rod and provides a resistance reading.

Figure 2–8 Using a clamp-on earth resistance meter is an easy way to measure ground rod resistance. *(Courtesy of Megger)*

Insulation resistance testers can have scales that read resistance values between megohms and infinity. They come in 250, 500, and 1,000 V models. For example, a 500 V insulation tester can be used at a meter base to check between two wires or between a wire and ground. A zero reading on the meter would indicate a problem such as a short circuit. The meter can also be used to check for continuity. **See Figure 2–9.**

Figure 2–9 An insulation resistance tester is useful for troubleshooting. *(Courtesy of Megger)*

Ohm's Law

The relationship between volts, amperes, and ohms is expressed in an equation known as Ohm's law. **See Figure 2–10.** The current flowing in a circuit is equal to the circuit voltage divided by the resistance. This is the most used electrical theory formula for the line trade. The equation is generally shown in a pie format as a visual aid in remembering how it is used.

Where:

E represents voltage (**e**lectromotive force)

I represents current (***i**ntensité*)

R represents **r**esistance

Figure 2–10 Ohm's law is the most common electrical formula used by linemen. *(Delmar, Cengage Learning)*

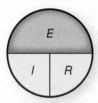   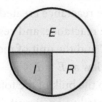   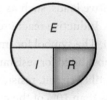

$E = I \times R$
The voltage is equal to the current times the resistance

$I = E \div R$
The current is equal to the voltage divided by the resistance

$R = E \div I$
The resistance is equal to the voltage divided by current

Examples Using Ohm's Law

1. The resistance of an electric heater is measured at 15 ohms. What is the voltage if the current through the heater is 8 amperes?

$$E = I \times R$$
$$= 8 \times 15$$
$$= 120 \text{ volts}$$

2. A clamp-on ammeter reads 10 amperes on a 120 V circuit in a home. What is the resistance of the circuit?

$$R = \frac{E}{I}$$

$$= \frac{120}{10}$$

$$= 12 \text{ ohms}$$

3. How much current would a 240 V heater with 30 ohms of resistance draw?

$$I = \frac{E}{R}$$

$$= \frac{240}{30}$$

$$= 8 \text{ amperes}$$

Conductance

Conductance is the opposite of resistance. The term conductance may not be used by the line trade. Technically, instead of discussing the resistance of various types of wood pole treatment, the conductivity could also be expressed as conductance.

The symbol for conductance is G. The unit of measure is mho (transposition of "ohm"). An mho is probably only discussed during training and has no real application in line work.

$$Conductance\ G = \frac{1}{R}\ resistance$$

$$Ohms\ resistance = \frac{1}{mhos\ conductance}$$

Impedance

In an AC circuit, the flow of current is impeded by reactance as well as resistance. The term for the combination of resistance and reactance is *impedance*.

In a heater or incandescent lamp, there is only resistance to the flow of electrical current. When AC current flows into a coil, there is additional opposition to the current flow caused by inductive reactance. Some electrical loads, such as motors, electronic equipment, and fluorescent lights, are inductive loads. A portion of the load current flows in and then back out of the load without doing any work. The inductive reactance created by these loads causes additional impedance to current flow.

Similarly, capacitive loads, such as the capacitive reactance created by long lengths of paralleling conductors, also oppose current flow. Inductive reactance, capacitive reactance, and resistance provide the total impedance to current flow in an AC circuit. The symbol for impedance is Z and the unit of measure is ohms. When working with AC circuits, impedance can be substituted for resistance in Ohm's law as $E = I \times Z$.

In an electrical utility circuit, resistance makes up the bulk of the total impedance in the circuit. The reactive current usually makes up less than 10% of the current in a distribution circuit. The opposition to current flow in an AC circuit, therefore, is resistance and reactance.

Conductor Resistance

In an electrical system, a short run of a large conductor provides the least resistance to electrical current. A wire table that shows the current-carrying capacity of different sizes of conductors is not used often in the utility business because there are other factors to be considered. The expected load current to be carried, the fault current the conductor can be exposed to for short periods of time, the length of the feeder, acceptable line loss, and expected voltage drop all have to be balanced with affordability when choosing a conductor size.

High Resistance and Insulation

Electrical current flows because there is a pressure (voltage) pushing it through any resistance toward an area of less pressure. It is like the common saying, "Electricity always tries to find a path to ground." For example, electrical current flows through the resistance of a heating element because the pressure from the supply voltage pushes the current through the element to the low or zero pressure of the neutral end. Current flows in the intended circuit as long as there is no low-resistance path to any other objects at a different pressure or voltage. High-resistance materials are used to insulate an electrical circuit from ground and other conductors.

Overhead conductors are generally bare and insulated from other objects by porcelain glass or polymer insulators. The longer the path through which the current has to flow and the higher the resistance of that path, the lower the current. Insulators have a number of curves or skirts along their surface. This increases the length of the path or creepage distance that current has to flow to ground. Increasing the size of an insulator increases its resistance and allows it to withstand a higher voltage.

Underground conductors are insulated by a rubber or polyethylene insulation. The insulation needs to be well protected because even a little damage will cause the voltage (pressure) to stress the damaged location and eventually cause a short circuit in the cable.

High voltage can stress any insulation and cause it to fail. For example, air is normally an insulator, but it can become a conductor when it is electrically stressed and becomes ionized. Air is ionized when electrons in orbit around the atoms are displaced because of being stressed by voltage. An electrical arc and lightning are visible examples of air that has become conductive.

The Effect of Rain on Insulators

Contrary to popular belief, water is not a good conductor of electricity. When it rains and water flows on clean insulators, the resistance of the insulators is not reduced substantially. Rain or water around electricity is very dangerous if an energized conductor has fallen to the ground. Water will absorb the salts in the earth and become highly conductive. Dry ground is not a good conductor, but when the same ground gets wet, it becomes a very good conductor. Water will greatly reduce the resistance of any dirty surface.

When it rains on contaminated insulators, the dirty and wet insulators are less resistive and can eventually short out the circuit. Insulators can become dirty when near industrial areas, saltwater, or roadways spread with salt in winter. The failure of contaminated insulators is delayed due to the irregular shape and skirts on insulators, which keep parts of the insulator dry and make a longer leakage path for current to flow.

Insulators are cleaned using high-pressure water or corncob and nutshell blasting to mechanically remove grime from the insulators. This work is done with the circuit energized. **See Figure 2–11.**

Figure 2–11 Using the right equipment and qualified workers, insulators can be washed with water while the circuit is energized. *(Delmar, Cengage Learning)*

Safety and Electrical Resistance

If an ohmmeter is used to measure a person's body from one dry hand to another dry hand, the resistance would be approximately 100,000 ohms. Most of this resistance comes from the skin. For some people, a 120 V source is not high enough to overcome the resistance of dry, calloused hands. If a person is perspiring, the resistance is reduced to approximately 35,000 ohms. Once contact is made, the skin resistance can break down in a very short time. After skin resistance breaks down, there will be much less resistance through the internal organs of the body. The internal resistance of the body ranges between 100 and 400 ohms.

There can be many variables involving a person's resistance when electrical contact is made. The clothing and gloves being worn can increase the contact resistance and make a difference during a time of electrical contact. When studies and calculations are made involving the resistance of the human body during an electrical contact, an average of 1,000 ohms is normally used.

Because people may have made contact with a 120 V source and were not hurt, some mistakenly think that 120 V is harmless; however, fatalities occur every year when contact is made with that voltage. The amount of resistance being imposed, often hand to hand, is not enough to prevent a small amount of current flow, considering that 100 mA can be fatal.

Second Point of Contact

A human body does not have enough resistance to prevent a fatal current flow through it when contact is made with a primary voltage. A more important consideration is, What other objects is a person touching when contact is made with an energized circuit?

Current must have a circuit before it can flow. Before current will flow through a human body, it must enter the body at one contact point and leave the body at another contact point that is at a different potential. An electrical accident usually occurs because one part of the body is in contact with an energized circuit and another part of the body is in contact with earth, a **neutral conductor**, a de-energized load, or another energized wire. If the body is not touching a second point of contact, no current can flow through it.

For example, some people claim they can touch a 120 V wire and feel nothing. The source may be 120 V, but a person is not exposed to all the available voltage unless another part of the person is well grounded. A person may not feel anything if standing on a wooden floor or wearing good boots. The same person standing barefoot on wet ground would probably receive a fatal shock.

Keeping away from a second point of contact is a fundamental rule for working on or near energized circuits. A body cannot become part of a circuit unless there is a potential difference across it. **See Figure 2–12.** If a power line technician makes contact with an energized conductor while not in contact with any second point of contact, no current can flow through the body. This is the basis for barehand work.

Figure 2–12 All the second points of contact is either covered up or out of reach providing a safe environment for rubber glove work. *(iStock.com/Kozmoat98)*

Electrical Power

Heat from Current Passing through a Resistance

When current passes through a resistance in a circuit, heat is released. The amount of heat produced varies with the amount of current squared (I^2). If the current doubles, the heat produced quadruples. A low resistance draws more current, and more heat is produced.

Joule's law of electric heating states that the amount of heat produced during each second by electrical current in a conductor is proportional to the resistance of the conductor and to the square of the current.

$$J = I^2 \times R \times t$$

Where:

J = the amount of heat in joules

I = the amount of current flowing

R = the amount of resistance in the circuit

t = the time the current is flowing

The same formula can be converted to the basic unit for electrical energy, which is a watt-second.

$$1 \text{ watt-second} = 1 \text{ joule}$$

When current travels through the resistance of a human body, heat is also released. Catastrophic electrical burns are immediately apparent as visibly blackened skin where contact is made and where current leaves the body.

The consequence of a less-serious electrical burn may appear on the body as a small entry burn and a small exit burn. However, the current flow that caused these burns flowed through the body along the lower-resistance blood vessels and nerves. The heat from the current flow will cause damage that can be much more serious than the external injuries might suggest. In other words, most of the damage is not visible, and even so-called minor electrical burns should have medical attention.

Electrical Power and Energy Basics

Electrical energy is the product measured at a customer's meter. Electrical energy is Electrical Power × Time. In other words, electrical energy and electrical power are not the same thing.

Recall that electrical power is the product of volts and amperes, and is measured in *watts* (P). The common equations for calculating power are:

$$P = E I$$
$$P = I^2 R$$
$$P = \frac{E^2}{R}$$

Where:

P = watts

E = voltage

I = current

Electrical energy is the product of watts and time. A watt-second is 1 watt × 1 second. A kilowatt-hour is 1 kilowatt × 1 hour.

Power Formulas Derived from Ohm's Law

There are many electrical equations based on Ohm's law. An equation wheel shows the interrelationship of these equations. **See Figure 2–13.**

Figure 2–13 The equation wheel can be used to calculate voltage, current, resistance and power in many different ways. *(Delmar, Cengage Learning)*

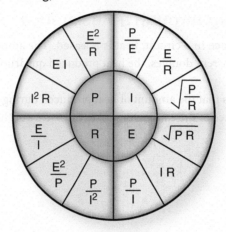

P = power in watts

I = intensity of current in amperes

R = resistance in ohms

E = electromotive force in volts

Large Units of Electrical Power

A watt is the basic unit of power, but it is very small. In the electrical utility business, kilowatts (kW) and megawatts (MW) are the most common units.

In the line trade, the terms kilovolt-ampere (kVA) and megavolt-ampere (MVA) are used when discussing power. To transmit AC power, more volts × amperes (VA) are needed to deliver the actual wattage used by a customer. In an AC circuit, a counterforce causes a reactance that impedes the flow of electrical current. In other words, slightly more than 1 kVA is needed to deliver 1 kW of power.

The amount of power transmitted by a transmission line is termed in megawatts. For example:

A 500 kV line can carry approximately 1,000 to 1,500 MW.

A 345 kV line can carry approximately 700 to 1,000 MW.

A 230 kV line can carry approximately 300 to 500 MW.

A 138 kV line can carry approximately 100 to 200 MW.

A 69 kV line can carry approximately 25 to 75 MW.

Putting Power into Perspective

1 kilowatt (kW) = 1,000 W	Hardware store portable generators have ratings such as 750 W, 2.5 kW, or 5 kW. A typical toaster oven is 1.5 kW.
1 horsepower (HP) = 746 W	With 100 HP = 74.6 kW, a 75 kVA transformer is needed to run a 100 HP motor. However, in reality, a motor is only about 80% efficient, and a 100 kVA transformer would be needed.
1 megawatt (MW) = 1,000 kW	Some of the older hydroelectric plants on small rivers have generating units rated at about 1 MW. The utilization station at a small factory is often rated at 1–3 MW. 1 MW = 1 million joules per second.
1 gigawatt (GW) = 1,000 MW	The largest generating units are about 1 GW and are found in newer thermal or nuclear plants. Hydroelectric plants such as the R. H. Saunders/Robert Moses generating station across the St. Lawrence River have 32 units: 16 in Canada and 16 in the United States. Each unit is rated at 57 MW, which means the total plant output is 57 MW × 32 units = 1.8 GW. The summer peak load of New York City is about 11 GW.

The Kilowatt-Hour

As a unit, the watt is the amount of power being used at a given instant. It is also necessary to know how long the power is used to determine the amount a customer is charged for energy. Customers are billed based on the kilowatt-hour (kWh), where the kilowatt is the rate at which energy is used, and the hour is the length of time the power is used.

$$kWh = kW \times hours$$

Kilowatt-hour meters are installed to measure the kilowatt-hours used by customers. A large variety of revenue metering is used to measure other variables, but the primary charge for power used is the kilowatt-hour.

Kilowatt-Hours in Perspective

One kilowatt-hour is a small unit. When utilities, and even countries, are compared, the amount of energy generated is shown in billions of kilowatt-hours.

Putting Kilowatt-Hours into Perspective

1 kWh	1 kWh will run a 1,000 W microwave oven or a 1,000 W hair dryer for 1 hour.
	1 liter of gasoline has the energy equivalent of approximately 10 kWh. (1 U.S. gallon of gasoline has the energy equivalent of approximately 37 kWh.)
1,000 kWh	An average household in North America is considered to use 1,000 kWh per month, a figure often used when comparing electric bills between utilities.
	One cord of dry hardwood has the energy of approximately 6,000 kWh.
	One barrel of crude oil has the energy equivalent of approximately 1,700 kWh.
1 billion kWh	Annual statistics for the sale or generation of electricity are expressed in billions of kWh. The total generating capacity of the world is approximately 12,000 billion kWh. Following are statistics from countries in the 1990s:
	United States 2,300 billion kWh
	Canada 500 billion kWh
	France 410 billion kWh

Review Questions

1. What physical matter is flowing in an electrical current?

2. How is high voltage, over 750 V, measured safely?

3. How can a person get into trouble when measuring voltage with a multimeter?

4. Why does a person normally not feel anything when hands are placed on each post of a car battery?

5. What means can be used to positively identify the nominal voltage of a primary circuit?

6. Do electrons travel at the speed of light in an electrical circuit?

7. Does electrical current stop once it gets to the load it is feeding?

8. Why is opening a ground wire or a neutral on an energized circuit dangerous?

9. If a circuit is carrying 100 A, what portion of that current could induce ventricular fibrillation and probably death?

10. What is the difference between load current and fault current?

11. How would a line crew normally restore power when the switchgear will not pick up the load because of cold load pickup?

12. Where along a line would an accidental short circuit be most explosive?

13. When an ohmmeter is hooked to two different wires and the meter reads zero, what does this indicate?

14. Why does keeping away from a second point of contact reduce the risk of an electrical accident?

15. How much current would a 240 V heater with 20 ohms of resistance draw?

Chapter 3

Electrical Power System Overview

Objectives

After completing this chapter, you should be able to:

1. Describe energy sources that spin turbines to generate electrical power.
2. Identify distributed electrical energy (DER) sources connected to the distribution system.
3. Explain how distributed electrical energy (DER) is a risk to power line technicians.
4. Visually identify the voltage on a transmission line.
5. Identify the electrical components in a substation.
6. List the segments that make up a distribution system.
7. Describe a distribution primary network.
8. Explain the advantages and disadvantages of overhead power lines versus underground power lines.

Introduction

Three Systems within the Power System

This chapter gives an overview of an electrical power system. There are three main systems within an electrical utility power system:

- The generation system converts other forms of energy into electrical energy.
- The transmission system transmits energy over long distances. It includes the rights-of-way, transmission lines, switching stations, and substations.
- The distribution system distributes the energy to industry, commercial customers, farms, and residences. It includes subtransmission lines, distribution substations, distribution feeders, transformers, and services.

Electrical Energy

Energy

To do any kind of work requires energy. Energy has the ability to produce change or exert a force on something. There are many forms of energy, including chemical, solar, potential, thermal, and electrical. Energy cannot be created or destroyed, but it can be converted from one form to another.

Many forms of energy can be converted to electrical energy. The generation of electricity is a process of converting other energy forms to electrical energy. Electrical energy is utilized when it is converted back to other forms of energy.

Utilization of Electrical Energy

Electrical energy is known as an energy source that is easily converted into power and light. The utilization of electrical energy comes from its four main effects:

- *Thermal effect:* The heat produced by electrical current is desirable for toasters, heaters, and ovens during the utilization stage, but it is wasted energy in the generation, transmission, and distribution stages.
- *Luminous effect:* Light is emitted when a filament is heated as in an incandescent or halogen light. Light is emitted when an arc is generated as in a fluorescent, sodium or UV lamp. Light is emitted when electric current passes through a light-emitting diode (LED).
- *Chemical effect:* Electrical current can break down certain chemical molecules into their component atoms. For example, water (H_2O) can be broken down into hydrogen and oxygen through a process called electrolysis. Electrolysis is used in industry for electroplating and the manufacture of aluminum. Hydrogen produced this way can be used as a source of energy.
- *Magnetic effect:* The magnetic field around a wire can be increased by winding the wire into a coil around a core of magnetic material. This effect is used by a utility for generators, transformers, and reactors. At the utilization stage, the magnetic effect is used for motors, solenoid switches, circuit breakers, telephones, and stereo speakers.

Generation of Electrical Energy

Generation Basics

When a wire is moved within a magnetic field, an electrical charge is induced into the wire. Almost all commercially generated electricity involves the movement of wire coils in a magnetic field. In practice, this normally means that many electromagnets are installed on a wheel or armature, which is turned inside a stator mounted with many wire coils. The armature is connected to a turbine. **See Figure 3–1.** The turbine is a wheel with blades mounted on it. The water, steam, or wind pushes against the blades and causes the turbine to turn.

Turning the Turbine

A surprising number of prime movers or sources of energy can be used to spin a turbine. The earliest energy forms used for this purpose were falling water and wind. Almost all of the suitable falling water or hydroelectric sites in the world have been harnessed, are spinning turbines economically, and are relatively pollution free. Harnessing the tides and winds to spin large turbines for commercial generation is a more recent development.

Figure 3–1 A force (water, steam, or wind) turns the turbine and the generator, which generates electric power. *(Delmar, Cengage Learning)*

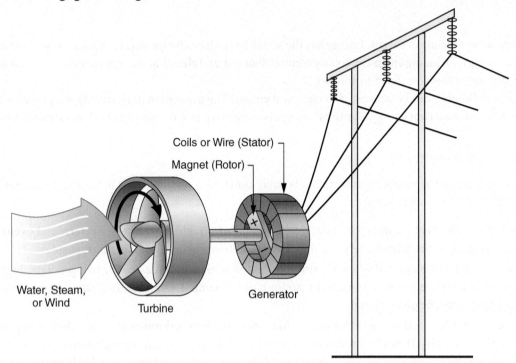

Most of the electrical energy produced in the world comes from the use of steam as a force to spin the turbines. The steam is converted from the heat energy of burning coal, oil, natural gas, wood chips, and garbage; or steam can come from the heat energy generated by a nuclear reactor or from geothermal (underground heat) sources.

Hydroelectric Generation

Hydroelectric stations are built in locations where water runs from a higher level to a lower level. **See Figure 3–2a.** This is normally accomplished by building a dam on a river with a suitable water flow and where a substantial difference in water level creates an advantage. The headwater formed by the dam is the potential energy that will be converted to electrical energy. **See Figure 3–2b.**

Figure 3–2 (a) A hydroelectric generator is turned by falling water. (b) The actual turbine used in a hydroelectric station is engineered to extract the most energy possible for the water flow available. *(iStock.com/Leschnyhan)*

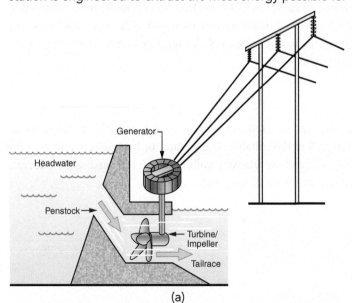

(a)

(b)

The headwater is funneled through a pipe called a penstock. Penstocks are mounted on the face of the dam into the powerhouse. **See Figure 3–3.** The water rushes down the penstock and hits the turbine blades with a force that equals the water's speed and weight. The turbine spins, which in turn spins the generator. The water continues out through the tailrace and back into the river.

Figure 3–3 The penstocks, step up transformers and tailrace are clearly visible in this hydroelectric station. Programs are in place to move transformers from the dams to reduce the risk of an oil spill into the waterway. *(iStock.com/Wolv)*

Generation from Steam

Generating heat from the burning of fossil fuels such as coal, oil, and natural gas, or from a nuclear reactor, is the most common commercial method of creating steam. **See Figure 3–4.** The steam expands and pushes against the turbine blades, causing the turbine to turn.

Figure 3–4 There is some movement to reduce dependence on coal-fired thermal generating stations such as this one because of global warming concerns. *(Courtesy of iStock Photo)*

The water in a steam plant is in a closed loop that continuously heats and cools. **See Figure 3–5.** A heat exchanger in the boiler heats the water in the closed loop, and another heat exchanger uses water from an ocean, lake, or river to cool the steam and condense it back into water. The water is then pumped back to the boiler to be reheated.

Figure 3–5 The water stays in a closed loop as it is heated and cooled in a thermal generating station. *(Delmar, Cengage Learning)*

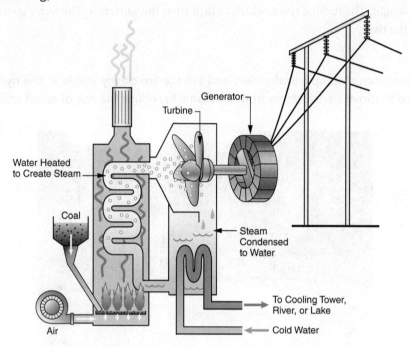

Nuclear Generation

A nuclear generating station is similar to a conventional steam plant except that it uses a nuclear reactor to create heat for making steam. The heat comes from uranium atoms splitting in a controlled reaction. **See Figure 3–6.**

Figure 3–6 A nuclear generating station is a thermal plant except the water is heated by a nuclear reactor. *(Delmar, Cengage Learning)*

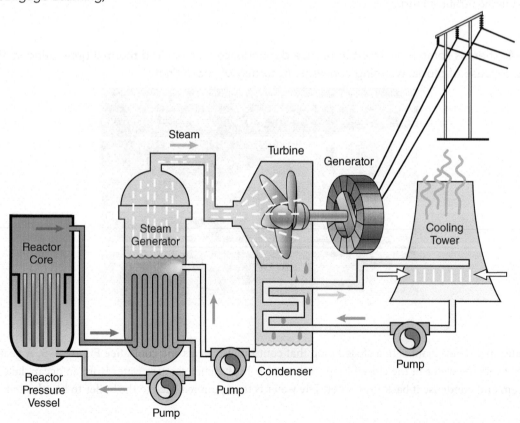

Uranium is a dense, unstable element. Neutrons, which are particles within the nucleus of an atom, are easily knocked free from a uranium atom's nucleus. A uranium atom splits if it is struck by a free neutron given up by another atom. When an atom splits, more neutrons are released that in turn hit other atoms, splitting them, and thereby causing a chain reaction. A nuclear fission (splitting) of atoms generates a huge amount of heat.

The same principles can be applied on a smaller scale. Nuclear reactors create heat to make steam used to generate electricity on some submarines and large ships. Small modular reactors (SMR) are also ideal for remote areas that typically rely on diesel units. SMR units can be manufactured, transported, and set up on site.

Gas Turbines

The hot exhaust gases from the burning of oil or natural gas in a high-pressure combustion chamber can spin a turbine when the exhaust gases expand through the turbine blades, much like the way a jet engine operates. High-pressure air is added to the combustion chamber to add more force to the escaping gases.

The most efficient way to use the gas turbine is in a combined-cycle system. **See Figure 3–7**. After the hot exhaust gases spin a gas turbine, the still hot gases heat water to make steam and spin a steam turbine. Usually, several gas turbines feed hot exhaust to one steam turbine.

Figure 3–7 Waste heat is used to turn another turbine in a combined-cycle generation plant. *(Delmar, Cengage Learning)*

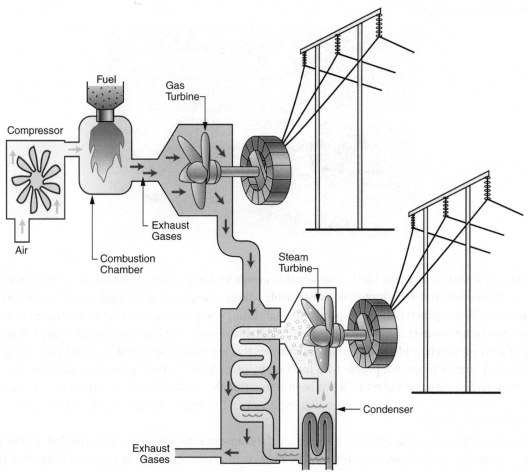

Cogeneration

Cogeneration plants are generating stations used to generate electric power and heat. The electric power can be sold to the grid, and the heat, which would otherwise be waste heat, is sold to a central heating plant or manufacturer. Because it is impractical to transport heat over any distance, cogeneration stations are built close to their heat users.

Cogeneration stations are fired by fuels such as natural gas, wood, agricultural waste, or peat moss. Steam pressure generated by burning the fuel turns the turbines and generates power. Normally, about one third of the energy in the original fuel can be converted to steam pressure to generate electricity. The excess heat supplied to the customer is usually in the form of relatively low-temperature steam exhausted from the turbines.

Generation from Wind

Wind power is solar energy because the sun creates the kinetic energy in wind. The wind is converted into mechanical energy when it spins a wind turbine and then converted into electrical energy when the turbine rotates an electrical generator.

Wind power varies directly with the cube of the wind speed. When the wind speed doubles, eight times more energy is generated. Wind turbines are, therefore, located at high altitudes, at seashores, at mountain or hill ridges, and offshore. Because so much power is generated by higher wind speed, much of the average power available to a wind turbine comes in short periods.

A large wind turbine can generate 10 MW. Considering that there are 900 MW thermal generating units, many wind turbines are needed to generate the equivalent amount of power. **Figure 3–8** puts the size of one blade of a Wind Turbine in perspective.

Figure 3–8 Modern wind turbine blades are very large and require expert riggers to raise them into position. *(iStock.com/Milehightraveler)*

Wind has the ability to meet the ideal of sustainable power with very little impact on the environment. The fuel (wind) for wind energy is free and clean. It is a sustainable energy source in that the fuel (wind) is not being depleted and lost forever to future generations such as oil and coal. Wind energy has a zero impact on the greenhouse effect. Large-scale wind farms connect to a transmission grid. Individual small wind turbines are also used in areas isolated from the grid and are especially ideal in remote communities that are otherwise served by diesel-powered generators.

When a wind farm is proposed, people who live nearby often have the perception that wind turbines are noisy, unsightly, and interfere with and kill birds. It has been stated that the most common objection to wind farms is probably the same objection that obstructs almost all generation and transmission line construction projects, which is NIMBY (not in my back yard).

Historically, the most recognizable wind turbines are the lattice steel towers with a multi-bladed windmill mounted on top. They were, and in some cases are still, used by farmers to generate electricity or pump water. **See Figure 3–9.** The many blades on the windmill provide good starting torque. When used to generate electricity, windmills had a DC generator that charged storage batteries. There was no need, therefore, to worry about AC frequency or voltage regulation.

Modern wind turbines tend to have two or three blades mounted on a horizontal axis. Tend to be propeller turbines with three-, two-, and even one-blade turbines. Turbine blades that rotate around a horizontal axis tend to be propeller

Figure 3–9 Old and new wind turbines use the same energy source. *(iStock.com/Twilightproductions)*

turbines with three-, two-, and even one-blade turbines. Large wind turbines typically have low-speed, large-diameter blades coupled to an electric generator by a high-ratio gear box. The individual blades are like helicopter blades that will turn in and out of the wind as electrical load and wind speed change. The low speed reduces maintenance requirements, and the inertia of the large blades helps maintain a more constant speed.

Power Line Techs are often involved in constructing wind farms. They have the ability and knowledge to work at heights and to work with heavy rigging. Constructing a wind turbine and mounting a generator on a structure are not unlike erecting a transmission tower. The footings are similar to those in a heavy anchor dead-end tower. **See Figure 3–10**. The structure is typically a large steel pole anchored in a concrete base.

Figure 3–10 The anchor bolts of a wind turbine structure can experience great amounts of stress. *(iStock.com/Andyqwe)*

The construction and maintenance of a wind farm collector grid and the grounding grid is also largely line/cable technician work. **Figure 3–11** illustrates the extensive cable network and the switchgear that make up a wind farm collector grid or system? with some typical voltages. The switchgear (circuit breakers), and its related relays, that connects a wind farm to a utility grid is designed to open when utility circuit goes out because of a fault. Power line technicians need to check that this switchgear is open before working on the line when isolated. However, protective grounding and bonding are the only true protection when working on a circuit.

Figure 3–11 Illustration of a wind farm collector grid.

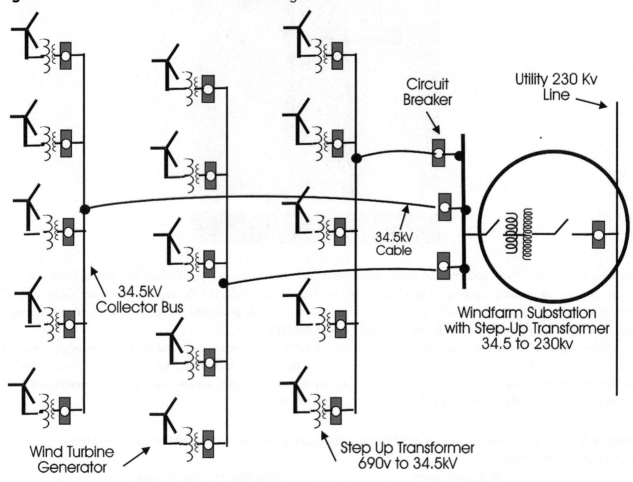

Solar Energy Converted to Electrical Energy

Solar Thermal Power Plants

Concentrating the sun's rays focused on a target can heat water and make steam to spin a turbine in a conventional thermal generating station.

In a *linear concentrator system*, long tubes (receivers) are placed in the focal point of a long run of curved (U-shaped) mirrors, as shown in **Figure 3–12**. The fluid flowing in the tubes is heated by the sun, and the fluid goes through a heat exchanger to boil water. Steam from the boiling water spins the turbine as in a conventional thermal plant to generate electricity.

Figure 3–12 Fluid is heated in tubes placed in the focal point of a long run of U-shaped mirrors. *(iStock.com /DougVonGausig)*

In a *power tower system*, sun-tracking parabolic mirrors (heliostats) concentrate the beams of light toward a receiver on the top of a tower, as illustrated in **Figure 3–13**. A fluid is heated goes through a heat exchanger to boil water. Steam from the boiling water turns the turbine in a conventional thermal plant to generate electricity.

Figure 3–13 Parabolic mirrors focus the sun's rays to a receiver on the top of a tower. *(iStock.com/Mlenny)*

Solar PV Utility Scale Power Plant

Solar energy can dislodge electrons and cause an electron flow in a photovoltaic (PV) cell made from material such as crystalline silicon. An individual PV cell generates only about 0.5 or 0.6 V and about 30 milliamps per square cm (0.16 square inches). However, when many PV cells are installed in a panel/module and connected electrically in series, a module can produce voltages such as 36 V. The solar panels are connected in groups or strings. For example, 42 of these modules in a string, connected in series could generate 1,500 V in the circuit. Multiple strings can be connected to a combine box where the output of all the strings becomes one direct-current (DC) circuit. This is fed to an inverter to transform the DC to alternating current (AC).

The output of the solar power plant is fed into a substation where there are more steps, including a feed into a step-up transformer to match the voltage of the utility grid.

The switchgear (circuit breakers) and its related relays that connects a solar farm to a utility grid is designed to open when the utility circuit goes out because of a fault. Power line technicians need to check that this switchgear is open before working on the line. However protective grounding and bonding are the only true protection when working on a circuit, see **Figure 3–14**.

Figure 3–14 A photovoltaic (PV) power plant used for commercial electrical generation. (iStock.com/Jeff_Hu)

Electrical Energy from Nontraditional Sources

- The burning of biomass can produce steam to spin steam turbines. Products such as sawdust and bark from the lumber industry, wood from fast-growing tree plantations, ethanol from corn, or methane from the decomposition of vegetation and garbage are burned in relatively small generating stations in many areas.
- Fuel cells generate electricity through an electrochemical process. The system converts the chemical energy of hydrogen or hydrocarbons and oxygen into electrical energy. In a fuel cell, hydrogen and oxygen are combined to form water and electricity (the opposite of the old experiment in which hydrogen and oxygen are produced when electricity is passed through water). The hydrogen needed for a fuel cell can be found in natural gas, coal-derived gas, ethanol, gasoline, and other fuels.
- *Tidal Power* generates electricity by placing the turbines under water. The surge of the tide's rise and fall spins the turbine. Tidal power is effective in locations known for unusual high tides, for example, the high tides in the Bay of Fundy in New Brunswick, Canada.
- *Geothermal* generating stations use steam that comes from hot water deep below the earth's surface. California and Hawaii have geothermal stations.
- *Engines*, such as locomotive or ship engines, are used to spin generators, especially in remote communities. On a smaller scale, there are many customer standby or portable generators found anywhere along a power line.
- *Hydrogen fuel cells* generate electricity through an electrochemical process. The system converts the chemical energy of hydrogen or hydrocarbons and oxygen into electrical energy. In a fuel cell, hydrogen and oxygen are combined to form water and electricity (the opposite of electrolysis, the making hydrogen and oxygen when electricity is passed through water). The hydrogen needed for a fuel cell can be found in natural gas, coal-derived gas, ethanol, gasoline, and other fuels.
- Stored *compressed air* can be used to drive a compressed air motor. It can also be injected in an internal combustion turbine, where it is burnt with fuel to provide mechanical energy which then powers a generator. Wind turbines can use excess power to compress air, this is usually stored in large aboveground tanks or in underground caverns.

Electrical Energy from Stored Sources

An electrical system needs to be able to meet the peak customer load, which tends to be the air-conditioning load on the hottest days. Customer load can change hour to hour. Electrical generation from wind and solar is not always available during peak load. This can cause problems with some traditional power generation because a thermal plant cannot be easily turned on and off. Ideally electrical energy can be stored for later use. Storing electrical energy can go a long way to stabilize the grid.

Pumped Storage

Electrical energy is being stored using pumped storage at some hydraulic generating stations. Water is pumped back up to the forebay during low electric power demand and then released during peak demand. On some rivers, water is held at the dam and released when needed, usually during the daily peak.

Battery Energy Storage Systems (BESS)

Large utility scale, battery energy storage system (BESS) has been installed connected to transmission and distribution lines and in substations as illustrate in **Figure 3–15**. Grid scale batteries give an operator more flexibility to meet peak customer demand. A battery energy storage system (BESS) is often installed along with wind and solar farms to store power until it is needed. A BESS system can also store energy from the grid during low customer demand, and low prices and then sold at a higher price during peak loads (arbitrage).

   Pole-mounted and underground batteries are also being installed. Batteries connected to the secondary can be charged by rooftop solar generation, which can be discharged to reduce peak demand, adding to stability in the whole grid. Customers can also install a battery backup uninterruptible power supply (UPS). A battery UPS is especially effect to store and release energy, especially those with solar panels.

Figure 3–15 An illustration of a battery energy storage system. *(petovarga/Shutterstock.com)*

BATTERY STORAGE

Distributed Energy Resources (DER) a Risk to Power Line Technicians

Electrical energy fed back from small electrical generation onto distribution lines (DER) can be found anywhere along a line. Power can be generated by gasoline or natural gas backup generation, from solar panels, as shown in **Figure 3–16** and stored energy battery packs. In other words, depending on how a customer makes their connections, it is possible for locally generated power to backfeed a distribution voltage to a high distribution voltage anywhere along the line.

   Checking that the switchgear is open at all possible electrical sources is not the best protection from backfeed. Apply protective grounds and bonds to protect power line workers from all generation sources.

   When working at a transformer remove or ground the secondary leads to prevent a possible energized secondary inducing a primary voltage at the high-voltage bushing.

Figure 3–16 Small-scale on-site generation can be found anywhere. Apply personal protective grounds. *(iStock.com/RoschetzkyIstockPhoto)*

Transmission of Electrical Energy

Transmission of Electricity

Electrical energy can be transported economically over long distances. Electricity is transmitted from the generating station to the customer load centers on high-voltage transmission lines. A transmission line can be compared to a water pipe: The higher the pressure and the larger the pipe, the more water will flow through the pipe. Similarly, the higher the voltage and the bigger the wire, the more electrical energy will be able to flow through the transmission line.

Typical Transmission Line Construction

The vast majority of transmission lines are overhead because underground lines are prohibitively expensive for long-distance transmission. Overhead conductors are suspended on structures such as lattice steel towers, wood poles, concrete poles, or steel poles. The purpose of a structure is to keep high-voltage conductors insulated from ground in all kinds of weather and out of reach of accidental contact. Tall structures allow long spans and, therefore, fewer structures.

The insulator length or size is dependent on the voltage: The higher the voltage, the longer the string of insulators. This is true in a transmission line corridor with lattice steel structures. **See Figure 3–17.** The tower in the foreground

Figure 3–17 The length of the insulator strings on the single-circuit transmission line on the left indicates that it has a higher voltage rating than the double-circuit line on the right. *(iStock.com/Yingxiaoming)*

is a single-circuit, **extra-high-voltage** (EHV) line. Notice that the tower in the background is a double circuit line with shorter insulator strings, which indicates that it is a lower voltage. Conductors are usually stranded aluminum with a steel core. Aluminum is a good conductor of electricity, and the steel core gives the conductor tensile strength. A strong, lightweight conductor can be strung with less sag over long spans.

Transmission Line Voltage

Commercial stations generate power at a voltage ranging from 13,800 to 24,000 V. A step-up transformer station next to a generation station boosts the voltage (pressure) so that it can be transmitted efficiently. Generation voltages are boosted up to common transmission line voltages such as 115,000, 230,000, 345,000, 500,000, and 765,000 V. The high voltages are normally expressed in kilovolts (kV) so that a 500,000 V line becomes a 500 kV line. As a rule of thumb, if the voltage is doubled, the energy that can be transmitted is quadrupled without an increase in line losses.

EHV lines, such as 500 kV circuits, use bundled conductors, which are two, three, four, six, or eight conductors tied together with spacer dampers. Conductors are bundled to counter certain problems caused by extra-high voltage; however, the increased conductor capacity plus the high voltage allows a single 500 kV circuit to carry the equivalent of eight 230 kV circuits.

Transmission System Substations

The terminals of transmission lines are at substations and switchyards. Substations are voltage-changing stations. Transformers can step up the voltage to allow for the efficient high-voltage transmission of power or can step down the voltage to allow for a more manageable voltage to distribute the power down roadways and streets.

A transmission substation with power flow from two 230 kV transmission circuits coming in from the left, through the transmission switches and breakers, and through the transformer, will step down the voltage to 46 kV subtransmission voltage. **See Figure 3–18.** Four subtransmission circuits exit this station on the right. A spare 46 kV breaker will be the foreground. In the layout for a small transmission substation, note that the switching arrangement for both the transmission line side and the **subtransmission line** side is designed so that the station can continue to feed out on each subtransmission line when any one component is out of service. **See Figure 3–19.**

Figure 3–18 This transmission substation transforms 230 kV to 46 kV. *(Delmar, Cengage Learning)*

Figure 3–19 A typical transmission substation layout has the ability to take any one element out of service and still provide power to each subtransmission circuit. *(Delmar, Cengage Learning)*

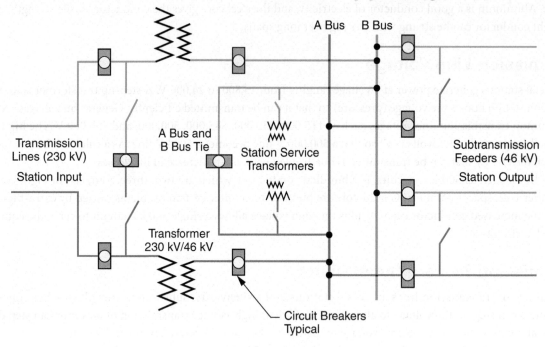

Switchyards

Switchyards are found at the terminals of transmission lines. A switchyard has disconnect switches, circuit breakers, relays, and communications systems to provide circuit protection. Switchyards allow the routing of power through various circuits to ensure that customers continue to receive service even when some parts of a power system fail.

A switchyard ties the many circuits coming into the yard to a common circuit called a bus. The term *bus* comes from the word omnibus, which means a collection of numerous objects or, in this case, a collection of numerous circuits. A bus must be able to carry very high current and, therefore, usually consists of large, rigid aluminum or copper pipe or very large conductors. Switchyards are usually within the same fenced area as the transformer and form part of the substation.

Communications between Stations

Operators in a control room monitor meters and alarms that indicate the condition of the substations and lines within their zone of control. An operator can open and close switchgear in remote generating stations and substations. This "supervisory control" of the system depends on communications systems between stations.

To transmit information and signals from station to station, utilities use telephone lines, wireless local area networks (LAN) utility-owned fiber-optic cable, power line carrier systems, microwave systems, or satellites. Because continuous communication is critical, usually more than one system is in place in case one fails. The use of the (Internet) for communications exposes the electrical system to hacking.

A power line carrier system uses the power line conductors to transmit information. The communication signals are sent on to and received from the power conductors by a device that looks like a potential transformer but is a coupling capacitive voltage transformer (CCVT). To keep the transmitted signals within the desired sections of the power line, wave traps are installed. The wave trap, which looks like a large cylindrical coil, stops the signals from continuing farther down the line.

Microwave communication between substations requires towers with microwave antennas in each station. Microwave sending and receiving antennas need a direct line of sight with no obstacles between them. Microwave towers are located on hills, where possible, and are about 35 to 60 miles (60 to 100 km) apart to relay the signals between towers.

Electrical Power Pools

Generating stations are interconnected by transmission lines into giant regional pools or grids that cross utility boundaries. The flow of electric power in these grids goes where it is needed. The flow could go south during a heat wave to feed peak air-conditioning loads or north during a cold snap to feed the peak heating loads.

Metering at line terminals or substations determines the amount of energy that crosses utility boundaries and what payment needs to be exchanged. Sometimes a utility's transmission line only transfers or wheels power from one neighboring utility to another. The utility receives payment for supplying this wheeling service.

Blackouts and Brownouts

There was a huge **blackout** in the central and northeastern United States and Canada on August 14, 2003. The failure in one element of the power pool started a chain reaction that led to a loss of most of the transmission grid. A fault on one transmission line caused another transmission line to overload and fail, which caused the generation that fed the line to be isolated from the grid, which overloaded neighboring lines and generation. Protection schemes are in place to isolate failures to the offending location; however, these protection schemes have been overwhelmed and failed.

Along with improved protection schemes, utilities have procedures that intentionally lower the voltage on or shed load from the system when the customer demand is greater than the system can supply.

When customer demand on the power pool is higher than available generation or transmission lines can provide, shedding or dropping load is a last resort. Before any load is shed, the voltage is lowered on the grid, which reduces the total energy supplied to the customers. Customers may notice that their lights become a little dimmer and their motors run hotter. About twice a year some utilities conduct tests by reducing the voltage on the system. These **brownouts** are generally only noticed by customers who are already receiving below-normal voltage in ordinary times.

After intentionally reducing the voltage on the system, and if there is still not enough to meet customer demand, some large industries have their loads dropped from the system first. These industries have a contract with the utility that allows the dropping of their loads in exchange for better rates.

In the unusually cold winter of 1993–1994, there was difficulty meeting the demand in Washington, D.C. Instead of implementing rotating blackouts, demand was reduced by closing federal buildings on the coldest days.

When all else fails, electrical load is shed on a rotational basis to the general population for a preset time. The deliberate dropping of load on a rotational basis results in rolling blackouts to specific geographical areas for specified periods of time, usually 30 or 60 minutes.

The Electric Power Grid

For a given large geographical area, generating stations, transmission lines and distribution systems are interconnected into one network. Power is bought and sold between electrical utilities through the grid. When generating stations, wind/solar farms are all tied together, utilities in the grid can shop for the cheapest power. When the grid crosses time zones, one utility can be under peak power demand and get relief from another utility where the demand is lower.

Smart technology is activated when a relay sees a fault in the system. Automatic switchgear opens to take a faulted generator or circuit out of service. Power to supply customers is automatically rerouted to other circuits, sometimes from other utilities within the grid.

Electrical Distribution

Distribution Basics

The transmission system brings electrical energy close to the load center and transforms the voltage down to a subtransmission voltage or directly to a **distribution voltage**. The distribution system consists of subtransmission lines feeding distribution substations, which transform the voltage down to distribution feeder voltage. **See Figure 3–20.** Distribution feeders deliver the energy to a transformer at the customer's premises and transform the voltage to a utilization level. By far, the biggest volume of line work involves the distribution system.

Figure 3–20 A distribution system feeds individual customers. *(Delmar, Cengage Learning)*

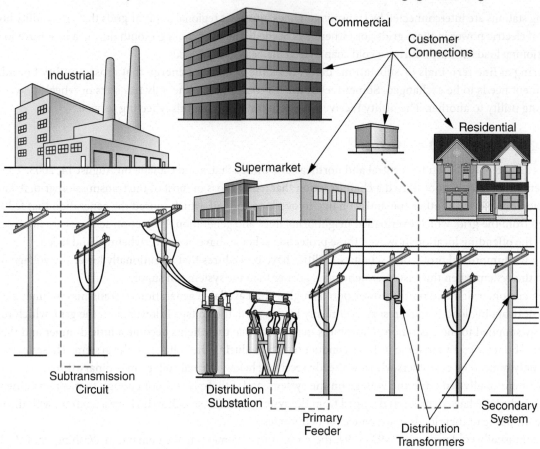

Six Main Segments of a Distribution System

Segment	Function
Subtransmission Circuits	Subtransmission circuits transmit power from the large transmission substations to the distribution substations. Examples of subtransmission voltages are 13.8, 27.6, 34.5, 46, and 69 kV. (**Note:** Voltages of 13.8, 27.6, and 34.5 kV are used in distribution in many places.) At these voltages, the structures and insulation are small enough to allow building lines along roadways. Some utilities may consider subtransmission lines to be part of the transmission system.
Distribution Substation	The transformer in the distribution substation steps down the subtransmission voltage to a distribution voltage. The substation consists of: • Switchgear on the subtransmission circuit • Transformer • Voltage regulation equipment • Distribution voltage bus • Multiple feeders connected to the bus • Distribution voltage switchgear for each feeder Many distribution substations are operated remotely from a central control room, which has access to substation data, such as feeder voltage and loading, and has the ability to operate the substation switchgear. Supervisory Control and Data Acquisition (SCADA) is the communications technology used to operate a distribution substation remotely.

(Continued)

Primary Feeders	The primary (**medium-voltage**) feeders leaving the station can be underground or overhead and are normally a three-phase type. A distribution feeder can be a radial feeder that branches off and ends at the end of a street or road, or can be networked into a grid with other feeders, allowing it to be fed from two directions. The loop between feeders can have a normally open switch keeping them separate or can be looped with automatic switchgear.
Distribution Transformer	The distribution transformer feeding the customer steps down the primary feeder voltage to a utilization voltage. Depending on the type of distribution system, the transformer can be overhead, on a concrete pad, or below grade in a vault.
Secondary Systems	A secondary system can range from a single service fed from one transformer to a secondary bus network fed from many transformers.
Customer Connections	The service to the customer can come directly from the transformer or from a secondary bus. The service can be overhead or underground. The responsibility for the utility service usually ends at the electric revenue meter.

Distribution System Designs

A distribution system can be laid out to give varying degrees of service continuity. A system with a high degree of service continuity is more expensive and is found where the customer density is high, namely in cities.

Radial System

The layout of a radial system is much like the design of a tree. **See Figure 3–21**. The main trunk is one of the three-phase feeders going out from a substation. Three-phase or single-phase branches or lateral taps feed customers along the circuit. The conductor in the main trunk carries the most load, and the branches get smaller as they feed out from the trunk. The length of a feeder is usually limited by the voltage and connected load.

Figure 3–21 Radial feeders are fed in one direction. *(Delmar, Cengage Learning)*

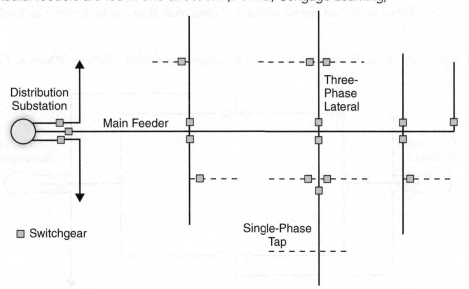

Loop Primary

A typical loop circuit starts at the distribution substation, makes a loop through the area to be served, and ends by returning to the substation. **See Figure 3–22**. It is similar to two radial circuits with their ends tied together. A loop primary keeps most customers' power on automatically when a fault occurs on the line.

Figure 3–22 A loop primary allows a system to be fed in more than one direction. *(Delmar, Cengage Learning)*

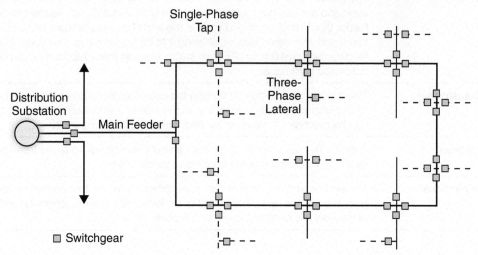

Circuit breakers are installed in the loop so that sections of the loop can be automatically isolated with the opening of any two breakers. Relays sense an overload situation and cause circuit breakers to open on each side of the fault. An underground tap is usually built in a loop, but the feed is kept radial by having an elbow removed and parked, kept as an open point between the two sources feeding the loop. On a switching diagram, this elbow would show as a normally open (NO) point.

Lateral taps from the loop are usually radial. An underground tap is usually in a loop, but an open switch within the loop keeps the two sides fed radially.

Primary Network

A primary network is used for heavily loaded downtown areas in a city. **See Figure 3–23**. It is similar to a loop primary except that the loop is fed from more than one substation and from more than one feeder from each substation.

Figure 3–23 A primary network can have complex automatic switching schemes. *(Delmar, Cengage Learning)*

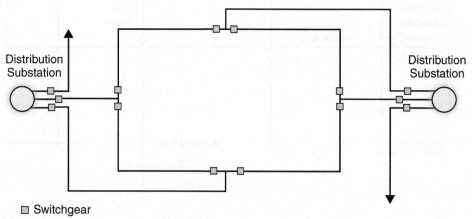

Overhead and Underground Systems

A distribution system is either overhead, underground, or a combination of both. Urban centers tend to have underground systems, and rural areas tend to have overhead systems.

Advantages of an overhead system include:

- Lower costs for conductor and associated switchgear and transformers
- Easier and quicker detection and repair of a breakdown within the system
- Much lower cost to upgrade an existing overhead system because of less need to dig up finished streets, curbs, and lawns

Advantages of an underground power system include:

- Almost no exposure to storms, trees, automobile accidents, insulator breakage, and insulator contamination
- More aesthetically acceptable to the public
- A necessity around airports, or where local laws require cable
- Long runs across water as submarine cable
- Less public exposure to the possibility of electric shock

Two Types of Underground Systems

Generally, there are two types of underground systems: the duct and maintenance vault system and the direct-bury system. The duct and maintenance hole system is used in cities where the presence of concrete and pavement would require expensive digging for maintenance or upgrading. The cables are in concrete or plastic ducts, and equipment such as transformers and switching units is below the surface in maintenance holes.

The direct-bury system is used primarily in residential subdivisions where most of the cable is buried under grass. The cable must have an envelope of sand around it to prevent any pressure points on the cable which are often sources of cable failure. Instead of using a sand envelope, most utilities put their cable into some kind of conduit, often PVC pipe, to provide better mechanical protection for the cable. The transformers and switchgear often have a "pad-mount design" and sit on the surface on a concrete pad. **See Figure 3–24.**

Figure 3–24 Pad-mount switchgear is commonly used in residential subdivisions. *(iStock.com/Benkrut)*

Managing a Distribution System

Most line work is in the distribution system. A variety of skilled people are needed to keep a distribution system functioning.

Skilled Workers in a Distribution System

People	Work
Design Engineers and Technicians	Set the design standards for structures, equipment poles, and so forth, considering their strength, electrical clearances, radio interference, lightning protection, and insulation.
Planning Engineers and Technicians	Monitor the system voltage and load. Decide on funding priorities for betterments. Carry out studies for fuse coordination, voltage regulation, voltage flicker, and load growth.
Customer Service Staff	Depending on the individual utility organization, customer service staff is the front-line customer contact regarding the following: • High bills • Meter reading • Conservation information • New service and service upgrade requirements • Collections for nonpayment of bills
Line Crews	Construct, maintain, and troubleshoot the overhead and underground distribution system. Numerous individual tasks could be listed here.
Tree Crews (arborists)	Keep lines clear of vegetation by trimming trees that are in close proximity to energized circuits, and keep customers relatively happy with the quality of their work.

Review Questions

1. The utilization of electrical energy comes from its four main effects. What are three of the four effects electrical energy can produce?

2. What physical act is required to generate power in almost all commercial generators?

3. What are three common energy sources used to drive a turbine?

4. Name four nontraditional energy sources.

5. If the voltage on a transmission line could be doubled, how much more energy would it be able to transmit?

6. What is the purpose of communications systems between substations?

7. Name the six main segments that make up a distribution system.

8. Name three advantages of an overhead distribution system versus an underground system.

9. Name three advantages of an underground distribution system versus an overhead system.

Chapter 4

Working in Substations

Objectives

After completing this chapter, you should be able to:

1. List the types of substations and their purpose.
2. Identify and describe the purpose of the elements found in a substation.
3. Trace the current flow through a substation.
4. Describe how communication is used to monitor a substation.
5. Operate a disconnect switch in an equipotential zone.
6. Perform an inspection of a substation.
7. Construct an equipotential ground grid in a substation.
8. Identify the hazards and the barriers to control the hazards in a substation.

What Is a Substation?

What Does a Substation Do?

A typical substation is a transformer station and a switching station. In substations, lines coming from different generating stations, voltage systems, and utilities are brought together, interconnected, and transformed into a needed voltage. A substation also has the equipment to provide protection for all the outgoing lines and for the transformers and switchgear inside the station yard.

The following equipment helps make the interconnections:

- Circuit breakers that control the lines going in and out
- Regulators that adjust the voltage
- Capacitors and reactors that adjust the volt-amperes reactance (VAR)
- Transformers that change the voltage of incoming lines to a different voltage for outgoing lines
- Auxiliary components such as current transformers, potential transformers, relay equipment, and metering provide information to SCADA (Supervisory Control and Data Acquisition); SCADA monitors the information and auto-matically activates switchgear during fault conditions
- Surge arrestors that shunt high-voltage lightning and other electrical surges to ground

Types of Substations

There is a tendency to call all stations downstream from a generating station a "substation." An electrical station can be a step-up transmission substation, a step-down transmission substation, an industrial substation, a distribution substation, a customer substation, a converter station, a mobile substation, a collector station, or a utility scale battery station.

A step-up transmission substation is located just outside the generating station. It looks the same as all substations except that the output voltage from the transformer is stepped up to a higher transmission line voltage. There is a practical limit to the voltage that a generator can generate, so lines (usually underground) from the generator go to the step-up transmission substation located nearby, where the voltage gets boosted to a more economical standard transmission line voltage. For example, a generating station generates three-phase power at a voltage between 11 and 22 kV. Three-phase cables rated at the generated voltage and large enough to carry the relatively high current go to the nearby step-up substation. At the substation, the cables are connected to circuit breakers, disconnect switches, and then to a transformer that steps up the voltage to a transmission voltage such as 230 kV. There may be autotransformers that then step up the voltage to extra high voltage such as 500 kV. The outgoing lines then go through a circuit breaker, which will protect the lines for a long distance.

A step-down transmission substation is usually located near a load center where the incoming transmission lines can be stepped down to a subtransmission line voltage, which is an easier voltage line to build along streets and highways. This substation is also a switching station, where lines from different parts of a grid come together and are switched to send power to the desired locations. The circuit breakers, instrument transformers, metering, and relays are set up to isolate any faulted equipment or line. A transmission substation is very similar to a distribution substation except that everything is bigger, and there is the added complexity of, for example, capacitor banks, reactors, and large instrument transformers.

Large industries often have their own step-down transmission substations that are fed directly by a transmission line. Because it costs less to supply a business that has its own substation, rates will be lowered accordingly.

Distribution substations are located near end users. Distribution substation transformers change a subtransmission voltage to one of a dozen standard lower-level voltages between 34.5/19.9 kV and 4.16/2.4 kV. This type of substation houses the source circuit breakers for the outgoing distribution feeders.

A rural substation typically consists of a single radial-feed subtransmission line, a single transformer, and three medium-voltage distribution feeders protected by reclosers. **See Figure 4–1**. To keep customers on during maintenance or an emergency, a mobile substation is necessary.

Figure 4–1 A typical layout for a small substation has distribution feeders fed through a transformer and a source subtransmission line. (*Delmar, Cengage Learning*)

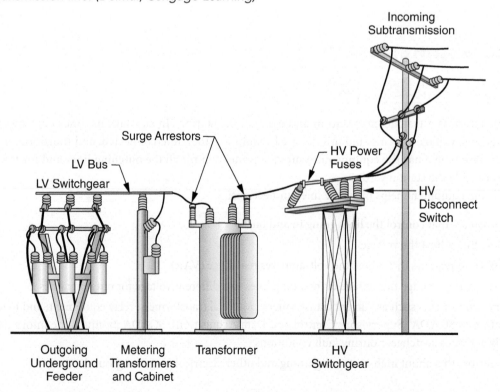

A larger substation can be similar but with a transmission line feeding through the high-voltage disconnect switches and fuses until it feeds into the high-voltage source side of the transformer. **See Figure 4–2.** The transformer in the same substation feeds the low voltage bus, where outgoing underground feeders are connected. **See Figure 4–3.** The feeders flow through circuit breakers that protect the outgoing distribution lines. An urban distribution substation has the same function and pattern but can be much more complex. It may have one or two overhead or underground incoming (e.g., 230, 115, or 69 kV) subtransmission lines, two or more transformers, and maybe eight or more outgoing overhead/underground distribution feeders. In this type of substation, one transformer can be taken out of service (depending on load); with some switching, the other transformer can pick up the load. Similarly, a distribution feeder breaker can be racked out for maintenance and the load switched to other breakers. There is a practical limit to the number of distribution feeders that can come out from one substation and distribute among nearby streets.

Figure 4–2 A transmission line is the source for the high-voltage side of a larger substation. *(Delmar, Cengage Learning)*

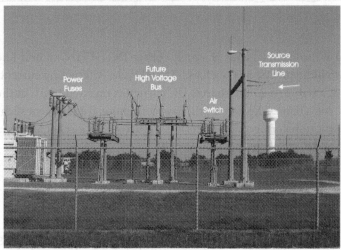

Figure 4–3 The distribution feeders exit underground from the low-voltage bay of the same substation. *(Delmar, Cengage Learning)*

Mobile Substations

If a major element in a distribution substation, such as a transformer, fails, customers can often be fed from a second bank within the substation or be backfed from other substations. In more remote locations, the economics of building a double-bank substation or having the ability to backfeed from other substations is not an option to provide power during a fault or during major maintenance. A mobile substation can be brought in to provide power to customers. **See Figure 4–4.** Ideally the faulted substation is designed and set up to accept the mobile substation.

Figure 4–4 A mobile substation can be moved in to provide power to customers.

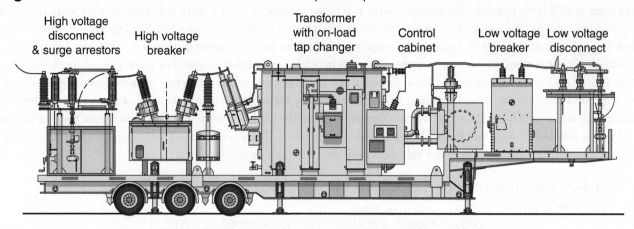

High voltage disconnect & surge arrestors — High voltage breaker — Transformer with on-load tap changer — Control cabinet — Low voltage breaker — Low voltage disconnect

Converter Stations

Converter stations are found at the terminal ends of a high-voltage DC (HVDC) power line, a submarine DC line, or a short DC tie line between two different systems, such as 50 to 60 cycle. Alternating current (AC) is converted, or rectified, to direct current (DC) at a station near the source. DC current is inverted to AC at the other terminal. Solid-state devices (thyristors) make the conversion. The station will have conventional equipment such as switchgear, transformers, reactors, capacitors, instrument transformers and metering. **Figure 4–5** is a schematic of a converter station with AC circuit coming from the left and DC circuit going out one right.

Figure 4–5 AC to DC converter station.

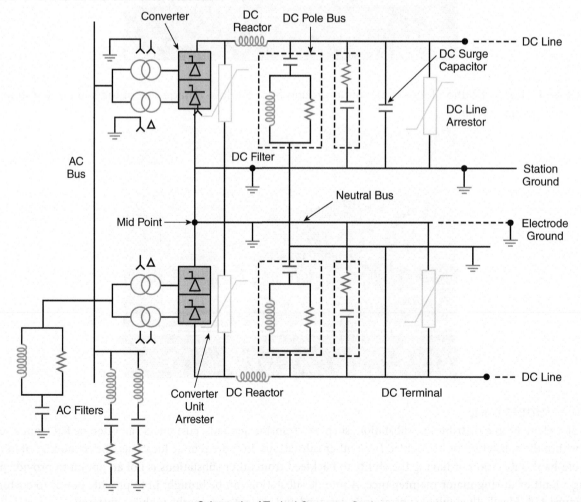

Schematic of Typical Converter Station

Collector Substation

A collector substation is typically found at a wind or solar farm. Multiple electric feeders from the individual wind turbines, for example, are combined into one. A transformer in the station boosts the voltage to match the electrical utility grid voltage. Equipment in a collector substation includes, metering, switchgear, transformer, and overhead or underground conductors. **Figure 4–6** shows the underground cables coming from the wind turbines gathered together and fed into a step-up transformer.

Figure 4–6 Cables from wind turbines come together in the collector substation. *(iStock.com/Pedrosala)*

Grid-Scale Battery Station

Utility grid-scale "battery energy storage system" (BESS) stations stabilize the grid by reducing peak loads and providing load during normally low customer demand. They are especially effective for integrating the energy from wind and solar farms. A utility can charge batteries in their battery station when electrical power prices are low and discharge them when prices rise (arbitrage). The battery station can provide power when the electrical system is down, as part of storm hardening.

The battery station will have typical substation equipment such as switchgear, an inverter/charger, a step-up transformer, protective switchgear, and metering. **See Figure 4–7.** The batteries will likely be monitored remotely by a battery management system (BMS), which will monitor voltage, current, and temperature.

Figure 4–7 Utility-scale battery station.

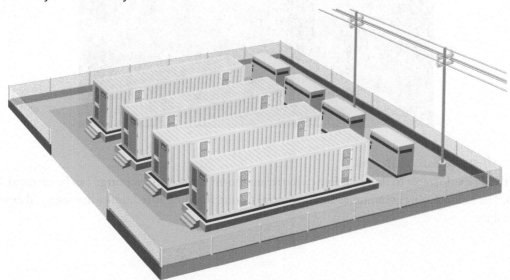

Types of Substation Construction

Substations can be either air insulated or gas insulated. Both of these can be outdoor or indoor, totally enclosed or partially outdoor, or completely underground.

An air-insulated substation refers to the classic, most common substation. Older substations typically have a lattice steel structure. **See Figure 4–8**. Air is the insulating medium around the bus and switchgear, but oil or gas is the insulating medium in the transformer, switchgear, and other equipment.

Figure 4–8 Except for the transformer and reclosers, air provides the insulation for this particular substation. *(Delmar, Cengage Learning)*

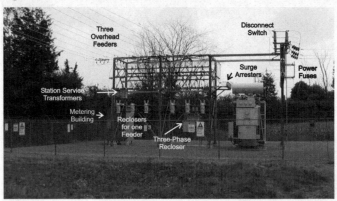

A gas-insulated substation can be indoors and totally enclosed or outdoors. Gas-insulated switchgear and transformers are very compact, and the substation resembles a network of pipes. Sulfur hexafluoride (SF_6) gas is both an insulating medium and a cooling medium for transformers, and an arc-quenching medium in switchgear.

In metal-clad distribution stations, the medium-voltage switchgear and bus are inside a metal cabinet. To make a substation virtually invisible to the public, the metal-clad switchgear can be in a building, the incoming and outgoing lines can be underground, and the transformers surrounded by a wall. **See Figure 4–9**. Variations of this configuration show overhead lines or transformers sitting in the open.

Figure 4–9 A substation can blend into the neighborhood and have the source and outgoing feeders underground. *(Delmar, Cengage Learning)*

Storm-Hardening Substations

Substations that are vulnerable to flooding from a hurricane storm surge or ocean surge can be elevated, as shown in **Figure 4–10**. The risk of losing communication or control during flooding is reduced when using fiber-optic instead of electrical cables.

Figure 4–10 Substation raised for flooding protection.

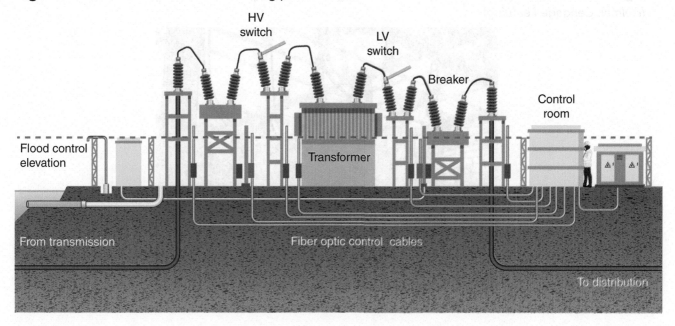

To protect substations vulnerable to wildfires, vegetation is kept as far away as feasible. Increasing the length of insulators will reduce the risk of flashover when insulators are contaminated.

Elements of a Substation

Substations have common elements, such as switchgear, transformers, and surge arrestors, but because of voltage, insulating medium, or inside/outside differences, each element can look quite different from substation to substation.

Station Yard

A station yard is one large, equipotential zone. There is a grounded wire-mesh network buried underneath the gravel or concrete of a substation. Each piece of equipment, the neutral of a transformer, the line neutrals, and the structure and fence are bonded to the grounding network. Numerous ground rods are driven to bring the grounding network resistance (impedance) as low as possible, or at least to a level to meet standards. A line-to-ground fault in the station or on a line results in a rise in potential on the total grounding network in the station. When everything is bonded together, there is less chance of a worker in the yard getting between two different potentials. The grid is also extended slightly beyond the fence perimeter to protect outsiders who may be in contact during a fault.

Switchgear

The most sophisticated switchgear in an electrical system is found in a substation. Switchgear can range from large oil, air, or vacuum circuit breakers that protect extra-high-voltage (EHV) lines to distribution line reclosers to protect a distribution feeder. Circuit breakers provide automatic opening and closing under fault conditions. Oil circuit breakers have been the most common, but SF_6 gas and vacuum breakers have advantages of less size and maintenance and are more suitable for indoors. **See Figure 4–11.**

A substation has load break and non-load-break switches that provide a visible open point to allow maintenance on, among other things, the circuit breaker and the transformer. For example, a 230 kV switch can be opened and have protective grounds installed to provide a visible open point to carry out maintenance in the substation. **See Figure 4–12.**

Figure 4–11 On a SF$_6$ breaker protective working grounds are in place for carrying out maintenance. *(Delmar, Cengage Learning)*

Figure 4–12 Protective working grounds are put in place on the deenergized side of a 230 kV air break disconnect switch. *(Delmar, Cengage Learning)*

High-Voltage and Medium-Voltage Bus

Every circuit coming out of or going into a substation is connected to a **bus**. A bus is like a line (conductor) except that it is inside the substation. It is often made of an aluminum-alloy pipe and sits on insulators, or it can be an ordinary power conductor supported by suspension insulators. Depending on the bus arrangement and naming convention, a bus can be a main bus, ring bus, transfer bus, A bus, or B bus. For example, if two 345 kV transmission lines feed the X bus and the Y bus and one circuit feeding the bus trips out, the other circuit can be switched over to feed the other transformer. **See Figure 4–13.**

Similarly, medium-voltage A and B buses are arranged so that the outgoing feeders can be switched to allow circuit breakers to be taken out of service while maintaining service, by supplying customers through other circuit breakers.

Figure 4–13 Some small substation layouts allow maintenance on any one element without needing to de-energize a 69 kV feeder. *(Delmar, Cengage Learning)*

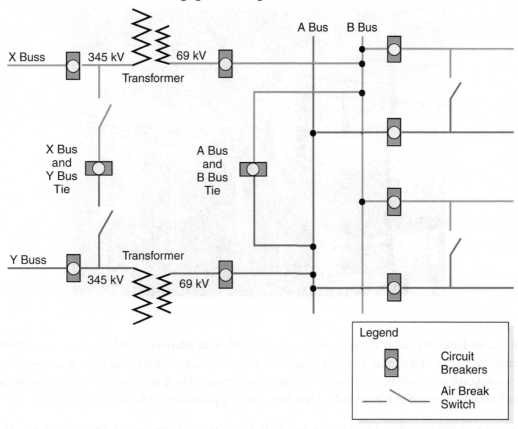

Power Cables

A substation can have underground transmission, subtransmission, distribution, and secondary cables below the ground grid in the substation yard. The cables can be direct buried or in conduit, raceways, or duct banks. The lengths of cables are often short, just long enough to get to an overhead structure or vault outside of the station fence. Work on the cables in the substation will be similar to work on cables outside of a substation, except that distribution feeder cables will likely be considerably larger and heavier.

Transformer

The transformer is the biggest and heaviest piece of equipment in the substation. The line voltage coming into the high side of the transformer is changed to a standard voltage on the low side of the transformer (which is opposite the setup of a step-up transformer). Transformers are most often oil insulated and cooled, but gas-insulated (SF_6) transformers are gaining popularity, especially in the city because they are noncombustible and more compact. The high-voltage side of the transformer will have the bigger insulators. **See Figure 4–14.** The high voltage coming in from the left is a 230 kV bus, and a 46 kV bus leaves the transformer on the right. This transformer uses larger radiators for cooling the oil.

Voltage Regulation Equipment

The voltage being supplied from a substation must be boosted or bucked on a regular basis as demand rises and falls. Many transformers have the ability to change the transformation ratio by using tap changers. Some transformers have an off-circuit tap changer, which means that the transformer has to be taken out of service to change the taps and the tap position is stationary. Other transformers have **on-load tap changers** that are much like a step-voltage regulator where the taps are changed automatically under energized conditions. Some substations will have a separate voltage regulator to boost and buck the voltage of the outgoing lines automatically under energized conditions.

Figure 4–14 The transformer steps down the incoming 230 kV circuit in order to supply multiple 46 kV subtransmission circuits. *(Delmar, Cengage Learning)*

Reactor

A substation reactor looks like and is about the same size as a station transformer, with large iron-core coils that generate inductive reactance. Each coil is connected phase to ground. Inductive reactance is needed in some substations to balance the capacitive reactance generated by very long EHV transmission lines. There are also air core dry-type reactors (ACRs) that resemble big coils of wire. **See Figure 4–15**. They have many applications, including the following:

Shunt reactors are connected to the circuit in parallel to serve the same purpose as the transformer-style substation reactor, that is, to introduce inductive reactance into the electrical system when needed.

Static VAR compensator reactors (SVCs) are teamed up with capacitors to form a complete piece of equipment that adjusts the reactive power (VARS) up and down as needed. In other words, the equipment can add capacitive or inductive reactance into the grid.

Figure 4–15 Protective working grounds are in place to allow maintenance on the air core reactor. *(Delmar, Cengage Learning)*

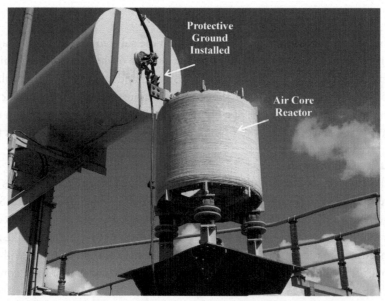

Current-limiting reactor (CLR) protects equipment by reducing any high fault current that would otherwise flow through it.

Damping reactors protect capacitor banks by reducing (damping) high inrush current when the capacitors are switched into service. The reactors are connected in series, whereas the capacitors are connected in parallel (shunt). **Figure 4–16** shows damping reactors mounted connected in series with a capacitor bank. Visible in the background are SVCs.

Power flow control reactors are switched by system control operators when the transmission power grid requires that the flow be adjusted in another direction. Generating stations and transmission lines are tied together in a grid. Power flows where there is least impedance. For example, power has been known to flow in an undesirable circle around and around Lake Erie, creating line loss. When the power in the grid will not flow in the desired direction, the power flow control reactor in a substation is turned on, the impedance is increased, and the current flow into the substation is reduced, causing current to flow in other directions within the transmission grid.

Figure 4–16 Damping reactors are installed to reduce high inrush current when the capacitors are switched into service. *(Courtesy of iStock Photo)*

Capacitor

Capacitor banks in substations generate capacitive reactance to counterbalance excessive inductive reactance produced by customer load and large transformers. Capacitors banks consist of lower-voltage capacitors put in series so that the series-connected capacitors can be connected to a high-voltage transmission line and neutral. Capacitors in a substation are isolated from ground by setting them on insulators. The capacitors are surrounded with danger signs to keep people from touching the framework above the insulators. A capacitor bank can be protected from a high inrush current by air core damping reactors. **See Figure 4–16.** Even though a substation yard is fenced, the capacitor bank may be surrounded by another fence or surrounded with danger signs to keep workers from touching the framework above the insulators.

Phase-Shifting Transformer

A phase-angle regulating transformer or phase-shifting transformer (PST) looks like a large substation transformer and is used to control the flow of power between different transmission lines and systems. It allows an operator to control the magnitude and direction of power flow by varying the phase shift between the input and the output of the phase-shifting transformer. Otherwise, when power is flowing in a grid or loop, the flow is in the easiest direction and not necessarily where it is needed.

Wave Trap (Line Trap)

Wave traps are seen in a substation when the power conductors are also being used as a communications line, called power line carrier communication (PLCC). **See Figure 4–17.** Communication between substations is critical in order to monitor the electric system and open and close switches. Typically, more than one form of communication is used in a substation, so in addition to telephone or fiber-optic communications, you may find PLCC. Wave traps look like ACRs, and are connected in series with the circuit to trap high-frequency signals that have been superimposed on the power conductor and to divert the signals to the control room.

Figure 4–17 Wave traps are used for power line carrier communication. *(Courtesy of iStock Photo)*

Inductor

During a fault, very high and damaging current can flow through transformers and breakers. An in-line, current-limiting inductor limits the current flow by creating a very high inductive reactance.

Station Service

In addition to having a transformer to supply the power needs for the station, a substation will have other devices, such as heaters, air conditioners, lighting, battery chargers, cooling fans, insulating oil pumps, air compressors, and equipment heaters. Secondary cables from the station service transformer are often underground. The cables are typically rated at 600 V. The station service transformer is an ordinary overhead distribution transformer used to feed the service in a smaller distribution substation. **See Figure 4–18.**

If you are called upon to restore a station service, remember that the fault current available in a substation is huge compared to a line outside of a station. Creating a short or an arc can generate a highly explosive, short-circuit current.

Potential and Current Transformers

Potential (voltage) transformers and current transformers are found connected to each source circuit and, in most cases, on each circuit leaving the substation. **See Figure 4–19.** The secondaries from potential and current transformers are a low-voltage and low-current representation of the actual voltage and current on the lines. These representative voltages and current go to meters and relays. For example, if the current goes higher than the values programmed into the relay, the relay sends a signal back to the circuit breaker to open the circuit.

Optical devices are installed as an alternative to potential and current transformers. These optical devices take advantage of the fact that the electromagnetic field of a conductor influences the polarity of light in an optical fiber. A computer in the control room interprets the effect on the light in the fiber and signals the meters and relays accordingly.

Figure 4–18 A standard distribution transformer is used to supply electrical service to a small substation. *(Delmar, Cengage Learning)*

Figure 4–19 Potential (voltage) transformers and current transformers are used to measure high voltage and current for relaying and metering purposes. *(Delmar, Cengage Learning)*

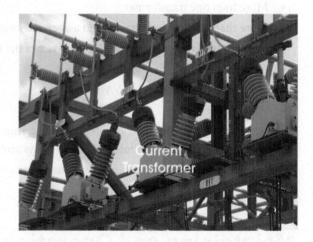

Control Cables

Most substations are operated and monitored from one central control center. Typically, there are thousands of control cables bringing in the necessary information to operate the electrical system. Control cables are communication cables between relays, instrument transformers, metering, and the control center. Substations communicate with the control center by radio, microwave, company-owned fiber-optic wire, and/or satellite. Inside the substation, the control cables can be power cables, telephone, fiber-optic, and/or coax. In the substation, yard control cables are underground or below the surface in accessible cable raceways, cable trays, and/or conduit.

Meter and Relay Equipment

Most substations are automated and remotely controlled. Equipment that is necessary to accomplish this task includes:

- Communications links and associated terminations and hardware for power line, telephone, fiber optics, and microwave carriers
- Control circuits, which are low-voltage lines that open and close circuit breakers
- Meters and relays, and all the wiring from the potential and current transformers to the control building
- Battery room and battery charger
- Control room or building, or cabinet, to house much of the auxiliary equipment

Power Flow through a Substation

Power flow through a step-down substation involves numerous steps: (1) A source or incoming line(s) flows through (2) a circuit breaker(s) that is capable of opening the circuit under a loaded or a fault condition, and then through (3) a switch on each side of the circuit breaker, so that the circuit breaker can be isolated from the circuit to carry out maintenance and to provide a visible open point in the circuit when required for work downstream, into (4) a high-voltage bus to which the incoming (and outgoing) line(s) is connected through (5) a transformer that drops the voltage (other than a step-up substation) into (6) a medium-voltage bus to which the outgoing lower-voltage lines are connected, through (7) a switch (fused) that can provide a visible open point for the outgoing circuit, into (8) a circuit breaker to provide protection and load break capability for the outgoing circuits.

The following complexities can make a substation look confusing (like a spaghetti jungle):

- Many high-voltage circuits, and associated switchgear, feeding into two high-side buses with switchgear to tie the two buses together
- More than one transformer
- More than one low-voltage bus and associated switchgear to tie the buses together
- Voltage-regulating equipment either as part of the transformer or separate
- Capacitor banks to provide leading VAR
- Reactors to provide lagging VAR
- Phase shifters
- Auxiliary equipment, such as potential transformers and current transformers on incoming and outgoing circuits to provide circuit condition information to relays and metering
- Incoming and outgoing communications circuits (telephone, microwave, fiber optics, power line carrier)

Maintaining and Operating a Substation

Entering a Substation

Most substations are remotely controlled. System control requires notification before anyone enters. If alarms go off in the control room, system control would like to know that people are there. System control will then have the option of making a call to the station before operating equipment.

Corporate safety rules and labor laws restrict entry to only people who have had electrical awareness training or are under escort by a qualified worker. One reason a substation has a fence, barbed wire, and warning signs is, unlike a power line, the clearance from live parts to ground or to a structure is easily reached with a boom, a ground rod, a ladder, or other equipment.

Substation Maintenance Programs

The ideal maintenance program is one in which work is carried out on equipment only when needed to keep it in optimal condition, and equipment that is in excellent condition is not overhauled. Reliability-centered maintenance (RCM) or predictive maintenance is preferred to maintenance based on a time schedule. There are good diagnostic tests available to reduce unnecessary maintenance and to focus more on priority needs.

Substation maintenance personnel can do testing to predict the need for maintenance. For example, they can take oil samples from in-service switchgear and transformers to analyze for levels of water, hydrogen, methane, ethane, ethylene, carbon monoxide, and carbon dioxide. A dielectric test of the oil will indicate the presence of contaminants such as water or particulates. A neutralization/acid test will measure the level of sludge-causing acid that is present in the oil. An interfacial tension test identifies the presence of polar compounds, which indicates oxidation contaminants or deterioration of the transformer materials, such as paint, varnish, or paper. A dissolved gas analysis (DGA) checks for any gas in the oil, which would indicate that arcing, corona, or overheated connections have occurred. Ultrasonic and vibration tests can detect arcing and corona discharge and pressure or vacuum leaks. Gas chromatography can determine the condition of insulating oil and insulating papers.

SF_6 gas-insulated equipment is monitored for gas pressure, temperature, and density. A reduced gas pressure means a loss of insulation that could cause a tripped circuit.

A thermographic (infrared) survey of a substation will point out any connections, splices, and switch components that may be loose or are getting hot and need attention.

Tests are also conducted before reenergizing a transformer if there are indications that it may have been subjected to a fault current. A test for each phase-to-ground resistance and each phase-to-phase resistance and or a DGA test can be done to ensure that a defective transformer is not being reenergized. Replacing all three fuses while a unit is out of service may prevent a weakened fuse from failing later.

Virtually every piece of equipment in a substation has a detailed written maintenance procedure, either the original equipment manufacturer's (OEM's) procedures or one prepared by the utility.

Detailed forms accompany each equipment maintenance procedure and are used to log all test results and repairs.

Inspecting a Substation

Compared to transmission and distribution lines, substations are inspected frequently. Transmission substations are inspected more frequently than distribution substations. A large urban distribution substation is inspected more often than a small, rural distribution station. Forms are used to provide a thorough checklist and log. The forms for a transmission station would be different than the forms for a rural distribution station. Some inspections are more thorough and require operational checks.

A walk-through inspection would allow a look at the following:

1. *Transformer:* Depending on transformer size, available gauges, and the like, look for damaged bushings, oil leaks, and the oil level in the main tank and tap-changer compartment. Under normal operating conditions, both the oil and winding temperatures should read below 160°F (70°C). Under loaded conditions, the winding temperature might read as high as 200°F (95°C).

2. *Tap changers and voltage regulators:* The operations on the load-tap changer are read and recorded from the counter indicator. Inspection of a voltage regulator is similar to that of a transformer as far as oil leaks are concerned. More detailed inspection may involve putting the regulator control in the manual position and operating it up or down over a small range and then returning it to automatic to watch it go back to the former position. Each of the three phases of the voltage regulator taps should be within four positions of each other.

3. *Oil, vacuum, SF_6, and air-blast circuit breakers:* Check for loose, contaminated, or damaged bushings, terminals, oil leaks, and proper gas pressures. Check oil level in bushings and main tank (as applicable). Check anticondensation heaters. Read and record the number of operations on the indicator.

4. *Disconnect switches:* Check for cracked, contaminated, or broken porcelain, loose connections, and corrosion to metal parts.

5. *Reclosers* are inspected for oil leaks. Recloser settings can be checked, where applicable, to ensure that the ground trip, reclosing, and supervisory settings are in the desired positions.

6. Check that the batteries are at about 125 V, check that the charger is working, and check for leaks.

7. Check for broken or contaminated insulators, surge arrestors, and bushings.

8. Check capacitor tanks for leakage or damage.

9. Check current and potential transformers for damage to cases, bushings, terminals, and fuses.

10. Take any requested meter readings.

11. Check visible ground connections at equipment, structures, switching platforms, and fences. Check continuity of fence grounds.

12. Check fencing for signs, gaps under the fence, third-party metallic fencing tied into the station fencing, and security of gates.

Operating in a Substation

Most substations are remotely controlled, although individual components may need to be operated by hand. Some distribution substations, especially rural ones, have no remote control capability.

A lineman may be asked to operate switchgear in a substation as an agent under the direction of system control. Ideally, you would have a written switching order with a device number and the operation requested with you.

It is essential that system control provide an operations drawing or a single-line diagram to follow as you work. A single-line diagram simplifies the physical arrangement, because a single line represents a circuit instead of the three phase conductors and the neutral. **See Figure 4–20.** Larger stations follow the same principle, but diagrams can look very complicated because of the quantity of transformers, buses, and tie switches.

Figure 4–20 Reference to a single line drawing is the best way to verify that a switching order is accurate. *(Delmar, Cengage Learning)*

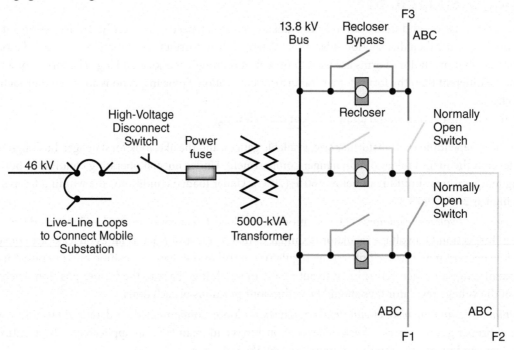

If asked to open or close a recloser or breaker from a control box mounted right at the equipment, be aware that a breakdown within the recloser or breaker can cause a fault to ground and energize the grounded network. Stand on a ground gradient grid or mat that is bonded to the steel structure. This equipotential arrangement ensures that an operator's feet and hands will be at the same potential if things go wrong. **See Figure 4–21.** Rubber gloves alone are not adequate protection.

Similarly, when a switch is being operated by someone standing on the ground, an insulator failure can cause the frame of the switch and structure to become energized. Standing on a ground gradient mat will protect the operator from any electrical potential. The grid may be buried under gravel, and some utilities insist on the use of a portable ground gradient mat if there is not a permanent one installed at the control box. **See Figure 4–21b.** If asked to open power fuses with a live-line tool, be aware that these devices are not designed to drop load.

Figure 4–21 Opening the switch handle while standing on ground gradient platforms ensures that there will be no voltage difference between the hands and the feet. *(Delmar, Cengage Learning)*

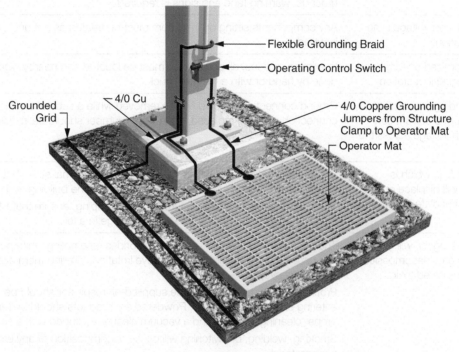

Hazards Working in Substations

The lineman can encounter numerous hazards in a substation. It's important to know the barriers that provide protection from these hazards.

Hazards	Barriers
Contact with energized overhead bus or leads going into equipment bushings and potheads	Live apparatus is much closer than apparatus encountered in line work. Minimum approach distances for truck booms and workers are easier to bridge. Long objects such as ladders, conduits, ground rods, and reinforcing steel should be carried by two people, horizontally, when in the live part of the station.
Identifying the wrong equipment as having been de-energized	Substations contain duplicate pieces of equipment. Accidents have occurred because people have climbed on the wrong breaker. Positive identification with grounds, warning tape and signs is required.
Contact with induced voltage in an energized substation	Any components sitting on insulators must be treated as hot until grounds are installed.
Contact with energized underground cables when digging in a station	In existing stations, all power lines must be located and nearby digging must be done by hand or with a "sucker" truck.
Opening a grounding connection, which can interrupt a current flow and create a high voltage across the open point	Ground connections should not be removed while a substation is energized. If a connection must be replaced, a temporary jumper should be installed.
The presence of SF_6, which is heavier than air, will displace oxygen and create an immediate suffocation hazard	SF_6 gas is odorless, tasteless, and nontoxic in its normal state, but an SF_6 gas leak in a basement will drift down to the lowest part of the building and displace air. To reduce the risk of a large volume of SF_6 escaping, ensure that SF_6 cylinders are stored outdoors and chained, to avoid damage in a fall.
Arcing inside switchgear, will decompose SF_6 gas; decomposition products are considered toxic	The decomposed products are metal fluorides resembling white powder. They can cause a rash, skin irritation, and eye irritation. Inhaling them can damage the respiratory system. Protective clothing and a full-face supplied-air respirator should be worn when entering electrical equipment. Powdered arc products should be removed with proper cleaning solvents and a vacuum cleaner equipped with a special in-line filter. Smoking, welding, or switching will cause decomposition of any escaped SF_6 into toxic material.
Damaged control cable, preventing a critical piece of equipment from operating	There may not be an immediate reaction to damage to a control cable. However, the damage will get worse over time and the control cable will eventually fail. This can result in equipment failing to operate at a critical time because it will not get the signal to operate. Call for repair if damaging contact is made with a cable.

Constructing a Substation

Locating a Substation

The ideal location for a substation is close to the load center but out of public view. The location must meet environmental and zoning regulations. Public meetings may be part of the approval process for locating a large transformer station. Large transmission substations with EHV lines going in and out are highly visible, no matter where they are located.

Some substations in a city center may not be noticeable to the public. Incoming and outgoing lines may be underground, and such stations may be situated inside large buildings.

The substation has to be built in a location where good grounding can be obtained. Good grounding is essential in order for relays to be activated to open switchgear for ground faults.

Work with the Substation Construction Drawings

The construction drawings for a substation are very detailed. They include measurements, wire sizes, types of electrical connections, torque required on bolts, pipe fittings, and more. The construction crew will have access to all this information, because installing control cables without the drawings would be impossible.

Construction in an Existing Station

Substation construction is often an addition to or a refurbishment of an existing substation where much of the existing station will stay energized. Barriers and warning signs are used to clearly mark out the energized part of the station. Equipment such as metal ladders must not be used. There can be energized underground high-voltage and distribution cables running through the construction zone that will have to be located. If digging is required around these cables, they must be hand dug or dug with a sucker truck. Typically, the first 5 ft (1.5 m) should be dug by hand or sucker truck before digging with a machine. The station yard grounding network has to stay intact or at least ensure that no open point is bridged elsewhere. Dangerous voltage and current can appear around an open point.

Where cables have been exposed, they must be supported to prevent stress on the cable. Mechanical protection over exposed cable is necessary to prevent accidental contact with tools and machinery.

All metal objects, such as excavation equipment, sheds, shoring, road plates, and temporary fencing, should be grounded.

Low-voltage and communications cables can also be present in a work zone. Contact with these cables may not lead to injury but can lead to a major outage.

Civil Work for a New Substation

There are many different types of substation structures, and the civil work varies accordingly. The description of the civil work in this section is for the more common air-insulated substation.

A foundation plan lays out the locations for the foundations of the station structure supports, the concrete pads that support equipment, and the control building. Concrete pads may need to support weights as heavy as 700 tons and to hold circuit breakers that can shake when operating under a fault condition. Structures need footings that will resist downthrust and uplift when tension is applied.

To reduce the risk of foundations shifting, topsoil and clay, which are compressible as well as affected by frost, are removed and replaced with a stabilizing backfill. The foundations are constructed with reinforced concrete. Oil retention systems are made to surround oil-filled equipment. Concrete is tested to ensure it is correct for its application.

Extensive trenching is required for power line cables, low-voltage circuits to equipment, and relay circuits from the potential transformers and current transformers. These cables can be in conduit or raceways covered by concrete slabs.

Constructing a Grounded, Equipotential Grid

A substation yard is constructed as one big, grounded, equipotential grid. The grounding grid is a mesh of wires and is bonded together at every crossing. The grid is connected and grounded so that electrical current, especially under fault conditions, will be carried into earth. The grid is equipotential, with all equipment, structures, and fencing bonded together to reduce touch and step potentials to safe levels. All electrical equipment and neutrals are connected to the grid.

When a fault occurs in a substation or on one of the circuits feeding into or out of the substation, a rise in voltage occurs in the earth and on the grounded grid. Having an extensive grid that bonds structures, equipment tanks, and fencing reduces the risk of dangerous potentials between different elements.

Many ground rods are driven in a station to carry electric currents into earth under normal and fault conditions without exceeding any operating and equipment limits or adversely affecting continuity of service. The grid network is typically made up from #4/0 bare copper in a grid pattern spaced about 10 ft (3 m) apart and about 1.5 ft (0.5 m) below grade. The grid is connected together electrically at each crossing and to each ground rod, usually with exothermic connections. The grid is also connected to an operating platform and the operating handles for switches. These connections must be in place to allow an operator to be in an equipotential zone to safely operate a switch.

The substation grid is connected to the fence and extended about 3 ft (1 m) outside the fence. **See Figure 4–22**. The intent is that anyone touching the fence during a fault condition will still be on the grid bonded to the fence. Step potential from the grid outside of the fence to earth beyond the grid is kept to a minimum by the very extensive grounding inside the substation. The fence has to meet regulatory standards for height, barbed wire along the top, and warning signs.

Figure 4–22 Connecting the fence to the substation grid ensures that the fence and the grid will be at the same potential during a ground fault. *(Delmar, Cengage Learning)*

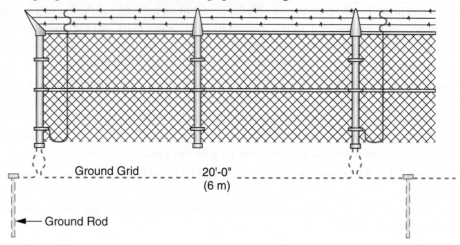

Erecting Station Structures

Most new substation structures are modular. Blueprints provide assembly and erection details. Bolts must be torqued to specifications. The aluminum-alloy pipe used for buses will need some rather exotic welding. Electrical connections that will be carrying very high fault and load current must be made to exacting standards. If installing long pieces of buses or structures, the use of tag lines is essential to keep the load under control, especially in substations with live sections.

Installing Heavy Substation Equipment

Transformers, circuit breakers, and other equipment are designed with provisions for lifting, jacking, and/or rolling. Units will have to be lifted using the base frame or from the lugs/clamps at the top of a unit. An experienced rigger needs to be in charge to lift and move heavy, yet delicate equipment. **See Figure 4–23**.

Figure 4–23 Positioning a large transformer requires skilled riggers. *(iStock.com/Fertnig)*

Wiring Auxiliary Equipment

There is a massive amount of low-voltage and relay-type wiring to do when constructing a substation, so it is essential to have the wiring blueprints. Each wire must be traced, color coded, and labeled, then tested to ensure that every piece of equipment and relay is connected to its intended terminal.

Review Questions

1. What is the application for a mobile distribution substation?

2. Why is the transformer in a collector substation a step-up transformer?

3. Which element in a substation will automatically open a line when there is a short circuit?

4. Why are disconnect switches in series with a circuit breaker?

5. What types of communication is used to monitor a substation?

6. How are the voltage and current of an incoming high-voltage transmission line measured?

7. What are the prerequisites before entering a substation?

8. Describe the construction required to make a substation yard into an equipotential zone.

9. How can opening a grounding connection in a substation be a hazard?

10. List four items in a substation design that are meant to protect the public.

11. Describe how SF_6 gas can be a hazard.

12. When digging in an existing substation, what kind of underground hazards must be avoided?

Chapter 5

Alternating or Direct Current

Objectives

After completing this chapter, you should be able to:

1. Explain why AC power was chosen for the distribution of electrical power.
2. Explain the importance of a constant AC frequency.
3. Determine the difference between measuring *instantaneous value*, *peak value*, *effective value*, *and average* voltage and current in a AC circuit.
4. Discover the effects of reactance in an electrical circuit.
5. Explain the advantages of 100% power factor.
6. Identify the characteristics of a DC electrical circuit.
7. Discover the applications for DC electrical power transmission.

Introduction

Why Alternating Current (AC)?

Thomas Edison originated the first investor-owned-electrical utility. The utility had a direct current (DC) distribution system that sold incandescent lighting. George Westinghouse and his partner Nickola Tesla built alternating current (AC) distribution systems to energize arc lighting. This resulted in the "Battle of the Currents" in the late 1800s.

Alternating current has one major advantage: the easy transformation from one voltage to another. Easy transformation allows the voltage to be stepped up for efficient transmission of electrical energy over long distances. AC dominates distribution and utilization by consumers.

Another advantage to AC power is that every time the voltage and current reverse direction, the magnitude of the voltage and current is zero. This assists in extinguishing arcs when opening switchgear.

AC does, however, introduce some complicating phenomena to electrical circuits that are not found in DC circuits. For very long distant transmission lines or long transmission cables DC current has some major advantages.

Characteristics of AC

AC Basics

Current flows in a circuit as long as a potential difference is present. To have a potential difference, one end of the circuit is at an opposite pole (polarity) to the other end. These polarities are labeled as positive and negative. The direction of the current flow in a circuit is determined by the polarity of the source terminals.

With DC, the polarity does not change and the current flows in one direction only. With AC, the polarity at the source alternates between positive and negative and the current direction changes with every change of the source polarity. **See Figure 5–1.**

Figure 5–1 Direction of current flow is dependent on the polarity of the source generator. *(Delmar, Cengage Learning)*

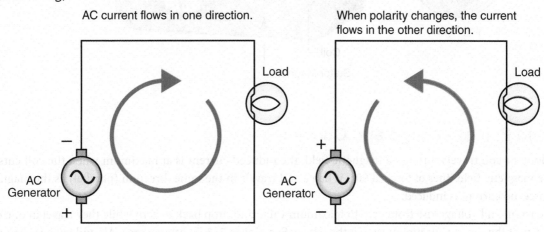

For example, the current in a single-phase circuit flows toward the load while the current in the neutral flows away from the load. In the next moment, the roles are reversed: The current in the neutral flows toward the load, and the current in the phase wire flows away from the load. The voltage on the phase wire is positive with respect to the neutral when it flows in one direction, and the voltage is negative when it flows in the other direction. The voltage on the neutral is unchanged and is close to zero with respect to a remote ground.

Frequency

In North America, alternating current supplied by electrical utilities travels 60 times in each direction per second. In certain parts of the world, 50 cycles per second is common. The term cycles per second has been replaced by the international standard term for frequency, which is **hertz** and is represented by the symbol *Hz*.

Unlike voltage or current, the **frequency** in a circuit stays constant right from the generator to the customer. When the frequency starts to drop, it is an indication that the generator supplying the electrical system is overloaded and slowing. A small reduction in frequency will trigger the electrical system to trip out of service. A typical range for frequency is 59.97 to 60.03 Hz. Some systems are set up to start load shedding when the frequency reaches 59.3 Hz.

Generation of AC Power

When a loop is rotated within a magnetic field, an electric current is induced in the loop. With AC generation, one half of a loop travels in one direction through the magnetic field while the other half travels in the opposite direction. The current flow induced in the two halves of the loop, therefore, also travels in opposite directions within the magnetic field but ends up traveling in the same direction within the loop. The maximum current is generated when the coil or loop is in a position to travel through the magnetic field at 90° to the magnetic lines of force. **See Figure 5–2**. In large commercial generators, many electric magnets are mounted on the rotating part (rotor), and many loops (coils) are mounted on the fixed part (stator).

Figure 5–2 The rotating magnet induces electrical current in the coil of this simple AC generator. *(Delmar, Cengage Learning)*

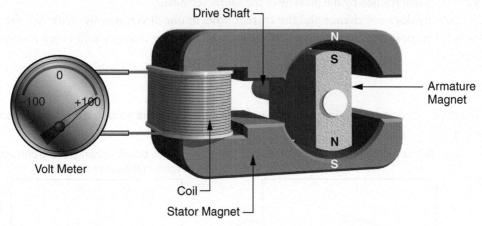

Rise and Fall of Voltage and Current

When a loop or coil travels through a magnetic field, the induced current is at maximum when the coil cuts straight across the magnetic field lines of force at 90°. When a coil travels in the same direction (parallel) as the magnetic field lines of force, no current is induced.

The current and voltage rise from zero to maximum value and drop back to zero while they travel in one direction and then repeat the zero-to-maximum rise on the return. **See Figure 5–3**. In other words, AC and voltage change in both polarity and magnitude.

AC Represented by a Wave

The rate of the rise and fall of the voltage and current and the direction of flow can be represented on paper by drawing a sine wave. **See Figure 5–4**. A sine wave is a wave shape that follows a mathematical form.

The wave above the zero current line represents the value of the current traveling in the positive direction, and the wave below the zero current line represents the value of the current traveling in the negative direction.

A Sine Wave as a Graphical Representation

A sine wave displays how AC power behaves over time. The sine wave is a graphical representation of the rate at which the current and voltage values rise and fall and of the direction the current travels during one cycle. The wave is about 3,100 miles (4,989 km) long on a 60 Hz line. Over the shorter distances that power line technicians do their work, the potential (voltage) on each part of a conductor is virtually the same at any instant in time. In the next instant, the potential along the wire is different, rising and/or falling as the wave rises and/or falls.

A sine wave is plotted against time, but it makes more sense to think about an electric charge on a wire in terms of distance. Except when speaking in terms of 10 miles or more, a typical conductor is at a virtual equipotential with the peaks and troughs of the sine wave occurring virtually simultaneously. Electric load and impedance of the circuit will reduce the magnitude of the peaks and troughs over distance. Transformers, voltage regulators, and other equipment can reduce or increase the magnitude of the peaks and the troughs so that customers receive the proper voltage at their place of business or home.

Figure 5–3 The changes in the magnitude and direction of electric current are dependent on the direction and angle of the conductor while passing through the magnetic field. *(Delmar, Cengage Learning)*

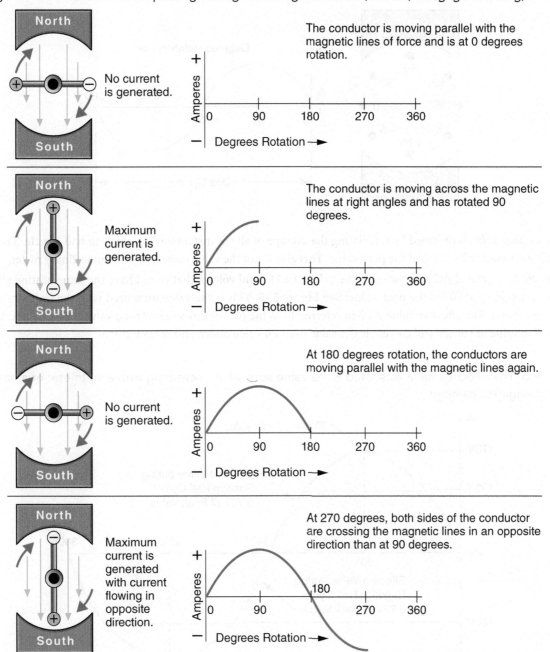

Values of AC and Voltage

With the values of AC and voltage continuously changing, what values do we actually read on an ordinary ammeter or voltmeter?

- The *instantaneous value* is the actual value of the voltage and current at each instant in the cycle. This value is not measured with an ordinary meter.
- The *peak value* is the highest instantaneous value that the voltage and current reach in both directions during the cycle. While this value is not measured by an ordinary meter, it must be considered by design engineers when planning the insulation needed for different voltage systems.

Figure 5–4 The magnitude and direction of AC electric current can be represented by a sine wave. *(Delmar, Cengage Learning)*

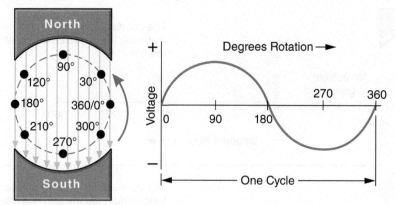

- The *average value* is obtained by calculating the average of all the instantaneous values in half a cycle. The average value works out to be 0.636 of the peak value. This also is not the value measured with an ordinary meter.
- The *effective value* of AC and voltage is the value of a DC and voltage that would have the same heating effect. The effective value is 0.707 of the peak value. **See Figure 5–5.** This is the value measured with the ordinary ammeter and voltmeter. The effective value is often referred to as the root-mean-square (rms) value. In AC applications, the effective value of voltage and current is the value used for calculations and measurement.

Figure 5–5 The effective value illustrated is the value seen when measuring with a voltmeter or ammeter. *(Delmar, Cengage Learning)*

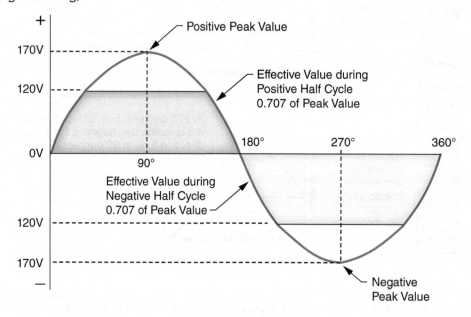

Voltage and Current in Phase with Each Other

When both the voltage wave and the current wave reach their maximum and zero values simultaneously, the voltage and current are considered to be "in phase" with each other. In an electrical system, the voltage and current are generally not exactly in phase.

When current is flowing into certain loads, such as a motor load, the current wave will peak after the voltage wave peaks. This current is out of phase and **lags** the voltage. When the current is flowing into a capacitor, the current wave will peak before the voltage wave peaks. This current is out of phase and **leads** the voltage.

Circuits in Phase with Each Other

An electrical power system of a utility is part of a large power pool or grid fed from many sources and locations. A generator must generate power so that the magnitude and the direction of voltage and current are synchronized with the other generators supplying power to the same grid. In other words, the voltage from all power sources must be in the same position on the sine wave before feeding power into the grid. All generators must be in phase with each other. If an out-of-phase generator is connected into the grid, it will act as a short circuit.

Reactance in AC Circuits

Loads Fed by AC Circuits

There are three kinds of loads fed by an AC circuit:

1. Heating and lighting are resistive loads. A resistive load can do work. The amount of work being done can be measured by a kilowatt-hour meter.
2. The energy used to magnetize a motor or a transformer is an inductive load. An inductive load does not generate heat or light.
3. The energy used to supply a capacitive effect at capacitors, paralleling conductors, or in cables is a capacitive load. A capacitive load does not generate heat or light.

Reactance in an AC Circuit

A resistive load opposes the flow of current, but so do inductive and capacitive loads. All three types of loads oppose the flow of current and add to the total opposition to current flow in a circuit. This disturbance imposed by inductive and capacitive loads in a circuit is called **reactance**. The symbol for reactance is X.

Reactance can be an inductive reactance (X_L) caused by loads, such as motors, transformers, fluorescent lights, and computers, or it can be a capacitive reactance (X_C) caused by capacitors or paralleling conductors.

Impedance in an AC Circuit

In an AC circuit, the opposition to current flow consists of resistance *and* reactance. This combination of resistance and reactance opposing the current flow is referred to as *impedance*. Impedance is measured in ohms and is represented by the symbol **Z**. Impedance can be used interchangeably with resistance in calculations using Ohm's law:

$$Z = IR \quad Z = \frac{E}{I} \quad I = \frac{E}{Z}$$

Where:

E = effective voltage in volts

I = effective current in amperes

Z = total impedance in ohms

Resistance in an AC Circuit

Light and heat are resistive loads and do not cause any other disturbance to the circuit. Either AC or DC can supply a resistive load. Ohm's law applies to a resistive AC circuit in the same way that it applies to a DC circuit. For practical applications in the line trade, resistance can be used for most calculations involving AC circuits, and the result will generally have less than a 10% error.

Induction in an AC Circuit

Voltage can be induced into a conductor:

- By moving a conductor within a magnetic field
- By having a conductor near a moving magnetic field

Voltage (electromotive force) is induced into a conductor when a conductor is moved through a magnetic field. There is no voltage generated unless the conductor is moving.

Voltage can also be induced into a conductor when a magnetic field from a nearby energized AC circuit moves through the conductor. Only an AC circuit would have a moving or fluctuating magnetic field. Induced voltage is always in a direction that opposes the direction of the current flow. Induced voltage is, therefore, the opposite polarity of the source circuit.

With AC, the magnetic field is moving (expanding and collapsing) and any nearby stationary conductor will have a voltage induced on it. This phenomenon is familiar to power line technicians because induction from energized circuits is a concern when grounds are installed on an isolated circuit. Some of the magnetic field around an energized conductor cuts through the energized conductor itself and induces some voltage onto itself. This self-inductance creates a counter-electromotive force (voltage) within the conductor.

When the conductor is part of a coil, such as in a motor or transformer, the magnetic field around the conductor cuts through adjacent wires in the coil, which increases the self-inductance in the conductors forming the coil. This opposition to the current flow delays the rise and fall of the current but not the voltage. The counter-electromotive force generated by the continuously changing voltage and current is called inductance.

Inductive current acts on a circuit as though it flows at 90° to the resistive current. **See Figure 5–6.** When the resistive current is added to the inductive current, the resultant current vector is longer, which means that the resultant current is higher.

Figure 5–6 To add the effect of inductive current to the total current in an AC circuit, the calculation must take into account that inductive current cannot simply be added to resistive current. (Delmar, Cengage Learning)

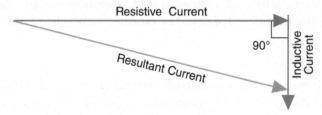

The symbol for inductance is *L*. The unit of measurement for inductance is a henry, but a power line technician would not have a reason to use this unit of measure.

Inductive Reactance

The opposition to current flow within a conductor or coil is called *inductive reactance,* represented by the symbol X_L and measured in ohms. The formula is shown only to illustrate the factors that influence the magnitude of inductive reactance:

$$X_L = 2\pi f L \text{ ohms}$$

Where:

X_L = inductive reactance in ohms

π = 3.14

f = frequency of the circuit in hertz

L = inductance in henrys

Amperage Lagging Voltage

Distribution feeders tend to have a combination of resistive load and inductive load (electric motors). The inductive load sets up an inductive reaction in the circuit, and this opposition to the current flow causes it to lag behind the rise and fall of the voltage. In other words, as represented by a sine wave, the voltage will peak before the current peaks. The formula for the power output of a circuit, power = volts × amperes, is valid only when the voltage and current in the circuit are working together, by being in phase with each other (e.g., with each reaching its peak value at exactly the same time). The power output of the circuit is reduced because of the formula.

When the current lags the voltage by 30°, it means the current is 30° out of phase with the voltage. **See Figure 5–7.**

Figure 5–7 When voltage and current are not working together, they do not reach their peak value at the same time and they are 30° out of phase with each other. *(Delmar, Cengage Learning)*

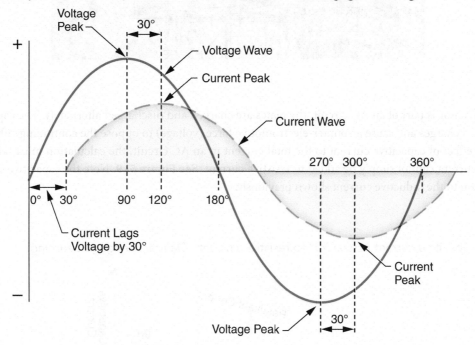

Applications of Inductive Reactance

Two applications of inductive reactance used in substations are the current-limiting reactor and the shunt reactor. A current-limiting reactor is a coil of cable that sets up an inductive reaction in a circuit. The current-limiting reactor is designed so that a large inductive reactance is created during a high fault current condition. The inductive reactance impedes the flow of damaging fault current in the substation.

To balance large capacitive loads caused by parallel conductors of a long transmission line, shunt reactors are installed in substations. The reactors produce an inductive reactance that cancels out an equal amount of capacitive reactance in the transmission line. A reactor looks like a power transformer and is connected into the circuit as a shunt (a parallel load). The only connections to the shunt reactor are to high-voltage bushings. There are no output connections, such as the secondary of a transformer. **See Figure 5–8.**

Pad mount reactors are installed on long rural underground distribution circuits to balance the capacitive reactance set up by the underground cable.

Capacitance

The major sources of capacitance in a system are paralleling conductors on long transmission lines, underground cable, and capacitors installed specifically to put more capacitance into the electrical system. The tendency of a circuit to store electricity when a potential difference exists between conductors is called *capacitance.* Capacitance occurs when an electrically charged conductor (plate) electrostatically induces an equal in magnitude but opposite polarity charge on a nearby conductor (plate). The two conductors are not in contact but are separated by some kind of insulation dielectric.

Figure 5–8 Reactors add inductive reactance to the circuit. *(iStock.com/BergmannD)*

When a capacitor is part of an AC circuit, the plates are charged and discharged alternately. The capacitor tends to store the acquired charges and cause a counter-electromotive force (voltage) to oppose the continuing voltage change.

To add the effect of capacitive current to the total current in an AC circuit, the calculation must take into account that capacitive current cannot simply be added to resistive current. **See Figure 5–9**. Note the capacitive current is in the opposite direction to the inductive current shown previously.

Figure 5–9 Capacitive current acts at 90° to resistive current. *(Delmar, Cengage Learning)*

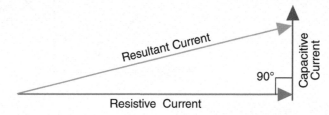

Capacitive current acts on a circuit as though it flows at 90° to the resistive current. When the resistive current is added to the inductive current, the resultant current is higher. When capacitive current is added to a circuit with inductive current, the two types of reactance cancel each other because they act in opposite directions. The symbol for capacitance is *C*. The unit of measurement for capacitance is a *farad*. However, farads are a very small unit of measurement, so line workers use kilovolt-amperes reactive **(kVAR)**.

Capacitors

Capacitance normally occurs in a circuit. A parallel conductor is like a capacitor where the two conductors are separated by air and the conductors electrostatically induce charges on each other. Underground cable is like a capacitor where the inner conductor is separated from the outer sheath by insulation and a charge is built up between the conductor and the sheath.

A capacitor is constructed to induce capacitance in a circuit. **See Figure 5–10**. A modern capacitor uses packs of foil plates in series for voltage drop and paralleled for capacity. As a capacitor, one electrically charged pack of foil plates will electrostatically induce an opposite polarity charge in the other pack of foil plates. The plates within the packs are separated by a dielectric material.

Figure 5–10 In a capacitor, an electrically charged metallic plate will electrostatically induce a charge in another metallic plate that is isolated from the first by insulation. *(Delmar, Cengage Learning)*

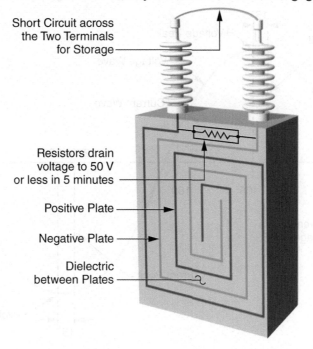

Short Circuit across the Two Terminals for Storage

Resistors drain voltage to 50 V or less in 5 minutes

Positive Plate

Negative Plate

Dielectric between Plates

Capacitive Reactance

When a capacitor is charged up, a counter-electromotive force (emf) equal to the source voltage is built up on the opposite plate. Current will flow into the capacitor only while the voltage is rising. When the voltage approaches peak value, the counter-electromotive force on the opposite plate is also approaching peak value, which causes the current flow to decrease. There is no current flow when the voltage is at its peak (90°). In a capacitor, the current reaches its peak before the voltage. A capacitor opposes a *change* in voltage, which is what the voltage is doing continually in an AC circuit.

This delaying or opposing of changes to the voltage is called *capacitive reactance.* Capacitive reactance (X_C) is measured in ohms. The formula is shown only to illustrate the factors that influence the magnitude of capacitive reactance:

$$X_C = \frac{1}{2\pi f C} \ ohms$$

Where:

X_C = capacitive reactance in ohms

π = 3.14

f = frequency

C = farads

Current Leads the Voltage

Because change to the voltage across the capacitor plates is delayed, this capacitive reaction causes the current wave to lead the voltage wave. **See Figure 5–11.**

Figure 5–11 Due to a capacitive reaction, the voltage and current are not working together and they do not reach their peak value at the same time. *(Delmar, Cengage Learning)*

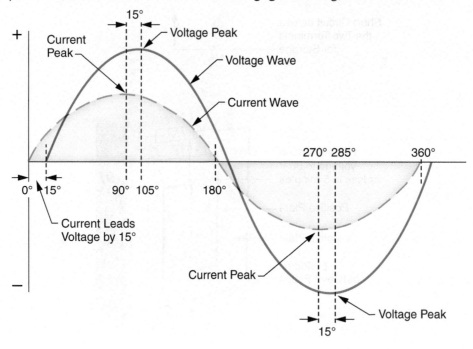

Applications of Capacitive Reactance

Capacitors are installed on electrical systems to draw a leading current from the circuit to counteract the more predominant lagging current in utility circuits. When the lagging current, due to motor load, is partially balanced by the leading current, due to the installed capacitor, there is less total reactive current. With less reactive current, there is less total current and, therefore, less voltage drop and line loss. The most common use of capacitors, therefore, is as a voltage booster.

Most customer load has a heavy inductive element to it, so an increased load on a transmission or distribution line means that there is an increase in the inductive load. As the voltage starts to decrease, an increase in capacitance is needed to keep the voltage up. Capacitors are installed to boost the voltage on the circuit. **See Figure 5–12**.

Figure 5–12 The installation of a capacitor bank will counter the inductive reactance in a distribution circuit, reduce the total current, and cause an increase in circuit voltage. *(Delmar, Cengage Learning)*

Effect of Frequency on Reactance

The formulas for inductive reactance and capacitive reactance show that the inductive reactance increases when the frequency goes up and capacitive reactance decreases:

$$X_L = 2\pi f L \ ohms$$

$$X_C = \frac{1}{2\pi f C} \ ohm$$

There is somewhat less inductive reactance in a 50 Hz circuit than in a 60 Hz circuit. A 60 Hz transformer has 10% to 15% less material than a 50 Hz transformer. A higher frequency sets up a greater magnetic flux linkage, and, therefore, a smaller iron core can be used. The electrical system in an aircraft is 400 Hz, which allows even less iron and less weight for electrical equipment using an iron core. Any power line technicians who has worked with transformers in the few locations that still have 25 Hz circuits can verify that a 25 Hz transformer is much heavier than the 60 Hz transformer of the same kilovolt-ampere rating.

A 50 Hz circuit would have more capacitive reactance than a 60 Hz circuit. The increased capacitance could be significant on long transmission lines.

Summary of Resistance, Inductance, and Capacitance

Resistance opposes the *flow* of current in an electric circuit. Inductance opposes a *change* in current. The nature of an AC circuit is that its current is always changing by rising and falling in magnitude and reversing direction. Loads containing wire coils are inductive and thus increase the amount of inductive reactance in a circuit.

Capacitance opposes any change in *voltage*. The nature of an AC circuit is its voltage is always changing by rising and falling in magnitude and its polarity is continually reversing. Capacitive loads such as capacitors increase the amount of capacitive reactance in a circuit.

AC Power

Active Power

Active power is referred to as effective power, true power, or **real power**, because it is the power that gives light, gives heat, and turns motors. Real power can also be expressed as voltage × resistive current.

In AC circuits, power alternates at the same rate as the voltage and current are alternating. The power measured at a customer's meter is referred to as active power. Therefore, the total active power supplied by a circuit is equal to total effective current × total effective voltage, or

$$P = IE$$

Active power is measured at a customer meter and is measured as *watts* or in blocks of 1,000 watts, which is *kilowatts*.

Apparent Power

The total power supplied by a circuit is called **apparent power** because in an AC circuit all the power does not perform actual work. Apparent power is a combination of *active* power and reactive power.

$$Apparent \ power = \sqrt{(active \ power)^2 + (reactive \ power)^2}$$

Apparent power is measured in volt-amperes or kilovolt-amperes (**kVA**). The term *kVA* is common terminology for the line trade when referring to transformer sizes. Using the term *kVA* for apparent power is more suitable than using the term watts power when referring to transformer size because a transformer has to supply the apparent power needed by the load.

Reactive Power

Reactive power is the element in the apparent power formula that does not do any work. In a circuit, reactive power is transferred back and forth between the reactive load and the circuit. Reactive power is not used up. Reactive power is equal to voltage × reactive current, or I^2X. It is measured in volt-amperes reactive (VAR) or kilovolt-amperes reactive (kVAR).

In most distribution circuits, the net reactive power is inductive due to a portion of the motor loads, fluorescent lighting, and electronic loads. Reactive power is normally very small compared to the active power available in the circuit.

Inductive reactive power is necessary to create electromagnetic fields in equipment such as transformers and motors. Capacitive reactive power is necessary to create electrostatic fields in capacitors. Volt-amperes reactive will increase current flow in a circuit but not represent energy consumption. Volt-amperes reactive store energy in one part of the cycle and return it in the next part of the cycle.

The Power Triangle

The mathematical relationship among the three kinds of AC power can be illustrated by a vector diagram. **See Figure 5–13**. In a vector diagram, the length of each side represents the magnitude of each type of power.

Figure 5–13 The relationship between active, apparent, and reactive power is illustrated in a power triangle. *(Delmar, Cengage Learning)*

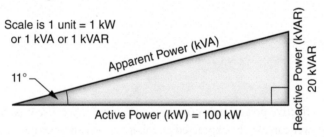

The sides of the triangle represent the magnitude of active power and reactive power. The hypotenuse represents the magnitude of apparent power. **See Figure 5–13**. If two sides are known, the remaining side can be measured or calculated. Active power is represented by 100 units, and reactive power is represented by 20 units. The resultant power can be calculated, as shown in the following equation, or measured from the vector drawing.

$$\text{Apparent power} = \sqrt{100^2 + 20^2} = 102 \ kVA$$

Power Factor

The ratio of active power to apparent power is called the **power factor** and is normally expressed in percentage. If there is no reactive power in the circuit, the power factor is 100%. Keeping the power factor high in a circuit reduces line loss and voltage drop, and results in even generation. The power factor is measured at some customers' locations by revenue meters, and customers are penalized for a low power factor.

The angle between apparent power and active power represents the amount of reactive power in a circuit. A power factor can be measured from the power triangle or calculated in one of two ways:

$$\text{Power factor} = \frac{\text{active power}}{\text{apparent power}}$$

or

$$\text{Power factor} = \frac{R}{Z}$$

In the example in Figure 5–13, the angle is 11°. The power factor is equal to the cosine of 11° = 0.98, or 98%, or, when using a formula:

$$\frac{\text{Active power}}{\text{Apparent power}} = \frac{100}{102} = 0.98 = 98\%$$

Power Factor Correction

In a circuit with a low power factor, the apparent power needed to supply the load on the circuit becomes unacceptably high. With too much reactive current in a circuit, more power must be generated to supply the load. More current must flow in the conductor, which causes more line loss (I^2R) and a greater voltage drop.

Because the reactance in most distribution circuits is due to inductive loads, the easiest way to correct the power factor is to install a nearly equal amount of capacitive VAR to balance the inductive VAR. Installing capacitors on the circuit would reduce the apparent power needed to supply power to the customer and, therefore, would reduce line loss and raise the voltage.

In industries with large inductive loads, the industry is billed for the apparent power used as well for the active power. These industries do their own power factor correction in the plant by installing capacitors.

Adding 5 kVAR capacitors to a circuit with 8 kVAR inductance will reduce the total inductive reactance in the circuit to 3 kVAR. **See Figure 5–14.**

Figure 5–14 Capacitors can reduce load current in a circuit. *(Delmar, Cengage Learning)*

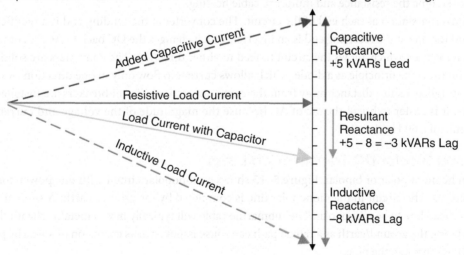

Additional Methods for Reactive Power Control

Controlling the reactance and thereby voltage in the grid is an ongoing task. The following methods are mentioned for information.

Synchronous Condenser

Some industries install a synchronous condenser, which is like an electric motor with an exciter system. The condenser adds reactive power/capacitance to improve the quality of power in their facility. Sometimes at generating stations, a generator will be run like an electric motor to add capacitance to the grid. The generator is not mechanically connected to anything but is allowed to spin freely, when powered from electrical the grid.

Static VAR Compensator (SVC)

A static VAR compensator uses thyristors (semiconductor rectifiers) to trigger switching between inductive reactors or capacitor banks in a substation. Balancing inductive and capacitive reactance is needed to regulate the voltage on the transmission system.

STATCOM (Static Synchronous Compensator)

A STATCOM is an improvement over the SVC because it uses a solid-state switching converter. Its function is to keep the reactance in a transmission line or a large industry balanced. By controlling reactance in a circuit, voltage disturbances are reduced.

Why Direct Current (DC)?

Characteristics of DC

The total power produced in a DC circuit is calculated by multiplying total voltage by total amperes. Because there is no reactance in a DC circuit, all the power in the circuit is active power/effective power/true power, or real power. The power factor of the DC circuit is 100%. The effective voltage in a DC circuit is equal to the peak voltage, whereas the peak voltage in an AC circuit is 40% higher than the effective voltage. Direct current in a DC circuit is surrounded by a fairly constant or stationary electric field. There is no electromagnetic energy and its associated induced current.

With reactance and skin effect eliminated, DC is a very efficient way to transmit power, and compensating shunt reactors are not needed. Resistance is the only impedance to current flow in a DC conductor. The cable size needs to be large enough to overcome the resistance and minimize cable heating.

There is a converter station at each end of the circuit. The converter at the sending end is a rectifier, which changes the AC to DC, and the one at the receiving end is an inverter, which changes the DC back to AC. A converter can be used as a rectifier or an inverter, which allows the circuit to feed in either direction. The converters are solid-state "thyristor" valves and work on the same principle as a diode, which allows current to flow only in one direction. A very large ground electrode grid is installed a short distance away from the converter station. Circuit breakers are installed on the AC side of the converters. It is easier to break the arc of AC because the magnitude of the voltage and current becomes zero, 120 times a second, on a 60 Hz circuit.

Monopolar and Bipolar DC Transmission Lines

A DC circuit can be monopolar or bipolar. **Figure 5–15** shows a monopolar circuit with one power conductor between the converter stations. The circuit in the monopolar line is completed by the ground/earth. A submarine cable is more likely than an overhead line to be monopolar. The submarine cable will typically have a metallic sheath that helps to complete the circuit. Using the ground/earth as a return path can cause issues such as corrosion of a nearby pipeline when the ground current flows through the pipe.

The more common bipolar DC circuit as shown in **Figure 5–15** has two power conductors (poles) and a grounded midpoint. One conductor is positive and the other is negative. The bipolar line is still just one circuit. With the midpoint ground/earth the circuit will continue to operate as monopolar when there is a fault on one of the conductors. Under the right circumstances, one conductor can be grounded to carry out maintenance while the circuit is energized.

Figure 5–15 Simplified schematic drawings of a monopolar circuit and a bipolar circuit. *(© Cengage Learning)*

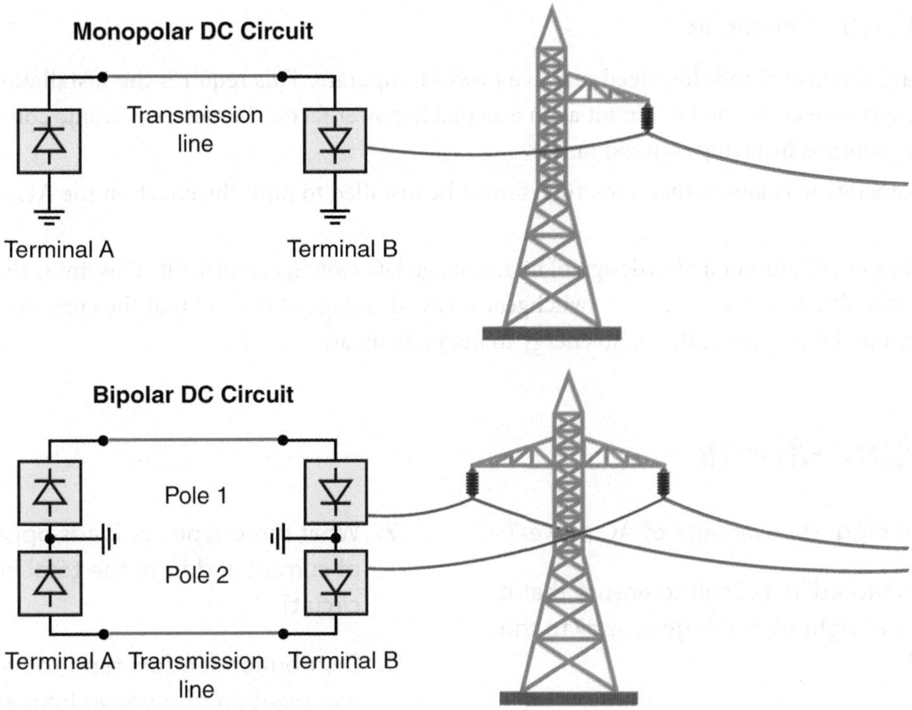

Applications for DC

1. *Very Long Overhead Lines*

    On long high-voltage DC transmission lines, there is no need to offset the effects of capacitive reactance that would add impedance and would reduce current flow. Only two power conductors are needed on a bipolar circuit. DC distribution has been used to feed a long distant remote load. This load would otherwise have needed its' own generator or a long subtransmission line and a substation.

2. *Underground or submarine Cable*

    DC is preferred for long lengths of submarine or underground cable, AC would generate a high capacitance reactance because of the power conductor, insulation and metallic sheath is close to each other and causes capacitance. There is a capacitance reactance in the DC cable caused by the charging current, only when the cable is first energized.

3. *Tying two incompatible AC systems*

    A DC circuit is ideal to tie two incompatible unsynchronized AC systems or to tie a 60 Hz and a 50 Hz system together. A DC circuit is independent of the frequency or phase angle of the AC circuits at each terminal. There are back-to-back converter stations, where the DC link and the source and load converters are in one station.

4. *DC circuits at wind and solar farms*

    The generator on a wind turbine is mostly AC because AC is more efficient than DC. The frequency of the generator at each wind turbine varies with wind speed. It would be difficult to collect the output of each wind turbine with its unique frequency together. A converter at each wind turbine rectifies the current, and the resulting DC circuits are compatible and tied together at the substation.

Disadvantages of DC Transmission

Disadvantages of DC transmission are:

1. Converters are expensive, and they need reactive power to operate. This requires the installation of capacitors to keep the AC system feeding the DC circuit at an acceptable power factor. Modern electronic converters solve many of the earlier problems involving conversion.

2. Converters generate harmonics; therefore, filters must be installed to limit the effect on the AC system feeding the converters.

3. The steady state of DC current makes design of high-voltage DC switchgear difficult. This limits the ability to operate between DC circuits. Alternating current switchgear takes advantage of the fact that the current wave goes through its zero point and there is momentarily no energy to maintain an arc.

Review Questions

1. What is the biggest advantage of AC power?

2. What is monitored in a circuit to ensure that it stays constant right from the generator to the customer?

3. What changes occur to AC as it goes through a cycle?

4. In relation to the peak value of AC power, what value does an ordinary ammeter and voltmeter measure?

5. What kind of AC load uses the energy needed to magnetize a motor or a transformer winding?

6. What kind of AC load uses the energy needed to supply a capacitive effect at capacitors, paralleling conductors, or in cables?

7. What three types of loads oppose the flow of current and form the total impedance to a circuit?

8. Distribution feeders tend to have a combination of resistive load and inductive load, creating a lagging power factor. What equipment is installed to counter a lagging power factor?

9. What are the three types of power in an AC circuit?

10. What are three applications of DC transmission lines?

Chapter 6

Three-Phase Circuits

After completing this chapter, you should be able to:

1. Describe the advantage of three-phase power over single-phase power.
2. Make three-phase connections considering three-phase rotation.
3. Differentiate between wye and delta three-phase circuits.
4. Explain that how the return flow in a delta system completes the circuit.
5. Explain the advantages of a wye system over a delta system.
6. Describe the effects of a phase to neutral fault on a wye system.
7. Make field calculations to aid in phase balancing.

Introduction

Three-Phase Basics

Electric current needs a circuit (a complete circle) before it can flow. A single-phase wye circuit can be a phase and a neutral with the current flowing through the "hot" wire and returning to the source through the neutral. Most utility circuits consist of three phases. A **three-phase** circuit is not three different single-phase circuits but one circuit with all three phases interconnected.

A live conductor in a circuit is called a *phase* because when three-phase voltages and currents are generated, each of the three conductors gets its voltage and current at a certain phase of a cycle. An electrical phase is often represented in drawings and text by the Greek letter phi (Φ).

Characteristics of Three-Phase Circuits

Why Three Phases?

The effect of three-phase power as compared to single-phase power is similar to the effect of a six-cylinder engine as compared to a single-cylinder engine. A six-cylinder engine produces six small, smoother pulses per cycle while the single-cylinder engine produces one large pulse per cycle.

The values of the voltage and current in each of the three phases overlap with the other phases; therefore, three interconnected phases provide a smoother power than the relatively more pulsating power of a single phase. Three-phase current supplies a rotating magnetic field. Even though the power on each individual AC phase pulsates when it goes through the AC cycle, the sum of the power in all three phases at any point is constant.

Large generators and large motors are more efficient and are considerably smaller as three-phase units compared to equivalently powered single-phase units. In a three-phase motor, the magnetic field automatically rotates, bringing the rotor along with it.

Generation of Three-Phase Power

A simplified three-phase generator has three coils mounted on the armature at 120° apart. **See Figure 6–1**. Each coil generates an AC and voltage, but the power generated in each coil reaches its peak and direction at 120° apart.

Figure 6–1 The rotating magnet induces electrical current in three sets of coils in a simple three-phase generator. *(Delmar, Cengage Learning)*

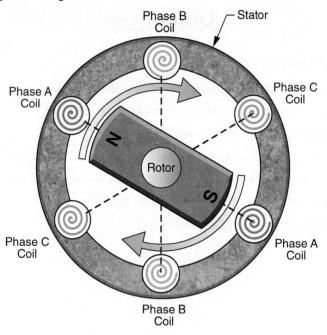

Commercial generators mount many coils on the stator and many magnets on the armature. The individual coils are wired so that they are connected together as three circuits, 120° apart. Each of the three circuits becomes a phase of a three-phase circuit.

Phases 120° Apart

The voltage and/or current in each of the three phases when 120° apart are not equal. **See Figure 6–2**. The vertical line shows the voltage and/or current values at one moment in time. Phase A, at 90°, is generating at the maximum value; phase B, at 210°, is climbing toward the zero value; and phase C, at 330°, is approaching the maximum return or negative value. The second vertical line shows what is happening to the power generated in each phase when phase A is 120° (one-third of the way) into the cycle, phase B is 240° (two-thirds of the way) into the cycle, and phase C is at 360° (at the end, which is also the beginning) of the cycle.

A three-phase circuit is like having three separate AC single-phase circuits with identical voltage that reach their peak values at a different time. At 60 Hz, the second phase reaches its positive peak at 1/180 (0.00556) second after the time the first phase reaches a positive peak, and the third phase reaches its positive peak 1/180 (0.00556) second later. The first phase again reaches a positive peak 1/180 (0.00556) second after the third phase, starting the next cycle. Even though each phase has the same voltage, they are out of phase with each other and there is a voltage difference between them.

Figure 6–2 The vertical line illustrates the relationship of three phases in one moment in time. *(Delmar, Cengage Learning)*

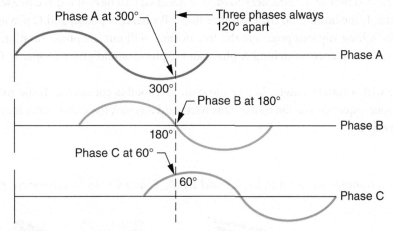

When the load on each phase is identical, the instantaneous power output of the three phases added together is constant. When one phase of a three-phase circuit reaches a peak voltage, another phase is close to zero volt and the third phase is on the return flow. Each phase is 120° out of phase with the other phase. The voltage and current in each phase are 1/180 second, or one-third of the distance in a cycle behind another phase.

Phase Designations

In the line trade, it is normal procedure to trace individual phases to ensure that the correct phase is connected to the correct terminal. Individual phases are named and marked at various locations on the system to keep them apart. **See Figure 6–3.** Utilities use various designations and markers, including the following:

1. Red phase, white phase, and blue phase
2. Red phase, yellow phase, and blue phase
3. A phase, B phase, and C phase
4. #1 phase, #2 phase, and #3 phase
5. X phase, Y phase, and Z phase

Figure 6–3 Phasing markers on underground cable ensures proper connection to overhead conductors also identified with phase markers. *(Delmar, Cengage Learning)*

Phase Rotation

When a three-phase motor is part of a customer's load, it is necessary to have the three phases in a sequence, or the motor will run backward. If the hookup to the customer had a B phase, A phase, and C phase sequence, the motors would run backward. Switching any two phases at the transformer will put the phase rotation back in sequence. For example, with a B, A, and C sequence, switching A phase and C phase would give a proper B, C, A (which is A, B, C) sequence.

A test can be made with a rotation meter before the customer load is connected. If the rotation meter shows the supply phases are in proper sequence and the motor runs backward anyway, the customer will need to reverse two leads on the motor. **See Figure 6–4.**

Figure 6–4 The rotation meter shows that L1, L2 and L3 (A, B and C) are in sequence. *(Courtesy of Fluke Corporation)*

(a)

(b)

Three-Phase Connections

A three-phase circuit starts at a generator. After generation, these three phases are connected to the input and output sides of transformers throughout the transmission and distribution system. A three-phase circuit is interconnected at the transformers in two ways. **See Figure 6–5.**

1. The three phases can be connected in parallel. One end of each of the three coils is connected together to a common point, and the other ends of the three coils are the three-phase connections. This is referred to as **wye**-connected (named after the letter *Y*). A vector drawing showing the three phases interconnected is in the shape of a Y.

2. The three phases can be connected in series, which is referred to as **delta**-connected (named after the Greek letter delta, Δ). A vector drawing showing the three phases interconnected is in the shape of a delta.

Figure 6–5 Wye phases are connected together at one point and are in parallel to each other. Delta phases are connected in series. *(Delmar, Cengage Learning)*

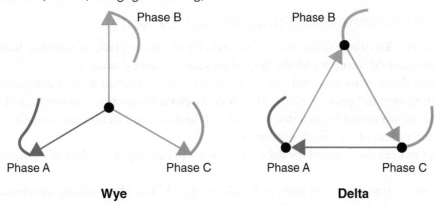

Wye and Delta Systems

All circuits worked on by the line trade are fed from a transformer bank somewhere in the system. When the circuit source is from a transformer bank with a delta-connected output, the circuit is a delta system. Similarly, a wye circuit comes from a transformer bank with a wye-connected output.

A three-phase delta circuit consists of three wires. Each wire is a phase 120° out of phase with the other wires. Transformers are connected phase to phase. Lateral taps branching off with two phases are used as single-phase circuits, which feed single-phase transformers and customers.

A three-phase wye circuit consists of four wires. Three are phase wires 120° out of phase with each other, and the fourth wire is the connection to the common point, which is the neutral. Lateral taps branching off, consisting of a phase and a neutral, are single-phase circuits, which feed single-phase transformers and customers.

Connections to the three coils of three-phase transformers or motors are always made based on the standard wye or delta shape. **See Figure 6–6.**

Figure 6–6 Three-phase transformer or motor coils will be connected together as wye or as delta. *(Delmar, Cengage Learning)*

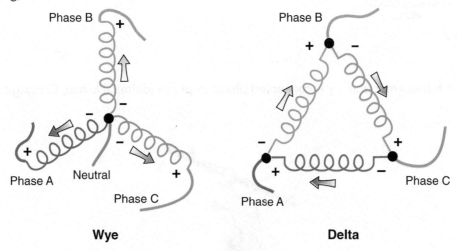

Delta-Connected Systems

Delta- or Series-Connected Three-Phase System

A delta system is a circuit fed from a delta-connected secondary of a three-phase transformer bank. The transformer bank provides a three-phase delta output with the three phases interconnected as one circuit.

To operate, each coil of a transformer must have a potential difference across it. On a delta connection, this can be achieved by connecting each coil phase to phase. The phase-to-phase connections are not at random, but each coil is connected so that the end of one coil is connected to the end of the other until all three are connected all the way around. In other words, the three phases are connected in series.

The current does not circulate around the delta. The current in each leg is traveling in a different direction at 120° out of phase with each other.

Delta connections are represented on paper in the shape of a "Δ" but when making transformer connections the transformer coils will typically be side by side. **See Figure 6–7**. The delta side by side connections are still in series with each other.

A single-phase transformer connected to a delta system must have two primary bushings because the transformer is connected phase to phase. **See Figure 6–8**. The transformer must have two primary bushings because the transformer is connected phase to phase. The transformer installation has two surge arrestors, and when fused cutouts are used, two are required.

Figure 6–7 The coils in both left and right illustrations are connected in a delta configuration. *(Delmar, Cengage Learning)*

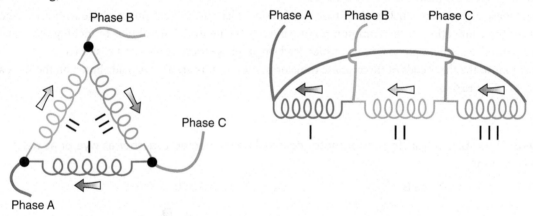

Figure 6–8 This transformer primary is connected phase to phase (delta). *(Delmar, Cengage Learning)*

Voltage in a Delta System

The voltage across each coil is the same as the voltage measured from phase to phase. There is a voltage across each coil or between any two phases because each phase is in a different part of the AC cycle. The three phases are, in fact, always 120° apart.

Current in a Delta System

For this discussion, the phase current is the current in each coil and the line current is the current in each conductor leaving the transformer bank. The current leaving each junction point is the resultant of the current in two phases 120° apart. The current through each coil is equal to the full-phase current, but the line current leaving each junction point is 1.73 × the phase current.

Return Flow in a Delta Circuit

To complete a circuit, electrical current must return to the source. In a delta circuit, the return flow to the source is in the opposing phase conductors. There is no return flow through the ground. There is a potential difference between the three phases of a delta circuit, but there is no "theoretical" potential between a phase and ground because going from phase to ground does not complete a delta circuit and there is no reference point between ground and any phase.

Using the same theory, a metal car body is the return path to the battery in an automobile. Theoretically, if a person avoids touching the car body, that person can touch the spark plug and not get a shock. If some of the return current takes a path to ground down one tire and up another tire, however, a person would still be exposed to a shock.

If one of the phases of a delta circuit became grounded, the voltage between the other two phases and ground becomes equal to phase-to-phase voltage. Unless a circuit breaker is equipped with a ground fault relay, a grounded conductor on a delta circuit will continue to work as a three-phase circuit. If any additional phase is grounded, the ground will create a short circuit.

Ground Fault Protection in a Delta Circuit

Many but not all delta circuits have protection against ground faults. A delta circuit would not notice a phase-to-ground fault; therefore, a grounding transformer and relays must be installed. A grounding transformer is installed in a substation and connected between one phase of a delta circuit and ground. This connection through the transformer incorporates the ground part of the delta circuit. There is normally no voltage difference between a delta phase and ground, so there would be no current flow through the transformer. During a phase-to-ground fault, the current flowing back to the source will flow through the transformer and send a signal to the ground fault relay. The relay will detect this current and trip out the circuit.

Wye-Connected Systems

Wye- or Parallel-Connected Three-Phase System

The source of a wye system is a three-phase transformer bank with a wye-connected secondary. A delta/wye or a wye/wye transformer bank provides a three-phase wye output with the three phases interconnected as one circuit. When connected in a wye configuration, each coil of a three-phase transformer is connected between a phase and the neutral. One end of each coil is connected together at a common point, which is the neutral. A voltage will be available at the other ends of the three coils when measured between any two phases. There is also a voltage between each phase and the common point. The three coils are connected in parallel.

Wye connections are represented on paper in the shape of a "Y" but when making transformer connections the transformer coils will typically be side by side. **See Figure 6–9**. The wye side by side connections still have a common point and three phases in parallel.

Figure 6–9 The coils in both illustrations are connected in a wye configuration. *(Delmar, Cengage Learning)*

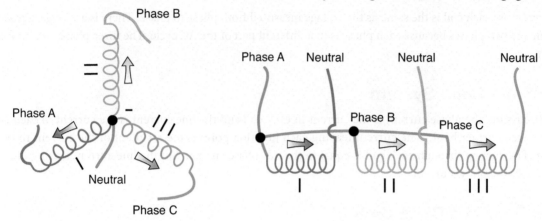

A single-phase transformer connected to a wye system has one connection to a primary phase and the other end to the common neutral point. **See Figure 6–10.** A phase to neutral connection is a parallel connection. The transformer installation has one surge arrestor, and when fused cutouts are used, one is required.

Figure 6–10 A transformer connected phase to neutral (wye). *(Delmar, Cengage Learning)*

Voltage and Current in a Wye System

Two different voltages are available in a wye system: phase-to-phase voltage and phase-to-neutral voltage. When the neutral is well grounded, the phase-to-neutral voltage is the same as the phase-to-ground/earth voltage. The phase-to-phase voltage is $\sqrt{3} \times$ the phase-to-neutral voltage. The phase-to-phase voltage is the resultant (vector or algebraic) sum of two phase-to-neutral voltages that are 120° out of phase with each other.

The phase current going through each coil in a transformer is the same as the line current that flows into each phase of the wye circuit. The current flow in a wye circuit is $1 \div \sqrt{3}$, or 58% of an equivalent delta circuit.

The Neutral in a Wye System

The neutral is the common point of each transformer winding in a wye system. If the loads on the three phases were equal to each other, there would be no current on the neutral. The neutral carries the sum (algebraic or a vector sum) of the three phase currents back to the source. With the three phases at 120°, the three currents cancel out each other. The neutral wire can be smaller than the phase wires because it is sized to carry the unbalanced load between phases.

The worst-case scenario for an unbalanced wye circuit would be a phase-to-ground fault. Most neutrals are multi-grounded, so fault current that enters the ground would return to the source through the ground and the neutral. For the duration of a phase-to-ground fault, the neutral current gets very high and the voltage between the neutral and ground also rises. On a single-phase line, the neutral—and to a lesser extent the ground—is the only path back to the source to complete the circuit.

Some three-wire circuits are actually wye circuits, which do not carry a neutral. The need for a neutral is minor when the load is balanced among the three phases. Transmission and subtransmission lines have relatively balanced loads and are often wye circuits without a neutral. In some utilities where there is good earth, the neutral is omitted on a distribution feeder because the earth is able to complete the circuit for unbalanced load and fault.

Single-Wire Earth Return

In some rural, thinly populated locations around the world (e.g., Australia or Saskatchewan, Canada), no neutral conductor is strung on the wye distribution system. A single-phase circuit would consist of only one conductor strung on a pole. This system uses the earth as the neutral and is called a single wire earth return (SWER) system. Earlier DC transmission lines were SWER, but the ground current caused corrosion to buried facilities; therefore, most new DC lines are bipolar. SWER systems are against the electrical code of many countries. An example of a SWER system is the electrical system of a car. The metal body of the car is used as the ground wire and is attached to the negative pole of the battery.

The advantage of the earth return system is that with fewer conductors, fewer pole-top fittings, and ease of construction, these lines require less capital to build. One disadvantage is that the down-ground wire on a distribution transformer pole is the transformer primary neutral. Anyone getting across a break in the down-ground wire is exposed to full primary voltage. It is important to ensure that the down-ground has good physical protection installed where it is accessible to the public.

Converting from Delta to Wye

Many delta circuits have been converted to wye circuits because of economical and safety reasons. When the output connections of a distribution substation transformer are converted from delta to wye, the phase-to-phase voltage will be increased 1.73 ($\sqrt{3}$) times. For example, a delta-connected transformer bank that has a phase-to-phase voltage of 2,400 V across each coil of the transformer secondary can be converted to a wye connection by having one end of each 2,400 V coil connected to a common point. Now, the phase-to-phase voltage of the newly connected wye circuit will be 4,160 V, and the phase-to-neutral voltage will be 2,400 V. (***Note:*** Usually a conversion involves going to an even higher voltage fed from a new substation transformer.) For the same load, the current in each wye-connected phase is reduced to 1 ÷ 1.73, or 58%. A lower current results in less line loss and less voltage drop.

During conversion, an existing distribution transformer that was connected as phase to phase on the delta circuit can now have the same primary voltage by connecting it phase to neutral on a wye circuit. (***Note:*** Usually a conversion involves going to an even higher voltage and installing dual-voltage transformers to reduce the time of customer outage during the cutover changeover.)

Single-phase wye-connected transformers require only one cutout and surge arrestor instead of the two needed on a delta transformer. Similarly, a single-phase lateral on a wye system needs only one line switch.

The multigrounded neutral on a wye system is an excellent ground and is available for transformers and secondary services regardless of possible poor local grounding conditions. On a delta system, the ground at a transformer is dependent on the driven ground rod at the transformer and at the customer.

Fuses and circuit breakers trip out quicker due to overcurrent when there is a good return path to the source. Phase-to-phase short circuits on both wye and delta circuits trip out a fuse quickly because of the good return path to the source through the other phases. Phase-to-ground faults have a good return path to the source on wye circuits through the multigrounded neutral. A tree contact would blow a fuse much quicker on a wye system than on a delta system.

Three-Phase Power

The Combined Power in Three Phases

When voltage and current are in each phase of a three-phase circuit, power is also being delivered. The power delivered by each phase will be 120° out of phase with the other phases. In other words, the power delivered by one phase could be at its peak when the power delivered by another phase is one-third farther into the cycle and the power delivered by the third phase is two-thirds farther into the cycle.

Therefore, the total power in a balanced three-phase system is not simply three times the power of one phase. The square root of three, which is 1.73, times the power in one phase will give the power in three phases. When the load is unbalanced, the load in each phase must be measured individually and the average value is put into the equation. The power factor must be known to calculate the true power delivered by a three-phase system.

Field Calculations

Converting amperes in a circuit to kilovolt-amperes (kVA) and converting kVA into amperes per phase are sometimes valuable tools in the field. Field calculations are valuable when balancing three-phase circuits. The ampere loading of a phase should be converted to kVA to determine which transformers to transfer over to another phase.

Before working energized line on a transmission circuit where jumpers are going to be installed, it is valuable to know the expected load on a phase. The controlling station can give the total load on the circuit, which can be converted to amperes per phase using field calculations.

Apparent power (kVA) is used in the field instead of true or real power (kilowatts) because apparent power is what a feeder carries. True power is measured by revenue meters. To determine the total load on a transformer, field calculations can convert amperes per phase to kVA.

Calculations Involving Power

It is important to familiarize yourself with the formulas used to calculate power in three-phase and single-phase circuits. **See Figure 6–11**. Calculations for field applications use the kVA formulas. Calculations involving three-phase power use line-to-line (phase-to-phase) voltage. The value of the square root of three is 1.73.

Figure 6–11 Power Formulas

To Find	Direct Current	Single-Phase AC	Three-Phase Wye AC
Kilowatts (kW)	$\dfrac{I \times E}{1,000}$	$\dfrac{I \times E \times PF^*}{1,000}$	$\dfrac{I \times E \times 1.73 \times PF}{1,000}$
Kilovolt-Amperes (kVA)		$\dfrac{I \times E}{1,000}$	$\dfrac{I \times E \times 1.73}{1,000}$
Amperes (when kW are known)	$kW \times \dfrac{1,000}{E}$	$kW \times \dfrac{1,000}{E \times PF}$	$\dfrac{kW \times 1,000}{1.73 \times E \times PF}$
Amperes (when kVA are known)		$\dfrac{kVA \times 1,000}{E}$	$\dfrac{kVA \times 1,000}{1.73 \times E}$

*PF = power factor

Calculations with "Handy Numbers"

To do the power calculations quickly in the field, a "Handy Number" can be used for approximations. **See Figure 6–12.**

Figure 6–12 Examples of "Handy Numbers" for Some Voltage Systems

Calculating Approximate Loads with "Handy Numbers"				
	Three-Phase kV	Handy #	Single-Phase kV	Handy #
kVA = Amperes × "Handy Number"	230	400		
or	115	200		
Amperes per Phase = $\dfrac{kVA}{\text{"Handy Number"}}$	69	120		
	46	80		
	25	40	14.4	14
	12.5	22	7.2	7
	8.32	14	4.8	5
	0.208	0.36	0.12	0.12

To calculate the "Handy Number" for other voltage systems:

    For three-phase lines, the "Handy Number" = $L - L$ voltage × 1.73 ÷ 1,000

    For single-phase lines, the "Handy Number" = $L - N$ voltage ÷ 1,000

Where:

    L = line

    N = neutral

$$\text{Approximate kVA} = \text{amperes} \times \text{"Handy Number"}$$
$$\text{Approximate amperes per phase} = \text{kVA} \div \text{"Handy Number"}$$

Examples Using "Handy Numbers"

Example One phase of an 8.3/4.8 kV feeder has 80 amperes (A) more load than the other two phases. To balance the feeder, how many kilovolt-amperes should be transferred to the other two phases?

Solution Using a "Handy Number," which is 5 for a 4.8 kV single-phase line, × 80 A = 400 kVA. Therefore, 400 kVA should be distributed between the three phases, and 400 ÷ 3 = 133 kVA should be transferred to each of the other two phases.

Example Load on each of the three phases of a 120/208 50 kVA pad mount transformer bank are 150 A, 170 A, and 140 A. Is the transformer overloaded?

Solution The average load on the three phases is 150 + 170 + 140 ÷ 3 = 153 A. Using the "Handy Number," which is 0.36 for 120/208 V service, × 153 = 55 kVA. The transformer is only slightly above its rating, but the load should be balanced more between the three phases to prevent one transformer winding from being overloaded.

Review Questions

1. What advantages are there for a three-phase versus a single-phase circuit?

2. What does it mean when the voltages of each phase of a three-phase circuit are 120° apart?

3. What can be done at a three-phase transformer bank if a customer's three-phase motor is running backward?

4. What determines whether a three-phase circuit is considered a wye or a delta circuit?

5. What is the voltage-to-ground on a 4,160 V delta circuit?

6. Why does a source breaker see a phase-to-ground fault on a delta circuit?

7. If the phase-to-phase voltage on a wye feeder is 25 kV, what is the phase-to-ground voltage?

8. What condition would put a very high current on the neutral and also raise the voltage between the neutral and ground?

9. Name two advantages for converting a delta circuit to a wye circuit.

10. A control room operator tells you a 115 kV circuit is carrying 70 MVA. How much current would a temporary jumper have to carry?

Chapter 7

Awareness in an Electrical Environment

Objectives

After completing this chapter, you should be able to:

1. Explain how a person can make electrical contact in series with a circuit and in parallel with a circuit.

2. Identify the work method needed for protection from electrical contact when working in series with a circuit.

3. Describe the need to install a jumper across an open point in a conductor exposed to electromagnetic induction.

4. Determine the work method needed for protection from electrical contact when working in parallel with a circuit.

5. Take steps to ensure that the leakage current along live line tools and rubber gloves are below the minimum threshold.

6. Identify the work method needed for protection from electromagnetic and/or electrostatic induction.

7. Take steps to prevent electrical contact from a potential ground fault.

8. Take steps to prevent electrical contact from step and touch potentials when setting a pole in an energized circuit.

9. Explain how grounding a truck to the neutral can prevent a more serious accident when a truck boom makes accidental electrical contact.

10. Describe electrical hazards when working on a neutral.

11. Identify hazards from the sources of electromagnetic fields.

12. Memorize the minimum approach distances for the circuit voltages being worked on.

Introduction

Working on or Near Electrical Circuits

When working on or around electrical circuits, people are exposed to electrical influences from sources that are not always obvious. The topics in this chapter apply knowledge of electrical theory to the various tasks involved when working in an electrical environment. An increased recognition of a potential electrical hazard will allow for safer work in an electrical environment. To be considered qualified to work closer than 10 ft (3 m) to a distribution circuit you must recognize these hazards and know the types of barriers to set in place to control them.

Making Contact with an Electrical Circuit

When working on or near energized circuits, a person can make two types of electrical contact. The person making contact is like an additional electric load to the circuit, and a load can be connected in series with the circuit or connected in parallel with the circuit. Some distinct differences exist in the way voltage and current act on a load connected in parallel versus a load connected in series.

Connecting a Load in Series with a Circuit

What Is a Series Circuit?

Two or more loads within a circuit are considered to be in series when one (common) current is flowing through all the loads. When the current has only one path to take through two or more electrical loads, then the loads are in series with each other. **See Figure 7–1.** In other words, when a component is connected end to end into a conductor, it is connected in series.

Figure 7–1 The electric lights have one continuous wire running from the generator and back and are therefore connected to the circuit in series. *(Delmar, Cengage Learning)*

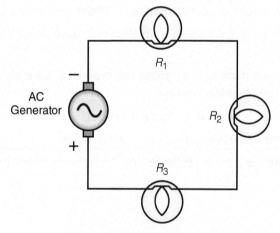

Current through each of R_1, R_2, and R_3 is equal to the current in the circuit.

AC Generator

R_1

R_2

R_3

Voltage across R_1, R_2, and R_3 varies, depending on the value of R.

Characteristics of a Series Circuit

In a series circuit, an equal amount of current runs through the complete circuit and all connected loads. An ammeter reading would show the same value any place in the circuit. If any change is made to the current flow in any part of the circuit, the change to the current applies throughout the circuit.

$$I_{Total} = \text{The current through any part of the circuit}$$

In a series circuit, a voltage drop occurs across each load. The voltage drop across each load is dependent on the resistance of the load and can be calculated using Ohm's law ($E = I \times R$). A large load or low resistance would have a large voltage drop across it. A break in the circuit would introduce an infinite high resistance to current flow and would result in a total voltage drop. This means that the total circuit voltage (recovery voltage) is available across the break. It is, therefore, a basic line-trade procedure to always jumper any conductor that is carrying current before it is cut.

The sum of all the individual voltage drops is equal to the applied voltage of the circuit. The voltage drops some as it goes through each load in the circuit and is "all used up" when it gets back to the source.

$$E_{Total} = IR_1 + IR_2 + IR_3 + \cdots IR_N$$

The total resistance in the circuit is equal to the sum of all the individual resistances within the circuit.

$$R_{Total} = R_1 + R_2 + R_3 + \cdots R_N$$

See Figure 7–2. Given a 120 V generator, feeding loads of $R_1 = 10$ ohms, $R_2 = 20$ ohms, and $R_3 = 30$ ohms, then,

$$R_{Total} = R_1 + R_2 + R_3$$

$$R_{Total} = 10 + 20 + 30 = 60 \text{ ohms}$$

$$I_{Total} = E \div R = 120/60 = 2 \text{ amperes}$$

The voltage drop across $R_1 = 2 \times 10 = 20$ volts

The voltage drop across $R_2 = 2 \times 20 = 40$ volts

The voltage drop across $R_3 = 2 \times 30 = 60$ volts

$$E_{Source} = IR_1 + IR_2 + IR_3$$

$$E_{Source} = 20 + 40 + 60 = 120 \text{ volts}$$

Loads Placed in Series

Few applications of series-connected loads are in an electrical utility. Series street-lighting circuits are still present in some utilities. The supply transformer keeps the current constant throughout the circuit, and the voltage varies depending on the number and size of the lights in the circuit. There are series-connected loads at the utilization level, but they are mostly within such electrical equipment as appliances and electronic devices. Some strings of Christmas tree lights are connected in series. In a series-connected string of lights, when the element of one lamp fails, it opens the circuit and the lights go out along the whole string. The input voltage is equally divided over all the lamps in the string. For example, strings have 10 lamps, 25 lamps, or 40 lamps; the voltage rating of the lamps from a 120 V source is 120 divided by the number of lamps in a string. Similarly, a person with a 240 V service could connect two 120 V lights in series to allow the use of standard 120 V lamps.

Making Series Connections

The line trade is involved in connecting equipment such as switches, reclosers, and voltage regulators in series within the circuit. Anytime a person is involved with cutting, splicing, or connecting or disconnecting a conductor, there is a risk of getting into series with the circuit. When a power conductor is cut, all the current flow is interrupted. The resistance across the open point is infinite; the voltage (IR) is at full-line voltage. This full-line voltage (recovery voltage) appears across the two ends of the cut conductor. Installing a jumper across an open point keeps the current flowing and keeps the potential across the cut at 0 V.

Figure 7–2 In a series circuit, the total voltage drops proportionately across each load while the current remains constant. *(Delmar, Cengage Learning)*

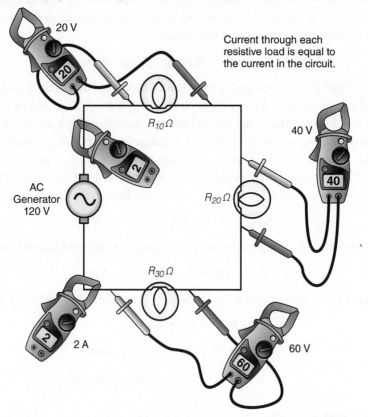

Current through each resistive load is equal to the current in the circuit.

The Human Body in a Series Circuit

A person's body can complete a circuit when it bridges an open point within the circuit. When a body completes a circuit, it is exposed to the full-line voltage of the circuit and all the current the voltage is able to drive through the resistance of the body. Every time a person makes a cut or splice in a current-carrying conductor, one can accidentally put oneself into series with the circuit. Even on an isolated and grounded circuit, enough current flow can be in the conductors to cause a lethal shock. Only body resistance, clothing, and gloves limit the current flow when a body is put into series with the circuit. If the voltage is high enough across the open point, this resistance will be overcome, a person's body will complete the circuit, and a current flow will be established through the body.

Examples of Accidents Involving Series Contact

Accidents occur when a person's body completes a circuit, as in the following examples.

- A power line technician received electrical burns when he opened the neutral of an energized circuit without installing a bypass jumper across the open point. His body was in series with the neutral conductor. This occurred even though the neutral was multigrounded.

- A power line technician received electrical burns when he removed a jumper on a grounded transmission line. Grounds were installed on each side of the jumper.

- Attaching to (getting in series with) an energized secondary bus with one hand and an isolated service drop with another hand caused a power line technician to be "frozen" on the line. Frozen means unable to release his hands. The voltage across the open point was equal to the service voltage.

- Forgetting to install a jumper before cutting a conductor while doing rubber-glove work or barehand work results in an electrical flash and exposure to a lethal voltage across the open point.

Protection from Series Contact

Every time a *cut* or *splice* is made to a current-carrying conductor, one can accidentally put oneself into series with the circuit. Even on an isolated and grounded circuit, current flow in the conductors can be sufficient to cause a lethal shock. Installing a jumper across an open point keep the current flowing and keeps the potential across the cut at 0 volts.

A power line technician should be careful to avoid being in series while making or breaking a connection. For example, when making a connection, the power line tech must have both hands on one wire until contact is made between the two wires. Even with a jumper in place and as a matter of habit, a power line tech should never place the hands across an open point in case an earlier step has been forgotten.

A piece of cover-up or an insulated aerial bucket provides protection from electric shock in cases of accidental contact. However, a bucket or cover-up will provide no protection from contact made by hands across an open point. Wearing rubber gloves provides protection, but no backup protection is available if the only thing preventing a current flow from hand to hand is the integrity of the rubber gloves.

Connecting a Load in Parallel with a Circuit

Making Parallel Contact

Power line technicians frequently put themselves in a parallel path to ground when working on a wye circuit. For example, when working with rubber gloves on a live conductor, a person is parallel to other loads on the circuit. The rubber gloves may prevent a dangerous current flow through the worker. On a delta circuit, a person must make phase-to-phase contact to make a parallel connection with other loads on the circuit.

What Is a Parallel Circuit?

When more than one path is available for current to flow through, the loads in each path are connected in parallel with each other. **See Figure 7–3.** The current divides into each branch of the circuit. The magnitude of the current in each branch depends on the resistance within that branch.

Figure 7–3 Each individual light is connected between the source and the neutral and is therefore connected in parallel. *(Delmar, Cengage Learning)*

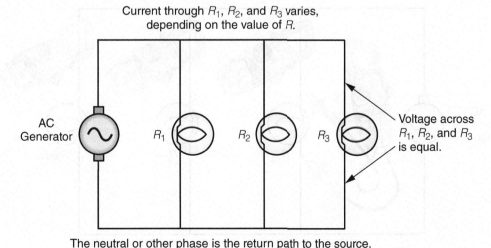

Current through R_1, R_2, and R_3 varies, depending on the value of R.

AC Generator

R_1 R_2 R_3

Voltage across R_1, R_2, and R_3 is equal.

The neutral or other phase is the return path to the source.

Characteristics of a Parallel Circuit

The same voltage appears across each path connected in parallel because each path is connected between the same two wires supplying the source voltage. Each branch is independent of the others. Each branch has the same voltage, but the current and resistance can change in a branch, which will not affect another branch.

$$E_{Source} = E_1 = E_2 = E_3 = E_N$$

The total current in a circuit with its loads connected in parallel is equal to the sum of the currents flowing into each path.

$$I_{Total} = I_1 + I_2 + I_3 + \cdots I_N$$

The current within each path is dependent on the resistance or load in each path. Most current takes the path of least resistance, but *each* path will have *some* current flow through it. The magnitude of the current through each path is inversely proportional to the resistance. The branch with the least resistance takes the most current. The total current in the circuit is always more than the current in the branch with the least resistance. This means, therefore, that the total resistance of the circuit is less than the smallest resistance in the circuit.

$$\frac{1}{R_{Total}} = \frac{1}{R_1} + \frac{1}{R_2} + \frac{1}{R_3} + \cdots \frac{1}{R_N}$$

A short-circuit fault is the lowest-resistance load in a circuit and, therefore, attracts the most current flow. However, every other element connected in parallel still takes a portion of the current. Because each parallel path is connected between the same two wires, the same source voltage (120 V) appears across each path.

In a parallel circuit, each branch has the same voltage, but the current through each path depends on the load (resistance). **See Figure 7–4.**

$$I_1 = E \div R_1 = 120 \div 10 = 12 \text{ amperes}$$

$$I_2 = E \div R_2 = 120 \div 20 = 6 \text{ amperes}$$

$$I_3 = E \div R_3 = 120 \div 30 = 4 \text{ amperes}$$

$$I_{Total} = I_1 + I_2 + I_3 = 22 \text{ amperes}$$

Figure 7–4 The sample calculations show that the current through each of the different sized loads varies but the voltage across each load is the same. *(Delmar, Cengage Learning)*

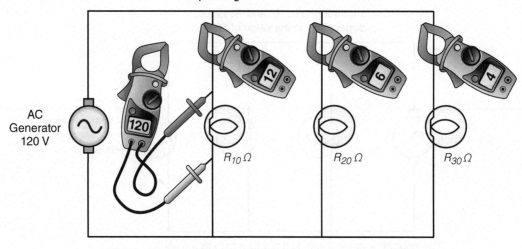

Current through R_1, R_2, and R_3 varies, depending on the value of R.

The neutral or other phase is the return path to the source.

Calculating the Total Resistance in a Parallel Circuit

The total resistance of a parallel circuit is not the sum of the resistance as in a series circuit. The total resistance is calculated using the following formula:

$$\frac{1}{R_{Total}} = \frac{1}{R_1} + \frac{1}{R_2} + \frac{1}{R_3} + \cdots \frac{1}{R_N}$$

Example Calculation

If three transformers are connected in parallel, calculate the equivalent or total resistance of the circuit, given a source of 5,000 V and three parallel branches with a resistance of 250, 500, and 100 ohms. **See Figure 7–5.**

Figure 7–5 Calculate the total resistance in the circuit. *(Delmar, Cengage Learning)*

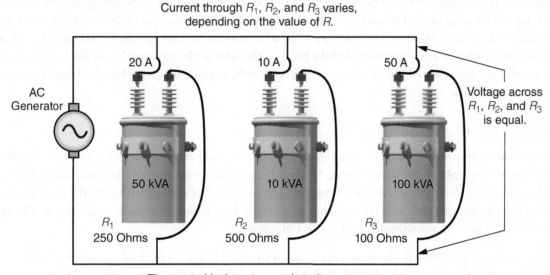

Current through R_1, R_2, and R_3 varies, depending on the value of R.

AC Generator

20 A   10 A   50 A

Voltage across R_1, R_2, and R_3 is equal.

50 kVA   10 kVA   100 kVA

R_1 250 Ohms R_2 500 Ohms R_3 100 Ohms

The neutral is the return path to the source.

The question in this sample requires finding the total resistance of a circuit that has three different sized (loads) transformers. Substituting the resistance values into the formula:

$$\frac{1}{R_{Total}} = \frac{1}{250} + \frac{1}{500} + \frac{1}{100}$$

$$\frac{1}{R_{Total}} = \frac{2}{500} + \frac{1}{500} + \frac{5}{500}$$

$$\frac{1}{R_{Total}} = \frac{8}{500}$$

$$R_{Total} = \frac{500}{8} = 62.5 \text{ ohms}$$

Note that the total resistance of this parallel circuit is smaller than the resistance in any one branch. The larger transformer has a large load current but a smaller resistance.

Loads Connected in Parallel

Almost all loads are connected to an electrical system as parallel-connected loads. Loads on a distribution feeder are fed through transformers. Almost all transformers on a feeder are connected in parallel to each other. All transformers on a feeder are also connected to the same voltage, and the current flowing into each transformer is inversely proportional to the load.

Loads at the utilization level in homes and factories are connected in parallel. When an appliance is plugged into a socket, it is connected to the same voltage as other appliances plugged into that electrical service.

Leakage Current

In a wye circuit, anything that is in contact with both the ground and an energized conductor is a parallel connection and has some current flow through it. Even high-resistance objects, such as insulators, have some current flow through or along them. This small leakage current is in inverse proportion to its resistance and is measured in microamperes.

Live-line tools, rubber gloves, and insulated booms are all subject to some leakage current. The care taken to keep these tools clean and in good condition is intended to keep the leakage current below a certain threshold.

Leakage Current Working Barehand

When the power line technician works barehand from an aerial basket, the truck boom itself is a parallel path to ground. The resistance of the boom is very high, but some current flow still travels through the boom to ground. A boom contamination meter is used to confirm that the leakage current is at an acceptable level. The leakage current is never at zero. A typical acceptable leakage current for an insulated boom is less than 1 microampere (μA) for each 1,000 V to ground. A power line technician in an aerial basket is exposed to even less current because the metal grid below the feet is bonded to the conductor. In other words, the potential of the hands is the same as the potential below the feet. On high-voltage circuits, a conductive suit is worn to reduce the uncomfortable voltage gradients around the body and to keep all of the body at the same potential.

The Human Body in a Parallel Circuit

Anytime a person makes contact with a hot conductor, whether with rubber gloves or with a live-line tool, that person will be exposed to some voltage and some current. In a wye circuit, high-resistance objects—such as insulators, live-line tools, rubber gloves, insulated booms, and anything that is in some kind of contact with both the ground and an energized conductor—create a parallel connection on the circuit and will be subject to some leakage current. The magnitude of the current will depend on the resistance of the object in contact.

- Because some current flow will always be present when a parallel contact is made, the power line technician is protected by keeping the current flow through the body to ground below a dangerous threshold. Current flow through a person will be kept below dangerous values when resistance is increased between an energized conductor and the second point of contact by the use of rubber gloves, live-line tools, cover-up, and insulated platforms.

- Current flow through a person's body will be kept below dangerous values when there is no potential difference between one part and another part of the body. When switching or operating stringing equipment, the use of a ground gradient control mat will ensure that hands and feet stay at the same potential. When there is no potential difference, there is no current flow.

- Even when working from an insulated basket, there is leakage current flowing down the rubber gloves and down the insulated boom. The current flow is very small, probably less than 10 μA, and it is below the threshold at which one would feel anything. If a person fails to wear rubber gloves or reduces the resistance of the electrical path through the body in any way, the body will take a greater share of the current flow.

Examples of Accidents Involving Parallel Contact

Accidents occur when a person's body is a parallel path to a current flow, as in the following examples.

- A power line technician received a lethal shock when he made an inadvertent contact with an energized circuit while standing on a pole waiting for the circuit to be isolated. The 4,800 V in the line were high enough to overcome the resistance of his leather work gloves and the wooden pole. The amount of current going through him, as a parallel path to ground, was well above the usually fatal 100 mA.

- An uninsulated vehicle boom made contact with the bottom of a transformer cutout while a power line tech was getting material from the truck bin. The truck was grounded to a temporary ground probe but the circuit did not trip out. He was a parallel path to ground and received fatal electrical burns.

- While a power line tech was opening a gang-operated switch, the operating handle became energized when an insulator broke and an energized lead dropped onto the steel framework of the switch. The power line tech was standing on a ground-gradient mat that was bonded to the operating handle and his back was in contact with some brush growing next to the mat. The brush was a parallel path to a remote ground, and he received electrical burns on his back.

Protection from Parallel Contact

Because there is always some current flow when a parallel contact is made, a person is protected by keeping the current flow through the body to ground below a dangerous threshold. Current cannot flow unless there is a circuit. A person must prevent a circuit from being completed through the body. For example, a person in contact with a single energized conductor, while working from an insulated aerial device, is part of an open circuit. Other than a small leakage current, the current has nowhere to flow because there is no second point of contact. Removing or keeping away from a second point of contact is a key element in all live-line work procedures.

Current flow through a person is kept below dangerous values when resistance is increased between an energized conductor and the second point of contact by using rubber gloves, live-line tools, cover-up, and insulated platforms. Current flow through a person's body is kept below dangerous values when there is no potential difference between any two parts of the body. The use of a ground-gradient control mat when switching or operating stringing equipment ensures that the hands will stay at the same potential. When there is no potential difference, there is no current flow.

Electrical and Magnetic Induction

Three Electrical Effects without Contact

A person can be affected in three ways by electrical phenomena without being in contact with an energized conductor. All people, not just utility workers, are exposed to static electricity. Static electricity is noticed when one touches a doorknob after walking across a carpet. The spark to the doorknob is a DC charge, which is quite different from the type of **induction** experienced working near energized AC circuits. An object or person near an energized AC circuit is exposed to electromagnetic induction. Electromagnetic induction is composed of electric field induction and magnetic field induction.

Static Electricity

Although static electricity is not the same as the induction experienced by electrical-utility personnel working near energized lines, it is a phenomenon that can be a hazard in some work settings. Static electricity is a DC charge. It is called static because it is a stationary electric charge. This charge can be generated by rubbing together two nonconducting substances. When a static charge is discharged, a current flow travels between a charged object and the body that makes contact. Lightning is a discharge of static charge. Static buildup occurs on fast-moving conveyor belts and when removing synthetic clothing, walking on carpets, and sliding out of a car seat while contacting the metal door. A spark from static electricity can cause a fire, especially when the spark occurs in the presence of flammable vapor, dust, or gas. Following are two examples of how to eliminate fires due to static.

- The hoses at gasoline pumps are conductive to prevent a static discharge.
- When fueling a helicopter, a bond is installed between the fuel pump and the helicopter.

Electric Field (Capacitive) Induction

An isolated, ungrounded conductor strung near an energized conductor has an **induced voltage** on it. It is called capacitive induction because, the energized conductor acts as one plate of a capacitor, the isolated conductor acts as the other plate of the capacitor, and the air between is the insulation between the two plates. **See Figure 7-6.** The isolated conductor has a voltage on it induced from the energized conductor. The isolated conductor is not part of a circuit if the switches are open at each end and, therefore, no current flows. If the isolated conductor is grounded to earth at one location, the voltage is

Figure 7–6 An energized conductor near an isolated conductor separated by air (insulation) acts as a capacitor and induces voltage on the isolated conductor. *(Delmar, Cengage Learning)*

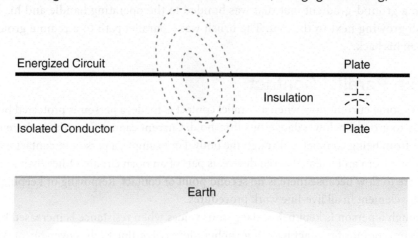

drained from the conductor and a minor amount of current flows down the ground. The magnitude of the induced voltage is higher in the following instances.

- The voltage on the energized conductor is high.
- The distance between the two conductors is decreased.

The parallel length between the energized conductor and the isolated conductor is not a big influence on the amount of voltage induced on the isolated conductor but is a big influence on the amount of current induced on the conductor.

Electric Field Effect on an Isolated Conductor

The voltage induced on an ungrounded isolated conductor from a neighboring high-voltage circuit can be very high. Contact with such a conductor can result in a lethal steady-state shock that continues as long as contact is maintained. Installing a set of grounds on the conductor discharges the induced voltage effectively. Bonds such as between vehicles, ground probes, and conductors eliminate exposure to potential differences among them.

Electric Field Effect Working in a High-Voltage Environment

When working in a high-voltage environment, such as climbing a transmission tower or working in a high-voltage substation, a person's body acts as the second plate of a capacitor even though the body is not necessarily well insulated from ground. An electric field induces a voltage on a body. The voltage becomes apparent every time a grounded object is touched and the voltage is discharged to ground. The resultant spark (spark-gap effect) consists of a high-voltage but low-current discharge.

One hazard from an electric field is an involuntary movement or a fall due to the surprise of a shock. Wearing conducting-sole boots reduces this hazard when working on transmission towers or working in high-voltage stations. When approaching an energized transmission line, the electric field gets more intense as a worker gets closer to an energized conductor; in other words, the kilovolts per inch (or per centimeter) increase as a person gets closer to the conductor. When working barehand, a power line technician is shielded from the high-intensity electric field with a conductive suit that includes a hood which fits over the hard hat, conducting-sole boots, and conducting work gloves. This shield or conductive "blanket" around the worker keeps all parts of the body at the same potential and forms a Faraday cage, which prevents current from the electric field from flowing through the conductive cover suit. Similarly, a person using rubber gloves on a medium-voltage distribution line, such as a 34.5 kV line, might feel a vibration or "bite" in the rubber gloves. The rubber glove acts as insulation between two conductive plates, the first being the 34.5 kV and the second the worker.

Magnetic Field Induction

An isolated and grounded conductor paralleling an energized current-carrying conductor has current induced on it. This occurs because each conductor acts like a coil in a transformer and the air between acts like the core of a transformer. Current flowing through an energized conductor acts as the primary coil. A nearby grounded conductor acts as the secondary coil. **Induced current** flows when a circuit has been created in the isolated conductor. The installation of two sets of grounds (bracket grounds) on the isolated conductor creates a circuit with current flowing along the conductor, down one set of grounds through earth (or neutral) and up the other set of grounds. **See Figure 7–7.** The magnitude of the induced current is higher in the following situations.

- The current on the energized conductor is increased.
- The distance between the two conductors is decreased.
- The length of parallel between the two conductors is increased.

Figure 7–7 Placing grounds on the isolated conductor creates a circuit and the magnetic field around the energized conductor induces current in the isolated conductor. *(Delmar, Cengage Learning)*

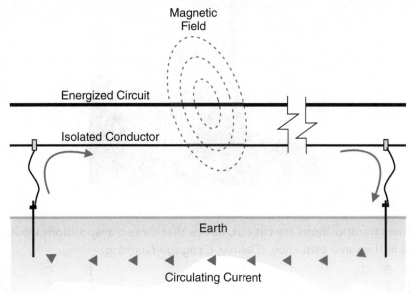

Because the amount of magnetic induction from a magnetic field depends on current and not voltage, this induction is also a hazard on lower-distribution voltages.

Working in an Environment with High Induction

A power line technician is exposed to induction from a magnetic field when working on an isolated and grounded circuit that is parallel to an energized circuit. When only one set of grounds is installed at the point of work, a circuit has not been created for circulating currents to flow. When the conductor is grounded in two locations, a circuit is created and current flows through the conductor and through the grounds. If this circuit is interrupted by removing a ground or cutting a conductor, a high voltage is available across the open point. That is why it is important to install and remove grounds with a live-line tool.

Transpositions

Many transmission and subtransmission circuits constructed before 1955 had transpositions installed in them. **See Figure 7–8.** A transposition involved changing the position of the three-phase conductors by crossing a conductor from one side of the structure to the other side. When the different phases crossed each other, the induced voltages on the phases canceled out each other. They were installed because large electromagnetic fields interfered with open-wire telephone circuits. With three transpositions, the A phase crosses under the other two phases to go from one side to the other. **See Figure 7–9**. Three of these transpositions (one barrel) are needed to bring the three conductors back to the original ABC configuration.

Figure 7–8 The transposition in this subtransmission circuit was originally installed to reduce the magnetic field and its effect on open wire telephone. The open wire telephone has since been removed. *(Delmar, Cengage Learning)*

Figure 7–9 Sometimes transpositions are cut out, notice that three transpositions would need to be cut out to keep the terminals in the same sequence. *(Delmar, Cengage Learning)*

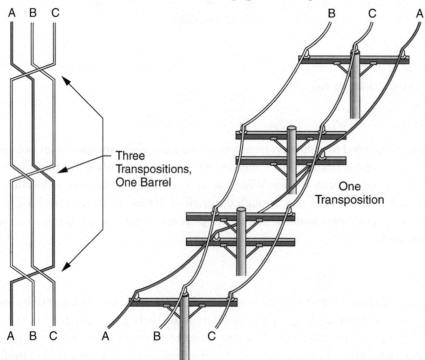

Voltage Gradients

Ground Faults

There is a hazard to anyone working at a location where a ground fault results in electrical energy flowing into the earth. A ground fault is a short circuit caused when an object in contact with earth contacts an energized conductor. The following are examples of ground faults that occur when electrical current flows into earth.

- A utility pole becomes energized due to contact with an energized conductor or because of a faulty insulator.
- An energized conductor comes in contact with a tree or a truck boom.
- A portable ground is accidentally installed on an energized conductor.
- An underground cable is dug up or punctured.
- An energized conductor is lying on the ground.
- The operating handle of a gang-operated switch becomes energized because of a switch insulator failure.
- The conductor being strung contacts an energized circuit which energizes the conductor at the reel stand or tensioning machine.

Voltage Gradients

At the point where the current enters earth, the current breaks up and flows in many paths, depending on the makeup and resistance of the earth. The voltage available is highest where the current enters earth. The earth acts as a network of resistors, and the voltage drops as the current flows through these resistors. The voltage at the current entry point is higher than the voltage one or two paces away from the entry point. Therefore, a difference of potential in the earth occurs around the current entry point.

It is easiest to visualize these voltage gradients as ripples in a pond emanating from where a stone has entered. The ripples are strongest at the center and get weaker as they radiate from the center. The difference in the intensity of the ripples represents the difference in the voltage levels. The voltage between the ripples lessens as the distance from the contact point is increased. **See Figure 7–10**. Voltage gradients are also known as *ground gradients,* potential gradients, and step potentials.

One form of protection from ground gradients is work from a ground-gradient mat (equipotential mat) that is bonded to the device that may become energized. A person standing on a fabric-style, ground-gradient mat that is electrically bonded to the switch being operated is protected. **See Figure 7–11**. If the cap and pin breaks off the post insulator and energizes the steel structure, the person will be safe because no voltage difference would exist between the hands and the feet.

A person working on a ground-gradient mat that is bonded to a pad-mount transformer, would also be protected. **See Figure 7–12**. If ground fault on a cable or elsewhere raises the potential on the transformer tank, the person would be safe because no potential difference would occur between the hands and the knees.

Voltage-Gradient Example

The line trade is exposed to a ground-gradient hazard when a pole is being installed in an energized circuit. If a pole contacts an energized circuit and also contacts the ground, a voltage gradient is set up at the base of the pole. The details involved with setting a pole in an energized line are used in the remainder of this section to explain the specific hazards involving ground gradients.

Setting a Pole in an Energized Distribution Line

A work procedure to set a pole in an energized line requires insulated cover-up on the conductors and/or on the pole. With the use of cover-up and an observer, a modern vehicle can set poles in an energized line without making contact with a bare, energized circuit. This procedure requires that the boom-equipped vehicle setting the pole be grounded,

Figure 7–10 Stepping near a fallen conductor can cause a person to have one foot at a higher voltage than the other foot which has been the cause of fatalities. *(Delmar, Cengage Learning)*

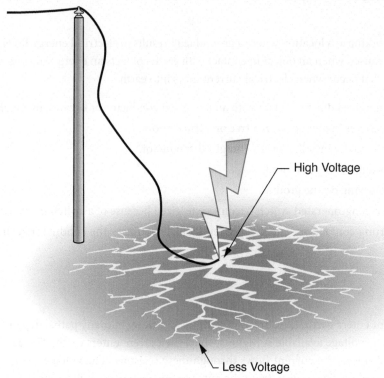

High Voltage

Less Voltage

Figure 7–11 The operator will not have a difference of potential from one foot to the other because he is standing on a ground mat. *(Courtesy of Kri-Tech Power Products, Ltd.)*

isolated, or barricaded so that the circuit trips quickly if the boom or its load makes accidental contact with the circuit. If the vehicle becomes energized, anyone near the truck is exposed to ground gradients. The boom operator stays on the operating platform or uses a ground-gradient mat. The circuit breaker or recloser can be put in a nonreclose position to ensure that the circuit remains isolated should it be tripped out due to an accidental contact.

Figure 7–12 If a ground fault raises the potential on this transformer cabinet, the worker will not be subjected to an electric shock because the ground mat is bonded to the transformer case. *(Courtesy of Kri-Tech Power Products Ltd.)*

A person controlling the butt of the pole wears rubber gloves to provide a barrier against touch potentials. **See Figure 7–13.** Pole tongs or cant hooks are also used to keep a person away from the base of the pole because the highest ground-gradient potentials are where the pole touches the ground. A person using rubber gloves to guide the pole into the hole could still have feet spanning two different voltage levels. At higher voltage levels, guide ropes tied to the butt of the pole can be used to get even farther away from the current entry point.

Figure 7–13 The pole tongs keep the worker farther away from the highest ground gradient voltage. *(Delmar, Cengage Learning)*

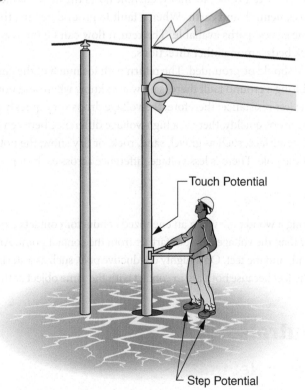

Touch Potential

Step Potential

Step and Touch Potentials

Electrical current flows into earth when an object such as a pole accidently makes contact with an energized conductor. Where the pole touches earth, a rise in voltage is relative to any earth farther away from the base of the pole. The current takes many paths as it flows away from the pole. The voltage at the base of the pole is higher than the voltage farther away (recall the ripple effect).

A step potential refers to having one foot on a high-voltage ring near the base of the pole and the other foot on a lower-voltage ring farther away from the pole. A step potential is defined as the voltage differential between two points on the ground separated by the distance of one pace or 1 yard (1 m). A touch potential refers to touching the pole where the hands are at one potential while the feet on the ground are at a different potential. A touch and step potential is defined as the voltage differential between both feet on the ground and an object being touched by the hand.

Magnitude of Step and Touch Potentials

The magnitude of the step and touch potentials depends on the voltage of the circuit, the conductivity of the pole, and the quality of the ground. The higher the voltage of the circuit, the easier it is for the current to overcome the resistance offered by the pole and the ground. A high voltage, therefore, is more likely to generate high touch and step potentials. The magnitude of the current flowing to ground depends on the type of earth at the base of the pole and can also be dependent on the contact with the grounded derrick handling the pole. The resistance of a wood pole varies with the moisture, weather, and pole treatment. On low-distribution voltages, a wood-pole contact does not normally trip out the circuit, and the wood resistance may reduce the step and touch potentials at the base of the pole to a minimal hazard. A steel pole, a concrete pole, or a wood pole with a pole/down-ground installed, on contact with an energized conductor, has a touch potential at the base of the pole at almost the full-line voltage. The current flow into earth depends on the resistance of the ground that the base of the pole is touching.

Size of the Step Potential Gradient

Electricity needs a circuit before current can flow. Normally, current flows through a conductor to the load and returns to the source through another phase, neutral, and earth. When a fault to ground occurs, the current flows through earth back toward the source through the easiest paths available. The return flow can be into earth, back up any ground wires to the neutral, along fences or creek beds, among other directions.

When setting a pole, the vehicle should be grounded. The return path for much of the current is back to the vehicle and up to the neutral. In other words, during a ground fault there is no way to know where and how far the ground gradients will travel. In good, moist earth, there is less resistance; therefore, the voltage drops very quickly before the current travels very far from the pole. When the voltage drops quickly, there is a high-voltage difference between the potential rings close to the base of the pole. On a high-resistance surface, such as gravel, sand, rock, or dry snow, the potential gradients drop off more slowly and farther from the base of the pole. There is less voltage difference across each step.

Voltage Gradients on a Pole

Voltage gradients can also occur along a wood pole when an energized conductor contacts a pole. For a person on a pole, the voltage at the contact point is higher than the voltage at a spot farther from the contact point. Anyone on the pole would have a potential difference between the hands and the feet. On a highly conductive pole, such as a steel structure, there is less potential difference between the hands and the feet because both are in contact with the same object at the same potential.

Working with Neutrals

Voltage and Current on a Neutral

A neutral or a metallic cable sheath tends to be treated as a non-energized conductor by the line trade. This is probably because the voltage on a neutral is normally below the threshold of sensation. The neutral voltage is usually below 10 V. There is, however, a current flow in the neutral, and when the current is interrupted, a recovery voltage appears across the open point.

Sources of Neutral Current

In an electrical circuit, all the current must return to the source. On a single-phase wye circuit, all the current returns by way of the neutral and the earth. The neutral and the earth are parallel paths for current to flow back to the source. Depending on the type of earth, about two thirds of the current flows through the neutral and the remaining one third flows through the earth. On a single-phase line, the amount of current on the neutral will depend on the load current being drawn.

On a three-phase wye circuit, all the current also returns to the source through the neutral and ground. On a balanced three-phase wye circuit, there would be no measurable current flow on the neutral because the return current from the three phases is 120° out of phase with one another and would cancel one another out. However, the loads on distribution circuits are not perfectly balanced between the three phases. The unbalanced portion will show up as a current flow through neutral and the earth back to the source. The more unequally the phases are loaded, the greater the current flow through the neutral. In an electrical system where feeders are tied between substations, the neutral can also carry current between substations.

The neutral on a three-wire 120/240 V secondary service also carries return current. When the load is balanced, there would be no measurable current flow on the neutral because the return current on one hot leg is equal to the current in the incoming hot leg and therefore cancel each other out. The load is typically not balanced equally between the two 120 V legs of the service, so there will be a current flow through the neutral and to a lesser extent the earth for the remaining current to flow back to the source.

Sources of Neutral Voltage

When there is a current flow, there also has to be some voltage pushing it. Therefore, if the current flow is reduced, the voltage is reduced. As the current gets higher, the voltage gets higher. Normally, on a neutral, the voltage is below 10 V and not felt by a worker.

- The voltage on a neutral increases if the neutral resistance is increased. Based on Ohm's law ($E = I \times R$), the voltage increases if the resistance or current is increased. A poor electrical connection in the neutral circuit raises the voltage.
- There can be a very high voltage on a neutral, in relation to earth, during a line-to-ground fault. A line-to-ground fault on one phase is like a large unbalanced load between the phases. A large portion of the return flow to the source is through the neutral and earth. A high-fault current will result in a high voltage on the neutral.
- There can be a very high voltage on a neutral when a phase is struck by lightning. The high voltage on the phase couples (is linked together electromagnetically) with other phases and the neutral. Electrical coupling with the neutral tends to bring the phase voltage down, but electrical coupling also raises the voltage on the neutral before the voltage is dissipated to ground on the multigrounded neutral.

Open Neutral

An open or broken neutral is a major hazard in the line trade. There is normally no apparent voltage on a neutral. A neutral is usually grounded on each side of a break at transformers or other equipment. There is, however, a high voltage across the break. On a multigrounded neutral, the voltage is kept low. If the current is interrupted by opening or cutting the neutral, all the voltage pushing the current appears across the open point. A break in the neutral introduces an infinite resistance to current flow. Applying Ohm's law shows that with an infinite resistance the total circuit voltage (recovery voltage) is available across the break. It is, therefore, a basic line procedure to always jumper any neutral before it is cut.

Precautions for Working with a Neutral

A power line technician must take special precautions that apply only to a neutral. The first precaution is to properly identify which conductor is the neutral. Placing a ground on the neutral is a positive identification method. If a neutral is to be cut or sliced, a bypass jumper must be installed, to avoid exposure to the voltage that is available across the open point.

On older constructions, the neutral position is not always standard. A power line tech who has had no exposure to delta circuits and working storm trouble in another utility could misidentify a neutral. For example, with some two-phase delta construction, one phase is in a position that would normally be in the neutral position in a wye system. Some three-phase delta circuits (a corner grounded delta circuit) have one phase conductor grounded that may or may not be in the neutral position. Stay away from all conductors, including the neutral, during a lightning storm.

Vehicle Grounding and Bonding

The Vehicle as an Electrical Hazard

Vehicles with booms are exposed to accidental or inadvertent contact when used in the vicinity of energized circuits. A vehicle can become an electrical hazard to people on the ground at a job site in the following three ways.

If	Then
A person on the ground is in contact with a utility vehicle while the uninsulated portion of a boom accidently contacts an energized circuit.	The person is a parallel path to ground and will take some share of the current flowing to ground.
A person on the ground is near a utility vehicle while the uninsulated portion of a boom accidently contacts an energized circuit.	The person is exposed to ground gradients at each location that current is entering the earth.
An uninsulated boom is used to lift or handle an isolated and grounded conductor in a high-induction area.	Unless the boom is bonded to the grounded conductors, the boom can be at a different potential to the conductors being handled.

How Vehicles Can Become Energized

Most utilities have procedures to protect people on the ground in case a vehicle should become energized due to inadvertent contact with an energized circuit. Many types of mishaps can cause a digger derrick or the uninsulated portion of an aerial device to become energized, such as the following:

- An energized conductor can fall onto a boom.
- A digger derrick can make contact while installing a pole or other equipment.
- The lower boom of an aerial device can inadvertently make contact with a tap running laterally from the circuit being worked. **See Figure 7–14.**

Figure 7–14 When the uninsulated portion of a boom makes contact with a lateral tap it can cause a truck to be energized. *(Delmar, Cengage Learning)*

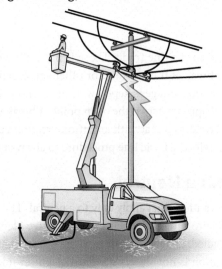

Ground-Gradient Hazards Near the Vehicle

When the vehicle becomes energized, there are ground gradients in the vicinity of the vehicle's outriggers, tires, attached trailers, and anything else in contact with the vehicle. **See Figure 7–15**. If the vehicle is grounded to the system neutral, there is a voltage rise on the neutral and a ground-gradient hazard at all the pole/down-ground locations along the line. If a ground rod or pole (down-ground) is used to ground the vehicle, there is a voltage gradient around them as well.

Figure 7–15 It is important to keep people away from a work site when a boom is in the vicinity of an energized circuit. When a pole makes contact with an energized line there are hazardous ground gradients in many places. *(Delmar, Cengage Learning)*

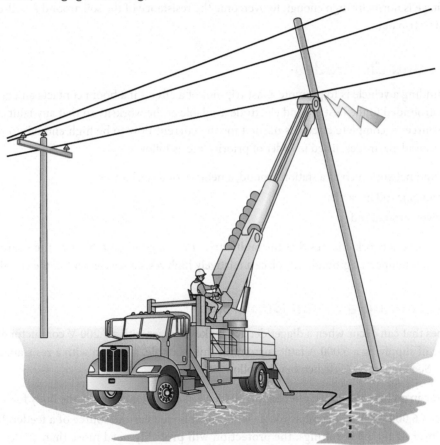

Protection from Shock

The only real protection available to people working around a vehicle when a boom makes contact with an energized conductor is to stay on the vehicle or stay away from the vehicle. Staying on the vehicle or on a ground-gradient mat bonded to the vehicle keeps a person at the same potential as the vehicle. As long as the person avoids contact with anything not connected to the vehicle, there will be no current flow through the person to another object. Keeping a safe distance from a vehicle that can become energized prevents contact with the vehicle and keeps one a safe distance away from high-ground gradients. Truck barricading promotes the need to stay away from the vehicle. High-resistance footwear provides additional protection by increasing the resistance of a person as a parallel path to ground. High-resistance footwear also provides protection from ground gradients.

Grounding the Vehicle

When there is contact between a utility vehicle's boom and an energized conductor, anyone touching the vehicle is a parallel path to ground and has some current flowing through the body. Even when the vehicle is grounded to an excellent ground, like a neutral, the amount of current going through the body could still be fatal. Grounding the vehicle to the neutral promotes a fast trip-out of the circuit should the vehicle become energized, and it also promotes the collapse of the voltage in the system. A fast trip-out reduces the exposure time to the hazard, and a collapse of the voltage reduces the voltage to which people are exposed.

On lower-voltage wye distribution systems, the vehicle must be grounded to a neutral to trip the circuit. The resistance of a ground rod or anchor rod is normally not low enough to generate a fault current high enough to trip out the circuit. On a delta circuit, the vehicle should be grounded to a ground rod. Most delta circuits are protected by ground fault relays that trip out a circuit when there is a line-to-ground fault. On higher-voltage distribution and subtransmission circuits, the voltage is normally high enough to overcome the resistance of the soil around a well-driven ground rod and cause the circuit to trip out.

Choice of Ground Electrodes

The purpose of grounding a vehicle is to promote a fast trip-out of a circuit if a boom contacts an energized conductor. A vehicle should be grounded to the best ground electrode available at the worksite so that any fault current has a good return path to the source. A complete circuit is needed for the current flow to be high enough to trip the protective switchgear. Typical ground electrodes, listed in order of priority, are as follows:

- Permanent ground network such as a station ground, a neutral, or a steel tower
- Ground rod or anchor rod in earth
- Temporarily driven ground rod

A permanent ground network ensures that the fault current has a good path back to the source to complete the circuit. Ground rods and temporary ground rods place a relatively high resistance element in the circuit.

Typical Values during a Boom Contact

Typical ampere values that can occur when a digger derrick makes contact with a 7,200 V conductor at a location where the circuit is capable of supplying a 6,000 A fault current. In this example, a person with a resistance of 1,000 ohms is touching the vehicle. **See Figure 7–16.** Remember the following:

- A person's heart can go into fibrillation after 50 mA (0.05 A) of current go through the body for a very short time.
- A location with a 6,000 A fault current available will probably be near the source of a feeder. Depending on the size of the recloser or the relay settings, the protection will probably need more than a 700 A fault current to operate.

Figure 7–16 Boom Contact with Energized 7.2 kV Conductor

Vehicle Ground	Fault Current Generated	Truck-to-Ground Voltage	Current through Person
Vehicle Not Grounded	200 A	5,500 V	6 A
Vehicle Connected to a Ground Rod	700 A	5,000 V	6 A
Vehicle Connected to a Neutral	5,000 A	200 V	0.2 A

Examples of Using a Ground Rod to Ground a Vehicle

A temporarily driven ground rod may not always trip out a circuit.

Transmission Lines Example

An accidental boom contact is made with a transmission line or subtransmission line. The vehicle is grounded to a temporary ground rod. The rod will probably provide a resistance low enough to trip out the circuit. The resistance of a temporary ground rod varies with the soil conditions but would seldom be less than 25 ohms and frequently is 100 ohms. Using Ohm's law and an example calculation of a 50 ohm ground rod for working on a 230 kV circuit, the fault current generated would be:

$$I = \frac{E}{R} = \frac{133,000}{50} = 2,660 \text{ amperes}$$

Where:

$R = 50$ ohms

$E = 230 \div 1.73 = 133$ kV phase to ground

The fault current generated by the ground rod in this example is more than enough to trip out the circuit, especially a circuit protected by a breaker with relays that sense phase differential and ground currents.

Distribution Lines Example

An accidental boom contact is made with a distribution circuit. The vehicle is grounded to a temporary ground rod. There is the likelihood that the resistance of the rod is too high to trip the circuit quickly or at all. Using Ohm's law and 50 ohms of resistance for the temporary ground rod on a 4,800 V circuit, the fault current generated would be as follows:

$$I = \frac{E}{R} = \frac{4,800}{50} = 96 \text{ amperes}$$

In this example, a fault current of 96 A will not trip out a circuit protected by a fuse larger than 50 A. The reason is that it takes 100 A to blow a 50 A fuse.

Boom in Contact with an Isolated and Grounded Conductor

A circuit is *not dead* when it has been isolated and grounded. There is often current flowing in the grounds, and there is often a voltage difference between the circuit and a remote ground. When a truck boom or crane is used on a job where a circuit is isolated and grounded, the boom should be bonded to the portable line grounds. On a right-of-way that has high induction from energized circuits in parallel or in the rare case of accidental reenergization, bonding ensures that there will be no potential difference between the vehicle, boom, winch, and conductors.

Examples of Hazards Using a Boom around Isolated Conductors

The following examples of hazardous incidents are rare occurrences, but they are all real.

- A person working on a conductor handles the winch of a crane and bridges the grounded conductors and an unbonded crane. **See Figure 7–17.** An unbonded vehicle is a remote ground in relation to the grounded conductors. In an area with high induction, there is a high voltage between the grounded conductors and the winch.

- When a vehicle is bonded to grounded conductors, the vehicle and the grounded conductors are at the same voltage. In a high-induction area, the bond sets up a voltage between the vehicle and any remote ground. A person standing on the ground and touching the vehicle could receive a shock. Both grounding and bonding the vehicle reduce the voltage difference between the ground and the vehicle.

- When using an aerial device to install grounds, especially on a multicircuit line, the vehicle is often grounded/bonded to the same ground electrode as the portable line grounds. If the grounds are mistakenly installed on an energized circuit, the truck also becomes energized, endangering those on the ground near the vehicle. People on the ground should stand clear of the vehicle until the grounds are installed.

Figure 7–17 The vehicle boom in this accident scenario is the source of a remote ground. *(Delmar, Cengage Learning)*

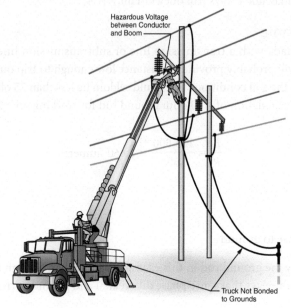

Electromagnetic Fields

Why the Interest in Electromagnetic Fields?

Before the 1970s, the only risks thought to be associated with electricity were electric shock and flash burns. There is now a question about the possible relationship between electromagnetic fields (EMFs) and some types of cancer. Most of the evidence comes from epidemiological studies. Epidemiology is a science that looks for statistical evidence of health patterns in people and the factors that may be responsible.

What Are EMFs?

Wherever there is electricity, magnetism appears, and vice versa. Electricity and magnetism both travel in the form of waves and in fields (in the same way that gravitation has a field), and they spread everywhere. EMFs are energy waves with both an electric component and a magnetic component. (EMFs should not be confused with *emf*, which refers to electromotive force.) Energy radiates from a source and can travel without the need of any material to conduct it. Electromagnetic waves can travel through a vacuum at the speed of light, and have a wavelength, frequency, and amplitude (or field strength). A wavelength is the distance between one peak on the wave and the next peak. The frequency, measured in hertz, is the number of wave peaks that pass by in 1 second.

Electromagnetic waves from power frequencies of 50 or 60 Hz are just one portion of a large band or spectrum of electromagnetic waves. The 50 or 60 Hz waves from power lines are very long waves (3,107 miles [5,000 km]) and are referred to as extremely low frequency (ELF) waves. When EMFs are discussed in the media, the term usually refers to electric and magnetic fields in the power frequency range that has a power line as its source. The term EMF, however, is also used for electromagnetic fields with higher frequencies, which radiate much farther than power frequency waves.

Electromagnetic Spectrum

All rays—including X-rays, ultraviolet light, visible light, infrared light, microwaves, radio waves, heat, and electrical power line—are electromagnetic waves. The properties that differentiate these various sources are the frequency and the wavelength. Some of the types of radiation that make up an electromagnetic spectrum. **See Figure 7–18.** The lines dividing the different types of radiation overlap because the division between them is not accurately defined.

Figure 7–18 The Electromagnetic Spectrum

Source	Effects	Frequency (Hz)	Wavelength
ELF (extremely low frequency)	Low energy	3 to 300	Power lines (60 Hz) are 5,000 km.
DC is 0 Hz Power lines are 50 to 60 Hz	No thermal effects		
VLF (very low frequency) AM radio	No proven health effects	3,000 to 30,000	AM radio waves are about 328 yards (300 m).
VHF (very high frequency) FM and television	Can cause heating	10^7 to 10^9	Television waves are 1 to 18 ft (0.3 to 5.5 m).
Radar microwave	High-induced currents	10^9 to 10^{12}	Microwaves are about 4.7 in. (12 cm).
Infrared visible light Ultraviolet	Energy waves can be seen Photochemical effects	10^{12} to 10^{15}	Visible light waves are from 0.75 to 0.04 micrometers/micron.
Ionizing radiation Ultraviolet radiation	High energy	10^{15} to 10^{22}	X-rays are about 0.03 to 0.00001 micrometers/micron.
X-rays	Radiation		
Gamma rays	Burns DNA damage		

Sources of Power Frequency EMF

All energized electrical wires, equipment, and appliances have an electric field and a magnetic field around them. Combined, these two fields comprise the EMF. The electric and magnetic fields behave differently and are measured separately, as seen in the following table.

Comparison of Electric and Magnetic Fields

Electric Fields	Magnetic Fields
The electric field is a voltage field. The higher the voltage source, the higher the electric field. An electric field is present around any energized conductor and is independent of the current flow in the circuit.	The magnetic field is a field of magnetic lines of energy. The higher the current in a source conductor, the higher the magnetic field. If there is no load on the circuit, then there will be no magnetic field.
An electric field is measured in volts per meter (V/m).	A magnetic field is measured in microteslas (µT) or milligauss (mG).
The strength of the electric field drops off with the inverse of the distance from the source squared $\left(E = \dfrac{I}{r^2} \right)$, where r is the radius from the source.	The greater the distance from an energized source, the less the strength of an EMF. The strength of the magnetic field drops off with the inverse of the distance from the source $\left(H = \dfrac{I}{r} \right)$, where r is the radius from the source.
Shielding from electric fields occurs if there are trees or walls, for example, blocking the way and draining the electric current to ground.	Magnetic fields are only weakened, but not stopped, by barriers such as trees. They are similar to the effect of a magnet where the lines of force can go through material.
The electric field from a power line would not normally penetrate into a building.	Magnetic fields from a power line can penetrate a building.

(Continued)

When a person is within an electric field, there is a voltage induced on the body. Workers feel the existence of electric fields when they are working in a substation or on a high-voltage transmission line. Hair tends to stand up as a person gets close to a high-voltage source. A small arc can occur when the voltage on a person's body is discharged to a grounded object.

An electric field can induce a weak electric current flow in the body by moving charges in the body. The redistribution of charges causes small currents, but these currents are typically much smaller than those produced naturally by the brain, nerves, and heart.

A magnetic field can induce a current flow in lengths of conducting material such as in a wire fence. A current, however, can flow only if there is a circuit. A fence kept insulated from the ground would have no current flow on it unless it was grounded in two places to form a circuit between the two grounded locations. An isolated fence would have a voltage on it.

A magnetic field will pass through a body but, because a person's body is not normally in series with a circuit, there should be no current induced in it from magnetic waves.

Methods of Reducing the Strength of EMF

There are ways to reduce the strength of EMF coming from energized power conductors.

1. Increasing the heights of poles and towers increases the distance from the ground to the power line.
2. A circuit converted to a higher voltage, given the same amount of load, would carry less current and, therefore, have a lower-strength EMF.
3. There is a canceling effect from three-phase lines and multicircuit configurations. With neighboring phases being out of phase with each other, there is a canceling effect that reduces the EMF. A closer spacing between conductors reduces the EMF. For example, a triplex service or house wiring has a reduced EMF due to the closeness and canceling effect of the current in the two legs flowing in opposite directions.
4. Underground power lines have a reduced electric field because of the shielding from the cable sheath and the surrounding earth. There is also a reduction in magnetic fields for underground lines, due not to shielding but to the EMF-canceling effect of the neutral and other phases in the closer conductor spacings allowed underground.
5. Transpositions in a power line reduce the EMF. Transpositions were originally put into circuits to reduce the effect of EMF on open-wire communications circuits. As a prudent avoidance measure, the EMF from electric blankets and waterbed heaters is reduced with transpositions when they are manufactured because the wires are continually crossing over each other.
6. Methods using metal shielding to reduce magnetic fields from power lines are available, but they are considered an expensive method for shielding large areas.

Measurement of EMF

To measure the strength of EMF, two instruments are needed—one for the electric field and another for the magnetic field. As noted in the previous table, the strength of an electric field is measured as volts per meter (V/m) or in larger units of kilovolts per meter (kV/m). The strength of the electric field depends on the voltage of the source conductor and the distance at which the measurement is taken. The magnitude of the electric field strength decreases as the distance from a location to the source increases inversely by the square of the distance. For example, if the distance away from the source is doubled, the electric field strength is reduced by a factor of 4 (2^2). If the distance away from the source is quadrupled, the electric field strength is reduced by a factor of 16 (4^2). Similarly, if the distance to the source is reduced by a factor of 2, then the electric field strength increases by a factor of 4 (2^2). The test instrument must be held away from the body because the body acts as a shield and distorts the readings.

The strength of a magnetic field is measured in teslas (T) or gauss (Gs). The gauss is common in the United States, and the tesla is used internationally. A tesla is a measurement of the magnetic field intensity or the density of magnetic lines. The strength of the magnetic field depends on the amount of current flowing in the source conductor and the distance at which the measurement is being taken. Measurements must be done in different locations over a period of time, because the load current of the circuit changes during the day. Tesla units are too large to measure common exposure; therefore, the microtesla is used.

$$1 \text{ tesla} = 1{,}000{,}000 \text{ microteslas}$$
$$1 \text{ tesla} = 10{,}000 \text{ gauss}$$
$$1 \text{ gauss} = 1{,}000 \text{ milligauss}$$
$$1 \text{ microtesla} = 10 \text{ milligauss}$$

When taking measurements, it is important to understand the difference between an emission and exposure. A measurement taken at a conductor will give the emission, but exposure to the emission is normally much farther away. Exposure must be measured at a location where people would be normally. The measurements will not mean much to the average person unless the numbers are used to compare the EMF strength from various sources. **See Figure 7–19**.

Figure 7–19 Sample Measurements of Electric Field Strength

Location of Measurement	Typical Electric Field Values	Typical Magnetic Field Values
Under a transmission line, the field strength depends on the voltage of the line and the height of the conductors	1 to 10 kV/m (1,000 to 10,000 V/m)	10 µT
On the edge of a transmission-line corridor	100 to 1,000 V/m	0.1 to 1 µT
Near an overhead distribution line	2 to 20 V/m	0.2 to 1 µT
In homes	200 V/m close to an appliance, to less than 2 V/m in other locations in the home	150 µT near appliances, to less than 0.02 µT in other areas in the home
Exposure to linemen, cable splicers, and substation workers	Typical 100 to 2,000 V/m with peaks as high as 5,000 V/m	Average 0.5 to 4 µT with as high as 100 µT

Dose

One problem in researching the health effects of EMF is to determine which dose to measure. The word *dose* means exposure that produces an effect. For example, if a risk is due to EMF exposure, what type of dose would produce harmful effects?

- Is a weak field safer than a strong field?
- Is it exposure to the peak electric field or exposure to a constant electric field that is harmful?
- Is it the going into and coming out of an electric field?
- Is it exposure to the peak magnetic field or exposure to a constant magnetic field?
- Is it the going into and coming out of a magnetic field?

Natural Levels of Electric and Magnetic Fields

Electric and magnetic fields exist naturally on earth. This EMF source is obviously not at 50 or 60 Hz, but it is a DC source with a fluctuating voltage and current level. It is the fluctuations that cause electric and magnetic fields to occur. Earth's atmosphere has a naturally occurring electric field that fluctuates on the surface around 130 V/m. Large electric currents through the earth's core creates a magnetic field on the surface that can range from 30 to 60 µT. The magnetic field is stronger at the North Pole and South Pole than at the equator.

A thunderstorm is the ultimate demonstration of naturally occurring electric and magnetic fields in action. The electric field strength is high enough to break down the insulation value of air and discharge as lightning. The high current within the electric arc generates a large EMF that can be noticed as interference on radio and television reception.

Limits of Exposure

The World Health Organization, the International Radiation Protection Organization, and other organizations have published threshold limits of exposure to EMF from power frequency sources (50 and 60 Hz). There are limits for the general public and for workers. For example, one organization has a threshold limit for workday exposure to EMF as 10 kV/m for electric fields and 5 Gs for magnetic fields. Utility workers should check for the limits their own utilities have adopted for exposure to EMF.

Working Near a Cellular Antenna or Microwave Dish

Power line technicians can be exposed to electromagnetic microwaves (MW) or radio frequency (RF) waves radiating from personal communications services (PCS), such as digital cellular phone service antennas and microwave dishes that are installed on utility-owned structures. It is most hazardous in the area immediately in front of a transmitting antenna at distances of less than 10 ft (3 m). RF/MW frequencies are at different points on the electromagnetic field spectrum. **See Figure 7–20.**

Figure 7–20 The radio and microwave frequencies on the electromagnetic spectrum are hazardous when a person is too close to the source. *(Delmar, Cengage Learning)*

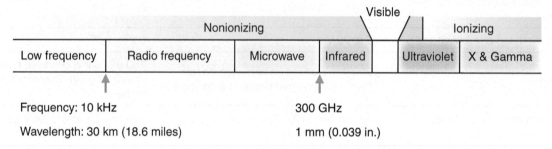

An immediate health effect to a high exposure to RF energy is heating. Like a microwave oven, RF energy has the ability to excite molecules and create heat in any exposed tissue. An early symptom of overexposure is similar to a person showing signs of heat exhaustion. The heating effect may not be noticed by a person until damage is done. Damage can range from raising body temperature to burned tissue and organs. Injuries from RF exposures are not cumulative as they would be from an ionizing radiation source.

When there is work to be done on a structure with an RF antenna or microwave dish, the employer must calculate the minimum safe distance for that antenna. Calculations are required to determine the distance required to receive less than the maximum permissible exposure (MPE), which is measured in milliwatts per squared centimeter (mW/cm^2) of tissue. Health effects of overexposure to RF/MW fields depend on the frequency and intensity of the fields, the duration of exposure, the distance from the source, and any shielding that is used. If a minimum safe distance cannot be maintained to do the required work, then other safeguards, such as a shutdown, may be required.

First aid is similar to that for heat stroke: Remove the casualty from the exposure area, provide cool drinking water, and apply cold water to burned areas. There could be internal tissue damage without visible skin injury, so it is advisable to seek medical attention.

Induced Current from an RF/MW Transmitter

An RF transmitter can also induce electromagnetic induction on nearby objects. The effect is the same as that from induction from a power line. A voltage can be induced on a person, and a shock can result when touching a grounded object. Sparking may occur across a gap between two metal objects that are close together as the induced current is trying to flow to ground.

Ultraviolet Radiation

Ultraviolet (UV) radiation is a form of electromagnetic radiation with wavelengths measured in nanometers (1 nanometer = 10^{-9} **meters**. On the electromagnetic spectrum, UV radiation comes between visible light and X-rays. It is divided into three wavelength bands according to its effects on living tissue: UV-A, UV-B, and UV-C. **See Figure 7–21.**

Figure 7–21 Ultraviolet light from switching arcs and sunlight are high frequency electromagnetic waves that can be hazardous depending on exposure. *(Delmar, Cengage Learning)*

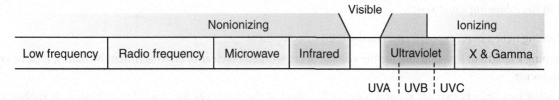

The sun is the main source of UV radiation, and it is known that in the short term it can cause sunburn and in the long term, skin cancer. An electric arc is a UV hazard where the eyes are most vulnerable because of the high intensity and short time exposure.

Reduce exposure through clothing and apply sunscreen with a sun protection factor (SPF) of 15 or higher of the type that is effective for both UV-A and UV-B rays. Plastic safety glasses are also good UV filters for protecting the eyes.

Minimum Approach Distance

Minimum Approach Distance as a Barrier

Maintaining a minimum approach distance to exposed, energized conductors is the most common barrier a power line technician uses to avoid electrical contact. Government regulators and utilities have tables listing the minimum approach distance from various voltages for different levels of qualified people and equipment. Minimum approach distances apply to work in the vicinity of an energized circuit, and they also apply to live-line work. **See Figure 7–22.**

Figure 7–22 Typical Minimum Approach Distances

Maximum Phase-to-Phase Voltage (Max. Phase-to-Ground Voltage)	Minimum Approach Distance Phase-to-Ground Exposure	Minimum Approach Distance Phase-to-Phase Exposure
0.05 to 1 kV	Avoid contact	Avoid contact
Up to 15 kV (8.7 kV)	2′-1″ (64 cm)	2′-2″ (66 cm)
Up to 36 kV (20.8 kV)	2′-4″ (72 cm)	2′-7″ (77 cm)
Up to 46 kV (26.6 kV)	2′-7″ (77 cm)	2′-10″ (85 cm)
Up to 72.5 kV	3′ (80 cm)	3′-6″ (1.05 m)
Up to 121 kV	3′-2″ (95 cm)	4′-3″ (1.29 m)
Up to 145 kV	3′-7″ (1.09 m)	4′-11″ (1.5 m)
Up to 169 kV	4′ (1.22 m)	5′-8″ (1.71 m)
Up to 242 kV	5′-3″ (1.59 m)	7′-6″ (2.27 m)
Up to 362 kV	8′-6″ (2.59 m)	12′-6″ (3.8 m)
Up to 550 kV	11′-3″ (3.42 m)	18′-1″ (5.5 m)
Up to 800 kV	14′-11″ (4.53 m)	26′ (7.91 m)

The flashover voltage for a live-line tool is the same as it is for air. A fiberglass live-line tool may be better insulation than air, but the distance needed on a tool is an air gap between the hands on a live-line tool and the energized conductor. Rubber glove work and barehand work may appear to be exceptions to the minimum approach table, but the table still applies, in reverse. Rubber glove and barehand procedures depend on the power line tech being insulated from ground or other phases.

The table should apply to the distance from the power line tech to any exposed objects that are second points of contact. Unless they are covered with applicable rubber or fiber barriers, a minimum approach distance is kept from other phases, the neutral, a grounded structure, or any other grounded objects. The distances apply to a qualified, competent worker, in the following situations.

- Working near a bare, exposed energized circuit or equipment
- Keeping a length of live-line tool for the voltage being worked on clear of encroachment by hands, tool, or energized conductor
- Keeping any objects that are second points of contact at the distances listed while working with rubber gloves or barehand

Factors Considered for a Minimum Approach Distance

The minimum approach distance tables consider the hazard of an electric discharge from an energized conductor to a person, and they consider an inadvertent movement by a worker in the vicinity of an energized conductor. A minimum approach distance, therefore, consists of an electrical factor distance and an ergonomic or human factor distance. (*Note:* The formula shown is an overview of the factors calculated, as the actual formula used is more complex.)

$$\text{Minimum approach distance} = A + (F \times B)$$

Where:

A = a human factor distance that takes into account a momentary inadvertent reaching into the prohibited zone

F = the minimum electrical clearance that would prevent a flashover from the highest transient voltage from an internal source

B = a safety factor to ensure there is no chance for a flashover across the minimum electrical clearance

Electrical Factor ($F \times B$) in a Minimum Approach Distance

Design engineers use standard electrical clearances when determining the distance allowed between a phase and a grounded object for a pole framing or transmission tower design. An electrical clearance distance is considerably less than the minimum approach distance a person must maintain. A minimum electrical clearance, which is represented by F in the preceding formula, is just greater than the flashover distance from the highest possible internal transient voltage. An internal source for a transient overvoltage would be a switching surge. An external overvoltage surge, such as lightning, could flash over the minimum electrical clearance distance; therefore, live-line work is not carried out with a lightning storm in the vicinity. In the formula, the electrical clearance distance F is multiplied by a safety factor B to provide extra confidence that no flashover will occur across an electrical clearance distance between the circuit and ground. The safety factor used tends to be approximately 1.25 times the electrical clearance distance. A nearby surge arrestor should shunt a transient overvoltage to ground to provide an additional assurance of no flashover across a normal electrical clearance distance.

Human Factor (*A*) in a Minimum Approach Distance

The human factor minimum approach distance is a space needed to prevent a person from encroaching into the electrical factor ($F \times B$) distance. A human factor distance is subjective and has not necessarily been studied scientifically. It could be said that it is unlikely that a power line tech would inadvertently encroach more than 3 ft (1 m) toward an energized conductor. If 3 ft (1 m), representing *A,* is added to the electrical factor ($F \times B$), in the formula, a minimum approach distance can be established for each voltage.

At transmission-line voltages, 3 ft (1 m) added to the electrical factor distance would establish a realistic minimum approach distance table and allow transmission-line work procedures to be carried out. At transmission-line voltages, no cover-up is available, and maintaining a safe approach distance is a necessary barrier for work on or near energized transmission lines.

At distribution voltages, the human factor distance is larger than the electrical factor distance. It is often necessary to go closer than 3 ft (1 m) to an exposed conductor. A human factor distance of less than 3 ft (1 m) requires extra barriers, such as the use of cover-up, an insulated platform, or a dedicated observer to reduce the risk of inadvertent contact.

Review Questions

1. What are the two types of electrical contact that a person can make?

2. Why does a full-line voltage (recovery voltage) appear across the two ends of a cut conductor?

3. When making a cut to an isolated and grounded conductor, is it necessary to install a jumper across the cut?

4. *True or False:* Anytime a person is in parallel with any load in a circuit, that person is exposed to some voltage and some current regardless of the amount of resistance being offered to the current.

5. Name the two forms of induction that make up electromagnetic induction.

6. Give an example of static electricity.

7. When working in a high-voltage environment, what form of induction causes a spark from one's body every time a grounded object is touched? What type of personal protective equipment will reduce this hazard?

8. Why is induction from a magnetic field a hazard on distribution lines as well as a hazard on high-voltage transmission lines?

9. What is the hazard when working at a location where a ground fault results in electrical energy flowing into earth?

10. Which ground gradient would spread out farthest: one in moist earth or one in rocky ground?

11. How do power line technicians protect themselves from a ground gradient when working at a pad mount transformer?

12. When an energized power line drops on a car, why does a person in the car not get an electric shock?

13. Name three ways that a power line tech can positively identify which conductor on a pole is the neutral.

14. Name three ways that the voltage on a neutral is increased.

15. Why is there a voltage across a break in a neutral?

16. When a truck boom is used in the vicinity of an energized circuit, what two ways can a power line tech prevent an electric shock should the truck become energized?

17. Why is a person not safe when touching a vehicle that is making contact with an energized circuit, even though the vehicle is grounded?

18. What is the purpose for grounding a line truck?

19. When a truck boom or crane is used for transmission-line work where a circuit is isolated and grounded, should the truck be bonded to the portable line grounds?

20. Which electric magnetic fields (EMF) can cause an immediate health effect?
    ( ) working close to a 50 or 60 cycle power line (very low frequency)
    ( ) working close to a cellular antenna (RF frequency)
    ( ) working close to a microwave tower (MW frequency)
    ( ) working in sunlight (ultraviolet frequency)

21. What two factors are used to make up minimum approach distance tables?

22. How can the minimum approach distance be maintained when working with rubber gloves?

Chapter 8

Constructing Overhead Power Lines

Objectives

After completing this chapter, you should be able to:

1. Perform an inspection of an overhead power line.
2. Evaluate required electrical clearances from the ground, from other utilities and for safe work.
3. Identify that the correct pole was specified for a job.
4. Take steps to ensure a pole or structure is set to the specified foundation.
5. Identify the hazards and the protective barriers needed to install a pole in an energized circuit.
6. Identify the type of guying needed to support a pole.
7. Take steps to take protection from any hazards when installing a powered screw anchor.
8. Identify the strength and voltage of suspension insulators.
9. Identify the numerous kinds of material that can make up the framing of a pole.
10. Explain the steps to safely perform tension stringing in the vicinity of energized circuits.
11. Identify the procedural barriers to protect traffic when stringing over a roadway.
12. Identify the construction and erection of various transmission line structures.
13. Explain the reasons to sag conductor to the specifications provided for the project.

Before a Work Order Is Issued

Planning for Maintenance and Capital Projects

Work in a utility is generally divided into two main categories: capital projects and maintenance. Maintenance is usually part of operations and maintenance (O&M) and is funded from revenue. O&M includes, for example, billing, trouble calls, low-voltage complaints, inspections, and pole-top maintenance. Capital work tends to be funded through borrowing. The philosophy behind borrowing the funds for a capital project is that the company or owners benefit from the project for a given number of future years and the future rates should pay for it.

A large part of both O&M and capital cost is the overhead that is applied to each job. The overhead pays for things such as engineering, accounting, safety professionals, and supervision.

Utilities set priorities for maintenance and capital programs needed to keep the electrical system operating and to ensure that the quality of power delivered to the customer service entrance is at an acceptable level and meets standards.

Maintenance and capital projects are prompted by circumstances or data such as the following:

1. Peak load voltage and current surveys at substations and on individual feeders
2. Problems with power quality, such as voltage flicker
3. Customer demand, such as subdivision layouts and road-widening projects
4. Nonstandard clearance between the conductor and the ground
5. Underground cable replacement
6. Rehabilitation projects, such as deteriorated pole replacement, conductor prone to breakage replacement, defective component replacement (e.g., aluminum capped dead-end insulators, certain brand porcelain cutouts prone to failure)

Types of Line Inspections

Maintenance work must be prioritized so that the allocated funding can be put where it is most needed. Inspections, along with trouble calls, provide information used by engineering to prepare the maintenance program. Inspection of the distribution and transmission are carried out because it is a good utility practice, but also to comply with applicable government regulations.

An inspection can vary from a simple foot patrol to the use of drones and helicopters. Specific effective inspections include an infrared survey, vegetation survey, insulator testing, corrosion survey, pole testing, and ground resistance testing.

Drones are used to have a close look at specific issues on a line as shown in **Figure 8–1**, and also to conduct a general line condition and vegetation survey. There is software available to let a drone fly independently, compile data linked to global information system (GIS) technology.

Figure 8–1 Drones can focus in on specific maintenance issues. *(istock.com/PJ_joe)*

Thorough helicopter inspections can use binoculars and cameras with a vibration reduction feature that can zoom in very close to individual hardware components. A foot patrol can pick up items such as erosion, loose or missing nuts on anchor bolts, corrosion 1 ft below ground line at suspect locations, broken counterpoise, and ground wire. Measurements can be taken if a ground clearance appears low. Dangerous trees can be identified.

Many utilities also own communications systems used for telemetering and operating their systems that are inspected and maintained.

Sometimes problems are found after they occur and a program is launched to correct it. For example, the jumper clamps on a transmission line were wearing through and one fell open, causing a line outage. The replacement was in rough terrain and linemen, belted into a hook ladder, were placed on each side of the jumper by helicopter. **See Figure 8–2.**

Figure 8–2 (a) Linemen reinforcing jumper clamps on 230 kV dead-end structures. (b) The worker is being lowered into position by helicopter while belted into the hook ladder. *(Courtesy of Wilson Construction Company)*

(a)

(b)

Types of Maintenance

Utilities are constantly looking for the most effective way to carry out their planned maintenance programs. The type of maintenance a utility wants to avoid is corrective maintenance (CM), which is unscheduled and requires immediate attention. Lack of scheduled maintenance can mean a disconnect switch will not open, a voltage regulator is stuck in one position, or electrical connectors have failed. A utility also wants to avoid unnecessary maintenance, such as over-hauling a recloser that when taken apart looks as good as when it was installed.

Preventative maintenance (PM) on substation equipment and line equipment such as voltage regulators and reclosers is often based on maintenance schedules either recommended by the manufacturer or "time based." Many utilities use loading history and/or number of operations as an indicator of the condition of their equipment.

Fixing hot connections and splices found by an infrared survey could be called predictive maintenance, because it allows a utility to do repairs before a complete failure occurs.

Reliability-centered maintenance (RCM) focuses on preserving the purpose of the line or equipment. Maintenance is prioritized so that the devices and lines with safety or potential outage problems that affect the most people get priority over devices and lines with less impact. A line or device that services customers that can be backfed from another line or device gets less priority than a radial feed to a customer. RCM also concentrates on doing maintenance on specific failure modes of lines and equipment; for example, on distribution overhead, prioritizing on failure modes means changing connectors that are prone to failure before replacing deteriorated guy guards.

A utility must set priorities when repairing deficiencies found during patrols. For example, a broken insulator or eight flashed insulators in a string of fourteen insulators on a 230 kV line would normally not require an immediate repair, but eight broken insulators would. Similarly, during an infrared survey, standards must be followed regarding when a hot connector or splice needs immediate replacement or when it can be scheduled at a convenient time. Chronic problems, such as corrosion at the ground line for many structures, would require a major maintenance project that is planned and budgeted in advance.

One example of a maintenance program is the inspection and maintenance of distribution lines in a given geographic area on a 5- to 10-year cycle. A utility may have a list of typical deficiencies that is prioritized based on the need to repair. Decisions regarding which prioritized deficiencies are repaired are based on the funding available in the program.

A prioritized list of items to report may look similar to the following:

1. Public hazards, such as inadequate clearance to flag poles, antennae, and buildings
2. Defective components prone to failure, such as brands of dead-end insulators known to be defective, porcelain in-span switches and/or cutouts, porcelain post insulators, and wood pins
3. Bad crossarms
4. Connectors prone to failure
5. Broken strands on conductors
6. Defective surge arrestors
7. Open wire services
8. Broken ground wires and exposed ground rods
9. Slack down guys and missing guy guards
10. Leaning poles
11. Nonstandard transformer installations

A pole may be identified for maintenance. **See Figure 8–3.** The single-phase lateral to the left is a nonstandard dead-end to the end of a crossarm, probably to save changing out the pole to gain height. The conductor tied in on the double crossarm has an old aluminum twist sleeve, which is vulnerable to failure. The neutral connection is made with a split bolt connector, also prone to failure.

Figure 8–3 A pole which has been pole identified for maintenance work. *(Delmar, Cengage Learning)*

Capital Project Planning

For line work, capital projects tend to be jobs where "plant" is installed. *Plant* refers to items such as structures, anchors, and conductors. Capital projects can be initiated by customer requests, such as for service, low-voltage and/or overloaded feeders, road moves, and the need for rehabilitation.

Justifying Capital Projects

Typically, a utility has a capital budget without sufficient funding for all possible projects. Projects are identified and given a preliminary estimate and are typically put into a 5-year capital projects plan. A priority is established for each project based on the justification prepared by a planning engineer. A planning engineer would provide details, cost benefits, present-day value, and other necessary information. Typically, a justification based on correcting low voltage or overloaded circuits gets priority because it would be intolerable to provide voltage below standard. The next level of priority includes new industrial customers needing service, new subdivisions requiring service, and lines that must be moved for road projects. Rehabilitation projects tend to be very low priorities, because the justification is subjective, and it is usually possible to get another year of use out of a line.

Estimating a Project

When a budget is prepared, the cost of each project is estimated so that decisions can be made as to which projects can be funded. Generally, each needed project is identified, estimated, and prioritized on a list. The amount of funds that will be allotted to the budget will determine how many projects will be planned for the year.

Estimates can be initial ballpark figures used to study two or three alternatives when justifying a project. Initial budget estimates are also ballpark figures that might come from cost-per-mile numbers that a utility uses for different types of line construction. When more accurate estimates are needed, a project may be planned on paper in more detail by an engineer, including drawings and estimates made from such drawings.

Before a work order is issued, a more accurate estimate is prepared based on actual staking data and field input. An estimate for an overhead distribution line, for example, would be prepared as follows:

- *Calculation of the material cost for conductor for the length of the line.* Generally, an estimator would have per unit costs for slack stringing and for tension stringing. Per unit costs would be included if existing conductors must be set out on auxiliary arms.
- *Calculation of the material costs for all the poles.* Generally, an estimator would calculate the per unit costs of setting poles based on the type of digging and whether the pole is to be set in an energized circuit.
- *Calculation of the material costs of the pole framing.* Typically, a construction specifications book will contain a material and labor cost for each drawing. Using the staking data, the framing for each pole can be totaled and the costs for each will be related to the construction specifications. On a new transmission line, each structure is likely to have individual specifications.
- *Calculation of transport and work equipment costs.* Each truck or piece of equipment, such as a tension machine, will have a cost figure (usually cost per hour) attached to it.
- *Calculation of administration costs,* including overhead and board and lodging costs if required.

Designing a Line

The design of structures, framing, vaults, and other components of the line must meet applicable regulatory standards (such as the National Electrical Safety Code [NESC]). A line design consists mainly of electrical and mechanical engineering considerations.

In terms of electricity, a line design is all about keeping the power conductor, whether overhead or underground, insulated from electrical short circuits and isolated from public contact. Three types of electrical clearances are considered when designing a line.

1. *Clearances from the ground, buildings, highways, railways, other electrical circuits, and communications cables to meet regulatory standards.* The maximum conductor **sag** that can occur during high temperatures due to weather and electrical load or during heavy ice conditions is included in the specified height. In other words, final sags and clearances must be calculated. Field measurements during stringing are not adequate to determine final sag.

2. *Design standards for minimum electrical clearances between phases and between energized parts, the structure, and down guys.* These clearances are much smaller than the minimum approach distances used for safe work. For example, on a three-phase 25 kV transformer installation, a standard may show that the minimum clearance between the riser drop wires to the three cutouts is 15 in. (40 cm) and the minimum distance between a phase and the transformer tank is 9 in. (13 cm). Metal hardware on a wood distribution pole must be at least 3 in. (8 cm) apart or bonded together to reduce the risk of sparking and radio interference.

3. *Design standards for clearances to allow safer work.* Climbing clearances have been part of design for many years. Other examples of when utilities must consider worker safety in design include lowering the neutral 10 ft (3 m), allowing clearance for a digger derrick to hang a transformer, using saddle clamps that can easily be clipped in with live-line tools, using hardware that can be installed easily with bulky rubber gloves or live-line tools, and using armless framing for easier lifting.

Mechanically, the design for structure and conductor strength is dependent on whether the line is in a heavy, medium, or light loading district. Loading district standards take into account the potential for extreme ice, wind, and combinations of ice and wind. In the United States, the NESC specifies the grades of construction; Grade B is used for most power lines.

For new transmission lines, typically each structure is engineered for height, strength, and foundation requirements. Each span is calculated for maximum sag and detailed drawings based on surveys are prepared. **See Figure 8–4**. Manuals or drawings for common structures—such as twin-pole H-frame, wishbone framing, spar structures, common towers, and steel/concrete poles—are available and are used for work on existing lines. Specifications books are used for most distribution work, including framing a pole, installing a cable, and connecting a transformer. Specifications are written as they are for good reason, but it is good practice to provide field input to design engineers. Linemen who choose not to follow specifications are often in violation of regulatory laws, as well as engineering practices.

Figure 8–4 A power line profile survey is necessary to establish the height of the structures to be installed. *(Courtesy of Power line Systems, Inc. Madison WI USA, http://www.powline.com)*

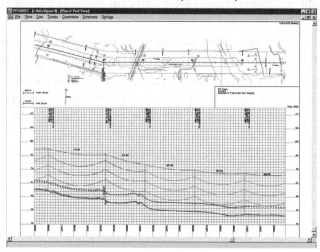

Design should consider workers so that all parts of a structure, insulator, and hardware assemblies can be reached for maintenance. Climbing space, ladders, where applicable and even permanent walkways or handrails may be designed to provide access for construction and maintenance.

Staking a Project

A layout, staking data, or survey is prepared for most line crew assignments. Different utilities have different job titles for the staking technician who does this work. Staking is specialized work. A person who lays out service to a new customer is normally not the same one who stakes out a line for a road-widening project or the same person or crew who surveys for a new transmission line.

Before setting up a new or changed service, a technician decides whether the feed is a primary or secondary service, what size transformer is appropriate, what new poles or cable are needed, where to locate the meter base, and what size the service conductors should be. Before a technician goes out on large and more complex distribution-line projects, decisions such as conductor size, general route, and type of pole or cable will have been made by a planning engineer. Ideally, the field supervisor, who will be doing the job, will go out to the site with the staking technician and engineer to provide input. For overhead lines, the technician must decide on pole heights, class of pole, anchor locations, and more. For underground lines, the route that the cable follows will be based on existing underground plant and disturbance to surface areas, such as streets and sidewalks. Obtaining property easements may also be part of a technician's responsibility.

Staking out a line using global positioning system (GPS) receivers and inputting the data into an automated geographic information system (GIS) helps to produce construction drawings, specifications, and maps. Once the information is in the system, it is used to easily make changes, prepare plant inventory, make material lists, and prepare as-built drawings.

The construction drawings used by field staff will show details such as pole locations, pole heights, pole class, trenching locations, transformer locations, conductor/cable size, secondary bus, and services. Any plant that must be removed (salvaged) is also shown on the drawings. Drawings will refer to the applicable specification drawings for each pole framing, equipment structure, pad-mount construction, and other details needed to carry out a project.

Joint Use with Other Utilities

There are some major cost advantages in going with joint-use structures and trenches where costs are shared among power, telephone, community antennae television (CATV) system coax cable, and in some cases gas utilities. Conductors are placed on poles such that the higher-voltage conductors are at the higher levels, and telephone and CATV are at a lower level separated by standard measurements. One standard measurement is 40 in. between the lowest power line conductor (often the neutral or secondary bus) and the communications cable. **See Figure 8–5.** The same spacing should exist midspan. Because the strand (messenger cable) is strung tightly before the cable is lashed to it, it is easy for the strand to contact the lower power company wires during stringing.

Figure 8–5 Minimum clearances between different utilities have been established to provide safe working space for each utility. *(Delmar, Cengage Learning)*

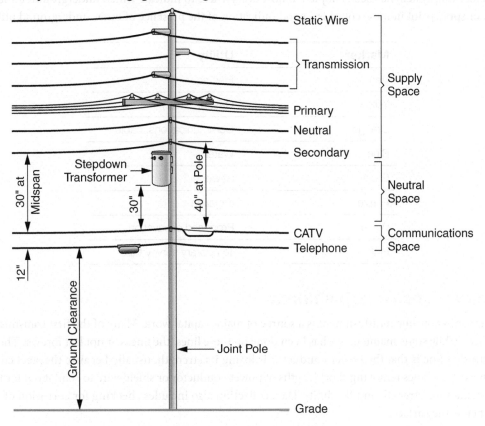

Often a higher grade of construction is necessary for a joint-use pole line. Anchoring is critical as can often be seen when a heavy telephone cable is added to a power line after the power line is in place. Poor anchoring or construction practices can cause a small movement of a dead-end or corner pole, which can cause a large change in power line sag.

Power line structures have become highly desirable locations on which to place new cellular antennas. Because of the energized conductors on power line structures, linemen are usually involved in installation of these antennas as well, including any time when work on them is required by a communications company. **See Figure 8–6.**

Figure 8–6 The telecom antennae on this power line structure was installed by linemen. *(Courtesy of iStock Photo)*

Color-Coded Marking System

A color-coded marking system has been adopted almost everywhere to indicate which underground utilities are present. Any flag, tape, or spray paint in these colors is a good indication of the presence of other underground utilities.

Marker	Utility
Red	Electric
Yellow	Gas, oil, steam
Orange	Communications, cable
Blue	Water
Green	Sewer
Purple	Irrigation
White	Proposed excavation
Pink	Temporary survey tape

Transmission-Line Refurbishment

The need for transmission-line refurbishment is a source of major capital work. Many of the first transmission lines built are still in service. While some maintenance has been done on these lines, the lines cannot last forever. The most pressing reason to refurbish a line is that the power conductor is losing its strength, usually because the steel core is corroded. Data collection work includes removing short lengths of power conductor or shield wire to send it out for testing that can determine the remaining strength and flexibility. Data collection also includes checking for corrosion of structure steel, especially just below the surface.

Major refurbishment projects include structure reinforcement, footing work, pole replacement, restringing, and grounding improvement. In many cases, a new higher-voltage line will be constructed on the existing right-of-way.

Planning a New Transmission Line

A new transmission line requires studies to determine the need for the line, the chosen route, and why alternative solutions and routes are unacceptable. A new transmission line involves a lot of property, as well as public meetings, environmental assessments, and government approvals.

A very large financial investment is involved in a new line. Typically, newly designed structures and new drawings are issued for such a project. Lines traditionally have been plotted on long roll plans that show the exact location of each structure, span lengths, angles, property lines, environmentally sensitive areas, and access roads. Profile drawings that show structure heights and conductor sag in relation to the terrain are prepared by using applicable software. A critical component of the design is obtaining assurance that the line-to-ground clearance meets standards regardless of temperature, load, and ice. It is not uncommon for a conductor on a line feeding an air-conditioning load to sag an extra 20 ft on a hot day.

Emergency Standards

Utilities have plans for emergencies such as restoration work after a major storm. Typically there would be emergency stock owned by one or more utilities. A minimum number of emergency restoration towers are kept in central locations. These towers can be installed temporarily to allow the stringing of a bypass around the downed structures. **See Figure 8–7.** Power is restored on the bypass while repairs are made to the original line. Written procedures are in place to carry out emergency work.

Figure 8–7 Temporary emergency bypass structures are installed to provide power while new permanent structures are installed. *(Courtesy of Lindsey Manufacturing Co.)*

Utilities cooperate with other utilities and emergency service agencies to form plans and carry out mock disaster exercises. Utilities register the minimum acceptable levels of standby and repair personnel, construction crews, and available equipment. This information is available to other utilities and is used when extra help is needed after major hurricanes, ice storms, and other disasters.

Constructing a Pole Line

Why Pole Lines?

Overhead lines are kept out of harm's way by stringing circuits high up on structures such as poles and towers. By far, most overhead distribution power line are pole lines. Poles can be wood, concrete, steel, laminated, aluminum, or fiber-reinforced polymer composite. Wood poles are most common because of their relatively low cost and ease of use. Poles support transmission lines, subtransmission lines, distribution lines, and outdoor lighting.

Choosing the Right Pole

Poles are chosen based on their strength, available heights, life cycle, cost, and availability. Other considerations include climbability, field drilling, type of foundation needed, and equipment needed for pole set-up.

Pole Strength

The pole strength specified for a job is critical, as can be seen after a storm. The appropriate pole strength will be based on number of conductors, height of conductors, guying, equipment to be installed, and whether the placement is a tangent, corner, or dead end. Pole strength also has to meet regulatory standards for the loading district and construction grade needed to withstand wind and ice loading. The critical point of strength (moment) for an unguyed pole is near the ground line. However, as seen after a windstorm, the moment on a pole with a double circuit of heavy conductors on top and a large telephone cable below tends to be just above the telephone cable. The circumference of a wood pole determines its strength and resistance to bending. The strength of steel, concrete, and other manufactured poles is engineered and more predictable than the variable strengths of a natural product such as wood.

Wood poles are divided into several classes, according to top circumference and circumference 6 ft (2 m) from the butt: classes 1 through 10 for normal strengths and H1 through H6 for higher strengths. A utility would typically specify a minimum pole class for their loading district and then use a bigger pole class for railway crossings, heavy dead-ends, corners, or equipment poles.

Steel, concrete, and composite poles are graded similarly for strength. Some utilities use tables that compare the class equivalents to wood class numbers.

Pole Length

Pole length is selected based on the number of circuits, equipment, communications cables, ground clearance, and possible future needs. A pole that is too long is better than too short because an extra 5 ft (1.5 m) is much more economical than having to change out poles later to add a transformer or to make room for telephone or television cable. Pole length is also selected so that the grade of a line will be relatively even, the intent being that there will be no uplift on a pole placed in a hollow. In mountainous country, back-to-back dead-ends are used to prevent uplift on a pole.

Wood Poles

Historically, wood poles have been the most common type of utility pole in North America. Wood poles have been plentiful and economical, are climbable with climbers, have moderate weight, and are easy to bore holes into for mounting equipment. Some linemen prefer wood poles and see them as less conductive and therefore safer to set in energized lines or when working with rubber gloves. Historically, some dry, untreated cedar poles have been quite nonconductive, but from a working perspective all poles should be treated as though they are very conductive.

Laminated wood poles compete with regular wood poles. Laminated wood poles can be made from old, defective poles, thus eliminating the environmental problems associated with the disposal of treated poles. From a climbing and working point of view, they can be framed the same way as regular wood poles. Because they are manufactured, they can be made 150 ft (46 m) long and strong enough to be used in unguyed corners and dead-ends.

To prevent or delay biological decay, many types of treatment have been developed, including tar, creosote (1900s), pentachlorophenol (1930s to 1960s), chromated copper arsenate (CCA), and chromated copper arsenate/ oil emulsion (CCA-ET). Some pole treatments have become environmentally unacceptable, some cause skin rash, and some have hard pole surfaces that climber gaffs cannot penetrate. These poles are treated with an oil emulsion (CCA-ET) in the outer layer so that they are easier to climb. There is also a chromated copper arsenate/polymer additive (CCA-PA) process that calls for injecting a polymer-based additive that creates a softer outer shell for better climbability. A complicating factor is that many locations consider treated poles as hazardous waste after they have been removed.

The most common location for pole decay is about 1 ft (0.3 m) below the ground line, which is where a pole is most likely to break when a lineman fails to support the pole when the work involves changing strain. The life expectancy of a wood pole is about 35 years, but a pole test and treat program is needed to ensure that no poles are decaying quicker. Structures can deteriorate to below the required strength level.

The decision to repair, treat, or replace a wood pole is made by a person trained to use one of the many instruments made for testing or by a boring/auger drilling inspection. The testing involves the process of boring a hole just below the ground line using an increment borer. **See Figure 8–8.** When the bit is withdrawn, the wood inside the increment borer can be examined for the thickness of sound wood left in the shell. One criterion is to replace poles when their strength falls to two thirds of the required strength. Others will replace poles based on a specified amount of shell that is still sound. Retreatment is an economical way to allow poles to be in service longer. Because testing a pole for possible replacement does not account for the work that may be done on a pole, a tested pole is not necessarily safe to climb. Wood poles are classed by their diameter which also translates into strength. In different species of wood poles, the weight of a wood pole is directly related with the strength and class of a pole. **See Figure 8–9.**

Figure 8–8 Testing with an increment borer will show the thickness of sound wood still available in the shell of the pole. *(Delmar, Cengage Learning)*

Figure 8–9 Weight of Douglas Fir Poles in Pounds

Length (ft)	40	45	50	55	60	65	70	80	90	100	110	120	125
Class H6	—	—	4,370	5,010	5,750	6,490	7,270	8,830	10,580	12,420	14,350	16,330	16,970
Class H5	—	—	3,910	4,550	5,150	5,840	6,530	7,960	9,520	11,180	12,880	14,670	15,640
Class H4		2,990	3,540	4,050	4,650	5,240	5,890	7,180	8,600	10,030	11,640	13,290	14,120
Class H3		2,710	3,170	3,680	4,190	4,740	5,240	6,490	7,730	9,060	10,490	11,960	12,700
Class H2	2,020	2,440	2,850	3,270	3,770	4,230	4,780	5,840	6,950	8,140	9,480	10,760	11,450
Class H1	1,840	2,160	2,570	2,990	3,400	3,820	4,280	5,240	6,300	7,360	8,510	9,710	10,350
Class 1	1,540	1,930	2,220	2,480	2,720	3,060	3,480	4,410	5,140	6,140	7,290	8,440	—
Class 2	1,310	1,610	1,870	2,130	2,430	2,750	3,080	3,960	4,740	5,680	6,660	—	
Class 3	1,160	1,410	1,620	1,850	2,110	2,430	2,690	3,520	4,170	—	—	—	
Class 4	1,160	1,410	1,620	1,850	2,110	—	—	—	—	—	—	—	

Concrete Poles

Concrete poles historically have been used in urban areas within North America but are found almost everywhere in many other countries. This concrete pole is framed with a set of switches. **See Figure 8–10**. The hollow center of concrete poles is often used as conduit to carry underground cables down a pole. Concrete poles have a long life span with known strength characteristics, although along roadways where salt is used in winter it is not unusual to see the steel reinforcement rods exposed and rusting.

Figure 8–10 Concrete poles in an urban area have the advantage of being clean and protecting the public from contact with wood preservatives. *(Delmar, Cengage Learning)*

Concrete poles are classed alphabetically. Each class identify which poles have a capacity for bending that is appropriate to the design requirements. **See Figure 8–11**. The minimum ultimate transverse load shown in the table is applied 2 ft (0.6 m) below the top of the pole.

Because manufactured poles can be engineered to exact standards, poles can be ordered with very precise requirements. **See Figure 8–12**.

Figure 8–11 Concrete Pole Classes

Class	Minimum Ultimate Transverse Load		Class	Minimum Ultimate Transverse Load	
	pound (lb)	kilonewton (KN)		pound (lb)	kilonewton (KN)
C	1,200	5.3	J	4,500	20.0
D	1,500	6.7	K	5,400	24.0
E	1,900	8.5	L	6,400	28.5
F	2,400	10.7	M	7,500	33.4
G	3,000	13.3	N	8,700	38.7
H	3,700	16.5	O	10,000	44.5

Figure 8–12 A load tree is used by engineering to determine the class and height needed for the expected load. *(Delmar, Cengage Learning)*

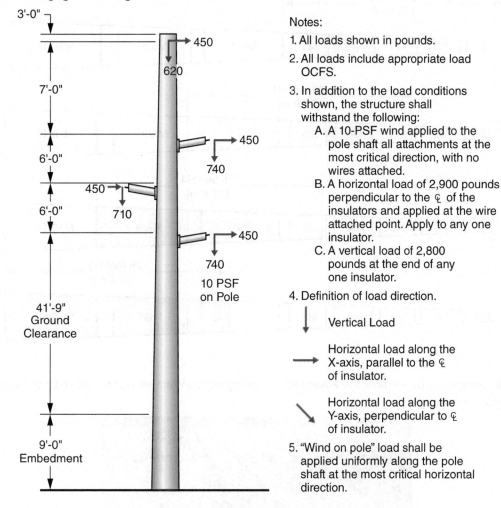

3'-0"

450

620

7'-0"

450

740

6'-0"

450

710

6'-0"

450

740

10 PSF
on Pole

41'-9"
Ground
Clearance

9'-0"
Embedment

Notes:

1. All loads shown in pounds.

2. All loads include appropriate load OCFS.

3. In addition to the load conditions shown, the structure shall withstand the following:
   A. A 10-PSF wind applied to the pole shaft all attachments at the most critical direction, with no wires attached.
   B. A horizontal load of 2,900 pounds perpendicular to the ℄ of the insulators and applied at the wire attached point. Apply to any one insulator.
   C. A vertical load of 2,800 pounds at the end of any one insulator.

4. Definition of load direction.

   Vertical Load

   Horizontal load along the X-axis, parallel to the ℄ of insulator.

   Horizontal load along the Y-axis, perpendicular to ℄ of insulator.

5. "Wind on pole" load shall be applied uniformly along the pole shaft at the most critical horizontal direction.

One advantage to concrete and steel poles is that pole-top fires and direct lightning strikes are not as damaging as they would be to wood.

Composite Poles

Composite poles can be made any length and strength required by the customer because they are available in modular form where the smaller sections (**Figure 8–13**) can be assembled on site. They are made with materials such as polyurethane, resins, and fiberglass. The advantages of fiberglass composite poles are advertised as having the highest strength-to-weight ratio, being nontoxic to the environment, nonconductive, and holes can be drilled in them to accept regular framing hardware. **See Figure 8–14**.

Steel Poles

Steel poles, like other manufactured poles, have known strength characteristics and are classified the same as wood poles, classes 1, 2 and 3. Stronger steel poles are available in classes H1 to H9. The galvanized tubular steel poles increasingly common in distribution lines are lighter weight than wood poles of the same strength and can be set as a direct bury similar to a wood pole or concrete pole direct bury. Steel poles also have a smaller diameter pole butt relative to a wood pole of similar strength, which is advantageous when installing in a drilled rock pole hole.

Figure 8–13 Modular composite poles can be used for many different lengths.

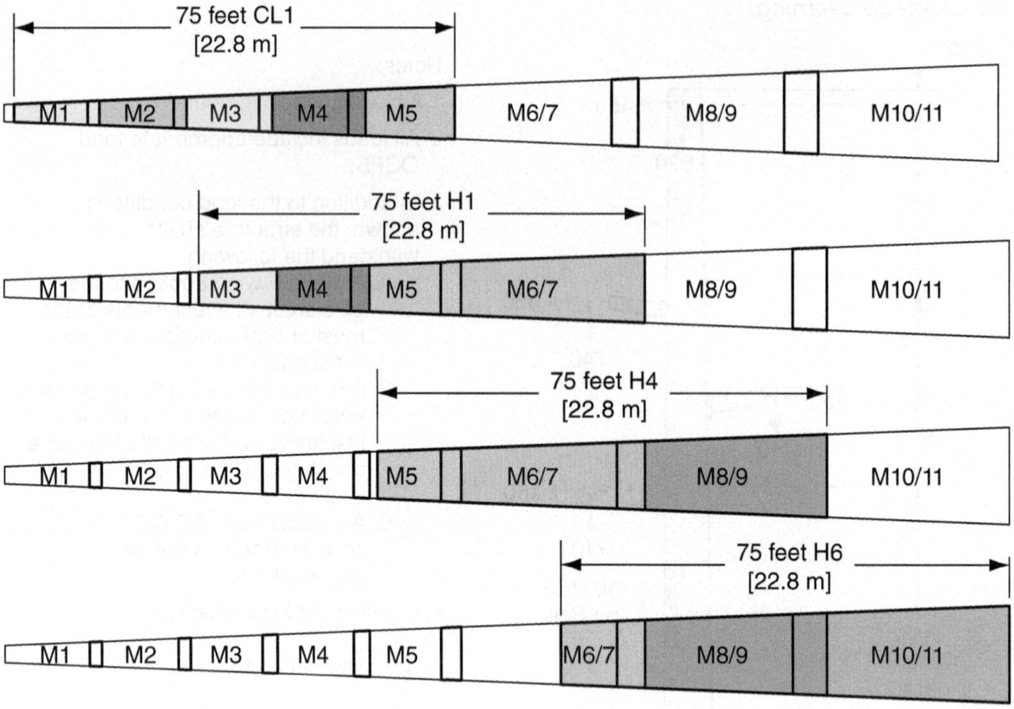

75 feet CL1
[22.8 m]

| M1 | M2 | M3 | M4 | M5 | M6/7 | | M8/9 | | M10/11 |

75 feet H1
[22.8 m]

| M1 | M2 | M3 | M4 | M5 | M6/7 | M8/9 | M10/11 |

75 feet H4
[22.8 m]

| M1 | M2 | M3 | M4 | M5 | M6/7 | M8/9 | M10/11 |

75 feet H6
[22.8 m]

| M1 | M2 | M3 | M4 | M5 | M6/7 | M8/9 | M10/11 |

Figure 8–14 Composite poles have the advantage of being nonconductive and nontoxic to the environment. *(Delmar, Cengage Learning)*

Steel poles are subject to corrosion, especially when placed in areas with a history of corrosion with galvanized anchor rods, for example. Extra protective coatings are needed to provide additional below-grade protection. Retreatment is also carried out by identifying any "rust" and painting on a zinc-rich paint.

Transmission-Line Steel Poles

Very high transmission-line poles are not direct buried but are bolted to anchor bolts that are previously installed in large engineered concrete footings. **See Figure 8–15.** Steel transmission-line poles are modular; manufactured in sections that are set into each other. They are often designed to be self-supporting at corners and dead-ends; therefore, they need extensive excavation, concrete, steel reinforcing, and long, large-diameter anchor bolts. Any required length of steel pole can be manufactured. **See Figure 8–16.**

Figure 8–15 Anchor bolts are used to set a transmission-line pole, this one carrying a 230 kV line. *(Delmar, Cengage Learning)*

Figure 8–16 A transmission line steel pole can be manufactured strong enough to be used to carry four transmissions circuits in a corner with no guys. *(iStock.com/Wesvandinter)*

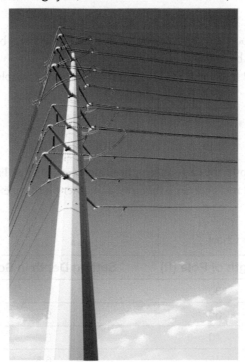

Pole Identification

A pole brand specifies the supplier, pole treatment, length, and class. (Other means are used to provide this information on steel and concrete poles.) Some suppliers bore a "through bolt" hole through the top of a wood pole, to indicate alignment of the pole face. Many utilities place a nail of soft aluminum 10 or 15 ft up the pole, depending on the pole length, upon which are stamped the date of installation and the owner's name. The bar code contains data regarding ownership, pole size, and GPS location. The next number sets indicate transformer location and telecom numbers, respectively. **See Figure 8–17**.

Figure 8–17 The bar code identifies all the needed pole data a utility needs. *(Delmar, Cengage Learning)*

Pole Foundation

The most common foundation for wood, concrete, and some steel poles is a direct bury into a pole hole. Tables specifying depth settings for poles are provided by utilities. Typically, the setting depth for a wood pole in earth is equal to the pole height in feet (m) divided by 10 plus 2 ft (0.5 m) (X ÷ 10) + 2 ft (0.5 m). For example, the following is the setting depth for a 40-ft (12 m) pole:

$$(40 \text{ ft} \div 10) + 2 \text{ ft} = 6 \text{ ft}$$

$$\text{or in metric } (12 \text{ m} \div 10) + 0.5 = 1.7 \text{ m}$$

There are established setting depths for various lengths of poles in soil. **See Figure 8–18.** A utility may have tables with different depth settings for solid rock, corner structures, and other points of extra strain.

Figure 8–18 Pole Depth Setting Chart

Length of Pole (ft)	Setting Depth in Soil (ft)
30	5.5
35	6
40	6
45	6.5
50	7
55	7
60	7.5
65	8
70	8
75	8.5
80	9

Exceptions to the hole depth are included in specifications books, because poles are set in earth as well as in hard rock, shale, sandstone, sand, gravel, swamps, and marshes. One typical exception allows a pole set in solid rock to be set 1 foot shallower. However, one interpretation states that this applies to drilled holes in solid rock, not to blasted holes. When a pole set in a marsh is cribbed and the crib is filled with rock or gravel, the above-ground height of the crib is not to be considered as additional depth of the pole setting.

Most holes are dug with machinery such as a digger derrick auger, backhoe, or sucker truck. Obtaining and paying attention to utility locations is essential because an auger has no mercy on buried gas, power, water, sewer, or communications lines. Occasionally, when all else fails, a crew must resort to "three-phase" equipment: the digging bar, spoon, and shovel; and in areas where there is less rock, "three-phase equipment" are shovel, spoon, and spade.

There can be a lot of continuous transverse wind load on a line, and it is all too common to see a pole leaning on a recently built line. A well-known rule of thumb states that three ambitious people with tampers are needed for each lazy person on a shovel when backfilling a pole hole. A hydraulic tamper would be used more often if it was mounted on the truck deck for easy retrieval. When earth conditions are not ideal, use rock or gravel as backfill. Composite backfill can also be used, and there is no need for tamping. It expands and forms a very solid foundation for the pole. Composite backfill will reduce leaching of pole preservative, reduce electrolysis that can cause corrosion of steel poles, and prevent marring the protective surface of composite poles. **See Figure 8–19.**

Some steel poles are intended for setting into a dug hole, and other types of steel poles are bolted onto bolts imbedded into a concrete foundation. **See Figure 8–20.**

Figure 8–19 Composite backfill provides a very firm foundation while preventing leaching of pole preservatives. *(Courtesy of Chemque, Inc.)*

Figure 8–20 Anchor bolts are ready for a steel pole installation. *(Delmar, Cengage Learning)*

Overhead Lines Design for Storm Hardening

Locations with higher risk of natural disasters such as hurricanes, tornadoes, flooding, ice storms, and wildfires, include storm hardening in their power line design. Stronger poles/structures, shorter spans, installing storm guys, inserting dead-ends in the line strengthen overhead power lines. Special smaller conductor that has more strength and similar conductivity reduces the vulnerability of the line to storms. Existing poles can be strengthened with the installation of a truss, as shown in **Figure 8–21**.

Figure 8–21 Storm hardening a pole with a truss.

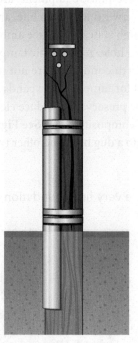

Constructing Overhead Lines in Wildfire Vulnerable Areas

Hardening Activity Examples

Note: when there is already a fire under a transmission line, hot gases and smoke from the fire can be conductive and cause a flash to ground.

There are three issues when an overhead power line is in a region vulnerable to wildfires.

1. The power line can be severely damaged by a fire.
2. A power line can be the ignition source of a fire.
3. Maintenance work activity can reduce or increase the risk of initiating a fire.

Damage to a line is reduced when using noncombustible hardware such as steel or concrete poles, or installing a fire-resistant wrap on composite poles and wooden poles. Specify ceramic or glass insulators, steel brackets or crossarms. More insulation will reduce the risk of flashover when the insulators are contaminated. Keeping only one circuit per right of way will keep damage to a minimum, and other circuits in other right of ways may survive.

To reduce the risk of a power line being the source of a fire it can be designed with shorter spans to reduce risk of conductors slapping together. The use of covered conductor such as the Hendrix covered conductor system reduce the risk of power outages or starting a fire when tree contact or arcing because conductor to conductor contact is almost eliminated. Improve lightning protection with more surge arrestors and improved grounding along the line. Switch gear such as non-expulsion fusing should be standard. Install protective wildlife deterrent and diverter hardware. Expanded well maintained vegetation clearances and the removal of trees that could hit a line is critical in wildfire regions. System operators at the control center can reduce the risk of initiating a fire by blocking the automatic reclose to limit a flashover to one occurrence. During extremely high wind conditions, proactive de-energization of the circuit is an option.

Maintenance activities, such as a patrol, may find weaknesses that require repair. Infrared surveys will find hot splices or connections. When using trucks "very high fire hazard severity zone" heat shields installed on the exhaust will reduce risk of a fire when parking in off-road. Some onboard firefighting equipment could put out a small fire preventing a new wildfire.

Setting Poles

Poles are set most often with a digger derrick, although they can also be set with pike poles, a gin pole, a helicopter, or a backhoe.

The oldest method for setting poles was with pike poles. Depending on the length of pole, the process will require six to ten people. The pole is laid down with the butt next to the hole. Two or three digging bars are set in the hole as a backstop. The top of the pole is lifted high enough to get the "raising horse," an H-frame device, underneath it. The raising horse is moved toward the hole as the pole is raised. A lot of coordination is required as the highest pike poles are removed and inserted at a lower position on the pole while the lower pike poles and the raising horse hold the weight. Anyone removing a high pike pole has to hold it tight to avoid dropping the pike onto someone holding a lower pike pole. **Figure 8–22** illustrates the method, but are not a substitute for a written procedure if a crew must use this method in a back lot or island situation.

Figure 8–22 (a) The raising horse supports the pole each time it is raised a few feet at a time. (b) Pike poles take over when the pole is raised to a height where the raising horse is no longer effective. *(Delmar, Cengage Learning)*

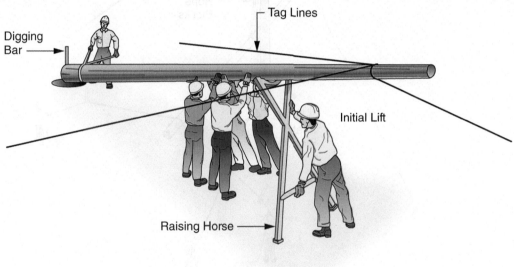

(a)

(b)

Another method for setting poles uses a temporary pole or existing pole as a gin. **See Figure 8–23**. Rope blocks or a winch are hung on the pole used as a gin and the new pole is raised into position. Sometimes the pole is tipped in and sometimes it is picked up just above the balance point, then lowered into the hole. There may still be applications of this method for pole replacement in subdivisions with back lot construction.

Figure 8–23 The pole is being set by using an existing pole as a gin pole. *(Delmar, Cengage Learning)*

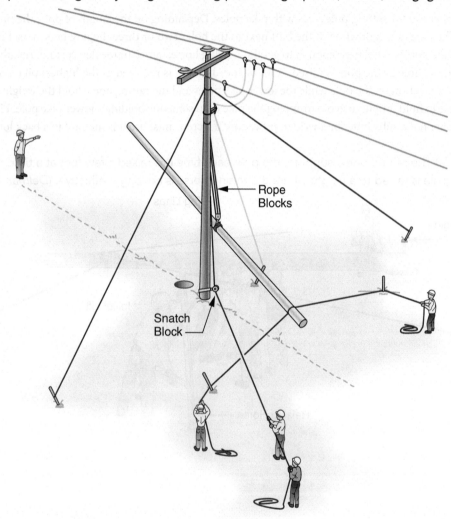

Rope Blocks

Snatch Block

Setting a pole with a derrick is a straightforward process, unless it is being set in an energized circuit. The winch is attached just above the balance point, and the pole is maneuvered into the hole while someone controls the butt.

The safest way to set a pole with a helicopter is to have the site rigged so that the structure can be set with no one in position under or within striking distance of the structure. Use of an eyebolt in the pole allows for the pole to be tied securely to the sling connected to the helicopter load hook. When setting a new pole next to an existing pole, a specially designed bracket must be installed on the old pole to hold the new pole after it is lowered into position. The new pole can also be held in place if a rope arranged with a loop is ready to receive the pole. **See Figure 8–24**. The hole is usually dug in advance, although it is possible to stand the pole first, dig the hole, and then jockey the pole into the hole. If the pole is to be set on a new right-of-way, it is usually set up with rope guys/tag lines (typically about ⅝ in. [1.5 cm] rope) already tied at the top of the pole and the other end of the rope coiled and tied just above the ground line. After the pole is lowered into the hole, workers on the ground take the rope guys and tie them to predetermined anchor points or bull pins before the pole is released by the helicopter.

Figure 8–24 Shown here is one method for receiving and securing a pole set by a helicopter in a 120 kV line. *(Delmar, Cengage Learning)*

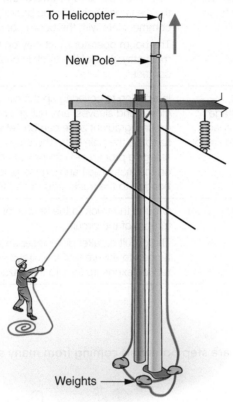

To Helicopter

New Pole

Weights

Problematic Pole Settings

The winch attachment just above balance point on a tall pole set into an existing line will be above the lower circuits or the communications cable when setting the pole. A lower attachment point will cause the pole to be top heavy and trying to hold the bottom of the pole has a high risk for losing control of the pole. For a production job, using a crane to drop the pole in from above works very effectively.

A crane can also be used effectively to drop long transmission poles into place. Inserting a link stick in series with a winch reduces the risk of the crane becoming energized in a mishap. No one has to be anywhere near the pole as it is being lowered.

Setting a Pole in an Energized Line

In addition to lifting a pole and setting it in a hole, two additional major steps are needed to set a pole safely near an energized circuit.

1. Use the equipment and procedures necessary to prevent the pole and equipment from making accidental electrical contact.

2. Use safety controls to reduce the risk of injury if that rare but dangerous contact is made.

Summary of Hazards while Setting Poles in an Energized Circuit

Hazards	Procedural Barriers
If a pole (wood, concrete, or steel) makes contact with an energized circuit, there will be touch potentials between the pole and earth. If the pole is also in contact with the earth, there could be high step potentials at the base of the pole. **See Figure 8–25.**	On distribution voltages, use rubber gloves along with pole tongs (cant hooks, pole handlers). The pole tongs keep a person away from the highest ground-gradient potentials where the pole touches the ground. On transmission-line voltages, use butt ropes and rubber gloves to guide the pole into the hole.

There will be touch and step potentials around the pole, truck, attached trailer, and ground rod when a bare pole contacts an exposed, energized conductor.	On distribution voltages, install protective cover-up on the conductors and/or on the pole. **See Figure 8–26**. Use a dedicated observer to monitor the clearance between the boom pole and energized conductors and to communicate with the boom operator. The boom operator must stay on the operating platform and other workers must stay clear of the truck in case the boom or its load makes accidental contact.
Setting a very conductive pole (steel, concrete, or wet wood) in a high-voltage distribution circuit with limited open space in which to set the pole increases the risk of contact and a high fault current.	In addition to covering up the circuit, covering up the pole, using rubber gloves and sleeves, and using a pole handler and/or butt ropes, a protective ground-gradient zone can be set up to provide protection for workers handling the pole butt. The ground-gradient matting (similar to tension stringing procedure) can provide an equipotential zone when a worker stays on the mat while handling the pole. The ground-gradient matting has to be bonded to the truck ground and to the pole.
A circuit may not trip out when a truck boom contacts an energized circuit.	Ground the truck to the neutral or other good ground to promote a quick trip-out of the circuit. The circuit breaker or recloser should be put in a nonreclose position and tagged to ensure that the circuit will remain isolated after it trips out. This will reduce exposure time to the hazard and facilitate a rescue if needed.

Figure 8–25 Notice that there are step potentials coming from many sources when contact is made. *(Delmar, Cengage Learning)*

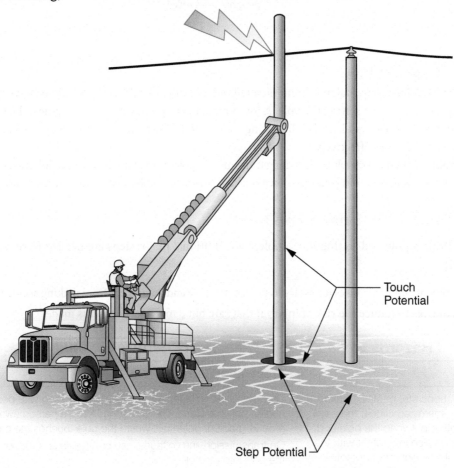

Touch Potential

Step Potential

Figure 8–26 A pole guard is installed for protection from a brush electrical contact. *(Delmar, Cengage Learning)*

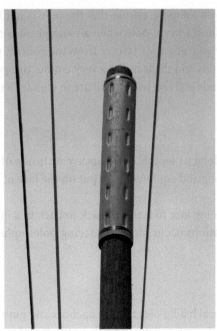

Preventing Contact

To prevent contact, insulated cover-up is put on the conductors and/or on the pole. Even though cover-up is installed, the pole must be controlled. With a modern derrick, this is relatively easy when the winch can be attached to the pole just above the balance point. Rope guying/tag lines may be needed with other installation methods. A dedicated observer positioned well for judging approach distances is standard, especially when installing a pole in three-phase lines.

No cover-up is available for transmission-line voltages. Larger clearances between conductors can allow a pole to be set in an energized circuit. Depending on the installation method, extra rope guying/tag lines to improve control may be needed.

Safety Controls

Safety controls refer to the extra steps taken to reduce the risk of injury if an electrical contact is made.

1. Ground the boom-equipped vehicle, setting the pole so that the circuit trips quickly if the boom or its load makes accidental contact with the circuit. If the vehicle becomes energized, anyone near the truck is exposed to ground gradients. The boom operator stays on the operating platform or uses a ground-gradient mat.

2. Have the circuit breaker or recloser put into the nonreclose position to ensure that the circuit remains isolated should it be tripped out due to accidental contact.

3. The workers most at risk, while setting a pole in an energized circuit, are the workers controlling the pole butt. The designated observer can stand clear of any vehicles or objects that may become energized, and the boom operator can stay on the operating platform. The person controlling the butt is exposed to touch and step potentials. Rubber gloves can provide protection from distribution voltage touch potential. Step potentials are generated where the pole touches earth. There is a rise in voltage relative to any earth farther away from the base of the pole. To keep a worker farther away from the high step potentials near the pole butt, tongs or butt ropes should be used to control the butt. Rubber gloves always should be worn while handling tongs because tongs are not usually electrically tested live-line tools. Ground-gradient mats bonded to the pole and derrick would provide even more protection from step potentials.

4. At transmission-line voltage levels, guide ropes tied to the pole butt can be used to get even farther away from the current entry point.

5. If climbing to remove the winch from the pole, ensure that the pole is not in contact with an energized conductor. Voltage gradients can also occur along a wood pole when an energized conductor is in contact. The voltage at the contact point is higher than the voltage at a spot farther from the contact point. Anyone on the pole could have a potential difference between the hands and the feet. On a very conductive pole, such as a steel structure, there is less potential difference between the hands and feet because all are in contact with the same object at the same potential.

Facing a Wood Pole

Every wood pole has a natural sweep along its length. The concave or inside of the sweep or bow is called the *face* and the convex side is called the *back*. Framing and equipment are put on the face of the pole to take advantage of the natural strength of the wood.

Poles are installed so they end up being face to face and back to back in a line. Corner poles and dead-end poles are installed to face the anchor. Some variations occur including facing poles uphill and facing the last two or three poles toward a dead-end.

Guying a Pole

Generally poles can support only a vertical load. Poles require anchors and guys to counter the pull of conductors in the opposite direction. Proper anchoring and guying allows the pole position and wire sag to remain constant throughout the life of the line and is a mark of good workmanship.

Types of Guys

When reading staking data, terms such as head guy, back guy, or side guy may be used. **See Figure 8–27.**

- A head guy is the guy used for the dead-end at the beginning of a job, at a change in conductor size, or before the start of some spans of secondary bus.
- A back guy is the guy holding the dead-end in the other direction.
- Side guys are down guys used at corners and to support lateral taps or services.
- A span guy is a horizontal guy, usually across a road to a pole (stub pole), anchor, and down guy.
- In some situations, conductors are dead-ended on crossarms on one side of the pole only. An arm guy is a span guy that is attached to the loaded side of the crossarm to prevent the cross arm(s) from twisting around.
- Storm guys are side guys installed at intervals down a line to help keep the poles straight during high winds. Storm guys can also be installed in four directions at intervals along the line to prevent a cascading when a pole breaks during an ice storm or hurricane.
- A strut guy is used with special hardware to span over a sidewalk.
- A push-pull brace is used where a down guy or span guy is not feasible.

Guy Wire

Choosing the type of guy wire, strength, minimum lead length, and total strength for a guy/anchor assembly needed for a job requires more information than a lineman would typically have available and is left to engineering. The type and strength of steel are also dependent on the specified ice and wind loading for the region (loading district) in which the line is to be constructed.

Five grades of guy wire are available: Siemens Martin, high strength, extra high strength, common, and utilities. The flexibility of the guy wire is the most common noticeable difference. Steel has a rated breaking strength (RBS) as well as allowable tension (AT) strength. For example, ⅜ in. (9.5 mm) high-strength steel has an RBS of 10,800 lbs (5,000 kg) and AT of 5,400 lbs (2,500 kg). Most guy wire is galvanized, but it is not uncommon to use aluminum weld in corrosive atmospheres such as seashores.

Figure 8–27 The method chosen to support a pole will be dependent on the availability of finding a location to install an anchor. *(Delmar, Cengage Learning)*

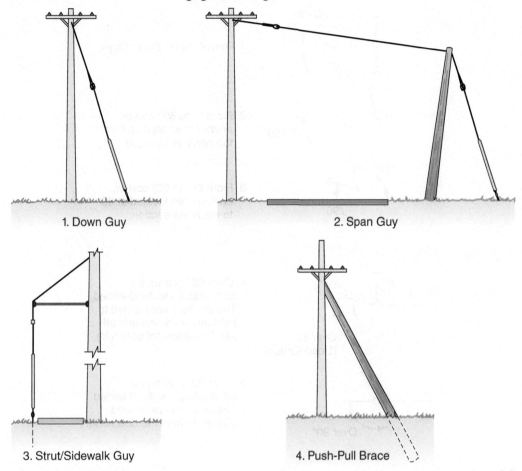

1. Down Guy

2. Span Guy

3. Strut/Sidewalk Guy

4. Push-Pull Brace

Guying Arrangements

There are various guying arrangements, a lineman must know how to decide which is best for each situation. **See Figure 8–28.** To determine the size and number of guys and anchors to be installed requires a calculation of the expected load. The load on the guy is dependent on the following:

1. Dead-end or bisect tension of each conductor (under ice loading conditions)
2. Height of the guy(s) attachment above the ground
3. Distance from the pole to the anchor

Typically, generic tables and graphs prepared by design engineers are found in specifications manuals so that calculating the load for each guy is unnecessary.

The tension on a down guy multiplies quickly as the distance between the pole and the anchor decreases. **See Figure 8–29.** The tension on a down guy is much higher than the conductor tension, and in most cases a guy already in service and under load would overload a standard chain hoist, grip, and pulling eye.

Guy wire tends to be 5/16 in. (8 mm), ⅜ in. (9.5 mm), and ½ in. (13 mm) galvanized steel or aluminum-clad steel. The guy holding a dead-end or corner transmission-line structure is considerably larger. It is important for the size of the guy wire to be properly identified. When a ⅜ in. (9.5 mm) preformed grip is used on 5/16 in. (8 mm) guy wire, it will not hold very much strain and may let go while someone is on the pole dead-ending conductor. It can result in a thrilling ride for anyone up the pole.

Figure 8–28 The anchor location is critical to ensure that corner poles will be properly supported. *(Delmar, Cengage Learning)*

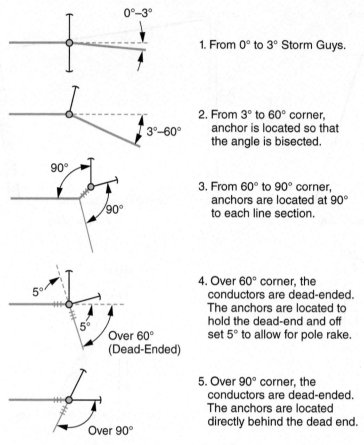

1. From 0° to 3° Storm Guys.

2. From 3° to 60° corner, anchor is located so that the angle is bisected.

3. From 60° to 90° corner, anchors are located at 90° to each line section.

4. Over 60° corner, the conductors are dead-ended. The anchors are located to hold the dead-end and off set 5° to allow for pole rake.

5. Over 90° corner, the conductors are dead-ended. The anchors are located directly behind the dead end.

Figure 8–29 Notice the increase in guy tension when the distance between the pole and the anchor is decreased. *(Delmar, Cengage Learning)*

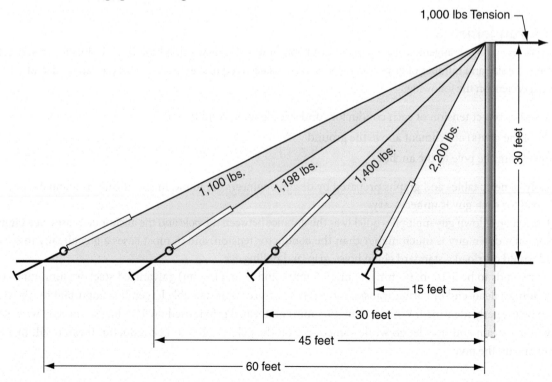

A line staker has to locate an anchor very carefully for guying corners on multicircuit poles. If the anchor is not located properly, then the guys coming from the top of the pole may be in conflict with conductors on lower circuits. **See Figure 8–30.**

Figure 8–30 The anchor is located so that guy wire will not interfere with the bottom circuit. *(Delmar, Cengage Learning)*

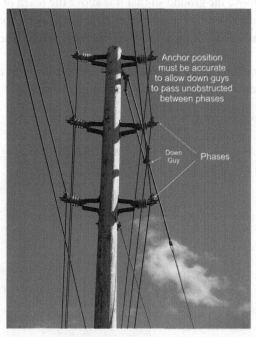

A separate guy is needed to offset the tension for each circuit on a double- or triple-circuit pole. In many cases, a double guy is needed in a location where there are large conductors under a high tension. Typically, two guys are allowed on one anchor rod. Multiple guys need to be pulled so that they are equally tight, which is relatively easy when using three-bolt clamps, but requires some guesswork when finishing off guys with preforms or automatic dead-ends. **See Figure 8–31.**

Figure 8–31 A preform and three-bolt clamps are shown as two ways to fasten a guy wire to an anchor. *(Delmar, Cengage Learning)*

(a)

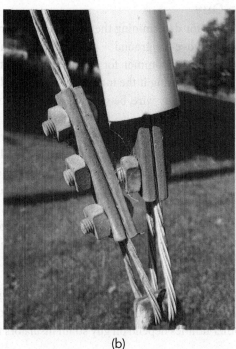

(b)

Guy Strain Insulators

A down guy can be energized accidentally when an insulator breaks down electrically, a conductor becomes dislodged, or because of a ground fault. Guy strain insulators protect the public when a guy becomes energized. Guy strain insulators can be made of wood, porcelain, or a fiber rod. **See Figure 8–32.** Note that with a porcelain guy strain insulator, the porcelain is under compression instead of tension. A strain insulator is located in a guy so that it will be below the lowest energized circuit and high enough above the ground so that it is above the reach of people, about 8 ft (2.5 m). There are differences of opinion among utilities regarding the use of guy strain insulators. Some utilities do not use any, some use them on distribution and not on transmission lines, and others use them on all structures.

Figure 8–32 A fiber rod or porcelain guy strain insulator is strong enough to withstand the tension in a down guy and provide an insulator into a guy. *(Delmar, Cengage Learning)*

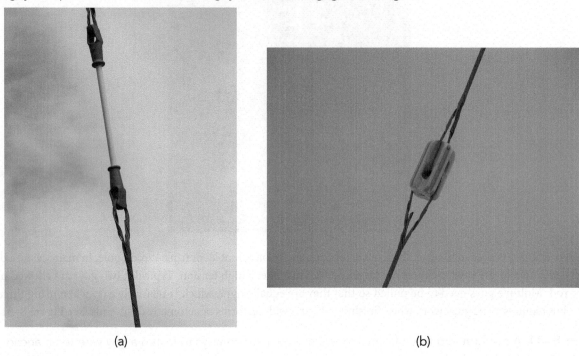

(a)   (b)

Pulling a Guy

A rule of thumb for determining the length of guy wire needed for a guy is, length of guy wire needed = height of the guy attachment above the ground + one half the distance from the pole to the anchor. Protect yourself when cutting guy wire. It has not been uncommon for an end of the steel to spring back and strike a person in the eye. Guy wire can be very springy, and when cutting it the individual strands will fly out and the end may come back and strike someone. When cutting springy-type guy wire, be sure to wrap tape around the guy wire on each side of the cutting point, and hold the wire on both sides of the cut.

When a guy is installed, the dead-end and corner poles are typically pulled over (raked) 1 to 1.5 ft (about 0.5 m) from plumb. If left straight and the anchor and guy settle a little, the pole will lean toward the line or into the corner. This arrangement will put a lot of extra strain on the guy assembly.

Except for ones very lightly loaded, a loaded guy should not be retensioned without a digger derrick supporting the pole. A loaded guy will typically overload a chain hoist, the guy wire grips, and the pulling eye. Guy guards or markers are required to keep down guys visible, especially where all-terrain vehicles and snowmobiles have access to a power line right-of-way. **Figure 8–33** shows an anchor rod preferred by linemen. This anchor does not need a separate anchor rod pulling eye.

Grounding and Bonding Down Guys

Many utilities require guy wires to be grounded by being bonded to each other and bonded to the system neutral to reduce stray electrical current and corrosion. There are anchor-bonding clamps that are installed between all down guys and anchors. Other utilities depend on guy strain insulators to prevent stray currents from flowing and to protect the public.

Figure 8–33 An anchor with built in pulling eye. *(© Van Soelen)*

Choosing and Installing an Anchor

The type of anchor specified for a job will be dependent on the soil conditions and on the type of load to be anchored. Historically, a log dead-man (slug) was used where a log was buried into the soil and the anchor rod cut into the soil to angle it toward the pole. Other anchors that were used where there was good digging, but are now used less frequently, were plate anchors and expanding anchors. **See Figure 8–34.**

Figure 8–34 These anchors were common before power installed anchors became the standard. *(Delmar, Cengage Learning)*

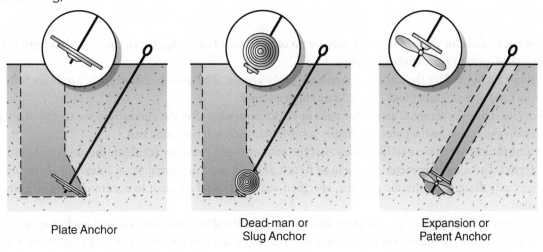

Plate Anchor   Dead-man or Slug Anchor   Expansion or Patent Anchor

Power-installed screw anchors are the preferred anchor where the soil conditions will hold the anchor and rocks do not prevent the installation. Different types are available for almost every type of load. They are installed by machine, and the result is an anchor that creeps up very little and causes very little soil disturbance. Utilities tend to use their local knowledge and experience to specify the type of anchor needed for local conditions. In sand or in swampy areas, anchor rods can be lengthened until good solid earth is reached. For heavy, critical anchoring such as transmission-line dead-ends, a more extensive study of soil conditions using a soil test probe should be carried out.

To install a power-installed screw anchor, the auger on the digger derrick is removed and a specially manufactured wrench is installed on the end of the Kelly bar. To install the anchor properly requires skill with a digger derrick so that proper alignment and down pressure are maintained as the anchor is screwed into the soil.

If a rock anchor is to be used, the rock must be solid. There are expanding rock anchors, rebar anchors, and rock eyes. The rock holes for expanding and rebar anchors are drilled in line with the intended guy. The rod is turned to expand the bottom of the expanding element in the bottom of the hole until it is tight. The quality of the rebar anchor depends on grout being applied around the rod in the rock hole. A rock eye is installed at 60° to the line, and the wedge in the bottom of the hole is expanded by hitting the top of the anchor with a sledge hammer. **See Figure 8–35.** Grouting keeps out the elements. Where there is overburden, an extension rod is installed on the anchor. Guy wire should not be buried.

Figure 8–35 The wedge type rock anchor requires solid rock to be effective. *(Delmar, Cengage Learning)*

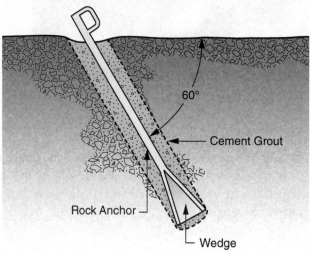

In poor rock, shale, or sandstone, there may be no other choice but to dig, break rock, or blast until solid rock is reached or a dead-man or plate anchor is installed.

The proper anchor rod for the application must always be chosen. The most common anchor rods are ⅝ in. (16 mm), ¾ in. (20 mm) and 1 in. (25 mm), and 1 in. (25 mm) high strength.

Anchors must be very secure. If an anchor pulls or settles somewhat after the conductors are sagged or after the telephone company puts cable on the poles, a formerly good-looking job will now look bad.

In critical situations where a sudden release of the anchor could cause an accident, the anchor should be tested. A digger derrick and a dynamometer can be used to test anchors used on distribution lines. For more sophisticated testing, a specially designed hydraulic pull test, specified wait times, and a transit to accurately measure any creep are recommended.

Ideally, when locating an anchor, adequate distance is maintained between the pole and anchor to reduce extreme tensions on a down guy. The location of an anchor is more critical when a guy must go between conductors of underbuild circuits to reach the top of a pole. A graph drawn to scale showing the pole height and framing is the easiest way to locate options for anchors on multicircuit poles.

Choosing the Proper Framing

A majority of the material in a utility storeroom is kept on hand to meet the many specifications for framing a pole. **See Figure 8–36.** Framing for tangent, various degrees of corners and dead-end poles using crossarms, armless framing, and aerial spacer cable framing are shown in specifications books. Drawings show the standards for installing equipment such as transformers, switchgear, regulators, capacitors, and riser poles.

Linemen should not deviate from specifications, because the specifications drawings are engineered and there are usually good reasons for the very specific details. A bill of material is part of most specifications drawings, typically

Figure 8–36 A utility warehouse needs a large inventory of many different kinds of distribution line hardware. *(Reprinted with permission of Hubbell Power Systems, Inc., Centralia MO USA)*

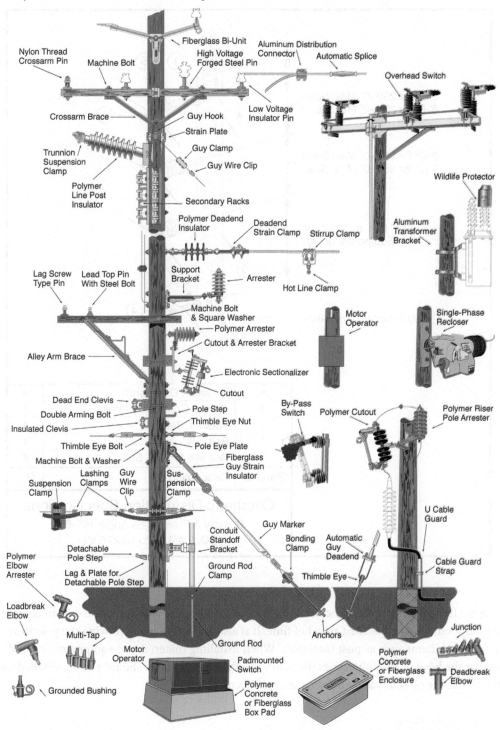

including stock numbers. A technician often will enter the drawing numbers for a particular job into a computerized system that will, in turn, apply the costs, tabulate the materials needed, and then order them. Stock keepers then issue material, and stockroom inventory numbers are automatically kept up to date.

Drawings in the specifications books are created with software that breaks down the standards to the subassembly level. Subassembles are then mixed and matched and the various standard drawings are created. **See Figure 8–37.**

Figure 8–37 A typical specification (Standard) drawing for a vertical corner on a distribution line. *(Delmar, Cengage Learning)*

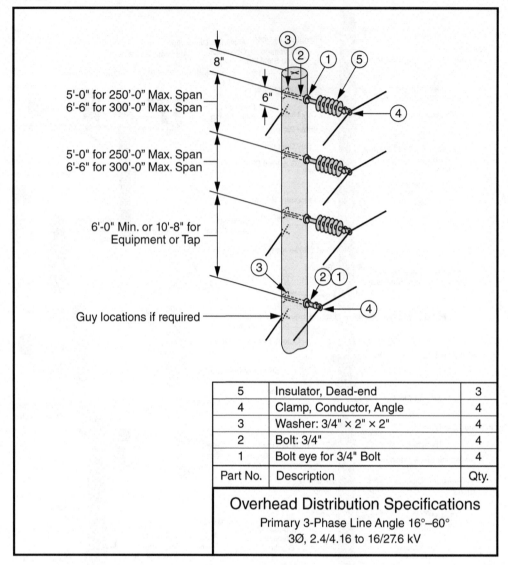

5	Insulator, Dead-end	3
4	Clamp, Conductor, Angle	4
3	Washer: 3/4" × 2" × 2"	4
2	Bolt: 3/4"	4
1	Bolt eye for 3/4" Bolt	4
Part No.	Description	Qty.

Overhead Distribution Specifications
Primary 3-Phase Line Angle 16°–60°
3Ø, 2.4/4.16 to 16/27.6 kV

Often, poles are framed on the ground before being installed. If the pole is to be installed in an energized circuit, the pole/down-ground should not be installed ahead of time. It is also important when moving the pole to avoid overstressing individual hardware items such as post insulators. When installing material and equipment on a pole, bolts should be installed so that the head of the bolt carries the heavy weight. Transformers should not be installed until the pole is vertical. Air pockets in the oil could cause a premature transformer failure.

Equipment on Poles

Transformers, regulators, capacitors, and different types of switchgear are mounted on poles. The pole framing dimensions and electrical connections should be available in a specifications manual. As a general rule of thumb, the "head" end of the bolt—not the threaded end—should carry the majority of the weight of any attachment. Heavy equipment, such as some regulators require a better-than-average class of pole. Some locations will construct a platform between two poles to carry a large piece of equipment.

Every piece of equipment has a number assigned to it, and an installation is not complete until a number is attached to the pole. Some utilities use GPS numbers to identify the locations of their equipment. Numbers are used to help identify locations for troubleshooting and for routine work. Numbers are assigned to other equipment, such as voltage regulators and capacitors, for similar reasons.

The numbers on switchgear are especially important when preparing a switching order using a formal lockout/tagout procedure. Switchgear numbers can be used to accurately relate field devices with the devices shown on operating drawings.

Installing Crossarms

Wooden crossarms and pin insulators were the first types of construction used to carry conductors, starting with telegraph lines. Double or even triple wooden crossarms have been used for dead-ends, corners, or long heavy spans. **See Figure 8–38**. Steel and fiberglass crossarms are designed to hold conductor tension at dead-ends and are now generally preferred to double wooded crossarms. **See Figure 8–39.** Twin-pole transmission lines also use poles, wooden timbers, or steel as crossarms. The "crossarm" on a steel tower is called a bridge.

Figure 8–38 Double wooden cross arms are used for added strength at dead-ends and corners. *(iStock.com/Westhoff)*

Figure 8–39 A single steel cross arm is often used for dead ends and corners. *(Delmar, Cengage Learning)*

Wood and fiberglass crossarms have the advantage of keeping the "ground plane" farther away from energized conductors. Extra **insulation** is needed when a ground plane, in the form of a grounded steel crossarm, is brought to the top of a pole. Design engineers have differences of opinion as to whether a steel crossarm should be grounded or left ungrounded (floating).

The use of wooden braces for wood crossarms is also meant to keep the ground plane away from energized phases. Some utilities use wood braces because the insulating value of the wood is considered safer for linemen working near them. Steel braces can be flat-strap, angle-iron, or **alley-arm** braces. Angle-iron braces are fitted to the bottom of the crossarm. The braces are attached to the pole with a bolt or a lag screw. A lag screw is to be driven in with a hammer to about 0.25 in. from the tight position and then seated with three or four turns with a wrench. Before fastening the braces, someone on the ground will sight in the crossarm until it is square with the pole (not square to the ground). Steel and fiberglass crossarms tend to have a gain brace as part of their construction and other braces are not needed. There are also gain braces available for mounting wood crossarms.

The maximum span length is based on the length of the crossarm and conductor spacing of the adjacent poles. The framing measurements also consider climbing space for linemen, for example to climb through underbuild to reach higher circuits.

Side or alley arms are used to clear obstructions such as buildings or trees. An alley arm is a regular crossarm that is mounted on one side of the pole and a special alley-arm brace is used to support it. The brace is long, and a step is provided for linemen to reach the outside conductors. Today, alternatives such as covered tree wire or extra long post insulators and brackets tend to be used. **See Figure 8–40.**

Armless framing with post insulators mounted on brackets is becoming standard in many locations, but span length options are more limited. **See Figure 8–41.**

Figure 8–40 A set of double alley cross arms are installed to support a slight corner. *(Delmar, Cengage Learning)*

Figure 8–41 The new pole has armless framing for a double circuit subtransmission and a distribution under build. *(Delmar, Cengage Learning)*

There is a wide variety of other framing hardware, such as secondary single-point and three-point racks, neutral spool bolts or clamps, different types of insulator brackets, pin insulators, post insulators, and strain insulators.

Installing Insulators

Conductors on overhead lines are electrically insulated from ground by either sitting on insulators or being suspended from them.

Insulation Materials

Insulators used on overhead circuits are typically porcelain, glass, or polymeric. Porcelain (ceramic) has a proven record of reliability. There are millions of porcelain pin insulators in service. "Cement growth" has been a problem with some identified insulators. Cement growth refers to the cement used between the porcelain and metal components, which have been found to expand or contract over time. **See Figure 8–42.** Cement growth in porcelain strain (dead-end insulators) has been a cause of mechanical failures, and utilities have developed programs to change out these insulators. Cement growth has also caused failures on two- and three-piece insulators.

Figure 8–42 An insulating cement is used between the steel and insulation. *(Delmar, Cengage Learning)*

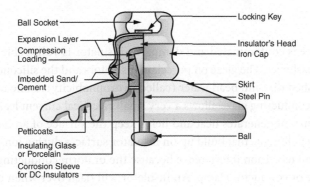

Glass insulators have a higher dielectric strength for a given thickness than porcelain insulators. Glass expands and contacts less with temperature change and maintains its insulation value over time because glass does not deteriorate. Glass is commonly specified for transmission lines. **See Figure 8–43.** Glass insulators are made of toughened glass. The glass does not chip but can shatter when struck by a lineman's pliers. Some utilities will use a glass insulator to distinguish the neutral from the phases, especially when the neutral is on a crossarm.

Figure 8–43 Glass insulators are often the preferred option on transmission lines. *(iStock.com/AWelshLad)*

Polymeric (composite) post and suspension insulators are popular with linemen because of their light weight. They are increasingly more popular, but some utilities are staying with insulators that have a proven 40- to 50-year life. Polymer insulators have a central fiber-reinforced plastic rod that provides the mechanical strength for the insulator. The rod is designed for applications such as compression cantilever and suspension. **See Figure 8–44.** The rod is covered with an outer insulating weathershed (housing) made of polymer material to keep water away from the rod and protect it from ultraviolet radiation. The sheds or ribs on the cover provide an increased electrical leakage distance.

Figure 8–44 This is one type of polymeric suspension insulator which can be used to dead-end conductors. *(Courtesy of Salisbury by Honeywell)*

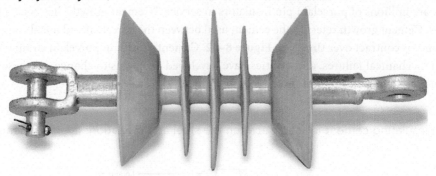

When an insulation surface is contaminated and damp, the insulator is vulnerable to leakage current, which will increase to cause an insulator flashover. The glaze on porcelain insulators and the silicone cover on polymeric insulators help the insulator's surface to shed dirt and bead water (called hydrophobicity). Some utilities specify semiconducting glaze (SCG) insulators. A semiconducting glaze allows a very small electrical current leakage (normally less than 1 mA) through the SCG layer. This current generates heat and helps keep the surface of an insulator dry and minimizes the problem of flashover caused by pollutants that build up on insulator surfaces during long dry spells. The SCG layer also minimizes corona and radio and television interference because the uniform conducting surface reduces voltage stress in any one point of the tie wire or conductor clamp. An insulator will flash over when the basic insulation level of the insulator is exceeded by lightning.

Insulators have a "Basic Insulation Level" (BIL) rating, which is a measure of the insulator to withstand a temporary high-voltage surge. For example, an insulator intended for a 115 kV circuit might have a BIL of 550 kV, and an insulator intended for a 69 kV circuit might have a 350 kV BIL. There will always be some minor electrical current leakage down an insulator. The skirts or sheds on an insulator creates a longer tracking distance and therefore a higher BIL.

Types of Insulators

1. *Pin-type insulators* are capable of being screwed onto threaded steel pins. One-piece porcelain pin insulators are used where voltages are less than 23 kV. For higher voltages, pin-type insulators are two or more pieces cemented together. **See Figure 8–45.** Pin insulators are no longer used on transmission lines. They are often selected over the suspension type because of cost, and because with the conductor tied in on the top, a shorter pole is needed to keep the conductor at a specified height.

Figure 8–45 Pin insulators, mounted on steel pins can be porcelain, glass, or polymeric. *(Delmar, Cengage Learning)*

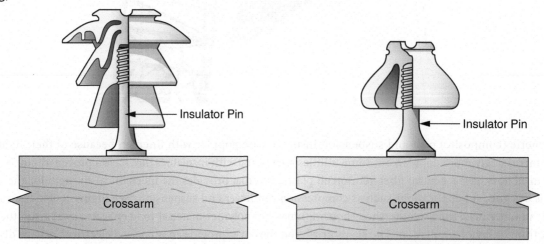

The neck size and design of a pin insulator are classified as having either a C, F, K, or J side and top grooves. The neck size must be known when choosing the correct size and type of factory-formed tie. When installing a pin insulator, it is not uncommon that it becomes tight on the pin before it is orientated properly to put the conductor in the groove. A light tapping on the steel pin while trying to turn it will usually gain a little more turn.

2. *Suspension-type insulators* can withstand the suspended weight of conductors or the tension (strain) of a dead-end conductor. Polymer suspension insulators can be one piece long, designed for any specified voltage. Porcelain (ceramic) and glass insulators come as individual disks (cap and pin type) that are normally rated at 11 kV each and are coupled with a split pin (cotter key) to the length needed for the line voltage. To determine the number of insulators needed in a string requires much more than dividing the voltage of the line by 11 kV. It is an engineering task that factors the capacitance because of the insulation and steel combination across the length of the insulator string, flashover voltage, impulse voltage, and withstand voltage. For example,

Nominal Line Voltage (kV)	Number of Insulator Disks
13.2	2 units
69	4 or 5 units
115	7 or 8 units
138	9 or 10 units
230	14 units
345	18 or 20 units
765	35 units

Suspension insulators are rated for strength in kilopounds (kip) (1 kip = 1,000 lbs) or kilonewtons (kN), typically 25 kip (111 kN), 30 kip (134 kN), 50 kip (222 kN), and 80 kip (356 kN). **See Figure 8–46.**

Figure 8–46 The suspension-type insulator is a ball-and-socket design that puts the insulating material under compression to create strength. *(Delmar, Cengage Learning)*

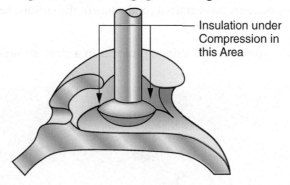

Insulation under Compression in this Area

An insulator string of cap and pin insulators must be handled in such a way as to prevent damage to the ball-and-socket cotter keys. When hoisting up a string of more than six insulators, a cradle or similar device should be used. A composite insulator must be handled carefully because a single cut, puncture, or tear in the weathershed material that covers the rod will expose the rod to moisture and eventual failure. The weathersheds are vulnerable to damage unless long suspension insulators are carried by two workers and the insulators are hoisted by the end fittings.

3. *Post insulators* are porcelain or polymer insulators that are positioned on a pole or structure to support conductors in a vertical or horizontal cantilevered position. **See Figure 8–47**. Post insulators come in all lengths and strengths. The insulators come with tops where the conductor can be tied in or clamped. Insulators are often installed before poles are set. Post insulators are especially vulnerable to damage because the insulators get dragged and bumped against objects as the pole is moved around.

Figure 8–47 Post insulators are designed to support conductors in the vertical and horizontal position. *(Delmar, Cengage Learning)*

Stringing Conductor

Stringing conductor can be one of the more complex jobs in power line work. Taking time to document a job plan may help you to recognize the many hazards that are likely to be present. Complexities include road and railroad crossings, an extra-long span, a heavy corner, energized underbuild, induction from nearby transmission lines, and deciding on the best locations to ground.

Slack Stringing

Slack stringing is appropriate when there are no energized conductors in the vicinity. **See Figure 8–48**.

Figure 8–48 Conductor is being run through stringing blocks in a slack stringing operation. *(Delmar, Cengage Learning)*

Slack Stringing or Removing Distribution Conductors Near an Energized Circuit

Hazards	Controls
While stringing a conductor in the vicinity of an energized circuit, there is a chance of the conductor making an inadvertent contact.	Tension stringing techniques should be used when stringing over or at the same level and on the same side of the structure as an energized circuit. If slack stringing underneath an energized circuit, use tie-down ropes midspan to prevent a snagged conductor from whipping up and making contact with the energized circuit.
Workers on the ground are exposed to possible electric shock while stringing in the vicinity of an energized circuit.	If slack stringing, the reel stands should be grounded and put on a ground-gradient matting or on a grounded trailer. Anyone tending the reels must remain on the ground-gradient matting or trailer. An operator of a reel trailer, tensioner, or puller must stay on the operating platform or on a ground-gradient mat that is bonded to the machine. The conductor must be held with grips and protective grounds installed before changing reels or doing other work at the tensioner or puller end.
A conductor makes contact with an energized circuit while a worker is tying or clamping it to an insulator.	Rubber gloves and sleeves must be worn by anyone handling the conductor unless they are in a bonded zone.
The energized circuit may not trip out when a conductor contacts the energized circuit.	A stringing procedure and precautions taken should recognize that a conductor can become energized and that it may not trip out the circuit. Ground the **stringing blocks** to the system neutral at regular intervals. Use traveling grounds at the reel stand or tensioner end.

Stringing across Roadways

Reduce the risk of a vehicle making contact with conductors being strung across a roadway by considering the following:

- Road-crossing spans are often longer than other spans in the line being strung. A lot of conductor will gather in an extra-long span and cause excessive sag. Consider stringing the road-crossing span by itself instead of it being just another span in a long pull.
- On high-volume roads, consider using road authorities and their equipment to channel and control traffic. The presence of a highly visible police vehicle may add to the effectiveness of traffic control.
- If tension stringing, use backup barriers such as guard structures/rider poles or a rope basket to catch conductors before they drop onto the road. Slow traffic by channeling vehicles into a single lane so that they can be stopped quickly if a conductor drops too low.
- If slack stringing, stop traffic before running a rope or wire across the road. Stopped traffic in each direction can serve as a barrier for traffic entering the work zone.
- A pole can break when a passing vehicle catches onto a conductor or when a running board catches on a stringing block. No one should be on a pole when the conductor is in motion. To free a running board from a stringing block, move the conductor only far enough for a person to go aloft to fix it.

Sagging Conductor

When the sag for a line is specified, the maximum tension on the conductor is in relation to its rated tensile strength (RTS) and its potential vibration, potential galloping, elasticity, tension, clearance from the ground, and ruling span, as well as temperatures, wind, and ice conditions (the loading district) that are likely to be encountered. Conductor creep, which is the final stretching and strand settling (up to a period of 10 years), is also a factor in specifying sag.

Sag charts are found in design standards for distribution lines and are usually part of the design specifications for a new transmission line. To measure the sag in a line, reference is made to the sag chart for the day's temperature, span length where the measurement is to take place, and the ruling span. The ruling span (also called the equivalent span) is an assumed uniform span that most closely represents the various lengths of spans that are in the section of line being sagged. Using the ruling span model to calculate sag will allow the best average tension throughout a line with varying span lengths and changes in temperature and load. The calculations for a ruling span are complex because the conductor sag takes the form of a catenary curve. The ruling span is given as part of the specifications for a transmission line. For less critical lengths of distribution line, an estimate of a ruling span can be made with the following formula:

$$\text{Ruling span} = \text{Average span} + \frac{2}{3}(\text{maximum span} - \text{average span})$$

Lines must be sagged to the specifications in the "initial sag" or "stringing sag chart." Sagging must be done accurately. If the section being sagged is long, measurements are needed in more than one span. The conductor can get quite tight where it is being pulled up before it starts to move at the end of the line or out of a long span. When the conductor gets close to sag, it should be taken up very slowly because if the conductor gets too high a lot of slack must be put back to get it to come down again near the end of the line.

Note: A lot of the distribution conductors are not sagged to specifications. Line crews tend to sag until it looks good (cornfield sag), which is usually too tight. Conductors that are sagged too tight in the summer may lead to structural failures (dead-ends, corners, guying) in winter and, in other cases, to excessive vibration. The preceding information should indicate that there is a lot of science involved in determining the proper sag—and that looking good is not one of the parameters.

The most common method of sagging has been the target method, that is, using sag boards. Once the temperature and the length of the span chosen for the measurement are known, sag charts will give the specified sag for the conductor size. The sag measurement is taken from the conductor support to a spot down on the structure where a board (a rolled-up flag or other target) is placed. The same is done at the next pole in the measured span, and the two boards can be used to see when the bottom of the sag lines up with the boards. On long transmission lines, a scope—instead of a board—makes it much easier to see and be more accurate, especially for the shield wire. **See Figure 8–49**.

The sag on transmission lines is often measured with surveying equipment where no climbing is involved.

Figure 8–49 The target method is being used to measure sag. *(Delmar, Cengage Learning)*

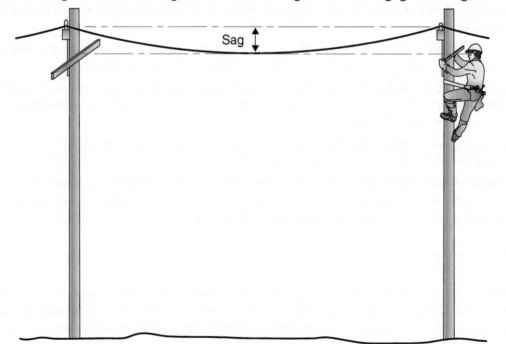

Measuring tension with a dynamometer is the best method for sagging messenger/strand cable, which is strung at a high tension. It is probably not very accurate for power conductors unless the tension more than the span measured where the conductor is pulled up. Depending on the length of the pull, the tension must be checked at multiple spans.

The return-wave method can also be used to check the sag on an existing line. This method is applicable regardless of the span length, tension, size, or temperature or the type of conductor. Start a wave in a conductor by pulling on it with a telescopic stick or jerking down on a rope over the conductor. The wave travels to the next structure and is reflected back and forth until the wave is eventually damped out. Record the time it takes (there is a special stopwatch available for this purpose) for three, five, or ten return waves.

The waves on long spans and large conductors can be more accurately counted than short spans and small conductors. The largest number of return waves minimizes errors in recording time. The initial impulse does not count as a return wave. Getting two equal results will give some assurance of accuracy. **See Figure 8–50**. The return-wave method is probably not the best while initially sagging because it does not give a continuous indication of where the sag is and because pulling has to be stopped each time a measurement is made. It is a very good method for checking sag on an existing line.

Figure 8–50 Return-Wave Method Chart

Sag in Inches (cm)	Return of Wave in Seconds		
	3rd Time	5th Time	10th Time
24 (0.61)	4.2	7	14.1
30 (0.76)	4.7	7.9	15.8
36 (0.91)	5.2	8.6	17.3
42 (1.07)	5.6	9.3	18.7
48 (1.22)	6	10	19.9
54 (1.37)	6.3	10.6	21.1
60 (1.52)	6.7	11.1	22.3
66 (1.68)	7	11.7	23.4
72 (1.83)	7.3	12.2	24.4
78 (1.98)	7.6	12.7	25.4
84 (2.13)	7.9	13.2	26.2
90 (2.29)	8.2	13.7	27.3
96 (2.44)	8.5	14.1	28.2
102 (2.59)	8.7	14.5	29.1
108 (2.74)	9	15	29.9
114 (2.90)	9.2	15.4	30.7
120 (3.05)	9.5	15.8	31.5
126 (3.20)	9.7	16.2	32.3
132 (3.35)	9.9	16.5	33.1
138 (3.51)	10.1	16.9	33.8
144 (3.66)	10.4	17.3	34.5
150 (3.81)	10.6	17.6	35.2
156 (3.96)	10.8	18	35.9
162 (4.11)	11	18.3	36.6
168 (4.27)	11.2	18.7	37.3
174 (4.42)	11.4	19	38
180 (4.57)	11.6	19.3	38.6
186 (4.72)	11.8	19.6	39.2
192 (4.88)	12	19.9	39.9

Tying or Clipping-In on Distribution Lines

After a conductor is sagged, it must be lifted out of the stringing blocks and tied in or clipped into the insulator(s). On distribution lines, the conductor is often lifted out of the stringing block and onto the insulator by hand. When working from the pole, lifting a heavy conductor by hand can put the body in awkward positions and result in musculoskeletal injuries. Gins are available to make this lift using mechanical aids.

Conductors are connected to insulators, using hand wire ties, preformed ties, or saddle clamps. Hand wire ties and preformed ties are used with top tie pin insulators and top tie post insulators. Aluminum tie wire is used on aluminum wire and copper on copper. Aluminum tie wire tends to be #4 soft drawn. Ties for a conductor smaller than 3/0 tend to be #6 aluminum. Set patterns are specified regarding how a conductor is to be tied in. **See Figure 8–51**. Other ties are specified for corners and spool insulators. Tie wire or preformed ties are intended to keep the conductor in the groove position of the insulator without having any constant strain on the tie wire. If a conductor will not stay in the insulator groove, such as at a corner, then the conductor should be tied on the side of the insulator. The tie should be secure and hand tight. If a pigtail is used to finish a tie, excessive twisting with pliers will damage the tie wire and over time it will break. Specifications books often give the length of tie wire needed for each size of conductor.

Figure 8–51 Two typical insulator ties, a pigtail tie which is used for conductors over 3/0, and a long tie for smaller conductors. *(Delmar, Cengage Learning)*

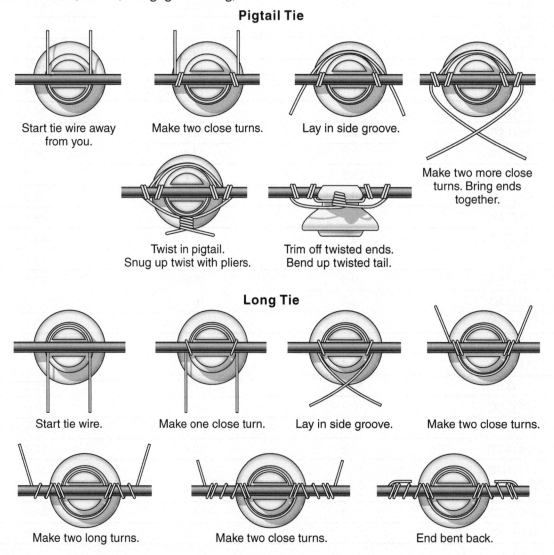

Some locations will prefer preformed or factory-formed ties, which are designed to tie the conductor in very securely. Preformed ties are less dependent on a worker always making a tie correctly. Factory-formed ties also serve as **armor rod** and help protect the conductor from broken strands due to vibration.

The conductor is subject to vibration. It can be damaged by abrasion where the outside surface of the conductor will wear away, by "fretting," where rubbing action between the strands causes internal damage, or by fatigue, where individual strands break. An armor rod over the conductor may be specified to reduce vibration and reduce the risk of broken or worn conductor strands at the insulator. Armor rods form another layer over the conductor and are like preformed ties where the rods are wrapped around the conductor. There is a rule of thumb to determine the length of tie wire needed to tie in a conductor over the armor rod. Cut the tie wire in feet equal to the number of rods in a set; for example, #4 armor rods have seven rods per set, so cut 7 ft 2 m of tie wire, and 1/0 rods have nine rods per set, so cut 9 ft 2.7 m of tie wire.

Saddle clamps are used on suspension insulators and can be found on most post insulators. Instead of a design with two U bolts, four washers, and four nuts, there are saddle clamps designed to easily accommodate a live-line tool wrench stick with one or two convenient bolts, where the nuts are captured in the conductor clamp. **See Figure 8–52.**

Figure 8–52 A single-bolt saddle clamp designed for easy maintenance with a live-line tool. *(Delmar, Cengage Learning)*

Tension Stringing

The probability of making contact with an energized circuit is relatively high when stringing in the vicinity of an energized distribution circuit. When using tension stringing techniques, the consequence of a worker having an electrical contact is usually low, because the set-up of the stringing equipment, grounds, and ground-gradient matting will prevent injuries when accidental contact is made. The tension stringing principles for transmission lines are similar to this distribution procedure. **See Figure 8–53.**

Figure 8–53 A view of tension stringing from the puller end; the energized conductors are set out on extension arms. The tensioners and conductor are set up at the other end. *(Courtesy of Wilson Construction Company)*

Typical Distribution Tension Stringing Procedure

Step	Action	Details
1.	Ensure that there is adequate spacing to allow stringing a new conductor.	**1.** To install a new circuit above an existing circuit, higher poles must be installed. The space above the energized underbuild should be at least 5 ft (1.5 m) and ideally 10 ft (3 m). See **Figure 8–54**. **Figure 8–54** Energized conductor on auxiliary arms. *(© Van Soelen)*  **2.** To replace existing conductors, install auxiliary arms and set out the existing energized conductors. The closest conductors should be at a minimum of 3 ft (1 m) to the side. **3.** To cross over energized circuits mounted on the same pole, the necessary clearance can be smaller because the energized circuit can be covered up. Crossing energized circuits in-span requires more clearance, but additional protection can be obtained by installing cover-up.
2.	Install stringing blocks.	The finger line (throw line) is a small rope put through the stringing blocks and tied off down the pole. The finger line is used later to pull the **pilot line** or **pulling rope** through the stringing block. **1.** Stringing blocks are normally installed when any necessary framing is done. Finger lines installed through the stringing blocks during installation will save additional set-up. The finger lines must be tied off above public reach and clear of energized underbuild. A clean finger line, in combination with rubber gloves, will tolerate momentary brush contact with energized conductors. **2.** On multisheave stringing blocks, keep the finger lines in the center sheave. **See Figure 8–55**. A pilot line has already been strung through the center sheave. **3.** Corner stringing blocks should, if possible, be set out so that the conductors do not have to be moved into place before sagging. **4.** Install grounds, as required by your utility, on the first and last stringing block and at regular intervals. **5.** *Note:* Friction from stringing blocks is a large factor in determining the pulling tension. Pulling Tension = Line Tension + (% Friction of Block Line Tension × Number of Structures) For example, pulling tension for a 20-span pull, with a 3% friction of blocks and line tension of 500 lbs = 500 + (0.03 × 500 × 20) = 800 lbs.

2.	(Continued)	**Figure 8–55** A three-sheave gang stringing block with a pilot line in position to pull in the bull rope. *(Delmar, Cengage Learning)* 
3.	Set up the puller.	1. To reduce the down weight on a structure, keep the puller back at a distance of at least one third of the span length. 2. Set up the puller so that it is in line with the pull or so that the installation of snatch blocks will allow the pull to be in line. 3. Ensure that the operator and anyone else near the machine will be on an operating platform or on a ground-gradient mat bonded to the puller. 4. Install a barrier around the puller to keep people away during the stringing operation. The electrical hazard at the puller increases when the conductor comes into the puller. 5. Use a reinforced steel reel made specifically for winding up the pulling rope under tension. Other reels can be crushed by the elastic properties of the rope. 6. Have protective grounds ready to install on the conductor when it comes to the puller.
4.	Set up the tensioner.	The tensioner is the reel carrier for the conductor and holds tension on the conductors by pulling back the conductors hydraulically or by applying brakes. 1. Set up the tensioners so that the conductors will be in line with the first pole or structure. **See Figure 8–56**. **Figure 8–56** The view from the controls of one type of three drum tension machine. *(Delmar, Cengage Learning)* 

Step	Action	Details
4.	(Continued)	2. Place ground-gradient mats so that the machine operator and people involved in changing reels are working from the mat bonded to the tensioner. 3. Install a barrier around the tension machine(s) to keep people away during the stringing operations. **See Figure 8–57**. **Figure 8–57** A ground grid and temporary fence is installed at this tension machine. *(Delmar, Cengage Learning)*  4. Have protective grounds ready to install on the conductors when the pull is stopped. For additional grounding during stringing, install traveling grounds. 5. Wooden conductor reels can collapse and cause a sudden loss of tension. It is generally acceptable to tension directly from a wooden or metal reel at tensions less than 1,000 lbs (450 kg). Use a metal mandrel all the way through a wooden reel to prevent a total collapse. Use a bull-wheel tensioner for higher tensions.
5.	String in the pilot line.	A pilot line is a small rope strung from the tensioner, through the stringing blocks to the puller, and used to pull back the heavier pulling rope. **See Figure 8–58**. (**Note:** The pulling rope may be pulled out directly without the use of a pilot line.) If stringing wire through multisheave stringing block (e.g., three sheaves), only one pilot line is needed to pull in one pull rope. 1. As the pilot line is strung and fed through the stringing blocks, it will likely make occasional contact with any energized underbuild. The first 150 ft (50 m) of the pilot line should consist of live-line rope or clean, dry, synthetic rope. Rubber gloves must be worn by the handlers of the rope. 2. Guard all road crossings during the stringing of the pilot line. **Figure 8–58** The different colors of the ropes on the pilot line winder reels help to keep the ropes in the proper position when stringing. *(Delmar, Cengage Learning)* 

6.	Pull the pulling rope back to the tensioner with the pilot line.	The *pulling rope* is used to pull in the conductor. It must be strong enough for the job, have very little stretch, and be clean enough to withstand brush contact with energized conductors. **1.** Connect the pilot line to the pulling rope. **2.** Using a tensioner, pull the pilot line from the puller back to the tensioner.
7.	Pull in the conductor with the pulling rope.	**1.** Use a minimum length of pulling rope to make the pull. Work as close to the drum core as possible because, unless it is a bull wheel puller, a large-diameter reel of pulling rope will require more pulling tension. **2.** The pull rope will be connected to one conductor or to a running board that may have more than one conductor attached to it when used on a transmission line. The connection to the pulling rope is made with a swivel joint. **See Figure 8–59**. If the swivel is not performing well, a running board will start flipping over in a span, often more than once, which creates a lot of work to unwind. **Figure 8–59** The swivel joint must be able to swivel under high tension. *(Delmar, Cengage Learning)* **3.** Establish radio communications between the puller, the tensioner, observers at road crossings, and workers following the running board or rope-to-conductor joint. If possible, use an exclusive radio channel for stringing. Only the puller can stop the pull, and only the tensioner can adjust the stringing tension or sag. **4.** Apply tension at the tensioners. Hydraulic tensioners are put in the take-up mode to tension the conductors. Tensioners that use braking to hold the conductors must have brakes that are designed for continuous braking and that will not overheat and fade. **5.** Pull in the pulling rope on the puller. **6.** When the conductor arrives at the puller, install grounds at both ends.
8.	You may experience problems while pulling in the conductor(s).	**1.** If the conductor sag is fluctuating and hard to control, the design of the pulling rope has too much stretch. The elastic-like action is causing the tension and sag to change constantly. Use a rope specifically designed for tension stringing. **2.** On a multiconductor pull, the running board can flip over between spans even when a swivel is used. The pulling rope should be designed not to rotate. **3.** If there is a long span within the pull, such as a wide road crossing or river crossing, and it is difficult to maintain the high tensions needed to maintain clearance from underbuild, that span should be strung separately. The weight of the long span will keep the sag in shorter spans excessively tight. **4.** A conductor smaller than 3/0 will cut into the lower layers of a conductor reel. Unless a tensioner is equipped with a bull wheel to remove the tension from the reel, small conductor should not be strung. **5.** If a tensioner is tensioning the conductor directly from a reel, the braking action must be reduced as the reel empties and the circumference gets smaller. **6.** For a heavy conductor on a long pull, tension can be lost when brakes heat and fade on the tensioner, so a tensioner with a hydraulic retarding system should be used. **7.** If the puller seems to be pulling at a very high tension and it is not tight at the tensioner: • The stringing blocks may be too small in diameter and cause extra loss due to the bending and straightening of the conductor as it passes over each stringing block. • There may be too much friction on the stringing blocks, too many corner structures or changes in elevation for the length of the pull. Losses in efficiency are additive. Even a good stringing block with a 2% loss at each suspension point can, after 15 structures, increase the total tension of the pull approximately 30%. • A flexible mesh pulling grip on the pulling rope can let go when it is not properly installed. The diameter of the rope will decrease considerably under high tension. Use a fiber or wood plug in the core of the rope to maintain the rope diameter.

(Continued)

Step	Action	Details
9.	Change conductor reels at the tensioner as required.	All workers must be completely in the bonded work zone or completely out of the bonded work zone. 1. To allow reels to be changed at the tensioner, the ground-gradient mat set-up should be big enough to accommodate the work. Protective grounds are installed on the conductor and bonded to the ground-gradient mat. If a vehicle boom is used from outside the bonded zone, the vehicle must be bonded to the work zone. The boom operator must stay on the platform, and all others must stay away from the vehicle. 2. However, *if* the pull is stopped, protective grounds are installed at the tensioner end, the conductors are temporarily dead-ended in grips, and the length of the pull is patrolled for any potential electric contact hazards, *then* the reels at the tensioner could be changed with a reasonable assurance that the conductors will not become energized.
10.	Free the running board or conductor when it snags onto a stringing block.	1. While the conductor or pulling line is being pulled (in motion), no one should be aloft on a pole or structure or aloft in the vicinity (in a bucket), except during a small amount of adjustment needed to guide the stringing sock or board through a stringing sheave. 2. When handling conductors during a tension stringing operation, use rubber gloves or apply protective grounds and/or bonds to the conductors and structure at the point of work.
11.	Sag the conductor.	1. Set the corner conductors in stringing blocks as close as possible to their final positions. 2. Sag the conductors using the tensioners, or bring the conductors in close to sag using the tensioners and finish sagging with chain hoists or a derrick winch.
12.	For long pulls that require more than one set-up, dead end the conductors temporarily.	1. Temporary dead-ends require standard anchoring and guy wire to hold the conductor tension. The use of rope guys or unsupported pole butts can lead to sagging into energized underbuild. 2. Energized underbuild conductors will need cover-up and/or setting out to provide clearance for the guying. 3. The slope of the guys must be adequate to prevent unacceptable downward stresses (e.g., on crossarms, insulators).
13.	Tie in or clamp in the conductor.	Use rubber gloves or apply protective grounds and/or bonds to the conductors and structure at the point of work.

Stringing Secondary Bus

Secondary bus is strung where it is more economical to install one transformer and feed up to 15 customers. Secondary bus is common on urban streets and in rural areas where more than one customer is located in a group.

Open wire bus can be found strung with bare wire (weatherproof) or insulated wire. Open bare wire bus is the most economical to install and can be installed vertically on three-point racks or on crossarms. It is still found strung on crossarms in some cities in North America and is very common in countries such as Australia. Open bare wire bus takes up a lot more space on a pole than open insulated wire bus because it requires good spacing to prevent conductors from slapping together midspan. **See Figure 8–60**. Dead ending a service midspan from an open wire bus works in locations with short spans and small, short services. More often all services from open wire bus are strung from a pole.

Aerial cable bus has insulated secondary wires lashed to a messenger (strand). The messenger wire can be 3/0 AASCR, which can serve as the neutral and has the strength to be tensioned very tightly. Because the messenger wire is tight, services can be strung to customers from midspan, and side guys are not needed. **See Figure 8–61**. It is important to prestress and string the messenger cable to the specified tension. If a messenger cable loses tension after the secondary cables are lashed to it, the tension on the secondary cables becomes excessive, making it very difficult to get enough slack to make service connections. It is also a good idea to test the anchors to ensure that one will not let go while someone is aloft as well as to remove some of the creep that would develop later. Preassembled aerial cable, which has the cable wrapped around the messenger instead of underneath, is also available. Terminating a service midspan requires terminating the lashing

Figure 8–60 The triplex service is dead-ended midspan on an open wire bus. *(Delmar, Cengage Learning)*

Figure 8–61 The triplex service is dead-ended midspan on aerial cable bus. *(Delmar, Cengage Learning)*

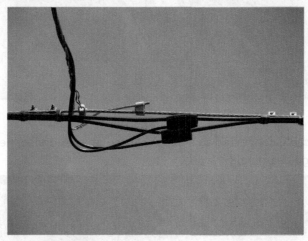

wire on each side of a new service, by dead-ending and pulling slack into the bus wires. If working on a hot secondary, it is wise to use cover-up, because all the connections are close and the neutral and lashing wires are bare.

Triplex and quadruplex cable is also used as secondary bus. Secondary bus can be 120/240 V, 120/208 V three-phase, and, in industrial areas, 277 V or 480 V three-phase. **See Figure 8–62.**

Figure 8–62 The three-phase quadruplex cable is dead-ended mid span to three-phase 120/208 V aerial cable bus. *(Delmar, Cengage Learning)*

Stringing Service to the Customer

A customer's electric service can be strung and connected directly from a transformer pole, strung and connected to a bus, or run down a pole to an underground service. **See Figure 8–63**. At the customer end, an overhead service is connected to a service stack (**see Figure 8–64**), and an underground service is brought up directly into a meter base.

Figure 8–63 This transformer is feeding one service and the service is connected directly to the transformer pole. *(Delmar, Cengage Learning)*

Figure 8–64 A 1/0 triplex cable connected to a 200 A service at a residence, nearby are the telephone and CATV wires. *(Delmar, Cengage Learning)*

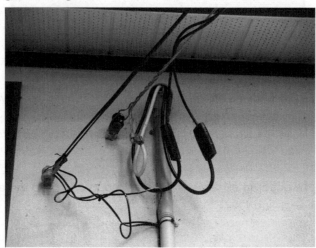

The most common overhead service conductors are a neutral supported cable, such as triplex and quadruplex. Insulated conductors are wrapped around a bare steel reinforced (ACSR) neutral wire. Because the service cable is accessible to the customer, the connectors at the customer end must be taped or covered well.

The size of wire used for a service is somewhat determined by the size of the electrical load. The size of wire also depends on the length of the service and the potential voltage drop over that distance. Utilities like to standardize on only two or three sizes of triplex with a tendency to provide a size that will supply a minimum 200 A service. Underground service wires tend to be larger to accommodate future load growth without the need to dig up the service. For very large services, a double run of service wires can be installed. When installing a double run, it is fairly critical to have both wires exactly the same length; otherwise, the shorter wire with slightly less impedance will carry a larger share of the load.

The minimum ground clearance for a secondary service is specified by government codes. The U.S. National Electrical Safety Code requires a service cable to be 12 ft (4 m) over sidewalks, 12 ft (4 m) from the ground line to the lowest point of building attachment, and 16 ft (5 m) over streets and roads. A service strung across a street with less than 16 ft (5 m) ground clearance may sag below the maximum height of a truck, at 13 ft 6 in. (4.2 m), during an ice storm or on a hot day with a heavy air-conditioning load. In other words, clearances over areas, such as roadways, are given the clearances with reference to loaded conditions.

When dead-ending a service at the utility pole, the tail of the service wire is often long enough to reach and contact energized primary conductors on the pole. Tie a rope to the end of the service wire to maintain control. Triplex service wires larger than #2 are typically pulled up with a set of two sheave blocks or with someone on the ground using a handline.

The most common connectors are insulated compression sleeves (insulinks). Split bolt connectors are notorious for being an eventual source of trouble calls. If a triplex service is to be spliced, the neutral will require a proper sleeve that can withstand full-line tension, not as seen occasionally where an insulink is used. An insulink is an electrical connection and is not designed for line tension.

There are a variety of tests that should be carried out depending on the situation.

1. Always do a voltage test, usually at the meter base, to ensure that the correct voltage is being supplied to the customer.
2. If a socket-based meter is being installed at an energized meter base, test the load side of the meter base to ensure that there are no short circuits. Installing a meter on a short circuit can be explosive. Use a meter puller/installer.
3. The integrity of the service can be tested with a 500 V insulation tester.
4. It is not always clear on some old services to identify which wire in the stack is the neutral. Connecting a hot leg to the customer neutral can and has resulted in burning down the building or burning out all the customer motors. With a building where the neutral is not identified, use a voltmeter connected from a hot secondary and test to each of the three wires in the service stack. With the customer breaker open, there should be voltage to the grounded neutral only.

Some older existing services (open wire service) consist of single wires with each wire terminated on a separate rack, keeping the wires separated. These wires are often just weatherproof wires, and if they slap together in the wind they result in a short circuit. The terminations at the customer end are somewhat of an electric shock hazard to a customer working on the house near the rack.

Constructing an Overhead Transmission Line

Deciding the Route

The decision making to establish a route for a new transmission line involves technical, environmental, and political issues. Technically, the route must go from one terminal to another. The most direct route, with a minimum number of corners and dead-ends, would be the most economical, but parks, homes, highways, pipelines, wetlands, and more affect the route.

Environmentally, a route must avoid wetlands, certain other habitats, and densely populated areas. Permission is needed to clear all trees and brush and to remove from the right-of-way all trees that could contact the line if they were to fall.

Many people will object to any route chosen. They will express concerns through the political process about health effects of the magnetic fields, visual pollution, and impact on property values. Political backing is needed when property is bought or easements obtained for a line and for access to each structure.

Designing the Line

Whenever a line is designed, a roll plan is produced. This plan shows the right-of-way measured from one end to the other (chainage) and the location of each structure. Another roll plan illustrates the profile of the line and shows the terrain, the height of each structure, and the sag in each span. A multitude of subassembly drawings and close-ups show all assembly methods, including foundations, structures, insulator strings, vibration dampers, shield wire attachments, and electrical connections. Tower loading diagrams will help when planning how to rig a tower for erection and for raising conductors.

An important consideration when designing a line is how to reduce the risk of the domino effect (cascading) occurring when one transmission-line structure goes down. Some types of structures, such as a wood-pole "H" frame, have very little ability to resist a longitudinal force; thus, additional dead-end or anchor structures are strategically inserted into the line. Some types of towers are deemed to have acceptable longitudinal strength to avoid cascading, such as a rigid, square-based latticed tower; a V, Y, or delta tower; and some types of steel poles. The swing of a suspension insulator string or the conductor slipping through a suspension clamp also reduces force on adjacent structures.

Clearing the Right-of-Way

Transmission lines transverse almost every kind of terrain, including mountains, deserts, forests, lakes, farm fields, and cities. The right-of-way must be cleared for construction and made suitable for maintenance for years to come. Temporary roads must be built into individual structures, although the use of helicopters allows structures to be built in all kinds of terrain with no need for access roads. Trees must be cleared and, if practical, the right-of-way should be grubbed and compatible ground cover should be planted.

Line crews must be informed about any special arrangements made with property owners. It has not been unusual for a line crew to drive through or over areas where agreement had been reached with the property owner that the area would be avoided.

Right-of-ways can cross valuable property and have compatible secondary uses. Parks, vehicle parking, and certain building are encouraged as sources of revenue for utilities. It is fairly common for power line up to and including 230 kV to be strung over industrial or commercial buildings, but not higher-voltage lines such as 345 kV, 500 kV, and 745 kV. Utilities have staff dedicated to handling requests for secondary use of transmission-line corridors.

Constructing the Structure Foundations

The foundation types for transmission towers include grillage, pad and pier, and augered. **See Figure 8–65**. In addition to the foundation needed to provide a solid and level base, it also must counter large uplift and down-thrust loads, especially dead-ends, corners, and towers subjected to wind and ice. The civil work involved with putting in a foundation has to be very precise and is normally carried out on construction projects by specialized crews.

Formerly, most steel-lattice structures had grillage footings, each with a good base width and a hole filled with the same material that was excavated. Today, the footing almost always involves excavating a hole for each leg, installing caissons, forming reinforced steel, putting in bolts to which the structure will be attached, and pouring concrete. The layout has to be very precise, and the portion above ground must be formed to make a reasonably attractive finish. **See Figure 8–66**.

The foundations of steel towers are grounded electrically to either counterpoise (wire such as 1/0 copper buried the length of the line) or crow's foot counterpoise (copper buried around the base of the tower in a crow's-foot pattern). The pole/down-grounds on wood poles are grounded to ground rods at the pole base. Other methods of grounding wood poles, such as a "hot plate" (wire fastened to the bottom of the pole) or a "butt wrap" (wire wrapped around the buried portion of the pole), are not nearly as effective as a ground rod.

Assembling and Erecting Structures

There are many different transmission structure designs for each transmission voltage. Pole structures include twin-pole wood (H-frame), twin-pole gulf port, single-pole wishbone, and single pole with a variety of post insulator arrangements. Steel-lattice structures include single circuit with many different designs, double circuit with many designs, and guyed V structures. Each design also has heavy anchor structures for dead-ends, heavy angle structures, and light angle structures.

Assembling the Structure

The work involved in assembling structures depends on the type of structure and on the method of structure erection. Steel-lattice structures are assembled in sections with a small crane used to lift and move the sections around as needed. When the structure is to be set by crane, the lowest section is placed on the foundation and the other sections

Figure 8–65 The type of foundation chosen to support a transmission structure is dependent on the soil conditions and the size of the intended structure. *(Delmar, Cengage Learning)*

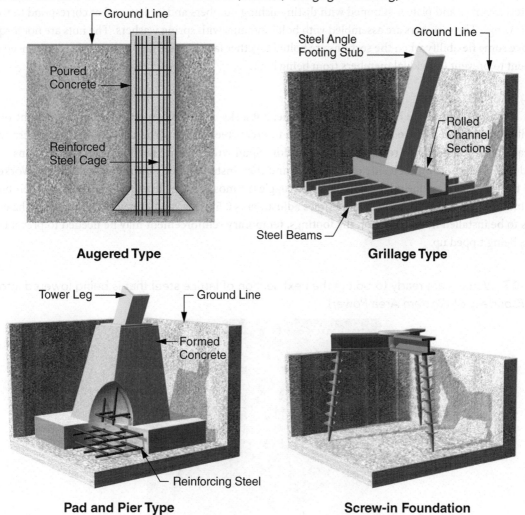

Augered Type

Grillage Type

Pad and Pier Type

Screw-in Foundation

Figure 8–66 The foundation for a large unguyed transmission line steel pole must be engineered to withstand huge stresses. *(iStock.com/Steverts)*

are assembled nearby. If the structure is to be set by helicopter, assembly yards (fly yards) are usually strategically placed where the structures are assembled to be flown to the structure site later.

Each steel member and plate is stamped with distinguishing numbers and/or letters that correspond to those on the structure drawings. The members are assembled with bolts and nuts with spring washers. The nuts are not torqued at this stage, to leave some flexibility when the sections are bolted together later. Sometimes individual sections need temporary reinforcement to prevent individual members from being bent.

Erecting the Structure

Transmission-line structures are most often set by crane, but a sky crane helicopter is common for right-of-ways that are difficult to access. **See Figure 8–67.** It is common to erect steel-lattice structures in sections where workers are placed at each leg of the tower to receive the next section. Spud wrenches are used to align two sections by inserting the pointed end into a hole on each side of the corner and then installing bolts into the vacant holes. Workers quickly realize that the last person to line up the holes has to struggle the most, so they usually compete to get their bolts in first. Good communication is needed because sometimes adjustments 0.5 in. or smaller have to be made by the crane. If the structure is to be installed in one piece on the footings, temporary reinforcement may be needed to prevent damage as the tower is being tipped up.

Figure 8–67 Workers are ready to bolt in the next section of lattice steel that is being lowered into position by crane. *(Courtesy of Western Area Power)*

If steel-lattice towers are erected in sections by helicopter, it is best *not* to have workers on the structure. Ideally, temporary brackets have been installed in each section, so each section can be set up while no one is on the tower. **See Figure 8–68.** Wood and steel poles can be set into pole holes with a helicopter, much like the distribution poles described earlier in this chapter.

Large steel-pole structures are visually more acceptable to the public than steel-lattice structures. The very large steel poles come in sections with a slip joint between each section. The sections must be pulled together to form a specified overlap. Steel poles are heavy, and lining up each section requires great accuracy. Each male and female section will have a mating mark to help with proper orientation and splice overlap. Such poles need more engineering than steel lattice because, like distribution poles, they flex and can be damaged very easily by improper rigging. If a circuit is to be strung on only one side, steel poles are sometimes set to counter the eventual deflection that would cause an unsightly lean

Figure 8–68 No one is needed on the structure as the skycrane helicopter is setting the top section on a steel transmission line pole. *(Courtesy of POWER Engineers)*

toward the loaded side. The pole is actually bent (or cambered) to offset the calculated expected deflection so that it will look straight after the wire is strung. Wood poles can be framed in advance. In some cases, especially when erecting by helicopter, even twin-pole structures can be assembled with the crossarms in place while leaving one cross brace loose to allow plumbing the structure after it is set in the holes.

Other methods are used to erect steel-lattice structures because they are manufactured in many small pieces. In the past, poles or steel-lattice structures have been set with a gin pole. **See Figure 8–69.** Today that would be a rare event, and the knowledge and skill needed to erect a steel-lattice tower with a gin pole are probably lost in most locations.

In brief, a gin pole (wood or steel lattice) is set up in the middle of a future tower. Four rope blocks are attached to the butt and tied to each tower leg. Four rope blocks are attached to the top of the pole, and the top of the pole can be moved to a location within the tower so that it is above the point for a lift of steel.

In places where fall arrest is used, the rope used by the fall arrest system is attached to the top of the structure before it is erected.

Hang Insulators and Stringing Blocks

Transmission-line insulators are most often porcelain (ceramic), glass, and polymer in suspension. Post insulators and braced post insulators are common when a line is built on a narrow right-of-way or along a highway. When braced posts are used, a post and suspension insulator supports heavier conductor in a fixed position and because the insulators do not swing, the braced posts can be used on smaller structures which would have less visual impact. **See Figure 8–70.**

Some structures are designed with a V arrangement of the insulator strings. A V arrangement is more likely to self-clean because the contaminants on both sides of each string are likely to be washed off by rain.

Insulators, stringing blocks, and, in some cases, ropes through the stringing blocks are sometimes installed before a tower is erected, but more often they are installed after the tower is erected. When sending up a string of insulators, a cradle or other device that can send the string up straight will prevent excessive bending of the string and damage to the ball, socket, and cotter key between the insulators. Polymer insulators are much lighter and easier to work with, but can be abused easily with rough handling.

Figure 8–69 When there is no crane or helicopter available, a tower can be raised with a gin pole. *(Delmar, Cengage Learning)*

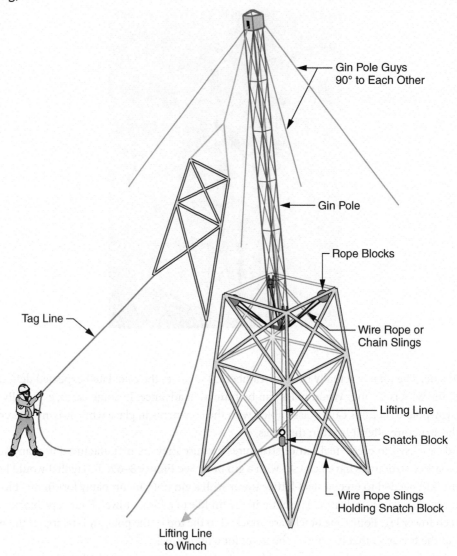

Gin Pole Guys
90° to Each Other

Gin Pole

Rope Blocks

Wire Rope or
Chain Slings

Tag Line

Lifting Line

Snatch Block

Wire Rope Slings
Holding Snatch Block

Lifting Line
to Winch

Figure 8–70 The use of brace and post insulators allows the double circuit 230 kV to be constructed on a narrower right-of-way. *(Delmar, Cengage Learning)*

Climbing up and down insulator strings when constructing a new line has been common in the past and may have been acceptable with porcelain insulation, especially if the feet were crowded in as close to the steel hub as possible. OSHA does not allow anyone to climb on suspension insulators for fear that a skirt will break and cause a fall or severe cut. Toughened glass insulators shatter completely when broken, porcelain insulators typically break off in chunks, and ceramic insulators also tend to shatter. Climbing on polymer insulators can damage the skirts, and the damage can be overlooked easily.

Stringing blocks for stringing bundle conductor come in single, double, and triple sheaves. **Figure 8–71** shows a multi-sheave stringing block designed to accept a pilot line dropped in by helicopter.

Figure 8–71 A multi-sheave stringing block is designed to accept a pilot line dropped in by helicopter. *(Courtesy of Sherman-Reilly)*

The diameter of a stringing block must be such that it does not damage the conductor. The strands of a large conductor will flex and loosen excessively as it rolls through the block. Smaller-diameter stringing blocks may be acceptable for a resagging job where the conductor is under line tension and will not travel very far.

Install Rider (Guard) Poles

On new lines, virtually all conductor stringing is done under tension. In a perfect world, the conductor could be strung and kept clear of highways and other critical spans. However, it is essential *not* to lose control of a conductor, so a backup system is necessary for preventing conductors from dropping down on roadways, railways, and circuits crossing below.

Guard structures or nets are installed at roadways, railroad crossings, and at locations where crossing circuits lie below. The guard structures are generally two- or three-pole structures with rider poles suspended horizontally between them. **See Figure 8–72.** Single poles under each conductor with crossarms mounted on top in V formation are also common (V poles). Another type of rider that is effective for shield wire stringing or restringing (when power conductors are in place) is a length of ⅜ in. (9.5 mm) or 5/16 in. (8 mm) steel strung across the three power conductors and tied to the ground with rope on each end. Guard poles are installed to protect traffic and a single-phase distribution line. Having flaggers as a backup at all such sites is common practice.

String Conductor

Historically, conductors were strung along the ground and pulled up through stringing blocks at every structure. Relatively short pulls and logs laid down in strategic locations prevented too much damage to conductors. Today,

Figure 8–72 Guard structures are installed to protect traffic and underbuild distribution lines. *(Delmar, Cengage Learning)*

almost all conductors are strung under tension and, thus, never intentionally touch the ground. Pilot lines (ropes) are dropped, usually from a helicopter, and strung into the stringing blocks that are designed to accept rope as it is dropped from above. When the pilot line reaches its target, a pull rope (sometimes a steel winch line) is attached and pulled back by the tensioner. The pilot line can be reeled back to the conductor/tensioner end with a pilot line winder machine.

The pull rope (or wire rope) is attached to the power conductor or running board. Running boards are available to string two, three, or four conductors with one pull. Ideally, the rope pulling the running board will be in the center sheave when it reaches the insulator string. If it isn't, however, the running board will center itself and each conductor will end up in its own sheave after the board has passed. The swivel between the pulling rope and the running board takes out the torsional forces that could flip the running board. **See Figure 8–73.**

Figure 8–73 The running board is designed to travel through a stringing block while pulling multiple conductors. *(Delmar, Cengage Learning)*

The puller pulls in the conductor while the tensioners hold enough tension to keep the conductors in the air. **See Figure 8–74**.

Figure 8–74 Conductor reels are set up in line with the tensioner. The tensioner will keep the conductors clear of the ground and other obstructions as the conductor is pulled out. *(Courtesy of Western Area Power)*

When working with conductor splices and connectors, sharp corners and projections capable of producing a corona discharge must be made smooth.

Shield wire is strung in a similar manner. It is common to take advantage of optical ground wire (OPGW) when stringing in new shield wire. OPGW is fiber optic cable surrounded by strands of steel shield wire, so placing it in the shield wire position is ideal and the fiber optic strand is not effected by the electric or magnetic field. The fiber-optic cable inside the shield wire can be damaged if it is bent too sharply or twisted during stringing. Larger-diameter stringing blocks and special hardware for dead ending and clipping are used to avoid excessive bending, and a swivel and weights hung near the pull rope prevent twisting. **See Figure 8–75**.

Figure 8–75 The glass fibers in OPGW cable are delicate and the cable needs to be prevented from twisting as it is being pulled. *(Delmar, Cengage Learning)*

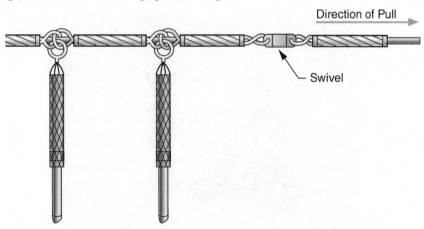

Sag and Clip-In Conductor

Normally, transmission lines are sagged more precisely than distribution lines. Scopes and transits are used to make precise measurements. More checks are made, especially in extra-long spans and spans on hillsides. After the conductors have been sagged, they are clipped in (clamped in) on the light angle and tangent towers between the dead-ends. **See Figure 8–76.** The conductor (or conductors, in the case of a bundled conductor) is removed from the stringing block and transferred to the saddle clamps. Vibration dampers (spacer dampers) are installed.

Figure 8–76 A hook ladder is used to reach out far enough to attach the hoist and grip when dead-ending the shield wire. *(Courtesy of Western Area Power)*

To lift the conductor out of the stringing block, a hoist is hung directly above the conductor on the tower arm. Nylon slings are generally used at the tower end, and unless a shoe is made specifically to hook onto the conductor(s), a nylon sling is used at the conductor end. To get down to the conductor level, the lineman—while wearing a harness and fall arrest system—a rope ladder or a rigid live-line ladder. If the hoist is hung upside down, all the up and down adjustments can be made at the conductor level. Because sometimes the chain will jam when lowering the conductor and can cause hoist failures, some locations do not allow the hoist to be hung upside down.

Vibration dampers must be installed at specified distances from the saddle clamp. The frequency of conductor vibration is highest near the saddle clamp, and it is important to place the vibration damper at the point and frequency for which it is designed. Spacer dampers are installed from carts ridden from tower to tower on the conductor. The carts are usually motorized and have hydraulic power for lifting the cart past the insulator strings and for operating impact wrenches. **See Figure 8–77.**

Figure 8–77 A cart (often motorized) is used to install spacer dampers in every span in the line. *(Courtesy of Western Area Power)*

On transmission lines, clipping in includes working from a hook ladder or rope ladder, lifting the conductor with a chain hoist, and installing the conductor into a saddle clamp. A nylon sling or special conductor shoes are used to prevent damage to the conductor stands. Bundled conductors have spacer dampers that are installed while clipping in, usually about four to a span. When the two, three, or four subconductors are fastened together they become a virtual single conductor. Apply the specified torque on the clamps because spacer dampers are subject to a lot of vibration and sometimes galloping. Improper torque during construction is a contributor to the conductor damage by improperly installed spacer dampers. Corona shields at suspension points and at dead-ends are also required on extra-high-voltage lines. These shields prevent the magnetic field from concentrating in any given location.

Dead Ending a Conductor

The highest tensions worked with in-line work involve dead ending transmission lines. The rigging components and rigging configurations require planning and engineering. Calculations must be made of the expected load on each tower component, snatch block, sling, hoist, and winch cable. The configuration of the rigging should be such that the angle through any snatch block will not increase the bisect tension of the anchor point beyond its load rating. For example, a winch line that changes direction through a block at 90° will put 1.5 times the line tension on the anchor point, sling, and snatch block.

The conductor is dead ended with compression or implosive dead-end terminals. It is common practice to sag conductors using pullers and tensioners and then to transfer the conductor to temporary anchors (tie-downs or helpers) to free up the machines and to allow tangent structures to be clipped in. Dead-end work involves working with one span between the dead-end tower and the tie-downs at the next tower. **See Figure 8–78.**

Figure 8–78 Heavy-duty rigging and scaffolding is needed to dead-end power conductors. *(Courtesy of Western Area Power)*

Major Transmission Projects with Existing Lines

To delay the need to build a new circuit, utilities are carrying out projects that will maximize the current-carrying capacity of existing lines. Work includes restringing or resagging power conductors and refurbishing existing lines.

Resagging a circuit improves the ground clearance and allows a line to carry more load. Reinforcement of structures, changing dead-end insulators, and changing saddle clamps and vibration dampers may be necessary. This is a job that can be carried out hot. The conductor is transferred to stringing blocks using live-line tools or barehand techniques from a tower or buckets. A hydraulic winch, much like the boom-tip winch of a digger derrick, is hung on the conductor using barehand techniques from buckets. Dielectric hydraulic hoses are connected to the tool circuit of a digger derrick that recently has had the oil electrically tested. The winch may have to be strung out to provide a two-, three-, or four-part line. Linemen working together in two bucket trucks install the grips and spread out the winch cable. When the conductor is pulled up, the slack between the grips is coiled up using barehand techniques. When the conductor is at the new sag, it is spliced and the conductors are clipped in. When required, dead-end insulators are changed out before resagging using live-line techniques.

Restringing usually involves putting the existing conductor in stringing blocks and pulling in the new with the old. The splice between the old and the new is the weak link that will require more than just boring out half of the small sleeve to accommodate the larger conductor. Shield wire and power conductor are restrung in this manner. Broken strands are not uncommon on old conductor, especially on old steel shield wire. A strand will break and unravel until a length of it is dropping on and tripping out circuits that are being crossed over, despite the existence of guard structures. In another case, a broken strand will bunch up at the front of a stringing block to the extent that it will stop the puller, damage the structure, or both. Constant helicopter patrol is needed when stringing and restringing.

Changing spacer dampers on existing lines seems to be an ongoing major maintenance project on lines with bundled conductors. Typically, the line is taken out of service, and the spacer dampers are replaced using a conductor cart in a similar manner as during construction. Spacer dampers are also changed out with the line energized using barehand methods via helicopter. **See Figure 8–79**.

Figure 8–79 Spacer dampers are being replaced using barehand techniques on an energized conductor. *(Courtesy of Western Area Power)*

Commissioning

Before a new line is put into service, it goes through a commissioning process to ensure that the line was built to specifications. Because a new transmission line is a huge investment, commissioning is actually continuous as the line is constructed. Linemen are used for commissioning because they can closely examine almost every part of certain structures. Detailed data about all structures are recorded on required forms. Utility representatives (inspectors) perform many tasks, including the following:

- Watch the foundations being installed and ensure, among other things, that any grounding connections are made.
- Torque a representative number of bolts on each tower to ensure that they meet specifications.
- Observe splicing and dead-end operations to ensure that the conductor has been cleaned and the work is done to specifications.
- Change as-built drawings as needed.

Sometimes, a commissioning team must number the structures. Numbers or labels are placed at the bottoms of structures and often include the circuit name or number. Numbers are also installed at the tops of structures to assist helicopter patrols. **See Figure 8-80.** The GPS is used to determine the latitude and longitude of a structure and to use their intersection as the structure number. For example, a structure number might be 41° 08′ N × 76° 15′ W. Warning signs for helicopters are also put on top of structures two or three spans before a circuit that is crossing over or under that circuit.

Figure 8-80 Structure and circuit numbers can be used to accurately identify the location of a structure when preparing a patrol report. *(Delmar, Cengage Learning)*

Installing Wind Turbines

The installation of a wind turbine needs people who can work with rigging, work with cranes, work at heights, make electrical connections, and install underground cable. In other words, linemen are a natural choice for wind turbine construction.

Most wind turbines are mounted on steel poles, not unlike steel poles used on transmission lines. The foundation can be huge. **See Figure 8-81.**

Figure 8-81 The foundation for a wind turbine is immense as the structure will be subject to high stresses. *(iStock.com/Andyqwe)*

Like the transmission-line steel poles, the steel poles for wind turbines come in sections where one section slips over the next section. **See Figure 8-82.**

Figure 8–82 The upper section of the structure is designed to slip over the lower section. *(iStock.com/Vinzo)*

The generator housing is mounted on top and then the generator components are lowered into the housing from the top. **See Figure 8-83.**

Figure 8–83 Generator housing is installed on top of the structure and then the generator is lowered into the housing. *(iStock.com/Vinzo)*

Due to the size turbine blades are assembled on the ground using cranes to position them. **See Figure 8–84**.

Figure 8–84 Rigging to install the large turbine blades in one piece requires an engineered lift. *(Baxtar/Shutterstock.com)*

The turbine is raised with the crane, and workers inside the generator housing bolt the turbine into position. **See Figure 8–85**.

Figure 8–85 The workers attach the turbine blades from inside the generator housing. *(iStock.com/MichaelUtech)*

The electric cables are installed inside the steel pole. In the case of a wind farm, the cables run underground to a substation where the voltage may be stepped up, and where it is connected to the transmission grid.

Review Questions

1. What are three electrical clearances to be considered when designing a line?

2. Before digging, what other utilities should be located?

3. Which wood pole has the larger diameter: class 1 or class 5?

4. What depth hole is needed in earth to set a 60-foot wood pole?

5. What measures are taken to avoid a pole or truck boom contact when setting a pole in an energized circuit?

6. What measures are taken to prevent injury to workers on the ground in case a pole or truck boom makes contact with an energized circuit?

7. What two factors determine the size or number of guys and anchors needed to hold a pole?

8. Name three precautions needed to install a power-installed screw anchor.

9. What kinds of protection can be taken to prevent injury when stringing a conductor near energized circuits?

10. During a stringing operation, how is a person on a tensioner protected from electric shock if the conductor being strung makes electrical contact?

11. What procedures are needed to prevent a vehicle from contacting a conductor while stringing over a highway?

12. Describe two types of transmission tower foundations.

13. What causes an uplift on a transmission tower foundation?

14. When assembling a steel-lattice tower, what two references are needed to determine the location of each individual piece of steel?

15. How can broken strands cause problems when using an old conductor to pull in a new conductor?

16. *True or False:* The strength of a polymeric suspension insulator is in the core.

17. What is the purpose of a semiconducting glaze that is found on some insulators?

18. *True or False:* When a conductor is sagged too tight, the conductor can be subject to too much tension.

19. Name one advantage of using *a preformed tie over a hand-formed tie when tying a conductor on the top of top-tie insulators.*

20. What is the purpose of rider poles/guard structures when stringing a conductor?

Chapter 9

Working on Underground Power Lines

After completing this chapter, you should be able to:

1. Describe the two main types of underground electrical distribution systems.
2. Identify the complications that can be encountered when doing civil work.
3. Determine the spacing needed for underground electric cable from other utility cables and pipes.
4. Describe the methods used to protect direct buried cable.
5. Describe the method used to put the pulling rope into the duct.
6. Take steps to reduce the tension on a cable when pulling it through a duct.
7. Identify cable terminations and install identifying labels.
8. Recognize the tools and hardware used in underground electrical distribution.
9. Describe a device used to test for shorted, high resistive and open conductors.
10. Identify types and design of transmission underground cables.

Underground Distribution Lines

Why Underground

The public always seem to be asking why utilities do not bury their power lines. The public tolerates extended outages less and less. The only obstacle has been the additional cost. However, the additional cost is being balanced by underground distribution tolerance to the extensive damage done by hurricanes, tornadoes, ice storms, thunderstorms, tree contact, animal contact, and wildfires. A cost/benefit study may show that with some projects underground is more economic.

Some municipalities legislate underground distribution for select parts of their jurisdictions. The difference in cost between overhead and underground is usually covered by the municipality or developer and passed on to the new tenants. Underground distribution is common in densely populated urban areas, and most residential subdivisions built in the past 40 years have been supplied underground.

Underground is more expensive than overhead, not so much because of the cost of cable compared to a pole line but because of equipment, such as pad-mount and submersible transformers and switching equipment. Conversion from overhead to underground is expensive because it also includes converting overhead services to underground at the customer's service entrance.

Underground/submarine lines are used for water crossings, especially when the span is too long for an overhead crossing. Usually, the only practical way to span even a short water crossing with a distribution line is with a submarine cable because of the height needed for an overhead line to avoid sailboats.

Transmission lines feeding city substations are underground because an overhead line is usually not feasible. The only feasible way for a transmission line to span a long waterway such as the English Channel is using DC (HVDC) submarine cables.

Two Types of Underground Distribution Systems

The two main types of underground distribution are the duct and **vault system** and the **direct bury system**.

The duct and vault system is used in cities where streets, sidewalks, and lack of space would be impractical for a direct bury system. In such systems, duct banks are installed between vaults and utility manholes and encased in concrete. **See Figure 9–1**. Switchgear and transformers are placed either under the street in large subterranean vaults or in buildings along the street. Duct banks are accessible at every utility manhole and vault. Sections of cable can be installed or replaced for years to come without having to dig up streets.

Figure 9–1 (a) A duct and vault system is found under urban streets and cable can be repaired, replaced and spliced without the need for civil work. *(Delmar, Cengage Learning)* (b) Cables entering a duct bank. *(© ELECTRIC CONDUIT CONSTRUCTION)*

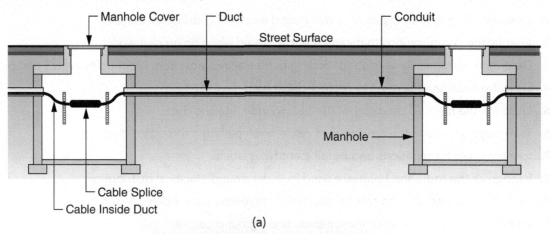

(a)

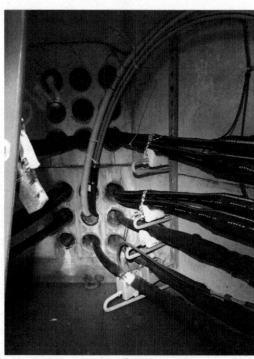

(b)

The *direct bury system* is more economical to install than a duct and vault system and is used mostly in **underground residential distribution** (URD). For our discussion, the direct bury system includes cable buried in sand, in big "O" pipe or in PVC pipe. In other words, cable still needs extra protection when not put in concrete duct banks. For example, cable on a reel can be purchased preassembled in a polyethylene **conduit** (cable-in-conduit system). **See Figure 9–2.** In other instances, the cable is buried directly in the ground and will have only an envelope of sand around it. With flexible duct such as PVC pipe, cable will be pulled in after the pipe is laid.

Figure 9–2 When cable in conduit is buried directly in the ground the conduit provides good protection for the cable. *(Delmar, Cengage Learning)*

Transformers and switchgear are either pad-mount design and sit on the surface of a concrete pad that sits over a well where cables come up to feed the equipment or are submersible where the equipment is installed below grade level. **See Figure 9–3**.

Figure 9–3 Utility enclosures used in underground systems are designed to provide maintenance access and protection for the public. *(Reprinted with permission of Hubbell Power Systems, Inc., Centralia, MO USA)*

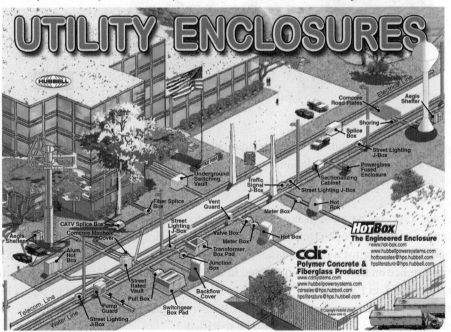

Civil Work: Trenching, Vaults, and Pads

The trenching and the digging needed to put a new duct system in a city require extensive civil work and will cause great disruption to the street. Locations for water, sewer, gas, other electrical circuits, communications cables, and even underground transportation systems must be considered when digging. Civil work involves breaking concrete/asphalt, extensive digging, setting concrete forms, placing conduit, pouring concrete, installing vaults, backfilling, and placing new street/sidewalk surfaces. **See Figure 9–4**. Ducts are mostly made of polyvinyl chloride (PVC) plastic or fiberglass, but existing ducts are also made of tile, concrete, or steel. Any bends or curves must be as large in radius as possible to keep the sidewall pressure on the cable to a minimum and to reduce the pulling tension on the cable during the cable installation.

Figure 9–4 The power cable ducts in a typical duct bank provides excellent protection for the cable. *(Delmar, Cengage Learning)*

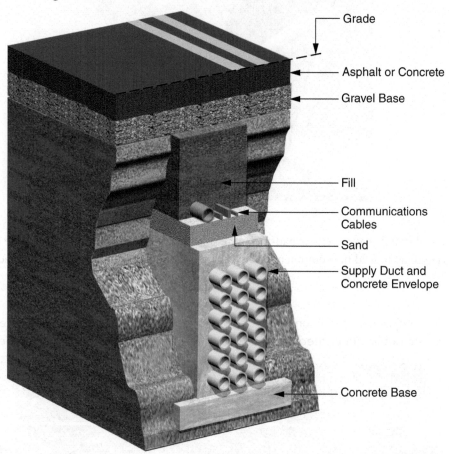

Very large, specially designed vaults are constructed much like basements, with elevations that are perfectly aligned with the street grade above and with foundations and covers that can support heavy overhead traffic. Typically, many circuits run in and out of a transformer vault, and switchgear may be in the vault as well, but this is not the case with all vaults. **See Figure 9–5**.

The civil work for direct bury systems is not as technical as for a duct and vault system. In a new subdivision, direct bury can be as simple as digging an open trench and laying cable with proper separation, and ensuring that the cable is protected. One rock pressing against a cable can cause a concentration of the electrical field at that point and eventual cable failure. The preference on how a cable is protected in a trench depends on the utility. Choices to protect the cable include enveloping it in sand, in PVC conduit, in flexible polyethylene (PE) conduit, and in pipe. Guided underground boring is also carried out where it is more economical or customer friendly to avoid disrupting streets and lawns. Smaller cable that is available preassembled in conduit can also be specified.

Figure 9–5 This is an economical transformer vault which supplies power to a condominium building. It uses above ground transformers in a concrete block vault secured with a wire mesh roof and a locked steel door. *(Delmar, Cengage Learning)*

Sharing a trench with other utilities is common with direct bury systems. Each organization must follow all specifications such as for depth, vault size, duct size, and minimum bend radius. **See Figure 9–6**. Watch for any color-coded marking, flag, tape, or spray paint that indicates the presence of other underground utilities, as shown below.

Marker	Utility
Red	Electric
Yellow	Gas, oil, steam
Orange	Communications, cable
Blue	Water
Green	Sewer
Purple	Irrigation
White	Proposed excavation
Pink	Temporary survey tape

The civil work for installing such equipment as transformers and switchgear on direct bury systems consists of installing and leveling prefabricated fiberglass, concrete, or polymer-concrete pads for pad-mount equipment and vaults for submersible (under grade level) equipment.

Some utilities serve each customer with a service straight from the transformer, while others run secondary bus to secondary pedestals or handholes/junction boxes, which are typically fiberglass boxes where services are connected to the bus. For secondary bus, conduit is installed between the transformer pads/vaults to secondary pedestals or handholes. The pedestals and/or handholes are installed, depending on the type, such that the top of the pedestal is at final grade or partly underground with the connections above ground.

Red warning tape is laid out about 1 ft (0.3 m) below the surface and above the underground cable to warn people digging above the cable. A bare ground wire strung along the top of the cable or duct bank can be used to more accurately locate cable in the future, as well as for gathering stray ground current.

Pulling and Laying Cable

As discussed elsewhere in this book, underground cables are delicate and any excessive bending or tension will damage them. A damaged cable may be in service for a while, but electrical stresses at locations where there are insulation voids,

Figure 9–6 Proper spacing between utilities in a joint use trench. *(Delmar, Cengage Learning)*

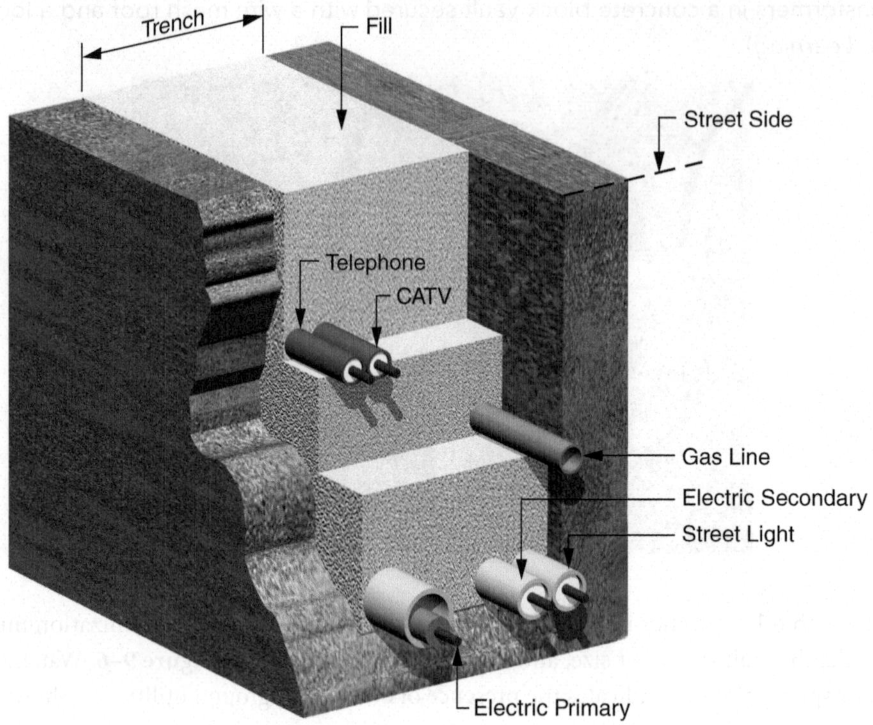

a damaged semiconducting layer, or other problem will cause premature failure. Therefore, it is essential when pulling in a cable to protect it by not pulling beyond the maximum allowable tension and not flexing it beyond the maximum allowable radius.

Maximum Allowable Cable Tension and Bending Radius

The maximum allowable tension for a cable is usually part of the manufacturer's cable specifications. The maximum tension allowed is dependent upon the conductor size and on how a pulling eye is attached to the cable. The core conductor(s) is the only part of a power cable that can take a fairly high pulling tension. Therefore, a crimp-on pulling eye is the best choice. A crimp-on pulling eye can be factory or field installed and is made to seal the cable against contaminants. If using a wire-mesh grip with a swivel eye, the pull will be on the sheath only, which can tear it away from the core. Therefore, the allowable pulling tension is very low, usually less than 1,000 lbs (500 kg). The manufacturer will also specify a minimum cable bending radius for the type of cable being installed.

Estimated Pulling Tension

An estimated pulling tension can be calculated before starting a pull or before the cable is purchased by entering certain known data into a computer program. The data will include the weight of the cable, number of cables in the duct, details about the length and clearances in the conduit, and sidewall pressure. Of course, the estimated pulling tension has to be less than the maximum allowable tension of the conductor. If not, a revision of the pulling layout is needed.

Preparing the Duct

Typically, a duct will contain a pulling wire or rope. If not, an air compressor can be used to blow a line carrier (bird) with a small nylon cord attached to it through the duct. **See Figure 9–7.** The cord is used to pull in a bigger rope, which in turn brings in a bigger pulling rope or winch line. Another method pushes a duct rodder from one vault to another. **See Figure 9–8.** One way a duct can be checked for proper diameter throughout and for obstructions is to pull a flexible steel mandrel of the correct diameter through the length of the conduit. **See Figure 9–9.**

Rigging the Pull

To make a pull within the tension limits of the cable, the vault at the cable reel end must be rigged so that there is a minimum amount of cable bends. A variety of cable guides are made specifically for keeping the cable running

Figure 9–7 Air pressure can blow this line carrier through the duct with a line tied to it. *(Courtesy of Comstar Supply, www.comstarsupply.com)*

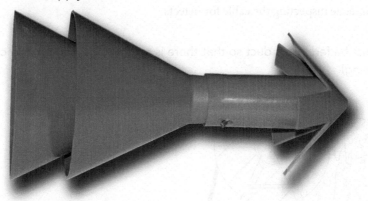

Figure 9–8 A duct rodder is pushed through the length of the duct where a line is attached and pulled back. *(Delmar, Cengage Learning)*

Figure 9–9 A flexible mandrel can be pulled through a duct to check for obstructions. *(Courtesy of Greenlee, A Textron Company)*

smoothly with a minimum amount of tension as it comes off the reel. Vaults are manufactured with embedded eyes for use as anchoring points for snatch blocks. A cable guide can be plugged into duct at the face of the vault wall to provide a smooth and low-friction change of angle. **See Figure 9–10.** A quadrant block allows the cable to make a smooth, low-friction, 90° turn. **See Figure 9–11.**

Figure 9–10 A cable feeding sheave (guide) provides a smooth change of angle for the cable. *(Courtesy of Greenlee, A Textron Company)*

Figure 9–11 A quadrant or lip-roller block allow the cable to make a smooth 90° turn. *(Courtesy of Condux International, Inc.)*

If the reel is set up, no extra rigging is needed, only muscle. **See Figure 9–12.** If this method is used, at least one worker, and probably more, should help to turn the reel, with someone in the vault to guide the cable, someone applying lubricant, and possibly someone inspecting the cable for defects.

Figure 9–12 Cable must be fed into a duct so that there is no sharp bending of the cable. *(Delmar, Cengage Learning)*

Making the Pull

After the puller and reel are set up the winch line can be attached to the end of the cable. **See Figure 9–13.** The pulling machines have gauges or monitors to indicate the tension on the cable. An electric puller monitors pull tension by translating the number of amps the motor is drawing into tension. **See Figure 9–14.** Applying a lubricant specifically made to be compatible with the cable to reduce friction is essential as the cable is pulled through the duct. A weak link can be installed at the pulling eye to act as a mechanical fuse and to break apart before a cable is overstretched.

Figure 9–13 Underground cable is pulled into a duct. *(Delmar, Cengage Learning)*

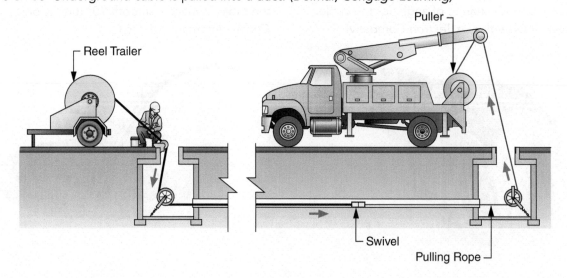

Figure 9–14 Underground cable pullers are designed to pull a long length of cable while monitoring and keeping the tension below the maximum allowable. *(a, Delmar, Cengage Learning; b, Courtesy of John Bellows)*

(a)

(b)

Installing Equipment

Equipment such as transformers, switchgear, reactors, and capacitors are set on pads, lowered into vaults, or rolled into buildings. Pad-mount equipment is very accessible to the public and is placed over the hole to the cable compartment below so as not to leave a space for vandalism. Before the equipment is installed, the primary and secondary cables are trained with the ends of the cable up through the hole.

Tools and Hardware for Underground Distribution

Working in underground distribution requires the use of some tools and hardware not found in any other sphere of work. It is important that a power line technician is trained on the proper use of the various tools and hardware that are found in underground distribution. **See Figure 9–15.**

Figure 9–15 Hardware and tools used when working on underground systems require skills that are not found in any other trade. *(Reprinted with permission of Hubbell Power Systems, Inc., Centralia, MO USA)*

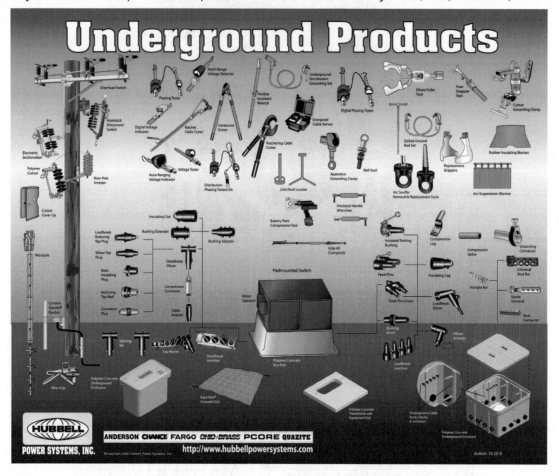

Terminations and Splices

Electrically, the most technical part of underground work is making **terminations** and splices. Some terminations are made at substations, some into equipment such as transformers and switchgear, and some on **riser poles** for a transition to overhead. **See Figure 9–16.** Terminations and splicing underground cable have to be done to exact standards. There are many cable types and termination types and unless a person is lucky enough to work only with one kind all the time, it is probably best to have a written procedure on hand. Each cable must be traced and labeled. Each switching device must be labeled with a corresponding number on an operating drawing.

Most utilities require electrical testing of new cables, splices, and terminations before energizing. Use the specified voltage for testing because it is very easy to damage a cable by testing it with too high voltage.

Underground Secondary

Underground secondary is mostly triplex and quadruplex. Troubleshooting, pinpointing faults and splicing are common tasks. There are many splicing options; taped splices, heat shrink, cold shrink, and gel-filled splices.

Figure 9–16 The underground riser provides fused protection and an isolating device for the underground cable. *(Delmar, Cengage Learning)*

There are some useful tools available to help trouble shooting underground secondary. A tester such as the Mega Beast, **Figure 9–17**, can test for a shorted, open, or high-resistive fault in phase conductors or neutral. It is effective to find a conductor that has a poor connection or other high-resistive fault that shows up only when the customer is drawing a heavy load.

Figure 9–17 A tester for shorted, high-resistive, and open conductors. *(Source: Arnett Industries, LLC)*

When one phase or the neutral is found to be defective a device such as the Restore-A-Phase, **Figure 9–18**, can temporarily supply regular 120/240 volt power until permanent repairs are made. It is a transformer with the neutral in the center; 120-0-120 volt. At the meter base volt the defective conductor is removed. The three wires of the device, which is like an autotransformer, is connected into the meter base. The missing phase is energized by the device restoring a regular 120/240 volt service.

Figure 9–18 A portable device to temporarily restorer power when one conductor is defective. *(Source: Arnett Industries, LLC)*

Storm Hardening Underground

An underground electrical system resists, hurricanes, tornadoes, ice storms, thunderstorms, tree contact, and wildfires reasonably well. Flooding, especially with salt water, can cause damage and operability problems in underground systems.

The duct and vault underground system is very vulnerable when the transformer and maintenance vaults fill with water. Submersible transformers and SCADA operated switchgear can relieve some of the effects of flooding.

Submersible transformers and switch gear in a direct bury underground system is designed to withstand flooding.

Maintenance on Underground Lines

Underground transmission cables are patrolled, sometimes weekly, to check for potential construction involving digging taking place along the cable route. Less frequently (i.e., annually) items such as, maintenance vaults, cable tunnels, cables attached to bridges, alarm systems, earth resistance, and insulating oil are inspected. Underground distribution systems have inspections where visible equipment such as vaults, pad-mount equipment, and risers are inspected. Many utilities also own communications systems that need to be inspected and maintained. The communication cables are needed for telemetering and operating the electrical system.

Underground Transmission Lines

Why Underground Transmission Lines?

Generally, underground transmission lines are used where there is no practical alternative, such as to feed the downtown core of a city, to build a line near an airport, to cross a river or channel, or to feed from a generating station to a step-up substation. Underground transmission lines have the following advantages.

1. Where it is impossible to build overhead, underground can be laid under streets, under water, through high-density residential areas, along very narrow right-of-ways, and through some types of environmentally sensitive areas.

2. Electromagnetic field (EMF) concerns are diminished because underground conductors are laid together closely with a greater canceling effect.

3. There are fewer visibility concerns.

4. Underground lines are not affected by the scourges of wind and ice.

Description of Transmission Underground Systems

Underground transmission cables are classified according to their insulation and whether or not they are in pipes. The main types are as follows:

1. *High-pressure, fluid-filled (HPFF) pipe type.* A HPFF type of cable consists of three high-voltage cables in a steel pipe filled with an insulating fluid (oil). Each cable is made of copper or aluminum, insulated with oil-impregnated paper insulation, and covered with metal shielding (usually lead). The steel pipe provides mechanical protection, is a container for the oil, and helps cool the cables. HPFF cable is, historically, the most common underground transmission cable.

2. *High-pressure, gas-filled (HPGF) pipe type.* A HPGF type of pipe uses pressurized nitrogen gas instead of oil. Because oil is a better insulator, each cable laid in nitrogen-filled pipe will be made with about 20% more insulation. **See Figure 9–19**. The cable is more difficult to cool and will have a lower current-carrying rating. Because there is no danger of oil leaks, this cable may be preferred for sensitive areas.

Figure 9–19 A high-pressure gas-filled (HPGF) pipe is typically pressurized with nitrogen. *(Delmar, Cengage Learning)*

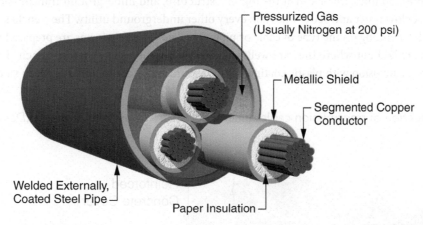

3. *Self-contained, fluid-filled (SCFF) pipe type.* In SCFF cable, one hollow cable is in each fluid-filled pipe; three pipes comprise a circuit. Each cable has paper insulation, a lead-bronze (or aluminum) **sheath**, and a plastic jacket. This cable is often used for water crossings.

4. *Extruded dielectric, polyethylene (XLPE) pipe type.* An **XLPE** (solid dielectric) cable is like distribution cable, except it is single cable, with each cable put in pipe or concrete duct for extra mechanical protection; three pipes comprise a circuit. Each cable has a copper or aluminum conductor, a semiconducting shield, cross-linked polyethylene insulation, an outer semiconducting shield, a metallic sheath, and a plastic jacket. The insulation is much thicker than oil-impregnated paper insulation. This type of cable is the choice for new underground transmission lines. Such cables were limited to 69 kV and 138 kV circuits, but higher-voltage XLPE is becoming available.

5. *Gas-insulated lines (GIL).* GILs are large, gas-filled (with sulfur hexafluoride [SF_6]) pipes with bare conductors suspended on insulators. A GIL is a good choice when very large conductors are needed, such as between a generating station and the step-up transformer substation.

Splices

Because of the practical limit to the amount of cable a reel can carry, splicing locations must be planned. Splices for oil-filled, pipe-type cables are in vaults constructed along the length of a line. The vault length is typically 15 ft (4 m) with one or two vents to the surface, and it is strong enough to withstand the weight of overhead traffic. XLPE splices can be in vaults, but some are put in temporary vaults filled with backfill.

Terminations

Transitions (risers or potheads) are made from underground to overhead lines or to substation potheads. The cable is terminated so that the high-voltage conductor is brought out from the cable in such a way that the cable is protected from moisture/contamination and, at the same time, has electrical clearance from the grounded shield, pipe, and structure. Porcelain or polymer housings or potheads are used to make these terminations.

Cable Current-Carrying Capacity

The rating or ampacity of the cable system is dependent on the size of the conductor and how well heat can be dissipated. Backfill is depended upon to do a majority of the cable cooling and, therefore, it plays a large part in determining the cable load-carrying capacity. Corrective thermal backfill material is designed to move heat away from the line, and is typically graded sand that is compacted or a thermal backfill made from a weak concrete solution (slurry).

Civil Work: Trenching and Vaults for a Transmission Line

The civil work required to install transmission lines is similar to distribution except that the transmission line is typically deeper, the pipes or conduit are larger, and more concrete is used to protect the cable. Because of the cost and consequences of an error when installing a transmission-line cable, the project is a team effort, with on-site engineering, on-site utility commissioning staff, and cable manufacturer representatives.

Overhead transmission lines are placed at the top of a structure, and underground transmission lines are typically placed at the bottom, below water and sewer lines and every other underground utility. The trenches are dug with gradual slopes and turns, and depending on the type of cable or pipe to be laid, concrete forms are prepared to build duct banks.

Oil-filled cables are laid out where they are well protected in a reinforced concrete trench. The pipes are welded, X-rayed, protected from corrosion with plastic coatings, and pressure and vacuum tested between vaults. **See Figure 9–20.**

Figure 9–20 Protection of transmission cable ducts is necessary to prevent outages. *(Delmar, Cengage Learning)*

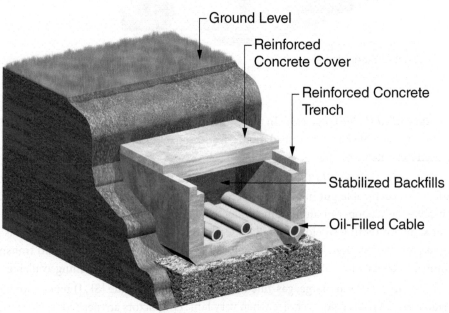

XLPE transmission cable duct banks are constructed very similar to distribution duct banks. **See Figure 9–21**. Extra ducts and communications cable are commonly installed as a prebuild for potential future use. Above the concrete duct bank, a ground wire is sometimes strung to carry stray ground current. In an urban area it is difficult to bring out more than two or three overhead feeders. Massive underground duct banks are needed to bring feeders out of large substations. **See Figure 9–22**.

Figure 9–21 An XLPE cable duct bank is very similar but larger than a distribution duct bank. *(Delmar, Cengage Learning)*

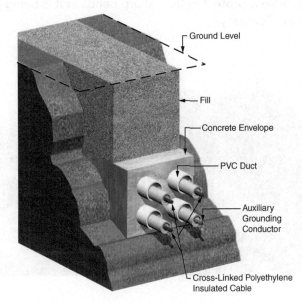

Figure 9–22 A duct bank constructed in conjunction with a new substation where all feeders exit underground. *(Courtesy of Wilson Construction Company)*

Underground transmission lines, other than some submarine cables, are generally less than 10 miles (15 km) in length. In that length, vaults are needed for locating splices and accessing the duct banks.

The trench is closed with thermal backfill and soil and the streets and sidewalks are resurfaced.

Pulling Cable

Cables are pulled between vaults in similar fashion to distribution except that the cable will be much heavier and the consequences of an error will be much greater. **See Figure 9-23.** When a transmission cable is pulled into a duct, the length of each pull is limited to the length of cable on a reel. As for distribution cable, there will be a maximum allowable cable tension. The calculations will determine the rigging needed to keep the tension below the maximum allowed. Some pullers have a port where a laptop loaded with a pulling program shows every aspect of a wire pull.

Figure 9-23 Notice how the cable is protected from sharp bends as it is being pulled into the duct. *(Courtesy of Wilson Construction Company)*

Terminations

At each end of a cable there will be a transition to either three separate overhead phases or three separate potheads. **See Figure 9-24.** The termination for underground transmission line is at a pothead on a transition structure, or at a pothead in the wall of an underground vault or substation.

Figure 9-24 These 230 kV underground to overhead terminations are for a water crossing. Extra protection is required because the cables are oil filled. *(Delmar, Cengage Learning)*

Making a splice or termination on XLPE transmission-line cable is not unlike working on distribution. Making a splice or termination on pipe-type cable is usually done by people who specialize in such work.

Extensive electrical testing is carried out before a cable is energized.

Auxiliary Equipment

Underground transmission cable is closely monitored for temperature, contaminants, pipe pressure, corrosion, and potential dig-ins.

Pipe-type cables will have oil-pumping stations that keep oil flowing in force-cooled cable systems. Oil samples are routinely scheduled for analysis, especially for the presence of air or moisture. Alarm systems will indicate low oil or low oil pressure. Cables in pressurized gas-filled pipes are similarly monitored (e.g., for pressure and contaminants).

Review Questions

1. What are the two main types of underground distribution?

2. What kind of maintenance is carried out on underground systems?

3. Name three other types of underground utilities that can be encountered when digging a new trench for a power cable.

4. Which of the two main types of underground systems require the most civil work to install?

5. How does a rope used to pull in a cable get installed into a duct?

6. What methods are used to ensure that the maximum cable tension is not exceeded when pulling it into a duct?

7. Name three types of underground transmission cables.

8. How does the backfill for an underground cable affect the current-carrying capacity of the cable?

9. Name three methods used to protect direct buried cable.

10. Name three advantages of a vault and duct system for an underground system over a direct bury system.

11. *True or False:* In a joint-use trench, power cables are laid in the top position.

12. Identify the cables in this joint use trench.

13. Which type of connection to a cable allows the highest tension for a pull, a crimp-on pulling eye or a wire-mesh grip with a swivel eye?

14. Before making a pull, where does a person find the maximum allowable tension for a given type of underground cable?

15. Why do pipe-type transmission-line cables have oil-pumping stations?

16. *True or False:* All transformers used on underground systems are installed below the surface of the ground.

17. How is a concrete duct in a vault and duct system prepared before a cable is pulled in?

18. How are cables identified at the terminations?

Chapter 10

Working with Conductors and Cable

Objectives

After completing this chapter, you should be able to:

1. Explain the three basic effects that occur when an AC electric current flows through a conductor.
2. Name two factors that influence the specification for the size of conductor needed for a project.
3. Describe the factors to consider choosing the type of conductor needed for a project.
4. Describe the factors to consider choosing a conductor size.
5. Explain methods used to reduce vibration and possible galloping of a conductor.
6. Recognize the types of bare overhead conductor and the tools needed to splice or make connections.
7. Explain the formation of underground distribution and transmission cable.
8. Perform a splice on a single-conductor concentric neutral underground cable.
9. Describe methods used to test, locate and trace an underground cable.
10. Describe work methods used to work with fiber-optic cable.

Electrical Properties of a Conductor

Conductors and Cable Work

Line work, whether overhead or underground, is all about working with conductors. Working with conductors involves stringing, splicing, making connections, and tensioning. Stringing can be some of the most complex work done by a line crew because it includes stringing near other energized circuits, working around road crossings, and rigging for heavy pull and high tensions.

Underground line work includes laying of cable, working in confined spaces, pulling cable through conduit, and locating faults.

Electric Current in a Conductor

When an alternating electric current flows in a conductor, three basic effects occur.

1. A magnetic field is set up around the conductor.
2. Heat is generated to some extent.
3. A drop in voltage occurs to some extent.

Some kind of conductor is used in every electrical circuit. A conductor can be wound into coils, as in a transformer coil; it can be a large aluminum pipe, such as in a station bus; or it can be in an electronic circuit board of a protective relay switch. This chapter describes conductors used in overhead and underground transmission and distribution circuits.

Conductor Selection

From an electrical perspective, the selection of a conductor is based on the ampacity and voltage requirements of the circuit. The larger the conductor, the greater the capacity to carry large amounts of energy with the least amount of line loss. However, some compromise has to be made between a large conductor and the mechanical properties of a conductor.

A larger conductor increases overhead line tension, adds to sag, allows for more ice buildup, and provides more surface for wind. However, while a large conductor is sagged lower, it will not heat up as much for a given load and will, therefore, have less sag. The bare conductor chosen should typically have a maximum thermal load of 100°F (38°C). Depending on the length of a span and the expected electrical load on a conductor, the mechanical strength requirements can be more critical than the electrical properties of a conductor.

On underground circuits, the selection is simpler. Current-carrying capacity and voltage regulation are the dominant factors involved in specifying a conductor.

Ampacity of a Conductor

Ampacity (ampere-capacity) is *the current in amperes that a conductor can carry continuously without exceeding its temperature rating*. When a conductor's temperature reaches the annealing point, its strength, brittleness, and elasticity are permanently changed.

Three factors affect the amount of heat in a conductor.

1. There is a resistance in the conductor itself. This resistance impedes the current flowing in the conductor and causes heat to be produced based on this formula:

$$\text{Watts} = I^2R$$

   In other words, a large conductor has less resistance to electrical current, which means that less heat is being generated within the conductor.
2. Heat generated in a conductor can transfer by convection or conduction to the surrounding environment. With an overhead conductor, heat transfers to the air; therefore, on a cold day, a conductor can carry more current than it can on a hot day. Underground cable has thermal barriers that slow the cooling of a conductor, including, for example, conductor insulation, the soil surrounding the cable, or poor air circulation in a duct.
3. The ampacity of a conductor is not normally the governing factor for choosing a conductor size. The conductor size chosen is normally much larger to reduce the voltage drop to a distant load.

Capacity Rating of a Transmission Circuit

Each transmission circuit has a continuous current rating. A normal electrical load causes some heating of the power conductors, and the conductor sag increases as the load increases. Overloaded conductors heat and sag into crossing circuits or trees, depending on the span length and conductor type. The circuit is rated to a maximum permissible loading based on conductor size and sag.

As the load on a circuit and the ambient temperature change, a control room operator uses precalculated tables to prevent the line from being overloaded and the conductor sag from dropping below regulatory standards. Circuit data is input into computer programs give a real-time thermal rating for a circuit. These programs continually calculate the maximum permissible loading as the load, temperature, and wind change. Some utilities have sensors on transmission lines that will give the actual conductor temperature.

A utility also has transmission line capacity ratings for emergency loading when other circuits are lost. An emergency rating allows additional sag as a temporary measure.

Capacity Rating of a Distribution Circuit

It is not normal to have a capacity rating for each distribution circuit. On rare occasions, an overloaded distribution circuit may sag into a neutral or secondary, but normally a switchgear tripping out a voltage problem would indicate a possible overloaded conductor.

Secondary buses and services are not as closely monitored as a distribution feeder, and a trouble crew will be alerted to problems during a low-voltage trouble call. An underground system should be monitored more closely because an overloaded cable will heat up, shorten the life of the cable, and eventually lead to failure.

Customer load determines the requirements for the capacity of a distribution circuit. Smart metering will give a distribution planning engineer good load data to decide when upgrades are required. Other indicators are switchgear tripping out, voltage problems at the end of the line, or on rare occasions, an overloaded distribution circuit may sag into a neutral or secondary.

Voltage Rating of a Conductor

The voltage to be used on a bare overhead conductor is not a factor in specifying conductor size until it is used on high-voltage transmission lines. On high-voltage lines, a minimum diameter is needed to avoid **corona loss**. Extra-high-voltage circuits use two, three, or four cables **bundled** together with spacer dampers to form a group, which creates a virtual large conductor.

On underground cable, the voltage rating is based on the property of the insulating material. The cable insulation, as well as other design considerations, must be suitable for the voltage of the circuit. Voltage induces electrical stresses on the insulating material; and an inadequate amount of insulation, a sharp bend in the cable, or external pressure from a rock will eventually cause a rupture in the insulation. Insulation will break down faster because of AC voltage stress versus DC voltage stress. The DC voltage rating of a specific cable insulation is three to four times higher than the AC voltage rating. The stresses induced on the insulation by DC voltage are constant, while the stresses induced by AC are fluctuating.

Voltage Drop in a Conductor

When current passes through a conductor, resistance opposes the flow and a voltage drop results. The amount of voltage drop can be calculated using Ohm's law ($E = IR$). The amount of resistance offered by a conductor depends on the conductor's size and length. Voltage drop is normally calculated by a computer program or voltage regulation tables. **See Figure 10–1** for a sample voltage regulation table.

Figure 10–1 Voltage Regulation Table

Percent Voltage Drop per Mile for 4.8 kV, 50 A Load, at 90% Power Factor					
	1 Mile	**2 Miles**	**3 Miles**	**4 Miles**	**5 Miles**
#4 ACSR	5.6%	11.2%	16.8%	—	—
#2 ACSR	3.8%	7.6%	11.4%	15.2%	—
1/0 ACSR	2.7%	5.4%	8.1%	10.8%	13.5%
3/0 ACSR	1.9%	3.9%	5.8%	7.8%	9.7%

This voltage regulation table illustrates the impact of conductor size and length in a single-phase line with a 50 A load. It also illustrates why voltage regulation is a bigger consideration than ampacity when choosing a conductor.

For example, the secondary voltage at an unloaded distribution transformer is 120 V. What would be the secondary voltage of an unloaded transformer 3 miles downstream if the primary conductor is 3/0 ACSR (aluminum conductor, steel-reinforced)?

Figure 10.1 shows that there would be 5.8% voltage drop over a 3 mile length of 3/0 ACSR, operating at 4,800 V which results in the following total voltage drop:

$$\text{The voltage drop is } 0.058 \times 120 = 6.96 \text{ V}$$
$$\text{The voltage at 3 miles is } 120 - 6.96 = 113.04 \text{ V}$$

Voltage Drop in a Three-Phase System

To use Figure 10–1 to calculate the voltage drop for a *balanced* three-phase circuit, the voltage drop for a single-phase circuit is divided by 2, because the flow back to the source is no longer on the neutral. The result is then multiplied by the square root of 3, which is the voltage drop per phase. Because the square root of 3 divided by 2 is 0.866, the voltage drop on one phase of a balanced three-phase system can be calculated by multiplying a single-phase voltage drop by 0.866. For example, the voltage at an unloaded three-phase transformer bank is 120 V per phase. What would be the voltage of an unloaded transformer bank 3 miles downstream if the primary conductor is 3/0 ACSR?

Figure 10–1 shows that there would be a 5.8% voltage drop over a 3 mile length of 3/0 ACSR. The voltage drop for a single-phase line as calculated earlier was:

$$0.058 \times 120 = 6.96 \text{ V}$$

For a three-phase line, the voltage drop would be:

$$6.96 \times 0.866 = 6.03 \text{ V}$$

Specifying Conductor Size

The current-carrying capacity of a conductor is not normally the controlling factor when specifying a conductor for an electrical utility circuit. Other factors include the following:

- A conductor must be large enough to keep the voltage drop to an acceptable limit.
- A conductor must be large enough to limit line loss and keep the fault current available at the end of the line high enough for the protective switchgear to see a fault.
- A conductor must be large enough to accept future load growth.
- A conductor on transmission lines must be large enough to limit corona loss.

On a very short length of line, ampacity can be a limiting factor because voltage drop or line loss will not be as noticeable. For example, a main line consisting of 336,000 **circular mils (kcmil)** aluminum (AL) has a voltage regulator with #2 copper input and output leads. This is acceptable because the #2 copper has the ampacity to carry the load current, and the short length will not affect the voltage regulation appreciably.

Conductor Sizes

Learning to recognize a conductor size on sight is a common skill for power line technicians. When in doubt, a conductor size should be confirmed with a gauge to avoid using the wrong sleeve or **connector**. The numerical systems used to size conductors can be quite confusing. The numbers used to indicate conductor size have no practical application to the line trade. Fortunately, most utilities standardize on a relatively small number of conductors and provide tables for their weight, die sizes, and grip sizes.

The **American Wire Gauge (AWG)** system is formed by defining a 4/0 conductor as 0.46 in. in diameter and a #38 wire as 0.005 in. in diameter. There are 38 sizes of wire, spaced in a geometric progression, between these two sizes. Conductor sizes are expressed by numbers. Common AWG stranded and solid sizes used in the line trade are #4, #2, #1/0, and #3/0.

Conductors larger than 4/0 are referenced to their cross-sectional area (CSA), expressed in circular mils. The size of a conductor is not calculated by using the formula for determining the area of a circle: πr^2. One circular mil is defined as a circle with a diameter of 1/1,000 or 0.001 in. Based on this definition, a solid conductor size can be determined by measuring the diameter of a conductor in mils (1/1,000 or 0.001 in.) and then squaring that number. The size of a conductor is normally expressed in thousands of circular mils, or kcmils.

However, only the conductive material is used when discussing conductor size, so type of stranding and steel wires in a conductor will change the calculated diameter. For example, a 336.4 kcmil ACSR conductor with 26 strands of aluminum and 7 strands of steel is measured as 0.720 in. in diameter, which when squared would indicate that this conductor should be a 518.4 kcmil conductor, which is incorrect.

In the metric system, a conductor size is referenced to the area of its cross-sectional area, expressed in square millimeters. The conductor size can be calculated using the formula for the area of a circle (πr^2). The aluminum and steel cores of ACSR conductors are measured separately and shown as 40/20 mm^2 for 40 mm^2 aluminum and 20 mm^2 steel core. **Figure 10–2** is a table comparing AWG aluminum conductor sizes with equivalent metric conductor sizes.

Figure 10–2 Metric conductor size equivalents

AWG Conductor	Standard International Conductor Size (mm²)
6	10
4	25
2	35
1/0	50
3/0	95
350 kcmil	185
500 kcmil	240
750	400

Corona Loss

An electric field around a conductor can be strong enough to break down or ionize the molecules in the air and cause sparks or a corona around the conductor. Under certain weather conditions, corona can be seen around a conductor on a dark night as a purple glow or as sparking. Corona can also be heard as a crackling or hissing sound, especially on a foggy or damp morning.

When the air between phases, or between phase and ground, is electrically stressed, the air becomes ionized (the air becomes conductive) near the surface of the conductor. The electric field generates a cloud of tiny electric discharges into the air surrounding the conductor.

Corona generates ozone gas (O_3), which is an ionized form of oxygen and is recognized by a distinct caustic smell. Ozone decomposes organic materials, such as rubber, and affects materials that are subject to oxidization. Older rubber cover-up was subject to ozone damage when the cover-up was left on a line for any length of time.

It takes energy to make corona, and corona loss results in power loss and causes radio and television interference. The design of a circuit, its conductor, and its hardware minimize corona loss, as follows:

- Voltage stress of the air between phases or between phase and ground is reduced by designing the circuit to have enough spacing between conductors. The higher the voltage, the greater the spacing required.

- Voltage stress is reduced at the surface of the conductor by spreading the stress over a larger area. High-voltage circuits, therefore, need a larger-diameter conductor. Extra-high-voltage circuits use bundled conductors to spread the voltage stress over a greater air space. Corona rings or grading rings used on extra-high-voltage circuits also spread the stress over a larger area.

- Voltage stress is reduced at the surface of conductors by smoothing any rough or sharp points where voltage stress of the air concentrates.

Skin Effect

Alternating current does not travel equally distributed throughout the conductor. Electrical current interacts with its own magnetic field. The electromagnetic field around the conductor also cuts through the conductor itself, and the created inductive reactance causes a counter-electromotive force (cemf). This force is greater at the center of the conductor than along the outside. The main current flow becomes concentrated along the outer surface of the conductor.

Large-diameter stranded conductors or tubular conductors are used to overcome the higher impedance to current flow caused by the **skin effect**. Skin effect increases with an increase in frequency. Although the frequency of a power line is a standard 50 or 60 Hz, the skin effect is increased when the circuit has high-frequency harmonics superimposed on it. Higher-frequency harmonics increase the amount of induction created by the electromagnetic field and increase the skin effect. For the same reason, harmonic distortion causes an increase in the heating of transformer and motor windings.

Vibration and Galloping of a Conductor

Conductors that are strung to high tension are subject to aeolian (caused by wind) vibration. Unless transmission lines have special vibration-resistant conductors, vibration dampers are installed in each span. **See Figure 10–3**. Vibration dampers can be armor rod wrapped around the conductor at the suspension clamp, stock bridge dampers, torsional dampers, spacer dampers, and fiber helix dampers. One method of reducing vibration in a span is to have torsional or stockbridge dampers installed at a specified distance relatively close to the suspension clamp to reduce high-frequency vibration and to have another damper farther out from the suspension clamp to reduce lower frequency vibrations.

Figure 10–3 Vibration dampers are installed in every span at specified distances from the suspension point. *(iStock.com/VladimirB)*

Typically, when conductors on distribution lines are strung tighter than 20% of the rated tensile strength (RTS) they are subject to aeolian vibration. Preformed spiral plastic dampers are often installed at the suspension points of a distribution line affected by aeolian vibration. Strand or messenger cable that is tensioned properly but has no cable lashed to it is very vulnerable to vibration.

Under conditions such as ice loading and a moderate wind, conductors can "gallop" or "dance," where the power conductors or shield wire in a span rises up and drops down violently. The wires rise and fall from a few feet to a distance equal to full sag. Severe galloping can have the conductor go up the same distance as the full sag and has in the past destroyed supporting structures. Galloping is usually vertical where shield wire and power conductor can clash, and on some occasions, galloping can also result in phase-to-phase contact. Subspan galloping can also occur where multiple oscillations occur in a single span. **See Figure 10–4**.

Figure 10–4 The differences between galloping, vibration, and subspan galloping conductors. *(Delmar, Cengage Learning)*

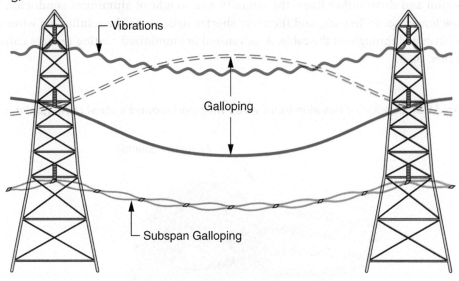

There continues to be much study on ways to reduce or eliminate galloping. Some solutions have been air-flow spoilers, midspan dampers, spacer dampers, detuning pendulums, and such circuit design changes as shorter spans.

Overhead Conductors

Types of Overhead Conductor

Overhead transmission line conductors are always bare. They are made up mostly of aluminum stranding with some kind of steel reinforcement. An overhead distribution conductor can be bare, covered with weatherproofing, or insulated. Weatherproof covering is apparently a carryover from earlier days when local or municipal electric codes called for conductor insulation. The weatherproof covering provides some protection from low voltages, but today it only provides some additional electrical protection from tree contact. A power line technician should treat weatherproof-covered conductors as though they are bare.

Secondary buses and services are mostly insulated. The live legs and neutral are often wrapped together to form a spun **aerial cable**, quadruplex or triplex cable.

Advantages of Aluminum

Almost all overhead power conductors strung today are made from aluminum, aluminum alloy, or aluminum with reinforcing. The conductivity of aluminum is about 62% that of copper, but the weight of an aluminum conductor is about half of a corresponding copper conductor with an equal resistance and length; in other words, aluminum has a conductivity-to-weight ratio that is twice that of copper. The light weight and cost advantage of aluminum have made aluminum the preferred conductor for overhead applications.

Types of Bare Conductors

The bare conductors available are a compromise between a conductor's tensile strength and its conductivity. Other properties considered in a conductor design are conductor weight per unit length, thermal expansion, elasticity, surface shape drag, fatigue resistance, and ability to dampen vibration and resist galloping.

Examples of some types of conductors include the following:

1. For transmission and distribution lines, the strength and weight of aluminum conductor, steel-reinforced (ACSR), allow longer spans, less sag, and therefore shorter structures. The reinforcing wires may be in a central core or distributed throughout the cable. A galvanized or aluminized coating reduces corrosion of the steel wires. **See Figure 10–5.**

Figure 10–5 An ACSR conductor has aluminum wires wrapped around a steel core. *(Delmar, Cengage Learning)*

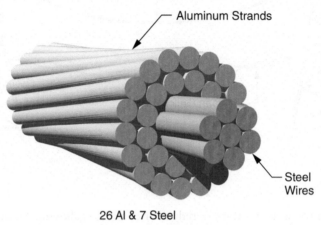

Aluminum Strands

Steel Wires

26 Al & 7 Steel

A large variety of stranding arrangements are available. For example, a 336.4 kcmil ACSR can have a ratio of aluminum strands to steel strands of 6/1, 18/1, 20/7, 24/7, 26/7, or 30/7. Each conductor would have a slightly different diameter. A design consideration is that a larger conductor would not only provide increased ampacity but unfortunately would also cause a greater resistance to wind and ice.

2. All-aluminum conductor (AAC), or aluminum-stranded conductor (ASC), has high corrosion resistance but relatively poor tensile strength. It is used in distribution in which shorter spans do not require steel reinforcement for strength. Corrosion resistance of aluminum has made AAC a conductor of choice in coastal areas where even the best galvanized steel can start corroding in two years.

3. All-aluminum-alloy conductor (AAAC) looks like an AAC, but it has higher-strength, individual, aluminum-alloy strands. The individual strands will be much stiffer to bend than the aluminum in AAC. Compared to an ACSR of the same diameter, AAAC has lighter weight, comparable strength, better current-carrying capacity, and better corrosion resistance. **See Figure 10–6.**

Figure 10–6 The stranding in an all-aluminum-alloy conductor (AAAC) has more strength than aluminum stranding. *(Delmar, Cengage Learning)*

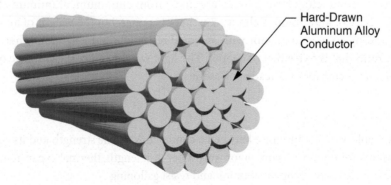

Hard-Drawn Aluminum Alloy Conductor

4. Aluminum conductor, aluminum-alloy-reinforced **(ACAR),** is composed of aluminum stranding and aluminum-alloy stranding. The aluminum-alloy strands provide a conductor with a balance of electrical and mechanical properties. ACAR can have any combination of the two types of strands to provide the best choice between mechanical and electrical characteristics for each application. In appearance, ACAR will look like AAC and AAAC, but ACAR is a combination of the two and will have both easy-to-bend strands and hard-to-bend strands. **See Figure 10–7.**

Figure 10–7 The aluminum alloy provides extra strength in an aluminum conductor, aluminum-alloy-reinforced (ACAR) conductor. *(Delmar, Cengage Learning)*

5. Aluminum-alloy conductor, steel-reinforced **(AACSR),** has steel strands for added strength. AACSRs have approximately 40% to 60% more strength than comparable ACSRs of equivalent stranding and only an 8% to 10% decrease in conductivity. It is a very high-strength conductor used for extra-long spans or for use as a messenger/neutral cable for spun secondary.

6. An ACSR with trapezoid-shaped aluminum wires **(ACSR/TW)** is a compact conductor. With a conductor diameter equivalent to conventional ACSR, there is a 20% to 25% increase of aluminum area. This provides a significant decrease in the resistance and an increase in the current-carrying capacity of the conductor, as well as a reduction in vibration to levels where vibration dampers are not needed. Less diameter for a given load allows using less sag and longer spans. It also means smaller ice and wind loads. **See Figure 10–8.**

Figure 10–8 The stranding in a compact trapezoidal conductor allows more aluminum for a given diameter conductor. *(Delmar, Cengage Learning)*

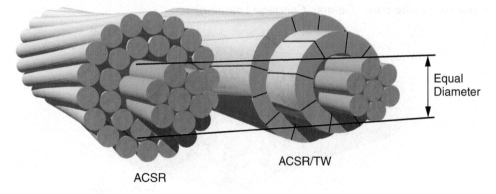

7. Aluminum composite conductor-reinforced **(ACCR)** is similar to ACSR except that the core is a composite material instead of steel. The core consists of a number of individual composite wires that look like the steel wires in ACSR, but each composite wire provides strength at about half the weight of steel, good conductivity, and corrosion resistance. Each composite wire contains many thousands of ultra-high-strength, micrometer-size fibers. **See Figure 10–9.**

An existing line can be restrung to approximately double its ampacity with no increase in conductor diameter, less weight, and no requirement to upgrade the structures.

Figure 10–9 The strength of aluminum composite conductor-reinforced (ACCR) comes from the composite core. *(Delmar, Cengage Learning)*

8. Aluminum composite concentric-lay stranded **(ACSS)** conductor is used for overhead distribution and transmission lines. Steel strands form the central core of the conductor with one or more layers of aluminum 1350-0 wire"0" (fully annealed or soft) temper aluminum. It is designed to operate continuously at elevated temperatures up to 250°C without loss of strength; it sags less under emergency electrical loadings than ACSR; it is self-damping if prestretched during installation; and its final sags are not affected by long-term creep of aluminum. These advantages make ACSS especially useful in reconductoring applications requiring increased current with existing tensions and clearances, new line applications where structures can be economized because of reduced conductor sag, new line applications requiring high-emergency loadings, and lines where aeolian vibration is a problem.

9. Aluminum conductor composite-reinforced **(ACCR)** core is composed of aluminum oxide (alumina) fibers embedded in high-purity aluminum. The core material is called composite but has no polymers; it is a fiber-reinforced metal. Similarly, aluminum conductor composite core **(ACCC)** uses a carbon glass fiber mixed into a thermoset resin. (**See Figure 10–10**). Because these conductors can withstand higher temperatures, it can be loaded with up to twice the equivalent size of ACSR conductor, making it an ideal replacement for existing ACSR using the same tension and ground clearance without the need to change the existing structures.

Figure 10–10 The aluminum conductor composite core, trapezoid wires (ACCC/TW) is similar to the ACSR/TW except for the composite core. *(Delmar, Cengage Learning)*

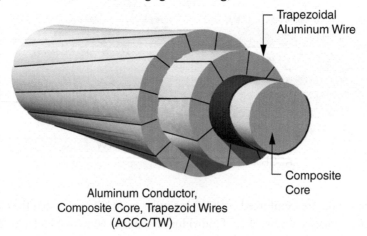

Trapezoidal
Aluminum Wire

Composite
Core

Aluminum Conductor,
Composite Core, Trapezoid Wires
(ACCC/TW)

10. An ACSR self-dampening conductor is also vibration resistant. It is constructed of steel wires in the core surrounded by layers of trapezoid-shaped aluminum wires. It is designed to keep a small gap between the layers of wire while under tension. The interaction of the different natural vibration frequencies of the steel core and aluminum layers provides an internal dampening effect. **See Figure 10–11.**

Figure 10–11 The small gap between the layers of wire creates a self-dampening conductor. *(Delmar, Cengage Learning)*

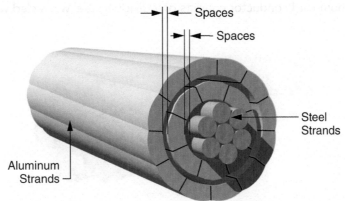

11. Vibration-resistant (VR) or twisted (T-2) conductor composed of two identical conductors twisted together, giving the conductor a spiraling figure-eight shape. VR conductor is used in areas subject to vibration and galloping due to wind or ice. The spiral shape presents a continuously changing conductor diameter to the wind, which disrupts the force of the wind on the conductor. This type of conductor can be sagged to full allowable tension without the need for additional vibration protection. During repair or splicing, each of these two conductors must have its own grips and hoist installed. Any splices should be about 15 feet (3 m) apart. **See Figure 10–12.**

Figure 10–12 The continuously changing diameter of the two conductors in the wind creates a vibration-resistant conductor. *(© Van Soelen)*

12. Aluminum-weld **(AW)**, or aluminum-clad, conductor has aluminum cladding bonded onto each individual strand of high-strength steel wire. It has the strength of steel with the conductivity and corrosion resistance of aluminum. A wire size of "7 No. 5" means there are seven strands of #5 wire. Aluminum-weld conductor is often used as overhead ground wire (shield wire), as a messenger (strand) wire, and even as guy wire in corrosive areas. Similarly constructed is copper-weld (CW), or copper-clad, conductor, which is used as a buried ground wire (counterpoise) along a transmission line. **See Figure 10–13.**

Figure 10–13 An aluminum-clad conductor consists of individual steel wires clad with aluminum. *(Delmar, Cengage Learning)*

Covered Conductor

Covered conductor prevents an immediate short circuit when a conductor makes contact with a tree or other grounded objects. Tree wire/aerial spacer cable as seen in **Figure 10–14**, usually has one layer of insulating material. One layer could prevent an immediate short circuit when there is contact with trees or other grounded objects. Covered conductor is not shielded with a grounded sheath. The insulation for covered conductor cannot be depended on for personal safety. Installing grounds and making connections may require stripping the insulation off. Insulation piercing connectors (IPCs) are available to ease making connections to covered conductor.

In fire-prone areas where a downed wire or arcing to trees is likely to start a wildfire, double- or triple-covered conductor is strung. This conductor is not shielded and is not treated like a shielded underground cable when work is to be done on an energized conductor.

When a covered conductor falls to the ground it may not trip out the protective switchgear. It could be a high-impedance fault. The only indication of a problem might be a customer calling in, or central control could be notified when smart meters indicate an outage.

Figure 10–14 Primary tree wire/aerial spacer cable is installed in the primary position. *(Delmar, Cengage Learning)*

Bundled Conductors

A bundled conductor is an arrangement of conductors in which each phase has two or more conductors in parallel. **See Figure 10–15**. The conductors are held a short distance apart by spacer dampers. Bundled conductors are frequently used for high-voltage and extra-high-voltage transmission lines. From an electrical point of view, a bundle of conductors is one very large conductor and has all the advantages of a very large conductor. **See Figure 10–16**.

Figure 10–15 The four conductors in each phase on this 500 kV double circuit have spacer dampers installed along the length of the line. *(Delmar, Cengage Learning)*

Figure 10–16 The spacer damper provides support and dampens the vibration on a conductor bundle. *(Delmar, Cengage Learning)*

A bundled conductor operates at a lower temperature, lower resistance, and lower line loss than does a single conductor with the same total amount of material. The inductive reactance in the circuit is reduced. For example, a two-conductor bundle has only about 50% of the reactance of a single conductor having the same circular mil area as a bundled pair. The greater the spacing between each conductor in the bundle, the lower the reactance. There are less corona and radio noise from a bundled conductor because corona loss from a conductor is related to the voltage gradient at the conductor surface.

Secondary Bus and Service Drops

Overhead secondary bus and service drops under 600 V can be aluminum or copper, many sizes, open wire, aerial cable, and neutral-supported aerial cables.

Open-wire bus and service drops are typically three wires strung separately and spaced apart. They can be bare, have a weatherproof covering, or be insulated. Open-wire bus is still being strung, generally with insulated wire. Except for very big industrial service drops, almost all service drops are now triplex or quadruplex. Open-wire service drops with weatherproof conductor are electrical hazards to the public because these services often have exposed, energized wires at the customer service rack. **See Figure 10–17.**

Figure 10–17 Open-wire secondary bus is strung in four directions from this pole. *(iStock.com/Flander)*

Aerial cable lashed to a messenger (strand) is popular in some areas. One major advantage is that side guying the poles for service drops is unnecessary. The tension on the messenger is tight enough to keep the poles straight. The messenger can also serve as a neutral and is typically 3/0 AACSR. Terminating a service midspan requires terminating the lashing wire on each side of a new service drop's dead-end and pulling slack into the service wires. If working on a hot secondary, it is wise to use cover-up because all the connections are close and the neutral and lashing wires are bare. **See Figure 10–18.**

Figure 10–18 The transformer secondary is connected to an aerial cable bus. *(Delmar, Cengage Learning)*

Neutral supported cable, such as triplex and quadruplex, is the most common service drop and is used as bus in many locations. When used as bus, a lot of sag occurs in the span, which may encroach on the space reserved for communications cables. As a service drop, the connectors at the customer end must be taped or covered so that energized connections are not exposed.

Working with Overhead Conductors

Making Splices and Connections

Work with conductors normally involves splicing and making connections. To avoid unnecessary line loss and a possible burn-off, splices and connections must be made on a clean conductor to ensure a low-resistance connection.

High-resistance oxides form very quickly on aluminum conductor. Aluminum oxide is a high-resistant transparent film that forms immediately on the surface of aluminum when exposed to air. Even an aluminum conductor that looks clean and bright must be cleaned. Use oxidation-inhibiting joint compounds to prevent reoxidation in the connections. The abrading action of the fired wedge connector will remove some of the oxide from the conductor during installation, but utilities will insist that the conductor be cleaned before any connection is made.

Over time, conductors are prone to creep or flow out (called cold flow) of a tight connector and eventually will cause the connection to be loose. A connector is designed so that it maintains a relatively low average stress, along with sufficient elasticity or spring-type loading, to provide a constant pressure even when there is some conductor metal flow out of the connector. Many old bolted connectors, such as parallel groove clamps and split bolts that are good for copper, will eventually fail on aluminum. Some examples of connectors are compression connectors, fired wedge connectors, and compression terminals with bolted pads.

Copper oxide, the green coating on the wire, is somewhat conductive, which is why copper connections rarely burn off. Approved connectors that join aluminum and copper together have a divider (often with a cadmium surface) between the two metals. The copper should be on the bottom to prevent the copper oxide from leaching over and corroding the aluminum. Copper and aluminum can be kept apart with an H-type connector. **See Figure 10–19.**

Figure 10–19 An H-type connector has a partition that can keep aluminum and copper separated. *(Courtesy of Burndy)*

Making a Tap Connection

A huge variety of connectors and lugs are available for transmission and distribution, ranging from split bolt connectors to large compression connections.

Connections on transmission lines, especially at terminals and for loops (jumpers) at dead-ends, tend to be compression fittings with lugs or pads. **See Figure 10–20.** When the pads are put together, they are cleaned and then bolted together. Others, such as Belleville (spring) washers, are beveled and are used to keep constant pressure on the connection by compensating for the expansion and contraction of the bolts and connection. It is important that they be properly torqued so that they are compressed to only 80% of their height; if no bevel is left in the washer, the connection is too tight and expansion could cause a failure. **See Figure 10–21.**

Figure 10–20 The use of a compression fitting and pad has proven to be reliable for many years. *(Courtesy of Burndy)*

Figure 10–21 Connecting lugs and pads are torqued to a specified value. *(Delmar, Cengage Learning)*

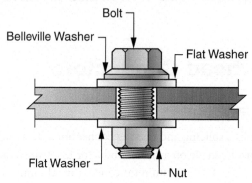

Fired-on wedge connectors are available in sizes suitable for a service entrance, up to and including relatively large conductors on transmission lines. Once the correct-size installation tool (gun), correct-size cartridge, and correct-size connector are chosen, the connection has a relatively fail-free history. The tool is powder actuated, and incorrect use can and has resulted in injuries. The short discussion in this text is not a substitute for the training that comes with use of the tool. A worker must know how and when to use either of the two platforms, choose the correct color shell, use the take-off clip, and clean and lubricate the tool. **See Figure 10–22.**

On distribution lines, there are split bolt, parallel groove, squeeze-on, compression, and many other types of connectors. To make a good connection with these connectors, it is necessary to use special tools, proper torque, and proper die sizes.

Connections to equipment, such as to a transformer or regulator, are bolted, and conductors typically are not installed directly into equipment lugs or pads. Transformers with secondary spade connectors are connected with wires terminated with compression lug and pads. Connections are similar to those made for a transmission line pad, with cleaning, no oxide compound, bolts, and Belleville washers. Wire connected to the bolted lugs of older transformers tends to be copper with an aluminum-to-copper transition elsewhere. **See Figure 10–23.**

Making a Connection to an Energized Circuit

Equipment or another line often must be connected by making a hot connection. The most common connector is a hot-line connector, which should be installed on a stirrup (bale) and not directly on the conductor. A poor connection directly on the conductor can cause the connector to heat up and the conductor to burn apart.

Figure 10–22 Follow the proper job steps to fire-on wedge connections to prevent injury. *(Delmar, Cengage Learning)*

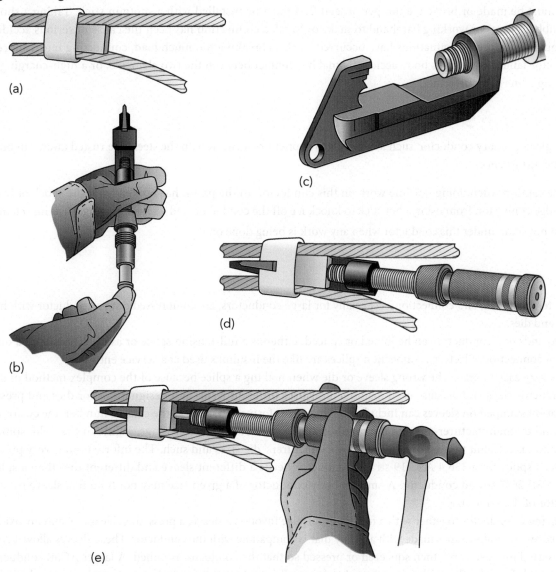

(a)

(b)

(c)

(d)

(e)

Figure 10–23 Compression fittings with bolted pad connections are used on this transformer secondary. *(Delmar, Cengage Learning)*

Fixed connectors, such as fired wedge connectors or compression connectors, are also connected to energized circuits, usually with rubber gloves or with live-line tools, or barehand methods. Anytime a connection to an energized circuit must be made or broken, a jumper or lead first must be installed with a shotgun stick. Taking a shortcut by using rubber gloves or working barehand to make or break a connection has been the cause of serious accidents. All too frequently unforeseen situations have occurred, such as breaking too much load, connecting into a short circuit, a connector failure, a worker's body accidently making contact between the two elements, or a high-charging current energizing a line or transformer.

Working with Small Conductor

An energized primary conductor, such as a brittle #6 copper or #4 ACSR with the steel core rusted away, can break with very little disturbance.

- Use caution when doing hot-line work on this conductor. In the past it has broken from the shock of firing on a wedge connector, from using a hot stick to knock ice off the conductor, and while removing a live-line clamp.
- Do not stand under this conductor when any work is being done on it.

Making a Splice

Splices, terminations, and connections, especially for large conductors, are compressed onto a conductor with hydraulic presses and dies.

Two ends of a conductor can be joined or spliced, either as a full-tension splice or as a splice intended only as an electrical connection. Electrical connection splices are like the insulinks used at a service entrance.

It is very easy to select the wrong sleeve or die when making a splice because of the complex method of designating conductor sizes and because each manufacturer uses different terms and designations for dies and presses. The information stamped on sleeves can include very specific information, such as the catalog number, the customer stock code number, manufacturer-specific die numbers, and conductor-diameter range. While in the field, some of this information is difficult to translate into other manufacturers' die sizes and such. The information is very precise—for example, a splice for a 336.4 ASC 19 strand conductor needs a different sleeve and different dies than a splice for a 336.4 ACSR 26/7 strand conductor. A smooth body conductor of a given size may not hold in a sleeve for a regular conductor of the same size.

Splicing a conductor together with a full-tension splice involves a sleeve, a press, dies, cleaning, and no-oxide inhibitors. Sleeves are hollow tubes made with an alloy that is compatible with the conductor. These sleeves allow a conductor to be inserted into each end, then squeezed or pressed so that the conductor is joined. A larger ACSR conductor has a separate steel sleeve for the steel core and an aluminum-alloy body that is pressed on to provide electrical conductivity. The procedure described for making a two-piece compression splice is generic, and splicing one-piece sleeves, dead-ends, and terminals is similar.

Procedure for Making a Two-Piece Compression Splice

Step	Action	Details
1.	Choose the correct sleeves and dies.	The material used to splice a large ACSR conductor consists of a steel sleeve for strength and an aluminum splice body to carry the electrical current. Refer to company documents or manufacturers' catalogs to ensure that the correct size and type of sleeves are chosen for the conductor size and type.
		Each steel and aluminum sleeve is stamped with the die size and usually the conductor type and size, along with information related to the manufacturer. It is important to use the dies specified on the sleeve because no standard is interchangeable among manufacturers.

2.	Slide aluminum sleeve onto one end of the conductor.	Slide the aluminum splice body over one end of the conductor far enough to allow the installation of the steel sleeve. This is the most embarrassing step to forget, especially when working hot on a tensioned conductor and not noticing the absence of the aluminum splice body until after the steel sleeve is pressed. To fix the situation will require a piece of conductor and two splices.
3.	Expose the steel core.	The aluminum strands on each end of the conductor being spliced are cut back, square with the conductor, half the length of the steel sleeve plus at least another ½ in. (1 cm) because the steel sleeve will expand lengthwise as it is compressed. Ideally, a cable trimmer with a cable trimmer bushing is used to make a square cut. If using a hacksaw, it is critical *not* to nick the steel wires. Taping the aluminum where the cut is to be made will prevent the individual strands from bending while being cut. After the steel wire is exposed, make or keep it straight and clean it. A small piece of wire wrapped near the end of the steel will prevent the strands from unraveling.
4.	Insert the steel core into the steel sleeve.	The steel must be measured and marked, often with tape to indicate when the steel core reaches the middle of the sleeve. If the steel sleeve has a crimp in the center indicating the middle, it will be fairly obvious when the center is reached. The rated strength of the splice will not be achieved if the steel ends are not in the center. The wire wrapped around the end of the steel stranding should slide back as the steel is inserted into the sleeve.
5.	Press the steel sleeve.	Use one of the many suitable compression tools and the die size and type that are stamped on the sleeve. Make the first compression directly over the center, or if present each side of the center indentation, capturing both ends of the steel core. Press from the center toward each end and overlap each compression by about 10%. Depending on the type of die and other factors, the sharp edges may have to be filed off the completed sleeve and the sleeve may need straightening. If the sleeve is being installed on tensioned conductor, back off the hoist and remove the grips so that you have more room to work with the aluminum, and birdcaging will be less likely when pressing the aluminum.
6.	Clean the aluminum stranding.	If the conductor is very black, it can be cleaned using a caustic sodalike lye. Before the steel sleeve is installed, each end of the wire should be inserted into hot water and lye. Eye protection must be worn. Ideally, a special "lye pot" made from a heavy-gauge steel is used for this method of cleaning conductor. The conductor should be rinsed after the cleaning. **See Figure 10–24**.

Figure 10–24 The use of a lye pot can be very effective to clean off excessive aluminum oxide. *(Delmar, Cengage Learning)*

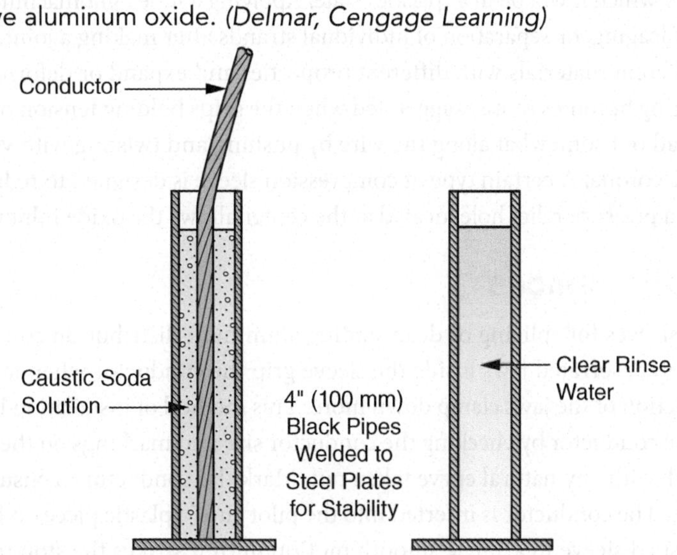

Conductor

Caustic Soda Solution

4" (100 mm) Black Pipes Welded to Steel Plates for Stability

Clear Rinse Water

(Continued)

Step	Action	Details
6.	(Continued)	The standard method of cleaning is to unravel (without bending) the aluminum strands and to wipe each strand with an abrasive cleaning pad or fine steel wool coated with an oxide inhibitor. This can be very dirty work, especially when working hot barehand or with rubber gloves. Remember that the no-oxide inhibitor is conductive.
		It is not normally necessary to unravel the last layer of aluminum strands next to the steel core.
7.	Slide the aluminum body over the sleeve.	Wrap the aluminum strands back into place so that it can be inserted into the sleeve.
		Measure and mark or tape the aluminum conductor so that when the aluminum body is slid into place it is directly over the center of the steel sleeve.
8.	Insert the joint into the filler hole.	One or two filler holes are typically in the aluminum body of a two-piece sleeve so that the cavity around the steel sleeve can be filled with joint compound. Joint compound is conductive and helps prevent oxidation. Joint compound is added with a grease gun or caulking gun to supply pressure. The compound is added until it begins to flow out between the aluminum body and the conductor. An aluminum plug or pin is then hammered into the filler hole until it is flush with the sleeve body.
		Use one of the many suitable compression tools and the die size and type that are stamped on the splice body. The starting point and direction are important for preventing high-stress points in the joint.
9.	Press the aluminum sleeve.	Make the first compressions on each side, just off center of the steel sleeve and move out toward the ends, overlapping each compression. Make no compressions over the top of the steel sleeve. The sleeve will expand out as the compressions are made, and any tape that marked the center should be removed.
		Pressing a one-piece sleeve is done in similar fashion except that the press can start on either side of the middle and can press full length each way. A terminal is pressed by starting at a point closest to the pad (tongue).
		Joints, especially on high-voltage lines, must have sharp edges filed off to make the joint smooth.

The Banana and Birdcaged Splice

A finished splice, especially on smaller conductor, may have bowed (and may now look like a banana). Such splices can be avoided by rotating the compression tool by 90° at each compression location. If you do end up with a curved sleeve, it can be straightened on a flat surface with a hammer. Surprisingly, little force can straighten a compression sleeve while it is being pressed, after which it will be in a "plastic" state. Applying a little light machine oil to the die faces also reduces bowing.

The birdcaging, or separation of individual strands after making a joint, occurs because the sleeve and conductor are made of different materials with different properties and expand or deform at different rates when under compression. The birdcaging becomes more exaggerated when the grips holding tension on the conductor are close. Birdcaging strands can be spread out somewhat along the wire by pushing and twisting with your hands. On high-voltage lines birdcaging can produce corona. A certain type of compression sleeve is designed to reduce birdcaging—as it is pressed first from the outer ends, a pressure relief hole located at the center allows the oxide inhibitor to bleed out while pressing.

Automatic Splices

Automatic sleeves for splicing or dead-ending aluminum distribution conductors are popular because they are easy to install. Tapered, serrated jaws inside the sleeve grip the conductor when tension is applied. As more tension is applied, the wedge action of the jaws clamp down more. This method of installation has to be quite precise. First, select the proper splice for the conductor by checking the conductor size and markings on the splice. The conductor must be straight when it is inserted, with any natural curve taken out. Mark the conductor to ensure that the conductor has reached the center of the sleeve. The conductor is inserted into the pilot cup (a plastic piece on the end of the sleeve to ensure centering) and into the pushed sleeve in a single smooth motion until it strikes the stop in the center. If the conductor does not go in smoothly, conductor strands are probably getting caught between the jaws inside the sleeve, which will prevent the jaws from working properly. Forcing and twisting the conductor will not work. Remove the sleeve and start again with a new one. Set the jaws initially by applying tension before letting off the hoist.

Automatic splices should not be used in some places and for some modes such as in a slack span. They are intended to be installed on conductor under tension and are not suitable for being strung and run through stringing blocks. They also are not suitable as an electrical connection in a nontension mode, as a compression sleeve is sometimes used. A utility may restrict use of automatic splices in coastal areas because of corrosion concerns, especially if it also does not use ACSR because the steel core corrodes.

Note: Automatic splices are not to be reused.

Implosive Sleeves

Implosive sleeves, dead-end hardware, and connectors have an implosive charge wrapped around them. After the usual preparations and instead of using a press and dies to compress the sleeve, the implosive (not explosive) charge is set off and virtually crushes the sleeve and conductor into a solid piece of metal. The result is an amazingly smooth, compressed sleeve. A separate steel sleeve is used on ACSR conductor. The implosive sleeve does not expand outward, so there is no birdcaging. The sleeves can be used during stringing and will go through stringing blocks.

The actual implosion has a lot of energy and is very loud. People living nearby are usually alerted to what is taking place. On a construction project, all the splices prepared on a given day are imploded at one time. On a dead-end tower with three bundled conductors of four each, 12 implosions may be set off simultaneously.

Exothermic Connections

Exothermic welding (manufacturer trade names such as Cadweld, Thermoweld, and Techweld Cadweld may be more familiar) bonds two materials into an electrical connection by melting them together. An electrical connection is made by pouring superheated, molten-copper alloy into a mold that contains the conductors being joined. The heating is due to a chemical reaction in which temperatures can reach 4,000°F (2,200°C).

Exothermic connections are common for electrical connections to ground rods or to fencing in a substation. The connector joins together materials such as copper, steel (plain or galvanized), clad steel, bronze/brass, and stainless steel.

Note: A wet mold can lead to an explosion caused by steam, and hot molds can be fire hazards.

Tools for Making a Splice or Connection

A large variety of tools crimp or compress sleeves and connectors. There are hand-ratchet crimpers and hydraulic presses. Hydraulic presses are available for almost all types of compression splices and connectors. The hydraulic pump can be powered by hand, foot, a regular electric motor, a gasoline engine, or a battery, as well as through an intensifier from a truck hydraulic system.

Generally, the dies will fit only the press for one manufacturer, so using the correct combination of tool, die, and connector is important for ensuring that the compression is made to the intended specification. A learning curve characterizes the process of mastering the terms used for the different presses—terms such as the Burndy designations for the OH25, MD6-8, and Y35. **See Figure 10–25.**

Figure 10–25 The hand ratchet crimper is used on INSULINK™ sleeves. *(Courtesy of Burndy)*

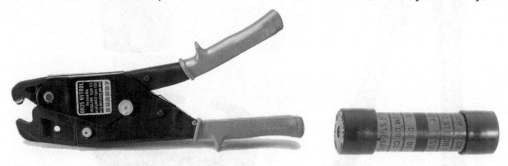

The OH25 hand ratchet crimper (linkit press) allows a one-hand operation for compressing insulink and linkit connectors commonly used at service entrances. One compression on each side of the connector does the job. It works with a ratchet mechanism that cannot be reversed once it has started, so the wire has to be held in the center until started.

The MD7-6 is a hand-operated press that can take considerable muscle. It is commonly used for splices and connectors on smaller conductors. **See Figure 10–26.**

Figure 10–26 An MD7-6 press is typically used on #2 ACSR and smaller conductors. *(Courtesy of Burndy)*

A Y35 is an intermediate hydraulic press. The varied types of hydraulic presses have different power sources. One type is pumped by hand, another type is battery operated, and yet another type is operated from an intensifier on a truck hydraulic system. A lighter hydraulic oil will allow it to be operated more easily on a cold day. A heavier oil is recommended for regular situations. **See Figure 10–27.**

Figure 10–27 A crimper can be pumped by (a) hand, (b) battery-operated, or (c) from an intensifier on a truck hydraulic system. *(Courtesy of Burndy)*

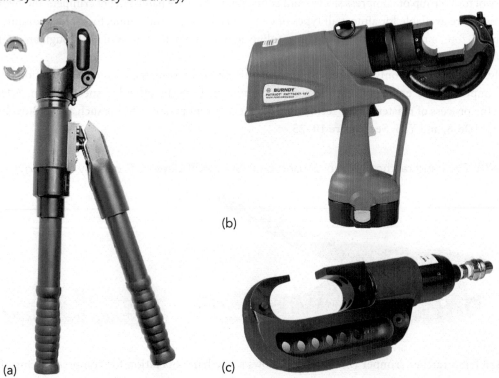

(a) (b) (c)

The Y60BHU, a 60 ton HYPRESS™ head, can come with a hand pump, foot pump, and all the other power systems previously described. Such a press can apply 40 tons of force. **See Figure 10–28.**

Figure 10–28 A Y60BHU, 60 ton HYPRESS™ head is rarely pumped by hand or foot. *(Courtesy of Burndy)*

Underground Cable

Types of Underground Cable

There are high-voltage transmission line cables, distribution voltage cables, and secondary voltage underground cables. Cables are generally identified or described according to insulation and shielding type, **see Figure 10–29.** For example, there are paper-insulated lead-covered (PILC) cables, ethylene propylene rubber **(EPR)**, and **cross-linked polyethylene** (XLPE) cables.

Transmission Cable

Early transmission cables were commonly pipe-type oil-paper insulated cables. The insulated cable is put in a pipe, and the pipe is filled with an insulating oil. Oil-filled cables are referred to as high-pressure fluid-filled (HPFF) cable. Older cables tended to be low-pressure oil-filled (LPOF) cables, now called self-contained fluid-filled (SCFF) cable. Oil has prevented voids (a common cause of failures in other types of cables) in the insulation and allowed continuous checking of the cable condition, as well as monitoring of the pressure and condition of the oil at the cable terminal oil-pumping stations. Environmental concern about oil leaks has resulted in some of these cables being converted to high-pressure gas-filled (HPGF) cable.

Gas-insulated lines use a metal busbar, supported on insulated spacers in the middle of the pipe, and the pipe is filled with an insulated gas such as SF_6.

Working with transmission cable tends to be a specialized cable trade. A power line technician's involvement may be limited to placing grounds at **potheads** or doing other work where lines equipment is needed. **See Figure 10–30.**

Figure 10–29 Some types of underground cable.

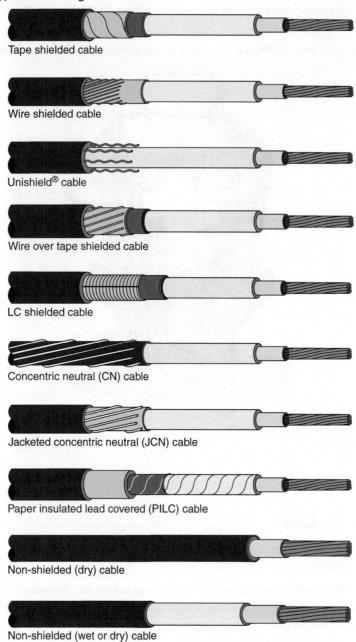

Tape shielded cable

Wire shielded cable

Unishield® cable

Wire over tape shielded cable

LC shielded cable

Concentric neutral (CN) cable

Jacketed concentric neutral (JCN) cable

Paper insulated lead covered (PILC) cable

Non-shielded (dry) cable

Non-shielded (wet or dry) cable

Figure 10–30 Work on oil-filled 230 kV overhead to underground transition potheads is specialized work. *(Delmar, Cengage Learning)*

Modern underground transmission lines are cross-linked polyethylene-insulated (XLPE) power cable, which is similar in design to XLPE distribution cable and much easier to splice and terminate. **See Figure 10–31.**

Figure 10–31 A transmission XLPE underground cable design is very similar to a distribution cable. *(Delmar, Cengage Learning)*

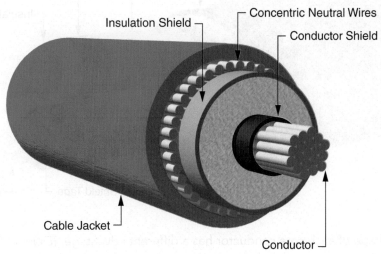

Insulation Shield — / Concentric Neutral Wires / Conductor Shield

Cable Jacket — / Conductor —

Distribution Cable

Earlier distribution voltage cables were PILC cables. The paper insulation was impregnated with oil, and the lead covering acted as a protective sheath. A lead-covered cable can withstand rainfall runoff or oil from the streets better than modern polymer cables. There is, however, a concern about the environmental effects of lead and oil. A substitute copper-alloy sheath is available for this type of cable.

Polymeric insulation has replaced the PILC, with each manufacturer having its own additives and process to improve upon the insulation and cable design. A dictionary definition of polymer is not particularly helpful, but it is worth citing: "a synthetic compound consisting of large molecules made up of a linked series of repeated simple monomers." Common polymeric cables are cross-linked polyethylene (XLPE), tree-retardant cross-linked polyethylene (TR-XLPE), or ethylene propylene rubber (EPR).

Submarine Cable

Submarine cable is like underground cable but with an added protective shield around it.

Secondary Cables

Underground secondary cables are typically three or four insulated conductors bundled together. Each conductor is insulated, generally with a 600 V cross-linked polyethylene (XLPE) insulation or jacket.

Cable Design

Underground cables have a common design requirement, which is to protect and ensure the continued integrity of the insulation. The electric field surrounding the conductor must be kept uniform so as not to become concentrated in any part of the insulation. If the electric field becomes concentrated in one area, the electrical stress in that location will lead to eventual insulation failure.

Cables at each voltage can be single phase or three phase bundled into one cable. **See Figure 10–32.**

The conductor inside all that insulation and protection can be aluminum or copper and of various shapes. The concentric round conductor is the most common. A cable with a compact round conductor, or the compact sector conductor of the same size as a cable with a concentric round conductor, has more conductive material and more ampacity. The annular and segmental conductors were designed to reduce the skin effect of the cable. The center of the annular and the space between the segments of the segmental conductors are nonconductive. **See Figure 10–33.**

Figure 10–32 Only one cable needs to be strung to provide a three-phase service when three phases are bundled into one. *(Delmar, Cengage Learning)*

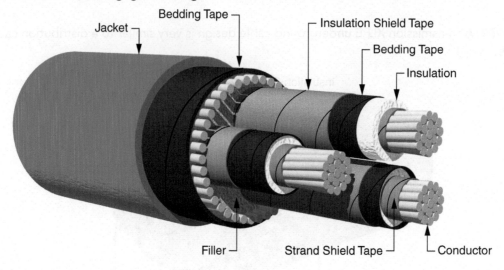

Figure 10–33 The shape of each core conductor has a different advantage. *(Delmar, Cengage Learning)*

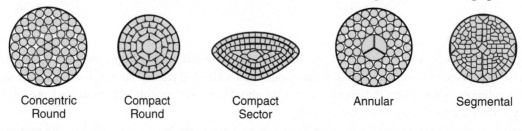

Concentric Round   Compact Round   Compact Sector   Annular   Segmental

Cable Shielding

The insulation of a cable has a semiconducting shield at both the inside and the outside of the insulation. A semiconducting layer on each side of the insulation spreads the electric field uniformly. **See Figure 10–34.**

Figure 10–34 The conductor and insulation shields protect the insulation from concentrated electrical stresses. *(Delmar, Cengage Learning)*

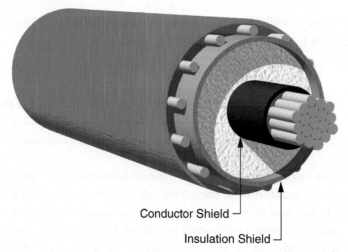

Conductor Shield

Insulation Shield

The **semiconductor** layer next to the conductor is referred to as the conductor or strand shield, which spreads an irregular electric field uniformly throughout the insulation. Electric stress would otherwise concentrate in certain areas because of the irregular shape of stranded conductor or bends in the cable.

The semiconducting layer on the outer surface of the insulation is the insulation shield, and it prevents a concentration of any electric field induced by the concentric neutral wires. The insulation shield also spreads the electric field uniformly over any minor cable damage, thereby reducing any concentration of an electric field at the damaged point.

When making splices or terminations, the insulation shield must be stripped without damaging the vulnerable insulation. Any voids or air pockets when applying the insulation shield (semicon) could result in a phenomenon called partial discharge and could eventually damage the cable. There can be a difference of potential between the energized cable core and the grounded insulation shield across any air trapped in the cable. Ozone produced in the air pockets breaks down insulation.

A metallic screen or armor is applied over the insulation shield. This metallic armor, typically called a concentric neutral consists of a screen or spiraled wires and is designed so that it serves as a path for the return current (neutral). The metallic armor acts as a path to ground to relieve electrical stress from the insulation. The metallic armor acts as a ground wire where it can provide a path for fault current and reduce any touch potentials when contact is made with the cable. The metallic armor (concentric neutral) can be bare or have another protective jacket over it.

Cause of Cable Failure

One cause of a cable failure is when electric stress is allowed to concentrate in one area. The semiconducting cable shield and insulation shield normally spread out any electrical stress, but any sharp bends, voids, or sharp pressure against the cable concentrate the electric stress at that point. When removing semiconducting and insulation shielding material it is critical to make an even square cut. Jagged edges become high stress areas when the electric field concentrates at the ends of a jagged cut. Avoid lifting the semicon off of the insulation during peeling of the semicon. Air can become trapped when it is pressed back down.

Water **treeing** is also a cause of premature failure. Cable can have some water or moisture in it. The moisture enters the ends where the cable is cut during storage or installation and is absorbed into the insulation filling any voids. Cable insulation breaks down to form stress lines in the shape or appearance of a tree; the root of the tree is at the cable, and the stress lines spread out from there toward the outer edge of the cable. Distribution submarine cable usually has a solid conductor instead of a stranded conductor to reduce the probability of water migrating between the strands. In TR-XLPE cable, a tree-retardant semiconduction compound is used that fills the area between the conductor strands to prevent moisture migration into the cable.

Cable Splicing and Terminations

Although there are a lot of cable types and a lot of different splicing kits and tools, the principles applied are the same for all. For a splice, this means exposing the inner core so that the core conductor(s) can be spliced and then rebuilding the cable so that a **conductor shield,** an insulation shield, a neutral, and a jacket are installed in such a way that there will be no added concentration of electrical stress on the insulation. **See Figure 10–35**.

Figure 10–35 A cable splice is made so that there will be no excessive electrical stress on the insulation. *(Delmar, Cengage Learning)*

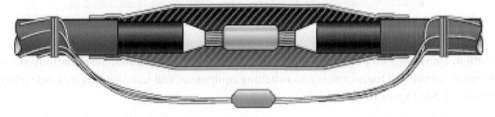

Making a termination means that the cable's shielding is being terminated. The core is exposed, and then the cable is rebuilt so that the exposed end will have a long tracking distance back to the cable shielding and neutral. **See Figure 10–36.**

Figure 10–36 A taped cable termination provides an insulated distance between the energized core and the cable. *(Delmar, Cengage Learning)*

Because there are so many different types of cable and splicing and terminating methods, a cable splicer must follow the manufacturer's or owner's specifications to the letter. Failure to use the proper splice for the type of cable, to handle the cable without adding undue bending stress, to keep the cable clean and dry, and to use exact measurements will lead to eventual failure. Although there are splices that depend less on tape, the generic taped splicing procedure that follows illustrates a rationale that applies to almost all splices.

The same principles apply to making a termination. There are many ways of terminating cable, depending on the voltage, the size of cable, the manufacturer's products, and the purpose. Terminations are made at transitions from underground to overhead. On higher voltage, the transitions tend to be called potheads. **See Figure 10–37.**

Figure 10–37 Three potheads on this pole are fed from a 46 kV subtransmission line. *(Delmar, Cengage Learning)*

On distribution, transitions on poles tend to be called riser poles or dip poles. Terminations are also made into elbows for dead front operating at transformers and switching equipment, and terminations are made into the switchgear at live front equipment. **See Figure 10–38.**

Figure 10–38 An elbow termination such as shown in this cutaway can be used as a switching point. *(Delmar, Cengage Learning)*

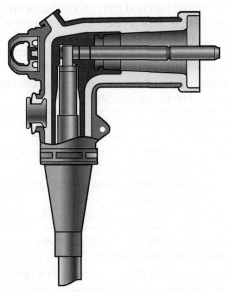

Procedure for Splicing a Single-Conductor Concentric Neutral

Step	Action	Details
1.	Train the cable into position.	The cable must be trained into position in such a way that it curves with the natural curve in the cable. A sharp bend can leave voids between the insulation and the conductor shield and also can leave electrical stress points at a sharp bend.
2.	Cut the cable.	Cut the cable where the splice is to be made. When the cable is cut leave enough length to jumper the metallic shield/concentric neutral wires. Because the core is being exposed, it is at this stage where the tools, hands, and environment must be kept clean and dry.
3.	Lay out the dimensions.	Using the manufacturer's or owner's specifications, prepare to remove the jacket at the exact location on the cable.
4.	Remove the cable jacket or sheath.	There are many types of cables and jackets. The surface of the cable jacket or sheath should be cleaned to remove all pulling lubricants, oils, and dirt that can later contaminate the splice. Cleaning a lead sheath can mean using a shave hook or rasp until it is shiny, and a polymer jacket can be cleaned by using an abrasive cloth. The finished splice will be overlapping onto the cleaned jacket. Use the special tools available for removing the jacket for the distance specified, with the understanding that the cable shield and the insulation must not be cut or damaged. Clean the removed jacket and sheath material from the work area.
5.	Remove the metallic shielding.	The metallic shielding is often composed of wires that must be controlled. Tape hose clamps; other methods can be used. Depending on the location of the cable and other energized cables, the metallic sheaths of the two cables being joined should be jumpered. Start each strand of the wire sheath at the end of the cable, usually with a knife, while being careful not to nick the cable insulation.
6.	Remove the semiconducting insulation shield.	The semiconducting material that makes up the conductor shield must be removed completely from the insulation. It is a good practice to leave 0.25 to 0.75 in. of the material to allow the semiconductor to connect both cables being spliced. The semicon material must be scraped off to leave the insulation clean. If cleaning with solvents or other cleaners, ensure that the cleaner is compatible with the cable insulation. When cleaning, wipe away from the end and toward the semicon.

(Continued)

Step	Action	Details
7.	Remove and pencil insulation.	The insulation is removed from the conductor core for a distance of half the sleeve or connector length, plus an allowance for a short length of exposed conductor between the insulation and the sleeve. Some materials are very hard to cut, and the application of heat from a hair dryer will ease the task. The insulation at the sleeve end is penciled to a specified distance with a knife or, preferably, a penciling tool.
		The space between the sleeve and the penciled end of the conductor is used to make an easier transition for the shielding tape.
8.	Splice or connect the conductor.	When using a sleeve to splice the conductors, use the sleeve, dies, and press specified for the type of conductor and in the splicing kit specifications. When splices are made between two different sizes or between copper and aluminum, sleeves made intentionally for that purpose must be used.
		If using a soldered sleeve, follow the instructions to prevent overheating and to allow for the proper cleaning of the joint.
		Make the joint smooth. Be sure to remove any sharp edges on the joint.
9.	Apply new conductor shield—semiconducting tape.	The conductor shield is conductive, prevents any electric stress from concentrating in one area, creating a faraday cage and fills any small indent at the sleeve or connector. Start the semiconductive tape at about ⅛ in. (3 mm) up the penciled slope, and tape a smooth layer back across the splice to the same position on the other slope.
10.	Clean the exposed cable insulation.	Remove all dust and contamination with an abrasive cloth or an approved solvent.
11.	Apply insulating tape.	Use calipers to measure the diameter of the sleeve/connector on the core, then double the width of the caliper. The doubled distance will be the thickness that the insulated tape will be applied over the connector. Apply tape to build up the insulation while stretching it to three quarters width. Apply tape half-lapped over the sleeve so that it is built up to the thickness set by the calipers while sloping the tape back to the original cables.
12.	Apply the insulation shield—semiconducting tape.	Apply the semiconducting tape, stretching to half of original width, and half-lapping for the full length of the splice until it covers the existing semiconducting layer on each end of the splice by 1 in. (2.5 cm).
13.	Apply a protective jacket.	On a taped splice, the protective jacket would be two layers of PVC tape, half overlapped over the full length of the splice.
14.	Connect the two ends of the metallic concentric neutral.	The neutral is joined with a specified sleeve or connector. If a jumper had been installed earlier it can be removed after the splice is made.

Limits to Ampacity

The limiting factor for the ampacity of cable is heat. When the cable is direct buried, the ampacity depends on the thermal resistivity of the earth. If the thermal resistivity is low, the heat is carried away from the cable and the ampacity of the cable is higher.

A cable in a duct does not have direct contact with earth; therefore, the cable will heat up quicker and have a lower ampacity. The center cable in a trench is not able to dissipate the heat as well as the outside cables. The section of cable going up a riser pole, inside a metal cover and exposed to the sun, may limit the ampacity of the whole circuit.

Electrical Current in a Cable Sheath

The electromagnetic field in an underground cable can induce unacceptable levels of current and voltage on the metal sheath and concentric neutral wires. The magnetic field from the current-carrying conductor induces a current in the metal sheath or concentric neutral wires when the sheath is grounded at both ends of the cable. When the sheath is grounded at both ends, a circuit is created for the induced current. The induced current in the sheath flows back to the source through earth. The current on the sheath can be high enough to cause heating and derate the ampacity of the cable. **See Figure 10–39**.

Figure 10–39 Grounds at both ends of a cable create a circuit for circulating current to flow through the sheath. *(Delmar, Cengage Learning)*

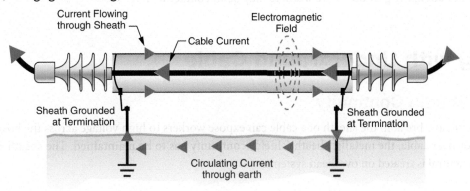

Voltage on a Cable Sheath

Planning engineers sometimes specify grounding a metal sheath at only one end to reduce heating caused by induced current. When only one end of the sheath is grounded, the circuit is open and no current can flow in the sheath. An additional conductor must be strung to serve as a neutral for the circuit.

At the end where the sheath is not grounded is a potentially high voltage between the sheath and the neutral. The voltage is generally less than 40 V. Under fault conditions, the voltage across the open point between the sheath and the neutral could be several thousand volts and hazardous to anyone working at the termination. **See Figure 10–40**.

Figure 10–40 When only one end of the cable is grounded, there is a voltage rise in the concentric neutral at the other end. *(Delmar, Cengage Learning)*

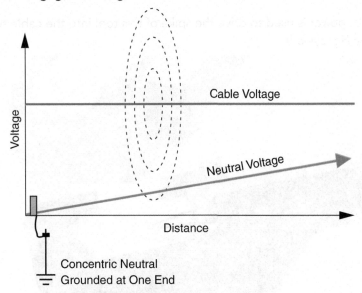

Engineering specifications should include a requirement to flag the open circuit location as a hazard. A surge arrestor could also be specified to limit the voltage across the open point.

Conductors in SF$_6$

In some station designs where space is limited, bare conductors are suspended inside pipes that are filled with an insulating gas, sulfur-hexafluoride (SF$_6$). The pipe containing the conductor can be above ground and is used as a bus in substations. This design has advantages similar to those of underground cable, because less electrical clearance is needed between phases, between phase and ground, and between circuits. The pipes are grounded; therefore, the bus can be laid out along the ground where workers may be in contact with it.

Working with Underground Cable

Maintain Sheath Continuity

In a looped circuit, any break in the sheath of a cable can expose workers to high voltage across the break. When work is performed on or near cable, the metallic sheath (shield) continuity has to be maintained. The sheath should be treated the same way a neutral is treated on overhead systems.

Identifying an Isolated Cable in a Trench

When underground cables are exposed for work in a trench, it is necessary to distinguish between the cable to be worked on and other cables buried in the same trench. A specialized instrument with a transmitter and a receiver is used. The transmitter sends a pulsing DC voltage down the sheath of the cable. The pulses can be varied with different time delays that help to ensure that the signal received is the signal being sent. The receiver is a DC voltmeter that is connected between the cable sheath and a remote ground probe.

Even with this reasonable assurance that the proper cable has been identified, a cable is "spiked" before it is worked on. A spike (spear) is forced into the cable using a spiking tool fired by a cartridge, or the spike is forced in with a hydraulic tool or turned in with a live-line tool. A spiked cable will cause a short circuit if the cable is energized, and it provides a positive physical identification of a desired cable. **See Figure 10–41**.

Figure 10–41 Hydraulic power is used to drive the spike of this tool into the cable to prove it is isolated. *(Courtesy of Salisbury by Honeywell)*

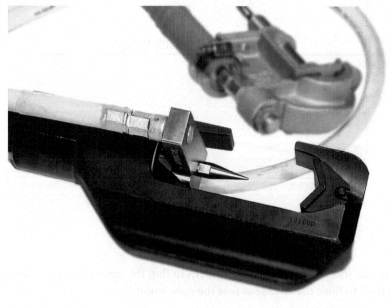

A cable-penetrating ground clamp can be used either as a tool for spiking jacketed cable or as part of a grounding jumper used to bond both ends of an opened cable during work operations.

Do Not Energize a Cable as a Test for a Fault
Very few transient faults occur on an underground cable. A blown fuse generally indicates a cable fault. Test for the existence of a fault before reenergizing.

When energizing a faulted cable, the high fault current generated can become either a ground-gradient hazard near the fault or an explosive hazard in a confined space. The high fault current can also damage an entire length of cable beyond repair.

Locating and Tracing an Underground Cable

Generally, it is a regulatory requirement to test for underground utilities before digging. Call before you dig; call 811 is advertised for public use. Underground Utility Locator Services use ground-penetrating radar (GPR) which sends a radio signal into the ground and the instrument detects any variations reflected back to the surface.

The receiver, a directional antenna, is swept from side to side. The strength of the electromagnetic field increases as the receiver gets closer to the cable. The electromagnetic field can be distorted by other cables, a turn, a change of depth in the target cable, or other metal, such as property stakes.

Three Locating Methods by Putting an Active Signal on a Cable
An underground cable is only visible at the two ends that come out of the ground. Maps and dowsing are unreliable tools for locating cable. Cable locators consist of a transmitter, which induces an electromagnetic field (radio frequency or audio frequency) on the target conductor, and a receiver that traces the field. The methods presented in this section do not apply well to cables in duct banks under city streets. However, cables in duct banks tend to be more protected from faults and better data are available on their routes and locations.

Consider the following cautions when putting an active signal on a cable.

- Ensure that the circuit to be tested is isolated. Reduce the risk of applying leads to an energized circuit. Take out an isolation certification, test and ground, apply the test leads, and then remove the grounds.
- A cable is a capacitor, and it will have a voltage on it long after a high-voltage test is completed. Ground while making the connections for a test and ground the cable before removing the test equipment. After a DC test is applied, it takes a long time to drain the charge off the cable. Wait 15 minutes, then ground it.
- Put up a barrier at both ends of a cable while it is being tested.

Method	Step	Additional Information
1. The transmitter is attached right onto the sheath or to an isolated core conductor with the direct-connect method. **See Figure 10–42.**	**1.** The transmitter can be connected to the sheath if the cable has an insulated cover over it. To trace a cable with a sheath that is continuously grounded (e.g., lead cable), use the lowest possible frequency. **2.** Ground the far end of the cable to complete a circuit so that the signal current returns to the transmitter ground through earth. **3.** Set to a low frequency (less than 10 kHz). A lower frequency is less likely to spill over (couple) to other buried utilities and the signal travels farther.	**1.** This is the most accurate method, because the signal (tone) is isolated to one cable. **2.** Can be connected to the sheath of an energized cable, *but not to an energized cable conductor.* **3.** When attaching the transmitter to a sheath at a riser pole with many cables, the signal current will take the easiest path, which may not be the target cable.

(Continued)

Method	Step	Additional Information

Figure 10–42 To trace the cable, a transmitter is connected directly to the cable or sheath when using the direct-connect method. *(Delmar, Cengage Learning)*

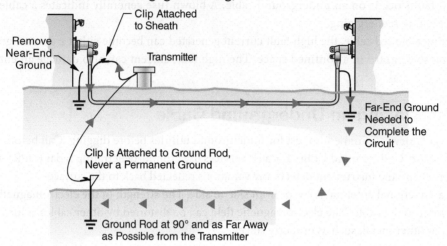

Method	Step	Additional Information
2. The general-induction method induces a signal into a cable from a transmitter set on the ground directly over the cable. As an operator with a receiver passes over the cable, the receiver picks up the tone so its route can be traced. **See Figure 10–43.**	1. The sheath must be grounded at both ends. 2. The receiver should be at least 15 paces away from the transmitter to avoid receiving a signal directly through the air. 3. Try a medium frequency (30 to 90 kHz) to reduce interference from other sources. A high frequency can be used to sweep a large area when looking for all buried utilities.	1. The transmitter can also induce a signal on other cables, pipes, etc., causing the receiver to pick up false signals. This is not a good method for congested areas. 2. For this method, the cable sheath has to have an insulated cover over it to prevent the signal current from following other ground paths. 3. This method is not suitable for cable deeper than 6 ft (2 m).

Figure 10–43 To trace the cable, a signal is induced into the cable with the general-induction method. *(Delmar, Cengage Learning)*

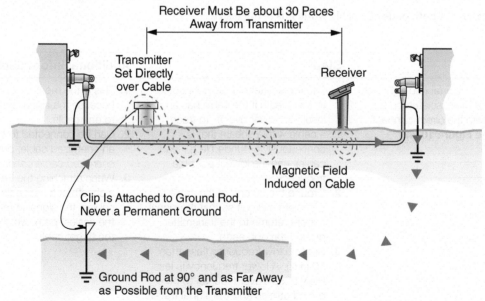

3. The inductive-coupling method uses a C-clamp (coupler) to concentrate the magnetic field signal onto the sheath or conductor. **See Figure 10–44**.

1. To use the inductive-coupling method, the cable sheath must be grounded at both ends to form a complete circuit for the signal current. For secondary cable without a sheath, a ground is placed on each end.

2. Use radio frequencies. Audio frequencies will not work.

1. The C-clamp can be connected over the sheath of energized cables or over energized secondary cables.

2. The concentration of the signal on a cable ensures less spillover to other cables on a riser pole or in a trench.

Figure 10–44 To trace the cable, a C-clamp concentrates the magnetic field signal on the cable with the inductive coupling method. *(Delmar, Cengage Learning)*

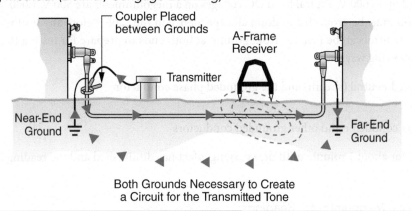

Coupler Placed between Grounds

A-Frame Receiver

Transmitter

Near-End Ground

Far-End Ground

Both Grounds Necessary to Create a Circuit for the Transmitted Tone

Locating a Faulted Section of Cable

When one section of cable has a fault, a fuse or breaker somewhere in the system can trip out many sections of cable. To find the faulted section of cable, a line crew has to progressively open, test, and close each section of cable until they find the fault, unless **fault indicators** have been installed.

A fault indicator will give a visual "flag" after the high magnetic field from a fault current triggers the unit. The fault indicators are put on the cable at various strategic locations. The system can be traced from the source through each fault indicator until one is found in the "no-trip" position. The fault indicator in the no-trip position did not have a fault go through it; therefore, the fault is upstream. **See Figure 10–45**.

Figure 10–45 Fault indicators will give a visual "flag" after the cable at this location has been exposed to a fault current. *(Reprinted with permission of Hubbell Power Systems, Inc., Centralia, MO USA)*

After a fault indicator has flagged a fault, it must be reset to the "no-trip" position. Some types are reset automatically by a normal electrical current, others are reset automatically after a specified time, and still others have to be reset manually.

Using an Insulation Megohmmeter to Verify the Existence of a Cable Fault

A cable fault will be caused either by some kind of insulation breakdown (anywhere from a dead short to a high-resistance fault) or a break in the conductor (open circuit). A megohmmeter (insulation megger, *not* a ground megger) can be used to test the insulation resistance of the cable and to verify if there is a fault that cannot be pinpointed. **See Figure 10–46.** This is a nondestructive test if the proper test voltage for the cable insulation is used—for example, use a 500 V megger on a secondary cable rated up to 600 V. Typical fixed DC voltages on a megohmmeter are 500 V, 1,000 V, 2,500 V, and 5,000 V.

The faulted section must be grounded to drain all capacitance in the cable before any testing is done. Using rubber gloves, the concentric neutral can be isolated from ground at both ends to prepare for testing the cable. The insulation can be tested in the following ways.

1. Between the isolated neutral (sheath) and the grounded phase conductor
2. Phase to phase
3. Between the target conductor and other grounded conductors

The voltage is applied for about 1 minute until the charging effect has diminished and the reading has stabilized.

Figure 10–46 Typical Megohmmeter Readings

Meter Reading	Type of Fault	Notes
0	A shunt fault, where the fault path is from the high-voltage conductor to ground or to the sheath (a short circuit or dead short).	Look for sources of potential dig ins.
More than 0 (hundreds of ohms resistance)	A low-resistance shunt fault allows current to flow to ground with very little resistance. Overcurrent protection will trip out the circuit.	The location of this type of fault is relatively easy to pinpoint with standard instrumentation. This is the most common fault because very fast circuit protection tripped the circuit before the cable was damaged more. Circuit protection may hold for a while when the cable is re-energized.
Less than infinity ∞ (thousands of ohms resistance)	A high-resistance shunt fault, where often both the center conductor and neutral are still intact. (A cable is considered good if it reads 1 megohm (MΩ) for each kilovolt of cable rating, that is, a 5 MΩ for a 5 kV cable.)	
Infinity ∞	Indicates a series fault where the conductor and/or neutral are burned open, or it indicates a cable without a fault.	Ground at the far end of the cable and measure again. A "0" resistance would indicate that it is not an open circuit.

An Aid to Locate a Cable Fault

The flow chart shown in **Figure 10–47** is a guide to locating a cable fault.

Figure 10–47 An easy to use device to test for a cable fault.

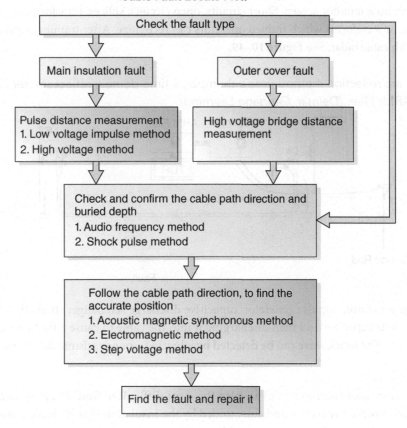

Cable Fault Locate Flow

Check the fault type

Main insulation fault

Outer cover fault

Pulse distance measurement
1. Low voltage impulse method
2. High voltage method

High voltage bridge distance measurement

Check and confirm the cable path direction and buried depth
1. Audio frequency method
2. Shock pulse method

Follow the cable path direction, to find the accurate position
1. Acoustic magnetic synchronous method
2. Electromagnetic method
3. Step voltage method

Find the fault and repair it

Pinpointing the Location of a Cable Fault

After a megohmmeter reveals the most likely type of fault in a cable, choose the instrument suitable for finding that type of fault. The methods described here are most suited to direct bury cables and generally not suitable for cable faults in duct banks under city streets.

Short-Circuit or Low-Resistance Fault

The short-circuit or low-resistance fault is the easiest fault to find. The cable tracing equipment described earlier may be used. After all the cable grounds have been removed, a signal current from a transmitter is forced out where the cable insulation has failed. The voltage gradient above the fault is picked up by a probe or A-frame connected to the receiver. The A-frame is moved and turned in the vicinity of the fault until the signal strength at each leg of the A-frame is equal (a "null" signal), indicating that the A-frame is directly over the fault. **See Figure 10–48.**

Figure 10–48 An A-frame is moved along above the cable to pinpoint a fault. *(Delmar, Cengage Learning)*

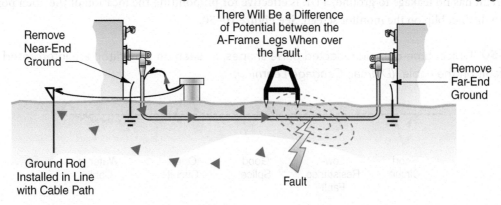

There Will Be a Difference of Potential between the A-Frame Legs When over the Fault.

Remove Near-End Ground

Remove Far-End Ground

Ground Rod Installed in Line with Cable Path

Fault

Cable radar (a time-domain reflectometer, or TDR) can be very effective as it sends a high-frequency pulse along the cable and measures the time it takes for the signal to be reflected back. Irregularities in the cable and their locations can be shown in graph form on a monitor screen. Short circuits, open circuits, splices, transformers, and other irregularities will show up on screen in waveforms which trained operators can recognize. After training, a person can determine the distance to the fault with cable radar. **See Figure 10–49.**

Figure 10–49 The arc reflection method uses a thumper, a time domain reflectometer (TDR), and an arc reflection method (ARM) filter. *(Delmar, Cengage Learning)*

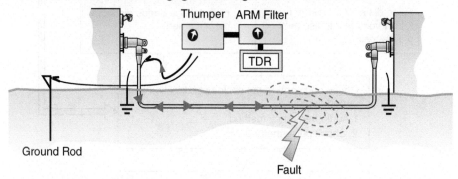

A **thumper** (surge generator, impulse generator, capacitive discharge set, banger) finds the short circuit by storing a high-voltage charge in its capacitor and then discharging it into the cable to cause a flashover and an acoustic shock wave at the fault location. The shock wave can be detected by vibration, sound, seismic detectors, or microphones.

High-Resistance Fault

High-resistance faults (the most common type) are also the most difficult to find. They are often caused by a faulted conductor burning back toward the source and surrounded by the insulation that is "healed over." It may be easier to find a high-resistance fault from the load end of the cable.

A high-resistance fault can be broken down into a low-resistance fault by a thumper or burner.

Caution: Aged cables that have been thumped excessively are likely candidates for a failure within the next year. The voltage chosen should be below the rated cable voltage and generated for as short a time as possible to reduce the risk of damaging the cable at other weak spots. A thumper should not be used on secondary cable in case the signal finds its way into the customer's wiring and ground. A very-low-frequency (VLF) AC high potential (**hipot**) burns much quicker than conventional DC burners.

The least damaging and most effective approach to finding a high-resistance fault would be to connect a thumper, cable radar (TDR), and an arc reflection method (ARM). The thumper "turns on" the fault momentarily as the cable radar captures the image of the brief flashover at the fault location. The ARM filter is needed with this method to prevent the thumper output from damaging the cable radar instrument. Damage to the cable is limited because frequent thumping is not needed and the voltage will not be much beyond the breakdown level as the fault breaks down before the voltage rises too high.

Open Circuit

If an open circuit has no leakage to ground, TDR is effective for pinpointing the location of the open point. An open circuit leaves a distinct blip on the monitor screen. **See Figure 10–50.**

Figure 10–50 Typical time-domain reflected pulse shapes, as seen on a monitor, are interpreted to identify the kind of fault on the cable. *(Delmar, Cengage Learning)*

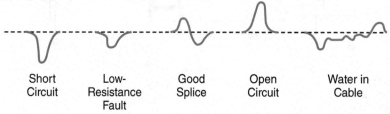

One Source of an Erratic Secondary Cable Fault

A fault on a secondary underground cable often starts out as a complaint of erratic power. A voltage reading at the customer's panel or service location may show a normal voltage.

A break in the insulation of a secondary cable is often a high-resistance fault. When water gets into an aluminum conductor, it slowly corrodes the aluminum and turns it into an aluminum hydroxide, which is a high-resistance, white, powdery material. The voltage may be normal until a load is put on the conductor because the corroded higher-resistance cable will cause an increased voltage drop as the load increases.

Use a 500 V insulation tester to test the service. A reading of anything less than 1 megohm is an indication of a faulty cable. Some insulation testers have the ability to send a 250 V, 500 V, or 1,000 V test voltage for insulation resistance testing. **See Figure 10–51**.

Figure 10–51 An insulation tester such as this one can be used to test a secondary service. *(Courtesy of Megger. The word "Megger" is a registered trademark.)*

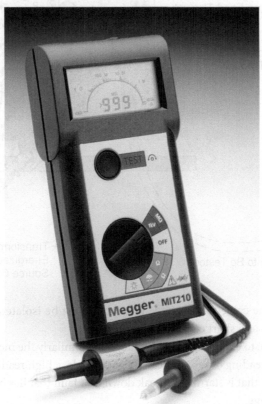

Testing Cables

A new or repaired cable installation is often tested before it is put into service. One method is to apply an overvoltage (hipot) test. Check with your utility for the proper test voltage because a hipot test is destructive. There is a concern about shortening the life of the cable with this test method, and some utilities will not use it.

A DC or the preferred VLF hipot test of a cable measures any leakage current in milliamperes between the phase conductor and the sheath or earth. The level of leakage current is not as critical if it remains at a steady level. If the leakage current increases rapidly during the test, it will indicate a potential fault. Stop the test if it begins to increase, to avoid damaging the cable. Of course you will now have a high-resistance fault to pinpoint.

For distribution voltage cables, an insulation tester (megger) or phasing sticks are common devices used to test after repair or maintenance of a cable.

Using a Hipot Adapter on Phasing Sticks

A DC hipot adapter attached to a phasing tool can test a cable with an AC voltage when using an existing energized AC source and converting it to DC. The DC voltage impressed on the cable will be at about the peak value of the source AC voltage.

The hipot adapter is installed on the meter stick. This stick is connected to an energized source, for example, a transformer bushing or an open cutout at a riser pole. Some bushing adapters and elbow adapters allow making the connection. **See Figure 10–52.**

Figure 10–52 Phasing sticks can be used as a handy hipot tester. *(Delmar, Cengage Learning)*

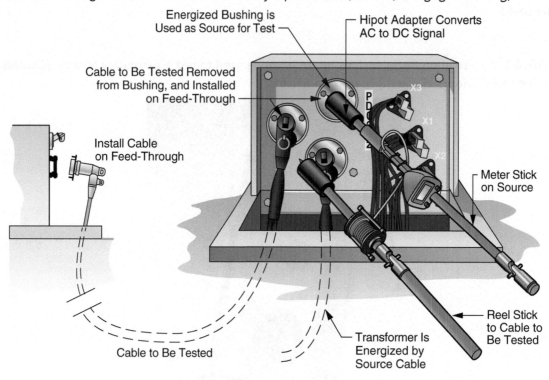

The reel stick is attached to the cable to be tested. The cable must be isolated at both ends because a DC signal through a transformer will show up as a short circuit.

A long cable will take some time to charge up to peak voltage and similarly the meter will show up to a full line voltage until the cable is charged. A meter reading of zero indicates a good cable, a high reading indicates a defective cable, and a fluctuating reading indicates a cable that is starting to break down or a fault that has burned back but will flash over when the cable gets up to or near full voltage.

Rejuvenating Cable

Cable is rejuvenated with the injection of a specific silicone fluid between the wire strands. Among other restorations, it repairs water treeing damage. The location of each splice needs to be found and dug up. The cable is cut and the injection continues at that point and new cable splices are made. The dielectric strength and cable life is increased.

Working with Fiber-Optic Cable

Why Fiber Optics?

Fiber-optic cable is an alternative to copper telephone, microwave, and radio communications between substations and switchgear. It is a superior communications cable because it offers a huge capacity in a very small cable. The optic fibers are made of glass and are nonconductive, so the communications signals are not affected by electric magnetic induction.

Utilities have the infrastructure, skilled staff, right-of-ways, structures, and underground duct banks that make them ideal vehicles for getting into the fiber-optic business.

Fiber-optic cable is also used on hydraulic boom trucks. On very long insulated booms used for barehand work, lightweight fiber-optic cables can replace the hydraulic hoses that are used as a "communication" to change a boom position. These long hydraulic hoses present a hazard should the oil level in the hose fall, creating a conductive partial vacuum.

Because fiber-optic cable can carry an immense amount of communications, utilities are able to sell some of their overcapacity to telephone companies, cable companies, and corporations for their internal intranet communications. Some utilities are running fiber directly to their customers to compete with suppliers of high-speed communications.

What Is Fiber Optics?

Fiber optics use light pulses to transmit information down glass fibers. Digital information, which is composed of a series of 0s and 1s, is transformed into a series of light pulses that are either on or off. Fiber-optic cable is a "light guide" in which the source of light pulses and a light-emitting diode (LED), or laser, is transmitted at one end and received at the other end. **See Figure 10–53.**

Figure 10–53 Electronic signals are transformed to light pulses and back again. *(Delmar, Cengage Learning)*

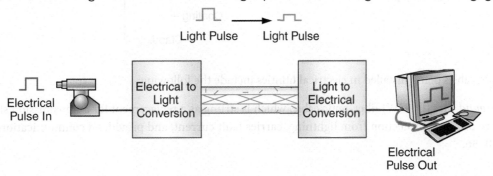

The inside of a fiber cable is coated with a mirror. The light pulses in one end can be seen at the other end because the mirror allows the light signals to bend around humps, bumps, and corners. When a light pulse strikes the mirror surface, it keeps bouncing back into the cable. **See Figure 10–54.**

Figure 10–54 Light pulses are transmitted through a mirror-lined fiber-optic cable. *(Delmar, Cengage Learning)*

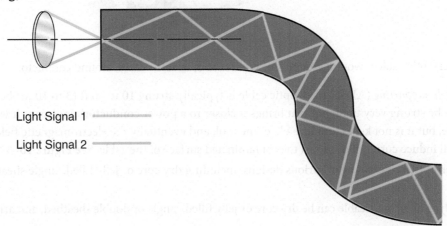

Types of Fiber-Optic Cable

A cable consists of the core, cladding, strengthening fibers, and the cable jacket, which is a coating or buffer. The cable is designed to protect the glass-fiber core and cladding. The optical fiber is a flexible filament of very clear glass that is just a little thicker than a human hair. The core is the light-transmission area of the fiber. The cladding is the layer that surrounds the core. The light in the core strikes the interface with the cladding at a bouncing angle and is trapped in the core. **See Figure 10–55.**

Figure 10–55 Fiber-optic cable construction varies depending on strength requirements and exposure to the elements. *(Delmar, Cengage Learning)*

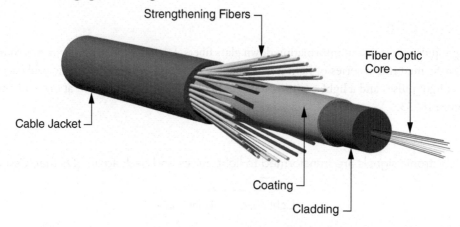

Fiber-optic cables being installed in electrical utilities include the following:

1. *Optical ground wire* (OPGW) fiber-optic cable, which is strung as a replacement for a transmission shield wire. The shield wire provides protection from lightning, carries fault current, and provides a communications channel. **See Figure 10–56.**

Figure 10–56 An optical ground wire (OPGW) is strung as a shield wire on transmission lines. *(Delmar, Cengage Learning)*

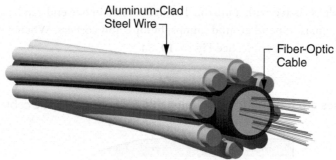

2. *Wrap-type* fiber-optic cable is wound around shield wires and, in some cases, around conductor.

3. *All-dielectric self-supporting* (ADSS) fiber-optic cable is typically strung 10 to 30 ft (3 to 10 m) below existing conductors. It can be strung very tight, but that brings it closer to a power conductor's midspan. It was intended to be dielectric cable, but it is not kept clean like a live-line tool, and eventually the electromagnetic field from the power conductors will induce current flow down the contaminated surface of the cable. **See Figure 10–57.**

4. *Duct-type* fiber-optic cable can be of various designs, including dry core or jelly filled, single sheathed, unarmored, and armored.

5. *Direct burial–type* fiber-optic cable can be dry core or jelly filled, single or double sheathed, and armored.

Figure 10–57 All-dielectric, self-supporting (ADSS) fiber-optic cable is strung near energized circuits. *(Delmar, Cengage Learning)*

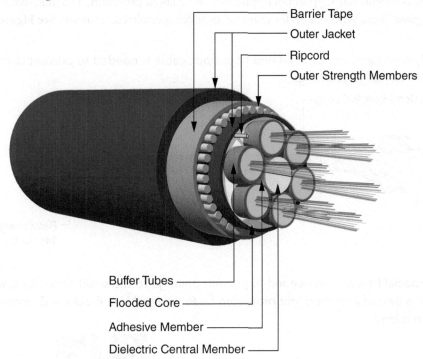

The fiber-optic cables typically used by utilities are loose-tube cables containing several individual cables (tubes), each with a fiber core within the main cable. The tubes are wound around a central member composed of aramid yarn (product similar to Nomex and Kevlar), which provides strength. The area around the loose tubes is filled with gel or water-absorbent powder. The loose-tubes design allows an individual tube to be accessed at intermediate locations without bothering other tubes and fibers. The outer buffer and outer jackets provide protection and strength.

Working with Fiber-Optic Cable

Working with fiber-optic cable is a very specialized undertaking. Some skills, such as stringing, climbing, and underground work, are common to both copper and fiber. This section provides only a brief overview. The specifications for dead-ending, stringing, and splicing will be included among a project's documents. Most of the technical documentation will involve care of the cable. Fiber-optic cable requires a lot of care. ADSS cable, for example, includes hardware such as special dielectric dead-ends, dielectric suspensions, abrasion protectors, a dielectric vibration damper, a dielectric corona coil, and download cushion.

Installation methods are similar for both wire cables and fiber-optic cables. Fiber-optic cable has more strength than copper wire if pulled straight, but the fibers break when bent too far.

A swivel pulling eye is necessary because any pulling tension will cause twisting forces on the cable. Twisting will stress the glass fibers.

Most cables cannot be pulled by the jacket, unless a special cable grip or special-strength members made of Kevlar or aramid yarn, a synthetic polymer that has great strength and resistance to high temperature, are in place and are used.

Splicing requires specialized tools and training that make the work easier than splicing together each tiny glass fiber.

Electrical Tracking on ADSS Cable

Fiber-optic cable was intended to be dielectric, but if it is not kept clean like a live-line tool, eventually the electromagnetic field from the power conductors will induce current flow (tracking) down the contaminated surface of the cable. Dead-end and suspension hardware must be designed to drain tracking without damaging the cable.

Hardware

The hardware used to dead-end and support fiber-optic cable is all about preventing twisting, sharp bends, and vibration that can damage the glass fibers. ADSS cable, for example, includes specialized hardware. **See Figures 10–58a, b,** and **c.**

Figure 10–58a Special hardware to dead-end fiber-optic cable is needed to prevent damage to the small optic fibers. *(Delmar, Cengage Learning)*

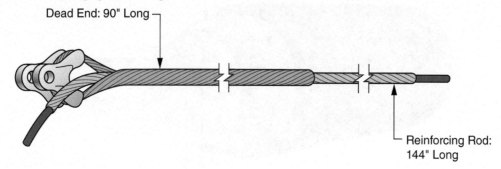

Figure 10–58b Special hardware is needed to prevent damage to the small optic fibers where the cable is suspended. *(Image is owned and copyright protected by Preformed Line Products Company, and reprinted herein with its permission.)*

Figure 10–58c Vibration dampers and armor wrap protects the small optic fibers. *(Delmar, Cengage Learning)*

Splicing Fiber-Optic Cable

Recall that splicing is carried out with specialized tools because of the small size of the individual glass fibers, which are not much larger than a hair. Splicing can be mechanical or by fusion.

With a mechanical splice the ends are joined together inside either the mechanical connector or an adhesive cover. A fusion splice uses an electric arc to fuse the fibers together. Neither splice can be put under tension, and therefore an opti-loop is formed at the splice. An opti-loop is a device that allows the storage of a reserve length cable and the splice without having any sharp bends in the cable. **See Figure 10–59.** The canister may contain a splice or a tap. **See Figure 10–60.**

Figure 10–59 An opti-loop in fiber-optic cable stores a reserve length of cable. *(Delmar, Cengage Learning)*

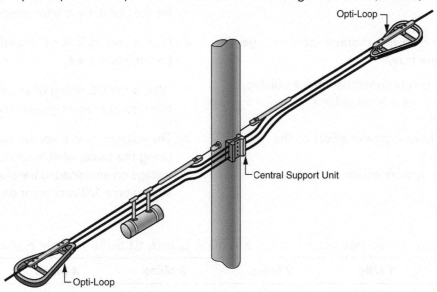

Figure 10–60 Fiber-optic cable splices are found in a canister. *(Image is owned and copyright protected by Preformed Line Products Company, and reprinted herein with its permission.)*

Other canisters are mounted on a tower. The coil bracket has enough cable stored on it to bring the canister to the ground so that splices can be made there with the special tools required.

Review Questions

1. What are three basic effects when an electric current flows in a conductor?

2. Why is the ampacity of a conductor higher in winter?

3. Which of the following characteristics of larger conductors are true?

   ( ) There is a greater potential for ice buildup.
   ( ) There is less heat buildup for a given load.
   ( ) Wind will have a greater effect on the conductor.
   ( ) Larger conductors are strung with less tension

4. How does the sag in a transmission line span affect the ampacity of a conductor?

5. Name two factors that influence the specification for the size of conductor needed.

6. How is corona loss minimized on a transmission line?

7. Why is the DC rating of an underground cable three to four times greater than AC rating?

8. The voltage on an unloaded transformer is 120 V. Using the table, what would be the secondary voltage on an unloaded transformer 2 miles down stream on a 1/0 conductor on a 4.8 kV circuit?

Percent Voltage Drop per Mile for 4.8 kV, 50 A Load, at 90% Power Factor

	1 Mile	2 Miles	3 Miles	4 Miles	5 Miles
#4 ACSR	5.6%	11.2%	16.8%	—	—
#2 ACSR	3.8%	7.6%	11.4%	15.2%	—
1/0 ACSR	2.7%	5.4%	8.1%	10.8%	13.5%
3/0 ACSR	1.9%	3.9%	5.8%	7.8%	9.7%

9. If a person working on a transmission line leaves any rough or sharp points on a conductor or other energized hardware, what effect will that cause?

10. Why is it important to clean aluminum conductor before making a connection or splice?

11. Identify the conductor shield, the insulation shield, the conductor, the cable jacket, and the concentric neutral wires in the drawing.

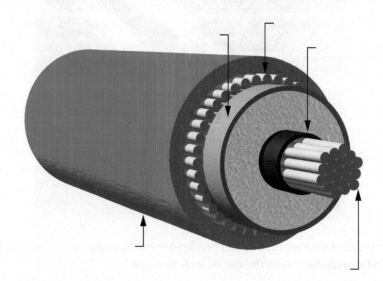

12. Can the insulation on unshielded cable, such as a covered conductor, be depended on for personal safety?

13. Why may sharp bends, voids, or sharp pressure against a cable cause an eventual failure?

14. Why should the cut end of an underground cable be sealed when it is stored on a reel in the yard?

15. What is the purpose of the conductor shield (semicon) and the insulation shield layers in an underground cable?

16. The magnetic field from the current-carrying conductor induces a current in the metal sheath or concentric neutral wires when the sheath is grounded at both ends of the cable. What can be done to reduce this current flow?

17. When the ground is removed from the sheath at one end to reduce heating in a cable, what hazard exists at the ungrounded end of the cable?

18. Before cutting a cable in a trench, how can the worker be sure the correct cable has been isolated?

19. A customer fed by a secondary underground cable complains about erratic power. A voltage reading at the customer may show a normal voltage. What else should be checked?

20. How can phasing sticks be adapted to carry out a hipot test on a distribution cable?

Chapter 11

Operating Switchgear

Objectives

After completing this chapter, you should be able to:

1. Differentiate between operating protective switchgear and isolating switchgear.

2. Describe the protection needed for protection from an electric arc.

3. Recognize a load interrupting switchgear.

4. Explain the verification needed before opening or closing a tie switch between two circuits from different sources.

5. Identify the limit to which a length of line, with no load, can be isolated with a disconnect switch.

6. Take steps to operate switchgear from a ground gradient grid in situations where there is exposure to touch and step potential.

7. Take steps to limit ground gradients when operating switchgear with a telescopic stick.

8. Explain the limits to opening a fused cutout to interrupt load.

9. Use a phasing tool to verify the phasing at open switchgear between two different sources before closing.

10. Identify the switchgear being operated on an operating schematic or drawing and prepare a "switching order" when needed.

11. Recognize the features of the many different types of overhead and underground switchgear that is available.

Switching Characteristics and Switching Hazards

Two Main Types of Switchgear

A lineman operates two main types of **switchgear**:

1. *Isolating switchgear,* such as a **disconnect switch**, an air break switch, or a load interrupter does not operate automatically during a fault but provides operating capability to isolate, sectionalize, or transfer loads at strategic locations. Some types can interrupt a load current, whereas other types have no ability to interrupt any load current.

2. *Protective switchgear,* such as a **circuit breaker**, **recloser**, or fuse, protects a circuit by providing automatic isolation when it is exposed to damaging faults. This type of switchgear can interrupt the extremely high current that occurs during a short circuit. Some protective switchgear, such as breakers and reclosers, will open and then close in again on a circuit for a prescribed number of times. If the fault is transient, customers will have power; if it is permanent, a line crew must be called out.

Arc Hazards

Excessive arcing can destroy a switch, trip out a circuit, and injure linemen. All switching must be carried out without arcing. If an excessive **arc** occurs and molten metal sprays out from the switchgear when switching, some kind of switching error has been made. **See Figure 11–1.**

Figure 11–1 All the properties of an arc can be hazardous. *(Delmar, Cengage Learning)*

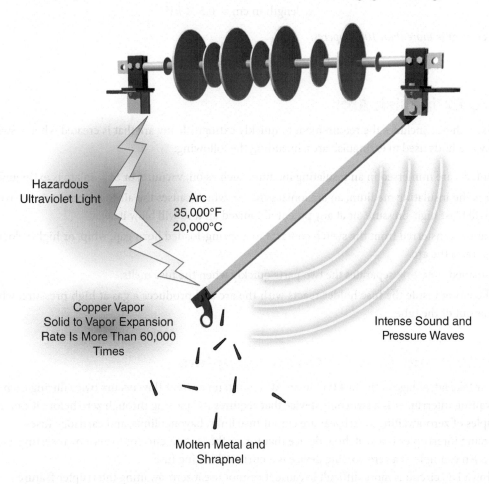

Hazardous
Ultraviolet Light

Arc
35,000°F
20,000°C

Copper Vapor
Solid to Vapor Expansion
Rate Is More Than 60,000
Times

Intense Sound and
Pressure Waves

Molten Metal and
Shrapnel

Many types of switchgear are designed to extinguish an arc, but many disconnect switches have no ability to extinguish an arc. When an arc is established, the heat from the arc causes the air to become ionized. The ionization of air means that electrons have been stripped from their atoms and that the air has become electrically charged. In other words, the air is changed from being an insulator to being a conductor. Once the air between the switch contacts becomes conductive, a follow-through current will flow across the switch gap.

An arc is more probable when opening than when closing a switch. Normally, a circuit can be closed without concern about an arc unless the device did not close the first time and a quick open and reclose (restrike) is attempted. It is the interruption of current that causes an arc.

Arcing generally occurs only when opening switchgear, especially opening switchgear that is not designed to open under load. Arcing can also occur when trying to close switchgear, for example a cutout. If the cutout does not close properly and opens again, a severe arc can occur.

An arc in the confined areas of underground switchgear can easily spill over to grounded cabinets or other phases, exposing a lineman to dangerous ground gradients and an explosive fault.

A potential arc length can be calculated. However, a lineman should not be asked to open a device unless the following conditions are met.

- The switchgear is designed to interrupt load current.
- The switchgear is capable of accepting a load bust tool.
- The switchgear will not be interrupting any load.

An arc length is dependent on the circuit voltage *plus* the current being interrupted.

Formulas that can be used as estimates for the expected arc length when opening a switch follow for when the current is *less than 100 amperes:*

$$\text{Arc length in cm} = 0.5 \times kV$$

and when the current is *more than 100 amperes:*

$$\text{Arc length in cm} = 50 \times kV$$

Designs to Extinguish Arcs

The design of switchgear includes the requirement to quickly extinguish any arc that is created when a switch is opened. There are many methods used to extinguish arcs, including the following:

- When contacts are immersed in an insulating medium such as oil, vacuum, or SF_6, an arc is extinguished quickly.
- When air is the insulating medium, an arc ionizes the air, which causes the air to become conductive. An **air-blast breaker** will blast high-pressure air at any potential ionized air and will blow it away.
- An arc can be transferred from the switch contacts to a spring-loaded horn gap, whip, or high-velocity interrupter, which separates the arc.
- A spring-loaded **fuse link** separates the two parts quickly when the fuse melts.
- A special coating inside the fuse holder reacts with the arc and produces a gas at high pressure, which expels the ionized gas out of the tube.

Zero-Awaiting and Zero-Forcing Interrupters

Most switchgear take advantage of the fact that in an AC circuit, no current flow occurs twice during each cycle.

A zero-awaiting interrupter is a switching device that requires AC passing through zero before it can interrupt a load current. Examples of zero-awaiting switchgear are cutout fuse links, bayonet links, and cartridge fuses.

A zero-forcing interrupter is a switching device that forces the system current to zero by inserting a large resistance into the circuit. An example of a zero-forcing device is a current-limiting fuse.

Interrupting a DC circuit is more difficult because it cannot use a zero-awaiting interrupter feature.

Remotely Controlled Switchgear

Remotely controlled switchgear keeps a safe distance between people and any switching hazards. When there is a choice, such as at a capacitor bank, it is always safer to operate from the control box. A ground-gradient mat should be used when operating at a control box when the box is on the same structure as the switchgear. A fault within the switchgear could short to the equipment grounding system, which will raise the potential on the control box.

Operating Three-Pole Switches

If only one phase of a three-phase circuit trips out, the protective switchgear and relays will experience a large unbalanced load with two phases loaded and one not loaded. The current and the voltage on the whole system will become unbalanced. The unbalance due to one phase being out of service is not as severe to distribution circuits as it is to transmission circuits. Three-phase motors will heat up when one phase is out, but the thermal protection should prevent them from being damaged. All transmission switchgear and a lot of distribution switchgear is gang operated. All three phases will trip out, even if only one phase is faulted.

SCADA Systems

A supervisory control and data acquisition (SCADA) system is an automated distribution system that brings needed information and remote control of switchgear into a central control room. It brings the operation of a distribution system to a level similar to a transmission system. SCADA systems are widely used in such other industries as telecommunications, water, pipelines, refineries, transportation, and nuclear-power plants.

A communications system, such as a telephone line, fiber-optic cable, radio, coaxial cable or, more likely, a secure private broadband network (Internet) brings information from *remote terminal units* (RTUs) in substations and from remote switchgear to a *central processing unit* (CPU) in the control room. The SCADA system gives an operator control room information such as system voltage, feeder loading, and the status of switchgear as open or closed.

A remote-controlled switchgear needs a remote-controlled motor that can open and close the circuit breaker or switch. The motor is set in motion by sending a signal to a motor-operating mechanism. The motor-operating mechanism has a low-voltage supply, a battery, RTU circuitry, and a communications connection to the control room. **See Figure 11–2**.

Maintaining and troubleshooting a SCADA-controlled switch includes bypassing the remote control and operating the switch manually. A line crew can be asked to go to a remote unit and operate it on-site during a breakdown.

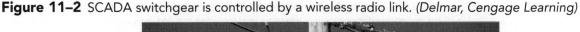

Figure 11–2 SCADA switchgear is controlled by a wireless radio link. *(Delmar, Cengage Learning)*

Understanding the Switchgear Being Operated

To avoid the creation of an excessive arc while switching, a person needs to know the purpose for operating the switchgear and whether the switchgear is capable of dropping load current.

Operating Situations

If	Then
A length of line, with a load on it, is to be isolated or interrupted.	A load current will be interrupted and an arc will have to be extinguished. Switchgear, such as a circuit breaker, a recloser, or a load interrupter disconnect switch, are capable of dropping load and quenching an arc. A disconnect switch without a load interrupter is not designed to interrupt any load. It may have a rating of 600 A, but that refers to the load it can carry through the switch continuously.
A length of line, with no load on it, is to be isolated.	The line can be isolated with an ordinary disconnect switch. The length of line that can be dropped with a disconnect switch depends on the voltage. For example, an air-break switch without a load interrupter can, with permission, drop up to 3 miles (5 km) of 230 kV or up to 16 miles (26 km) of 50 kV.
Switchgear is opened to break parallel in a circuit that is being fed from two directions.	Some current and voltage will be interrupted. If the switch breaks parallel between different stations or two unequal lengths of line, a voltage difference (recovery voltage) will occur and a current flow will be interrupted upon opening the switch.
Switchgear is opened to break parallel in a circuit that is being fed from two directions.	A control room operator has the needed information to calculate the amount of current to be interrupted. Switchgear capable of interrupting load current can break parallel easier than an ordinary disconnect switch can.
Switchgear is closed to make parallel between two sources.	There can be a voltage difference and current flow created when the switch is closed. Unless there is a restrike, all switches should be capable of picking up the rated load current. Closing switchgear between two circuits fed from two different substations that are fed from two different transmission systems can be dangerous because a large voltage difference may create a large current flow through the switch. A system control operator should have control of such switchgear.
A loop must be opened or closed. (Opening a circuit fed from two directions but fed from the same circuit breaker is breaking a loop.)	Unless a loop is excessively long it can be closed or opened with an ordinary disconnect switch or a hot-line clamp. An example of a loop is where a new line is constructed next to an old one, such as at a road-widening project. The new circuit can be energized in parallel with the old circuit by tying it in at both ends and forming a loop. Any load on the old circuit must be opened before breaking the loop and isolating the old circuit.
A switch is to be closed, as a test, to determine if a line is still faulted.	Switchgear can be closed without an arc. A major arc will occur if a switch is closed in on a fault and the lineman immediately reopens the switch. It is important to let the normal protective switchgear trip out the circuit automatically. Closing a switch on an underground cable as a test for a fault can cause a lot of damage and is not an approved procedure.

Operating a Disconnect Switch with an Operating Handle at Ground Level

Many disconnect switches are three-phase, gang-operated switches with operating handles that extend to ground level. If an insulator or other component breaks while operating the switch, the energized leads could contact the switch frame and energize the operating handle. Similarly, the control box at ground level of protective switch gear can be energized from a fault. It is critical, therefore, to stand on a ground-gradient mat or ground-gradient network that is bonded to the operating handle. **See Figure 11–3.** The specification for this type of switchgear, especially in substations, requires that a permanent ground-gradient mat be installed at the switch handle. **See Figure 11–4.** It is also a good practice to install a temporary mat each time, to provide good visible electrical connections between the mat and switch handle. To dodge falling porcelain, if necessary, jump clear either with both feet together or by hopping without ever having both feet on the ground at the same time.

Figure 11–3 A ground-gradient mat ensures that a person's feet and hands are at the same voltage. *(Delmar, Cengage Learning)*

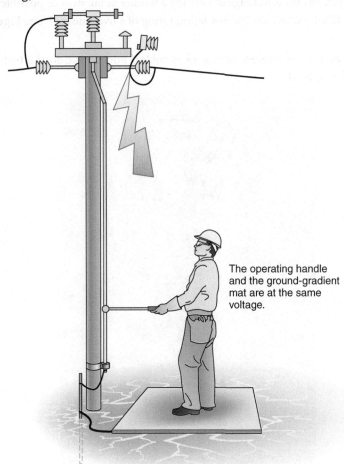

The operating handle and the ground-gradient mat are at the same voltage.

Figure 11–4 Permanent ground-gradient mats are installed at switches in a substation. *(Delmar, Cengage Learning)*

When the mat and operating handle are bonded together, the two will stay at the same voltage no matter how high the voltage becomes. If there is no voltage difference between the operating handle and the mat, there will be no voltage difference between the operator's hands and feet. If there is no voltage difference, there is no current flow.

Switching from the Ground Using a Telescopic Switch Stick

The use of a telescopic switch stick has the advantage of keeping a worker as far away as possible when energizing suspect equipment. Only the top foam-filled section has the insulation rating of a live-line tool. **See Figure 11–5.**

Figure 11–5 Only the top section of a telescopic stick is tested and considered a hot-line tool. *(Courtesy of HASTINGS Fiber Glass Products, Inc.)*

When using a telescopic switch stick, a person can still be exposed to an electrical shock when a high-potential ground gradient is created around the base of a pole during a switching operation. This has happened when a fault traveled down the ground wire because of a faulty surge arrestor or because an energized part contacted grounded hardware when a pedestal insulator came apart.

Keep some distance away from any ground rod at the base of the pole and keep your feet together during the switching operation.

Confirm Proper Phasing between Two Sources

It is often necessary to do some phasing checks before making connections or switching at the open point between two energized sources. The meter reading on the phasing tester should read near zero between phase 1 and phase A, for example, when the two sources are in phase. In the field, draw a table to record the results of your tests. **See Figure 11–6.**

Using a Phasing Tester

A phasing tester is like a high-voltage voltmeter. It has the ability to measure an approximate voltage between two different potentials on circuits over 750 V and, with the proper extensions or model of tester, can measure up to 161 kV. A phasing tester can be used as a potential tester, as a phasing tester where a reading near 0 V between two phases indicates they are the same phase, as an insulator tester, and as a hipot tester when using a special adapter.

At the contact end of each stick is a high-impedance resistor that lowers the voltage and current to a representative reading on the meter. From an operating perspective, keep a safe distance from the resistor sticks and cable. The resistors are protected by an epoxy covering, not a rated insulation; therefore, each stick should be in contact only with an

Figure 11–6 Use a form similar to this to perform a phasing test before making connections or switching. *(Delmar, Cengage Learning)*

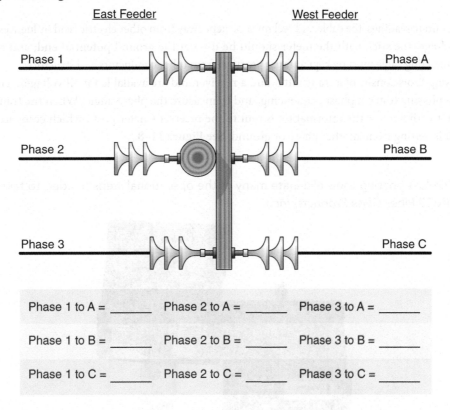

East Feeder   West Feeder

Phase 1   Phase A

Phase 2   Phase B

Phase 3   Phase C

Phase 1 to A = _____   Phase 2 to A = _____   Phase 3 to A = _____

Phase 1 to B = _____   Phase 2 to B = _____   Phase 3 to B = _____

Phase 1 to C = _____   Phase 2 to C = _____   Phase 3 to C = _____

individual phase. **See Figure 11–7**. Below the resistor are standard live-line tools on which a normal minimum approach distance should be kept from the splice of the stick. On higher voltages, additional resistors are needed and are available as extension sticks.

Figure 11–7 Phasing tools are capable of measuring high phase-to-phase and phase-to-neutral voltages. *(Reprinted with permission of Hubbell Power Systems, Inc., Centralia, MO USA)*

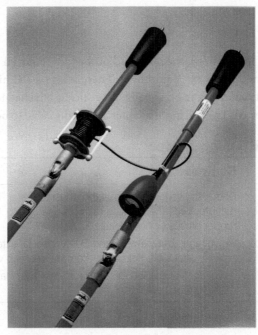

The cable between the sticks should be kept as short as practical to keep from sagging into structures or other conductors. The cable insulation is originally rated at 15 kV but should not be treated with the same confidence as a line hose.

To get more accurate readings, the cable and reel must be kept away from other electric field influences. When measuring line-to-ground voltage, the stick with the meter should be used at the ground potential end; and on phase-to-phase measurements, connecting cable must be kept as far as practical from other conductors and structures.

Wireless phasing tools consist of a transmitter and a receiver and are available for all voltages. The tool is used as a voltage detector, a phasing tester, a phase sequencing, and to measure the phase angle. When the transmitter (reference probe) is in contact with a phase the information is sent to the receiver (meter prob) which compares the information with the reading it is getting from another phase or ground. **See Figure 11–8.**

Figure 11–8 Wireless phasing tools eliminate many of the operational steps needed to take measurement. *(Courtesy HASTINGS Fiber Glass Products, Inc.)*

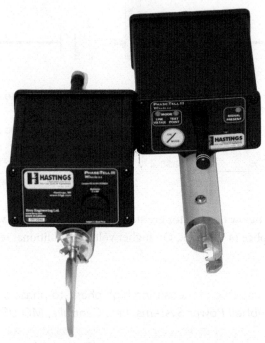

For measuring at underground installations, the overhead probes are removed and elbow adapters and/or bushing adapters are installed to allow testing **dead-front** equipment. There is also a DC hipot adapter that allows the tools to apply a DC voltage on a cable to test cables before energizing. One stick is attached to an energized source and the other stick is attached to the cable being tested. The DC hipot adapter allows the cable to be tested with a DC voltage.

Specific Hazards Using a Phasing Tester

Hazards	Controls
Creating a short circuit while using the phasing tester on a higher voltage than the tester is rated.	Check the voltage rating of the phasing tester. Typically, depending on the manufacturer, the phasing tester is used up to 15 kV and is handled directly with rubber gloves or hot sticks.
Note: These hazards do not apply to wireless phasing tools.	On voltages between 15 and 50 kV, special extension sticks are needed and the phasing sticks are held away from the lineman with attached universal live-line tools.
	At voltages above 50 kV and up to 161 kV, phasing testers specifically made for these voltages are used.

The insulated connecting cable between the cable reel and the other stick makes contact with a grounded object or person.	Typically, the connecting cable insulation is limited to 15 kV. Keep excess cable on the reel and use rubber gloves and/or universal sticks to avoid contact. In case of accidental contact, the resistors within the phasing tool should keep current flow at a safe level. Meter-reading accuracy will be affected when the connecting cable is influenced by other grounded objects or phases. When doing phase-to-ground tests, keep the reel stick on the grounded object.
The upper end of the stick, where the resistor is enclosed, makes contact with a grounded object or other phase and shorts out.	The voltage at the upper end of the resistor, enclosed inside the stick, is not intended for continuous contact. Keep the sticks away from other objects when they are in contact with an energized circuit.

Using Maps to Locate Switchgear

Circuit Identification and Structure Numbers

When preparing or applying for an isolation certification on a circuit, the circuit must be identified by a name, number, or acronym. Virtually all transmission and subtransmission circuits are identified with some kind of nomenclature. Transmission circuit nomenclature typically identifies the source substation and the destination substation. If there is more than one circuit, each must have a number. **See Figure 11–9**. Here, the source for the 230 kV circuits that feed the Benton Substation is the Latchford Substation. The circuits are therefore labeled L1B and L2B. The four 46 kV feeders coming out of the substation are labeled B1, B2, B3, and B4.

Figure 11–9 A typical schematic of a subtransmission system includes the nomenclature for all of the circuits. *(Delmar, Cengage Learning)*

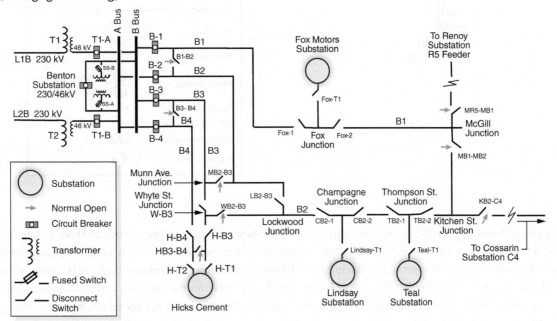

On distribution systems, often the source switch number identifies the circuit. Typically, the single-line drawing does not identify each circuit, but each line section can be identified by the source switch number. For example, a line crew would apply for an isolation certification on switch 390. **See Figure 11–10**.

Figure 11–10 A single-line drawing for distribution lines is needed to operate the system. *(Delmar, Cengage Learning)*

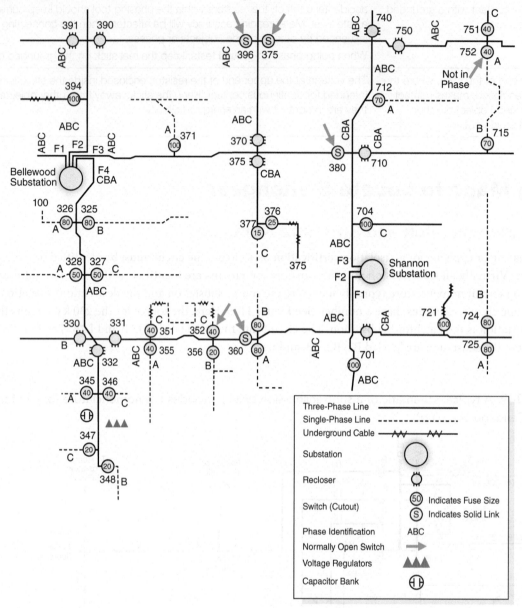

On a right-of-way, especially one with multiple circuits, the nomenclature on each structure should include the name of the circuit owner, the circuit name, and the individual structure number. Some system control people ask for the actual structure numbers that are to be worked on during an isolation certification. This information can be cross-referenced on maps and drawings for assurance that the proper location and switchgear for the isolation certification are identified and that no work on structures will be done outside of the isolation certification zone. Energized circuits have been grounded in the past because of two lines being similar in appearance or because of a double circuit.

Structure numbers are identifiers for recording data into computer-aided drafting (CAD) software that is tied into facility management (FM) software, which is tied into geographic information system (GIS) software and many evolving variations. The facility management database would typically have access to information for each structure, including the owner of property that must be crossed and the type of structure, height, and hardware manufacturer. Many utilities are using the latitude and longitude coordinates, as registered by their global positioning system (GPS) receivers, as their structure numbers.

Switchgear Nomenclature

When preparing or applying for an isolation certification on a circuit, the switchgear that feeds or is able to feed that circuit must be identified by a name, number, or acronym. Reliance on nomenclature and an operating drawing is especially critical on transmission lines and underground systems where it is not practical or possible to trace circuits physically. Relying solely on memory or physically tracing a circuit to identify isolation points on a distribution line to establish an isolation certification has a history of incorrect isolation and accidents.

Numbering and naming of switchgear are usually based on the circuit or source substation name. In many other locations, distribution switchgear numbers appear to be totally random. In a typical **schematic**, the switchgear numbers correspond to the feeder and/or junction names.

Electrical Schematic Drawings

The most efficient type of drawing for operating a system is a schematic drawing. Schematic drawings have virtually no relationship to the geography but make it much easier to trace a circuit to its terminal points, view the relationship of one circuit with another, and locate junctions and switchgear. Schematic drawings can be found for almost all electrical apparatus, from showing the electrical circuitry of a car to the circuitry of a transformer to the circuitry of a large transmission system at a system control center. The sample schematic drawing shows four 46 kV feeders coming from a substation. Each feeder feeds customers and the high side of distribution substations. The dark arrow at a switch indicates a normal open point. These open points can be closed to feed back on another circuit.

A system control center uses schematic drawings almost exclusively for transmission lines and substations. It is common in the control room to have the schematic drawing on a wall and also on a computer monitor. A line crew should also have access to system drawings to follow and check the switching order (sequence) issued by a system control center or for one prepared by themselves. When dispatchers or system control refer to a schematic drawing to discuss work, they do not necessarily understand exactly where a line crew is located.

Standard symbols are used to designate a breaker, disconnect switch, and load break switch, for example. Some utilities use different types of symbols for a line schematic drawing.

One-Line Distribution Drawings

One-line drawings are common for operating a distribution system. A one-line drawing shows the lines in their geographical location, but the background geography, such as roadways and lot lines, is either not present or very faint in the background. **See Figure 11–10.** Note that this is a typical one-line drawing with no geographic background, but the lines generally follow the roads. The drawings concentrate on what is needed to operate the system; therefore, distribution transformers and individual customers are not on these drawings. The symbols on the legend for single-line drawings are not consistent or standard among utilities. A legend will include enough information for understanding the drawing and identifying isolation points when needed.

Many utilities have one-line distribution drawings mounted on a wall. These provide the big picture of a system. These drawings are given priority for revisions. Changes occur daily, and line crews that take out isolation certifications must know about any change in normal feeds. **See Figure 11–11.**

Detailed Geographic-Based Maps

Maps that show distribution and/or transmission lines, transformers, and customers on a geographic background are detailed and would be too cluttered to use as operating drawings. Such maps are used to make drawings of new projects, to prepare a layout for a new subdivision, to record property easements, and to locate customers. Geographic maps, along with software such as automated mapping/facility management (AM/FM) software, are used to create an inventory of the locations and type of plant or facility the utility owns.

A transmission line map is a record of the type of structure, type of insulators, type of vibration dampers, among others. The data are used for such situations as identifying the location of hardware of a certain manufacturer or type when it has been found to be defective and a replacement program is needed. Transmission lines are also recorded in even more detail on roll plans that show each structure, span length, sag, ground clearance, access road, and property owner.

Figure 11–11 A single-line drawing is found in a dispatch office. *(Delmar, Cengage Learning)*

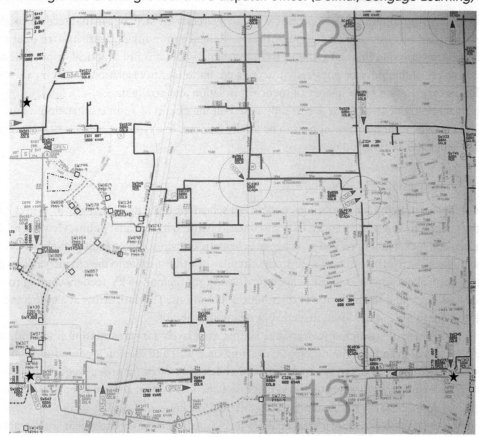

Planning Maps

Utilities also have planning maps. For distribution systems, a planning map shows the length of each line segment, conductor sizes in each segment, phasing of each line segment and lateral, voltage regulator location, capacitor location, transformer location, size, and connected phase. Engineering staff use these maps to prepare voltage and current surveys, short-circuit studies, protective device coordination studies, phase balancing studies, load management studies, and maintenance and capital programs.

Another type of drawing is the overview transmission line circuit diagram, which is similar to a distribution one-line drawing. The overview transmission line circuit diagram shows transmission circuits geographically placed but no background geography, and they are useful for planning maintenance programs.

Operating Isolating Switchgear

Operating a Nonload Break Disconnect

A disconnect switch is any solid-blade switch installed to provide a means to isolate or sectionalize a circuit. Unless a disconnect switch has a load interrupter, it has no capability to interrupt load current. A disconnect switch has no capability to open automatically under fault conditions.

Disconnect switches come in all voltages and can be single-phase bracket mounted, single-phase in-span, or three-phase gang operated. Depending on the type, disconnect switches are operated by hot stick, operating handle, or remotely by way of a low-voltage motor.

You may be asked to operate a three-phase gang-operated switch with no load break capability. This switch may show a rating of 600 A, but that means it is capable of carrying 600 A, not of breaking a 600 A load. The switch is operated with a handle at ground level, and the design of the switch installation should include a grid placed below and bonded to the handle. Some utilities require the use of a portable ground gradient mat. **See Figure 11–12.**

An in-span disconnect has no load break capability unless it is opened with a load break tool. **See Figure 11–13.** A fused in-span disconnect does not have load break capability either. The fuse can isolate a fault automatically, but when the device is opened with a switch stick, it operates as a nonload break disconnect. At distribution voltages, single-phase disconnect switches are normally equipped with hooks to allow a portable load break tool to be used.

Figure 11–12 A gang-operated air-break switch on a subtransmission line which has no load break capability. *(Delmar, Cengage Learning)*

Figure 11–13 A single-phase in-span disconnect switch which has no load break capability. *(Delmar, Cengage Learning)*

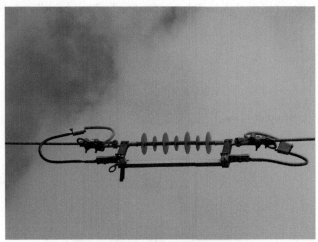

An ammeter check can be used to confirm that there is no load on a circuit before opening a nonload break disconnect switch. Some utilities suggest that a nonload break disconnect switch be operated by using the inching method. Because the switch should not be dropping any load, there would be no arc developing as the switch is opened slowly. If there is an unexpected load and an arc starts to develop, close the switch immediately. If a fairly long length of conductor (no load) is being de-energized by the disconnect switch, the switch should be opened in one fast operation.

Operating a Load Break Disconnect

Disconnect switches with load interrupters come in all voltages. A load break disconnect switch is designed to interrupt load current and can be operated without creating a dangerous arc. They can be three-phase, gang-operated switches or single-phase switches. Depending on the type, load interrupters are operated by hot stick, operating handle, or remotely by way of a low-voltage motor. Load interrupter switches can break load and are normally rated at 600 A or less. A load interrupter switch has no ability to open automatically under fault conditions. **See Figure 11–14.**

Figure 11–14 The combination of a load interrupter and power fuse is a common arrangement in a small substation. *(Delmar, Cengage Learning)*

Another design of the same arrangement has a power fuse to provide overcurrent protection but does not drop load current. The load interrupter switch is used to drop load but cannot interrupt a fault current automatically. This common type of load interrupter extinguishes the arc within an interrupter housing by deionizing gases that are exhausted flamelessly through a muffling device. **See Figure 11–15.**

Figure 11–15 The load interrupter of this switch is operated with a handle at the ground. *(Delmar, Cengage Learning)*

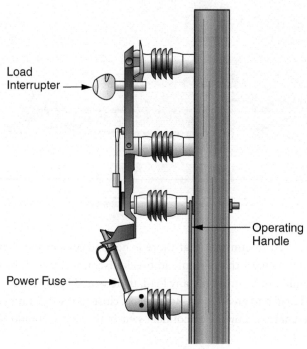

Operating a Hot-Line Clamp

Hot-line clamps are not switches, but they are used as a convenient way to disconnect and connect risers between equipment and circuits. There is, obviously, no arc-extinguishing capability with either a hot-line clamp or a duckbill clamp. When inadvertently dropping load with a hot-line clamp, a very long arc can be established. The air within the arc becomes ionized and, therefore, conductive so that as the hot-line clamp is pulled away the arc will follow. If, when the live-line clamp is pulled away it then touches a structure, an explosive fault current arc will be established.

To reduce the risk of an unmanageable arc, a hot-line clamp should not be used to drop any load. Hot-line clamps are sometimes used to drop a limited length of conductor (no load).

Operating Protective Switchgear

Circuit Breakers

Circuit breakers are the backbone of protective switchgear in an electrical power system. They are installed to protect generators, transformers, transmission lines, subtransmission lines, and distribution lines. They provide sophisticated circuit protection and switching capability. Circuit breakers, including distribution reclosers and metal-clad breakers, can be set to open and reclose a number of times before they will finally lock out on a permanent fault. On a transient fault, the breaker will stay closed once the fault has cleared from the circuit.

Circuit breakers open automatically for faults such as overcurrent, low voltage, or a drop in frequency. Potential transformers and/or current transformers on a circuit send representative voltage and current values to relays. Relays are programmed to send a signal when readings indicate a problem on a circuit. The signal can go to an alarm in a control room and simultaneously activate an electric motor that opens the circuit breaker. The motors of a circuit breaker are supplied by a low-voltage source, such as a substation service.

Circuit breakers are named according to their arc quenching medium. There are oil circuit breakers, air-blast circuit breakers, vacuum circuit breakers, and sulfur hexafluoride (SF_6) circuit breakers. Oil circuit breakers have been the most common circuit breakers and have been used for the longest time. The circuit breaker contacts are immersed in insulating oil, which quenches arcs. **See Figure 11–16.**

Figure 11–16 These 46 kV oil circuit breakers are operated remotely from a control room. *(Delmar, Cengage Learning)*

Air can be used as an insulation medium to quench an arc if the air is prevented from being ionized between contacts. Compressed air is piped to air-blast breakers and will "blast" an arc away from contacts. When these breakers are opened, the shotgun-type blast is extremely loud.

An arc occurs when the air is ionized and becomes conductive. When a vacuum is used as the insulating medium there is no air and therefore no arc. The contacts for a circuit breaker can be in a vacuum, and the remaining part of the breaker is immersed in oil. A separate **vacuum container** for the contacts prevents the oil in the rest of the circuit breaker from becoming contaminated.

Sulfur-hexafluoride (SF$_6$), an inert nonflammable gas with high insulating properties, is a common insulation for circuit breakers. Breakers using this gas are much smaller and can be used indoors. The operation is quieter than others and is ideal for substations in cities. SF$_6$ is not toxic, but under arcing, toxic by-products are produced that require special handling during maintenance. **See Figure 11–17.** Concerns about the effect of SF$_6$ on the environment have led to development of other insulation gases.

Figure 11–17 This 46 kV SF$_6$ circuit breaker is smaller than an equivalent oil circuit breaker. *(Delmar, Cengage Learning)*

Distribution Circuit Breakers/Reclosers

There are many types of circuit breakers/reclosers/interrupters found on distribution lines and in distribution substations. They can be overhead or underground (pad mount). They can be single-phase or three-phase units. Their contacts can be enclosed in vacuum bottles, SF$_6$ or oil. Most are multishot devices with the ability to automatically interrupt a circuit for a fault and reclose. Distribution circuit breakers can be operated or modified to be operated remotely through a supervisory control and data acquisition (SCADA) system.

A recloser can be set to open at various current levels and operating sequence. For example, a recloser is often set to trip out when the current going through the recloser coil is twice its rating. Therefore, it takes a minimum of 200 A to open a 100 A recloser. This ensures a trip-out is probably for a fault and not a temporary overload.

Reclosers can have their contacts enclosed in vacuum bottles, SF$_6$, or oil. They can be overhead or underground (pad mount). They can be single-phase or can be a three-phase unit, such as the one shown in **Figure 11–18**.

Metal-Clad Circuit Breakers

In urban distribution substations, a common circuit breaker is a metal-clad, draw-out breaker. This type is a multishot device with the ability to automatically interrupt a circuit for a fault and automatically reclose. It can be operated with a toggle switch or, more commonly, by remote control from a control room. The breakers are designed with various arc-quenching mediums, such as vacuum and air.

Typically, deep inside and along the length of a metal cabinet are bus bars that are fed from the secondary side of the substation transformer. Individual, metal-clad, draw-out circuit breakers are racked (pushed) into the substation bus bars. The output from each breaker feeds an underground distribution feeder. The breakers are mounted on a trolley or on wheels, or some other easy means that allows them to be racked out. A draw-out circuit breaker can be in the in-service position, in a test position, a disconnected position, or completely removed. Because of the extremely high-fault

Figure 11–18 Three-phase electronic reclosers can be bypassed by fused cutouts for maintenance. *(Delmar, Cengage Learning)*

current available and the limited space for operating as **seen in Figure 11–19**, full face protection and flame-retardant clothing should be worn when racking out a breaker.

Metal-clad circuit breakers can be operated remotely, keeping the operator out of the potential arc flash area. An operator using a remote racking system, such as shown in **Figure 11–20** can trip out or close the circuit breaker. Different designed remote racking systems are available to match other breaker manufacturers.

Figure 11–19 Metal-clad circuit breakers found indoors and outdoors. *(istock.com/Wisarut Pumipak)*

Figure 11–20 Breakers can be racked out remotely. *(© CBS ArcSafe)*

Pulse Closer Fault Interrupter

A typical circuit breaker opens and then recloses subjecting the line, connectors, splices and possibly downstream underground cable to a damaging high short circuit current. The interrupter shown in **Figure 11–21** is designed to test for a fault by sending a pulse limited to 5% of the available fault current when the switchgear is closed. The interrupter limits the current by closing on the sine wave point when it approaches zero. The interrupter can be operated using a hook stick on the "Open/Closed/Ready" lever or by SCADA.

Figure 11–21 A pulse closer fault interrupter. *(Source: S&C Electric Company)*

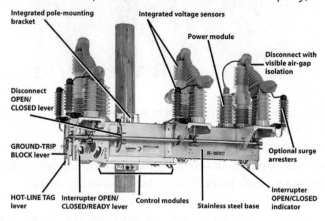

Types of Automatic Reclosers

There are many types of switchgear that a power line technician will be asked to operate. While there are similarities, it is required to familiarize with their operating characteristics. **Figure 11–22** shows an electric automatic circuit recloser.

Figure 11–22 An electric automatic circuit recloser. *(Van Soelen)*

Cutout Mounted Recloser

There are reclosers made to fit into a cutout holder, as shown in **Figure 11–23**. This recloser is applicable on laterals with load current 200 A or less. This would replace a fused cutout, which would typically need to coordinate with an upstream recloser. When the cutout mounted recloser senses a fault, it operates before the upstream protective switchgear opens. The mode selector lever can set the recloser into a nonreclose operation when working on the line. A lever on the other side of the cutout opens and closes the recloser.

Figure 11–23 S&C trip saver cutout mounted recloser. *(Source: S&C Electric Company)*

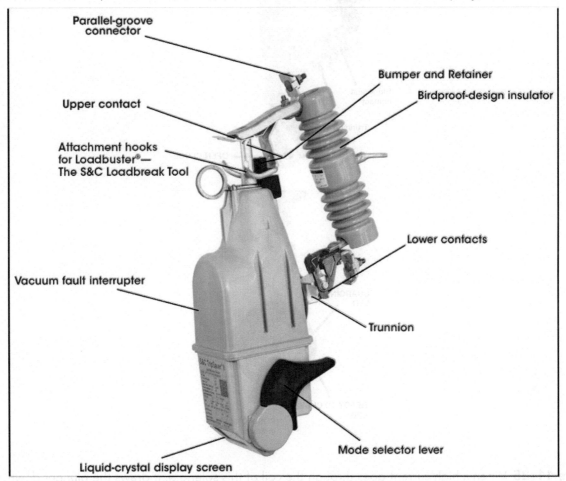

VacuFuse Self-Resetting Interrupter

The VacuFuse interrupter, shown in **Figure 11–24**, is similar to the cutout-mounted recloser and is applicable to laterals or a transformer with a load current of 199 A or less. It acts as a recloser and will operate before upstream protective switchgear. It would be safer to close than a fused cutout on a suspected faulted transformer.

Operating Hydraulic Reclosers

A hydraulic recloser is an oil-filled recloser that automatically opens when an overcurrent flows through its trip coil. A solenoid coil in the recloser is the fault sensor. When the current flowing through a coil is higher than the current rating of the coil, the electromagnetic action on a movable plunger inside the coil opens the contacts. Larger hydraulic reclosers have a low-voltage supply to an electric motor, which allows the heavy contacts to be opened and closed quickly. Any hydraulic reclosers that must be operated remotely (such as SCADA) would also have a low-voltage supply to operate the recloser. **See Figure 11–25.**

Figure 11–24 S&C VacuFuse self-resetting interrupter. *(Source: S&C Electric Company)*

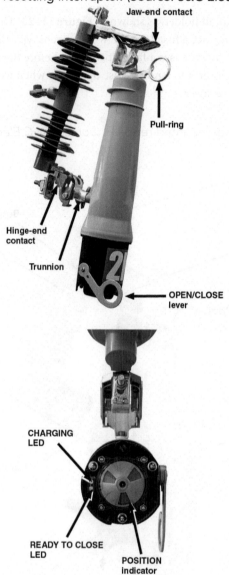

Jaw-end contact

Pull-ring

Hinge-end contact

Trunnion

OPEN/CLOSE lever

CHARGING LED

READY TO CLOSE LED

POSITION indicator

Figure 11–25 When a high current goes through the coil of this solenoid, it draws the plunger down, which in turn opens a circuit. *(Delmar, Cengage Learning)*

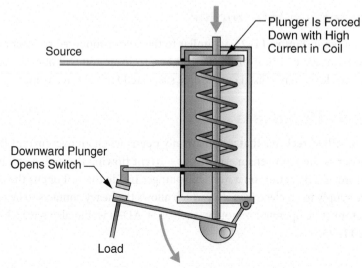

Source

Plunger Is Forced Down with High Current in Coil

Downward Plunger Opens Switch

Load

Like all reclosers, the recloser will close and reopen a specified number of times and at a specified speed. If the fault has cleared, the recloser will close and reset itself for the next time a fault occurs. On a permanent fault, the recloser will go through the full sequence of specified operations before staying open.

A hydraulic recloser can be used safely to interrupt load current by pulling down the operating lever under the sleet hood. There is also a nonreclose handle right beside the open-and-close handle. The handle that opens and closes the recloser is often painted yellow.

Hydraulic reclosers with a low-voltage supply can be opened with the handle, but some of these reclosers must be closed from the control box. The handle must be in the up position, or the remote control or SCADA control cannot close the recloser.

Operating Electronic Reclosers

Electronic reclosers are controlled by a relay. Relays can be programmed for a range of minimum trip values, a number of operations to lock out, and a range of minimum response times to coordinate with upstream and downstream protective switchgear. The size of the plug-in resistors in the control box determines the trip-current level. **See Figure 11–26**.

Figure 11–26 Familiarity with different control panels is needed for correct operation: (a) Cooper Form 6 control panel, and (b) SEL-351R recloser control panel. *(Delmar, Cengage Learning)*

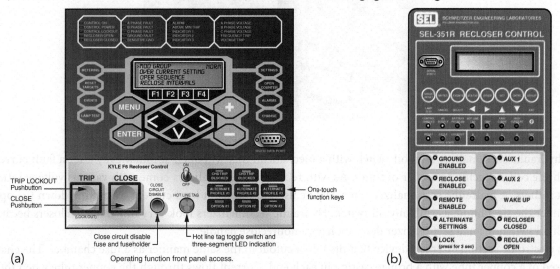

An electronic recloser has a low-voltage supply, which allows the recloser to be operated from a control box. A low-voltage supply can be from a regular distribution transformer on the supply side of the recloser or from a bushing current transformer mounted in the recloser. The low-voltage supply to the control box must also be in service to close the recloser.

The normal method of opening or closing these units is by the operating switch in the control box or remotely through the SCADA system. The main operating switch in a control box can open and close the recloser. The switch can be closed and held in the closed position to provide a cold-load pickup capability. Holding the switch blocks the instantaneous trip function. In other electronic reclosers, a cold-load pickup feature is programmed to prevent an instantaneous trip-out when closing after a lockout. The delayed trip will still open the circuit if an overcurrent situation lasts long enough. An electronic recloser has an open-and-close handle and a nonreclose handle under the sleet hood, the same as a hydraulic recloser.

An electronic recloser can detect a fairly small phase-to-ground fault. This sensitivity can prevent reenergizing at the recloser because the ground fault relay can interpret an unbalanced load as a phase-to-ground fault. Temporarily switching the *ground trip* toggle switch to the block position will prevent a small ground current from tripping out the circuit.

An operator should stand on a ground-gradient mat, or other grid, bonded to the control box when operating the main switch. The control box can become energized if the recloser malfunctions or the bushings flash over because the control cable bonds together the recloser and control box.

Operating Sectionalizers

A **sectionalizer** is a device that will isolate a faulted section of line when coordinated with an upstream multishot recloser or circuit breaker. A sectionalizer is a slave device that cannot interrupt a fault current but opens the circuit during a specified second or third interval while the upstream multishot device has the circuit de-energized (during dead time). **See Figure 11–27.**

Figure 11–27 This three-phase sectionalizer operates in conjunction with an upstream recloser. *(Photo courtesy of CAL FIRE)*

A hydraulic sectionalizer is an oil switch with a mechanism that will open automatically after a fault current goes through the coil a specified number of times. As with reclosers, sectionalizers come with various voltage ratings and current ratings. A hydraulic sectionalizer can safely be used to interrupt load current by pulling down the operating lever under the sleet hood, as with any oil switch. Hydraulic sectionalizers look very much like reclosers because they are both oil switches, but a sectionalizer does not have a nonreclose handle.

An electronic sectionalizer is a device that fits into a cutout in the same manner as a fuse chamber. The chamber is composed of a copper tube with a bronze casting in each end. Current flows through the copper tube when the unit is closed. A current transformer is mounted on the copper tube. The secondary of the current transformer is wired to an actuator on the bottom of the chamber that, when activated, drops the chamber and will look like an open cutout. An electronic sectionalizer is a slave device that opens during an interval when the upstream multishot device has isolated the circuit. An electronic sectionalizer is like any solid-blade switch: When it is to be operated manually, it requires a portable load break tool to interrupt load current. **See Figure 11–28.**

Operating a Distribution Cutout

The fused cutout is a common protective switchgear on distribution overhead lines to protect lines and equipment such as transformers, regulators, and capacitors. A fused cutout is a one-shot device that can interrupt a high-fault current.

An expulsion-cutout fuse link is the most common and economical fuse used in a distribution system. When the spring-loaded fuse melts, the fuse chamber drops open to provide a visible open switch. When a fuse element melts, the current continues to flow in the form of an arc through the particles of the vaporized fuse element and ionized gases. The heat from the arcing burns back the remaining element, and the heat generates the release of a large amount of gas from the inside wall of the fuse chamber. **See Figure 11–29.**

Figure 11–28 To open an electronic sectionalizer a portable load break tool is required. *(Reprinted with permission of Hubbell Power Systems, Inc., Centralia, MO USA)*

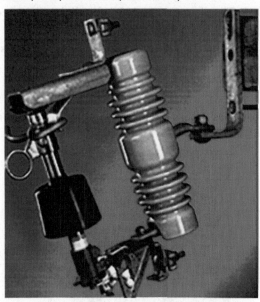

Figure 11–29 A fused cutout will extinguish an arc during a fault but not when it is opened with a hot stick. *(Photo courtesy of CAL FIRE)*

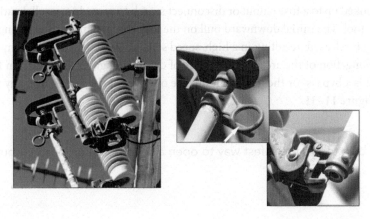

The resultant high-pressure gas and arc products are expelled from the tube. When a cutout is opened with a switch stick, the cutout is operated like a regular isolating disconnect switch that has no arc-extinguishing capability other than the air gap between the switch terminals. When opened with a switch stick, a cutout can only interrupt about 15 A safely. If a cutout is closed in on a faulted circuit during troubleshooting, it is important not to immediately open the cutout because it is likely to draw a large arc, possibly from terminal to terminal. The design of the fuse and fuse holder should be allowed to extinguish the arc.

It is not unusual to close a fused cutout and find that there is still a fault on the circuit. When that happens, exhaust gases and molten metal will be expelled and directed down from the vent end of the fuse holder. Use a long length of switch stick and stay to the side of the fuse tube alignment.

A fuse holder should not be left hanging in the open position in a cutout for long periods of time. Rain and ice will collect inside and eventually damage the fuse holder. Remove the fuse holder and hang it upright on the pole, usually on the neutral bracket.

A blown fuse in a cutout should be replaced with a fuse of the same size and speed to maintain coordination between other upstream and downstream fuses and switchgear. Feed the fuse cable clockwise around the stud and tighten it so that extra tension is not put on the link. The diameter of a 100 A fuse link is designed only for a fuse holder up to 100 A. A larger tube diameter is needed for larger fuses to operate properly.

A distribution cutout is a switch, not protective switchgear, when it is used with a solid blade as shown in **Figure 11–30**. It will not open for a fault.

Figure 11–30 A solid-blade cutout has no arc extinguishing capability and must be opened using a load break tool. *(Photo courtesy of CAL FIRE)*

Using the Load Bust Tool

A load bust tool can be hooked up to a fuse cutout or disconnect switch to provide a parallel path for load current through the tube of the load break tool. The initial downward pull on the tool charges an internal spring. At a certain point in the downward pull, the spring is released, resulting in a high-speed separation of the contacts. Any arc is extinguished inside the chamber by the fast elongation of the arc and the release of deionizing gases formed from the surrounding chamber material. A load bust tool is a bypass for the load current. The opening of the circuit and any resultant arc occur inside the load break tool. **See Figure 11–31**.

Figure 11–31 A load bust tool is the safest way to open a solid-blade or fused cutout. *(Courtesy of S&C Electric Company)*

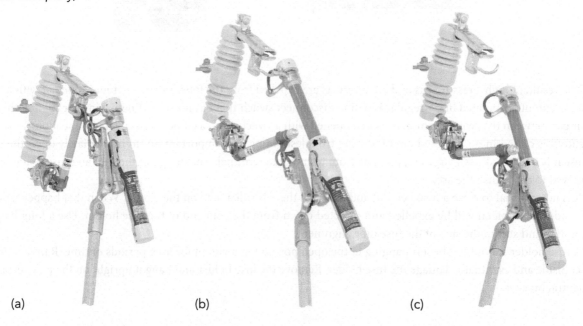

(a)   (b)   (c)

Fuses

The replaceable fuse link inside the tube of a cutout is a metal or alloy with silver, tin, lead, or copper that is designed to melt when a given amount of overcurrent for a given amount of time flows through it.

A fuse can carry load current without deteriorating, can carry some overload without immediate rupture, and must be able to interrupt a very high-fault current. The higher the current flowing through the fuse, the quicker it will blow. The amount of current a fuse is able to interrupt is based on its ability to extinguish an arc.

- An arc is extinguished by providing a fast separation of two parts of a melted fuse. One way to separate a blown fuse quickly is to have it spring loaded so that the two parts of the blown fuse will separate quickly.

- An arc inside a fuse chamber forms gases when it acts on a special coating on the inside walls of a chamber. The formation of gas forces the arc products out of the expulsion chamber and extinguishes the arc.

- An increase to the resistance of an arc path will extinguish an arc. Having a fuse immersed in oil or other insulation will cool and increase the resistance of the arc.

The melting and **clearing time** of a fuse should coordinate with upstream and downstream protective devices. Different speeds of fuses are available; for example, a K-link fuse is fast, and a T-link fuse is slower. The K-link and T-link fuses have specific time and current characteristics recognized by all manufacturers.

Solid-material fuses have a fusible element inside a heat-absorbing and arc-quenching material such as silica sand or borax. During a fault, the fuse vaporizes and the solid material cools the arc. These fuses can be fixed or dropout expulsion. The reduced expulsive emissions of boric acid fuses permit their use in enclosures, in substations, and in certain vulnerable overhead distribution applications.

The following list summarizes the types of fuses used in electrical utilities.
- Expulsion-cutout fuse links
- Under-oil expulsion fuses (bayonet style)
- Solid-material-filled power fuses (**See Figure 11–32.**)

Figure 11–32 A solid-material-filled power fuse has superior arc-quenching capability. *(Photo courtesy of CAL FIRE)*

- Nonexpulsion (NX) **current-limiting fuses**
- Under-oil backup current-limiting fuses
- Liquid-filled fuses (**See Figure 11–33.**)
- Fault-tamer fuses (**See Figure 11–34.**)

Figure 11–33 Liquid-filled fuses are replaceable, but installing a new one would be rare. *(Photo courtesy of CAL FIRE)*

Figure 11–34 A fault-tamer fuse reduces the possibility of an explosive opening during a fault. *(Photo courtesy of CAL FIRE)*

Current-Limiting Fuses

An ordinary expulsion fuse is not current limiting. It will limit the duration of an arc but not the magnitude. A current-limiting fuse will limit the magnitude of the current flow by introducing a high resistance after the fuse element melts. A current-limiting fuse is used in locations where a very high-fault current is available on the electrical system. A current-limiting fuse will limit the magnitude of a fault current and will reduce the risk of transformers and other equipment from failing explosively. A current-limiting fuse consists of a silver ribbon element wound around inside an insulated tube of silica sand. The silver ribbon is perforated with holes along its full length. During a fault, the silver ribbon element vaporizes along its length, starting at the perforations. The vaporized element is blown into the surrounding sand. The resulting arc heats the sand and turns it into a glasslike material that increases the resistance to current flow and chokes off the arc. **See Figure 11–35.**

Figure 11–35 A current-limiting fuse must be replaced after one operation. *(Delmar, Cengage Learning)*

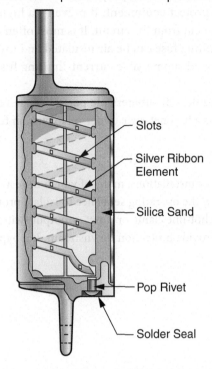

- Slots
- Silver Ribbon Element
- Silica Sand
- Pop Rivet
- Solder Seal

The two main types of current-limiting fuses are partial range and full range. A partial-range current-limiting fuse (or backup current-limiting fuse) is designed to limit only high-fault currents. It is used in series with a fused cutout to protect equipment from the high energy levels available in a high-fault-current location. It is available in different sizes and speeds to coordinate with a cutout fuse link. For example, a 25 K current-limiting fuse coordinates with a 25 K cutout fuse link. An air-insulated, partial-range current-limiting fuse is installed right on top of a fused cutout. **See Figure 11–36**.

On underground systems an under-oil submersible partial-range current-limiting fuse is used in series with low-current protective devices such as a bayonet-style under-oil expulsion fuse, cartridge fuse, or fuse link. If a transformer cutout fuse is found open, the current-limiting fuse (if properly coordinated) should not be damaged. Testing or replacing the current-limiting fuse before energizing the transformer with the cutout will reduce the risk of an explosive failure if the transformer is faulted.

Figure 11–36 A partial-range current-limiting fuse is placed in series with a regular fuse. *(Delmar, Cengage Learning)*

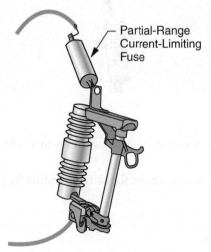

Partial-Range Current-Limiting Fuse

A full-range current-limiting fuse (or general-purpose current-limiting fuse) is not used in series with another fuse but is the only fuse needed to protect equipment. It prevents a high-fault current from going through equipment, and it isolates defective equipment from the circuit. It is most often used to protect underground or metal-clad equipment. A full-range current-limiting fuse can be air insulated and used for high-amperage-rated applications in live-front switchgear or as an under-oil submersible current-limiting fuse (bayonet-style mounted fuse) used with dead-front equipment.

On underground systems, an under-oil submersible partial range current-limiting fuse is used in series with low-current protective devices such as a bayonet-style under-oil expulsion fuse, cartridge fuse, or fuse link.

Bypassing Protective Switchgear

There are different designs for recloser installations, many of which have a means to bypass the recloser to allow maintenance. When bypassing a recloser, *if* the operating sequence to isolate protective switchgear is not followed, *then* load could be dropped accidentally with a hot-line clamp or input or output switch, resulting in an uncontrolled arc.

The bypass should be fused to provide protection downstream. If a bypass fuse size is not specified, use a fuse size equal to the recloser coil rating.

To isolate, do the following:

1. Close bypass.
2. Open recloser.
3. Remove hot-line clamps.

The framing of this single-phase recloser is designed for safer operation. A switch in series with the source lead allows a visible open point when using the recloser for an isolating point. The postinsulator on the other side of the pole is a clamp rest. **See Figure 11–37.**

Figure 11–37 The design of this single-phase recloser pole includes a bypass arrangement. *(Delmar, Cengage Learning)*

The design of this three-phase recloser should have a warning that it be operated from a bucket truck only. **See Figure 11–38.**

This three-phase recloser installation is designed for safer operation by providing clamp rests for the source and load leads. **See Figure 11–39.**

Figure 11–38 A three-phase recloser installation mounted vertically is designed to be operated from a bucket truck only. *(Delmar, Cengage Learning)*

Figure 11–39 A three-phase recloser is designed to be bypassed when working from the pole. *(Delmar, Cengage Learning)*

Nonreclose Feature on Breakers and Reclosers

Putting a breaker or recloser in a nonreclose position gives a crew working on a circuit the assurance that the circuit will not be reenergized automatically if an incident on the job trips out the circuit.

The nonreclose position does not guarantee that a circuit will trip out during an accidental contact; it only ensures that a circuit will stay out of service after it is tripped out. Placing a tag at the breaker or recloser will prevent other utility personnel from closing the recloser without first checking with the crew working on the circuit.

Underground Distribution Switchgear

Types of Underground Switchgear

Switchgear in an underground distribution system is operated in confined areas where any arc could easily spill over to grounded cabinets or other phases. Some types should not be operated under load, and others require special tools to operate safely. There are dead-front and live-front types of switchgear in pad-mounted equipment or as submersibles in below-grade vaults. There are also large vaults with overhead-style switchgear.

Switching distribution underground involves operating load break–separable connectors (elbows), arc strangler switches, bayonet-style under-oil expulsion fuses, vacuum switchgear, SF_6 switchgear, and other devices with arc-quenching capability. A portable load break tool is needed to operate some types of switchgear safely. Often switching an underground system involves opening and closing ordinary overhead switchgear at a riser pole where the overhead-to-underground transition exists. **See Figure 11–40**.

Figure 11–40 Cable can be isolated from this riser by opening the load break switch. *(Delmar, Cengage Learning)*

The original design of some types of underground switchgear may have allowed operation under energized conditions, but field experience, higher fault-current levels, and higher distribution voltage levels have changed this. Usually, an accident or incident investigation will restrict the manner in which some types of switchgear are operated. Some switching is carried out in a vault or manhole where there is the additional hazard of creating an arc in a confined space. A line crew always checks for the oxygen level and for the existence of flammable or toxic gas before and during switching inside a confined space. There are utilities that require all energized switching to be done from outside a vault, especially when the switching involves oil-filled equipment. **See Figure 11–41**.

Figure 11–41 These load break elbows on the red-phase side of a three-phase switching cabinet can be removed using a hot stick. *(Delmar, Cengage Learning)*

Operating Cable-Separable Connectors (Elbows)

Separable connectors are the live-line clamps of an underground system, although fused load break elbows are available. Load break elbows are used to isolate equipment or to sectionalize segments of cable. They are found at the input and output of a transformer and at other locations where separating a cable from equipment under energized conditions is necessary. **See Figure 11–42.**

Figure 11–42 Load break elbows are used to isolate cable in pad-mount switchgear. *(Courtesy of John Bellows)*

A load break elbow can be pulled from the load bushing with a live-line tool to interrupt a 200 A load current. The elbow must be pulled off in one quick motion, especially on high-voltage distribution. If the lubricant between the elbow and the bushing has dried, an elbow does not always come all the way off with the first pull. An arc between the load break pin and the bushing contacts can spill over to the outside of the bushing to a grounded tank or cabinet. Pulling elbows has a history of flashovers. The root of the problem is that while the load break elbow was being pulled from the bushing, a partial vacuum was formed across the mating interface. A partial vacuum reduces the dielectric strength across the mating interface. While a vacuum is a perfect insulator (as seen in vacuum-insulated switchgear), a partial vacuum is a poorer insulator than normal air pressure. Manufacturers have designed elbows with vent holes around the cuff of the elbow to prevent the partial vacuum from forming.

There are fused elbows available that can be installed to protect the system from down stream faults or to protect equipment. **See Figure 11–43**.

There are non-load-break elbows that should not be operated while the circuit is energized. A non-load-break elbow can have a current rating as high as 600 A and is often used to terminate a main feeder into a switching cabinet. **See Figure 11–44**.

Figure 11–43 A fused elbow connector can be installed to protect cable from downstream faults. *(Courtesy of Thomas and Betts)*

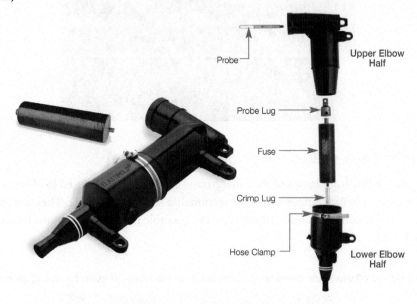

Figure 11–44 Recognize the difference between load break and non-load-break elbows. *(Delmar, Cengage Learning)*

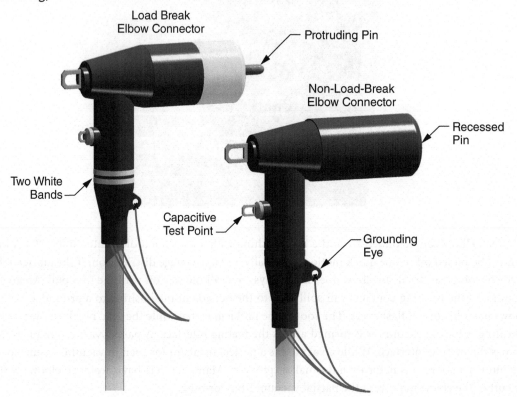

The capacitive test point on an elbow can be used to determine whether or not a circuit is energized. The cap over the test point is removed with a live-line tool, and a potential tester designed to pick up the capacitive voltage is used at the exposed test point. The test point is not a direct connection to the energized conductor inside the elbow but has a potential because of capacitance. An ordinary voltmeter will not pick up a potential at the capacitive test point. The test point can also be used for phasing checks when using a phasing tool that has the multiple-functions feature. **See Figure 11–45.**

Figure 11–45 A capacitive test point is used to test for de-energization. *(Delmar, Cengage Learning)*

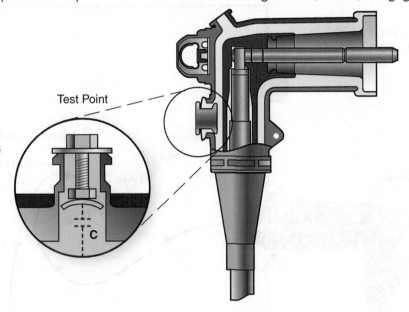

Test Point

Nx Switches

Nx (nonexpulsion) switchgear, also called an arc strangler switch, is found in live-front switching cabinets and transformer vaults. It is operated with a live-line tool in the same manner as operating a fused cutout. The Nx switchblade is an arc quenching, load break device that is available as a current-limiting fuse or as a solid-blade device. The load break works like a portable load break tool and must be cocked to operate. **See Figure 11–46.**

To cock the barrel, the arc strangler sleeve is pulled down and held in that position by a latch spring. When the switch is closed, the latch spring is depressed and the sleeve is held down by the switch itself. When the switch is opened, the arc-strangler sleeve is released and snaps up to activate the load break feature of the switch barrel.

Figure 11–46 An arc-strangler switch operates similarly to a load break tool. *(Delmar, Cengage Learning)*

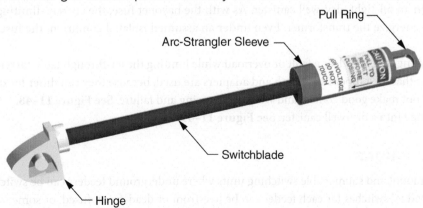

Pull Ring

Arc-Strangler Sleeve

Switchblade

Hinge

Operating Under-Oil Bayonet-Style Fuses

The elbow, plugged into a bushing, is a load break device, but it cannot interrupt a fault. Often, the elbow is in series with an under-oil bayonet-style fuse. The bayonet fuse is an under-oil expulsion fuse contained in a load break holder. The fuse provides protection to the transformer by interrupting a fault, but it is not necessarily a good device to drop load. **See Figure 11–47.**

Figure 11–47 An under-oil bayonet-style fuse must not be used to drop load. *(Delmar, Cengage Learning)*

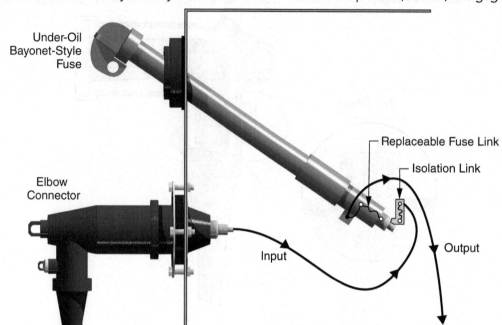

Some utilities require that the transformer be isolated before removing or inserting the fuse. To avoid oil from being expelled when the fuse is withdrawn, relieve the pressure in the transformer tank by operating the external relief valve. To prevent dropping oil on the rubber elbows, pull the fuse out about 3 in. (5 cm) and hold it to let the oil drain from it. When it is removed, wipe the oil off the fuse.

Dry-Well Canister Fuses

Three-phase, pad-mount transformers often have dry-well canister fuses. A current-limiting fuse that protects the transformer is mounted in an oil-tight, dry-well canister. As with the bayonet fuse, the current-limiting fuse should not be used to isolate or re-energize the transformer. Even under an assumed isolated condition, the fuse should be removed with a shotgun stick.

The current-limiting fuse will clear a fault or overload while limiting the let-through fault current to prevent damage to equipment. Ensure that the proper fuse length and adapters are used, because they can differ for different voltages. An incorrect length will not make good contact and will lead to arcing and failure. **See Figure 11–48.**

The fuse is inserted into a dry-well canister. **See Figure 11–49.**

Multipoint Junctions

There are large pad-mount and submersible switching units where underground feeders can be switched from one feeder to another. The individual switches for each feeder can be live-front or dead-front, fused, or some other kind of breaker. **See Figure 11–50.**

Figure 11–48 A current-limiting fuse will protect equipment by keeping out damaging fault current. *(Courtesy of Thomas and Betts)*

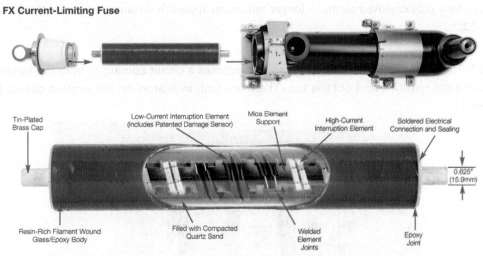

FX Current-Limiting Fuse

Tin-Plated Brass Cap

Low-Current Interruption Element (Includes Patented Damage Sensor)

Mica Element Support

High-Current Interruption Element

Soldered Electrical Connection and Sealing

0.625" (15.9mm)

Resin-Rich Filament Wound Glass/Epoxy Body

Filled with Compacted Quartz Sand

Welded Element Joints

Epoxy Joint

Figure 11–49 A molded dry-well canister contains the current-limiting fuse. *(Delmar, Cengage Learning)*

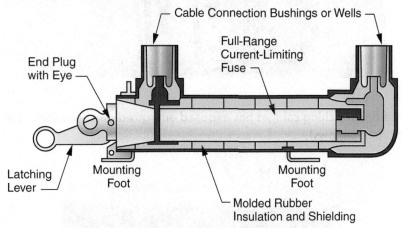

Cable Connection Bushings or Wells

Full-Range Current-Limiting Fuse

End Plug with Eye

Latching Lever

Mounting Foot

Mounting Foot

Molded Rubber Insulation and Shielding

Figure 11–50 Feeders can be connected together or separately to the center bus and in turn feed elsewhere; this is the top of a switching and connection arrangement with four feeders. *(Delmar, Cengage Learning)*

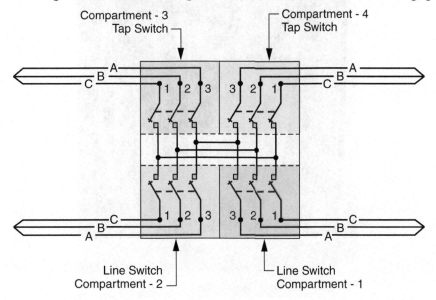

Compartment - 3 Tap Switch

Compartment - 4 Tap Switch

Line Switch Compartment - 2

Line Switch Compartment - 1

This S&C PME pad-mounted switching cabinet has similar switching arrangements. **See Figure 11–51.**

There are switching cabinets with live-front switchgear. An employer typically has additional work procedures, such as a lock-to-lock rubber glove rule and a longer minimum approach distance to operate this kind of switchgear. **See Figure 11–52.**

Figure 11–51 The open door of a switching cabinet exposes a circuit coming in from the bottom through the switch behind the partition and out the top. There are fault indicators on the bottom circuit. *(Courtesy of John Bellows)*

Figure 11–52 A live-front switching cabinet has exposed energized parts requiring a minimum approach distance. *(Courtesy of John Bellows)*

Undercover-Style Switchgear

Switchgear with features such as SCADA, electronically programmable fault interrupting is available in padmounts and submersibles, as shown in **Figure 11–53**. The switchgear can be motor or manually operated in open, closed or grounded position. Similar to overhead switches, the contacts can be in vacuum or SF$_6$ to allow for smaller tanks.

Figure 11–53 Undercover switchgear. *(© S&C Electric Company)*

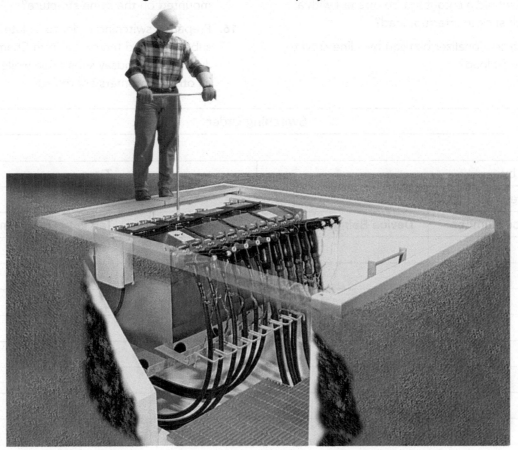

Review Questions

1. What two main types of switchgear are used in an electrical system?

2. What is the main potential hazard when opening any type of switchgear?

3. Name three methods used to extinguish an arc in various switchgear.

4. Can a load interrupter switch interrupt a fault current?

5. When planning to close a switch between two circuits fed from different substations, what check should be made?

6. *True or False*: A switch can open (isolate) any length of line that has no load on it.

7. When a disconnect switch is operated, a breakage may cause energized leads to contact the switch frame. What is the best protection for an operator when standing on the ground, to open a three-phase gang-operated switch with an operating handle?

8. Why is the use of a ground gradient mat required when operating switchgear from a control box?

9. What potential hazard exists when switching from the ground using a telescopic switch stick?

10. Why is it important to check that all three poles are open after opening a three-pole switch?

11. Why should a cutout not be opened with a switch stick to interrupt load?

12. May a sectionalizer be used by a line crew to interrupt load?

13. Why is it important to use a phasing tool before closing a switch between two sources?

14. Name four ways to verify a line is isolated when applying a lockout/tagout procedure.

15. What potential hazard exists when operating an electronic recloser from the control box mounted on the same structure?

16. Prepare a switching order to isolate Lindsay substation and the circuit from Champagne Junction to Lindsay substation while keeping all other customers energized.

Switching Order

Purpose				
Ordered by		Time		
Completed by		Time		
Sequence Number	**Device Being Operated**	**Operation**	**Tag #**	**Initials**

Chapter 12

Circuit Protection

Objectives

After completing this chapter, you should be able to:

1. Explain why circuit protection is needed to detect and clear abnormal voltages and currents.
2. Identify the types of faults circuit breakers can be programmed to detect.
3. Describe the damage that can take place when overcurrent protection does not operate.
4. Explain why a good multigrounded neutral is essential to activate circuit protection switchgear.
5. Explain why downstream protection is specified to coordinate with upstream breakers or fusing.
6. Explain the problem that is created when switchgear is bypassed with a solid jumper.
7. Describe the difference between connections to a neutral and connections to ground.
8. Explain why staying in a car is safe during a thunderstorm.

Introduction

The Purpose of Circuit Protection

An abnormal voltage or current is an indication of a problem somewhere in the circuit. Protective equipment is installed to detect and clear these abnormal voltages and currents.

- Circuit protection limits the time people are exposed to hazardous voltage and current due to situations such as a fallen conductor.
- Circuit protection limits the time equipment is exposed to damaging voltage and current.
- Circuit protection minimizes the number of customer outages during adverse conditions by automatically isolating and removing a faulted circuit from the system.
- Circuit protection is programmed to open and reenergize a circuit that was subjected to a transient fault.

A common task for a power line technician is tracing problems due to overcurrent, abnormal voltage, or poor system grounding. Troubleshooting will be easier with an understanding of how protective equipment is specified to operate.

Transmission System Protection

Circuit Protection

When a fault occurs in a transmission system, circuit breakers go into action and automatically separate the fault from the system. Circuit breakers are used to protect every part of an electrical system. **See Figure 12–1.**

Figure 12–1 Circuit breakers provide protection to the electrical system. *(Delmar, Cengage Learning)*

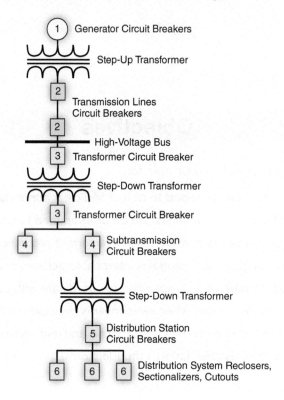

Other than a lightning surge, there are two main types of faults in an electrical system that cause a circuit breaker to operate: short circuits and open circuits. Relays can detect a fault only if measurable conditions show up at the terminals of the circuit. Three types of faults, or combinations of faults, can show up and are measured at a terminal.

1. The current flow in one or more phases becomes abnormally high.
2. The current in the three phases becomes abnormally unequal.
3. The voltage in one or more phases becomes abnormally low.

Protective Relaying

A relay is a low-voltage switch that is normally in either an open or a closed position. Depending on the design, the relay switch will operate either when electrical current passes through it or when electrical current stops passing through it. The relay can be electromechanical or solid state (electronic). Each relay is designed to operate when a certain amount of current is reached.

Potential transformers and current transformers installed on high-voltage circuits send representative low voltage and low current to the relays. Relays are designed to operate when the voltage and/or current is beyond the range of the specified relay settings.

A relay designed to detect overcurrent would operate if the representative current from the current transformer was beyond the specified setting. The relay switch would operate and send a low-voltage signal to the circuit breaker control mechanism and cause it to open.

Relays Controlling Circuit Breaker Operation

An electrical system can become unstable with adverse conditions ranging from a low overcurrent fault, such as a tree contact, up to and including a geomagnetically induced current (GIC) on long transmission lines, due to increased sunburst activity on the sun.

Relays can be designed to detect various conditions on a circuit as long as the voltage and current representing the condition of the circuit can be brought into the relay. Examples of some of the relays that can be installed to protect a circuit are the following:

- An **overcurrent relay** detects current when it exceeds predetermined limits.
- An undercurrent or underpower relay detects current when it has decreased beyond a predetermined limit.
- An over or undervoltage relay detects voltage change beyond predetermined limits.
- A differential relay detects current entering a protected zone that does not equal the current leaving the zone.
- A current or voltage balance relay can detect a predetermined difference between two circuits or between phases on a circuit.
- An under or overfrequency relay detects frequency when it changes abnormally.
- A thermal relay detects an abnormal rise in the temperature of a generator or transformer.
- A directional power relay detects a change in the direction the power is flowing on a circuit within the grid.
- A power factor relay detects changes to the reactance in a circuit beyond predetermined limits.

Circuit Breaker Operation

Relay settings specified by a planning engineer will determine the sequence speed and number of times when a breaker will open and reclose on a fault. A circuit breaker must open within a speed range that will prevent damage to equipment, such as a substation transformer, and still coordinate with other switchgear on the system so that only the faulted section will be isolated.

The tripping time for a breaker is very fast. The speed is normally expressed in cycles. For example, a breaker can open on as few as three cycles, which means that a circuit breaker on a 60 Hz line can operate in 0.05 second after sensing a fault.

Zone Protection

The **protection scheme** of the high-voltage system is divided into protective zones. When trouble occurs in a zone, relays sense the problem and send a signal to the appropriate circuit breakers to disconnect the zone from the system. The usual protective zones are generator, transformer, bus, and lines. **See Figure 12–2.**

Due to its length, the line zone has the greatest exposure to faults and is the most frequent zone to be disconnected from the system.

The Blackout of 2003

Northeastern and midwestern USA plus the Canadian province of Ontario suffered a major blackout in 2003. It is an example of how a protection system can fail. The complete cause of this particular blackout is complex. The system was already stressed with a high customer demand, voltage swings, and a need for more reactive power (capacitance) in the grid. Under these unstable conditions, a transmission line tripped out due to a short circuit caused by a brush fire under the line. Hot gases from the fire ionized the air and caused the air under the line to become conductive. Normally within a grid, other lines pick up the load dropped by the fault line. Extra load on another circuit caused the power conductors to warm up and sag into a tree and trip out. A generation plant tripped out when the transmission lines were no longer available. The resultant **overload**, voltage collapse, and current swings in the rest of the grid caused instability in the power system and circuits tripped out, one after the other.

System operators and the protective relays are supposed to isolate a fault to protect equipment from damage and prevent the fault from cascading into neighboring systems. However, unless there is intervention, the route that power will flow in an electrical grid is controlled by the laws of physics. The flows will go over many different lines, choosing

Figure 12–2 Circuit breakers protect individual protection zones. *(Delmar, Cengage Learning)*

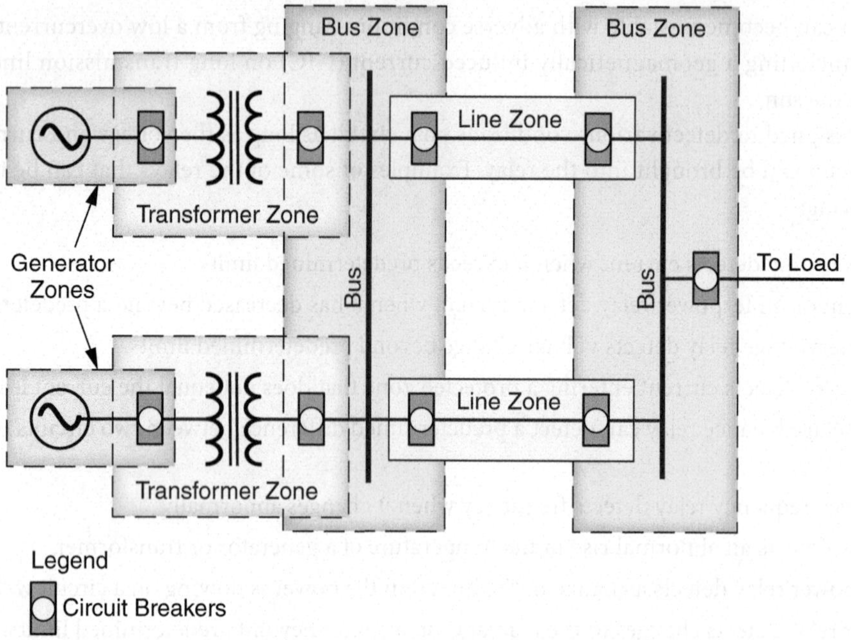

a path of least resistance to get to the load center; meanwhile, relays will sense any overload, a voltage collapse, low frequency, reactive power overload, and other factors and cause signal breakers to open.

System operators in different jurisdictions can reroute power to a different transmission line by turning down the generation in one location and turning up generation at a different location. This assumes that there is reserve generation capacity available and that there is excellent coordination and communication between jurisdictions. Meanwhile, under normal conditions, a system operator (market player, generator, or trader) considers the price of power from different generation sources. A system operator is supposed to ignore economics and markets to relieve a transmission line overload, but these could still be complicating factors when making quick decisions.

Distribution Protection

Distribution Protection Requirements

A distribution system is exposed to overcurrent, overvoltage, and open-circuit conditions. Protective equipment is strategically placed to limit and isolate the faults so that the remainder of the system is not affected.

In many cases, especially in urban areas, a circuit breaker in a substation can detect faults in the full length of the circuit. Sensing problems such as open circuits, abnormal ground current, or abnormal phase imbalance can be obtained by using circuit breakers or electronic reclosers and their associated relays.

Downstream equipment, such as fuses or hydraulic reclosers, will trip out and sectionalize a circuit automatically, without the need for relays or other power sources. Only an overcurrent condition will trip out these devices.

Protection in a Distribution Substation

At a distribution substation, normally high-voltage (HV) fuses or a circuit breaker are on the high-voltage source side of the substation transformer, and circuit breakers or reclosers on each distribution feeder are on the low-voltage (LV) load side. **See Figure 12–3**. A fault in the substation should trip out the HV fuses before the circuit breakers at the source of the incoming subtransmission line trip out. The LV reclosers should trip out a faulted feeder before there is any damaging current through the substation transformer. **See Figure 12–4**.

Figure 12–3 This distribution substation is protected by HV fuses and LV reclosers. *(Delmar, Cengage Learning)*

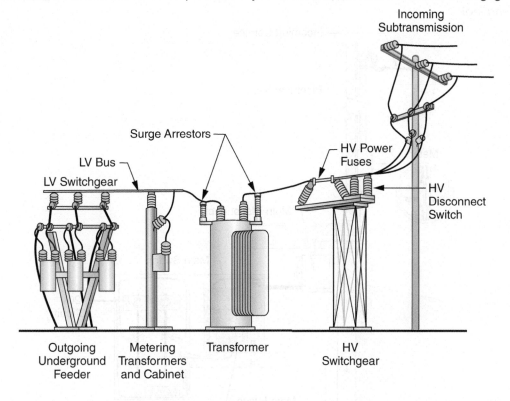

Figure 12–4 An example of a small rural distribution substation which is protected by HV fuses and LV reclosers. *(Delmar, Cengage Learning)*

Protection Using Home Wiring as an Example

Protection for the electric wiring in a home is not unlike protection of a radial distribution system. Home wiring for a 120/240 V service is fed to a main breaker (or main fuses). Many individual circuits, each with its own breaker (or fuse), go from the panel to various loads in the home. **See Figure 12–5.**

The breaker (or fuse) on an individual circuit trips out when it is exposed to overcurrent, such as an overload or a short circuit. The rating of the main breaker is coordinated so that the individual circuit breaker will trip before the main breaker. This protection scheme is similar to the protective setup with a distribution substation and its feeders.

Circuits feeding a bathroom or the outdoors need more sophisticated protection to protect people in these well-grounded locations. Protection that will open a circuit at a current level below the threshold of a person's sensation is available. A ground fault circuit interrupter (GFCI, or GFI) is used on these circuits where it is desirable for the circuit to trip out quickly for even a minor fault. The protection of a household circuit using a GFI can be compared to the ground

Figure 12–5 Circuit breakers and fuses provide and fuses protection for residential wiring. *(Delmar, Cengage Learning)*

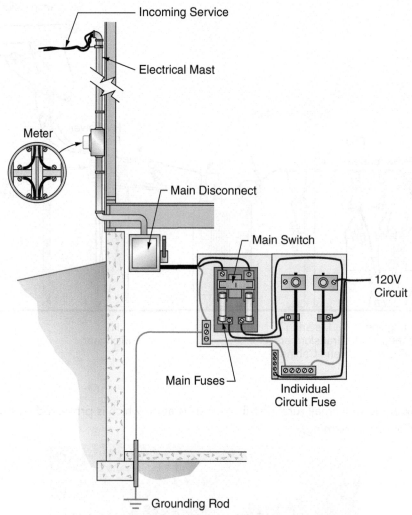

fault relay protection available on transmission and distribution circuits. A GFI monitors the current flowing to the load and the current returning to the source on the neutral. **See Figure 12–6.**

The energized and neutral wires pass through a coil. Under normal conditions, the current through each wire would be equal but would be flowing in opposite directions. No magnetic field would be induced into the coil because of the canceling effect of the two wires of opposite polarity. If some of the current flow returned to the source through ground, there would be more current flowing in the hot wire than in the neutral wire, and there would be a small magnetic field induced in the coil. This primary coil induces a voltage into a secondary coil whose output is amplified by an electronic amplifier. The output from the amplifier is applied to a relay coil, which opens the circuit. A typical GFI will trip out within 30 milliseconds when there is a current imbalance of only 5 mA. Faster and slower GFIs are available. A very low level of current will trip a GFI quick enough to prevent a person from feeling any electrical sensation.

Consequence of Poor Overcurrent Protection

Faults such as phase-to-phase contacts, phase-to-ground shorts, or a follow-through current after a lightning strike show up in a circuit as an overcurrent condition. *If* the circuit stays in service during an overcurrent condition, *then* the following will occur.

- Equipment such as transformers, regulators, and conductors will overheat and be damaged.
- An overcurrent condition will cause the voltage on the system to collapse, which if not corrected can damage customer equipment.

Figure 12–6 A ground fault circuit interrupter detects current in the ground wire and trips out the circuit. *(Delmar, Cengage Learning)*

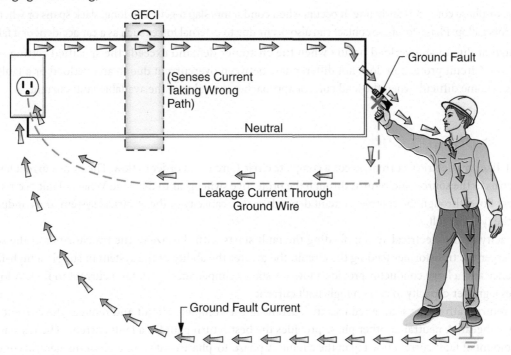

- There will be a greater risk to the public when they are involved with a conductor falling to the ground or when they have an accidental power line contact with a ladder, antenna, or tree.

Just as it is essential to have a fuse or breaker panel to protect household circuits, utility circuits also need protection. Having no protection is similar to a homeowner putting a coin behind a fuse to bypass the fuse protection.

Transient or Permanent Faults

Depending on the protection scheme and location, about 90% of overhead circuit faults are transient. A transient fault such as a lightning strike or a conductor contact with a tree in a windstorm can cause an initial trip of the circuit, but the circuit is automatically reclosed, and—if the fault is gone—the circuit stays energized. A line crew is needed to repair and reenergize the circuit for the 10% of faults that are permanent.

Transient faults are rare on underground circuits. Most underground faults are permanent and require a line crew to make repairs. Underground faults are not as obvious as overhead faults because faults on underground systems tend to be high-resistance faults. The protection opens the circuit before the fault becomes a dead short, thereby making the location of a fault more difficult to find.

Causes of Overcurrent

A circuit is faulted anytime an energized conductor makes contact with another element that is at a different potential. When the resistance or impedance between an energized conductor and another element is low enough, the fault becomes a short circuit.

- Lightning is the most frequent cause of transient faults. At the flashover point, a high-voltage arc establishes a path of ionized air to ground. A high follow-through current is established through the ionized air and causes a fault current to flow in the circuit.
- A phase-to-ground or a phase-to-neutral fault is the cause of about 70% of permanent overcurrent faults. A phase-to-ground fault could be caused by a broken insulator, a contaminated insulator, a tree contact, a broken conductor, an animal contact, a car accident, or an accidental public contact with a crane, sailboat, ladder, or antenna.

- A phase-to-ground fault on an underground cable can occur due to an insulation breakdown, a dig-in, or a driven fence post.
- A phase-to-phase contact is fairly rare. It occurs when conductors slap together in long, slack spans or when ice-covered conductors gallop. Phase-to-phase contact can also occur due to external forces such as a car accident or a falling tree.
- Overcurrent due to an overload occurs when the customer demand exceeds the specified setting of the circuit protection. Circuit protection does not differentiate between overcurrent due to an overload or a fault. Protection settings become difficult when the load current approaches the value of the available fault current.

Magnitude of Fault Current

Fault current, like any electrical current, needs a complete circle (circuit) in order to flow. The same current level flows into the fault, returns to the source, and returns through the protective switchgear to the fault. When a fault occurs, the amount of current that flows through the complete circuit depends on the capacity of the electrical system, the conductivity of the circuit, and the type of fault.

The capacity of the electrical system feeding the fault starts with the size of the transformer at the source of the feeder. The larger the transformer feeding the circuit, the greater the ability of the system to supply a high-fault current. A short distance and a large conductor provide a low-resistance (impedance) path for a current to flow. A low-resistance path provides a greater capacity to carry a high-fault current.

A poor return path to the source reduces the current in the complete circuit and reduces the current through the protective switchgear. A neutral or other phase provides the best return path for a fault current. The resistance of a fault affects the amount of fault current flowing in the circuit. A phase-to-phase contact or a phase-to-neutral contact is a dead short and has virtually no resistance to impede current flow. A conductor lying on the ground or contacting a tree has a relatively high resistance and limits the amount of fault current generated. A high-resistance fault is really just another load on the circuit if the current generated in the circuit is not high enough to trip any overcurrent protection switchgear. Likewise, a customer load is just like a high-resistance fault.

Impedance to Fault Current

A fault can be a dead short or a partial short circuit, depending on the impedance or resistance between an energized conductor and the object causing the short. A fault to ground, such as a dry tree limb or a broken conductor lying on dry or frozen ground, is a high-resistance fault and does not provide a good path for current to flow into earth. A high-voltage feeder is more likely to overcome any impedance and is able to generate a higher current flow back to the source. A phase-to-phase or a phase-to-neutral fault is a low-impedance fault. The circuit conductors provide a good return path to the source. The complete circuit has a high-fault current flowing through it and trips the protective switchgear quickly.

Distribution Protection Scheme

A protection scheme for a distribution feeder has a protective device at the source and protective devices downstream at junctions, lateral taps, and transformers.

Feeder Protection at the Substation

The source of a feeder can be protected by fuses, reclosers, or circuit breakers. A fuse is a one-shot device and is sometimes used in small, older stations as feeder protection. A recloser and a circuit breaker are multishot devices that provide a number of tripping and reclosing operations to give transient faults time to clear.

Downstream Multishot Protection

A substation recloser or circuit breaker is often able to provide multishot protection to the end of the feeder, especially on the shorter urban feeders or on high-voltage distribution feeders. Long rural feeders require downstream reclosers to maintain multishot protection to the end of the line.

Downstream Sectionalizers

A sectionalizer is a slave device to an upstream multishot device. A fault downstream from a sectionalizer will cause an upstream multishot device to trip out and reclose the circuit for a specified number of times. Depending on how the protection is specified, a sectionalizer will isolate a permanent fault during one of the intervals while the circuit is de-energized by the multishot device. A sectionalizer can interrupt a load current, but it does not have the capacity to interrupt a fault current.

Downstream Fuses

A fuse can be specified to isolate a permanent fault after the upstream multishot device has operated a specified number of times. Generally, the speed of the specified fuse will be set to melt before the third or fourth operation of the multishot device.

Specifying Protection for a Distribution Feeder

Planning a Protection Scheme

Planning a protection scheme is normally outside the scope of a power line technician's duties, but it is useful and interesting to have an understanding of the process. There are three types of protection schemes: radial, loop, and network. The radial protection scheme is based on a single source for a feeder. A permanent power failure interrupts everyone downstream of the automatically opened switchgear. The method to set up a radial protection scheme has four major steps.

1. Gather the data needed to calculate a feeder profile.
2. Calculate the voltage, load current, and fault current on the feeder.
3. Interpret the results of the study.
4. Specify the protective devices needed on the feeder.

Gather Feeder Data

A planning engineer prepares a map or a schematic drawing of a feeder and breaks the feeder into segments. Each line segment is assigned a numbered node or point. **See Figure 12–7.**

- Nodes are assigned to each junction, switch, conductor-size change, and load center.
- The length and conductor size in each line segment are recorded.

Figure 12–7 Distribution feeder data is collected from node to node. *(Delmar, Cengage Learning)*

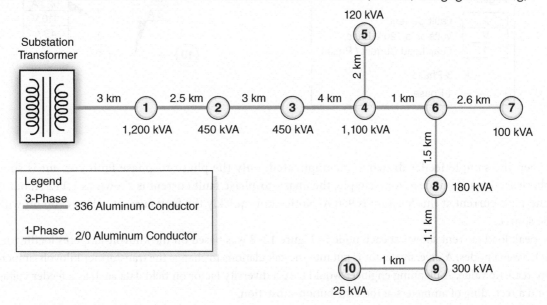

- The transformer load from node to node is added up and recorded. Transformers on a feeder are not always loaded to 100% capacity, so a diversity factor is used to approximate the actual loading. In the sample feeder shown a **balanced load** is assumed between three phases. In reality, many utilities would add up the load on each individual phase.

- The available fault current at the source of the feeder is needed to do the feeder calculations. This information is based on the impedance and size of the distribution substation transformer and the voltage, conductor, size, and length of the circuit feeding the substation. In the sample feeder shown, the available phase-to-phase fault current at the secondary of the substation transformer will be given as 5,000 A.

Calculating Feeder Information

Converting the field data to fault current, voltage, and peak load current at each node is not normally done manually. The feeder data are input into a computer program, and the output shows the fault current, voltage, and peak load current at each assigned node.

Calculations for fault current are based on a dead short, such as a phase-to-phase fault or a phase-to-neutral fault at each node. Most faults are, of course, not dead shorts, and this is taken into consideration when protective switchgear is specified. **See Figure 12–8.**

Figure 12–8 The field data is used to calculate the fault current, load current, and voltage at each node. *(Delmar, Cengage Learning)*

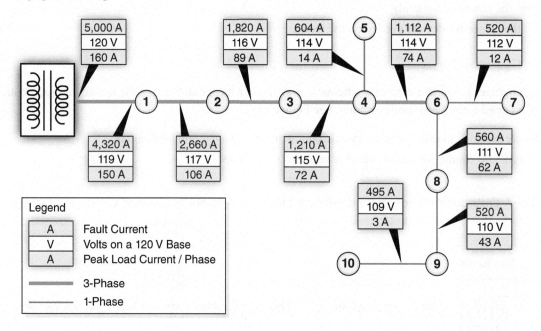

To keep the sample feeder drawing uncomplicated, only the phase-to-phase fault current is shown in the three-phase sections. At node 6, for example, the phase-to-phase fault current is shown as 1,112 A, but the phase-to-ground fault current at this location is 960 A. Notice how quickly the fault current values drop off with distance from the source.

The peak load current shown at each node in Figure 12–8 was based on the kilovolt-amperes loading in each line segment between nodes. A diversity factor is put into the calculations to change the transformer kilovolt-amperes loading to a more realistic value. A planning engineer would base a diversity factor on field data such as a feeder voltage/current survey and a recording of ammeters at the distribution substation.

The voltage at each node is also calculated by the computer program. The voltage in the sample feeder is shown to drop in proportion to the feeder load and the distance from the source. In this example, the voltage was put on a 120 V base. The voltage on the single-phase line farthest from the station is shown to be too low, and the planning engineer would probably specify a voltage regulator.

Protection Scheme Philosophy

A radial protective coordination scheme provides automatic isolation of faulted circuits from the system while leaving the rest of the system in service. Where two protective switchgear devices are in series with each other, the *protected* switching device is on the source side and the *protecting* switching device is on the load side. If a fault develops downstream from the protecting device, it should trip out before the protected device. For example, a transformer fuse should have a size and speed that will clear a faulted transformer from a circuit before a protected upstream line fuse is damaged.

Most distribution feeders have a recloser or a circuit breaker that provides multishot protection to the feeder. A multishot protective device can be set for a sequence of a specified speed and number of trips and reenergizations. A *fuse-saving* philosophy specifies that the first trip-out operation of a source-multishot device is set to be so fast that a downstream fuse does not even begin to melt. For lightning and other transient faults, this protection scheme reduces outages and trouble calls. Everyone on the feeder experiences at least one trip-out for each transient, which starts their digital clocks blinking. A permanent fault on a lateral will cause a multishot device to trip out and reenergize the circuit for the complete specified number of times. The downstream fuse will blow on the second or third re-energization, depending on how the protection is set up.

A *time-delayed-instant* philosophy specifies that the first trip-out operation of the source multishot device is delayed long enough for a downstream fuse to blow first. This prevents everyone on the feeder from being exposed to a momentary trip. This philosophy is often used with fuses protecting underground feeders, where most faults are permanent. The time-delayed-instant philosophy keeps a feeder exposed to a fault longer and increases the likelihood of damage, such as a conductor burning down.

Interpreting the Results of a Feeder Study

The results of a feeder study, similar to the sample, give a planning engineer all the information needed to specify, prepare, and justify betterment specifications for the feeder.

- The maximum and minimum fault current available along the feeder gives the necessary information to prepare a protective coordination scheme. For example, the sample feeder study shows that the reclosers in the distribution substation must be big enough to interrupt 5,000 A of fault current and carry a load current of approximately 200 A continuously.

- Normal data collection and calculations are done for each individual phase. The results show the load current on each phase and whether there is a need for some phase balancing work.

- The study results show where there is a need for voltage correction. A voltage profile of the feeder is available from node to node. Voltage regulators or capacitors can be located near the node where the equipment will provide maximum benefit.

- Justification for major betterment work is based on a feeder study. The cost of line losses can be worked out and compared to the cost of re-conductoring. A feeder study will show whether conversion of a single-phase circuit to three phases will solve phase balancing, voltage, or protection problems.

Specifying a Hydraulic Recloser Frame Size

Recloser types vary depending on the voltage rating, the interrupting capacity, and their mode of operation. There are a variety of recloser frame sizes, each with a voltage rating, maximum current rating, and maximum interruption rating. Each manufacturer provides a catalog designation to identify each type of recloser. The designations of recloser types help to avoid using a particular manufacturer's designation. **See Figure 12–9.**

Figure 12–9 Typical Recloser Ratings

Types of Reclosers	Types	Maximum Rating (kV)	Maximum Current (A)	Interruption Rating (A)
Hydraulic Reclosers				
The ratings of five typical sizes of *hydraulic* units are shown. The contacts on hydraulic reclosers open when overcurrent flows through a solenoid coil, which creates a magnetic force on a plunger and opens the contacts.	H_1	14.4	50	1,250
	H_2	14.4	100	3,000
	H_3	14.4	280	6,000
	H_4	24.9	100	2,000
	H_5	24.9	280	4,000
Electric Reclosers				
The two typical *electric* reclosers shown require a low-voltage supply to operate their heavy contacts.	EL_1	14.4	560	12,000
	EL_2	34.5	560	8,000
Electronic Reclosers				
The ratings of four typical three-phase relay-controlled *electronic* reclosers are shown. The relays can be set to detect specified overcurrent and ground current conditions.	ET_1	14.4	400	6,000
	ET_2	14.4	560	10,000
	ET_3	34.5	400	6,000
	ET_4	34.5	560	8,000

Identifying the types of reclosers as H_1, EL_1, ET_1, and so forth, simplifies reference back to this figure for the protection examples shown later. The figure does not give all the needed information about a recloser rating. For example, the interrupting capacity of the H_3 recloser is shown as 6,000 A, but its actual interruption rating is 6,000 A at 4.8 kV and only 4,000 A at 14.4 kV.

When a recloser is chosen for a specific location, it must have a suitable frame size, which means a unit must have its contacts and its trip mechanism big enough to interrupt the highest fault current that could occur at that location. In the sample feeder, a fault immediately in front of the substation recloser could generate 5,000 A. The recloser frame must have its contacts and a spring mechanism able to safely carry and interrupt 5,000 A.

Note in Figure 12–9 that the smallest recloser for a 12.5/7.2 kV system that can interrupt 5,000 A would be an H_3 recloser. This recloser can interrupt 6,000 A; however, this interruption rating applies to 8.32/4.8 kV. The interrupting capacity for the H_3 recloser on a 12.5/7.2 kV system is 5,000 A.

Specifying a Trip Coil Size

An overcurrent situation will cause a recloser or circuit breaker to trip out a circuit. Relay-controlled devices, such as a circuit breaker or an electronic recloser, will trip out when a relay senses the overcurrent and sends a signal to trip out the circuit. A hydraulic recloser opens when overcurrent flows through its trip coil. A trip coil installed in a recloser must be able to carry the expected peak load current. For example, if a sample feeder has a projected peak load of 160 A per phase, a 200 A trip coil would probably be specified to allow for load growth and unbalanced load. A 200 A recloser will carry 200 A continuously and will carry some overload without damage.

In general practice, an overcurrent device will not trip for a small or temporary overload. A trip coil in a hydraulic recloser is normally designed to trip a circuit when the current flowing through it is twice the rating of the coil; therefore, when 400 A flow through a 200 A coil, the circuit will trip out. An overcurrent of double the recloser setting would normally be a fault.

A Recloser Protection Zone

A recloser will provide multishot protection downstream to any location where the circuit has the capacity to generate the needed fault current to trip it out. For example, the lowest fault current available in the sample circuit shown in

Figure 12–8 is 495 A phase to ground at node 10. A dead short between phase and neutral at node 10 could generate 495 A, which would be enough to trip a 200 A recloser at the substation. A more common higher-impedance fault, such as a broken conductor lying on dry ground or a tree contact, will not generate the 400 A needed to trip the circuit.

In the sample feeder, the reclosers at the substation will likely "see" most faults just beyond node 6 where the circuit can generate a phase-to-phase fault current of 1,112 A and a phase-to-ground fault current of 960 A. At node 6, even a high-impedance fault is likely to generate the minimum 400 A needed to trip a 200 A recloser. Ideally, a safety factor is needed to provide a greater likelihood of a circuit tripping out for a high-impedance fault. A safety factor of two would require a minimum of an 800 A fault current to be available at the end of a zone protected by a 200 A recloser.

Specifying Recloser Speed and Operating Sequence

A planning engineer chooses a recloser with a sequence of opening and closing speeds based on coordination with upstream and downstream protective devices. When a fault is beyond a downstream fuse, the planning engineer would choose a fuse size and speed to blow at the desired time—for example, at the second or third closing of the recloser. The timing sequence of a recloser is coded as follows:

A = instantaneous

B = retarded

C = extra retarded

D = steep retarded

A typical speed and sequence may be A1B3, which means the recloser opens instantaneously on sensing a fault and recloses. If the fault persists on the line, the recloser opens and recloses after a retarded delay. For a permanent fault, the recloser will go through the complete sequence of one instantaneous trip and three retarded trips before remaining open. The speed and sequence chosen by the planning engineer depend on the protection scheme philosophy and on the need to solve particular protection problems.

Some utilities have the size of recloser, frame size, and operating sequence painted on the recloser (e.g., 70 A, 4H recloser with a B13 sequence). To change the operating sequence of a hydraulic recloser, the recloser needs to be removed from the pole and the solenoid trip coil replaced. **See Figure 12–10.**

Figure 12–10 The recloser type and sequence is painted on the tank. *(Delmar, Cengage Learning)*

The minimum trip setting and operating sequence on an electronic recloser can be reprogrammed in the field by switching the specified DIP switches on or off. A DIP switch is a "dual in-line package" that switches a "package" of switches in a circuit board. Certain reclosers can be reprogrammed in the field and also have the additional advantage of being able to change the rating of the unit, for example, from a 70 A unit to a 100 A unit. **See Figure 12–11.**

Figure 12–11 The rating of the Versa-Tech® vacuum interrupter recloser can be changed in the field. *(Reprinted with permission of Hubbell Power Systems, Inc., Centralia, MO USA)*

Typical TCC Curve for a Recloser

When specifying a protective device, the speed of operation must be known in order to be able to coordinate the device with other protective equipment in the circuit. A time current characteristic (TCC) curve is available for each size of fuse, sectionalizer, recloser, or relay. A TCC curve is plotted on graph paper and shows the response time, minimum damage time, and total clearing time. **See Figure 12–12.**

Note that the TCC for the 200 A recloser in Figure 12–12 shows that the minimum fault the recloser will see is 400 A. The A curve shows that 400 A will trip out the circuit in about 0.5 second. The recloser will clear a permanent fault of 400 A in about 15 seconds. A permanent 5,000 A fault will clear in about 0.2 second.

Specifying a Downstream Recloser

In a sample feeder, a downstream recloser is needed to provide multishot protection to the end of the feeder. A higher-impedance fault would not be "seen" by the source recloser. Ideally there should be a multishot protection to the end of every feeder. For economic reasons, multishot protection may be unavailable for higher-impedance faults near the end of a feeder. The end of the feeder can still be protected by a fuse, but without multishot protection a transient fault will blow a fuse and cause an easy callout for a line crew. To have multishot protection for a higher-impedance fault on the single-phase line past node 6, a downstream recloser should be installed.

Figure 12–12 A TCC is used to coordinate the tripping time of one protective device with another. *(Delmar, Cengage Learning)*

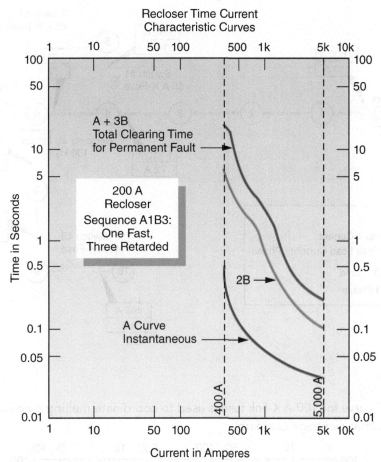

For example, a 100 A H_2 recloser on the single-phase line at node 6 will carry the 62 A of load current and should see most faults at node 10, which has 495 A of available fault current. To provide node 7 with multishot protection, another recloser would need to be installed in that line. **See Figure 12–13**.

Specifying Fuse Size and Speed

A line or a station fuse is chosen based on its current rating and speed. The speed of a fuse is indicated by the letter on the fuse link (e.g., K-link or T-link). In order to coordinate the speed of the fuse with upstream and downstream protective switchgear, the minimum damage time, minimum melting time, and total clearing time must be known. The TCC curve for a 100 A K-link fuse shows how quickly the fuse melts as the current going through it is increased. **See Figure 12–14**.

In the figure, 200 A will melt the fuse link after about 100 seconds of exposure, and 1,000 A will melt the fuse in about 0.2 second.

Specifying Downstream Fuses

A fuse size and speed can be chosen so that the fuse will melt in any of the energized intervals of an upstream multishot device as it goes through its sequence during a permanent fault. In the sample feeder, the lateral protected by a fuse in switch 1 is specified to coordinate with the 200 A substation reclosers. The lateral has a peak load current of only 14 A. The 25 A fuse would protect this lateral, but it would melt before the upstream recloser could go through the instantaneous trip. A 65 A fuse would stay intact as the substation recloser went through its instantaneous trip for a fault on the lateral. The 200 A reclosers can see to the end of the lateral because the available fault current is 890 A.

Figure 12–13 The protection needed to protect this sample feeder has been determined and is recorded on the feeder drawing. *(Delmar, Cengage Learning)*

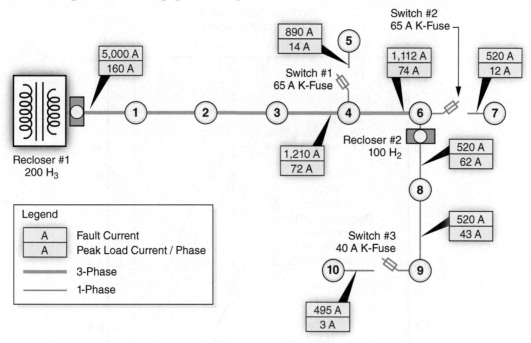

Figure 12–14 A TCC curve for 100 A K-link fuse is used to coordinate melting time with upstream protective devices. *(Delmar, Cengage Learning)*

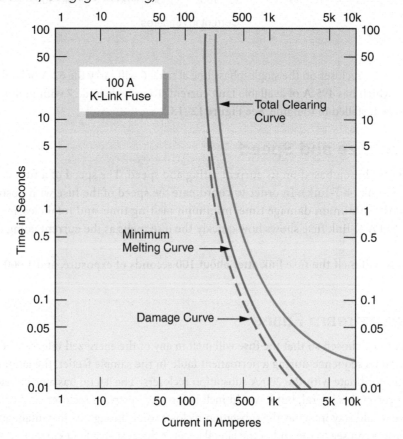

A TCC curve for a 65 A fuse superimposed on the TCC curve for a 200 A H_3 recloser can show that the two devices coordinate with each other. A quicker method is the use of a recloser-to-fuse coordination table. **See Figure 12–15**.

This data can be used with a 100 A H_2 recloser. In the sample feeder shown in Figure 12–13, the lateral tap protected by a fuse in switch 3 is specified to coordinate with recloser 2, which is a 100 A H_2 recloser. The lateral is lightly loaded with a peak load of 3 A, but if the circuit is exposed to transients from lightning or trees, it would save a callout to have the fuse in switch 3 coordinate with recloser 2. The fault current at the end of the lateral is 495 A, which means that the 100 A recloser should see most faults. The recloser-to-fuse coordination table shows that the lateral at switch 3 has a minimum of 495 A and a maximum of 520 A fault current and needs a 40 A fuse to coordinate with a 100 A recloser.

Figure 12–15 Recloser-to-Fuse Coordination Table

A 100 A H_2 Recloser Set at Sequence: One Fast, Three Retarded (A1B3) Type K Fuse Links								
	Recloser-to-Fuse Coordination Range in Amperes		Maximum Fuse-to-Fuse Coordination Current within the Recloser Zone of Protection					
Protecting Fuse Link	Min.	Max.	Protecting Fuse Link	Protected Link				
				50	65	80	100	140
40	200	690	40	600	690	690	690	
50	200	1,020	50		510	1,020	1,020	1,020
65	200	1,400	65			600	1,400	1,400
80	280	1,880	80				1,280	1,880
100	560	2,560	100					2,560
140	1,970	3,000						

In the sample feeder, the lateral protected by the fuse in switch 2 has multishot protection for low-impedance faults from recloser 1. The available fault current of 620 A at the end of the line may not generate enough current for a higher-impedance fault such as a tree limb. A judgment has to be made regarding what to do, as follows:

1. Install a recloser at switch 2.
2. Install a fuse that coordinates with recloser 1.
3. Install a small fuse that would blow quickly for any fault.

Option 2 is a compromise wherein recloser 1 sees most faults except for higher-impedance faults near the end of the line. A 200 A recloser table indicates a 65 A fuse coordinates with recloser 2.

Specifying Sectionalizers

Sectionalizers are specified in a way similar to how fuses are specified. TCC curves or tables determine the most appropriate size and speed sectionalizer to coordinate with an upstream multishot device. A sectionalizer must coordinate with a multishot device because it isolates or opens a faulted circuit in an interval when the multishot device has tripped out the circuit. A sectionalizer is more reliable than a fuse for specifying at which interval of the upstream multishot device sequence it will open during a permanent fault. A trip coil for a sectionalizer is specified for its current rating and has a two or three-shot sequence.

A Loop Protection Scheme

A loop protection scheme circles through the service area and is tied to another in-phase source. Unlike a radial system where the tie switches are left normally open, open automatic tie reclosers are placed in the loop in various locations. The tie switch closes automatically when power is lost on one side or the other and feeds as far as the location where another open point is created.

A Network Protection Scheme

In a network protection scheme called distribution automation, customers are supplied by more than one source in parallel with each other. When one source fails the any parallel sources keep the power on. During a fault on a line, distribution automation switchgear will communicate with each other and will isolate a faulted section from the grid and allow one of the parallel feeds to continue to supply the customer. A computerized distribution management system coordinates, controls, and monitors the automated distribution network.

Automated switchgear that can communicate back to the control center includes SCADA-controlled reclosers, remotely controlled circuit breakers and switches such as S&C IntelliTEAM II, S&C Remote Supervisory PME Pad-Mounted Switchgear, and Scada-Mate. **See Figure 12–16.**

Figure 12–16 (a) Remote Supervisory PME Pad-Mounted Switchgear and (b) Scada-Mate. *(Courtesy of S&C Electric Company)*

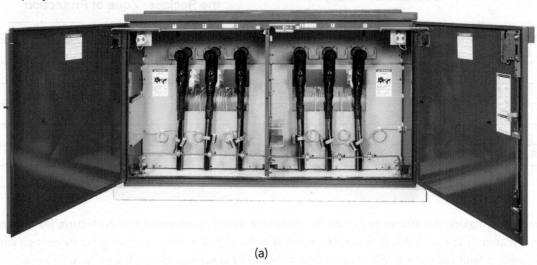

(a)

(b)

Overcurrent Trouble Calls

During an overcurrent situation, a line crew is called out after circuit breakers, reclosers, sectionalizers, or fuses have gone through their sequences and have isolated any circuit with a permanent fault.

If	Then
There has been wind, ice, wet snow, or lightning.	The problem is often the traditional broken conductors, broken poles, or fallen trees and branches.
Protection tripped out on a nice clear day.	The fault is sometimes due to public contact, such as a car accident, tree cutting, or a boom contact. An outage on a clear day suggests that a patrol should be carried out before energizing the circuit.
An overload is the likely cause of an outage.	The most common fix is phase balancing. The planning engineer often assumes that the load on the feeder is balanced among the three phases. When the load is not balanced, one phase of the three-phase system is carrying more than its share of the load and the protective device will trip out.
The circuit has been out for a while, especially during peak load periods.	There is a high initial in-rush current when the switch is closed, and the line trips out again. A heavily loaded circuit may have to be picked up a section at a time.

High-Voltage Distribution Conversion

Many utilities have voltage conversion programs to convert their systems from typical primary voltages such as 8.3/4.8 kV and 12.5/7.2 kV to higher distribution voltages such as 25/14.4 kV and 34.5/19.9 kV. Following are the advantages of a higher-voltage distribution system.

- The substation breaker or recloser can see out much farther because the available fault current level remains high much farther downstream. There is often no need for downstream reclosers or fusing to maintain protection at the end of the line. Sectionalizing devices are still installed for operating purposes.

- There are fewer substations needed and fewer feeders needed from the substation.

- For a given load, a higher-voltage feeder would have less load current and, therefore, less voltage drop along the feeder. Downstream voltage regulators are rare, except for a very long or very heavily loaded feeder. There is a rule of thumb that a distribution feeder with an average customer load can feed about 1 mile (1.6 km) per 1 kV (phase to neutral) before needing voltage regulation. Based on this rule of thumb, an 8.3/4.8 kV feeder would need voltage regulation about 5 miles (8 km) from the substation, and a 34.5/19.9 kV feeder would need voltage regulation at about 20 miles from the station.

- A higher-voltage feeder can feed much more load. For example, an 8.3/4.8 kV feeder typically can be considered heavily loaded when supplying about 3 MW, while a 34.5/19.9 kV feeder is considered heavily loaded when it is supplying about 15 MW.

- A utility would normally not be able to feed an individual customer that has a requirement for a 750 kW service on an 8.3/4.8 kV system. A subtransmission feed would be required, along with all the associated costs. Meanwhile, if a 25/14.4 kV or 34.5/19.9 kV feeder were available, a 750 kW load could be supplied on that system with standard distribution transformers without creating a problem for other customers on the feeder.

Overvoltage Protection

Causes of Overvoltage

Other than lightning, not many trouble calls are due to overvoltage. Lightning is the most common source of overvoltage and one of the most transient faults on an electrical system. The high voltage and current generated by lightning usually cause a momentary outage to customers because the automatic circuit protections open and reclose after the strike.

Switching operations can cause a transient overvoltage. The magnitude of the surge is much lower than lightning and normally is significant only on circuits of 230 kV and above. An overvoltage situation can occur when a voltage regulator is stuck in a maximum-boost position, a switched capacitor is left in service during an off-peak time period, a higher-voltage wire falls onto or makes contact with a lower-voltage wire, or, in rare cases, a transformer with a partially shorted coil results in a high secondary voltage.

Convection and Frontal Thunderstorms

The two types of thunderstorms are convection storms and frontal storms.

A convection thunderstorm is the most common thunderstorm, especially on a hot summer day. It is a localized storm that occurs when the hot air near earth's surface rises and meets the cold air at a higher altitude. A convection thunderstorm does not last long because it is usually accompanied by rain. The rain cools the earth, which removes the energy source for the storm.

A frontal thunderstorm occurs when a cold front meets a front of warm, moist air. The storm can stretch for hundreds of miles and last for hours. It can regenerate itself because air masses continue to move in as the fronts collide. A frontal storm is more severe than a convection storm.

Description of Lightning

Lightning is part of an overall electrical circuit between earth and atmosphere. Electrical charges are created in the atmosphere by the friction between particles of rapidly moving air. The cloud normally associated with a frontal storm or a convection storm is the cumulonimbus cloud, also called thunderhead. During rain or hail, negative charges fall to the bottom of the cloud. When these negative charges travel toward positive charges, the air becomes electrically stressed and breaks down, resulting in high voltage and high-current discharge (lightning) between the positive and negative charges. The electrical discharge can be within the same cloud, clouds, or to earth. **See Figure 12–17.**

Figure 12–17 An electrical circuit is created between the atmosphere and ground. *(Delmar, Cengage Learning)*

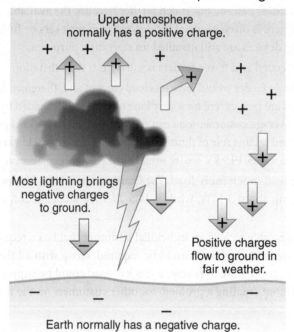

The direction or polarity of the current flow due to lightning is not important as far as protection is concerned. Both positively charged and negatively charged lightning strikes earth. Contrary to its appearance, the most common direction for the current flow is from cloud to earth. A rarely visible charged leader from the cloud reaches earth, and, as it starts to neutralize there, it becomes a visible arc from earth to cloud. The following are some characteristics of lightning.

- Lightning can have a charge of up to 100,000,000 V.
- Lightning establishes an electrical path through which the follow-through current flow can be from 2,000 to 200,000 A.

- Lightning can generate heat in a 1 in. (2.5 cm) path of approximately 60,000°F (35,000°C).
- Lightning generates heat, which causes an explosive expansion of air and is heard as thunder.

Estimating Distance from Lightning

The flash-bang method is a quick method for estimating the distance to lightning. Count the number of seconds between seeing the lightning and hearing the thunder:

Number of seconds divided by 3 = Number of kilometers away

Number of seconds divided by 5 = Number of miles away

Determining the direction in which a storm is moving is important for determining safe working conditions on distribution lines. If the storm is 5 miles (8 km) away but heading toward a work area, it is time to get clear of the circuit. The next lightning flash will be closer. As soon as lightning is seen or thunder is heard, work on power lines, especially transmission lines, should be suspended because some part of the line could reach into the storm area.

Some jurisdictions use a 30/30 safety lightning rule. If thunder is heard within 30 seconds of seeing a lightning flash, then get off the line and remain off for at least 30 minutes after the last lightning flash.

Effects of a Direct Stroke of Lightning

The effects of a direct stroke of lightning can be very dramatic. When lightning flashes, an electrical circuit is being completed. When lightning strikes earth, an electrical circuit is completed through the closest and lowest-resistance path to ground.

A tall, pointed object is the most attractive object for a lightning strike. A 1,000-foot (300 m) structure may be struck four times a year, while a 100-foot (30 m) structure is struck once every 25 years. Tall towers and buildings with lots of steel in them are struck by lightning but can carry the current to ground without overheating and causing damage.

Concrete footings of transmission line towers can be damaged because lightning will heat the moisture in the concrete and expand it explosively. Concrete footings are, therefore, usually bypassed with a heavy-gauge conductor to carry the current. Trees and wood poles are more attractive to lightning than earth is. Although trees and wood poles are higher-resistance paths than steel buildings or towers, they are still lower-resistance paths than air. When lightning strikes these objects, the moisture in the wood is suddenly vaporized and the resultant explosion will cause the wood to splinter.

Ground-Gradient Effect

There are step and touch potentials up to 70 kV per foot (200 kV per m) in the immediate area of a lightning strike. People standing near a tall object that is struck by lightning are injured because of the high step and touch potentials at the base.

Even when lightning discharges in the clouds and does not reach earth, voltage-gradient changes are taking place on the ground. Induction from the lightning discharge can cause gradients of 3 kV per foot (10 kV per m) on the ground. The induced gradients cause corona discharge from pointed, grounded objects and can cause hair to stand on end and skin to tingle. Sparks from induced gradients and corona discharge can cause fires around unprotected flammable liquids, even without a lightning strike.

An object isolated from the ground can have an induced potential up to 100 kV. When lightning is seen or thunder is heard, there may already be an induced potential buildup on isolated conductors.

Safety Tips for Working Near Lightning

The following safety tips for working near lightning can apply to on-the-job and off-the-job situations.

- The safest place to be during a thunderstorm is in a vehicle. If a vehicle is struck by lightning, all the metal in the vehicle will be at the same potential and anyone in the vehicle will not be exposed to any potential difference. The vehicle provides an equipotential zone and an effective Faraday cage.
- When your hair stands on end and your skin starts to tingle, there is a potential difference building up between the earth you are standing on and the clouds above. Lightning is about to strike. Drop into a fetal position.

- Lightning usually strikes tall objects such as trees. People nearby are injured due to the ground gradients of the lightning dissipating into earth. The ground gradients are like the rings formed when a stone is thrown into water. Each "ring" is a different voltage with the greatest potential differences being near the center. Staying away from tall objects reduces exposure to the highest-voltage gradients. Keeping your feet together reduces exposure to the different potentials of the rings, or to ground gradients. In the outdoors, with no nearby shelter, crouch low with feet together.
- In buildings, stay away from grounded objects. Grounded objects rise in potential when lightning strikes nearby.

Tracking Lightning Storms

Many electrical utilities have access to information from lightning detection systems, which detect the location and intensity of lightning discharges. This information is valuable for determining which transmission lines are at risk of a lightning strike and, therefore, at risk of being tripped out. When possible, system operators reduce the loads on vulnerable transmission lines and increase the loads on other generators and lines to make up the difference. Putting a storm limit on a key extra-high-voltage (EHV) circuit is critical because an unexpected loss of a large source of power could destabilize an entire electrical system.

Lightning-detection systems can also provide an early warning to line crews working on transmission lines. A voltage surge due to lightning is a hazard when working on an energized line, as well as when working on an isolated and grounded line.

Effect of Lightning on a Circuit

A voltage surge from lightning travels like an ocean wave in all directions, except that it travels at the speed of light. The wave is attenuated (diminished) when it encounters a location where it will flash over, preferably at a **surge arrestor** where a sparkover occurs and some or all of the energy is dissipated. A voltage surge, which causes a coupling effect between the phase conductor and the neutral or shield wire, can be attenuated in other ways. The coupling effect reduces the voltage on the phase conductor while it raises the voltage on the neutral or shield wire. When the neutral and shield wire are well grounded, the voltage is quickly lowered and the coupling effect, in turn, lowers the voltage on the phase conductors.

A high-voltage surge will be attenuated by corona discharge because distribution conductors are too small for such high voltages. The relatively small conductor also limits the current-carrying capacity of a conductor because of skin effect. The current flow is subject to skin effect in the conductor even though the current resulting from lightning is DC. However, it is not a steady-state current; it reaches a peak value and then diminishes. The effect of the change in current means it has a similar effect on conductors as AC. When an ocean wave hits a wall, it is reflected back double in size and then dissipates quickly. A voltage surge acts out similarly when it comes to a dead end. That is why transformers on dead-end poles are more vulnerable to lightning damage.

Underground cables are very sensitive to voltage surges and are exposed to surges when lightning strikes at or near a cable riser pole. While **arrestors** are installed at a riser pole, the cable shield and arrestor ground are tied together and the discharge voltages from arrestors will travel through the ground and cable sheath. The design of a riser pole and location of arrestors minimize the effect of lightning through factors such as arrestor lead length and location.

Switching Surges

Switching operations cause transient overvoltages. The magnitude of the voltage surge is much lower than lightning and is significant only on circuits 230 kV and higher. If a switch opens and is immediately reclosed, a brief overvoltage occurs when energizing a capacitor. A length of parallel conductors on a transmission line is like a big capacitor and arcing inside a circuit breaker can have the same effect as though the breaker is reclosing and a restrike is occurring. An arc of a restrike on a capacitor causes a **switching surge** in the line.

When carrying out live-line work, the risk of a restrike or switching surge is reduced if the circuit breaker is put into a nonreclose position.

Effects of a Voltage Surge

A voltage surge can cause a flashover, a sparkover, or a puncture.

- A *flashover* is a disruptive discharge along a solid material such as an insulator string or a live-line tool.
- A *sparkover* is a disruptive discharge through the air.
- A *puncture* is a disruptive discharge through a solid material such as rubber cover-up, rubber gloves, or a fiber conductor cover.

A voltage surge can be compared to the action of a wave of water. **See Figure 12–18.**

Figure 12–18 A voltage surge can be compared to wave action. *(Delmar, Cengage Learning)*

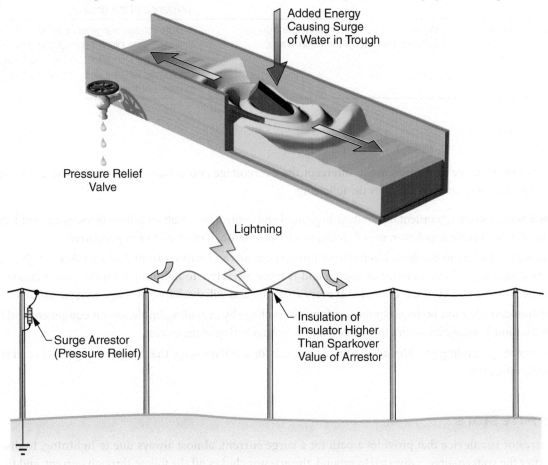

Comparison of Voltage Surge to Water Surge

Surge Activity	Water System	Electrical System
Initiation of surge	An outside energy source such as a brick is thrown into the water.	An outside energy source such as lightning strikes a line.
Speed of energy	The speed of a wave of water can be calculated with a relatively complicated formula. The speed of a wave of water is relatively slow.	The speed of a voltage surge through a power line is at the speed of light.
Surge or wave impeded	A wave of water is impeded by these: • The size and shape of the trough or conduit holding the water • The frictional resistance of material making up the trough	A large conductor offers less impedance than a small conductor.

(Continued)

Surge Activity	Water System	Electrical System
Surge or wave action at a dead-end	The wave is reflected off the dead-end and is doubled in magnitude.	The wave is reflected off the dead-end and is doubled in magnitude. It diminishes in a relatively short distance.
Uncontrolled surge discharge	Water spills over the edge of the trough.	Voltage surge spills over or sparks over at location with least insulation.
Controlled surge discharge	A pressure relief valve reduces the pressure of a surge in a water system. The amount of discharge is dependent on the spring pressure of the pressure relief valve and the size of the discharge pipe.	A surge arrestor with a voltage rating lower than the system insulation provides a path for a voltage surge to be dissipated into the ground. The amount of discharge depends on the resistance of the path to ground and the resistance of the ground.
Dissipation of surge	Wave action on the edge of the trough reduces the size of the main wave.	A high-voltage surge on a small conductor causes some dissipation of the curve due to corona and skin effect. A high-voltage surge dissipates when it couples with a nearby phase, neutral, or ground.

Circuit Protection from Overvoltage

The basics of overvoltage protection are to intercept the overvoltage and conduct it to earth, where it is dissipated. Circuits are protected from overvoltage by the following:

1. Surge arrestors shunt a transient overvoltage to ground and control the resultant follow-through current before insulation, such as porcelain, polymer, air, oil, cable, or wire insulation, flashes over or is punctured.

2. Shield wire, which is an overhead ground wire, protects circuits and stations from direct strokes of lightning. Shield wire is strung high above a circuit or station and is more attractive to lightning than the power conductors and equipment below. Shield wire is grounded at every structure and will discharge a voltage surge to ground.

3. Insulation coordination protects equipment from overvoltage by providing insulation on equipment and lines that can withstand a voltage considerably higher than the voltage rating of the circuit.

4. Good system grounding provides a low-impedance path for a voltage surge that has gone to ground at an arrestor to dissipate into earth.

Surge Arrestors

A surge arrestor is a device that provides a path for a surge current, almost always due to lightning, to discharge to ground. After the voltage surge is diverted to ground, the arrestor chokes off the follow-through current, and the arrestor is restored and ready for the next surge.

A normal voltage will not discharge to ground across a gap or many gaps inside the arrestor. Surge arrestors are designed to operate at a voltage higher than the circuit voltage but lower than the voltage that would cause an insulator to flash over or a transformer to be damaged. The arrestor must operate in time to avoid damage to equipment or, in other words, coordinate with the withstand curves of equipment needing protection.

Surge arrestors can fail violently and, if in a porcelain housing, the shrapnel coming from a violent failure can cause severe injuries. Some utilities require that all arrestors over a certain voltage, such as 15 kV, be tested before being re-energized. A tester applies a voltage across the arrestor. The arrestor conducting at lower voltages than specified indicates that it could fail while the power line tech is energizing it. A tester is used to test distribution class arrestors. The tester puts out 0 to 30 kV DC in 1 kV steps, and the current meter will indicate if there is any current leakage while the voltmeter will display the breakover voltage of the arrestor. **See Figure 12–19.**

Because of the high fault current available, arrestors in a substation should be energized remotely. Polymer housing of an arrestor will reduce the damaging effect of a failure.

Figure 12–19 The bottom of this arrestor has been blown off by a lightning surge. *(Delmar, Cengage Learning)*

Bottom of surge
arrestor blown off

Types of Surge Arrestors

An arrestor provides a sparkover location for a voltage surge, provides a path for a high-surge current flow to ground, and then stops the current flow to ground before it becomes a permanent **ground fault**. The following paragraphs present some types of surge arrestors. Improvements always are being made.

A spark gap or rod gap is installed between the line and ground to discharge a surge and protect insulators. This type of protection is found mostly on transmission lines. Once an arc is established, however, the ionized air becomes a conductor and the arc continues until the circuit trips out. This continuing arc is a fault current called the power-follow current. The spark gap will erode and be damaged by a prolonged arc. **See Figure 12–20.**

Figure 12–20 Spark gap surge protection on these 230 kV overhead to underground potheads includes a grounded rod between the potheads, which acts as a spark gap. *(Delmar, Cengage Learning)*

Spark
Gap

A valve-type arrestor is designed to limit a power-follow current after passing the voltage surge to ground. A valve-type arrestor is made of air gaps and a special resistor called a valve element. The valve element in the arrestor will pass a lightning surge to ground, but it has a high resistance to 60 Hz and stops any 60 Hz power-follow current.

In metal-oxide varistors (MOV), the gaps are not air but layers of highly resistive metal-oxide additives. They do not need air gaps to prevent normal voltage from going to ground. This material can pass a voltage surge without damage. The material is nonlinear, meaning that it sets up a reactance that prevents the voltage and current from being at their peak value at the same time, thereby limiting the energy being dissipated through the arrestor. **See Figure 12–21.**

Figure 12–21 MOV arrestors are installed to protect a live-front transformer. *(Delmar, Cengage Learning)*

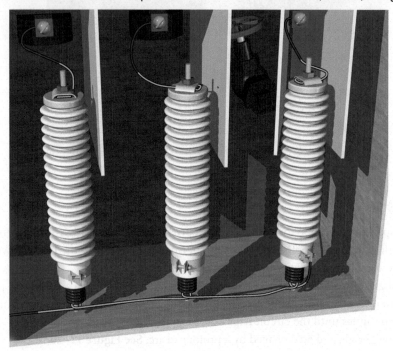

A dead-front arrestor is assembled in a shielded housing, such as an elbow, and installed for the protection of underground and pad-mounted distribution equipment and circuits. **See Figure 12–22.**

Arrestors should be located as close to equipment needing protection as possible. **See Figure 12–23.**

Having a longer than necessary length of wire to the arrestor is an example of poor workmanship. This compromises the effectiveness of surge protection. When lightning strikes, much of the high-current flow that is supposed to flow through the arrestor will be choked off by the coil. **See Figure 12–24.**

Classes of Surge Arrestors

Valve-type surge arrestors come in the following four classes.

1. A station class arrestor provides the highest degree of protection. It is used in transmission substations because it can withstand a very high surge current.

2. An intermediate class arrestor is used in distribution substations and on subtransmission circuits and is available in ratings up to 144 kV. It is used where the higher quality and cost of a station class arrestor are not justified. The difference between the circuit voltage and the sparkover voltage of the intermediate class arrestor is narrower than it is with the distribution class arrestor. This provides better protection for the equipment that the arrestor is protecting.

3. A distribution class arrestor is used on distribution transformers and other line equipment and is available in ratings up to 42 kV. It provides a reasonable balance between protection and cost. Distribution class arrestors are classified according to the following:

   - Heavy-duty class is used to protect overhead distribution systems exposed to severe lightning currents.

   - Light-duty class is used to protect underground distribution systems where the major portion of the lightning strike current is discharged by an arrestor located at the riser pole.

   - Normal-duty class is used to protect overhead distribution systems exposed to typical lightning currents.

4. Secondary class arrestors are used for the protection of secondary services and are available in ratings up to 650 V.

Figure 12–22 Elbow arrestors are used to protect dead-front equipment. *(Courtesy of Thomas and Betts Corporation)*

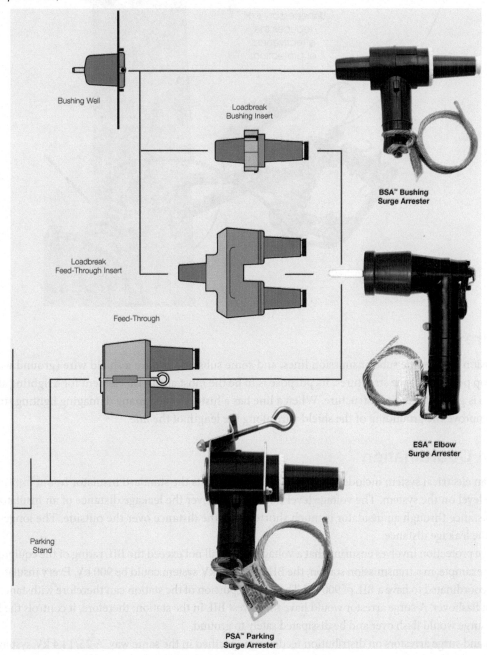

Bushing Well

Loadbreak
Bushing Insert

BSA™ Bushing
Surge Arrester

Loadbreak
Feed-Through Insert

Feed-Through

ESA™ Elbow
Surge Arrester

Parking
Stand

PSA™ Parking
Surge Arrester

Figure 12–23 Surge arrestors are most effective when mounted close to the equipment being protected. *(Delmar, Cengage Learning)*

Surge
Arrestors

Surge
Arrestors

Figure 12–24 A coil-in arrestor lead is poor workmanship. *(Delmar, Cengage Learning)*

Unnecessary coil
reduces the
effectiveness
of protection.

Shield Wire

Most transmission lines, some subtransmission lines, and some substations have a shield wire (ground wire, static wire) strung in the top position on the structures. Its purpose is to be the most attractive element for a lighting strike.

Shield wire is grounded at each structure. When a line has a history of too many damaging lighting strikes, often the solution is to improve the grounding of the shield wire along the length of the line.

Insulation Coordination

The design of an electrical system includes using insulation that meets the standard insulator **basic impulse level** (BIL) for the voltage level on the system. The voltage level that can flash over the leakage distance of an insulator is the BIL. A straight-line distance through an insulator is much shorter than the distance over the outside. The longer distance over the outside is the leakage distance.

Overvoltage protection involves ensuring that a voltage surge will not exceed the BIL rating of the equipment or line it is protecting. For example, in a transmission station, the BIL for a 230 kV system could be 900 kV. Every insulator, breaker, and transformer is coordinated to have a BIL of 900 kV. The 230 kV portion of the station can therefore withstand up to a 900 kV surge without a flashover. A surge arrestor would have the lowest BIL in the station; therefore, it controls the location where a high-voltage surge would flash over and be dissipated safely to ground.

Insulation and surge arrestors on distribution feeders are specified in the same way. A 25/14.4 kV system with a BIL of 95 kV means that all the insulation and equipment should withstand 95 kV. A surge arrestor would have a voltage rating of less than 95 kV but more than 14.4 kV. On a 25/14.4 kV system, an arrestor would have a voltage rating of 18 kV.

Equipment Voltage–Time Curves

Equipment such as a transformer has a BIL voltage–time curve, which represents the amount of voltage and time a transformer can withstand before it is damaged. Equipment is manufactured to meet a BIL voltage–time curve so that a standard surge arrestor can be used to protect it. **See Figure 12–25.**

Shorting Insulators for Live-Line Work

When working on an energized transmission line, the BIL at an insulator string will be reduced by tools, insulated strain links, and live-line rope used by the line crew. On a compact—designed transmission structure with smaller spacing

Figure 12–25 A transformer withstanding voltage–time curve shows the BIL voltage–time curve for a transformer. The curve of a voltage surge, shunted to ground by a surge arrestor, is well below the level that would damage a transformer. *(Delmar, Cengage Learning)*

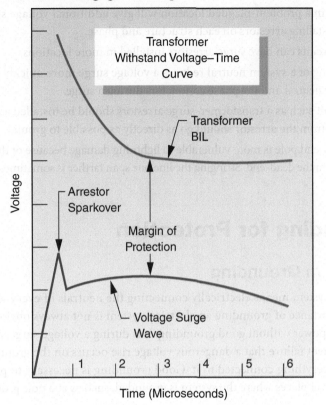

between conductor and ground, live-line tools have an even greater likelihood of lowering the BIL. A location with a lower BIL is more vulnerable to a flashover if a voltage surge occurs. Although a voltage surge due to switching or lightning is extremely unlikely to happen while live-line work is carried out, protection from a possible flashover can be taken.

Prevention of voltage surges at a job site is improved by having a lightning warning system in place. Getting clear of the line when thunderstorms are in the area eliminates exposure to the greatest cause of voltage surges. To avoid a switching surge, the reclose feature of the circuit breaker must be blocked. A blocked recloser avoids the switching surges involved with reenergizing the circuit if it is tripped out for some reason.

To reduce the likelihood of a flashover at a worksite, the BIL of an insulator string at an adjacent structure can be lowered. Any voltage surge that occurs while working on the circuit would then flash over at the adjacent structure and not at the work location.

Lowering the BIL of an insulator string at an adjacent structure can be accomplished by shorting out some insulators or by installing a portable protective gap (PPG) device. A PPG can be set to provide a spark gap for a given value of voltage surge and to shunt it to ground.

Protection at Problem Locations

Some circuits and locations are exposed to more lightning than others. The following steps can be taken to reduce outages to circuits prone to lightning strikes.

- On transmission lines where a shield wire is already in place, improvements to grounding of the shield wire dissipate the effects of lightning more quickly.

- On problem subtransmission and lower-voltage transmission lines, lightweight, polymer, and metal-oxide arrestors can be installed on all three phases at the most vulnerable locations. In some cases, these arrestors are installed on every pole or tower. One arrestor on each of the three phases is better than a single arrestor on the center phase because, when lightning strikes one phase, the three arrestors allow all three phases to go to the same voltage, preventing flashing (backflash) from one phase to another. These types of arrestors have a hot-line clamp on the top and a grounded isolator on the bottom and can be installed while the line is energized.

- Stringing a shield wire in a problem-plagued location will give additional voltage surge protection, although it is more expensive than installing arrestors on each structure and phase.
- Problem distribution circuits can have surge arrestors installed in more locations.
- Improving the grounding of a system neutral reduces a voltage surge more quickly because of the coupling effect that occurs between the neutral and the phase conductors during a surge.
- On individual equipment such as a transformer, surge arrestors should be installed as close as possible to the equipment. The ground wire from the arrestor should go as directly as possible to ground.
- A transformer on a dead-end pole is more vulnerable to lightning damage because of the doubling of voltage when the lightning comes back from the dead-end. Stringing the line one span farther is sometimes done in problem locations.

System Grounding for Protection

Purposes of System Grounding

System grounding a power system means electrically connecting the neutrals of every wye-connected transformer or generator to earth. The importance of grounding an electrical system is not always obvious. Under normal conditions, an electrical system delivers power without good grounding. It is during a voltage surge, an unbalanced load condition, a short circuit, or an equipment failure that a dangerous voltage rise occurs on the grounding system. The voltage rise occurs on the neutral and everything connected to it. Good grounding is necessary to protect the public and workers from a dangerous voltage rise at places where the system is grounded, such as at a pole ground/downground.

The Benefits of Good Grounding

Good grounding provides people protection, equipment protection, and circuit protection.

People Protection

Grounding an electrical system limits the rise in potential on the neutral, metal structures, noncurrent-carrying electrical equipment, and everything electrically connected to ground. When equipment is connected to a low-resistance ground, a voltage surge is kept to a lower level and will dissipate more quickly. These grounds are called protective earth (PE) connections in the international standard.

Circuit Protection

Good grounding improves the likelihood of a circuit breaker, recloser, or fuse tripping out the circuit for a phase-to-ground fault. Most faults on an electrical power system are line-to-ground faults. A well-grounded electrical system provides a better return path for the fault current to flow back to the source to complete the circuit for the fault current. These grounds are called functional earth connections in the international standard.

Importance of System Grounding

When a line or substation is built, the installation of a grounding and bonding network is essential for safe operation. The design of electrical installations includes system grounding so that a person touching any equipment during a fault condition is not subjected to a dangerous current or potential.

- The installation of a ground-gradient control grid reduces step and touch potentials around such equipment as a pad-mount transformer or a switching kiosk.
- The use of crushed rock around equipment, especially in a substation, increases electrical resistance under the feet of workers.

- Standing on a ground-gradient control mat that is bonded to a switch handle protects the switch operator from electric shock if the handle becomes energized during a switch failure. With the mat bonded to the switch handle, the operator's feet and hands are at the same potential, which means there will be no current flow.

- Grounding the system neutral at frequent intervals provides a good return path for current from a line-to-ground fault. The current from a fault would have a relatively short path back up to the neutral.

- During a fault condition, voltage rises in the ground grid and everything attached to it. Any railway climber that enters a station has one section of rail removed.

- When a new feeder is being strung from a substation and protective grounds are connected to the grid, the substation grid will be extended out of the substation along the conductors being strung. Ensure that the public has no access to the conductors especially if the conductor is being pulled out on the ground.

- To ground electrical equipment means to connect transformer tanks, metal-clad equipment, and more, to earth. The main purpose is to reduce shock hazards by limiting the potential difference between the grounded equipment and earth. It is also important to electrically bond together all tanks of electrical equipment to prevent someone from getting in between and becoming a conductive path from one piece of equipment to the other.

The International Grounding Standard

This section introduces the terms used in international standard (IEC 60364) for grounding (earthing). This information will lead to a better understanding of the importance of grounding.

The international standard defines three types of grounding/earthing arrangements and uses two-letter codes TN, TT, and IT to describe them.

1. The TN grounding/earthing system refers to the wye or star point being connected to ground. To complicate this standard further, the TN system has three variations.

    a. TN–S   The protective earth (PE) connection and the N (connection to earth via the supply network, in other words, the neutral) are separate conductors that are connected together only near the power source. A good example of this is that in North America, the neutral and ground are connected together in the customer's service panel and then kept separate in all the downstream connections.

    b. TN–C   A combined protective earth connection and neutral (PEN) conductor fulfills the functions of both a PE and an N conductor. In other words, a PEN conductor refers to a "multigrounded neutral." In England, the PEN conductor is called "protective multiple earthing" (PME) and in Australia a "multiple earthed neutral" (MEN).

    c. TN–C–S   A TN–C–S grounding system refers to a multigrounded neutral that somewhere downstream divides into two separate conductors: a neutral conductor and a ground wire. Most customer wiring around the world has the neutral and ground connected in the customer's service panel or at the service cutout (European) and then kept separate in all the downstream connections.

2. A TT grounding/earthing system means that a customer's ground connection is separate from the supply authority. In other words, the customer's connection to earth is to their own ground rod with no connection to a utility ground or neutral. This is apparently done to keep a customer free of high and low-frequency "noises" that come from the neutral. A service feeding a telecom customer may require a TT service connection.

3. In an IT grounding/earthing system, there is no connection to ground/earth in the distribution system except through a high-impedance connection. This would be typical in a delta distribution system.

Ground versus Neutral Connections

On an equipment installation, ground wires and their connections should not be confused with neutral wires and their connections. A *neutral* is part of the electrical circuit and is a current-carrying conductor that provides a path for current to flow back to the source. If a neutral connection on energized equipment is opened, voltage is equal to the circuit voltage across the open point.

A ground wire grounds the equipment tank and surge arrestor lead-to-earth potential. During a fault or a voltage surge, a good ground reduces the potential rise on the equipment and prevents equipment failure or flashover. An equipment installation may appear to have a ground wire and a neutral connection doing the same job, but skipping one of the connections can cause a dangerous potential rise or service problem. **See Figure 12–26.**

Figure 12–26 When a missed neutral connection causes the pole ground to become the primary neutral, it can cause a fatal transformer grounding accident. *(Delmar, Cengage Learning)*

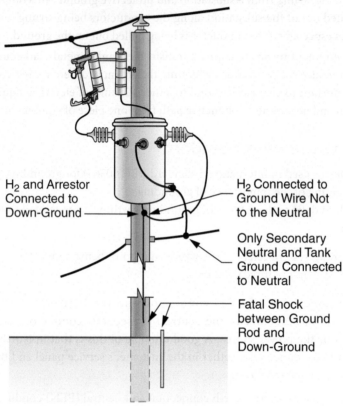

H₂ and Arrestor Connected to Down-Ground

H₂ Connected to Ground Wire Not to the Neutral

Only Secondary Neutral and Tank Ground Connected to Neutral

Fatal Shock between Ground Rod and Down-Ground

The figure shows a typical transformer-grounding installation with one major oversight: The primary neutral of the transformer is not connected to the system neutral. A person working on the ground would be electrocuted while trying to repair a damaged ground-rod connection. In this illustration, pole ground/downground is part of the primary circuit, and, when it is opened, a primary voltage difference appears across the opening.

The Neutral as Part of the Grounded System

For a circuit to be complete, all the current leaving the source must return to the source. In a balanced three-phase circuit, the return flow occurs in the phase conductors. Unbalanced three-phase and single-phase wye-distribution circuits use the neutral for current to find its way back to the source.

The neutral also provides a return path for a fault that goes to ground. On a well-grounded, multigrounded neutral, about two thirds of the current returns to the source through the neutral and the remaining current returns through ground.

Neutral Potential

From Ohm's law ($E = I \times R$), it can be seen that if there is a current flow, there is a voltage. There is always some current on a distribution neutral. A neutral potential, which is measured between the neutral and a remote ground, should be kept at less than 10 V.

A neutral potential will be lowered if the current flowing through it can be reduced. A neutral is the return path for any unbalanced current between phases. The current on the neutral can be reduced by converting heavily loaded single-phase circuits to three phases and balancing load between the three phases. A neutral potential can also be lowered by installing more grounds in good earth in many locations along its length.

If the resistance of a neutral is lowered, the potential is lowered. Ohm's law shows that less resistance means less voltage. Neutral resistance can be reduced by stringing a larger conductor and ensuring that all neutral connections are in good condition.

Note: Under a line-to-ground fault condition, the voltage on the neutral can rise to many kilovolts.

Grounding for Overcurrent Protection

A line-to-ground fault is by far the most frequent type of fault. When a line-to-ground fault occurs, the protective switch-gear trips out the circuit, but only if the current feeding the fault is high enough. To draw a fault current high enough to trip the circuit, a return path to the source must also be able to carry the same amount of current. The return path is the neutral and earth.

Frequent grounding of the neutral in good earth aids the fault current in finding its way back to the neutral and the source. When a conductor falls in a location that is not well grounded, such as rocky ground, sandy soil, or dry snow, insufficient fault current is generated to trip the circuit because the resistance of the return path through the earth is too high. Every power line technician, however, has seen circuits fail to trip out when they should have. In these cases, the culprit is poor grounding or an outdated protection scheme.

Inserting an Impedance in the Ground Return

At some substation transformers, the neutral point is not grounded directly to earth. A resistor or reactor is placed in series with the connection to earth. The fault current levels on high-voltage distribution feeders from some substations can be very high and very damaging. Inserting a resistor or reactor into the return circuit from ground has the effect of reducing the fault generated from a line-to-ground fault in the whole circuit.

Surge Protection

The dissipation of a voltage surge is greatly improved through good grounding. Circuits in a location with a poor ground are more vulnerable to outages due to lightning.

When a voltage surge occurs on a phase conductor, the neutral conductor acts as a coupling wire. The voltage on the neutral rises along with the voltage on the phase conductor. This coupling lowers the potential difference across insulators and equipment. The coupling of a well-grounded neutral or shield wire also helps lower the voltage of the surge more quickly. A good grounding design dissipates a voltage surge.

- The ground resistance of a driven ground rod should be low, preferably below 25 ohms.
- A grounding wire should be short because there is a big voltage drop during the high current due to a fault condition. The added resistance of a long lead will add substantially to the voltage drop.
- A ground lead should be as straight and direct as possible because a high-voltage surge will jump across sharp bends.
- A surge arrestor should be well grounded and as close as possible to the equipment being protected.

Choose the Best Ground Available

When you have a choice, use the best ground available. The resistance of earth varies with the type of earth, moisture content, and temperature. The resistivity of earth is measured in ohms per meter, which is equal to the resistance between the opposite faces of a cubic meter of soil. **See Figure 12–27.**

Soil with no moisture content would be an insulator. The mineral salts of earth, dissolved in water, give the soil low resistance. A ground rod should be driven straight down, where moisture is more likely to be found.

The temperature of earth affects its resistivity. When the temperature decreases, the resistivity of the earth increases. When the moisture in the earth freezes, the resistance increases dramatically. A ground rod must be driven below the frost line to be effective in winter. Frozen ground around part of a ground rod can double or triple the resistance.

Figure 12–27 Typical Soil Resistance

Earth Type	Resistivity (Ohms per Meter)
Sand saturated with seawater	1 to 2
Marsh	2 to 5
Loam	5 to 50
Clay	5 to 100
Sand/gravel	50 to 1,000
Sandstone	20 to 2,000
Granite	1,000 to 2,000
Limestone	5 to 10,000

Measuring Ground Rod Resistance

Measuring the resistance of a ground rod can confirm that an installation meets design requirements. For example, an acceptable resistance for a ground rod at a transformer installation is typically 25 ohms. Ground resistance measurements are also carried out when investigating calls such as those regarding tingle voltage, high-neutral voltage, excessive vulnerability to lightning, or failure of protective switchgear to operate.

Two different types of instruments are used to measure ground rod resistance.

1. The method that has been used for many years is the fall of potential test. The fall of potential test is done by applying a fairly high voltage to the ground rod and measuring the voltage at a potential probe and the current at a current probe that is installed far enough away to represent remote ground. This method is for a bare ground rod, with no connections. **See Figure 12–28.**

Figure 12–28 The leads of the earth resistance tester and probes are connected to measure ground rod resistance. *(Delmar, Cengage Learning)*

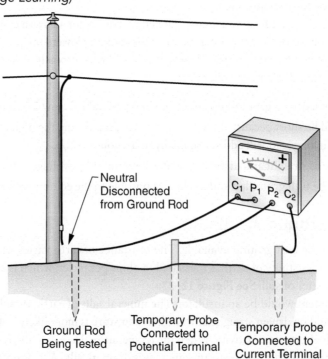

Neutral Disconnected from Ground Rod

C_1 P_1 P_2 C_2

Ground Rod Being Tested

Temporary Probe Connected to Potential Terminal

Temporary Probe Connected to Current Terminal

2. A simpler test is the clamp-on tester. It is clamped over a ground rod much like a clamp-on ammeter is put on a conductor. Instead of magnetic induction being imposed on the clamp by a current-carrying conductor, the clamp imposes induction on the ground rod. The unit then measures the current and voltage drop to measure resistance. This instrument tests a grounding system and does not measure the resistance of an unconnected single ground rod. It works best when the ground being measured is a ground rod in a system of many other ground rods. **See Figure 12–29.**

Figure 12–29 A clamp-on ground tester is the easy way to measure ground resistance in a ground rod. *(Courtesy of Fluke Corporation)*

Protection from Corrosion

Corrosion Hazards

Corrosion reduces the life of metal hardware used in electrical utilities. Corrosion due to road salt, fertilizers, seawater, and damaged paint surfaces is a normal occurrence. There is, however, an electrical component to understanding the cause of premature corrosion of metal hardware. Line materials such as tower footings, anchor rods, electrical connections, and underground/submarine cable sheath normally last a long time, but occasionally these items corrode prematurely and create severe hazards.

The most vulnerable locations for premature corrosion are near pipelines and in marshy areas. The corrosion is usually at its worst just below the ground line, so it is reasonably easy to inspect vulnerable locations. When a tower footing or anchor rod corrodes, an unseen hazard results that could cause a structural failure. When the metallic sheath of an underground/submarine cable is also the neutral for the circuit, an electrical hazard exists when the sheath has corroded away and a voltage difference occurs across the open point.

Dissimilar Metals in a Circuit

Two dissimilar metals can generate DC when placed in contact and connected together into a circuit. There is a small, self-generated electric current without any external source. For example, a lead-acid battery consists of two different metals (lead and lead oxide) in an electrolyte (a conductive solution, such as diluted sulfuric acid). The battery becomes part of a circuit when it is connected to a load. One post of the battery gives off ions, eventually deteriorates (corrodes), and disappears. Similarly, two dissimilar metals connected together with some kind of electrical bond and placed in a conductive solution are like a battery.

A circuit is created when a self-generated current leaves one metal (an **anode**), flows through the electrolyte, and enters another metal (the **cathode**). The self-generated current flows because of a natural difference between the potentials of the metals in the electrolyte. As current leaves the metallic anode and flows into the electrolyte, metal ions separate from the anode and cause the anode to be consumed (corrode).

Galvanic Series of Metals Used in Line Hardware

Metals have a tendency to dissolve into ionic form and lose electrons when they become part of a circuit capable of conducting electricity. Some metals (such as aluminum) have a strong tendency to lose electrons, while other metals (such as copper) have a weaker tendency. An active metal with a strong tendency to lose ions is anodic in an electrical circuit and will corrode. A less active "noble" metal with a weak tendency to lose ions is cathodic in an electrical circuit and does not corrode.

Below are the metals used in electrical utilities and the relative placement of their activity in relation to other metals. The closer to the top of this list the metals are, the greater the rate of corrosion.

Galvanic Series of Some Metals

Anodic End (Most Active)

Positive Polarity

Magnesium

Zinc (Galvanized Metal)

Aluminum

Cadmium

Steel or Iron

Lead

Brass

Copper

Graphite

Silver

Cathodic End (Least Active)

Negative Polarity

Note: Closer to the top of this list the metals are, the greater the rate of corrosion.

Galvanic Corrosion

Premature corrosion of metal hardware occurs when the metal becomes part of a corrosion cell. A typical corrosion cell involving line hardware must have the following conditions present.

1. There must be an anode and a cathode. The anode and the cathode must be connected to each other by a wire or some other kind of metallic contact.

2. An electric circuit must be completed between the anode and the cathode through an electrolyte. An electrolyte is a conductive solution, such as the earth in a wet, marshy area.

3. There must be a DC potential causing current to flow in the circuit. Dissimilar metals connected in a circuit generate a DC potential. When current leaves the anode, it takes minute particles of metal with it and the anode corrodes. A flow of DC through the ground can cause corrosion to vital components of an electrical system.

Copper-to-Aluminum Connections

It is well known in the line trade that when copper is connected directly to aluminum, the connection corrodes prematurely and fails. Aluminum is an active metal, while copper is less active. To prevent galvanic corrosion of the electrical connection, normally a spacer with another metal, such as cadmium, is placed between the copper and aluminum. A conductive grease can also be used to slow down any corrosion. With an aluminum-to-copper or aluminum-to-brass connection, the aluminum should always be placed on top to prevent copper salts from leaching and corroding the aluminum.

Cathodic Protection

A tower leg or anchor in wet soil becomes vulnerable to corrosion. Cathodic protection is installed at vulnerable locations. Cathodic protection consists of electrically connecting a sacrificial anode, such as magnesium, to a tower leg and burying it nearby. A circuit is set up between the sacrificial anode and the tower leg through the surrounding soil (electrolyte). The magnesium is a more active metal than the galvanized tower leg. Thus, the magnesium anode corrodes instead of the tower leg. The anodes must be replaced on a regular basis as they corrode away.

Stray-Current Corrosion

Stray-current corrosion is due to a DC of external origin flowing from a metal into an electrolyte. When a DC leaves a metal and enters an electrolyte, some metal leaves and the metal corrodes. Alternating current is not significant in producing corrosion because the reverse flow will build up the anode again. One source of stray DC can be a natural current due to the magnetic fields in earth. Long pipelines can be vulnerable to this current. Human-made DC that can stray into earth can come from a nearby DC transmission line, a mining operation, or a welding operation.

Pipeline Cathodic Protection Hazard

Buried electrical utility line materials are subject to corrosion when a pipeline is nearby. The cathodic protection used to protect a pipeline from corrosion accelerates the corrosion of nearby utility anchors, cable sheaths, and ground rods. To protect a pipeline from corrosion, a negative potential is applied to the pipeline. This prevents a current from flowing into the surrounding earth (electrolyte) from the pipeline and prevents any metal ions from leaving the metal pipe. Any stray current flows toward the pipeline.

Review Questions

1. Name three reasons why a circuit needs protection.

2. Name four types of faults that relays can be programmed to detect on a transmission line system.

3. If a recloser is bypassed with a solid jumper or switch, what is the consequence if a conductor falls to the ground somewhere downstream?

4. What are three influences that affect the magnitude of a fault current during a short circuit?

5. Why is a good multigrounded neutral important for circuit protection on a wye system?

6. Why does a dry tree limb or a broken conductor lying on dry or frozen ground not always trip out the circuit protection?

7. What is meant by the frame size of a recloser?

8. Name three advantages for converting to a higher distribution voltage.

9. What is the most common cause of overvoltage on a distribution system?

10. Why is a vehicle the safest place to be during a thunderstorm?

11. What is the difference between connections to a neutral and connections to ground?

12. How does good grounding of the neutral improve overcurrent protection on a circuit?

13. When making an aluminum-to-copper connection, why is the aluminum kept in the higher position?

Chapter 13

Installing Personal Protective Grounds

Objectives

After completing this chapter, you should be able to:

1. Identify the three reasons to place protective grounds before working on an isolated circuit.
2. Explain how protective grounds control the voltage.
3. Explain how protective grounds control the current.
4. Determine the size of the protective ground cable needed for the circuit to be worked on.
5. Choose the best ground electrode at the protective grounding site.
6. Describe how to place protective grounds to create an equipotential protective grounding zone.
7. Explain how a second set of grounds on a line can create a circuit for electromagnetic induction.
8. Describe locations where protective grounds can be installed on underground cable.

Reasons to Install Personal Protective Grounds

Reasons for Placing Protective Grounds

The main reason to ground a circuit is to make an electrical connection to Mother Earth, where we feel safe. Three more technical reasons to install grounds and bonds on an isolated circuit before work is started include the following:

1. *Install protective grounds to prove isolation.* After isolation and testing, the installation of grounds is the final proof that you are about to work on the correct circuit.
2. *Install protective grounds to have protection from accidental reenergization.* When working on a grounded circuit, it is necessary to have protection from accidental reenergization, which can come from operator error, contact with neighboring circuits, lightning, or backfeed.
3. *Install protective grounds to provide protection from induction.* Grounds are needed to provide protection from two kinds of induction hazards. Electrostatic induction will induce a voltage on an isolated circuit. Electromagnetic induction will induce a current flow in the conductor and grounds of an isolated and grounded circuit.

Protective Grounds Control Current and Voltage

Voltage and current on a circuit will likely not be zero after protective grounds are installed. There can be a voltage relative to a remote ground, and there can be a current flow in the conductors. The saying "It is not dead if it isn't grounded" is not entirely correct. "It is not safe to work on a circuit until protective grounds are installed" would be more accurate.

The discussion in this chapter is on line grounding. It does not include grounding for stringing operations or truck grounding, because most of the effort there is for touch and step potentials for working around equipment on the ground, which adds to the complexity of the line grounding discussion.

To ensure that a person will not be exposed to any hazardous current or voltage *after* the protective grounds are installed, adhere to the grounding principle, which will control the current around a worker and adhere to the bonding principle, which will limit the voltage exposure to a worker.

- *Grounding Principle:* Protective grounds are installed to reduce any current flow through a worker to an acceptable level by providing a low-resistance parallel shunt around the worker. At the same time, if the circuit is or becomes energized, the grounds must be big enough to withstand any fault current in the circuit.
- *Bonding Principle:* Bonds must be installed so that a worker is kept in an equipotential zone. A worker must not be able to bridge between a grounded circuit and any unbonded structure, vehicle, boom, wire, or other object not tied into the bonded network.

Applying the Grounding Principle to Control Current

Sources of Current in Protective Grounds

Applying the grounding principle means controlling the current after the grounds are applied. Even after protective grounds have been installed, a constant current could be due to induction from neighboring circuits, backfeed from a generator, or backfeed from within the electrical utility system. A very high current would occur if the line were to be accidentally reenergized. The protective grounds do not act as a gate that shunts everything to earth.

Use the Proper Size of Protective Grounds to Control the Current

When the size of protective grounds are specified for a circuit, they are sized to carry a fault current for enough time to trip out the circuit. During accidental reenergization, a set of grounds on a circuit provides a major short circuit. The ground wires and clamps will be subject to all the current that the circuit can deliver to that point.

Engineering at each utility will have data on the maximum fault current available on each circuit. Many utilities opt to standardize on one or two sizes of ground cable and clamps to cover all their circuits.

Promoting a Fast Trip-Out of Circuit Protection to Control Current

Circuit breakers, reclosers, and fuses open when a current feeding a fault goes through these devices for a specified period of time. In the case of reenergization, grounds on all three phases will provide an excellent phase-to-phase fault that will activate the protective switchgear. If one phase is energized, the protective grounds need to be connected to a good ground electrode. The best ground electrode is a permanent ground network such as a neutral, station ground, or tower steel.

Providing a Parallel Path around a Worker to Control Current

When a person on a structure is in contact with a circuit that is accidentally energized, the protective grounds and the person form two parallel paths to ground. Based on the parallel circuit theory, both paths will carry some current to ground. **See Figure 13–1**.

Figure 13–1 A worker is in a parallel path and is vulnerable to fatal electric shock if the circuit is re-energized. (*Delmar, Cengage Learning*)

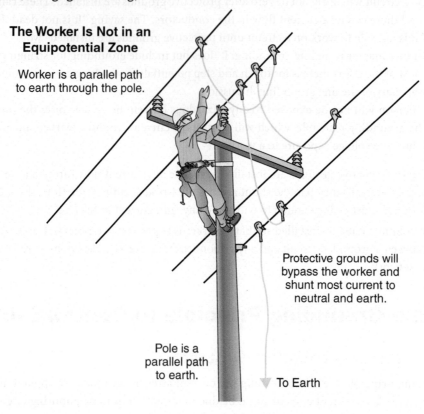

The Worker Is Not in an Equipotential Zone

Worker is a parallel path to earth through the pole.

Protective grounds will bypass the worker and shunt most current to neutral and earth.

Pole is a parallel path to earth.

To Earth

By far, most of the current will go through the low-resistance grounds. However, because a person working from a structure is a parallel path to earth, some current will go through the worker. To provide a good shunt around the worker, the protective grounds set must be kept in excellent condition. The current that flows through the worker must be kept well below a 10 mA threshold and, ideally, below the threshold of sensation.

When discussing controlling current flow, the worst case is when one phase is accidently reenergized. Protective grounds installed on three phases provide a balanced three-phase fault if the line should be accidently reenergized from a source three-phase breaker. The fault current would flow back to the source through the other phases and very little would flow back through the earth or neutral. However, grounding procedures and ground cable size, for example, must address worst-case single-phase reenergization, such as one phase contacting an overbuild or underbuild circuit, back-feed on one phase, and single-phase circuits.

If an operations error caused a three-phase line to be reenergized, the grounds will be a three-phase fault. Applying the grounds in a T pattern will reduce current flow in the cable connected to the grounding electrode and therefore reduce the proportion of current flow through a worker in parallel to ground. **See Figure 13–2.**

Grounding Differences of Wye and Delta

The phase-to-phase installation of grounds will short out both wye and delta circuits. On single-phase installations, a phase-to-neutral ground on a wye circuit and a phase-to-phase ground on a delta circuit will short out the circuit and trip out the protection.

On a wye system, the protective grounds connect the phases to the neutral. The connection to the neutral is made first, and an energized circuit mistakenly left so will trip out when the first phase is grounded. Grounding the phases to the neutral also lowers the potential of the grounded conductors at the worksite.

Figure 13–2 Protective grounds installed in a T pattern reduce the current flow to ground during a three-phase fault. (*Delmar, Cengage Learning*)

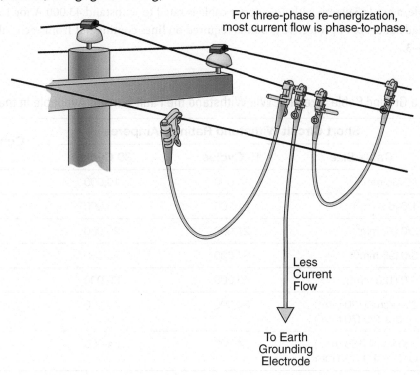

For three-phase re-energization, most current flow is phase-to-phase.

Less Current Flow

To Earth Grounding Electrode

On a delta system, the protective grounds connect the phase to a ground rod in earth. Because one purpose of protective grounds is to prove isolation, the first ground must be able to trip out a circuit that is mistakenly still energized. If the delta circuit protective switchgear is equipped with a ground fault relay, a connection to earth will trip out energized delta circuits if the ground rod is in good earth. The most positive way to ensure that a delta circuit will trip out is to apply a jumper ground from phase-to-phase first. In order to not be in contact with the grounds while installing the phase-to-phase jumper, use a grounding support stud as shown in **Figure 13–6**.

Protective Grounding Hardware

Protective grounds must carry very high fault current long enough to cause the protection for a circuit to trip out. Grounds also must be in excellent condition to provide a very-low-resistance shunt around a worker so that all but a harmless amount of current passes through the parallel path established by a worker. Virtually all utilities specify their grounding hardware to meet the American Society for Testing and Materials (ASTM) F855, "Standard Specification for Temporary Grounding Systems to be Used on De-energized Electric Power line and Equipment." This standard specifies the minimum standards for cables, clamps, and ferrules.

A typical jumper ground or ground set consists of a flexible copper stranded cable covered with a jacket rated at 600 V. The jacket is intended mostly for the mechanical protection of the small stranding of the cable. During an accidental reenergization, the jacket may provide protection for anyone near it because the voltage drop along the cable will reduce the voltage in the cable. A ferrule is used to make the cable connection to the clamp to reduce the problem of broken strands and corrosion at the clamp connection. The small stranding provides cable flexibility, but a regular inspection will find occasional broken strands where the cable enters the ferrule.

The clamp chosen for grounding has to be the right size, have the correct current-carrying capacity, and have the right shape for the object to which it is to be attached.

There are many styles and sizes of clamps designed to fit on conductor, bus, ground rod, tower steel, and underground hardware. ASTM F855 classifies grounding hardware into different grades based on maximum fault-current capability, for example, a grade 5 clamp with 4/0 copper cable is rated to withstand 43,000 A for 15 cycles. Engineering should specify the grade and size of protective grounds required on lines coming out from each substation within their utility. **See Figure 13–3.**

Figure 13–3 Use a Ground Cable Size That Will Withstand the Fault Current Available in the Circuit

ASTM Grade	Cable Size	Short Circuit Withstand Rating—Amperes		Continuous Current Rating (kA)
		15 Cycles	30 Cycles	
1	#2 (35 mm^2)	14,000	10,000	200
2	1/0 (50 mm^2)	21,000	15,000	250
3	2/0 (70 mm^2)	27,000	20,000	300
4	3/0 (95 mm^2)	34,000	25,000	350
5	4/0 (120 mm^2)	43,000	30,000	400
6	250 kcmil (120 mm^2) Or 2 × 2/0 (70 mm^2)	54,000	39,000	450
7	350 kcmil (185 mm^2) Or 2 × 4/0 (120 mm^2)	74,000	54,000	550

If two smaller sets of protective grounds are used (e.g., 2 × 2/0 or 70 mm^2 instead of one 250 kcmil or 120 mm^2) the first set of grounds may burn off if inadvertently installed on an energized circuit.

A typical protective jumper ground has a conductor clamp at each end and is used to go from one conductor to the other conductor. **See Figure 13–4.**

Figure 13–4 A jumper ground is designed to go from one conductor to another. (*Delmar, Cengage Learning*)

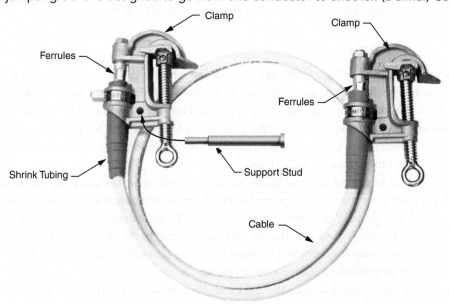

Another typical protective ground has a conductor clamp at one end and a clamp that fits on a ground rod, stud, or flat steel on the other end, and is typical for grounding on a steel transmission line. **See Figure 13–5.** Some utilities insist on installing a stud on steel for a ground attachment instead of attaching to flat steel. Ball-and-socket studs and clamps are used in many locations, especially in substations. The ball-and-socket studs are permanently installed where grounds would typically be installed. **See Figure 13–6.**

Figure 13–5 A ground cable with flat jaw clamp is designed to fit a ground rod or stud. *(Delmar, Cengage Learning)*

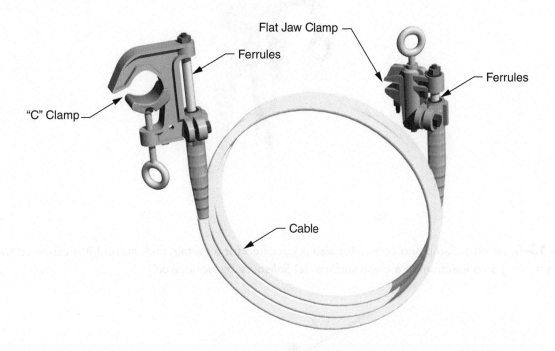

Figure 13–6 At locations where grounds are installed frequently, ball-and-socket studs are preferred. *(© Salisbury by Honeywell)*

Ground sets used for going from conductor to conductor can have a grounding support stud (piggyback stud), to allow an easier installation of grounds by one person from the end of the shotgun stick and to maintain clearance. **See Figure 13–7.**

To make a good, low-resistance connection between the protective ground clamp and the conductor, the contact surface of the clamp and the conductor must be cleaned. There is no easy way to clean a conductor at the end of a hot stick, so the step is often overlooked. A little bit of effort with a brush, however, can make a big difference. **See Figure 13–8.**

Figure 13–7 A jumper ground with piggyback clamp allows one person to safely install a jumper ground phase to phase. (© *Salisbury by Honeywell*)

Figure 13–8 When a clamp and conductor size is specified for a certain fault current it is assumed that the clamp is making a connection on a clean surface. (© *Salisbury by Honeywell*)

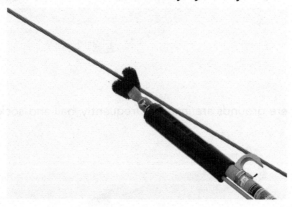

Inspect and Test Grounding Sets

Grounding sets must be maintained and tested on a scheduled basis. They are prone to broken strands—especially near the clamps—and loose and corroded cable connections at the clamps. A visual inspection will not necessarily pick up these defects because many are hidden under the jacket and heat shrink cover over the ferrule.

When dealing with parallel circuits, some current will go down each path. A few extra milliohms of resistance in the grounds means that the other parallel path—for example, a power line technician will carry more current. Therefore, use a grounding set tester to verify the condition of the grounds. **See Figure 13–9.**

Figure 13–9 A ground set tester will verify that the conductor under the ferrule is making a good connection.

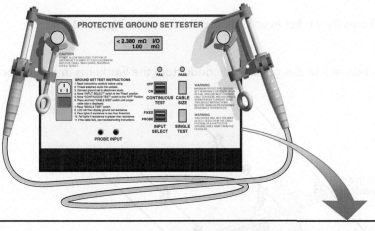

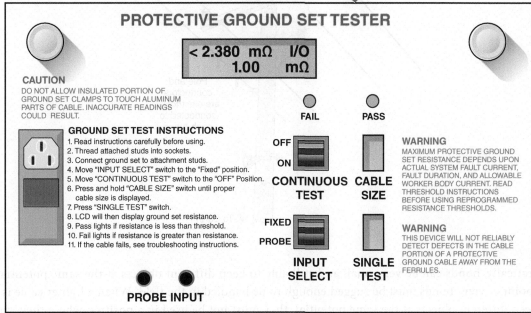

Enlarged View of Instrument Panel

Applying the Bonding Principle to Control Voltage

Sources of Voltage on a Grounded Circuit

Applying the bonding principle means controlling the voltage after the grounds are applied. Even after protective grounds have been installed, voltage may be present on the conductors in relation to a remote ground. Voltage can be from induction from neighboring circuits, backfeed from a generator, or backfeed from within the electrical utility system. Voltage on the line would be very high in relation to a remote ground if the line were to be accidentally reenergized.

Providing Equipotential Bonding to Control Voltage

When protective grounds are installed, ensure that everything a person is likely to touch will be at the same potential while working on the circuit. Protective grounds also act as bonds, keeping all grounded conductors at the same potential. If there is little or no potential difference across a person's body, there can be no current flow.

When working on a pole or steel structure, a power line tech is a parallel path to ground. When a structure is bonded to the grounded conductors in such a situation, there would be little or no voltage difference for a worker to bridge across.

Attaching the grounds to the pole keeps the worker in an equipotential zone. **See Figure 13–10.**

Figure 13–10 When a structure is bonded to the grounded conductors the worker is in an equipotential zone. (*Delmar, Cengage Learning*)

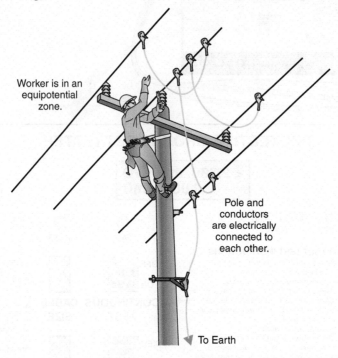

Worker is in an equipotential zone.

Pole and conductors are electrically connected to each other.

To Earth

Theoretically, bonds can be very small and still able to keep different objects at the same potential. From a practical point of view, bonds must be rugged enough to be handled in the field. When a lighter cable is used as a bond to keep different objects at the same potential, they must not be used in a position where they would carry fault current.

A common cause of accidents over time has been a failure to bond (ground) across an open point in a conductor. The assumption had been that once protective grounds were installed on each side of a break or open point, it was safe to go to work. Consider the following examples.

1. People have received serious and fatal electrical burns while removing or installing jumpers/loops on a transmission line, even with grounds installed on each side. Installing or removing loops can interrupt or complete a circuit and a lethal current flow.

2. People have been hurt when preparing to make a splice on a three-phase line that was completely separated by a tree or car accident. Even with grounds installed on each side of the break, when the first conductor ends were brought together they completed a circuit. **See Figure 13–11.**

3. In one accident scenario a connector on a 44 kV lateral feed to a distribution substation burns off. Protective grounds were placed on the three-wire circuit and down to a temporary ground rod in earth. The lateral was grounded to the substation grid. The second circuit on the 44 kV was left energized. When the power line tech tried to remake the connection at the tap pole, he received a severe shock. The ground electrode at the substation was much better than the ground rod, and a circuit was completed between the two different electrodes when the worker tried to make the connection. **See Figure 13–12.**

Figure 13–11 A hot stick and jumper should be used when making a connection across the break of the first conductor to be spliced. *(Delmar, Cengage Learning)*

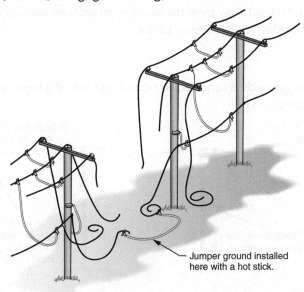

Jumper ground installed here with a hot stick.

Figure 13–12 To eliminate a difference of potential between two sets of grounds, the grounds need to be bonded together. *(Delmar, Cengage Learning)*

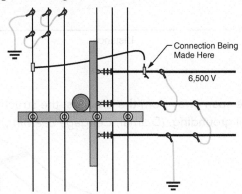

Connection Being Made Here

6,500 V

Bonding to Control Voltage Rise Due to Induction or Reenergization

If a circuit with protective grounds already installed is exposed to induction or accidental reenergization, the voltage will rise in the complete circuit, the grounds, the neutral, and the earth where a ground rod is installed. If a vehicle is connected to the grounding setup, the voltage will rise on it as well.

Any component not connected electrically to the bonded network will be at a different potential, until the circuit trips out or the source of induction is removed. A person working on a pole or structure that is not bonded to the grounds will be a parallel path to ground through that pole or structure.

At the site where a set of protective grounds is installed, the phase and the neutral or the phase and steel structure are tied together and will remain at an equal potential in case of induction or accidental reenergization. When working farther away from the installed set of grounds, the potential on the phase will remain high while the potential on a multigrounded neutral or a remote structure will drop quickly. The difference of potential between the phase and a neutral or structure can become dangerous when working more than 300 ft (100 m) from the installed set of grounds. **See Figure 13–13.**

When working up to 300 ft (100 m) away from the installed grounds on another structure, the structure must be bonded to one of the grounded conductors. **See Figure 13–14.**

Figure 13–13 Point-of-work grounds is required to reduce the risk of a high potential between the phase and neutral. *(Delmar, Cengage Learning)*

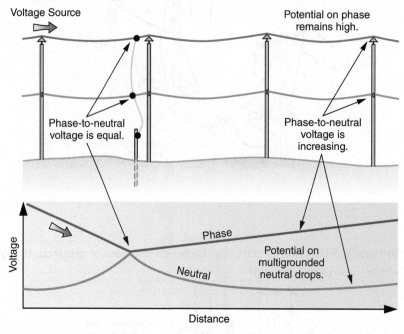

Figure 13–14 When working one span away from the grounds, the structure still needs to be bonded to a conductor to achieve equipotential grounding. *(Delmar, Cengage Learning)*

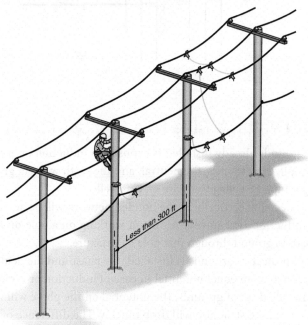

Bonding the Structure for Equipotential Grounding

When an isolated circuit becomes accidently reenergized or is exposed to high induction, the potential at the worksite will rise and be at a different potential than any other objects not bonded to the grounded conductors. The structure will be a parallel path to ground. To eliminate the potential difference between the structure and the circuit, the structure is bonded to the circuit.

Making a bond between the grounded conductors and a wood pole will not be a typical electrical connection. Two common methods to bond a pole to the grounded conductors is to install a pole band below the feet or attach a ground clamp to a bolt through the pole. **See Figure 13–15.** The connection made to a bolt through the pole or to a band around the pole lowers the potential difference to acceptable levels when the wood pole is bonded to the grounded conductors. **See Figure 13–16.**

Figure 13–15 A pole band bonds the pole to the protective grounds. *(Courtesy of Salisbury by Honeywell)*

Figure 13–16 Connecting to a bolt through a pole will bond the pole to the protective grounds. *(Delmar, Cengage Learning)*

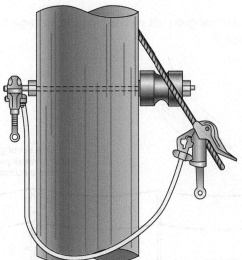

If working on a transformer pole or other equipment pole, the transformer is bonded to the pole, to ground, and to the neutral. A protective ground placed on the neutral, therefore, also bonds the pole and equipment to the power conductors. Even if working from a bucket, the protective grounds must be bonded to the structure in case a worker is in contact with both the structure and the conductors.

On a steel pole or tower, a connection between the grounded conductors and the structure will be a good electrical connection. The complete steel structure will be electrically uniform. A person working on the structure will not be exposed to a dangerous potential difference between the structure and the conductors.

Controlling Induced Voltage and Current from Electromagnetic Induction

The discussion on electromagnetic induction is a prerequisite to understanding our discussion here.

Applying Protective Grounds to Control Induction

The electromagnetic field from an AC power line causes two kinds of induction: an *electric field* (capacitive) induction and *magnetic field* (inductive) induction. Proper application of protective grounds/bonds will control, but will not eliminate, voltage and current induced on isolated circuits from nearby power line. Protective grounds/bonding will control voltage and current to a level that is safe for workers.

Reducing Voltage from Electric Field Induction

A power line will induce a *voltage* on a nearby isolated conductor. There is no current flowing in the isolated conductor if it is not part of a circuit (closed loop). The relatively small current that would be in the conductor is due to leakage current at insulators and surge arrestors. If a worker was to get between an isolated conductor and the earth, the exposure would be to a high induced voltage but a relatively low current. While in relative terms there is very little current, the small amount remaining is likely enough to cause a fatal shock.

One set of protective grounds on an isolated circuit will collapse the voltage due to electric field induction, but there will still be no circuit for current to flow. Note, while there is no circuit for large amount of current to flow, there is still enough leakage current down insulators that a person touching the circuit will be a parallel path that can draw a fatal current flow. **See Figure 13–17.**

Figure 13–17 One set of equipotential grounds will reduce the voltage from electric field induction to a safe level. *(Delmar, Cengage Learning)*

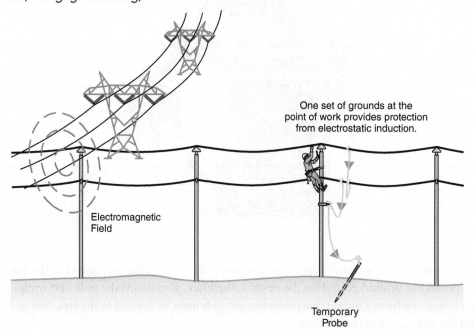

One set of grounds at the point of work provides protection from electrostatic induction.

Electromagnetic Field

Temporary Probe

Controlling the Current from Magnetic Field Induction

A power line will induce *current* on a nearby isolated conductor when that conductor forms part of a circuit. Grounding a conductor on each side of a work zone (bracket grounding) will create a circuit through the conductor, down one set of grounds, through the earth, and back up the other set of grounds. **See Figure 13–18.**

Figure 13–18 Magnetic field induction will set up a current flow when a circuit is created in the isolated conductor. *(Delmar, Cengage Learning)*

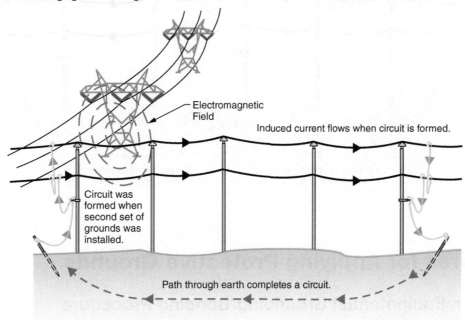

The first set of grounds installed on an isolated conductor will collapse the induced voltage from the electric field, and a second set of grounds will create a circuit for an induced current to flow from the magnetic field.

When bracket grounds are removed, the first set of grounds removed will open the circuit and interrupt the current flow. An immediate voltage will appear (recovery voltage) across the gap between the ground clamp and the conductor. In high-induction areas, a long arc could be drawn. When more sets of grounds are installed along the line, it will be the second-to-last ground removed where a long arc could be drawn.

Sometimes a ground clamp that is in the way of making a splice or other task must be removed, and because other grounds are on the circuit, a power line tech may be tempted to move the clamp by hand or using rubber gloves. However, it can be seen that the second ground can be quite dangerous to handle with anything but a hot-line tool. It is also against grounding principles to install or remove grounds at any time using rubber gloves instead of a hot stick.

Working and Grounding between Bracket Grounds

Once bracket grounds are installed, the conductor between the two sets of grounds will have very little induced voltage but could have a high electrical current flowing through it. When grounds are installed at a worksite (between the bracket grounds), the worksite set of grounds will have relatively very little current flowing in it.

The middle set of grounds creates two circuits for the ground current to flow where previously there was only one. The current from one circuit goes up the middle set of grounds, and the current from the other set of grounds goes down the set of grounds. The two currents tend to cancel each other out, so very little current flows in these grounds. **See Figure 13–19.**

Figure 13–19 Very little current flows through the middle protective grounds. *(Delmar, Cengage Learning)*

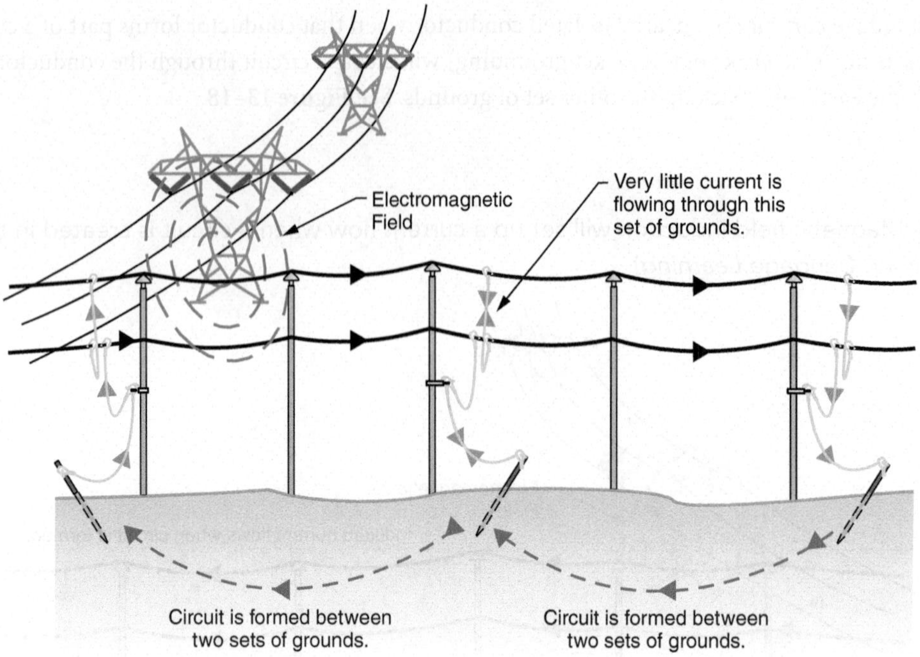

Procedures for Applying Protective Grounds

Distribution Equipotential Grounding/Bonding Procedure

A grounding procedure is probably the most sacred procedure in line work, and every utility/employer has some very rigid concept about how circuits are to be grounded. The grounding procedures described in this section would be superseded by an employer's procedure.

Equipotential grounding (also called single-point grounding) refers to grounding and bonding to a single point at a worksite so that everything that will be worked on or touched is tied together. Thus, everything a person is likely to touch will have the same potential and not have any current flow through them, even during an accidental reenergization.

This grounding and bonding procedure applies to all distribution overhead circuit applications where grounds can be physically accessed at the point of work.

Single-Point Grounding and Bonding Procedure

Step	Action	Details
1.	Arrange for an isolation certification of the circuit.	Use a lockout/tagout procedure to ensure that the line will be isolated and remain isolated. Apply tags and/or locks on isolation points.
2.	Choose suitable grounding location(s).	Ground at the point of work. Grounding on the structure being worked on or grounding within 300 ft (100 m) of the work location is considered grounding at the point of work. **Note:** When the grounds are not on the structure being worked on, a bond must be installed from the structure to one conductor (neutral).

3.	Determine the proper grounding cable size.	The grounding cable and clamp must be large enough to carry the short-circuit current available at the work location to ground and stay intact until the circuit protection opens. The level of short-circuit current available will be dependent on the ability of the system to feed the fault. If the worksite is close to a substation and the circuit conductor is large, the circuit can deliver a large amount of current. A utility planning engineer would have the fault current available for the circuits to be worked on. Many utilities merely specify a minimum cable size for the whole utility and, perhaps, designate some specific locations for larger ground cables.
4.	Choose the most effective ground electrode.	On a wye system, the system neutral is the best electrode. Some utilities/employers may insist on an additional cable going to a ground rod in earth. On a delta circuit, the ground electrode will be the best electrode found at the earth level—usually an anchor or a ground probe.
5.	Establish a bonded work zone at the worksite.	Install bonds between the ground electrode and the pole. If the pole is wood, this means making an electrical connection between wood and the ground electrode: • Install a ground support assembly around the pole. *or* • Install a jumper from the neutral to any bolt through the pole. Tightening the bolt first will improve the electrical connection. To keep everything in the work zone bonded, any wires or winch cables brought into the work zone must be bonded to the same ground electrode. Workers on the ground who are handling conductors are outside of the bonded zone and must use rubber gloves or, in more hazardous situations, ground-gradient mats.
6.	Test the circuit to be grounded to verify isolation.	Use an approved potential tester to verify that the circuit is isolated. Buzzing or teasing is not a reliable method. The hook of a **clamp stick** (shotgun) is the only metal in the head of the stick, and on distribution voltages it is not enough metal for a reliable indication. If more metal is used, buzzing still does not adequately distinguish between induction and an actual energized circuit. **See Figure 13–20.** **Figure 13–20** Test to ensure that the line is de-energized before installing grounds. (© *Salisbury by Honeywell*) 
7.	Install the ground clamps on the phase conductors.	Ensure that all personnel are either completely in or out of the bonded work zone before installing the protective grounds. Use a live-line tool (shotgun) that is at least 8 ft (2.5 m) and wear eye protection. Some utilities/employers will insist on rubber gloves also.

(Continued)

Step	Action	Details
7.	(Continued)	Put one grounding jumper clamp on the neutral. Then, in one firm movement, apply the other end to the closest phase conductor. Install ground sets from the grounded conductor (or neutral) to the other conductors. Using grounding jumpers with a support stud arrangement allows you to keep a distance from the cable while installing the ground. **Note:** If a ground is mistakenly installed on an energized circuit, *do not remove* the ground clamp. Removing the ground clamp could draw a very long and hazardous arc. Let the circuit protection take out the circuit.
8.	Remove grounds in reverse order.	If a ground rod in earth was used as an electrode, do not remove any clamps until all the ground clamps have been removed aloft in order to avoid accidently removing the wrong clamp. Use a clamp stick when removing ground. In high-induction areas, removing the grounds can interrupt current and drop voltage.

Distribution Bracket Grounding Procedure

Bracket grounding *does not protect* a worker from accidental reenergization. Bracket grounding *will protect* a worker from induction and backfeed from small portable generators.

Bracket grounding should be used only where it is very impractical to ground at the worksite *and* where accidental reenergization is impossible. An example of when it is impractical to apply equipotential grounding at a worksite involves a stringing operation where existing wire is being removed and new wire strung along the ground is being put up on structures.

Because bracket grounding does not protect a worker from accidental reenergization, it must be seen as an exception to equipotential grounding and require a supervisor's approval for each application. Every time a bracket grounding procedure is used, a written job plan for grounding, approved by a supervisor, should be created. The grounding plan should identify all potential sources of reenergization and the control barriers that are put in place.

This bracket grounding procedure should be considered as an exception to the single-point grounding/bonding procedure.

Distribution Bracket Grounding Procedure

Step	Action	Details
1.	Arrange for an isolation certification of the circuit.	Use a lockout/tagout procedure to ensure that the line will be isolated and remain isolated. Apply tags and/or locks on isolation points.
2.	Check for additional reenergization hazards.	• Isolation points with suspect insulation; flashed, broken, and suspect brands or type. • Suspect switchgear, SCADA operation disabled. • Potential for the work operation to cause contact with overbuild, underbuild, or crossing circuits. • Physical condition of pins, insulators, ties, crossarms, etc., of overbuild and crossing circuits.
3.	Choose the suitable grounding locations.	• Grounds must be placed on all sides of the work zone not more than 2 miles (3 km) apart.
4.	Determine the proper grounding cable size.	The grounding cable and clamp must be large enough to carry the short-circuit current available at the work location to ground and to stay intact until the circuit protection opens. Technically, bracket grounding is used only where it is virtually impossible for the circuit to be reenergized; however, it is still valid to use the ground cable size meant for the system in that location.
5.	Choose the most effective ground electrode.	On a wye system, the system neutral is the best electrode. Some utilities/employers may insist on an additional cable going to a ground rod in earth. On a delta circuit, the ground electrode will be the best electrode found at the earth level, usually an anchor or a ground probe.
6.	Test the circuit to be grounded to verify isolation.	Use an approved potential tester to verify that the circuit is isolated. Neither buzzing nor teasing is a reliable method. The hook of a clamp stick (shotgun) is the only metal in the head of the stick, and on distribution voltages it is not enough metal for a reliable indication. If more metal is used, buzzing still does not adequately distinguish between induction and an actual energized line.

7.	Install the ground clamps on the phase conductors.	Ensure that all personnel are either completely in or out of the bonded work zone before installing the protective grounds.
		Use a live-line tool (shotgun) that is at least 8 ft (2.5 m) and wear eye protection. Some utilities/employers will insist on rubber gloves also.
		Put one grounding jumper clamp on the neutral and then apply the other end to the closest phase conductor in one firm movement. Install ground sets from the grounded conductor (or neutral) to the other conductors.
		Note: If a ground is mistakenly installed on energized circuit, *do not remove* the ground clamp. Removing the ground clamp could draw a very long and hazardous arc. Let the circuit protection take out the circuit.
8.	For additional protection, establish a bonded work zone at the worksite.	When working between bracket grounds, the grounds will reduce potential on the conductors and trip out a circuit if it should become reenergized. However, in high-induction areas, the grounds are not on the structure being worked on, so a bond must be installed from the structure to one conductor (neutral).
		• In a high-induction area, the induction on the bracket grounded circuit will be low; the induction on work equipment, and on any wires or winch cables brought into the work zone, could be high and at a different potential than the grounded circuit.
		• Working in a bonded area will provide additional protection from accidental reenergization.
9.	Remove grounds in reverse order.	To avoid accidently removing the wrong clamp from the ground probe, do not remove any clamps until all the ground clamps have been removed aloft.
		• When the grounds are removed and circuit is being opened, use a hot stick and eye protection.

Transmission Equipotential Grounding/Bonding Procedure

The grounding and bonding procedure applies to all transmission overhead circuit applications where grounds can be physically accessed at the point of work.

Grounding/Bonding Procedure for Overhead Transmission-Circuit Applications

Step	Action	Details
1.	Arrange for an isolation certification of the circuit.	Use a lockout/tagout procedure to ensure that the line will be isolated and remain isolated. Verify that the control center has isolated the proper circuit and opened and locked/tagged the correct switchgear on an operating drawing. If there are grounding switches on the circuit, ensure that they are closed.
2.	Choose a suitable grounding location.	Ground at the point of work. Grounding on the structure being worked on or grounding within 300 ft (100 m) of the work location is considered grounding at the point of work.
		Note: When the grounds are not on the structure being worked on, a bond must be installed from the structure to one conductor or to the same ground electrode.
3.	Determine the proper grounding cable size.	The grounding cable and clamp must be large enough to carry the short-circuit current available at the work location to ground and stay intact until the circuit protection opens. The level of short-circuit current available will be dependent on the ability of the system to feed the fault. If the worksite is close to a substation and the circuit conductor is large, the circuit can deliver a large amount of current. The utility is responsible for specifying the fault-current availability for the circuits to be worked on.
		Note: If parallel grounds are needed, the first set of grounds installed on a circuit is supposedly not able to carry the fault current and will fail if the circuit being grounded is still hot.
4.	Choose the most effective ground electrode.	On steel poles and steel-lattice structures, install a ground stud on the structure. With the shield wire properly bonded to the steel structure, the structure becomes a good ground electrode. The structure is also well bonded to the conductor after the ground is installed.
		• On wood poles, use an anchor or a ground probe and install a ground clamp on the shield wire. If all the work is aloft and nothing on the ground must be bonded to the bonded network, only the shield wire must be used as a ground electrode.

(Continued)

Step	Action	Details
5.	Establish a bonded work zone at the worksite.	On wood poles, install a ground support assembly on the pole and bond it to the down leads, guys, and ground probes. • Steel structures provide a bonded area to workers aloft. • Install bonds on any wires, booms, or winch cables brought into the work zone for both wood pole and steel work. • Ensure that workers on the ground are protected by using ground-gradient mats or rubber gloves, as applicable.
6.	Test the circuit to be grounded to verify isolation.	Use an approved potential tester. • Teasing is not a reliable method to test for isolation because it does not adequately distinguish between induction and an actual energized line.
7.	In addition to installing single-point grounds at the worksite, install bracket grounds where necessary.	Install bracket grounds on each side of the worksite under the following conditions: • The worksite is in a high-induction area. • Poor soil conditions at the worksite reduce the effectiveness of the installed ground probe. Bracket grounds will reduce the voltage of the bonded work zone in relation to remote ground and will reduce the current in the on-site protective grounds.
8.	Install the ground clamps on the phase conductors.	• Ensure that all personnel are either completely in or out of the bonded work zone. • Use a live-line tool and eye protection. Some utilities/employers will insist on rubber gloves also. • Ground the closest phase conductor first. • On wood poles, install grounds on all three phases. On steel structures, only the phase being worked on must be grounded. ***Note:*** If the ground is installed on an energized circuit, *do not remove* the ground clamp. **See Figure 13–21**. **Figure 13–21** Protective grounds on an H-frame transmission line provide equipotential grounding for workers on the structure. *(Delmar, Cengage Learning)* 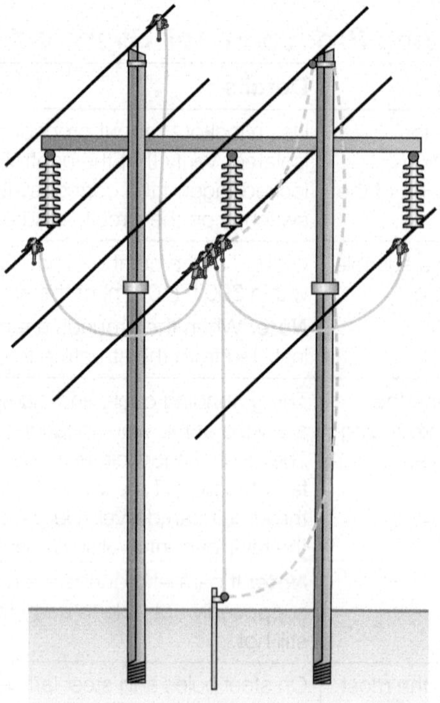
9.	Remove grounds in reverse order.	• Remove the clamp from the shield wire after the phase clamps are removed. There could be a high-potential difference between the grounded phases and the shield wire. • To avoid accidently removing the wrong clamp from the ground probe, do not remove any clamps until all the ground clamps have been removed aloft. • Use a clamp stick and eye protection when removing grounds.

Transmission Bracket Grounding Procedure

There are times when a transmission line bracket grounding procedure can be very practical, such as a project involving the setting of poles in a de-energized circuit. However, bracket grounding *does not protect* a worker from accidental reenergization, but *will* from induction. If bracket grounding is used, it must be considered as an exception to the single-point grounding/bonding that is planned, documented, and approved by a utility/employer.

Bracket grounding should be used only where it is impractical to ground at the worksite and where there is no possibility of accidental reenergization. A written job plan for grounding should include extra steps taken to disable any sources of reenergization, including measures such as removing a span or removing loops at a switch.

Grounding DC Transmission Lines

Grounding a high-voltage direct current (HVDC) power line has many similarities with grounding an AC power line. However, a standard potential tester used on AC circuits may not work on a DC circuit because potential testers measure the electric field around a conductor. There will be some electric fields because the actual DC current is not absolutely steady.

One scenario to grounding a DC circuit is placing grounds on a circuit that is in service. While a two-pole DC line is one circuit, one pole of the circuit can be grounded and worked on while the circuit stays in service. The other pole and the earth make up a circuit. The location and number of grounds must be such that the ground current does not flow back up the grounds through the circuit to get back to the source. In other words, there must be clear, written instructions before proceeding to install grounds.

Specific Grounding Hazards

Hazards and Procedural Controls

This section discusses specific grounding hazards, along with procedural controls to implement when appropriate.

Grounding an Energized Circuit

In addition to getting a work clearance on a circuit, a person must identify the correct circuit, conductor, or cable in the field with a potential tester suitable for the voltage rating of the circuit.

Teasing (buzzing) a conductor with the metal head of a live-line tool is not an effective method of checking for isolation. There is not enough metal on a stick attachment for low voltage to be heard, and it is difficult on high voltage to differentiate between dynamically energized and induction. *Applying protective grounds is the ultimate test for isolation.* **See Figure 13–22.**

Cutting or Connecting a Grounded Conductor

If a current flow is interrupted by a break in a circuit, an immediate high voltage appears across the break. Even if a set of protective grounds is installed on each side of a break in the conductor, it is still likely a potential difference might occur across the break, unless the two grounds are connected together electrically.

Use a live-line tool to install a jumper across an open point in a circuit. When planning the location of the protective grounds, any switchgear or fuse in the work zone that could operate during reenergization should be treated as a potential break or open point.

Electrical Hazards to Workers on the Ground

Accidental reenergization or induction will create touch and step potentials around any ground probes, vehicles, or wire that is bonded to the protective grounds. Workers on the ground are highly vulnerable unless they are in the bonded zone.

Figure 13–22 A voltage detector can be used to test before grounds are installed. (© *Salisbury by Honeywell*)

A worker on the ground must continuously apply the bonding principle for protection from electrical shock. Applications of the bonding principle while working on the ground include, but are not limited to, the following:

- Stay clear of everything attached to the bonded network.
- When operating a boom, stay on the truck platform.
- If tools are required from a bonded vehicle, use rubber gloves.
- If a ground wire must be sent aloft, use a ground-gradient mat.

In a station yard, the station ground network and the graveled yard provide some protection from ground gradients.

Wood Pole Work in Poor Soil or in High-Induction Areas

When working on a three-wire system, especially on transmission lines, protective grounds will not necessarily reduce the voltage of the grounded conductors to acceptable levels in relation to a remote ground. A truck boom or unbonded down guy being sent aloft could become a source of a remote ground if they are not bonded to the protective grounding setup.

When available, grounding to the shield wire (static wire, ground wire, or counterpoise) will lower the voltage on the bonded system. The shield wire can be the best ground electrode available at the worksite because it is multigrounded and a direct path to the station ground.

In addition to installing grounds at the worksite, installing bracket grounds (grounds somewhere on each side of the worksite) will lower the voltage at the worksite. The bracket grounds can be installed in any convenient location that has a good ground electrode.

Grounding Gradients When Only One Phase Has a Protective Ground Installed

Grounding all three phases results in fast clearing in the case of reenergization. When the three phases are tied together there is less current flow through the ground cable and there are less ground currents flowing to earth. An additional benefit is that there will be a reduced level of touch and step potentials for the workers on the ground.

On steel transmission lines, some utilities will allow the grounding of only the phase being worked on. The steel provides a good ground electrode, and grounding only one phase will trip the three-phase circuit breaker during an accidental reenergization. The workers on the steel structure will be working in a bonded area, but workers on the ground should stay clear of the tower.

Protective Grounds Whipping during a Fault

During a high-fault current condition, the portable ground cable can whip violently. The ground cable can also burn off if it is wrapped around a steel structure or is left coiled. The huge magnetic field around the ground cable when it is carrying a high-fault current results in major mechanical forces exerted on the cable.

Use a rope to tie off grounds where workers could be exposed to the whipping action. Keep grounds as short as practical and lay out the ground so it is not coiled or wrapped around steel.

Protective Grounding for Backfeed from a Portable Generator

Backfeed from a generator that was improperly connected at the customer can feed back through the secondary of the transformer and create a high potential on the primary side. Listening for a generator or patrolling for a generator before working at a transformer is not totally dependable; for example, a generator might be started after a person has started to work. The installation of equipotential grounds is the only sure way to know that a person will be protected from electric shock.

When working on an isolated circuit, protective grounds on the circuit will reduce the voltage due to backfeed from a portable generator to an acceptable level. The transformer impedance and the resistance of the conductor between the grounds and the generator will probably not create a fault current large enough to cause the generator to overload. Even with the protective grounds installed on the primary, the generator keeps running, and there will be a small voltage and a current flow in the conductor. Because there is always a chance of current flow in a conductor, a conductor must not be opened, cut, or spliced without first installing a jumper. The magnitude of the voltage and current from a large industrial or commercial generator would cause a higher voltage and current to appear on grounded conductors at the worksite, but because these generators are more likely to be installed and inspected properly, backfeed is very unlikely. The hazard from backfeed is from the unknown portable generator.

Working from a Transformer Pole or Lateral Tap Pole

If work is to be done on a transformer pole, and there is a source of unknown secondary backfeed, the protective grounds will protect a power line tech as long as the transformer fuse is intact. However, it is against protective grounding principles to rely on a fuse for continuity of grounds. The fuse could blow or the cutout could fall open, leaving the high-voltage transformer bushing at full line voltage. The secondary transformer leads should be removed before working on the power conductors.

Using Painted Steel, Slip Joint, or Weathered Steel Pole Structure as a Ground Electrode

When a painted steel tower is used as a ground electrode for a jumper ground, remove a bolt and install a grounding stud that will accept the ground clamp.

A temporary bonding cable may need to be installed across a slip joint on steel pole unless a permanent one is part of the design. The protective oxide on weathered steel is highly resistive. A permanent grounding stud should be installed (welded) at each location on the structure where grounds need to be installed.

Protective Grounding of Underground Cable

Reasons to Ground Underground Cable

The reasons to apply protective grounds on underground cable include the following:

1. *Install protective grounds to prove isolation.* Underground cable is difficult to trace physically, so it is critical to test the cable at a riser, to test at the capacitive test point at an elbow, or to spike the cable before placing grounds.

2. *Install protective grounds to have protection from accidental reenergization.* When working on a grounded cable, a set of protective grounds must always be in place to have protection from accidental reenergization.

3. *Install protective grounds to provide protection from induction.* Underground cable is a capacitor that can maintain a charge for a long time. A protective ground must be installed and maintained to drain these charges.

Applying the Grounding Principle

A protective ground, sized for the available fault current, is needed to trip out the protective switchgear in case of accidently grounding an energized circuit or in case of reenergization. Apply grounds to the cable wherever it is possible to get access—for example, at risers, at switching cabinets, at transformers, and at live-front switchgear cabinets.

Grounding at a Riser

Conventional protective ground sets can be applied at a riser pole (also called a dip pole or transition pole), as long as enough conductor is exposed somewhere on the structure. A method that allows a protective ground to be connected to the bottom of a cutout is to use a clamp similar to the device shown. **See Figure 13–23.**

Figure 13–23 Grounding the bottom of a cutout is more secure when using a cutout clamp ground. (© Salisbury by Honeywell)

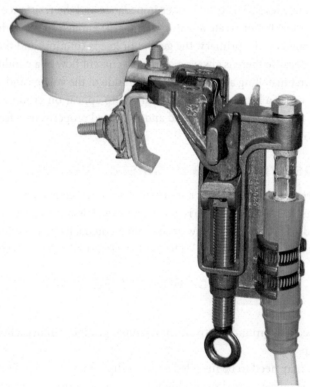

Grounding at a Dead-Front Transformer or Switchgear

A cable terminal that is a load-break elbow plugged into a transformer or other dead-front equipment can be grounded by removing the elbow from the transformer bushing and inserting it into a grounded bushing. A portable ground elbow can serve as the grounded bushing. **See Figure 13–24.**

Using a hot stick, a feed-through is placed on a parking stand in the transformer cabinet. Using an elbow puller, remove the elbow of the cable to be grounded and insert into one-half of the feed-through. If the other end of the cable terminates as a load-break elbow, the same actions are performed there. The cable is now in an isolated state but is not grounded.

Figure 13–24 A portable grounding elbow can be used to install a ground at a bushing of dead front equipment. (© *Salisbury by Honeywell*)

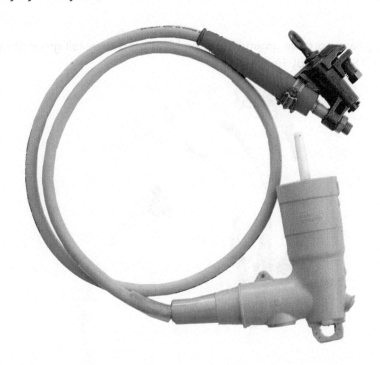

Verify that the cable is isolated by using a stick-mounted potential tester and testing at the capacitive test point. **See Figure 13–25**.

On confirmation that the cable is isolated, a portable grounding elbow is fastened to the grounding bus in the cabinet and the elbow end is inserted into the other half of the feed-through. When this is carried out at both ends of the cable, it effectively grounds the cable.

Figure 13–25 The capacitive test point at an elbow allows a test for potential without contacting bare energized apparatus. (*Delmar, Cengage Learning*)

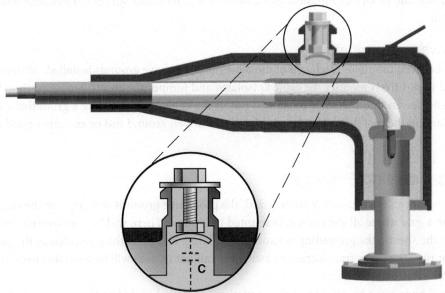

A protective ground set is typically used to ground a three-phase feeder at a dead-front switching cabinet. See Figure 13–26.

Figure 13–26 A three-phase portable grounding assembly is used to install grounds at three-phase dead-front equipment. (© *Salisbury by Honeywell*)

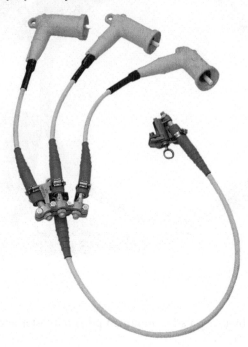

Grounding at a Live-Front Transformer or Switchgear

On live-front equipment, the load side of switchgear is exposed and somewhat accessible to protective grounds that have clamps suited to the equipment to be grounded. Isolating the cable to be grounded requires opening the appropriate switches at both cable terminals. One way to ground is to remove the fuse tube (door) and place a ground at the load side of the switch using the proper clamp, such as a cutout clamp. Ball-and-socket connections also are available for grounding at a switch.

The Cable between Terminals

To work on a cable between terminals, each end of the cable must first have grounds installed. At the work location, the cable is then spiked to positively identify the cable as isolated and jumper the sheath before cutting through the cable. A spiking tool can be driven into the cable to short it out. If the cable does not have a grounded concentric neutral, there is a "ground lug" on the tool that has to be connected to a driven ground rod or any other good ground electrode. See Figure 13–27.

Applying the Bonding Principle

If a protective grounded cable is accidently reenergized, the protective grounds will trip out the circuit. However, the grounds are not like a gate where all the current is shunted to ground. There will be a momentary rise in potential on the core conductor, the sheath, the grounding network in a cabinet, and everything bonded to the protective grounds. Anyone touching or near the cable, transformer, or switching cabinet (kiosk) will be a parallel path to a remote ground and subject to a lethal electrical burn.

An equipotential zone must be created so that a worker is totally bonded to the protective grounds. A ground-gradient mat bonded to the protective grounds will keep a worker within an equipotential zone. The mat serves the same

Figure 13–27 A spiking tool will confirm that a cable is isolated. *(© Salisbury by Honeywell)*

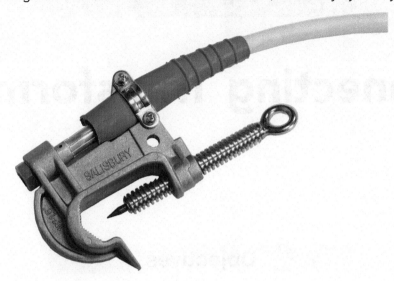

function as bonding a structure to the protective grounds in an overhead lines situation. When working on a cable in a trench, work from a ground-gradient mat bonded to the sheath/concentric neutral.

When the cable core is exposed it will not be in the established equipotential zone. When practicable, install a bond and/or use rubber gloves. If a cable is being cut in two, there could be a high potential between the two ends of the cable, even with protective grounds installed at each end. A jumper should be installed to bond the two ends before cutting to keep continuity between the grounds. Distance and objects between ground locations make this a real-world problem.

Review Questions

1. What are the three reasons to place protective grounds before working on an isolated circuit?

2. When using protective grounds, does the application of the grounding principle control current or voltage?

3. Does all the current in a grounded conductor take the easiest path to earth?

4. How does a power line tech determine the size of grounds needed in a given location?

5. Why is it important to test grounding sets?

6. What is the best ground electrode on a wye circuit?

7. What is the bonding principle?

8. How does bonding a grounding set to a structure provide an equipotential zone for a power line tech?

9. When removing protective grounds on an isolated circuit, which set of grounds is more likely to create an arc: the first set or the second set?

10. At what locations is an underground cable exposed enough to be able to apply grounds?

Connecting Transformers

After completing this chapter, you should be able to:

1. Describe the transformation of voltage through a transformer.
2. Explain the effect transformation has on current from the input to the output.
3. Perform a turns ratio test on a transformer.
4. Calculate the expected voltage on the secondary side of the transformer with a given input voltage and the transformer turns ratio.
5. Perform a test for the polarity of the transformer.
6. Install the proper fuse protection and surge protection at a transformer installation.
7. Make the connections to single phase overhead and underground wye and delta transformers.
8. Describe how the secondary fault current is very high close to the transformer.
9. Choose the correct transformers and connections for the required three phase transformer installation.
10. Explain the steps to troubleshoot a transformer installation.
11. Recognize special installations such as primary step down transformer, grounding transformer and constant current transformer.

Introduction

Transformers are used throughout an electrical system, first to raise voltage for efficient transmission of electrical power, then to reduce voltage to a manageable level for local distribution along roadways and streets, and eventually to reduce the voltage to a utilization level.

The power carried in a circuit is equal to volts × amperes. To transmit a large block of energy would require an extremely large conductor if the voltage were not stepped up. Likewise, it would be extremely costly to distribute power to customers along residential streets at high transmission voltages. The transformer is the link between the different voltage systems.

This chapter discusses the **distribution transformer**, which is used to reduce the voltage on local distribution lines to a utilization voltage. It is the most common piece of line equipment that the line trade installs and maintains.

Large transformers found in substations work on the same principle as distribution transformers. Substation transformers often have underload **tap changers**, additional cooling radiators, and fans. Transformer reliability affects many customers; therefore, most transformers are monitored for temperature and voltage output by control room operators.

Distribution transformers can be overhead, underground residential distribution (URD) as **pad-mount** or submersible, as network transformers found in the downtown core of some cities, and large surface-mount transformers found in small buildings (called substations) feeding long lengths of secondary bus in European systems.

Transformer Basics

Components of a Transformer

A transformer is an electromagnetic device that provides a magnetic linkage between two electrical circuits. Energy is transferred from one circuit to another by the magnetic field. A transformer consists of a tank, a steel or iron core, and one primary and one or two secondary coils of wire. **See Figure 14–1.**

Figure 14–1 An energized coil will induce a voltage on another coil wrapped around a common iron core. *(Delmar, Cengage Learning)*

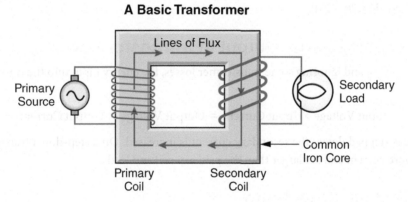

A Basic Transformer

The windings are placed in a common magnetic flux field which couples the coils into a position of mutual induction. On a single-phase transformer, two separate coils of wire are mounted on a common iron core. When a voltage and current are applied to one coil of wire, a strong electromagnetic field is created in the iron core. The electromagnetic field in the iron core induces a voltage and current into the second wire coil. Therefore, even though the two coils are not electrically connected to one another, voltage and current in one coil induce voltage and current into the second coil.

Two Types of Steel Core Configurations

The actual configuration of the steel core and the windings are of two main types. The steel cores are either shell type or core type. The primary and secondary coils are not necessarily wound around separate legs of the steel core but are wound separately around the same core. The shell-type design is more economical for low voltage and for a high kilovolt-ampere (kVA) rating, and the core-type design is more economical for high voltage and a low kVA rating. **See Figure 14–2.**

Turns Ratio

The same electromagnetic field that magnetizes the iron core cuts through both the primary (connected to the source) and the secondary (connected to the load) coils. The same amount of power is induced in each coil. The total voltage induced in each coil is proportional to the number of turns in the coil. When the secondary coil has fewer turns than the

Figure 14–2 The best design of a steel core is dependent upon the voltage and size of the transformer. *(Delmar, Cengage Learning)*

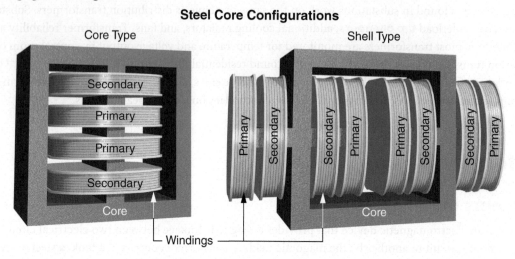

primary coil, the voltage output is reduced at the same ratio as the ratio of the number of turns of wire on the primary coil to the number of turns of wire on the secondary coil. This is called a step-down transformer. For example, a transformer selected to step down 4,800 V primary to 240 V secondary must have 20 times more turns on the primary or a 20-to-1 ratio, because 4,800 divided by 20 is 240.

Energy Input Equals Energy Output in a Transformer

Regardless of the turns ratio, and ignoring some transformer losses, the energy input into the transformer is equal to the energy output.

$$\text{Input Voltage} \times \text{Input Current} = \text{Output Voltage} \times \text{Output Current}$$

When the voltage is stepped down, the secondary current is increased. On a step-down transformer, the secondary coil and leads carry more current and are larger than the primary coil and leads.

Transformer Test for Turns Ratio

Every transformer has a nameplate showing the rated primary voltage and the expected secondary voltage. A field test can confirm that the actual turns ratio is equal to the ratio shown on the nameplate. A ratio test will ensure that there are no shorts between turns in the windings.

To conduct a turns-ratio field test on a transformer, energize the *high-voltage* coil with a low voltage, such as 120 V. **See Figure 14–3.**

Measure the input voltage and the output voltage with a voltmeter. The ratio is calculated with this equation:

$$\text{Input} \div \text{Output} = \text{Transformer Ratio}$$

Caution: Transformers work both ways! For this test, the low-input voltage must be connected to the *high*-voltage winding. If the input voltage was connected to the low-voltage winding, the transformer would be a step-up transformer and a lethal voltage would appear at the high-voltage terminal.

Transformer Taps

Some distribution transformers have off-load/no-load tap changers. By changing the position of the tap changer, additional or fewer turns are applied to the primary winding, which changes the turns ratio of the primary coil to the secondary coil. Increasing the tap adds to the number of turns of the primary coil; therefore, lowering the secondary voltage and lowering the taps reduces the number of turns of the primary coil raising the secondary voltage.

Figure 14–3 It is critical to apply voltage to the high-voltage terminals when conducting a turns-ratio field test. *(Delmar, Cengage Learning)*

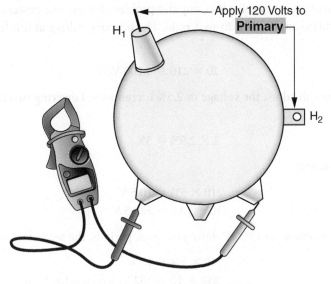

Apply 120 Volts to
Primary

H_1

H_2

Turns-Ratio Field Test

Distribution transformers with taps typically have tap settings that raise or lower the voltage 2.5% for each tap. There will be either two tap positions above and two below 100%, or four below 100%. **See Figure 14–4**.

Figure 14–4 Check the tap settings on the name plate to help calculate any voltage adjustment needed. *(b, © John Bellows)*

Taps	
%	
105	A
102.5	B
100	C
97.5	D
95	E

(a)

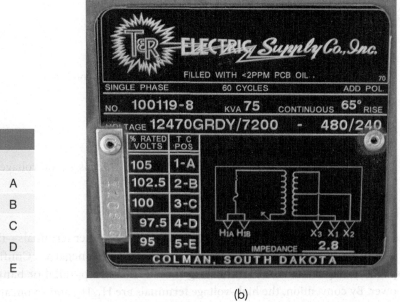

(b)

An external switch knob can be turned to the various positions to change the taps. *The transformer must be de-energized before turning the knob.* On some older transformers, the cover is removed to access the tap changer switch knob. Feeder voltage changes during the day and in different seasons. Therefore, changing the taps at the transformer could produce extreme voltages when the primary voltage returns to its normal level in off-peak periods. Consult with engineering before making a tap change that may give a customer an extreme voltage.

Example 1

A single customer on the end of a long line is continuously receiving about 210 V instead of the rated 240 V. The upstream regulators are difficult to set to solve this singular problem for the one customer at the end of the line. The 4,800/240 V transformer at this location has a 20-to-1 ratio. The primary voltage at this location must be very low and can be calculated:

$$20 \times 210 = 4,200 \text{ V}$$

The transformer has taps that will adjust the voltage in 2.5% increments. Lowering two tap settings would change the secondary voltage:

$$2 \times 2.5\% = 5\%$$

Therefore, 210 V can be boosted to:

$$210 \times 5\% = 10.5 \text{ V}$$
$$210 + 10.5 = 220.5 \text{ V}$$

Caution: If the primary voltage returns to 4,800 V during off-peak periods, the secondary voltage will increase:

$$240 \times 5\% = 12 \text{ V}$$
$$240 + 12 = 252 \text{ V, which is too high.}$$

This is an example of an attempt being made to solve a voltage problem by changing transformer taps when the problem could be a capacitor bank not switching on at the right time or a malfunctioning voltage regulator that is out of service or voltage.

Example 2

The voltage regulators on a line are set high in order to boost the voltage at the end of the line. The customers fed from a transformer just downstream from the regulators are receiving 260 V instead of the rated 240 V. The 14,400/240 V transformer has four, 2.5% tap settings. Which tap will give the customer 240 V?

Reducing the voltage with two tap settings, $2 \times 2.5 = 5\%$

$$\frac{260}{5\%} = 13 \text{ V}$$
$$260 - 13 = 247, \text{ which is too high.}$$

Reducing the voltage with four tap settings, $4 \times 2.5 = 10\%$

$$\frac{260}{10\%} = 26 \text{ V}$$
$$260 - 26 = 234 \text{ V, which is a good voltage.}$$

Transformer Polarity

The terminals of a transformer have a fixed **polarity** in relation to other terminals—that is, when the voltage at one end of a coil is positive, the voltage at the opposite end of the coil is negative. Confirming the actual transformer polarity is especially important when connecting transformers in parallel or banking transformers to provide three-phase power. By convention, the high-voltage terminals are H_1, H_2, and so on, and the secondary terminals are X_1, X_2, and so on.

The polarity of a single-phase transformer is considered to be either *additive* or *subtractive*. Whether the transformer is additive or subtractive is based on the direction of the coil winding. This direction determines the direction of the current flow at the primary terminals with respect to the direction of the current flow at the secondary terminals. **See Figure 14–5.**

From a power line technician's perspective, the polarity of a transformer is based on how the secondary leads are brought out of the transformer. In simpler terms, the X_1 is either on the left side or the right side when facing the

Figure 14–5 The position of the X_1 transformer terminal cannot be assumed. *(Delmar, Cengage Learning)*

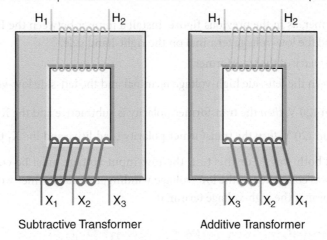

Subtractive Transformer   Additive Transformer

transformer. When the X_1 terminal is on the left side, the transformer is subtractive and when the X_1 terminal is on the right side, the transformer is additive. The H_1 terminal of all transformers will always be on the left when facing a transformer.

ANSI standard C57.12.00 states, "Single-phase transformers 200 kVA and below with high-voltage ratings of 8,660 V and below (winding voltage) shall have additive polarity. All other single-phase transformers shall have subtractive polarity." Therefore, a transformer smaller than 200 kVA, rated at 7.2 kV to ground, would be additive and a similar size 14.4 kV transformer would be subtractive. Dual-voltage transformers (e.g., 7.2 kV and 14.4 kV) would be subtractive.

The nameplate on an additive transformer will show the H_1 and the X_1 terminals diagonally opposite one another. The nameplate of a subtractive transformer will show the H_1 and the X_1 terminals directly opposite one another on the left side of the transformer. Large substation transformers and instrument transformers (PTs and CTs) have subtractive polarity.

Test for Transformer Polarity

A field test can determine whether a transformer is additive or subtractive. **See Figure 14–6.**

Figure 14–6 When conducting a transformer polarity test it is critical to apply the voltage to the high-voltage terminals. *(Delmar, Cengage Learning)*

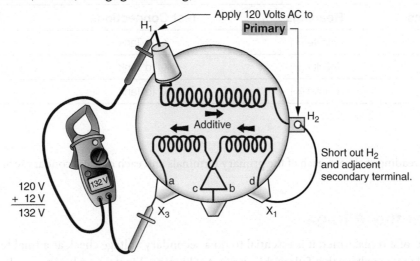

For a 10-to-1 additive transformer,
voltmeter will read 132 V.

Testing a Transformer for Additive or Subtractive Polarity

1. When facing the transformer, as in the previous figure, install a jumper between the H_2 terminal (or the right-side high-voltage terminal) and the low-voltage terminal on the right-hand side.
2. Apply 120 V across the *primary* of the transformer.
3. Measure the voltage between the left-side high-voltage terminal and the left-side low-voltage terminal.

   If the voltage is lower than 120 V, *then* the transformer polarity is subtractive and the X_1 terminal is on the left.

   If the voltage is higher than 120 V, *then* the transformer polarity is additive and the X_1 terminal is on the right.

Caution: Transformers work both ways! For this test, the low-input voltage must be connected to the *high*-voltage winding. If the input voltage was connected to the low-voltage winding, the transformer would be a step-up transformer and a lethal voltage would appear at the high-voltage terminal.

Connections for Transformers with Different Polarities

It is possible to use transformers with different polarities by ensuring that the X_1 of one transformer is connected to the X_1 of the other transformer (similarly with X_2 and X_3), regardless of the position of the terminals on the transformer tank.

When single-phase transformers are to be connected in parallel or connected into a three-phase **bank**, it is normal to select transformers with the same polarity. However, an additive and a subtractive transformer can be banked together by ensuring the X_1 of one transformer is connected to the X_1 of the other transformer (similarly with X_2 and X_3), regardless of the position of the terminals on the transformer tank.

Insulation Resistance and Continuity Testing

Testing a transformer for damaged insulation or coil continuity can be carried out using a 1,000 V insulation tester (megger). Remove any ground from the X_1 terminal for these tests.

Insulation Test

The readings between the high-voltage terminal and the low-voltage (X_1, X_2, or X_3) terminals should be infinite. Readings between the low-voltage terminals and the transformer tank should be infinite. **See Figure 14–7.**

Figure 14–7 Insulation Resistance Should Be Tested at All Six Terminals of a Three-Phase Transformer

HV to LV Insulation Resistance		LV Insulation Resistance to Tank	
Test Connections	**Readings**	**Test Connections**	**Readings**
H_1 to X_1, X_2, or X_3	Infinite (∞)	X_1 to Tank	Infinite (∞)
H_2 to X_1, X_2, or X_3	Infinite (∞)	X_2 to Tank	Infinite (∞)
H_3 to X_1, X_2, or X_3	Infinite (∞)	X_3 to Tank	Infinite (∞)

Continuity Test

In a **continuity test** readings between each of the primary terminals and each of the secondary terminals should be zero. **See Figure 14–8.**

Always Check the Voltage

After the installation of a transformer, it is essential to do a secondary voltage check as a final test to ensure that the customer is supplied with a voltage that falls within range A. Skipping this step can lead to much embarrassment later as well as damaged appliances and electronic equipment. Range A shows favorable voltages, while the voltages in

Figure 14–8 Continuity Should Be Tested at All Six Terminals of a Three-Phase Transformer

Continuity Tests of HV Windings		Continuity Tests of LV Windings	
Test Connections	Readings	Test Connections	Readings
H_1 to H_2	Zero	X_1 to X_2	Zero
H_2 to H_3	Zero	X_2 to X_3	Zero
H_3 to H_1	Zero	X_3 to X_1	Zero

range B are considered tolerable but plans should be made to upgrade the voltage to range A. The voltage readings are utilization voltages, which means that the voltage readings should be taken at the meter base, not at the transformer. **See Figure 14–9.**

Figure 14–9 Voltage at the Customer's Meter Base Should Be in Range A

Service Voltage	Range A Minimum (V)	Range A Maximum (V)	Range B Minimum (V)	Range A Maximum (V)
% of Nominal	95%	105%	91.7%	105.8%
120/240 3 wire	114/228	126/252	110/220	127/254
240/120 4 wire	228/114	252/126	220/110	254/127
208Y/120 4 wire	197/114	218/126	191/110	220/127
480Y/277 4 wire	456/263	504/291	440/254	508/293

Transformation Effect on Current

Load Current

When the voltage from one coil is being induced into a second coil, with an attached load, current is also induced into the second coil. The current, however, is transformed in the *inverse* ratio to the voltage transformation. For example, if the turns ratio of the transformer is 20 to 1, the current transformation is 1 to 20. The voltage is stepped down, but the current is stepped up. In a transformer:

$$\text{Input: Volts} \times \text{Amperes} = \text{Output: Volts} \times \text{Amperes}$$

Example: A 4,800 V transformer with a turns ratio of 20 to 1 has a 200 A load on the 240 V secondary. What current does the transformer draw on the primary side of the transformer?

$$\text{Input: 4,800 V} \times \text{A} = \text{Output: 240 V} \times \text{200 A}$$

$$4,800 \text{ V} \times 10 \text{ A} = 240 \text{ V} \times 200 \text{ A}$$

The primary coil on a 20-to-1 ratio transformer draws 10 A to produce 200 A on the secondary coil.

Calculating Load Current

The actual load on a transformer can be measured and calculated in the field by measuring V_1 and V_2 and I_1 and I_2, and applying it to the following formula. **See Figure 14–10.**

$$\frac{(I_1 \times V_1) + (I_2 \times V_2)}{1,000} = \text{Total kVA load}$$

If there is a need for a fairly high degree of accuracy, *then* the current and voltage on each leg should be taken at the same time to reduce the effects of a load shift from one side to the other.

Figure 14–10 An example of calculating the load on an in-service transformer. *(Delmar, Cengage Learning)*

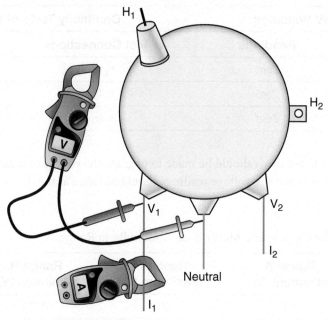

Why Is the Primary Coil of a Transformer Not a Short Circuit?

When one end of a transformer coil is connected to an energized primary source and the other end is connected to a grounded neutral, why is the coil not a dead short? The resistance of a transformer coil is low and a DC would, in fact, cause a dead short between the positive and negative ends of the coil. When AC is applied to the primary coil, there is an induced counter-electromotive force (cemf) that limits the amount of current that can flow in the primary coil and prevent a short circuit. This is explained by Lenz's law, which states: "An induced electromotive force (emf) always tends to oppose the force that causes the induction."

The changing current in the primary of the transformer induces a voltage on the secondary coil. When current flows in the secondary coil, a voltage is induced back to the primary coil from the secondary coil. When more load is connected to the secondary, more current will flow in the secondary. This additional current increases the strength of the secondary magnetic flux field which reduces the primary counter-electromotive force, allowing more current to flow into the primary coil. Because primary $(V \times I)$ = secondary $(V \times I)$, when more load is added to a transformer the changing magnetic fields automatically adjusts to allow more current to flow into the primary coil.

Secondary Fault Current

A step-down transformer can generate a very high current on the secondary side if the secondary wires are accidentally shorted. The magnitude of the fault current available to a secondary of a transformer is mostly dependent on the size of the transformer. The larger the transformer, the greater the capability to generate a large fault current. If the secondary wires are shorted, the transformer can briefly carry a load more than 10 times its rating and supply a high fault current to the faulted location. Depending on the impedance of the transformer, a short at the secondary terminals of a 100 kVA transformer can be as high as 18,000 A. The level of fault current at the location of the short depends on the impedance of the circuit back to the transformer—that is, it depends on the distance from the transformer and the conductor size. **See Figure 14–11.**

The level of fault current available on secondary wires drops quickly with distance away from the transformer. Secondary services are not always recognized as high-energy circuits by power line technicians. However, the magnitude of a fault on the secondary can be very explosive when a short circuit is close to the transformer. In addition to wearing regular PPE, ensure that safety glasses are worn when working on an energized secondary because an eye can be permanently damaged by a large flash.

Figure 14–11 The level of fault current available on a service varies with the distance from the transformer. *(Courtesy Delmar, Cengage Learning)*

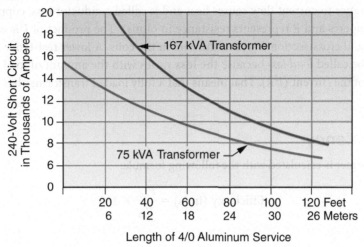

Length of 4/0 Aluminum Service

Transformer Losses and Impedance

Transformer Losses

It would appear from the following equation that there are no losses in a transformer:

$$\text{Input } V \times A = \text{Output } V \times A$$

This is essentially true because power transformers are more than 97% efficient; however, transformers do account for the largest proportion of energy loss in a distribution system. For example, a typical 25 kVA transformer could have losses of about 150 to 350 watts, which would add up, considering the number of transformers in an electrical system. The losses in a transformer are due to **core loss** and conductor loss.

Core Loss

Energy is consumed when the iron core is magnetized and demagnetized by AC power at 120 times a second. This energy loss is referred to as hysteresis loss or *excitation loss*. To reduce hysteresis loss, the core is made of more permeable iron, which means the core is made with various steel alloys that are easier to magnetize and demagnetize.

When the magnetic field induces a voltage into a transformer coil, a voltage is also induced into the iron core. This induced voltage causes current, known as eddy current, to flow in the iron core. Eddy currents cause heating and are a waste of energy. Eddy-current flow can be reduced by designing an iron core that raises the resistance of the core to current flow while not lowering the flow of magnetic flux. This is done by use of a laminated steel core that has thin strips of steel core material insulated from each other. Eddy currents are, therefore, kept smaller within each separate lamination.

Hysteresis loss and eddy-current loss together make up core loss or iron loss in the form of heat. Core loss can be referred to as no-load loss, because there is energy loss even when there is no load on the transformer. When load is added hysteresis and eddy currents increase and losses increase.

Relatively excessive core losses occur when a larger-than-necessary transformer is installed for the load to be served. If a 25 kW load were fed from a 100 kVA transformer, there would be considerably more core loss than if the load were fed from a 25 kVA transformer.

A high-permeability material called amorphous (having no regular form) steel offers 70% less power loss than the conventional iron core. It is also called metallic glass because of its atomic structure similar to glass. It is made when hot liquid iron is forced through super cooled rollers. The steel solidifies so fast that it forms a crystalline structure. The result is an amorphous or irregular structure that is more permeable.

Conductor Loss

Resistance to current flow occurs in any wire. A transformer coil is made from a long length of wire, and it offers resistance to current flow. The resistance to current flow causes heat and is called conductor loss, copper loss, or I^2R loss (where I represents current in amperes and R represents resistance in ohms). The product of I^2R = watts. Conductor loss is dependent on the length and cross-sectional area of the wire in the coils. A lower resistance results in a lower I^2R loss. Conductor loss can also be called *load loss* because the loss varies with the amount of load on the transformer. Load losses vary by the square of the current (I^2R). That means that a fully loaded transformer has four times the copper loss as one loaded to 50%.

Transformer Efficiency

The efficiency of a transformer is calculated with the following formula:

$$\text{Efficiency (in \%)} = \frac{P_{\text{out}}}{P_{\text{in}}} \times 100$$

Where:

P_{out} = total output power delivered to the load

P_{in} = total input power

For example, a fully loaded 100 kVA transformer that is about 98% efficient could have a 2,000 W conductor loss at full load, and a 500 W core loss at a 90% power factor.

Transformer Impedance

The turns ratio of a transformer determines the ratio between the primary voltage and the secondary voltage. When load is applied, the load current is impeded by the resistance and reactance in the windings. When under load, the voltage at the secondary terminals is lower than the voltage indicated by the turns ratio. The impedance of a transformer is expressed as the percentage of the voltage drop at a full load compared to the voltage drop at no load. For example, if a transformer with a 2.2% impedance delivers 240 V at no load, it will deliver 2.2% less than 240 V (234.7 V) at a full load.

There are practical limits to designing a transformer with lower impedance; and there is an advantage for a transformer to have some impedance because the impedance does limit current going through the transformer during a fault. If the current is limited by a higher impedance then there is a lower probability of a catastrophic failure when a downstream fault occurs.

Voltage-Survey Accuracy

When a planning engineer carries out a voltage survey, recording voltmeters are installed at the end of a feeder where a problem low voltage is likely to first show up. This voltage is used by the planning engineers to verify and update their computed feeder calculations.

To get an accurate line voltage, a recording voltmeter is installed on the secondary of an unloaded transformer. An unloaded transformer is used, because with a loaded transformer the internal impedance will cause a voltage drop and the actual feeder voltage would still be unknown.

Transformer Protection

Preventing Insulation Breakdown

Transformer problems are usually due to an internal insulation breakdown. Transformer protection is focused on preventing insulation breakdown because of heat or a voltage surge.

Transformer Overheating

Overheating of a transformer is usually due to overload, a short-circuited secondary, or the follow-through current initiated by lightning. The transformer is designed to be fairly tolerant to an overload for short duration. An oil-filled transformer, for example, may withstand an overload of 25 times the rated current for 2 seconds or two times the rated current for 30 seconds.

The kilovolt-ampere rating of a transformer is based on it being at 30°C. When the average temperature is higher or lower than the standard 30°C, the transformer rating can be changed. The kVA rating of a transformer can increase 1% for a decrease of each degree below 30°C and is decreased 1.5% for an increase of each degree above 30°C.

Larger transformers found in stations normally have a permissible rating that is calculated based on ambient temperature, load factor, and the existence of external cooling, such as by fans. **See Figure 14–12.**

Figure 14–12 This transformer rating sheet is for a distribution substation transformer in a more northern climate. *(Delmar, Cengage Learning)*

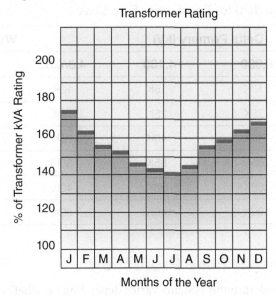

Transformer Loading

The heating effect of current flowing in a transformer coil determines the amount of energy a transformer can supply without causing damage to the insulation. The heat developed in the transformer winding is based on the following formula:

$$\text{Heating effect in watts} = I^2R$$

If a transformer were allowed to carry three times more than its rated load current, the heating effect would be nine times as great as with full-load current. The full-load current in the windings of a transformer can be calculated for either the primary or secondary coil.

$$\text{Full-load current} = \frac{\text{kVA} \times 1{,}000}{\text{Voltage across coil}}$$

Example: The following is the full-load current for a 25 kVA, 14,400/240 V transformer:

$$\text{Primary full-load current} = \frac{25 \times 1{,}000}{14{,}400} = 1.74 \text{ A}$$

Secondary full-load current:

$$\text{Secondary full-load current} = \frac{25 \times 1{,}000}{240} = 104 \text{ A}$$

Overhead Transformer Fuse Protection

The fuse in a transformer cutout melts when exposed to overcurrent. The fuse melts before the transformer is damaged from the heat generated by an overload or a secondary short circuit. The fuse is also coordinated so that it isolates a faulted transformer before any upstream protection opens the primary circuit.

A link fuse speed and size are specified according to the preferences of the utility. The speed of a transformer fuse is generally specified as a K-link (fast) or a T-link (slow) fuse. The slower T-link fuse reduces nuisance fuse blowing due to transients, such as lightning.

The kVA rating of the transformer and the primary voltage are two factors that govern fuse size. For example, a transformer on a 2.4 kV system has a fuse size of approximately 1 ampere per transformer kilovolt-ampere, and a transformer on a 14.4 kV system has a fuse size of approximately 0.2 ampere per transformer kilovolt-ampere. Always install the specified fuse, current-limiting fuse, and surge arrestor at a transformer. **See Figure 14–13.**

Figure 14–13 One Utility's Standard for Transformer Fuse Sizes

Transformer Size	Delta Primary (kV)		Wye Primary (kV)		
	8,320	4,160	4.8/8.32	8/13.8	14.4/25
10 kVA or 3% 10 kVA	10K	15K	10K	10K	10K
25 kVA or 3% 25 kVA	25K	25K	15K	10K	10K
50 kVA or 3% 50 kVA	40K	40K	25K	15K	10K
75 kVA or 3% 75 kVA	65K	65K	30K	20K	15K
100 kVA or 3% 100 kVA	65K	80K	40K	30K	20K
167 kVA or 3% 167 kVA	100K	140K	80K	50K	25K

Underground Transformer Fuse Protection

Protection for transformers on underground systems varies depending on whether the transformer is in a vault, is submersible, or is a pad mount. Transformers in a vault often have standard overhead protective switchgear. Energized leads, terminals, and switchgears are exposed and must be in locked enclosures to prevent accidental contact.

Pad-mount transformers are used on underground systems and sit on a concrete or fiberglass/epoxy pad above the surface of the ground. A pad-mount transformer can be live front or dead front. A live-front transformer, which is typically an older installation, has exposed energized switchgear when the metal enclosure is opened.

The switchgear in a dead-front transformer is insulated. Elbow connectors are used for switching, sectionalizing, or isolating the transformer. A bayonet-style fuse is an under-oil expulsion fuse cutout that has a stab-sheath arrangement to hold the fuse and is field replaceable by a line crew. **See Figure 14–14.**

An isolation link or a current-limiting fuse is used in series with a bayonet-style fuse. During a transformer failure, the isolation link or current-limiting fuse opens the primary lead to the faulted transformer. This safety feature prevents a line crew from re-energizing a faulted transformer. Because the isolation link or the current-limiting fuse will only operate during a transformer failure, it is not replaceable in the field.

Current-Limiting Fuse Protection

On overhead transformers, a current-limiting fuse is sometimes installed in series with a fuse cutout to limit the amount of current that can rush into a transformer during a fault. A current-limiting fuse protects a power line tech from a catastrophic transformer failure that could occur when a worker is trying to energize a defective transformer in a high-fault-current location. A transformer on an underground system often uses a current-limiting fuse to limit arcing, because an arc could easily spill over to a nearby grounded object.

Figure 14–14 This dead-front transformer is protected by a draw-out load break, bayonet-style expulsion fuse. *(Delmar, Cengage Learning)*

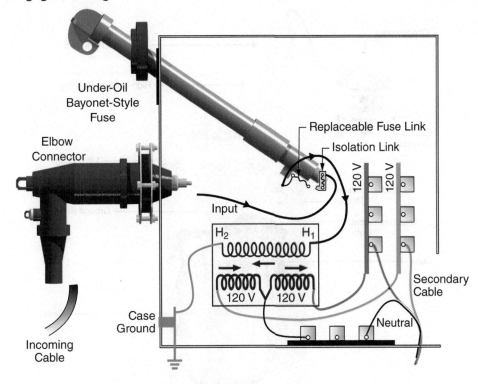

Completely Self-Protecting Transformer

In a completely self-protecting (CSP) transformer, the primary fuse is inside the transformer tank. The primary link blows when there is a fault inside the transformer. It is somewhat less vulnerable to lightning because the surge arrestor on a CSP transformer is mounted right on the tank to shunt away any follow-through current. Replacing a blown primary fuse is not considered a job to be carried out in the field. **See Figure 14–15**.

A CSP transformer also has an internally mounted secondary circuit breaker which protects the transformer from secondary short circuits and overload. During a no-power callout the secondary breaker may have tripped open. Using a switch stick, the external handle can be used to close the breaker. The breaker will reopen during a fault even if the external handle is held in a closed position. On some CSP transformers a small red light mounted near the breaker will come on when the transformer has been overloaded.

Some CSP transformers have a secondary breaker that has an emergency overload device. It is found just above the breaker operating handle. Moving the emergency overload device lever in a clockwise direction allows more load to be placed on the transformer. This is intended to be used just long enough until a new transformer is installed.

Power line technicians face a hazard when working on the secondary of a CSP transformer. When the primary is still energized and the secondary breaker is open there may still be a capacitive or magnetic coupling between the primary and the secondary, causing a high-voltage shock on the secondary. The primary lead should be removed for work on the secondary. Some utilities require a cutout to be installed on the primary even though there is an internal primary fuse link to make it safer to isolate the transformer primary.

Voltage Surge Protection

A surge arrestor is installed to channel any voltage surge and its associated current away from the transformer. On distribution circuits, the source of a voltage surge is almost always lightning. Occasionally, an accidental contact with a higher-voltage overbuilt circuit results in a surge.

Figure 14–15 A completely self-protecting (CSP) transformer does not necessarily need an additional cutout or surge arrestor for protection. (© *Cooper Power Systems, LLC*)

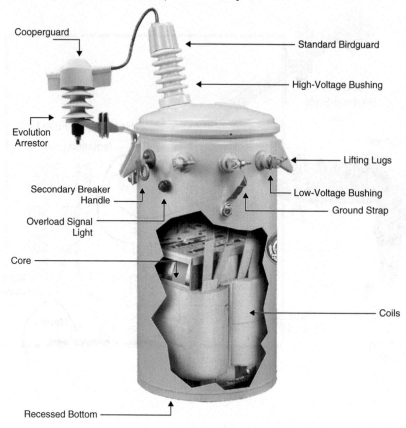

The surge arrestor is installed ahead of the primary transformer terminal; however, utilities differ on whether to install the arrestor ahead of or after a fuse cutout. An arrestor protects best when it is installed as close as possible to the equipment it is protecting, but there is a concern about nuisance fuse blowing when the arrestor is installed after the fuse cutout. There is usually a high follow-through current associated with a voltage surge, and a surge arrestor causes the current to bypass the cutout when it is installed before the fuse.

The rating of an arrestor is specified at a slightly higher voltage than the circuit. If the rating is too low, continuous exposure to small surges will cause the arrestor to deteriorate prematurely. If the voltage rating of the arrestor is too high, a damaging voltage and current may not be bypassed. The ground wire leading away from the arrestor must be as short as possible and at least the same size as the primary lead. A rule of thumb is that each foot of #6 copper wire that is added to a surge arrestor lead adds 1.6 kV to the discharge voltage of the surge arrestor.

Neutral Connections and Ground Connections

The neutral and ground connections on a transformer may appear to be interchangeable because the two are connected. For all transformer installations, the neutral connections have one purpose and the ground connections have another. A neutral is a grounded current-carrying conductor at ground potential. The neutral is part of the electrical circuit and during normal operation carries the current back to the source.

A ground wire is a *grounding* conductor. Ground-wire connections provide a path for current under abnormal conditions, such as during a lightning storm when an arrestor or an insulator sparks over. The ground wire also bonds the transformer tank and related equipment to keep them all at the same potential.

Single-Phase Transformer Connections

Single-Phase Transformers

Single-phase transformers have one high-voltage primary coil with two high-voltage terminals. The high-voltage terminals (**bushings**) can be vertically mounted on the transformer cover or horizontally side mounted. Only one high-voltage terminal needs to be insulated when the coil is connected between a phase and a neutral. Both high-voltage terminals must be insulated when the coil is to be connected phase-to-phase.

In North America, 90% of all transformers are installed as single-phase transformers, using a center-tapped secondary coil connected to supply a standard 120/240 V service. **See Figure 14–16**.

Figure 14–16 This single-phase transformer has a standard three-wire 120/240 V service. *(Delmar, Cengage Learning)*

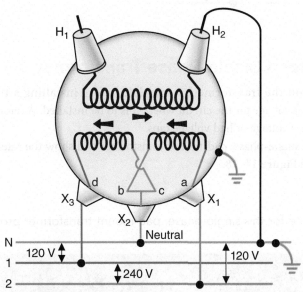

Note that the head of the vector representing one secondary coil is connected to the tail end of the vector representing the other secondary coil, which means the two secondary coils are connected in series. A customer receives a three-wire service consisting of two 120 V hot wires and a neutral. The two 120 V hot wires are at opposite polarity to each other and any contact between them will create a short circuit. The voltage between one 120 V leg to the other is 240 V. Because the two legs are not always equally loaded, the neutral carries the current difference of the two legs.

The two low-voltage coils can be connected in parallel or in series. **See Figure 14–17**.

Transformers with a center-tapped secondary coil (two secondary coils connected in series) have a certain voltage induced across the full length of the coil and half of that voltage on each side of the center tap. Power line technicians occasionally need to take the lid of the transformer and move the internal leads so that the secondary coils are either in series or in parallel. When the two secondary coils are connected in series (with b and c connected together) inside the tank, the secondary voltage is double the voltage of each individual coil.

Placing the two secondary coils in parallel (a is connected to c and b is connected to d) allows the complete coil and therefore the full kVA of the transformer to be available at 120 V. The X_3 terminal will only be acting as an oil seal.

In some other places in the world, where a standard 240 V (or 220 V) is used, the service is fed from a bus supplied by a 240/415 V, three-phase transformer or from a single-phase, 240/480 V transformer. A residential customer would have a three-wire service with one active (hot) wire, one neutral wire, and one safety ground wire. Higher voltage would be available with a four-wire service.

Figure 14–17 The coils of a center-tapped secondary can be connected in series or parallel. *(Delmar, Cengage Learning)*

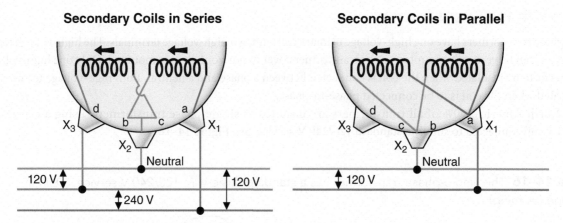

An overhead transformer is the same as an underground transformer except underground transformers are constructed

Typical Nameplate for a Single-Phase Transformer

Power line techs should consult the transformer's nameplate before installing a transformer, to ensure that the transformer voltage is the correct voltage for the circuit where it is to be installed. A nameplate would also show whether the transformer had dual primary voltage or had voltage taps.

A nameplate on a standard single-phase, pad-mount transformer will show the rated voltage in electrical schematic and the transformer weight. **See Figure 14–18.**

Figure 14–18 The nameplate for this single-phase, pad-mount transformer provides needed information to a lineman. *(© John Bellows)*

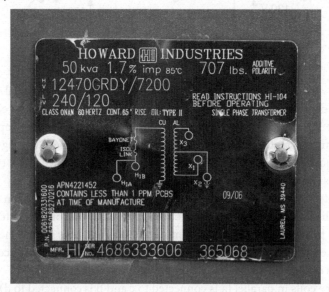

An overhead transformer is the same as an underground transformer except underground transformers are constructed for installation either in a vault, as a pad mount, as a submersible, or direct buried. **See Figures 14–19 and 14–20.**

Figure 14–19 A typical single-phase overhead transformer is connected between a phase and neutral. *(Delmar, Cengage Learning)*

Figure 14–20 A typical single-phase, dead-front, pad-mount transformer. Note that the secondary is exposed. *(© John Bellows)*

Connections to Delta or Wye Primary Systems

The primary of a single-phase transformer can be connected in two ways to create a high-voltage potential between the H_1 and H_2 terminals, either phase-to-phase (delta) or phase-to-neutral (wye).

Phase-to-Phase (Delta) Systems

For a single-phase connection to a delta circuit, connect one high-voltage terminal to one phase and the other high-voltage terminal to another phase. The transformer must have two insulated high-voltage terminals for a delta connection. To protect the transformer, a fused cutout and a surge arrestor are installed on both of the high-voltage primary leads. **See Figure 14–21.**

Phase-to-Neutral (Wye) Systems

For a single-phase connection to a wye system, one high-voltage terminal is connected to the phase and the other is connected to the system neutral. Transformers intended for connection to wye systems can be constructed with only one insulated high-voltage terminal. The primary neutral H_2 connection is not insulated and is connected directly to the transformer tank.

It is possible to connect a transformer with two high-voltage terminals phase-to-phase on a wye system, but the voltage rating of the transformer primary would have to be suitable. For example, on a 7.2/12.47 kV wye system, a transformer connected phase-to-phase (delta) would need to be rated as 12.47 kV, and a transformer connected phase-to-neutral (wye) would need to be rated as a 7.2 kV.

Figure 14–21 A single-phase transformer is connected phase-to-phase on a delta circuit. *(Courtesy of iStock Photo)*

Transformers Connected in Parallel

Occasionally a transformer is changed out, usually for a larger one; or, to increase capacity, a second transformer is installed and connected to the same bus with a new bus break installed between the transformers. To prevent an outage, the new transformer can be temporarily connected in parallel to the existing transformer where both will be temporarily energized.

The transformers must be connected to the same phase. Do not make any connections to the secondary of the new transformer before all work is completed on the primary. An energized secondary will cause backfeed to the H_1 bushing at primary voltage. To ensure that the X_1, X_2, and X_3 of the new transformer is connected to the same wires as X_1, X_2, and X_3 of the existing transformer, voltage checks should be made. After the new transformer is energized and the X_2 is connected to the secondary neutral, use a voltmeter to ensure a zero reading before connecting the X_1 and X_3 to the common secondary. **See Figure 14–22.**

Figure 14–22 A second transformer can be added to a common secondary if all connections are parallel and backfeed is avoided by connecting the secondary last. *(Delmar, Cengage Learning)*

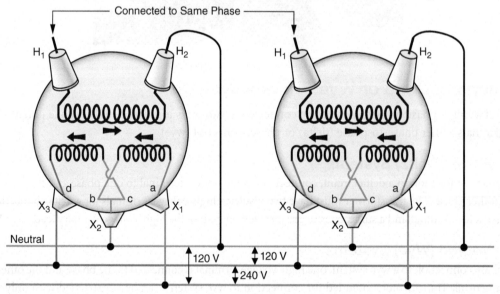

Although rare, two smaller transformers are sometimes connected in parallel to give the equivalent capacity of one large single-phase transformer, mainly as a means to use up older surplus small transformers. Two transformers would be connected together on one pole and the connection to the phase would be through one fused cutout.

Three-Phase Transformer Connections

A three-phase service can be supplied by one three-phase transformer unit or by banking three single-phase transformer units. One three-phase unit is smaller than an equivalent-size bank consisting of three single-phase units. **See Figure 14–23.**

Figure 14–23 Installing a single-unit, three-phase transformer is easier than banking three individual single-phase transformers. *(Delmar, Cengage Learning)*

Single-unit three-phase transformers tend to be used in underground vaults or as a pad-mount transformer. Most three-phase pad-mount transformers have the primary source connected to the transformer with elbow connectors in the left-hand primary cabinet and outgoing secondary cables in the right-hand-side secondary cabinet. **See Figure 14–24.**

Figure 14–24 A three-phase, pad-mount transformer like this allows work in the secondary cabinet with the door closed and locked on the primary side. *(© John Bellows)*

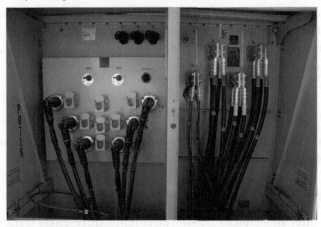

One three-phase unit is easier to install because the polarity and connections between the phases are fixed. **See Figure 14–25.**

The use of three single-phase units is common in overhead distribution. When single-phase transformers are banked together, they can be connected to supply more than one type of service. For example, three transformers with 120/240 V secondaries can supply a 120/208 V, 240/416 V, or a 240 V three-wire service. Fewer specialized spare emergency transformers are needed when single-phase transformers are used.

Figure 14–25 An illustration of the internal connections of a single-unit, three-phase wye-to-wye transformer. *(Delmar, Cengage Learning)*

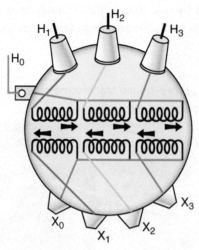

Choosing the Correct Transformers

1. The voltage rating of the transformer primary coil must be compatible with the applicable circuit. The voltage impressed across the primary coil will depend on whether the coil is connected in a wye (phase-to-neutral) or a delta (phase-to-phase) configuration.

2. The transformer must be able to deliver voltage needed by the customer. The supplied secondary voltage will be dependent on the following:
   - The voltage rating of the secondary coil
   - Whether the transformer secondaries are connected in a wye or delta configuration
   - Whether the secondary coils inside the transformer are connected together in series or in parallel

3. If equipped with tap changers, the transformers must be on the same voltage tap. Dual-voltage transformers must be set on the proper voltage.

4. The impedance of the transformers in the bank should be within 0.2% of each other to avoid having the transformer with the lowest impedance drawing the most current and because it is supplying more of the load than the other transformers in the transformer bank; and if the bank is heavily loaded, the transformer with the lower impedance will overheat and burn out prematurely. In other words, if one transformer has an impedance of 2%, then the impedance of the other transformer should be between 1.8% and 2.2%.

Typical Nameplate for Three-Phase Unit

A transformer nameplate should be checked to determine which three-phase configuration and voltage the transformer is able to supply. **See Figure 14–26.**

Using Vectors as an Aid in Banking Three Single-Phase Units

When connecting (banking) three single-phase transformers into a three-phase transformer bank, the connections between the transformers can get confusing, even when following a specifications drawing. **See Figure 14–27.**

The relationship between phases, series connections, parallel connections, and polarity can all be represented on paper as a vector drawing. For electrical drawings, the proper term is *phasor* instead of vector, but in line work, *vector* continues to be used. A vector representation of a wye and/or a delta configuration shows the three vectors, each representing a phase that will always be 120 degrees apart from the other. **See Figure 14–28.**

When the vectors are in a wye or parallel configuration, the tails will be tied together as a common neutral. If the head of the vector comes in contact with a tail of a vector, it will be a dead short.

Figure 14–26 The nameplate shows the transformer is an oil-filled, three-phase unit to be fed from a 12.47/7.2 kV primary. The secondary will feed a three-phase 208/120 V service. Both the primary and the secondary are wye connected. (© John Bellows)

Figure 14–27 The connections needed to bank three single-phase transformers must be done precisely and will vary with the type of service needed. (Delmar, Cengage Learning)

Figure 14–28 Notice that the tails of each vector go to a common point in the wye configuration and the head and tail are connected in the delta configuration. (Delmar, Cengage Learning)

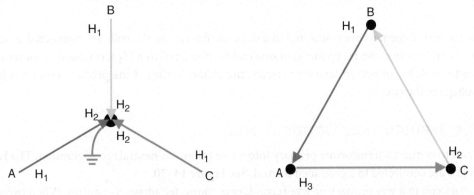

When the vectors are in a delta or series configuration, the head of one vector makes contact with the tail of the next vector. If a head-tail combination comes in contact with another head-tail combination, there will be a dead short.

Putting vectors in the drawing puts logic into the drawing. Completing a vector drawing before connecting up a three-phase bank makes it more likely that the work will be done in a logical fashion, rather than by rote.

Wye or Delta Connections

A transformer coil must have a potential difference matching the nameplate voltage across it to operate. Following are two ways to get a voltage across a transformer coil.

1. One way is to connect a coil between one phase and another phase. When each of the three transformers has its coils connected between the phases AB, BC, and CA, the transformers are connected in a delta configuration.

2. The second way is to connect a coil between one phase and the neutral. When each of the three transformers has its coils connected between each phase and a common neutral, the transformers are connected in a wye configuration.

Primary Delta Transformer Connections

There are three ways to connect a transformer primary into a delta (phase-to-phase) configuration. The direction of the vectors reduces confusion when making connections. Each transformer coil in a delta primary or delta secondary is connected phase-to-phase. **See Figure 14–29.**

Figure 14–29 Primary delta connections are phase-to-phase and can be made in three different situations. *(Delmar, Cengage Learning)*

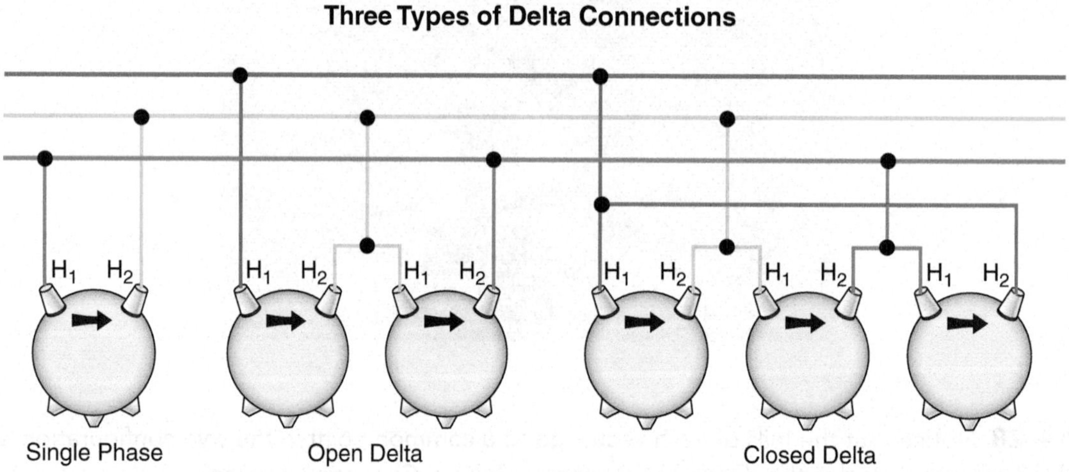

Three Types of Delta Connections

Single Phase   Open Delta   Closed Delta

When two or more transformers are connected in a delta configuration, the coils are connected in series with each other. To connect a coil in series, the H_1 terminal of one coil is connected to a H_2 terminal of another coil. If a transformer primary is to be delta connected on a wye circuit, the voltage rating of the primary coil must be equal to the phase-to-phase voltage of the circuit.

Primary Wye Transformer Connections

There are three ways to connect a transformer primary into a wye (phase-to-neutral) configuration. The H_1 is connected to a phase and all H_2s are connected to a grounded neutral. **See Figure 14–30.**

Each transformer coil in a wye primary or wye secondary is connected phase-to-neutral. When two or more transformers are connected in a wye configuration, the coils are connected in parallel with each other.

Figure 14–30 Primary wye connections are phase-to-neutral and can be made in three different situations. *(Delmar, Cengage Learning)*

Three Types of Wye Connections

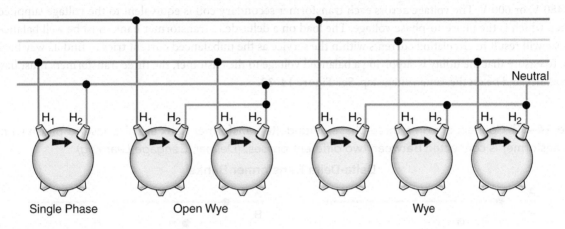

Single Phase   Open Wye   Wye

Wye-Wye Transformer Banks

A wye-primary–wye-secondary transformer bank can supply 120/208 V, 240/416 V, 277/480 V, or 347/600 V services. The phase-to-phase voltage is 1.73 times the phase-to-neutral voltage. The voltage across each transformer coil is equivalent to the phase-to-neutral voltage. **See Figure 14–31.**

The primary neutral must be connected to the secondary neutral and grounded in a wye-wye transformer bank. The expression used is, "Wye-wye you must tie." This neutral connection provides a path for any fault current or current from an unbalanced load to get back to the source. There is a potentially lethal voltage (third harmonic) between the primary and secondary neutrals if they are not connected together.

Figure 14–31 Connections on a three-phase wye-wye transformer bank are all phase-to-neutral, with the primary and the secondary neutrals grounded and tied together. *(Delmar, Cengage Learning)*

Wye-Wye Transformer Banks

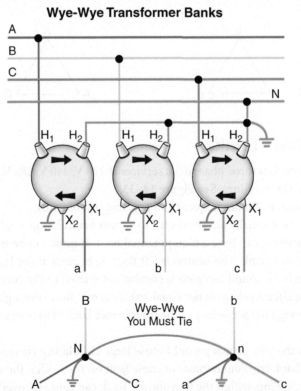

Delta-Delta Transformer Banks

The delta secondary can be connected in two ways, which changes the angular displacement of the secondary, as explained later in this chapter. A delta-primary–delta-secondary transformer bank supplies three-phase power at 120 V, 240 V, 480 V, or 600 V. The voltage across each transformer secondary coil is equivalent to the voltage supplied to the customer, which is the phase-to-phase voltage. The load on a delta-delta transformer bank must be well balanced. Any imbalance will result in circulating currents within the service as the unbalanced current tries to find its way back to the source. To ensure that the utility is supplying a balanced voltage to the customer, the three transformers must have similar impedance and be on the same voltage tap. **See Figure 14–32.**

Figure 14–32 Connections on a three-phase delta-delta transformer bank are all phase-to-phase so that each transformer is connected between two different phases. *(Delmar, Cengage Learning)*

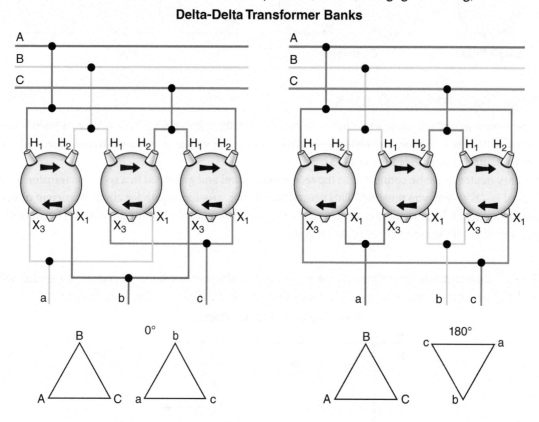

Delta-Delta Transformer Banks

Wye-Delta Transformer Banks

A wye-delta transformer bank supplies three-phase delta services at 120 V, 240 V, 480 V, or 600 V. A wye-delta bank must have transformers with insulated H_2 bushings. **See Figure 14–33.**

Wye-delta banks can be a source of more frequent and puzzling trouble calls.

The H_2 bushings are banked but usually not connected to system neutral or grounded. The neutral connection is left floating (ungrounded) and, therefore, can have a high potential on it. It must not be treated like a grounded neutral by anyone working on the transformer bank. The neutral is left floating because if the H_2 bushings were connected to the system neutral, the transformer bank would carry extra current not related to the current needed to supply the normal service load. If the primary wye circuit voltage is not equal, extra current flows through the delta secondary as it tries to balance the primary voltage through the secondary of the transformer bank. This extra current can overload the secondary and cause a burnout.

Any voltage unbalance on the wye primary would cause large circulating currents in the delta secondary. When one phase feeding a wye-delta bank trips out because of some fault on the feeder, the result would be a very large voltage unbalance; and circulating currents within the transformer bank can cause an overload to the transformer resulting

Figure 14–33 A wye-delta transformer bank has standard wye connections to the primary and standard delta connections to the secondary. *(Delmar, Cengage Learning)*

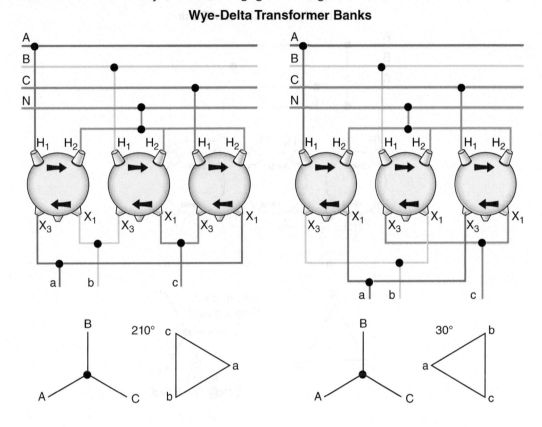

in a burnout and/or blown fuses. Because of backfeed other customers on the same primary feeder will experience low voltage, which can cause electric motors to burn out.

A wye-delta transformer bank can easily be converted to an open-wye/open-delta transformer bank. If one transformer is bad, it can be isolated and then by connecting the H$_2$ of the bank to the system neutral the bank becomes an open-wye/open-delta bank. The two energized transformers continue to provide three-phase power but are subject to burnout because of overload, as discussed later in the chapter.

Two transformers now carry the load normally supplied by three transformers. This arrangement has a capacity of 57.7% of the capacity of three transformers, or 86.6% of the two remaining transformers.

Certain utilities, however, require the primary neutral to be grounded. For example, at least one utility requires the primary neutral to be grounded if the single-phase load is more than two thirds of the three-phase load. This is because if one transformer is lost due to an open phase on the primary feeder or the fuse of one transformer fails, then the transformer bank automatically turns into an open-wye/open-delta transformer bank, keeping the power on as discussed earlier.

Delta-Wye Transformer Banks

A delta-primary–wye-secondary transformer bank can supply standard three-phase wye services such as 120/208 V and 277/480 V. The secondary neutral should be well grounded because a primary system neutral is unavailable. **See Figure 14–34**.

Open-Wye/Open-Delta and Open-Delta/Open-Delta Transformer Banks

A three-phase delta service can be supplied with two single-phase transformers, but three wires are needed: A wye primary would need two phases and a neutral, and a delta primary would need three phases to feed an open secondary service. This type of service is called an open delta because the delta configuration is missing one side—that is, using two transformers instead of three, preventing it from being a closed loop. An open delta transformer bank is typically used where there is a small three-phase load on an otherwise single-phase service. **See Figure 14–35**.

Figure 14–34 The secondary neutral on a delta-wye transformer bank must be connected to a good low-resistant ground. *(Delmar, Cengage Learning)*

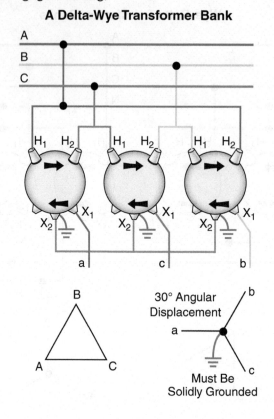

A Delta-Wye Transformer Bank

Figure 14–35 Two transformers can be used to supply a three-phase delta service using an open-wye/open-delta transformer bank. *(Delmar, Cengage Learning)*

Open-Wye/Open-Delta Transformer Connections

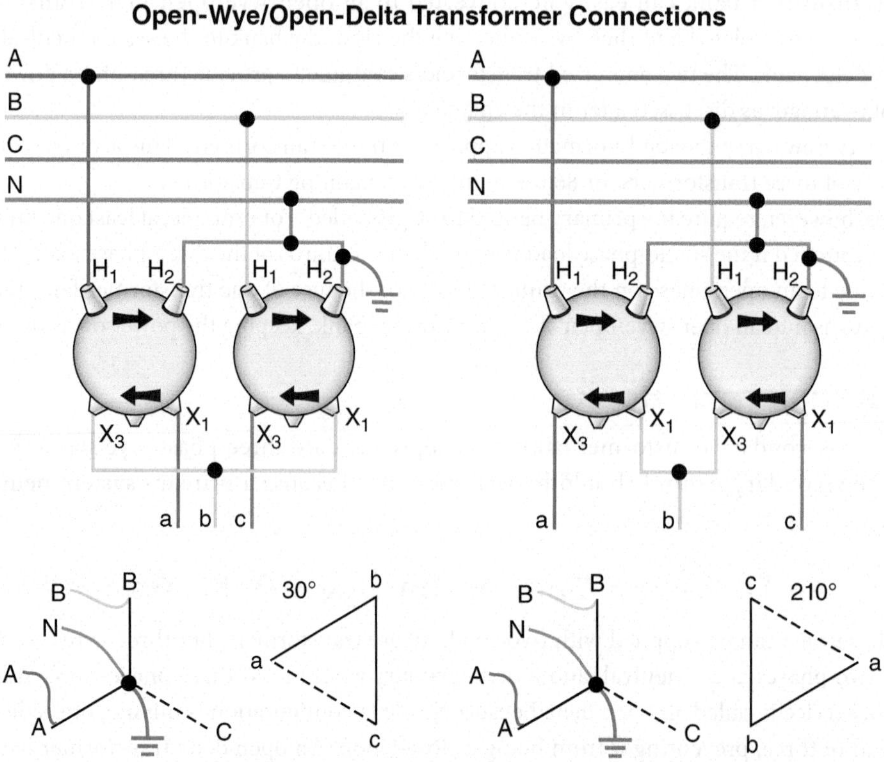

The figure shows two ways that the delta secondary can be connected, which changes the angular displacement as explained later in the chapter. The neutral must be grounded and will carry the full-load current. If the connection to the neutral is opened while the bank is energized, then a high voltage appears across the open and the customer's voltage is reduced to 86.6%. An open-delta/open-delta transformer bank is less common. The delta primary must be three phases. Notice that there are two ways that the delta secondary can be connected, which changes the angular displacement, as explained later in the chapter. **See Figure 14–36.**

Figure 14–36 Two transformers can be used to supply a three-phase delta service using an open-delta/open-delta transformer bank. *(Delmar, Cengage Learning)*

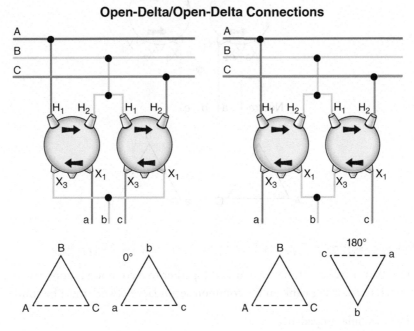

Open-Delta/Open-Delta Connections

Providing an open delta secondary is sometimes used as an economical way to feed a small three-phase delta load. One of the two transformers is often called the lighting (lighter) transformer and is sized larger to feed the single-phase portion of the load. The lighting transformer X_2 bushing is grounded to provide a 120 V phase-to-neutral voltage. **See Figure 14–37.**

The smaller transformer is often called the power transformer and is there to help provide the three-phase load, typically a three-phase motor supplying an air conditioner or water pump.

An open-delta secondary provides three-phase power but the capacity of the bank is reduced. Two transformers now carry the load normally supplied by three transformers. This arrangement has a capacity of 57.7% of the capacity of three transformers, or 86.6% of the two remaining transformers. For example, two 100 kVA transformers in an open delta bank are 100% loaded when they supply $0.866 \times 100 \times 2 = 173$ kVA.

When one transformer of a normal three-phase delta-delta or wye-delta transformer bank is found to be defective, the connections can be changed to an open-wye/open-delta or open-delta/open-delta configuration which restores the service as an open delta. The customer should be told to reduce demand on the service until the transformer is replaced because the two good transformers will now only have the capacity to supply 57.7% of the capacity of three transformers.

Figure 14–37 An open-delta/open-delta transformer bank can supply a 120 V lighting load. *(Delmar, Cengage Learning)*

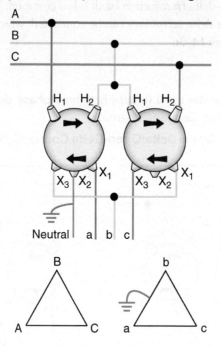

Open-Delta/Open-Delta with Lighter

Connecting Three-Phase Transformer Banks in Parallel

Three-phase transformer banks are sometimes networked together to a common secondary to add extra capacity and security to the service. Each transformer bank on the common secondary network must be similar.

- Each bank must have a similar impedance.
- Each bank must be on the same voltage tap setting.
- Each bank must have the same angular displacement or phase shift.

Angular Displacement of Wye-Delta and Delta-Wye Transformer Banks

A wye secondary from a wye-wye transformer bank cannot be connected in parallel with a wye secondary from a **delta-wye** transformer bank. Similarly, a delta secondary from a delta-delta transformer bank cannot be connected in parallel with a delta secondary from a wye-delta transformer bank. This is because there is an angular displacement or phase shift between the primary and secondary of some types of three-phase transformer banks. **See Figure 14–38.**

Notice that "A" primary phase and "a" secondary phase are both at 8 o'clock; "B" primary phase and "b" secondary phase are both at 12 o'clock; and "C" primary phase and "c" secondary phase are both at 4 o'clock. In other words, a wye-wye bank has 0° angular displacement between the primary and the secondary because all three phases are going through their voltage cycle at the same time. This is also true for a delta-delta transformer bank. However, the wye secondary of a wye-wye transformer bank and the wye secondary of a delta-wye transformer bank cannot be connected together in parallel. **See Figure 14–39.**

There is a 30° angular displacement or phase shift between the primary and the secondary of the delta-wye transformer bank. This means that a wye secondary of a wye-wye bank, which has a 0° angular displacement, cannot be connected in parallel with a wye secondary of a delta-wye bank, which has a 30° angular displacement.

Going back to the drawings of the different transformer banks in this chapter, the angular displacements between the primary and the secondary will be either 0°, 30°, 180°, or 210°.

Figure 14–38 This wye-wye transformer bank has a 0° angular displacement and can be connected in parallel with another wye-wye transformer bank. *(Delmar, Cengage Learning)*

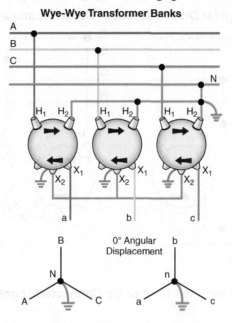

Figure 14–39 Although both secondaries are wye, they cannot be connected in parallel because they have a different angular displacement. *(Delmar, Cengage Learning)*

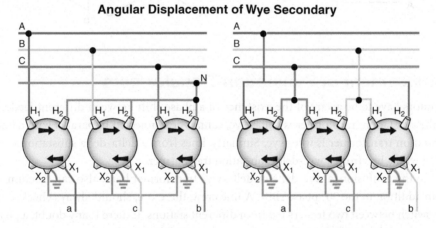

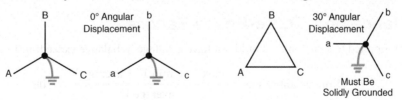

The angular displacement between the primary and the secondary of two different transformer banks must match in order to be able to parallel them together. For example, a delta-delta bank, which, depending on how the secondary is connected, can have a 0° or 180° angular displacement and cannot be connected in parallel with a delta secondary of a wye-delta bank, which has a 30° angular displacement. **See Figure 14–40.**

There is no way a line crew can switch secondary connections around to allow transformer banks with different angular displacements to be connected in parallel.

Figure 14–40 Although both secondaries are delta they cannot be connected together in parallel because they have a different angular displacement. *(Delmar, Cengage Learning)*

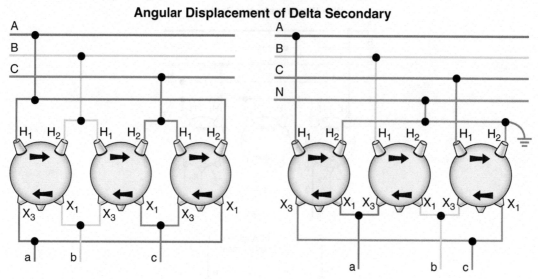

Angular Displacement of Delta Secondary

These Delta Secondaries Cannot Be Connected Together in Parallel

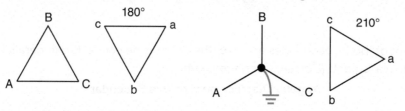

Angular Displacement of Substation Transformers

Control room operators have to be aware of the existence of a phase shift between different feeders. A line fed from a substation, where the substation transformer is delta-wye, cannot be connected in parallel with a line fed from a substation where the substation transformer is wye-wye. Similarly, lines from a delta-delta substation transformer cannot be connected in parallel with a line from a wye-delta substation transformer.

To add further to the confusion, a wye-delta or delta-wye transformer bank also can be connected so that there is a 180° phase shift in addition to the 30° phase shift. A line crew, therefore, should always check with operating control before closing a tie switch between two feeders fed from different stations. If there is any doubt, a phasing set can be used to verify that there is a no voltage difference across an open-tie switch before closing it.

Voltage Imbalance on a Three-Phase Service

A three-phase service supplied to a customer should not have a voltage imbalance exceeding 1%.

$$\%\text{Voltage unbalance} = \frac{\max V \div \min V - \text{average } V}{\text{average } V} \times 100$$

Where:

Average V is the average of the three voltages

Maximum and *minimum V* are the voltages with greatest difference from average

Any imbalance in a delta secondary will result in circulating currents within the service as the unbalanced current tries to find its way back to the source. To ensure that the utility is supplying a balanced voltage to the customer, the three transformers must have similar impedance and be on the same voltage tap.

An unbalanced secondary voltage can be caused by any of the following:

- An unbalanced customer load
- An unbalanced primary voltage
- Banked single-phase units with different voltage tap settings
- Banked single-phase units with different impedances

Three-Phase, Secondary-Voltage Arrangements

The secondary voltage from a three-phase transformer bank depends on more than just the transformer ratio. Three transformers with a given ratio can be connected to provide up to four different types of services. The secondary voltage is based on whether the transformer secondary is connected as wye or delta. The secondary phase-to-phase voltage is:

- Equal to the actual voltage across the transformer coil *with a delta connection*
- Equal to 1.73 times the voltage across the transformer coil *with a wye connection*

The secondary voltage is also dependent on whether a transformer with a center-tapped secondary coil has the two parts of the coil inside the tank arranged in parallel or in series. The output voltage of series-connected coils is double the output of two parallel-connected coils.

Services Available

There are many three-phase voltages available from common distribution transformer secondaries in North America. **See Figure 14–41.**

Figure 14–41 Standard North American Three-Phase Voltages

Transformer Secondary	Type of Three-Phase Service	Coil Arrangements Inside the Tank	External Configuration
120/240	120/208	Parallel	Wye
	240/416	Series	Wye
	120	Parallel	Delta
	240	Series	Delta
240/480	240/216	Parallel	Wye
	240	Parallel	Delta
	480	Series	Delta
277	277/480	NA	Wye
347	347/600	NA	Wye
600	600	NA	Delta

In much of the world, the three-phase voltages available to the customer equal the standard voltage in that country $\times \sqrt{3}$ or 1.732.

For example, the following are common three-phase voltages.

North America: $120 \times 1.732 = 208$ V

Europe: $220 \times 1.732 = 380$ V

Other countries: $240 \times 1.732 = 416$ V

Three-Phase Voltages from a 120/240 V Secondary

Three single-phase transformers with a 120/240 secondary can be banked together to supply a three-phase 120/208 or a three-phase 240 V service.

1. A *three-phase, 120/208 V service* has the secondary coils in each transformer internally connected in parallel to provide 120 V output. **See Figure 14–42.**

Figure 14–42 The secondary coils inside the transformer are connected in parallel to provide a three-phase 120/208 V wye service. *(Delmar, Cengage Learning)*

Three-Phase 120/208 Volt Wye Service

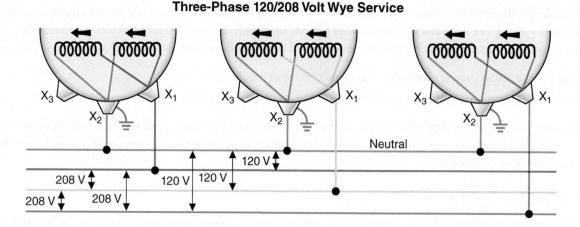

The external secondary leads of the three transformers are connected into a wye configuration. Each phase-to-neutral voltage is 120 V, and the phase-to-phase voltage is 120 × 1.732 = 208 V. Large residential and commercial buildings are often fed at 120/208 V.

2. A *three-phase 240 V* service has the internal secondary coils of each transformer connected in series to provide 240 V. **See Figure 14–43.**

The center tap (X_2) is left ungrounded. The external secondary leads of the three transformers are connected in a delta configuration, and 240 V are available from phase-to-phase. A phase-to-ground voltage reading would be 0 V

Figure 14–43 The secondary coils inside the transformer are connected in series and the X_2 terminal is not grounded to provide a three-phase 240 V delta service. *(Delmar, Cengage Learning)*

Three-Phase 120/208 Volt Delta Service

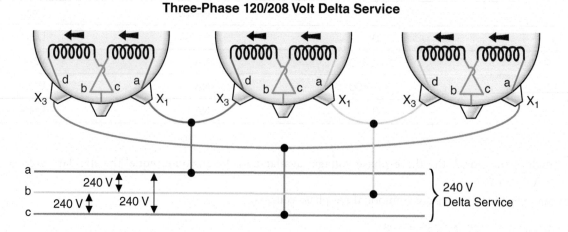

because there is no path or circuit back to the ungrounded delta. If there was a voltage from phase-to-ground, it would mean that there was a phase-to-ground fault somewhere or that there was a lighting service from one of the transformers. A phase-to-ground fault would mean that the earth was at 240 V in relation to each of the two other phases. One phase-to-ground fault would not normally blow a fuse because there is no path for current to flow through earth back to the source.

3. A *four-wire (lighting) service is available from a 240 V delta secondary.* Some customers fed from a 240 V three-phase delta service still need a 120 V single-phase lighting source. To get a 120/240 V lighting load, from a 240 delta secondary, the center tap (X_2) of one transformer secondary coil is grounded. The ground *must* be removed from the X_2 terminals of the other two transformers. **See Figure 14–44.**

Figure 14–44 Connecting the X_2 terminal to ground in one transformer provides a four-wire (lighting) service from a 240 V delta secondary. *(Delmar, Cengage Learning)*

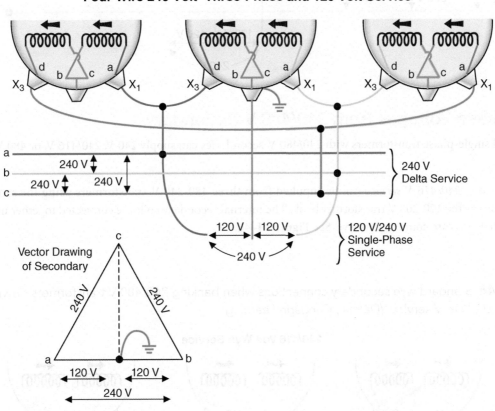

There will be 120 V phase to neutral available from each of the two phases connected to the grounded transformer. The remaining secondary phase is sometimes called a *wild phase* and has a phase-to-neutral voltage of about 208 V. The lighting transformer will have a higher kVA rating because it will have both the single-phase lighting load and the three-phase load.

4. A *five-wire service is available from a 120/208 V wye secondary.* A standard single-phase, 120/240 V lighting service can be provided from a three-phase, 120/208 V service. The internal secondary coils in the center transformer are left in series to provide a 120/240 V supply. **See Figure 14–45.**

The kilovolt-ampere rating of the center transformer is usually increased to handle its share of the three-phase power load and the single-phase 120/240 V lighting load.

Figure 14–45 Connecting the secondary coils in series in one transformer provides a five-wire service from a 120/208 V wye secondary. *(Delmar, Cengage Learning)*

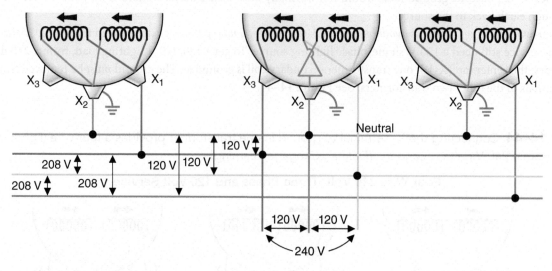

Three-Phase Voltages from 240/480 V Secondary

Three banked single-phase transformers with 240/480 V secondaries can supply 240 V, 240/416 V, or 480 V three-phase services.

1. A *three-phase, 240/416 V service* can be supplied from three 480/240 V transformers using the same secondary connections as the 120/208 V transformer bank. The internal secondary coils are connected together in parallel, and the external leads are connected as wye. **See Figure 14–46.**

Figure 14–46 Standard wye secondary connections when banking 240/480 V transformers provide a three-phase 240/416 V service. *(Delmar, Cengage Learning)*

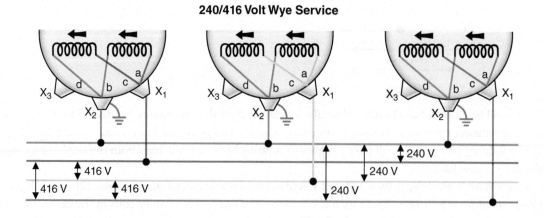

2. A *three-phase, 480 V service* is supplied from a 480 V secondary. The secondary coils are connected in series inside the tank, and the external leads are connected in a delta configuration. The phase-to-phase voltage is 480 V. A 240 V single-phase supply could be made available by grounding the center point on one transformer to create a neutral for the 240 V load. **See Figure 14–47.**

Figure 14–47 Standard delta secondary connections when banking 240/480 V transformers provide a three-phase 480 V delta service. *(Delmar, Cengage Learning)*

480 Volt Delta Service

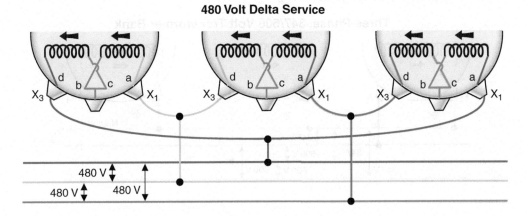

Three-Phase Voltages from a 277 V Secondary

Three single-phase transformers with a 277 V secondary coil can supply a 277/480 V three-phase service. These transformers have one secondary coil and two secondary bushings. A three-phase, 277/480 V transformer bank has the secondary connections in a wye configuration. Each phase-to-neutral voltage is 277 V, and each phase-to-phase voltage is $277 \times 1.73 = 480$ V. **See Figure 14–48.**

Figure 14–48 Wye secondary connections when banking 277 V transformers provide three-phase 480/77 V wye service. *(Delmar, Cengage Learning)*

277/480 Volt Wye Service

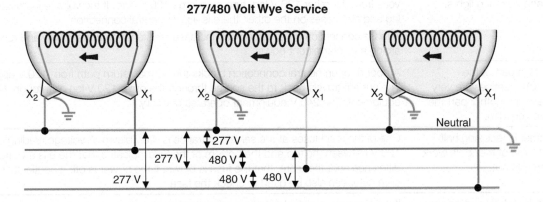

Three-Phase, 347/600 V Transformer Bank

Three single-phase transformers with a 347 V secondary coil can supply a 347/600 V three-phase service. The three-phase, 347/600 V transformer bank has the secondary connections in a wye configuration. Each phase-to-neutral voltage is 347 V, and the phase-to-phase voltage is 347 × 1.732 = 600 V. **See Figure 14–49.**

Figure 14–49 Wye secondary connections when banking 347/600 V transformers provide three-phase 347/600 V wye services. *(Delmar, Cengage Learning)*

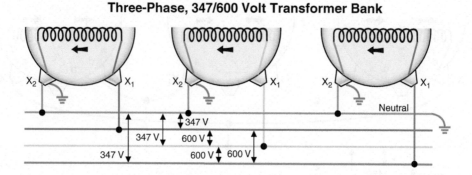

Three-Phase, 347/600 Volt Transformer Bank

Troubleshooting Transformers

Troubleshooting transformers mainly involves checking for problems on secondary services. Many *no-power* or *partial power* calls are due to internal customer problems. A utilities responsibility normally ends at the service entrance. Opening the main switch and taking a voltage reading will determine whether the utility is the cause of the problem.

Troubleshooting a Single-Phase, 120/240 V Service

If	Then
A customer has intermittent power and flickering lights.	Check for a voltage imbalance in the two 120 V legs. If one 120 V leg varies two or more volts from the other 120 V leg, turn on a large 120 V load. If the voltage increases on one leg and decreases on the other, there is a poor neutral connection.
	A loose connection on either leg can increase resistance to current flow and result in arcing and intermittent power.
The lights in part of the customer's premises are very bright, and in another part the lights are very dim.	A poor or open neutral connection blocks the normal return path from 120 V appliances. The current travels back to the source through the other 120 V leg or through 240 V equipment. The 240 V equipment operates normally.
A customer is receiving half power. Some of the lights work and some do not.	One of the main fuses at the service entrance is likely blown. A voltage reading of about 120 V between the top and the bottom of the fuse indicates that there is a voltage difference and, therefore, that the fuse is blown. If the power is off, a continuity tester or ohmmeter can also be used to check the fuse.
None of the 240 V appliances work.	If the fuses are good, check the connections back to the transformer.
A customer complains about erratic or low voltage.	A poor connection could cause low voltage at various times. Heavy-duty equipment used by a neighbor on the same bus can cause voltage problems for others on the bus.

Troubleshooting a Wye Secondary (120/208 V, 240/480 V, or 347/600 V) Service

If	Then
A customer has an abnormal voltage.	Open the customer's switch and check the voltage. If the voltage on a phase-to-neutral reads 0, then check for a defective transformer or an open phase on the primary feeder.
	If the voltages on all three phases are balanced, close the customer switch and check the voltage. An unbalanced voltage indicates the customer's load is unbalanced.
Three-phase motors are overheating, and the thermal overload protection trips out the motor.	The usual cause of trouble on one phase is an unbalanced load. This type of service has single-phase loads. Unbalance beyond 3% can cause the low-voltage leg to draw more current and create heating of equipment.
	A faulty winding in a three-phase motor can cause a high current on a phase.
The load is balanced at the service entrance, but there is still a problem.	Open the customer's main switch and verify that the utility-supply voltage is correct.
	Close the customer's circuit breakers one circuit at a time and take voltage and current readings. A suspect circuit will have an unbalanced voltage or abnormal current readings for the equipment being fed.
All three-phase equipment and some single-phase equipment will not operate.	One phase is probably out. A phase-out can be due to a feeder problem, a transformer problem, or a blown fuse at the service entrance.
The single-phase equipment has intermittent power. The three-phase load is operating normally.	A poor or broken neutral connection will cause problems with single-phase loads and can be very difficult to find.
	A customer-owned secondary (dry) transformer that steps down the higher 600 or 480 V supply to 120/208 V can be the cause of problems.

Troubleshooting a Delta Secondary (600 V, 480 V, or 240 V) Service

If	Then
The customer has abnormal voltage. With the customer's switch open, the voltage on the supply side shows one or more of the phase-to-phase voltages at 0 V.	One or more phases feeding the customer is out of service. Check the transformers and the primary feeder.
	The customer's switch must be open for this voltage check because the backfeed in a delta service can show a voltage at the service entrance.
At the main panel, the phase-to-phase voltage readings are lower than normal and one phase-to-phase voltage reading is 0 V. For example, on a 240 V service, the readings are 210 V, 210 V, and 0 V.	When one phase of a primary feeder loses power, the voltage at a wye-delta transformer bank will have a reduced voltage on two of the phase-to-phase readings and 0 V on the third phase-to-phase reading.
	On a wye-delta transformer bank, the primary neutral is not connected to the system neutral or the secondary neutral, but depending on the wishes of the utility may or may not be left ungrounded or floating. If the transformer neutral was grounded and one primary phase is opened, the transformer bank would become an open-delta bank and continue to supply three-phase power at a 57% reduced capacity, exposing the bank to a burnout because of an overload.
At the main panel, all three phase-to-phase voltage readings are normal, but one phase to ground is reading very low or 0 V.	A ground fault exists on a phase. When one phase of a delta circuit is faulted to ground, it does not blow a fuse because there is no path for a return current to the source. Both the phase and the ground are energized; therefore, a faulted phase-to-ground voltage reading will show a very low or 0 V difference.
	Both the faulted phase and earth have a difference in potential in relation to the other two phases.
	A customer with a delta service often has a ground fault indicating light connected between each phase and ground. The light will go out when there is a ground fault and the ground becomes energized.

Working on a Voltage Conversion

The earliest distribution system conversions in North America were converting the 2,400 V delta circuits to 4.16/2.4 kV wye (4,160 phase-to-phase and 2,300 phase-to-ground). As loads increased, the old 4.16/2.4 kV circuits were converted to 8.32/4.8 kV, 12.5/7.2 kV, and 13.8/8 kV. Since then these circuits are being converted to 25/14.4 kV and 34.5/19.9 kV.

Preparing the Circuit for Voltage Conversion

Conversion of distribution systems to higher primary voltages, such as 25 kV, 27.6 kV, and 34.5 kV, makes it possible to serve a lot more customers on existing lines. One substation with 34.5 kV feeders can make it possible *not* to have to build six or more substations with lower-voltage feeders. The higher-voltage distribution also saves on restringing, voltage regulators, capacitors, and downstream reclosers.

Work that can be done on the circuit *prior to the day of the voltage conversion* includes the following:

1. If going to a higher voltage, change out all insulators on the circuit where that has not been done previously during prebuilding projects.
2. Change out any switchgear to the type that can handle the duties for both voltages.
3. If going to a higher voltage, change out any surge arrestors that are on the circuit. The utility will have to decide if the risk of having the higher-voltage arrestors on the circuit for a period of time is acceptable.
4. Primary underground cable that cannot accommodate the new voltage must be replaced or fed at the original voltage using a primary step-down transformer.

Work to be done *on the day of the voltage conversion* includes the following:

1. If going to a higher voltage, existing voltage regulators will likely not be needed, especially in the same location and they can be removed from service.
2. If going to a higher voltage, existing capacitors will likely not be needed, especially at the same location and can be removed from service.

Working at Transformer Installations

Work that can be done at transformer installations *before the day of the voltage conversion* includes the following:

1. Install dual-voltage transformers. There may not be any dual-voltage transformers for 277/480 V or 347/600 V three-phase banks.
2. Install surge arrestors for the new voltage.
3. Install cutouts and fuses for the new voltage.
4. If needed, install current-limiting fuses for a higher-voltage distribution system.
5. Make preparations if existing wye-delta transformer banks have to be changed to wye-wye on conversion day to accommodate single primary bushing transformers. The customer service entrance must be upgraded to accept a neutral.

Work to be done *on voltage conversion day* includes the following:

1. Isolate all transformers from the circuit and energize the circuit at the new voltage.
2. Change out transformers that are not dual voltage.
3. Change the tap setting on the dual-voltage transformers to the new voltage.
4. If the circuit has been energized at the new voltage, energize the transformer and take voltage checks and phase rotation checks before the customer applies load (remove meter or open customer breaker).
5. If, however, it is preferred to energize the new higher-voltage surge arrestors from a remote location, continue to work under the general outage and attach the riser and arrestor to the circuit. Leave the customers disconnected and do voltage checks at each transformer after the circuit is energized.

6. If a transformer is to be fed from a primary step-down transformer, ensure there is proper voltage and phase rotation before connecting customers.

7. Using a checklist for these operations will reduce the risk of forgetting to change the tap on a transformer and leaving a customer with a very high and damaging voltage.

Working at Underground Installations

Work that can be done at underground installations *before the day of the voltage conversion* includes the following:

1. If using existing cable at the new voltage, cable size may not match the bushing inserts of new switching cabinets or transformers. Nonstandard equipment may be needed.

2. If a new higher-voltage cable is installed, larger elbows (separable connectors) before conversion day may not match existing equipment. Nonstandard equipment may be needed.

3. Install dual-voltage transformers or plan a change-out during the outage.

4. A higher-voltage distribution system may need current-limiting fuses.

5. Replace switchgear and fusing to accommodate the new voltage.

6. Change out surge arrestors at riser poles and any elbow arrestors installed at dual-voltage transformers.

Work to be done *on voltage conversion day* includes the following:

1. Change the tap setting on the dual-voltage transformers to the new voltage.

2. If the circuit has been energized at the new voltage, energize the transformer and take voltage checks and phase rotation checks before the customer applies load.

3. If the transformer is to be fed from a primary step-down transformer, ensure that there is proper voltage and phase rotation before connecting customers.

Working at Primary Step-Down Transformer Installations

During a voltage conversion project, primary step-down transformers are often installed in line locations to delay the conversion of some line sections, especially underground. **See Figure 14–50.**

Figure 14–50 Inserting a three-phase step-down transformer allows completion of the voltage conversion downstream at another time. *(Delmar, Cengage Learning)*

Work that can be done at underground installations *before the day of the voltage conversion* includes the following:

1. Do a turns ratio test on the primary step-down transformer because setting the external switch or tap does not always make the proper internal connection.

Work to be done *on voltage conversion day* includes the following:

1. Reduce the risk of causing damage to a customer's appliance by isolating all the transformers on the load side of the primary step-down transformer and doing a voltage test and phase rotation tests on an individual basis as the transformers are energized.
2. Make a clear separation at tie points between the two primary voltage levels.

Other Transformer Applications

In addition to transformers that are used for stepping up or stepping down voltage, some transformers in an electrical system are used for other purposes, including the following:

- A *neutralizing transformer* is one that is used where telecommunications (telephone) cable enters a substation. The grounded network in a substation is subject to a voltage rise when a fault occurs in the electrical system. Any communications cable coming in from outside the station would be seen as a remote ground, and there would be a dangerous potential difference between the communications cable and anything attached to the grounded network in the substation. The neutralizing transformer is a 1-to-1 ratio transformer that removes the direct connection between the cables in the substation from the cables leaving the substation. Fiber-optic communications cable will eliminate the need for these specialty transformers, because fiber-optic cable does not transmit electricity and will not propagate voltage and current from a station.

- A *grounding transformer* is used as an indirect way to ground one phase of a delta circuit. The grounding transformer provides the means for a phase-to-ground fault to get back to the source. A ground-fault relay senses this ground current and trips out the circuit.

- *Instrument transformers* are potential transformers and current transformers that reduce voltage and current to lower, manageable levels. The lower voltage and current represent the primary voltage and current at a given ratio for use in relays and metering.

- A *constant-current transformer* is used for applications such as series street lighting where the current must be kept constant and the voltage is allowed to fluctuate.

- A *voltage regulator* is a tapped **autotransformer** that can regulate voltage under loaded conditions.

- A *compensator starter* is a tapped autotransformer that is used to soft-start large induction motors. Soft-starting a large motor will prevent a voltage flicker for other customers on the circuit.

- An autotransformer is an exception to the transformer with two or more coils, as described earlier in the chapter. The autotransformer is a single winding transformer. The primary voltage is distributed over the length of the coil. Tapping into a correct portion of the coil provides a desired lower voltage.

- Autotransformers are most efficient when the high- to low-voltage ratio is 3 to 1 or less. Applications are found in high-voltage substations (e.g., to transform from 240 kV to 120 kV and 230 kV to 500 kV step up) and by industrial and commercial customers supplied at 600 or 480 V as a means to step the voltage down to utilization voltage such as 120 V. **See Figure 14–51.**

Autotransformers are not used on distribution voltages anymore because a breakdown of insulation can result in the customer receiving the full primary voltage.

Figure 14–51 The coil in this autotransformer has 480 V across it but a tap in the coil allows this autotransformer to feed a 120 V service. *(Delmar, Cengage Learning)*

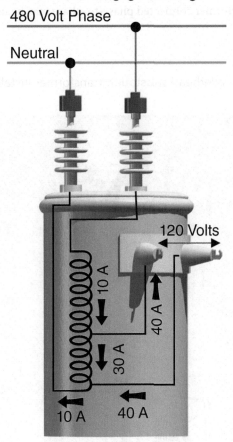

European Secondary Systems

Around the world there are two main ways that the utilization voltages are fed to the customer. The two systems tend to be called European or American, although most people worldwide use the European distribution system.

The European distribution layout uses a large 300 to 2,000 kVA transformer, typically a delta-wye transformer to feed a long length of four-wire three-phase 220/380 or 230/400 or 240/416 bus feeding many customers. There is a move to make a 230/400 V secondary standard. The transformer installation, often called a substation, can be overhead. **See Figure 14–52.**

On the primary side there is a three-phase gang-operated switch and three power fuses to protect the transformer. The transformer (substation) is more often found in small rectangular but tall buildings. **See Figure 14–53.**

One three-phase transformer is used to feed about eight times the length of typical single-phase secondary bus used in the American system for a given load and for the same voltage drop limitation. The bus can be up to 1 mile (1.5 km) long because a three-phase bus can feed twice as far as a single-phase bus for a given load and with a utilization voltage 230/400 V, double the American utilization voltage. The bus can feed four times the distance for a given load.

A service feeding a residence is typically two wires with a 230 V to neutral service. A service cutout is installed before the revenue meter. A three-phase service would be a four-wire 400 V phase-to-phase or 230 V phase-to-neutral.

In rural areas, a single-phase transformer connected phase-to-phase may be used to supply a 230 V phase-to-neutral service.

Figure 14–52 A European system overhead substation transformer installation feeding a 230/400 V bus. *(iStock.com/Linleo)*

Figure 14–53a A European system substation. Primary overhead is coming in from the right and a secondary circuit exiting on the left. The primary switchgear, transformer, and secondary breakers are found in the building. *(Delmar, Cengage Learning)*

Figure 14–53b A gang operated primary switch is used to isolate the transformer. There are indications on the wall that this solid blade switch was opened under load. *(Delmar, Cengage Learning)*

Figure 14–53c The transformer primary in the building is connected with elbows. (Delmar, Cengage Learning)

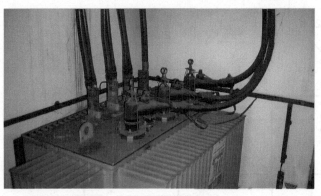

Figure 14–53d These secondary breakers in the European system substation feed a long length of secondary. *(Delmar, Cengage Learning)*

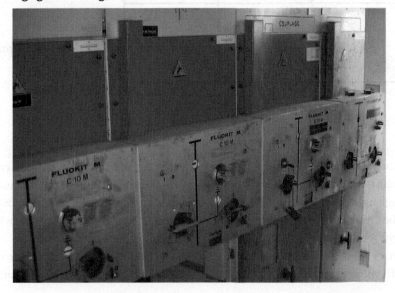

Urban Secondary Network Grid

Customers in the downtown core of many cities are supplied with an underground secondary network grid system. Over the years this has been the most reliable method to feed customers. A failure of a source primary circuit or a transformer does not cause customers to lose power. **See Figure 14–54.**

Figure 14–54 An urban secondary network can have many transformers feeding into one very large secondary bus. *(Delmar, Cengage Learning)*

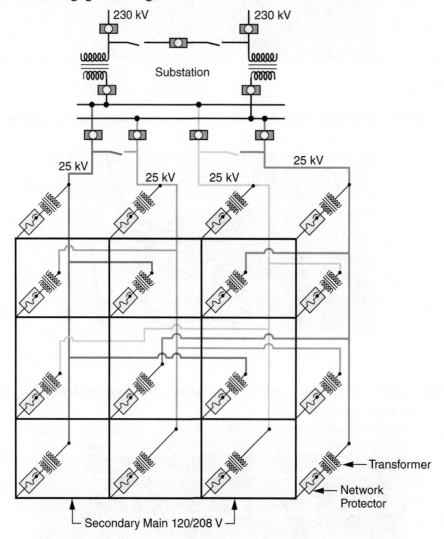

More than one primary feeder feeds a number of network transformers connected together in parallel to feed one large secondary bus (main). Typically the primary feeders come from the same substation, because when two substations are tied together through a secondary main there can be excessive current flow when the substation voltage or phase angle are not exactly the same. In case of a failure at the substation, a primary network will switch the network transformer to another supply. A city may have several separate secondary networks.

Network transformers are three-phase units and can be 300 to 2,500 kVA. The transformer secondary is most often 216Y/125 V (which with voltage drop supplies customers at approximately 208Y/120 V) but can be 480Y/277 V or any other voltage required by the planners. There can be as many as 1,000 transformers feeding one secondary bus. The network transformer has a primary breaker and fuse often monitored by supervisory control and data acquisition (SCADA). Transformers used in this system are designed with the primary circuit breaker connected to the tank on one side and the network protector connected to the tank on the other side. Network transformers are also intentionally designed with a higher impedance so that the explosiveness of a secondary fault will be reduced.

Network protectors are enclosed in a tank bolted to the network transformer tank. The network protector is an electrically operated low-voltage circuit breaker that is installed to protect the secondary network, not the transformer. A reverse current relay in the network protector is designed to detect the reverse current flow (backfeed) and then open. The network protector opens to protect the secondary (bus) from problems that might occur on the source side (network transformers and primary feeders). If, for example, a network transformer developed a short circuit and there was no network protector, then all the available current in the network bus fed from all the other transformers would feed into the fault.

The secondary bus (network main) conductor has to be very large and can be found with four paralleling cables of 500 kcmil copper per phase. The network cables are joined at each intersection. The fault current available on the secondary grid can be as high as 200,000 A, which explains why there are occasional large explosions under a city street. Limiters, which are high-capacity fuses, are installed on the cable at junction points. The limiters are designed and fused to blow before the cable insulation is damaged by the heat from excessive current feeding a fault. The limiter will blow and isolate a faulted section of cable from the network. There may be networks without limiters on the secondary grid because it is assumed that any short circuit would burn clear very quickly. Service connections from the main are made with a specially designed "crab joint," which allows submersible connections from the large 500 kcmil cable to the much smaller service cables such as #2.

Large customers such as an airport, mall, or any large building in a downtown may have their own secondary bus fed from multiple transformers connected in parallel from more than one primary feeder. When an individual customer used a secondary network to feed large areas, the network is called a spot network instead of a grid network.

Secondary Network (Banked Secondary)

Most utilities feed their secondary bus radially from one transformer. Secondary bus will have bus breaks installed between transformers to prevent electrical interconnection with other transformers. When work is to be done at a transformer, it is good practice to check that there are no jumpers across a bus break. Occasionally a bus break is jumpered out temporarily to prevent an outage to the customer while working on a transformer. On at least one occasion the bus break jumper was not removed after the work was completed. At a much later time a power line tech was replacing the hot-line clamp on the transformer primary and lost both of his arms. Backfeed from the neighboring transformer fed the bus and energized the transformer primary.

There are utilities where you will find more than one transformer feeding a common bus intentionally. **See Figure 14–55.**

Following are advantages to using banked secondaries.

- The load is divided among all the transformers connected to the bus.
- Individual customers with peak loads are supplied by the greater available reserve capacity of the multiple transformers.
- All of the transformers on a common bus are connected to the same phase and connected with the same polarity.
- Customers do not lose power if one transformer fails.
- *When the primary of the transformer is opened, the primary terminal will remain energized because of backfeed from the energized secondary. Remove secondary leads to isolate a transformer completely.*

Figure 14–55 Connecting transformers in parallel to a common secondary increases reliability to the customer but creates backfeed hazards to linemen. *(Delmar, Cengage Learning)*

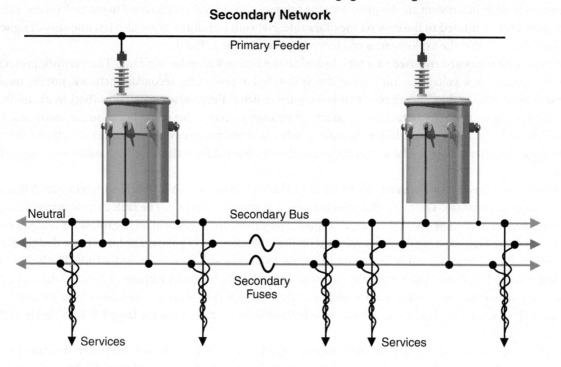

Specific Hazards Working with Transformers

Transformer Backfeed

A power line technician must always be alert to a transformer that is accidentally fed backward and to the fact that a secondary voltage can be stepped up to a full primary voltage at the transformer primary terminal.

Portable generators available at hardware stores are likely sources of transformer backfeed because unqualified people often connect the generators to the service without opening their main breaker. When working on a primary feeder the proper placement of personal protective grounds is the most positive protection. While it is a requirement to de-energize from all sources before working on the line, a generator could be missed or started up after work has started.

When grounds are applied to a primary circuit that is backfed from a portable generator, the generator will keep running and does not short out. The impedance of the circuit between the generator and the grounds is too high to overload or short out the generator.

A power line technician is protected when working on a feeder where protective grounds are installed because the voltage at the grounded location has been lowered to an acceptable level. However, current will be flowing in the conductor, which is why a conductor must always be jumpered before cutting or opening the circuit.

Large, permanently installed generators do not normally create a hazard because they are more likely to be properly installed with double-throw switches so that the generators cannot feed back into the electrical system.

Removal of the transformer secondary leads, when working on a transformer, will ensure that there is no possibility of backfeed.

Sources of secondary backfeed are not always obvious and include the following:

- A portable generator connected into the customer system without opening the customer main switch
- A nearby recreational vehicle (RV) with a generator connected into a home wiring
- An extension cord from a neighboring house
- When working on a transformer that is networked with other transformers to a common secondary bus, and the primary terminal remains energized (if there are no network protectors opened) after the transformer is disconnected from the primary

Following are ways to be protected from transformer backfeed.

- When working on a transformer, remove the secondary leads to ensure that there is no possibility of backfeed into the transformer secondary terminals. Listening or patrolling for any generators is not 100% effective, and generators can be started after work begins.
- Shorting out the secondary with approved grounds can be effective.
- Removing meters from all customers fed from the transformer will eliminate backfeed from customer operations but not from other transformers connected to the same bus.
- A small (<5 kW) portable generator feeding back into a grounded circuit will keep running and not short out. The impedance of the circuit between the generator and the grounds is too high to short out the generator. When working on a primary circuit apply protective grounds. It can still be safe to work on the grounded conductors because the voltage at the grounded location has been lowered to an acceptable level. However, there will be current flowing in the conductor and the conductor must be jumpered before cutting or opening the circuit.

Backfeed from Three-Phase Transformers

When one or two phases trip out on a three-phase line, the tripped-out phases can still be energized because of backfeed from a downstream three-phase transformer. The three phases of a single unit three-phase transformer share a common iron core and any energized phase will induce voltage onto the other phases.

When single-phase units form a three-phase bank then any wye-delta or delta-delta bank will feed back on a tripped-out phase.

All three phases of a faulted circuit should be isolated and grounded to carry out repairs when one or two phases are down.

Potential Explosive Flash at a Transformer Secondary

A step-down transformer can generate a very high current on the secondary side if the secondary wires are accidentally shorted. The magnitude of the fault current available on a transformer secondary is mostly dependent on the size of the transformer. The larger the transformer, the greater the capability to generate a large fault current.

If the secondary wires are shorted, the transformer can briefly carry a load more than 10 times its rating and supply a high fault current to the faulted location. Depending on the impedance of the transformer, a short at the secondary terminals of a 100 kVA can be as high as 18,000 A.

The level of fault current at the location of the short will depend on the distance from the transformer and the conductor size. For work on secondary close to a large transformer, consideration should be given to open the transformer.

Specific Hazards When Working with Transformers

Hazards	Controls
Reenergizing a defective overhead transformer can result in a violent expulsion of molten products from the fuse chamber or, in rare cases, a transformer explosion.	Test the transformer before reenergizing or use precautions such as staying out from under the cutout, using a stick with an attached shield, and/or using an extra length of hot stick. A current-limiting fuse in series with the cutout fuse will reduce the risk of a violent transformer failure in locations where there is a high-fault current capability.
Bayonet-style fuses and current-limiting fuses in a dry-well canister have been known to fail explosively when using them to energize a faulted transformer.	Energize the transformer with a load break elbow or from a remote location to reduce risk.
The magnitude of a fault on the secondary is very explosive when a short circuit is close to the transformer.	When work is required directly at the secondary of a large transformer, consider isolating the transformer. Safety glasses should be worn when working on an energized secondary because an eye can be permanently damaged by a large flash.

(Continued)

Hazards	Controls
Ferroresonance can cause the voltage on a circuit to increase from two to nine times, causing equipment damage. The most common occurrence of ferro-resonance involves a three-phase wye-delta or delta-delta transformer bank fed with a length of underground cable.	Ferroresonance can be prevented by using a three-phase gang-operated switch, or by keeping a load on the transformer when it is being energized or de-energized, or temporarily grounding the floating neutral of the wye primary, which will short out a possible series circuit.
On a wye-delta transformer bank, the H_2 bushings are connected but not connected to system neutral or grounded. The neutral connection is left floating (ungrounded) and, therefore, can have a high potential on it.	It must not be treated like a grounded neutral by anyone working on the transformer bank.
When a single-wire earth return (SWER) system is feeding a transformer, then the pole/down ground is the primary neutral. Opening or cutting the pole/down ground while the transformer is energized can have a primary voltage across the opening.	Because the pole/down ground is a primary wire, always de-energize the transformer to do any work on it.

Review Questions

1. A transformer with a 240 V secondary has a 60-to-1 turns ratio. What is the primary voltage feeding this transformer?

2. To conduct a turns ratio test on a transformer, why is it necessary to energize the high-voltage coil with a low voltage such as 120 V?

3. What two kinds of tests can be carried out on a transformer using a 1,000 V megger?

4. Why is there a greater hazard working on an energized secondary close to a transformer rather than farther away?

5. Can a single-bushing transformer be installed on a delta circuit?

6. A transformer coil must have a potential difference across it to operate. What two types of connections are possible to get a voltage across a transformer coil?

7. What would be the expected full-load current (100% loaded) on the 240 V secondary of a 100 kVA, 7,200/240 V transformer?

8. When a secondary bus is fed from many transformers connected in parallel, what kind of secondary system is it?

9. When two or more transformers are inter connected with the positive terminal of one transformer connected to the negative terminal of another transformer, what kind of interconnection is it?

10. When two or more transformers are connected with the positive terminal of each transformer connected to a phase and all the negative terminals connected together, what kind of connection is it?

11. Why should the primary neutral of a wye-delta transformer bank be left floating (ungrounded)?

12. Can two phases of a delta primary supply a three-phase, open-delta service?

13. Can the wye secondary of a wye-wye bank be connected to the same bus as the wye secondary of a delta-wye bank of the same secondary voltage?

14. A center-tapped 120/240 V secondary coil has the two parts of the coil inside the tank arranged in parallel. What would be the voltage between the X_1 and the X_3 terminals?

15. Name three types of three-phase services that three single-phase units with a 120/240 V secondary can serve.

16. The lights in part of a customer's premises are very bright, and in another part the lights are very dim. What is the likely cause?

17. At the main panel, the phase-to-phase voltage readings of a delta 240 V service are lower than normal and one phase-to-phase voltage reading is 0 V. For example, the readings are 210 V, 210 V, and 0 V. What is the likely cause?

Chapter 15

Supplying Quality Power

After completing this chapter, you should be able to:

1. Define and recognize power quality.

2. Identify signs of disturbances to quality power.

3. Recognize that engineering solutions such as the installation of voltage regulators and capacitors may be needed.

4. Install voltage regulators and perform applicable operations.

5. Take steps to troubleshoot a voltage regulator.

6. Install capacitors and perform applicable operations.

7. Troubleshoot voltage disturbances at the customer's premises.

8. Troubleshoot "no power call" on overhead and underground systems.

9. Troubleshoot high and low voltage problems.

10. Recognize power quality issues such as harmonics, voltage flicker and ferroresonance.

11. Troubleshoot customer complaints experiencing tingle voltage.

12. Troubleshoot customer complaints where power lines causing radio and television interference.

Introduction

More Sensitivity to Power Quality

Electrical disturbances in power lines and on customer premises are to the result of disturbance-producing electrical equipment. The increased sensitivity of certain customer loads is less tolerable to customers. Other than power outages, a variation in voltage is the most noticeable and obvious power-quality problem. Some power-quality problems are not picked up by a voltmeter. It is worthwhile for power line technicians to recognize the signs of poor power quality when checking customer complaints of erratic power problems.

What Is Power Quality?

Definition of Power Quality

A high-quality power supply is one where the AC and voltage rise and fall at a rate that can be represented by a sine wave. Any deviation in the magnitude or frequency of the 60 Hz sine wave is considered a power-quality disturbance. Poor power quality affects the performance of electrical equipment adversely.

Power Supply Disturbances

Electrical disturbances have always occurred in the supply of power. The increased sensitivity of certain loads causes these disturbances to be unacceptable to the customer. Momentary disturbances that affect the quality of power can be due to any of the following:

- Switching surges, fault clearing, and capacitor switching
- Voltage flicker from starting large motors or arc welders
- Transient faults and operation of surge arrestors

Continuous disturbances that affect the quality of power can be due to any of the following:

- An unacceptable range of voltage rise and fall
- Intentional voltage reductions (brownouts) during periods of peak load
- Unbalanced voltages between phases
- Harmonic distortion
- Tingle voltage
- Electrical noise (radio and television interference)

Modes from Where Disturbances Are Measured

There are two modes, or means, where the voltage can be erratic. These modes refer to the points where the unwanted voltage is measured. The differential mode (also called the normal mode) refers to disturbances between phases or between the phase and the neutral. Most voltage problems on the supply system would be differential mode and the source of trouble calls needing investigation by the line trade.

The common mode refers to disturbances between the neutral and the ground. This is also called noisy ground. Most common-mode problems would be on a customer's premises. For example, tingle voltage involves a voltage between the neutral and the ground. Trouble in this mode would normally involve engineering staff.

Corrective Measures Available

Customers understand that a utility cannot completely eliminate power outages, but some types of customers are becoming less tolerant to momentary outages and other disturbances. Often, a power-quality problem suffered by a customer comes from equipment on the customer's own service, such as large motors or arc-welding equipment.

A utility can improve power quality on a feeder by installing voltage regulators, capacitors, and surge arrestors. It can even build a dedicated substation or dedicated feeder to a customer who would be willing to pay more for a quality power supply with fewer disturbances.

Some customers own an uninterruptible power supply (UPS), which is a backup power supply, to protect them even from momentary outage. Batteries and/or some kind of generator is used to maintain service. Surge arrestors can be installed on customer equipment for extra protection from voltage surges. Filters are available for installation on customer equipment to limit harmonic interference.

A Frequent Fix for Power-Quality Problems

Poor grounding is often a cause of power-quality problems. In most cases, the poor or improper grounding is within the customer's premises. Good grounding is important to the customer because grounding balances the electrical system by bleeding off overvoltage or overcurrent.

There is a difference between a neutral and a safety ground. The neutral is intended to carry current. A safety ground is usually a small wire intended to drain away voltage that occurs during a failure of equipment, such as appliances and tools.

The neutral and the safety ground are normally tied together both at the transformer supplying the service and at the service panel. The neutral and the ground should not be tied together anywhere else on the load side of the main service disconnect because a circuit will be formed through the ground and the neutral. The safety ground wire will then share with the neutral the job of carrying current back to the source. The resulting ground currents can cause tingle voltage as well as disturbances to the normal operation of electronic equipment in the customer's premises.

Supplying "Custom Power"

Custom power is the ultimate solution to power-quality problems. Instantaneous **voltage regulation**, voltage flicker control, reduced harmonics, and no momentary outages are possible with specialized equipment installed on the utility system. The equipment uses the technology developed for high-voltage DC (HVDC) transmission and flexible AC transmission systems (FACTS). An electronic controller can convert DC from a backup source to AC with any wave shape needed.

Electronic rectifiers and inverters convert AC to DC and vice versa. High-speed switching (less than 10 milliseconds) is possible with solid-state breakers (SSB), which are practically instantaneous. A backup DC source provides power to an electronic controller, which can counter voltage dips, momentary outages, voltage flicker, and changes in reactive power.

For example, a dynamic voltage restorer (DVR) is an injection-transformer device that is installed in series with a circuit. Capacitors in the DVR maintain an internal bus, which is a DC power source that an electronic controller can draw on to supply AC power. During a disturbance, the electronic controller reshapes the power wave back to a proper sine wave by injecting real or reactive power as needed.

Similar technology is used in a distribution static compensator (DSTATCOM), which is connected into the system like a shunt capacitor. An electronic controller will put in and take out real power and reactive power as needed from a rechargeable energy-storage system.

Other devices are static VAR compensators (SVC) and adaptive VAR compensators (AVC), which are electronic devices that instantaneously input reactive power to counter changes in a supply system.

Factors Affecting Voltage in a Circuit

Voltage as a Measure of Power Quality

There is a voltage drop along every circuit and through every transformer. The extent of the voltage drop depends on how much the current flow is impeded by the device through which the current flows. For example, more current flow is impeded in a long length of a conductor than in a short length of a conductor, and more current flow is impeded by a small-diameter conductor than by a large-diameter conductor.

Voltage fluctuates throughout the day and throughout the seasons in proportion to fluctuation of the electrical load (current flow). Almost all circuits need some kind of additional voltage regulation and control. Some loads are very sensitive to voltage variations. It is important to keep the voltage to a customer within an acceptable range. High or low voltage is noticed by a customer, and both high and low voltage can damage customer equipment.

Customers and their motors, lights, and computers require a voltage that falls within a standard range. A standard would apply to steady-state voltages, not to a fluctuation caused by switching or motor starting.

A utility has to meet a voltage standard when supplying power to a service. Specific voltage ranges must be met and measured at the point of delivery, usually at the meter base. Standard range A refers to a favorable voltage level, and the utility planning engineer designs the distribution system to supply to this voltage standard. Standard range B refers to voltages outside of this range and, although tolerable for a time, corrective action to fix the situation should be carried out. Once voltage falls outside of standard range B, the customer equipment will not operate properly. See Figure 15–1.

Figure 15–1 Voltage Standards

Service Voltage	Range A Minimum	Range A Maximum	Range B Minimum	Range B Maximum
% of Nominal	95%	105%	91.7%	105.8%
120/240 3-wire	114/228	126/252	110/220	127/254
240/120 4-wire	228/114	252/126	220/110	254/127
208Y/120 4-wire	197/114	218/126	191/110	220/127
480Y/277 4-wire	456/263	504/291	440/254	508/293

Voltage Control and Voltage Regulation

The terms voltage regulation and voltage control tend to be used interchangeably.

Voltage control refers to the direct method of voltage change, such as changing a transformer output with transformer taps or changing the feeder voltage with line voltage regulators.

Voltage regulation refers to the indirect method of keeping voltage at a proper level. Voltage is regulated by ensuring that the conductor size and distance and the power factor of a circuit are adequate. Improving the power factor in a circuit reduces the amount of apparent power or current needed to supply the load. Less current in the circuit reduces the voltage drop.

Voltage Drop

When the load in an electrical system increases, the voltage at the load decreases. The amount of current flow affects the voltage drop. The following equations show that line loss and voltage drop are related to current flow.

$$\text{Power loss} = I^2R$$

$$\text{Voltage drop} = IR$$

Current flow varies according to the amount of customer load and the impedance offered by the power line and transformers feeding the load. Resistance is the largest component of the total impedance of current flow in a circuit. The design of the electrical power system includes keeping losses as low as practical because line loss in a power line or transformer is wasted energy.

Voltage Regulation

Voltage regulation is the difference between no-load voltage and full-load voltage, expressed as the percentage of no-load voltage.

$$\%V\ regulation = \frac{(no\text{-}load\ V) - (full\text{-}load\ V)}{no\text{-}load\ V} \times 100$$

For example, if a station transformer delivers 4,900 V at no load and 4,800 V at full load, the voltage regulation on the transformer is this:

$$\frac{4,900 - 4,800}{4,900} \times 100 = 2\%$$

If there is no load on a transformer, it would have a near-perfect voltage regulation of 0%. However, because load changes constantly, an electrical system should be designed so that the voltage regulation does not exceed a range of 2% to 3%.

Conductor Size and Length

Conductor size and circuit length affect the magnitude of the voltage drop in a circuit. A large-diameter conductor offers less resistance to current flow than a small-diameter conductor. Less resistance reduces line loss and, therefore, reduces voltage drop. The distance to a customer affects the voltage, because a longer conductor imposes more resistance to current flow than a shorter conductor.

Daily Changes

Because a load increase can result in increased voltage drop, voltage control equipment must adjust the voltage to reflect the load changes during the day. **See Figure 15–2.**

Figure 15–2 This graph displays a daily voltage profile of a customer at the end of a residential feeder. *(Delmar, Cengage Learning)*

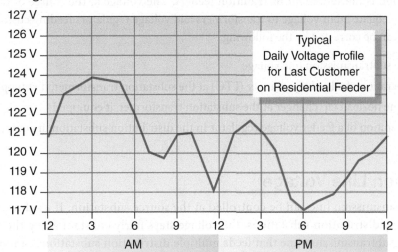

Seasonal Changes

The heat in summer and the cold in winter produce the peak-load periods for an electrical system. Voltage regulation studies by a planning engineer consider these peak loads as the benchmarks for worst-case, low-voltage conditions.

Peak-load statistics are used to determine where an upgrade is needed to substation transformers, the conductor size of feeders, or the number of feeders, voltage regulators, and capacitors.

Reactance in a Circuit

The amount of reactance in a circuit adds to the total impedance of the circuit. If the current was kept in phase with the voltage, very little voltage drop would be due to reactance. Transformers, motors, and fluorescent lighting have an inductive component in their energy demand, and, therefore, an inductive reactance is set up in the distribution feeder. Additional impedance caused by reactance in a circuit results in the need for additional current to feed the load. A higher current will cause a greater voltage drop.

Voltage on the Transmission Lines System

Voltage Control of Transmission Lines

Transmission line voltage is regulated and controlled by equipment at substations. The voltage at the source of a transmission line can be boosted so that the voltage at the load end of the line is at the required level. This method would not work on a distribution line because customers are usually spread out along the whole circuit. Voltage can be boosted by changing the transformer voltage taps to alter the transformer ratio. Transformer taps must change position regularly to keep the voltage constant as the load changes.

At the end of the line, voltage can be regulated by installing capacitors. Capacitors improve the power factor on a circuit. An improved power factor results in a lower current and, therefore, a boost in voltage. Relays sense the amount of inductive reactance and control the amount of **capacitance** needed to maintain a relatively constant voltage.

Parallel conductors on very long transmission lines increase capacitance on a circuit to a level where the capacitive reactance causes a high impedance to current flow. At transmission substations, reactors are installed for input of inductive reactance, balancing the capacitive reactance of the long line.

Distribution Substation Voltage

Voltage at the Distribution Substation

A distribution substation is the source for distribution feeders. The voltage at the source of the feeder must be high enough to provide an adequate input voltage to transformers and voltage regulators feeding customers downstream. The voltage at a substation can be corrected by the following:

- Adjustment of the subtransmission line voltage
- The automatic operation of a load tap changer (LTC) at the substation transformer, if equipped
- The changing of the no-load tap changer at the substation transformer, if equipped
- The automatic operation of a feeder voltage regulator in the distribution substation, if equipped

Subtransmission Line Voltage

The voltage of a subtransmission line can be controlled at the source substation. If a subtransmission line is short and does not feed many distribution substations, the voltage stays fairly constant along the full length of the line. The voltage on a long subtransmission line that feeds multiple distribution substations cannot be controlled to suit the needs of each substation. Each distribution substation will need its own voltage control to supply the distribution feeders. Voltage regulators can be installed along a subtransmission line to ensure that the distribution substation receives an acceptable voltage.

Voltage Control at a Distribution Substation

A distribution substation transformer with underload tap changers (ULTC) can adjust the voltage by changing the ratio between the primary and the secondary coils of the transformer. Similar to a line voltage regulator, a transformer ULTC can boost or buck voltage as needed while the transformer is in service.

A distribution substation transformer with a no-load tap changer requires all the load to be dropped from the transformer while the tap change is made. A transformer with a no-load tap changer cannot make regular adjustments during the day and would depend on the subtransmission line source or line step-voltage regulators to regulate the voltage for daily adjustments.

Feeder Voltage Regulator in a Substation

A feeder voltage regulator is sometimes used in small or lightly loaded substations where the substation transformer is not equipped with an LTC. A feeder voltage regulator is more commonly installed downstream from the substation on long, individual feeders. The voltage is boosted or bucked as needed to ensure that the customers receive a voltage within the standard range.

Distribution Feeder Voltage

Voltage Profile of a Feeder

The design of a distribution feeder facilitates keeping the voltage drop along every element of the circuit to a minimum. To provide an acceptable voltage to the customer, each part of the distribution system must also have an acceptable voltage. **See Figure 15–3.**

Figure 15–3 A feeder voltage profile illustrates how the voltage can drop below acceptable levels unless a correction is made. *(Delmar, Cengage Learning)*

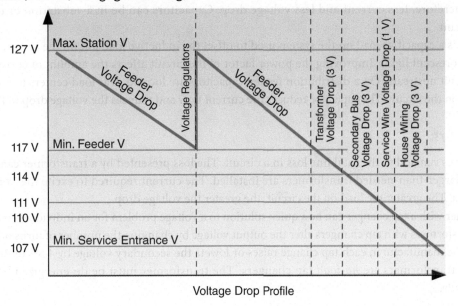

Voltage Drop Profile

Conductor Size

The amount of current in a circuit is a prominent factor affecting voltage drop. The less resistance there is to the current flow, the less line loss or voltage drop there will be. A large conductor over a short length offers the least resistance.

An ampacity chart to determine conductor size for a utility circuit is not very useful. For example, the ampacity chart for conductors may show that a 3/0 aluminum conductor has a capacity to carry 255 A. However, at 255 A, the voltage would drop about 1% every 40 ft (10 m), which means it can carry a 120 V service at that current for about 200 ft (60 m) before the voltage is below standard.

Distribution circuits have larger conductors than are needed for ampacity. A larger conductor is used to reduce the voltage drop and line loss on a circuit. Less line loss also keeps the available fault current higher for a longer distance and allows circuit breakers or fuses to "see" a short circuit farther downstream. For example, a short lead in and out of a set of voltage regulators would need to be sized to carry the current but does not need to be the same size as the main line conductors. The use of a smaller conductor for the short distance involved in and out of the voltage regulator would not reduce the voltage significantly.

The cost of conductors affects the size of a conductor chosen for distribution feeders. Usually, a utility decides on a few standard sizes for their feeders and selects conductors large enough to keep the voltage drop within practical and economical limits.

Locating Feeder Voltage Regulators

To improve the voltage on a feeder, it is usually more economical to install voltage regulators than to carry out a major improvement—such as stringing larger conductors, converting single phase to three phase, or installing additional feeders. Feeder voltage regulators should be installed in a location where the voltage is still high enough that the voltage is boosted from an adequate base. Voltage regulators step-up voltage in percentages. For example, boosting 2% of 117 V boosts the voltage to about 119.3 V, which stays at a higher level for a longer distance than boosting 2% of 110 V to about 112.2 V.

As a point of interest, a rule of thumb states that it requires 1,000 V to feed 1 mile of a normally loaded circuit before the voltage needs boosting. A 4,800 V circuit should be able to feed about 5 miles before needing a boost.

Locating Capacitors

Installing a capacitor is the most economical way to improve the voltage on a feeder and also benefit the whole electrical system. When the power factor of the circuit improves, there is less apparent power needed to feed the load and, therefore, less current and less voltage drop. Capacitors can be in a substation or downstream on a distribution circuit.

A capacitor is a capacitive load on the system used to offset the inductive load put on the system by motors, transformers, and fluorescent lights. Improving the power factor of the circuit affects the amount of current and the power factor of the circuit upstream. On a distribution feeder, capacitors are located near load centers to reduce the need for apparent power on the circuit. The capacitor reduces the current flow and reduces the voltage drop in the circuit.

Transformers

Transformer losses can add to the total line loss in a circuit. The loss presented by a transformer can be unnecessarily excessive when larger-than-needed transformers are installed. The current required to excite the iron core is constant regardless of load. The greater the load on the circuit, the greater the voltage drop.

A transformer with a tap changer can be a quick solution to a voltage problem for an individual customer. Individual distribution transformers with tap changers alter the output voltage by changing the number of turns on the primary coil. Depending on the manufacturer, each tap change raises or lowers the secondary voltage by 4.5% or 2.5%. Tap changers on distribution transformers are *no-load* tap changers. The transformer must be de-energized before turning the tap-changer handle.

Feeder Voltage Regulators

Voltage Regulator Operating Principle

The most common feeder voltage regulator used on a distribution feeder is a step-voltage regulator. A step-voltage regulator corrects excessive voltage variation and raises or lowers the voltage during low- or high-voltage conditions. Step-voltage regulators are normally used on long rural feeders or at small, older distribution substations where the transformer is not equipped with an automatic underload tap changer.

A step-voltage regulator works on the same principle as an autotransformer. On an ordinary distribution transformer, the primary and secondary coils are coupled magnetically; on an autotransformer, the primary and secondary coils are connected magnetically and electrically. The secondary coil is connected in series with the primary. If the transformer ratio was 10 to 1, then the voltage on the load side would be 10% higher than the source voltage. **See Figure 15–4.**

Figure 15–4 An autotransformer is designed to boost the primary voltage at a constant rate. *(Delmar, Cengage Learning)*

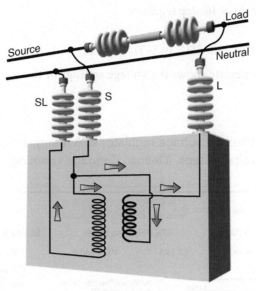

Step-Voltage Regulator

A voltage regulator is an autotransformer with an underload tap changer on the secondary coil. It has the ability to step the voltage up or down in small, incremental steps by moving a contact along the series-connected secondary coil sections. If the ratio of the primary to the secondary is 10 to 1, the secondary voltage at the highest tap would be 110%. **See Figure 15–5.**

Figure 15–5 A step-voltage regulator is designed to step up and step down the voltage automatically. *(Delmar, Cengage Learning)*

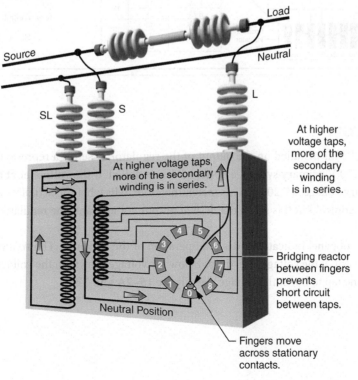

As the figure shows, the 10-to-1 ratio regulator has the secondary coil divided into eight equal sections. Each tap change would change the voltage by $10 \div 8 = 1.25\%$. A reversing switch allows the regulator to either boost or buck the voltage eight steps each way to make this a 16-step regulator.

Typical Step-Voltage Regulator Nameplate

A typical step-voltage regulator nameplate shows the voltage settings of the regulator and the types of connections available. **See Figure 15–6.**

Figure 15–6 A nameplate identifies the voltage regulator kVA rating, voltage rating, weight, and technical details related to the size of the voltage steps. *(Delmar, Cengage Learning)*

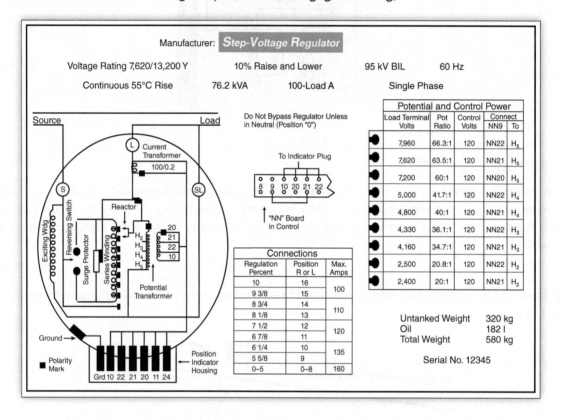

Regulator Controls

A potential transformer (PT) is installed on the output of the regulator. It sends a representative voltage to the control box. For example, on a 7,200 V primary system, a 60-to-1 PT is installed. If the voltmeter at the control box reads 120 V, it is known that the primary voltage is 7,200 V. This meter, therefore, tells what the *actual* voltage level is on the regulator output. The voltage-level control knob is set to the *desired* output voltage, and the regulator raises or lowers the actual output to the desired level.

The settings of a control panel indicate the desired operation of the regulator. The *voltage-level* setting is the desired voltage output of the regulator. If the voltage level drops below the voltage shown on the voltage-level setting, the regulator automatically moves up one tap to boost the voltage. **See Figure 15–7A and 7B.**

Figure 15–7A The settings made at the regulator control panel are specified by engineering but familiarity is needed to make adjustments during troubleshooting. *(Delmar, Cengage Learning)*

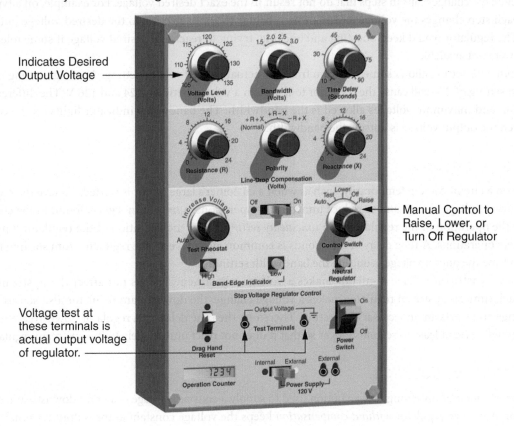

Figure 15–7B Each manufacturer's control panel may appear different, but the operations are similar. *(Source: Eaton)*

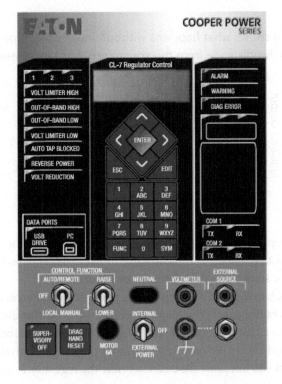

Bandwidth Setting

A voltage regulator changes taps in steps that do not result in the exact desired voltage. For example, on a typical voltage regulator, each step changes the voltage 0.625%, which brings the voltage close to the desired voltage but is probably not exact. The regulator would keep boosting and bucking, trying to reach the desired voltage if some tolerance for an inaccuracy were not available.

The *bandwidth* setting allows some variation from the actual desired voltage setting. A voltage setting of 125 V and a bandwidth setting of 2 V will cause the regulator to maintain a voltage between 124 and 126 V. The difference between the minimum and maximum voltages allowed is the bandwidth. The bandwidth indicator lights on the control panel indicate when the output voltage is outside the bandwidth.

Time Delay Setting

The voltage on a circuit can dip temporarily, such as when a customer's large motor is started. To save the regulator from reacting to every voltage change and to reduce unnecessary operations, a *time delay switch,* found in the control box, is set to delay the operation of the regulator. The *time delay setting* delays the operation of the regulator long enough to avoid needless operations. A time delay of 30 seconds is common and prevents the regulator from starting to adjust the voltage each time the output voltage is outside the bandwidth setting.

When a downstream voltage regulator makes a voltage adjustment, it does not affect the upstream regulator. However, each time an upstream regulator makes a voltage change, the downstream regulator also senses a need for a voltage change and starts an unnecessary operation. Therefore, the time delay on a regulator downstream from another regulator should be set at least 10 seconds longer so that it does not react immediately to the upstream regulator.

Compensation Settings

The line drop compensation settings are an option used to supply a constant voltage at a point downstream, remote from the regulator. A voltage regulator *without compensation* keeps the voltage constant at the output terminal. As the load current changes, the voltage at the output terminal stays constant, while the voltage drops at the end of the line. To supply the end of the line with an adequate voltage, the voltage setting at the regulator must be increased. Customers close to the regulator would then have a constant high voltage. **See Figure 15–8.**

Figure 15–8 A voltage profile shows what happens without line drop compensation. The voltage starts to drop and continues to drop as the distance from the source regulator increases. *(Delmar, Cengage Learning)*

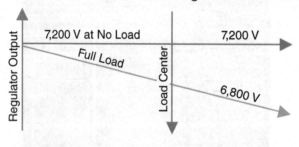

A regulator with *line drop compensation* keeps the voltage swings on the circuit to a minimum when the load current changes. Instead of keeping a constant voltage at the output terminal, the compensation settings give the regulator the ability to keep a constant voltage at some point downstream. **See Figure 15–9.**

The resistance and reactance between the regulator and a point downstream are calculated. These values become the compensation settings in the control panel. The projected voltage drop for that distance is automatically added to the voltage output by the compensator circuit in the regulator. When the load current increases, the voltage at the regulator terminal increases so that the load center downstream continues to have a relatively constant voltage. The planning engineer must choose settings that will not cause the customer near the regulator to get a voltage that is too high.

Figure 15–9 A voltage profile shows what happens with line drop compensation. The voltage regulator is set to provide proper voltage at the load center. *(Delmar, Cengage Learning)*

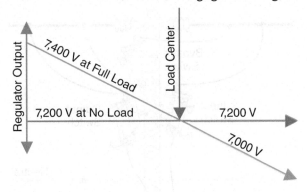

Operating a Regulator

Operating a regulator normally involves putting the regulator into service or taking it out of service. The operation involves a bypass, a source (S) switch, and a load (L) switch. **See Figure 15–10.**

Figure 15–10 Each of the three single-phase voltage regulators on the platform has its own control panel and bypass switch. These regulators are connected without an input and output switch. *(Delmar, Cengage Learning)*

The most critical operation involving a regulator is ensuring that the source and load voltages are equal when the bypass is about to be closed. If the source and load voltages are not equal when the bypass is closed, the regulator will be subjected to a short circuit.

Note: A bypassed regulator is damaged more quickly when it is one step up or down from the neutral tap (shorting 45 V) than when it is at full boost or buck (shorting 720 V). When the series winding of the regulator is shorted out, there is more impedance to the current flow when it travels through all of the windings. With the regulator at the number-one tap, there is not enough impedance to reduce the current and prevent the series winding from burning out. The *control switch* is used to manually raise or lower the regulator to the neutral position. The *neutral indicator* light should come on when the regulator is in neutral position. To equalize the voltage, the auto-manual switch in the control box is turned to manual operation. The voltage then can be raised or lowered until it reaches the zero or neutral tap. When the neutral tap is reached, the switch is turned to the off position. The bypass switch can then be closed, and the input and output cutouts can be opened to isolate the regulator. Some utilities require that a test be made to prove that the electric neutral of the regulator coincides with the neutral position indicator before carrying out any switching. **See Figure 15–11.**

Figure 15–11 At this regulator the switching arrangement includes an input, output, and bypass switch. *(Delmar, Cengage Learning)*

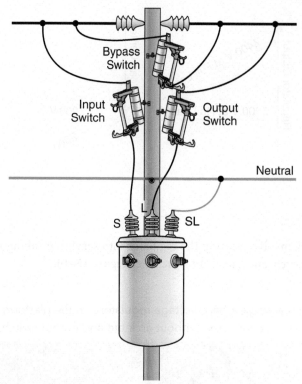

Troubleshooting a Regulator

A troubleshooting guide assumes that the trouble crew does not actually maintain the regulator or calculate the required setting for bandwidth, time delay, or compensation. A temporary fix means that the regulator problem will be reported to have the unit fixed as soon as practical.

Guide for Troubleshooting a Regulator

If	Then
There are excessive regulator operations.	The regulator could be overloaded. Phase balancing may unload the affected regulator.
There is a voltage complaint on a circuit where there is a regulator.	Often, the tap changing mechanism gets stuck on a tap. Raise and lower the voltage using the auto-manual switch, and then return the switch to automatic to see if the tap changer will move of its own accord.
The voltage problem was not due to a sticking tap changer.	The compensation settings may be out of date due to changes to the circuit, such as the installation of a capacitor, an upstream regulator, new conductors, or an increase in load. As a *temporary* measure, adjust the voltage using the auto-manual switch and then turn the switch off. Test the voltage at the customer and at the voltage test studs in the regulator control box. Now customers close to the regulator and/or at the end of the line will be exposed to more extreme voltages when the load current changes.
The regulator is chattering or hunting. The regulator is changing taps continuously.	The bandwidth or time delay settings are incorrect. As a *temporary* measure, adjust the voltage using the auto-manual switch as before. This fix is temporary because customers may be exposed to extreme voltages when the load current changes.
The regulator does not respond to the previous solutions.	The regulator may be damaged. Take the regulator out of service. If the tap changer cannot be put in the zero or neutral position, arrange to take the circuit out of service by opening the source switch with a load-break tool, then open load switches and close the bypass.

The regulator is at the maximum boost position but is not able to supply an acceptable voltage.	1. An increased load in the circuit may have reduced the input voltage to the regulator. 2. Phase balancing could unload the affected phase. 3. An upstream or downstream regulator may be needed.
The actual position of the zero or neutral tap is not certain.	The pointer on the tap position indicator can be broken and/or the neutral indicating light is not working. Sometimes the specifications for the control settings do not have the zero tap at the center. If in doubt, take the regulator out of service by opening the source switch with a load-break tool, as before.

Reverse Feed through a Voltage Regulator

When a circuit is temporarily fed in reverse through a regulator, the regulator continues trying to adjust the voltage on its load side, which is now the source. The voltage sensor measures the input instead of the output voltage. The regulator tries to change the input voltage, though it is not able to do so.

Typically, the regulator goes to the maximum boost or to the maximum buck position. To avoid these problems, the regulator should be bypassed and removed from service before the circuit is fed in reverse.

The S terminal must always be connected to the source and the L terminal always connected to the load. It is possible to set up a *reverse-power-flow* switching arrangement, which swings the input and output around so that the new source goes into the S terminal. This fairly complex switching arrangement is useful in cases when reverse feeding is a common requirement.

Capacitors

The Purpose of Capacitors

Most capacitors are installed to provide power factor correction on an electrical system, which in turn boosts the voltage. An electrical power system must supply the apparent power needed to meet customer needs. Customers tend to have motors that cause inductive reactance in the circuit, which increases the overall impedance of the circuit.

When capacitors are installed, a capacitive reactance is introduced into the circuit, which neutralizes the inductive reactance. Therefore, the overall impedance of the circuit is reduced. Often, with less impedance, less current is needed to supply the load. Less current results in less voltage drop. To improve the voltage on an electrical system, installing capacitors is more economical than installing voltage regulators, stringing larger conductors, or adding more feeders. **See Figure 15–12**.

Figure 15–12 A capacitor installation neutralizes inductive reactance in the circuit which reduces total impedance; therefore, load current is reduced which results in less voltage drop. *(Delmar, Cengage Learning)*

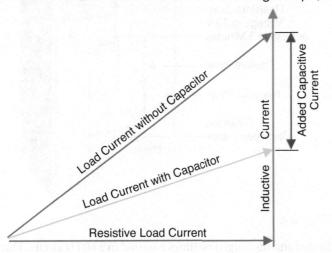

Power Factor Improvement or Voltage Booster?

Sometimes capacitors are installed on a distribution system to help improve the power factor on the transmission system right back to the generator. There is little need for system capacitors today because most utilities have capacitors on their high-voltage systems within the transmission substation yards. There is more control over station capacitors because operators can install a variable amount of capacitance as required by the system.

From a system point of view, there is very little control over capacitors on distribution circuits. Most are installed to boost distribution voltage and to reduce line loss. The location of the capacitors on a feeder is important because the voltage boost is upstream toward the source. While the best location can be calculated, a typical rule for placement of distribution lines is to place the capacitor bank about two thirds of the distance from where the voltage has dropped by two thirds (the "2/3–2/3 location rule"). The rule is intended to prevent the customers closest to the source from exposure to overvoltage and to improve the voltage at the end of the line.

Construction of Capacitors

Capacitors consist of two plates with insulation between them. The larger the plates, and the closer the two plates are to each other, the greater the capacitance. A typical distribution capacitor consists of two plates made up of two long sheets of aluminum. The long sheets of aluminum are rolled up with insulation, such as oiled paper or polyethylene film, between them. Each aluminum sheet or plate is connected electrically to a terminal. The rolls are made up flat so that they are more compact and can be stacked with other rolls in the capacitor tank. Multiple rolls within a unit can be connected in series or parallel depending on the capacitor kilovolt-ampere reactive rating and voltage rating. The capacitor case is filled with an insulating oil. **See Figure 15–13.**

Capacitors store a charge. Inside the capacitor unit, between the two terminals, are discharge resistors designed to drain the electric charge from the capacitor after a capacitor is isolated. It is normal to wait about 5 minutes after isolating a capacitor to let the resistors drain the charge before a power line technician applies grounds to the capacitor. The terminals of a capacitor should always be left shorted out when the unit is exposed to contact by people.

Figure 15–13 A capacitor is constructed with two plates insulated from each other so that when one plate is energized it induces an electric field onto the other plate. *(Delmar, Cengage Learning)*

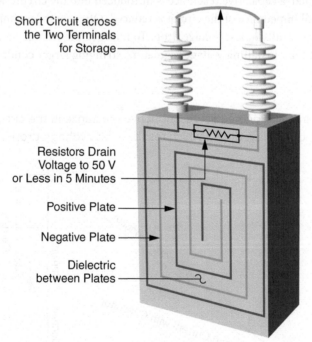

How a Capacitor Works

The plates of a capacitor are charged and discharged 60 times a second in a 60 Hz circuit. During the first half of a cycle, one plate is positively charged, which causes a negative charge of equal voltage to be electrostatically attracted to the other plate.

Current flows into the capacitor only while the voltage is rising. When the voltage approaches peak value, the counter-electromotive force is also approaching peak value, which causes the current flow to decrease. There is no current flow when the voltage is at its peak (90°). In a capacitor, the current reaches its peak before the voltage reaches its peak. A capacitor opposes the *change* in voltage, which in an AC circuit is constantly changing.

Because change to the voltage across the capacitor plates is delayed, this capacitive reaction causes the voltage wave to lag behind the current wave. In a circuit where the current lags behind the voltage, the capacitor effect of the current leading the voltage helps to cancel the two effects and to bring the circuit closer to a unity power factor.

Shunt Capacitors versus Series Capacitors

In an electrical power system, most capacitors are connected in parallel and are often called shunt capacitors. Unfortunately the term "shunt" is sometimes used by power line technicians to mean a short circuit. In the case of shunt capacitors it means that the circuit does not go through the capacitors (as in a series capacitor) but the circuit bypasses or shunts past the capacitor bank. A shunt capacitor has one capacitor plate connected to a phase and the other to the neutral—that is, there is only one connection to each phase. On a distribution system, shunt capacitors are connected near the load center to help reduce voltage drop. The voltage drop is reduced because the capacitors reduce the line loss in the complete circuit back to the source.

Series-connected capacitors are used to reduce a severe voltage flicker on radial circuits where frequent motor starting, electric welders, or electrical arc furnaces affect other customers on the circuit. Construction-wise, shunt and series capacitors are the same. Series capacitors are connected in series and, therefore, conduct the full-line current through them. The voltage drop across a series capacitor changes instantly when the load changes, and, therefore, reduces the effect of a voltage flicker.

Switched Capacitors

On distribution lines, shunt capacitors are installed downstream closer to the load center. During peak-load periods, the capacitors are needed to reduce line loss and keep the voltage at the proper level. During light-load periods, the capacitors can improve the power factor to the point where the voltage will be too high.

Capacitors must be switched off before damaging high voltage occurs. Capacitors can be switched on or off with electrically operated oil switches or solenoid driven vacuum switch. Control of the switches can be by a time control, radio control, voltage control, power-factor control, or kilovolt-ampere reactance (VAR) power control.

Daily switching of capacitors in and out of service, especially large capacitor banks in substations, causes a **transient overvoltage** in the circuit. Sensitive customers, such as those with variable-speed-drive motors, may find these fairly large voltage fluctuations unacceptable. **See Figure 15–14**.

Figure 15–14 This three-phase capacitor bank installed on a distribution feeder is switched in and out with solenoid-driven vacuum switches. *(Delmar, Cengage Learning)*

System Capacitors in Substations

A capacitor installation affects the complete system right back to the generator. To provide capacitance to the total electrical system, it is often more economical to install capacitors on distribution lines than on transmission lines. However, these systems could be a problem to distribution workers trying to solve a voltage problem on a feeder.

System control operators have more control of the power factor of the transmission system when capacitor banks are installed in substations. On distribution substations, a capacitor is either all in service or all out of service. Capacitor banks in transmission substations are more sophisticated, and there are choices as to how many VARs an operator wants to put into the system. Because capacitors constructed for high transmission voltages would cause operating problems, lower-voltage capacitors are put in series, for example, two 12 V batteries are put in series to produce 24 V. These capacitors are connected between the transmission line and neutral, set up on insulators and surrounded with danger signs.

Operating Capacitors

Unlike transformers, shunt capacitors draw a constant current regardless of the customer load. The current drawn by the capacitor is due to the energization of the relatively large amount of "metal" of the capacitor plates. Large capacitors have oil switches to allow safe energization or de-energization. Large capacitors should not be energized or de-energized with a cutout unless a load break tool can be used.

The current a capacitor bank draws can be measured or calculated. The following are some typical current readings for a 25 kV system:

600 kVAR bank 13.9 A

450 kVAR bank 10.4 A

300 kVAR bank 7.0 A

225 kVAR bank 5.2 A

Use the following equation to calculate the expected current for other voltages on wye systems:

$$I = \frac{\text{kVAR of capacitor bank}}{3 \times \text{phase-to-ground kV}}$$

Where:

I = current in one phase

kVAR = rating (of capacitor bank)

kV = phase-to-ground voltage

For example, how much current can be expected in the leads feeding a 600 kVAR capacitor bank on a 12.5/7.2 kV system?

$$I = \frac{600}{3 \times 7.2} = 27.8 \text{ amperes}$$

Operation of the Discharge Resistors

The discharge resistors in a capacitor drain the voltage of an isolated capacitor to a level below 50 V in about 5 minutes. Draining the voltage allows for safer installation of portable grounds.

When a capacitor is isolated with the intent to reenergize it, it is important to wait 5 minutes to allow the voltage on the capacitor to drain before reenergizing. If a *charged* capacitor is returned to service, the line voltage builds up well above normal.

There is a danger when energizing capacitors with a fused cutout because, if proper contact is not made the first time, any immediate second strike to retry closing the cutout results in double the line voltage across the cutout. Wait 5 minutes before trying to reclose a capacitor.

When available, reenergize a capacitor bank with the manual settings of the automatic capacitor control, as shown in **Figure 15–15**.

Figure 15–15 A capacitor control found on SCADA-operated capacitors. *(© S&C Electric Company)*

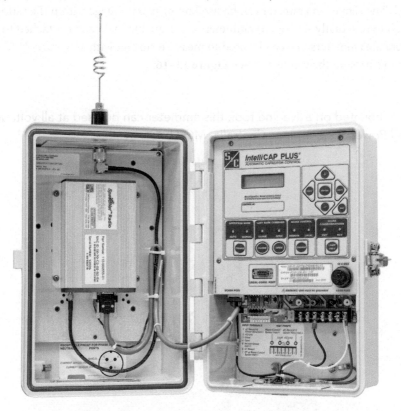

Troubleshooting No Power, High Voltage, or Low Voltage

Troubleshooting an Individual Customer's Service

Usually the first thing a power line technician does is check for voltage somewhere at the service. The following are typical items to check when troubleshooting an individual service.

- *At the customer's breaker.* If there is normal voltage at the top of the breaker, look for an open or faulty breaker or for blown fuses (breakers) in the panel. Some utilities will not check beyond the meter.
- *At the meter base.* If there is power to the top of the meter base, check for a faulty meter or bad connections in the meter base.
- *At the transformer.* If the primary circuit is supplying normal power to the transformer, check for a blown primary fuse, loose connections, bad neutral connections, a blown surge arrestor, an open current-limiting fuse, or a faulty transformer.

Using an Ammeter/Voltmeter

Checking or testing voltage is a task that is carried out often. Testing for the existence of any voltage can be carried out on transmission and distribution lines with a potential tester, and for circuits under 750 V with a voltmeter. Phasing sticks can be used on subtransmission and distribution circuits to measure voltage, but the most accurate voltage readings are made at the customer service entrance (meter base). Typically, the voltmeters used by power line technicians are clamp-on ammeters with a voltage testing function.

Literature for the safe use of a voltmeter and ammeter will state that to read current with an ammeter the leads must be connected in series, and that the leads must be connected in parallel to read voltage with a voltmeter. However, to measure current, the clamp-on ammeter feature is much safer than having to break into a circuit to connect an ammeter in series to measure current while maintaining service to the customer and ensuring that the ammeter is rated for the current to be measured. The clamp-on ammeter can be used on primary voltage within the rating of the rubber gloves, and clamp-on (although they actually stay open) ammeters are available that can be attached to a hot-line tool to take ammeter readings. There are ammeters that can be used to measure current with a **live-line tool** where there is no need for the ammeter clamp to surround the conductor. **See Figure 15–16.**

Figure 15–16 When mounted on a live-line tool, this ammeter can be used at all voltages. *(Reprinted with permission of Hubbell Power Systems, Inc., Centralia, MO USA)*

Voltage measurements are actually taken in parallel or in series. For example, power line technicians sometimes test for voltage between the source side and load side of a meter base, and they test between any two terminals, objects, or wires within the voltage rating of the voltmeter. A voltmeter has a high resistance compared to the circuit being tested because there should be very little current flow through the meter. An extra safety measure is to use fused leads or probes with high interrupting capacity (HRC) fuses that will operate quickly under high fault current conditions.

If voltage tests are needed to investigate tingle voltage, a more accurate voltmeter with a capability to read very low voltage will be needed.

If there is an ohmmeter feature on the multimeter, ensure the circuit is de-energized before taking resistance readings. Also ensure the multimeter settings are not in the ohmmeter setting when taking a voltage reading. (The multimeter will blow up unless the fused leads protect it.)

Troubleshooting a No-Power Call on an Overhead System

From a power-quality perspective, having no power is the ultimate in poor-quality power. A no-power call is always a priority call. The following table outlines some common causes of outages.

Troubleshooting a No-Power Call on Overhead Distribution

Common Problems	Restoration
Lightning is the most frequent cause of a transient fault. At the flashover point, a high-voltage arc establishes a path of ionized air to ground. A high follow-through current is established through the ionized air and causes an overcurrent fault.	In an urban area, a patrol on the circuit is always wise before closing in the circuit. In a rural setting, lightning without high winds should allow reenergization without a patrol.
When *wind, ice, or wet snow* have been present, the cause of an outage is often a phase-to-ground fault. A phase-to-ground fault is the cause of about 70% of permanent faults. Probable causes are a bad insulator, tree contact, broken conductor, or animal contact. Accidental contact by the public includes car accident, crane contact, sailboat contact, ladder contact, or antenna contact.	In a populated area, patrol the line before reenergization. In a rural setting, patrol the line unless special circumstances indicate that local knowledge would not allow reenergization.
On *very cold or very hot days,* suspect an overload problem. Overcurrent due to an overload occurs when the customer demand exceeds the specified setting of the circuit protection. Circuit protection does not differentiate between overcurrent due to an overload or a fault.	The most common fix to an overload problem is phase balancing. When the load is not balanced between phases, one phase of the three-phase system is carrying more than its share of the load and the protective device may trip out. A low-resistance tree contact on a tap protected by a fuse can be seen as an additional load by the upstream three-phase device.
On an underground cable, a phase-to-ground fault can occur due to an *insulation breakdown, dig in, or a driven fence post.*	Other than an overload, a blown fuse on an underground cable normally calls for checking out the cable before reenergizing.
The *circuit has been out for a while,* especially during peak load periods.	A heavily loaded circuit may have to be picked up one section at a time.
There is a high initial in-rush current when the switch is closed and the line trips out again.	On an electronic recloser, the handle can be held closed momentarily until the in-rush current drops. On a hydraulic recloser, holding the handle closed will not prevent a trip-out.
After patrolling, there is *no apparent cause* for a permanent line outage, but the protective device trips out each time the line is reenergized.	• The cause may be a faulty surge arrestor. • There may be a fault downstream past the section that was patrolled. When the downstream protective devices do not trip out in proper sequence, it could be that incorrect fuses were installed during previous work. • A punctured dead-end insulator could also be the cause.

Troubleshooting a No-Power Call on an Underground System

Troubleshooting power outages on underground systems—whether duct bank/vault systems or direct bury/pad-mount/submersible systems—relies much more on testing and technology. An initial patrol can determine if there is an obvious dig-in, failed or charred elbows, dead animals, a burnt smell, or other problem. Locals may be aware of cables in specific locations that are vulnerable due to aging, overloading, improper backfill, or stresses from settling soil.

The outage will be either with the system at large, with the transformer, or with the individual customer.

A no-power call from an individual customer starts with a voltmeter check at the meter base. For no voltage, check out other customers fed from the same transformer. The voltage checks will show if the problems are a faulted service cable or a transformer problem.

If a larger section of the underground system is out, it is not a good idea to use an approach common in overhead lines—that is, close in on the circuit and see if the fault is still there. Very few transient faults are found on an underground system. Energizing the circuit for testing or sectionalizing will stress the cable, elbows, and more, each time the high fault current generated by the fault goes through it. Using a smaller fuse does not change the fact that the cable is subjected to the full fault current generated by the fault.

If fault indicators have been installed on the system, the faulted section can be found by observing the fault indicators. In this relatively simple example a line crew would find the switch at the riser pole open, check the fault indicators at each transformer, and find the fault indicators at the first three transformers showing that the fault passed through there and that the fault did not go through the indicator at transformer 126. The conclusion would be that the fault lies somewhere between transformers 125 and 126. **See Figure 15–17.**

Figure 15–17 A fault indicator will flag that a fault current has passed through it. Patrol a faulted circuit until reaching the first fault indicator that has not been activated to determine that the fault location is between it and the last flagged fault indicator. *(Delmar, Cengage Learning)*

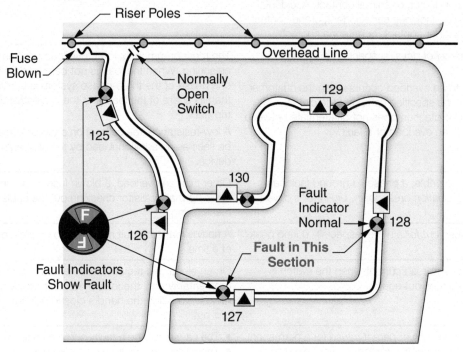

Without fault indicators, sectionalize the system and test the cable sections with instrumentation and a voltage that will not damage the cable.

If the riser pole is in the section that has faulted, the cable there should be the first suspect. The cable is more vulnerable because the cable on the pole is not able to dissipate heat, so as the cable enters the ground it is vulnerable to frost and settling.

Troubleshooting Low-Voltage Problems

Low-voltage problems usually show up during peak demand times, such as during cold and hot weather and at dinnertime. A customer may notice low voltage when an electric motor is running hotter than it should. During low voltage, the motor draws more amperes and the torque that the motor produces is reduced. The torque is inversely proportional to the square of the voltage. For example, even though most motors are designed to run at 90% of their voltage ratings, a motor running at 90% of voltage produces 81% torque and runs hotter.

Low voltage has many causes. Finding the precise cause is a step-by-step process that starts at the customer and works toward upstream equipment.

Finding the Cause of Low Voltage

Step	Action	Details
1.	Check the voltage at the customer meter base.	If the voltage is low, check for any recent increase in the load a customer is drawing. An increase in load could make the length or size of secondary conductors inadequate for the additional load. The planning engineer or engineering standards books have voltage regulation charts for secondary bus and services.
2.	Remove all load from the transformer and take a voltage reading at the transformer. If the voltage reading is low at the unloaded transformer, the primary voltage is low and the cause of the problem is upstream. Other customers should be affected by upstream problems. A quick fix to an individual customer is to make a change to the transformer tap if equipped.	If the voltage reading is normal at the unloaded transformer, the low-voltage problem must be due to the length and/or size of the secondary or possibly an overloaded transformer. A recording voltmeter may need to be installed if the problem is intermittent.
3.	Check out any upstream voltage regulator or transformer with a load tap changer (LTC).	A voltage regulator could be stuck in a low position. Put the regulator in "manual" position and raise the voltage. Similarly, the LTC at the substation transformer could be malfunctioning.
4.	Check out any capacitors on the line.	Switched capacitors should be in service during daily peak load times and switched out of service during lightly loaded times. Check to ensure that the time clock or other control is working. Check that the motor-operated oil switches or solenoid switches are closed during peak load periods.
5.	On a heavily loaded circuit, check the phase balance.	The planning engineer may show that the circuit is able to carry the load, but if one phase is carrying more than its share, it could be overloaded and cause an excessive voltage drop on that phase.

Troubleshooting a High-Voltage Problem

A high-voltage problem becomes noticeable to a customer when lights seem brighter and light bulbs, motors, and electronic equipment burn out prematurely. High voltage has many causes. Finding the cause is a step-by-step process that starts at the customer and works toward upstream equipment.

Finding the Cause of High Voltage

Step	Action	Details
1.	Check the voltage at the customer's meter base.	If the voltage is high and the customer is close to a distribution substation or close to a voltage regulator, the voltage is often relatively high at these locations to provide good voltage farther downstream. Check the voltage of a neighbor fed from a different transformer to find out if the high voltage is unique to the one customer that is complaining.
2.	If the neighbors have normal voltage, do a ratio test on the transformer causing the high voltage.	A transformer can occasionally suffer some shorted-out turns in the coil and cause high voltage to customers.
3.	Check out the substation load tap changer (LTC) or upstream regulator.	A voltage regulator could be stuck in a high position. Put the regulator in "manual" position and lower the voltage. Similarly, the LTC at the substation transformer could be malfunctioning.

(Continued)

Step	Action	Details
4.	Check out any nearby capacitors.	Capacitors that have not automatically switched out of service when the circuit is lightly loaded can raise the voltage during off-peak times.
		Check to ensure that the controller (load controller, VAR controller, time clock, etc.) of the capacitor switch is working. In other words, check that the motor-operated oil switches are open during off-peak load periods.
5.	If all other conditions are normal, change the voltage taps only on the transformers feeding customers with the high voltage.	Depending on the manufacturer, each tap will raise or lower the secondary voltage by 4.5% or 2.5%. The nameplate on the transformer must be consulted. *The transformer must be isolated before changing the voltage tap.*
		The feeder voltage will change during the day and at different seasons. Therefore, changing the taps at the transformer could produce extreme voltages if the primary voltage returns to a normal level.

Harmonic Interference

What Are Harmonics?

With AC power, the magnitude and rate of current and voltage rise and fall are represented graphically by a sine wave. Harmonics are AC and voltage disturbances that can also be represented by sine waves but are at different frequencies than the fundamental 60 Hz.

The harmonic frequency is a multiple of the fundamental 60 Hz frequency. Harmonic frequencies include 30 Hz, 120 Hz, and 180 Hz. A 120 Hz harmonic is twice the fundamental frequency and is called the second harmonic. Similarly, 180 Hz is the third harmonic.

Many of the various harmonic waveforms exist in the power supply at one time. These waveforms are out of phase with each other. The many differing waveforms tend to sum and cancel each other to form a distorted, but predominantly 60 Hz sine wave.

Typically there are many harmonics, but sometimes there is only one harmonic. The 60 Hz power wave stays predominant in a normal distribution system. The resultant 60 Hz wave becomes distorted and, when it becomes severely distorted, it affects sensitive electronic equipment. **See Figure 15–18.**

Figure 15–18 Quality power can be represented with a sine wave. Harmonic disturbance will distort quality power which is represented by a distorted sine wave and typically causes electronic equipment to malfunction. *(Delmar, Cengage Learning)*

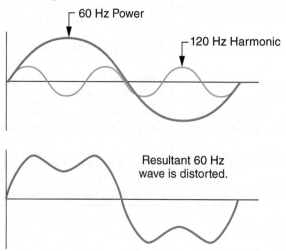

Sources of Harmonics

Electric loads that are nonlinear propagate harmonics back into the electrical supply system. Nonlinear loads are not purely resistive but have a capacitive or inductive component in them.

There are many kinds of nonlinear loads, and all distort the "pure" 60 Hz waveform to some degree. Examples of nonlinear loads are transformers, arc furnaces, arc welders, adjustable-speed and variable-frequency motor drives, electronic lighting ballasts, converters, rectifiers, and large computer systems.

Harmonics are steady-state disturbances that exist in the system as long as the equipment generating the harmonics is in operation.

Conductive Harmonic Interference

Harmonic interference can be conducted back into the power supply from the direct connection to the harmonic-producing customer load. The harmonic currents conducted back into the power source add additional current to conductors, transformers, and switchgear. Harmonic currents can also interfere with protective relays, meters, and induction motors.

Capacitors are exposed to overloading because they act as a sink for the higher-frequency harmonics. When capacitors are on the system, there is also a possibility for resonance to occur at one of the other frequencies.

Inductive Harmonic Interference

Just as voltage and current can be induced on adjacent circuits, harmonic voltage and current can be induced on a neighboring power supply or telephone circuit. Unshielded telephone cable running parallel to power circuits is especially vulnerable to induced harmonics from the power line.

Signs of Harmonic Problems

One of the first signs of a problem due to harmonics is from customers with sensitive electronic equipment. Harmonics distort the needed voltage and current magnitude, which can show up when a computer acts erratically or loses data.

Conductors, transformers, and motors are subject to heating because the high-frequency component of a harmonic disturbance causes an increase to the skin effect of conductors. A larger proportion of current is carried on the outer edge of a conductor, which limits its current-carrying capacity. Harmonics reduce the reliability of electric signals that activate meters, relays, power line carriers, and equipment such as robots.

Testing for Harmonics

Before remedial action is taken for harmonic interference, tests should be carried out to confirm its presence. A specialist uses a power harmonic analyzer to make tests, interpret the waveform, and predict the most likely cause of the disturbance.

Harmonic Interference Solutions

Once a problem has been diagnosed as a harmonic problem, filters can be installed at the customer to limit the effect on their sensitive loads. The filter consists of an inductor, a capacitor, and a resistor, which are tuned to provide a low impedance to ground for specific harmonic frequencies. For large customers with big power-quality problems, arrangements can be made to supply "custom" power.

Voltage Flicker

The Voltage Flicker Problem

Erratic fluctuation in voltage shows up as blinking lights or intermittent shrinking of a television screen. For a customer with sensitive electronic equipment, such as a computer or an electronic cash register, the problem becomes unacceptable.

Normally, calculations for potential voltage flicker problems are carried out by the planning engineer when a customer with large motors or welders applies for service.

Sources of Voltage Flicker

A loose neutral or other bad connections are possible sources of flickering lights, especially during windy or heavy load conditions. When a customer has a very noticeable dip in voltage on a regular basis, other sources must be investigated. Large motors, arc welders, X-ray machines, and electrical arc furnaces have loads with varying demand, are mostly unbalanced, and have a poor power factor. Under starting conditions, these loads draw considerably more current than when operating. A voltage flicker is noticed by other customers served by the same feeder as the offending motors and welders, for example. The extent of the voltage flicker depends on the capacity of the feeder supplying the load to the customer.

When Does a Voltage Flicker Become Objectionable?

There are standards for an acceptable or unacceptable voltage dip. These standards may change as more sensitive electronic devices come on the market. The standards combine the percentage of voltage dip with the frequency of its occurrence. **See Figure 15–19.**

Figure 15–19 A threshold of objectionable voltage flicker is shown in this graph. This typical standard shows that a 3% voltage dip five times per minute would be objectionable. *(Delmar, Cengage Learning)*

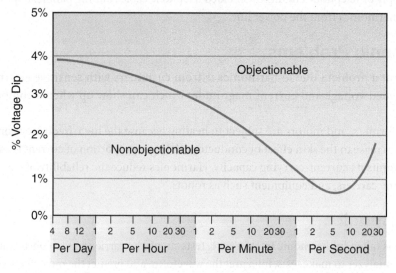

Frequency of Voltage Fluctuation or Number of Starts

Factors Affecting Voltage Flicker

The size of voltage flicker is dependent on the type of load and the capacity of the feeder supplying the load. The load itself can have equipment that reduces the start-up in-rush current. In-rush current information for equipment such as motors and arc welders is normally found on the equipment nameplate. An electric motor has an in-rush current of about six times its load current. This can be reduced by using motors with soft-start capability.

The greater the capacity of the feeder supplying the load, the less the feeder is affected by voltage flicker. Voltage flicker is reduced in the following circumstances.

- The source transformer is large and able to carry a high peak load for a short period of time.
- The distribution feeder conductor is large.
- The distribution feeder distance to the offending load is short.

Calculating Predicted Voltage Dip

To predict a voltage dip, detailed calculations can be made. However, the planning engineer normally enters the needed information into a computer program or uses an estimating chart. The information needed is the in-rush current for the load to be fed, the capacity of the power supply system (which is the phase-to-phase fault current availability of the feeder at the source substation transformer), and the length and size of the conductor from the station to the load.

Ferroresonance

What Is Resonance?

Objects have their own natural frequencies. For example, a rattle in the dashboard of a car can vibrate annoyingly when a car travels at a certain speed. The frequency of the engine matches the frequency of the dashboard and sets up a sympathetic vibration in the dashboard. Soldiers are told to march out of step when they cross a bridge because when they march in step they sometimes match the natural frequency of the bridge; the sympathetic vibration set up in the bridge can cause severe damage. When the vibration in one object matches the natural frequency of another object, the two objects vibrate together in resonance.

Three effects of ferroresonance are as follows:

- Loud hum, growl, or rumble from transformer when closing in or opening the first two cutouts, normal hum when all cutouts are energized
- Flashovers at terminals of transformers or cutouts on the open phase(s) during single-pole switching
- Failure of lightning arrestors due to overvoltages on the open phase(s) during single-pole switching

Resonance in an Electrical Circuit

In any series RLC circuit (circuits that have resistance, inductive reactance, and capacitive reactance), resonance occurs at some frequency. When the frequency rises, the inductive reactance increases and capacitive reactance decreases. As the frequency rises, the increasing inductive reactance will at some point be equal to the decreasing capacitive reactance and vice versa. When this happens, the inductive reactance cancels out the capacitive reactance, which means there is no reactive load impeding the current flow in the circuit.

If the resistance in the circuit happens to be low, there is very little to impede current flow and the current increases. As the current increases, the voltage also increases and rises above the source voltage.

The tuner on a radio varies the capacitance in a circuit to match the inductance at a desired frequency. The signal is amplified when it resonates at the desired frequency. Of course, the frequency on an electrical power system is constant. Resonance only occurs by coincidence when the capacitive reactance is in series with and happens to match the inductive reactance at the standard 50 or 60 Hz. When resonance occurs, the voltage can build up from 5 to 15 times the normal phase-to-ground voltage. The increase in voltage is evidenced by a loud hum, whine, or rumble from a transformer during switching; flashovers at terminals insulators, switchgear; and failure of lighting arrestors.

Resonance with Harmonic Frequencies

The frequency of an electrical power system may be a constant 50 or 60 Hz; however, there are frequencies superimposed into the system from certain loads. The power at these "harmonic" frequency waves sometimes resonates.

Causes of Ferroresonance

Resonance is a rare occurrence in a circuit because *all* of the following factors must be in place.

- The inductive reactance X_L is equal to the capacitive reactance X_C.
- The inductive load and the capacitive load must be in series with each other.
- There must be virtually no resistive load on the circuit.

When these factors are in place, there is practically no impedance to current flow in the circuit. Ferroresonance usually occurs when one or two phases are disconnected from the source by a fault; or, by switching a single-pole device, the transformer windings connected to the open phases are excited through the capacitance to ground in the cables and between phases.

The Source for Inductive Reactance in Series

Most electrical system circuits have loads that are a source of inductive reactance, but these loads are normally connected in parallel. The most common and possibly the only way to get an inductive load in series with a circuit is to open one or two phases feeding a three-phase transformer bank.

Often, the windings in the transformer with a delta primary are in a series configuration when one phase is open. The windings in a transformer with a wye primary do not become a series load when one switch is open. This only applies when the wye point is grounded. However, if the wye point on a wye-delta transformer is left ungrounded or floated, which is typical for a wye-delta transformer, then when switching the transformer bank in or out of service a ferroresonant condition can occur. **See Figure 15–20.**

Figure 15–20 When one phase is open, the inductive reactance in a transformer coil can be in a series circuit. *(Delmar, Cengage Learning)*

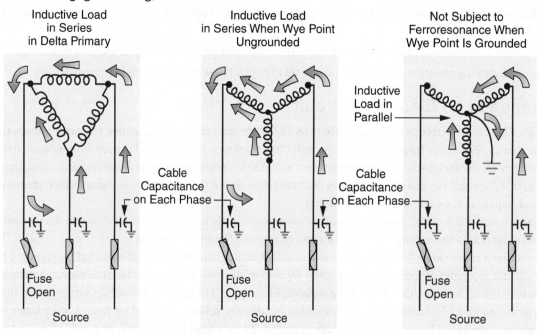

The inductive reactance of a transformer changes as the magnetic field from the coil magnetizes and eventually saturates the iron core. This gives a range of inductive-reactance values, which increases the possibility of matching the capacitive reactance in the circuit. Smaller transformers have less inductance and are more likely to match the typically small capacitance set up in an overhead or underground line feeding a transformer bank. The process of saturating the iron (ferrous) core is where the term ferroresonance originates.

The Source of Capacitive Reactance in Series

There is some capacitance on any circuit because an energized conductor acts as one plate; the air or cable insulation is the **dielectric** between the plates, and any conductive material near the conductor acts as the second plate. Paralleling overhead conductors causes a capacitive reactance in a circuit. The capacitive reactance from long transmission lines is often countered by the installation of reactors at stations. Underground cable is a natural capacitor with the energized conductor separated from another conductor, the grounded sheath, by relatively thin insulation. Long lengths of underground cable need reactors to cancel some of the capacitive reactance. The right amount of capacitance to cause resonance is more likely to occur on higher voltage distribution cables. The capacitance generated on 20 to 35 kV cable is greater than on a 12 kV cable and is more likely to match the reactance of a transformer bank.

Field Examples of Ferroresonance

The most common occurrence of ferroresonance involves a three-phase transformer bank fed with a length of underground cable. A certain critical length of underground cable provides the critical amount of capacitive reactance in the circuit. A transformer bank can provide the inductive reactance matching the capacitive reactance of the cable. The inductive load and the capacitive load must be in series with each other, and this occurs when only one or two phases are energized. Resonance is most likely to occur when remotely switching single-pole devices on a high-voltage distribution underground cable feeding a transformer with a delta primary. **See Figure 15–21.**

Figure 15–21 Ferroresonance is set up on this circuit when one switch is opened and the other two phases stay energized. *(Delmar, Cengage Learning)*

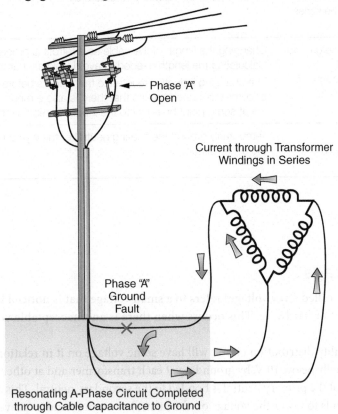

Phase "A"
Open

Current through Transformer
Windings in Series

Phase "A"
Ground
Fault

Resonating A-Phase Circuit Completed
through Cable Capacitance to Ground

Another example of ferroresonance occurred on an overhead 115 kV line when one of the blades of a gang-operated air-break switch failed to open during the isolation of a substation transformer. With one phase closed, the unloaded transformer became an inductive load connected in series with the 115 kV line. The paralleling 115 kV conductors supplied the capacitive load. A corona discharge was seen on the 115 kV conductors, and surgearrestors operated as the voltage climbed higher.

Prevention of Ferroresonance

Ferroresonance can be avoided by removing one of the causes of the phenomenon. By changing the design of the installation or changing the switching procedure, the capacitance, inductance, series connection, or loading can be changed.

Prevention of Ferroresonance

Causal Factor	Action
Operating single-pole switches to energize or de-energize an underground cable feeding a three-phase transformer allows the transformer coils to be an inductive load in series with the circuit.	Install a three-phase gang-operated switch.
If there is no load on the transformers when they are energized, the low resistance in the circuit causes a higher voltage when resonance occurs.	Keep a load on the transformer when it is being energized or de-energized. The increased resistance in the circuit will lower the effects of resonance.
A three-phase bank with a neutral grounded wye primary shorts out the two windings that are part of the series circuit, causing induction.	At the planning stage, where possible, use a transformer with a wye primary and the wye point connected to the neutral.
Changing the length of the underground cable feeding the three-phase transformer will change the amount of capacitive reactance in the circuit.	Changing the length of the cable as a retrofit is probably an expensive option. Calculating the length needed to avoid ferroresonance is complex. The changing inductive reactance that occurs before the transformer core becomes saturated means that there is also a range of capacitive reactance that will at some point be equal to the inductive reactance.
A wye primary with a floating (ungrounded) neutral is susceptible to ferroresonance.	Temporarily ground the floating neutral of the wye primary, which will short out the series circuit.

Tingle Voltage

What Is Tingle Voltage?

The term tingle voltage (also called stray voltage) refers to a small voltage that is noticed by people or animals while contacting certain equipment or hardware. This occurs when there is an unacceptable voltage between the neutral and earth.

A system neutral on a utility distribution circuit will have some voltage on it in relation to a remote ground. This voltage is kept very low, normally below 10 V, by grounding at each transformer and at other locations. At transformers, most specifications require that the primary neutral is bonded to the secondary neutral. This connection takes advantage of the customer service grounds to lower the voltage on the neutral even more and to prevent a potentially dangerous open circuit in the grounding network. However, depending on the quality of grounds and other factors, there will still be some voltage left on the neutral.

With the ground bonded to the neutral at the service entrance, the safety ground will have some voltage on it because it is connected to appliances, equipment, and fixtures by the many electrical circuits coming out of the customer panel. Tingle voltage occurs between a grounded object and a remote ground. There will always be some current flow back to the circuit's source through earth. Where there is current flow, there has to be some voltage. **See Figure 15–22.**

Figure 15–22 In this example, tingle voltage can be felt when a circuit is established through a person between a shower control and a main drain. *(Delmar, Cengage Learning)*

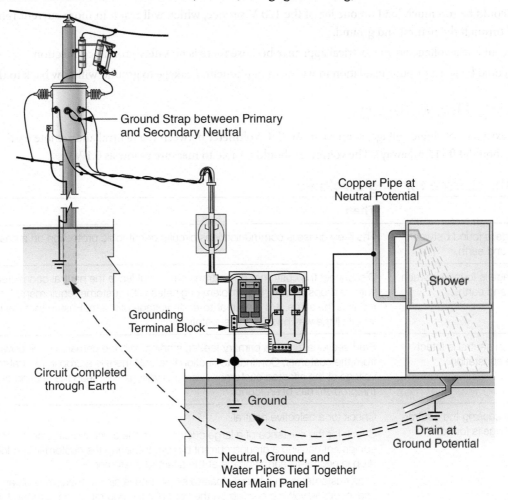

Ground Strap between Primary and Secondary Neutral

Copper Pipe at Neutral Potential

Shower

Grounding Terminal Block

Circuit Completed through Earth

Ground

Drain at Ground Potential

Neutral, Ground, and Water Pipes Tied Together Near Main Panel

Examples of Tingle Voltage Problems

The voltage between the neutral and earth is normally well below the threshold of sensation for the vast majority of customers. Vulnerable customers, such as dairy farms, require additional efforts to reduce tingle voltage. Milk cows have been the most vulnerable to tingle voltage. A small voltage between the stanchions, water bowl, or milking machine in relation to the floor that the cow is standing on results in a current flow that turns the animal into a "dancing cow." In these conditions, a cow can feel a potential difference as low as 0.5 V.

Normally, people would not feel tingle voltage because 10 V or less is not enough to overcome a person's skin resistance. However, skin resistance is reduced when it gets wet. People have noticed tingle voltage in a shower, kitchen, wet basement, or barn. In a shower, a person can feel a potential difference as low as 2 V.

Investigating a Tingle Voltage Problem

Although it is not usually part of a power line technician's job to investigate a service past the meter, some typical customer-generated tingle voltage is listed here for information.

- There should be only one electrical connection between the neutral and earth: at the service entrance. A connection between the neutral and the safety ground wire at an appliance some distance from the main panel will mean that the neutral and safety ground will be parallel paths. This will encourage more current flow through the safety ground wire. More current in the safety ground wire will mean more voltage in the safety ground network.

- There could be a poor neutral or ground connection at the service entrance or at junctions.

- There could be ineffective grounding due to poor earth, insufficient number of ground rods, or poor ground rod connections.
- There could be too much load on one leg of the 120 V service, which will result in more current returning to the source through the neutral and ground.
- There could be a voltage on an electrical appliance because there is no safety ground connection.
- There could be worn or poor insulation in wiring or equipment. Leakage to ground will flow back to the source.

Measuring Tingle Voltage

To find the existence of tingle voltage, connect an AC/DC voltmeter between the neutral and remote earth (a temporary ground rod about 50 ft [15 m] away). The voltmeter should be able to measure as low as 0.1 V.

Finding the Source for Tingle Voltage

If	Then
A DC voltage is found between the neutral and earth.	The likely cause is communications circuits or cathodic protection on a nearby pipeline.
An AC voltage is found between the neutral and earth.	Disconnect the customer at the transformer but leave the neutral connected. If the voltage drops to zero, the problem originates with customer equipment. If the voltage remains the same, it is a neutral-to-earth voltage problem. Ensure that the neutral is intact all the way back to the substation.
The common neutral is intact back to the substation.	Start sectionalizing the primary feeder, starting with the primary circuit downstream from the customer. Continue to sectionalize until there is a drop in neutral-to-earth voltage at the affected customer. A drop in voltage will mean the section of line causing the problem has been found.
The section causing the neutral-to-earth voltage is found.	Check for a defective neutral. Check for the existence of a large customer on the same circuit. Check for defective equipment causing a ground current by disconnecting the customer and looking for a drop in neutral-to-earth voltage at the affected customer. If the customer is on a heavily loaded single-phase circuit, there may be an unacceptable voltage buildup on the neutral if the neutral is poorly grounded.
The source of the complaint is a high system neutral voltage and there is a need for an engineering solution.	Engineering solutions include the following: • Additional ground rods are driven to lower the voltage on the neutral. • A larger neutral conductor can be strung to promote more current flow through the neutral and less through earth. • A conversion to three phase will promote more current returning to the source in the other phases and less current flow in the neutral.
An immediate solution is required.	Have the customer install a tingle voltage filter. In an emergency, with concurrence from the supervisor, split the secondary neutral from the primary at the transformer. Install a warning sign on the pole to warn others about the split neutral.

The Utility as a Source of Tingle Voltage

Single-phase circuits have current flowing back to the source through the neutral, and a portion of this current flows through the parallel path in earth. When there is a current flow, there is some voltage. A heavily loaded single-phase circuit will have an unacceptable voltage buildup on the neutral if the neutral is poorly grounded. Normally, the voltage on the neutral should not exceed 10 V.

Delta circuits and three-phase circuits have considerably less current flowing back to the source through earth because the phase conductors carry current back to the source. The more evenly the load is balanced on the three phases, the less current there is in the neutral.

If the voltage on the neutral has to be lowered, *then* the following actions are taken.

- Neutral connections are checked.
- Additional ground rods are driven to lower the voltage on the neutral.
- A larger neutral conductor can be strung to promote more current flow through the neutral and less through earth.
- Conversion to a three-phase circuit is done to promote more current returning to the source in the other phases and less current flow in the neutral.

Neutral Separation at the Transformer

In extreme cases, a high neutral voltage on the utility supply system will be isolated from the customer's neutral by removing the connection between the primary neutral and the secondary neutral at the transformer. A separate ground is installed on the secondary service. The separation of the neutral and the two separate pole/down grounds are hazards to utility personnel because there could be a potential difference across the open circuit between the primary and secondary neutral and the associated pole/down grounds. Under fault conditions, the voltage on the transformer tank or on the pole/down ground could be lethal. A caution sign is normally placed on the pole to alert power line technicians to the hazard of the separated neutrals. **See Figure 15–23.**

A gas-tube type of surge protector can be installed between the secondary neutral and the transformer tank. During a voltage surge, the protector sparks over internally and temporarily connects the secondary neutral to the transformer tank. With the transformer tank connected to the pole/down ground and primary neutral, the primary and secondary neutrals are temporarily connected together, thereby preventing a secondary insulation failure in the transformer.

Figure 15–23 Neutral separation at transformer will eliminate tingle voltage for the customer but because it is a hazard to linemen it should be temporary until a permanent fix is found. *(Delmar, Cengage Learning)*

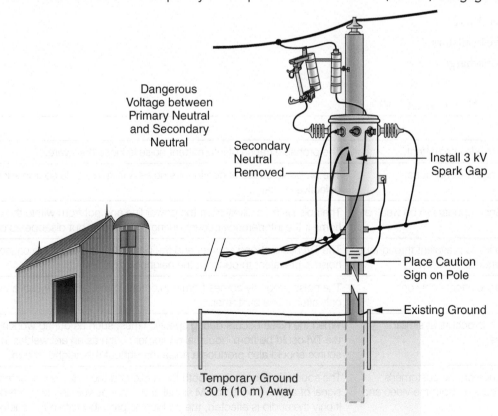

Customer Remedial Action

The usual first step to reduce the voltage on the neutral is to improve the grounding of the neutral at the service entrance. An inspection of the insulation of the wiring, neutral connections, and safety ground connections is also carried out.

The neutral-to-earth voltage can be reduced to acceptable levels by installing a tingle voltage filter, which is designed for this purpose. It is possible to install a grid bonded to the source of the tingle voltage, using the same principle as the grid attached to tension machines when stringing near energized circuits. A grid on the stable floor bonded to the stanchion and water bowl would ensure that there is no exposure to any difference of potential. This is not always practical as a retrofit.

Investigating a Radio and Television Interference (TVI) Complaint

TVI Trouble Calls

Trouble calls involving finding and fixing the source of radio and television interference (TVI) can be a frustrating experience for the lines trade. Customers who have interference with their reception are quite often on the fringe of being too far from the radio or television transmitter. The source of the interference is not always obvious and is not always due to the electrical utility facilities. Having a technician who uses specialized instrumentation is the best way to find difficult TVI sources. While checking the instrument or listening to an AM radio, the source pole of TVI can often be found by giving a pole a bump with a sledge hammer.

Three Main Sources of TVI on Utility Facilities

The three main sources of TVI on utility-owned electrical circuits are as follows:

- Loose hardware
- Defective insulators
- Corona discharge

Finding the Source of TVI

If	Then
The interference is intermittent.	The probable cause of intermittent noise is loose hardware.
The interference is continuous.	The probable cause of continuous noise is from a stable source such as a defective insulator.
The interference appears to be weather dependent.	The noise is most likely from the power line and not from within the customer's home if the interference occurs during dry weather but disappears during a rain.
The interference is prevalent throughout the neighborhood.	The TVI has a strong source, and the power line is a probable cause when the noise is prevalent throughout the neighborhood.
The interference affects only one customer.	The noise probably comes from equipment in the building when a weak source only affects one customer.
The interference occurs at similar specific times.	When the noise occurs during specific times, such as during working hours, the TVI could be from industrial equipment such as an arc welder. This type of source should also produce a noise throughout the neighborhood.
The interference on the customer's television occurs on both the video and the audio.	The source is strong when both the audio and the video are affected. The audio signal of a television is an FM signal and is not as vulnerable to interference. If only the audio is affected, the problem is probably within the customer's equipment.

The interference is from devices in the customer's premises.	Devices in the home that have been known to cause noise are an electric motor, a fluorescent light, electronic equipment, a doorbell transformer, flashing decorative lighting, an aquarium pump, heating pad, or dimmer switch.
The interference is from an outdoor source other than a power line.	Check for potential sources such as radio and television transmitters or two-way radio transmitters used by police, utilities, and other businesses.
The noise is the same across all television channels.	Depending on the strength of the noise, channels 2 to 6 are affected first, channels 7 to 13 are affected next, and UHF is almost never affected. Suspect the television itself if the noise is constant for all channels.

Loose Hardware

Loose hardware is the most common and the most intense source of TVI. Sparking occurs between loose pieces of metal that are not electrically bonded to each other. For example, a washer does not make a good electric connection to a bolt unless the bolt is tight.

When the wood on a wood pole or crossarm dries and shrinks, the hardware loosens. Dielectric gaps build up where metal pieces are not forming a tight bond with each other, and sparking occurs across the gap. The sparking occurs in repetitive bursts, which results in a radiated noise that interferes with radio and television reception.

Defective Insulators

Pin-type insulators on subtransmission lines have been a common source of TVI. The problems typically occurred with older insulators or poor conductor ties. Newer pin-type insulators have a semiconducting glaze (Q glaze) on the top surface of the insulator. The semiconducting glaze provides an equipotential area that prevents sparking between the conductor, the conductor ties, and the insulator.

Corona Discharge

Corona discharge as a source of TVI is most likely from higher-voltage circuits. The discharge emanates from sharp points on energized hardware. It is most likely to show up when a new line is first put into service or after maintenance work is carried out on the line. The design of the energized hardware on a high-voltage circuit requires everything to be smooth or rounded off. The usual cause of a corona discharge is a connection that is not properly smoothed off. Radio noise because of corona discharge should be a rare occurrence.

Review Questions

1. What does high-quality power look like?

2. Name four types of disturbances to power quality that can be continuous.

3. What is a common fix for poor-quality power?

4. Which of the following are considered measures used by a utility to reduce disturbances to a power system?

    (   ) Install voltage regulators
    (   ) Install a recording voltmeter at the customer.
    (   ) Install capacitors
    (   ) Install surge arrestors
    (   ) Install a new revenue kwh meter
    (   ) Construct a dedicated feeder

5. The neutral and the safety ground are normally tied together at the transformer supplying the service and also at the service panel. What happens when the safety ground is tied together downstream on the customer's premises?

6. Identify the formula for voltage drop and power loss.
   $IR = ?$
   $I^2R = ?$

7. What influences the amount of voltage drop in a circuit?

8. Why do distribution circuits have larger conductors than what is needed for ampacity?

9. *True or False:* Capacitors improve the voltage upstream from the capacitor installation.

10. The leads going in and out of a set of voltage regulators are very small compared to the main line conductors. Will the relatively small leads affect the capacity of the feeder to supply the load?

11. Why would a time delay of 10 seconds on a voltage regulator not be considered a good practice?

12. *True or False:* A bypass switch at a voltage regulator can be closed after the regulator has been zeroed.

13. When a circuit is temporarily fed in reverse through a regulator, what adjustments should be made at the regulator?

14. How does a set of capacitors installed on a distribution feeder benefit the electrical system?

15. How much current can be expected in the leads feeding a 450 kVAR capacitor bank on a 25/14.4 kV system?

16. When a capacitor is isolated with the intent to reenergize, why is it necessary to wait 5 minutes before re-energizing?

17. Name three possible causes of a low-voltage trouble call.

18. Name three possible causes of a high-voltage trouble call.

19. What can be seen or heard when a line is subject to harmonic interference?

20. Name three possible sources for flickering lights.

21. How can ferroresonance be prevented when energizing a three-phase transformer bank?

22. What can a utility do to reduce the risk of being the cause of tingle voltage at a customer?

23. Why is separating the primary neutral from the secondary neutral a hazard for power line technicians?

24. What are the three main sources of TVI complaints caused by utility equipment?

25. What solution is available to a utility to correct a corona problem that is causing TVI interference?

Chapter 16

Working with Aerial Devices and Digger Derricks

Objectives

After completing this chapter, you should be able to:

1. Perform a pretrip vehicle inspection.
2. Take the necessary steps to inspect the air brake system on a truck.
3. Perform a pre-use hydraulic system inspection.
4. Troubleshoot a hydraulic system.
5. Explain the method used to emergency lower an aerial device boom.
6. Take the steps needed to stabilize an aerial device and/or digger derrick truck.
7. Take steps to reduce the risk of electrical shock to personnel when a noninsulated boom is being used in the vicinity of energized lines.
8. Use hand signals when communicating with a derrick or crane operator.
9. Take steps to reduce the risk of electrical contact when working with an insulated boom.
10. Explain the lifting capacity limits of a digger derrick and an insulated aerial device.

Checking Out the Truck

Typical Daily Pre Trip Inspection

A daily pre trip inspection is required by law for large trucks. Most utilities/employers require a pre trip inspection for all other fleet vehicles. Under the law, the driver is totally responsible for the vehicle. If a truck has poor brakes, is overloaded, or has defective tires, any charges laid will be to the driver, not the mechanic or the utility/employer. Most utilities have a list or illustration showing the items to be checked in a logical order and a requirement to record the results in a logbook. When a different truck is introduced to your crew, make sure you ask what inspection points are unique to that truck.

To reduce the possibility of a high-risk failure, prioritize on specific items that can lead to a catastrophic failure.

- Check for signs that indicate lug nuts are loose. An off-center valve stem in a truck wheel opening is an indicator of a loose wheel. Visual indicators can also be installed. **See Figure 16–1**.
- Too much travel (should be less than 2.5 in./6 cm) in the slack adjusters of an air brake is a common fault leading to poor braking performance. The angle between the push rod and the adjuster arm should not be more than 90° when the brakes are applied. **See Figure 16–2**.
- Check for a leaking hub or axle seal. If equipped with a sight glass on the wheel, check the oil level.
- Adjust trailer brakes so that they will not lock up when a brake pedal is applied. A trailer with locked-up brakes can slide sideways/jackknife and strike other vehicles.

Figure 16–1 If a lug nut indicator has turned it means that the nut needs to be tightened. (*Delmar, Cengage Learning*)

Figure 16–2 If the slack adjusters are over 90° they will not meet regulatory compliance. (*Delmar, Cengage Learning*)

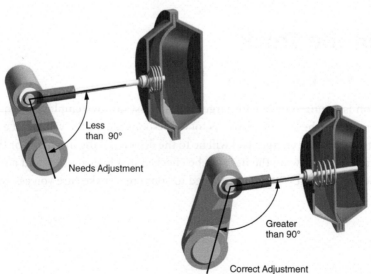

Summary of Pre Trip Air-Brake Inspection

The following summary of a pre trip air-brake inspection is a memory device only. Presumably the driver has checked under the hood and will inspect the physical condition of the braking system and slack adjusters.

1. Drain air tanks to zero pressure, then do the following:
   1.1. Build up pressure; low-air warning signal cuts out at 60 psi (415 kPa) minimum.
   1.2. Pressure buildup between 85 to 100 psi (590 to 690 kPa) must be less than 2 minutes.
   1.3. Governor cuts out between 100 and 135 psi (690 to 930 kPa).
2. At full pressure, fan/pump the brake pedal, then check the following:
   2.1. After a drop of about 25 psi (170 kPa) from the maximum pressure, the governor should cut in.
   2.2. At about 60 psi (415 kPa) minimum, the low-air warning cuts in.
   2.3. Between 45 and 20 psi (300 to 140 kPa), the parking brakes (spring brakes) should come on automatically.
3. Build back to full pressure, then do the following:
   3.1. With engine stopped, fully apply the foot brake and hold for 1 minute. Pressure drop must be less than 2 psi (15 kPa). With a trailer, the drop must be less than 4 psi (30 kPa).
   3.2. With the engine at idle speed and the transmission in first gear, check the parking brake adjustment by trying to move forward.
   3.3. With the parking brakes off, drive ahead and apply the service brake as a check for response.

With the truck in motion, activate the trailer brakes (spike) as a check for response.

Jump-Starting

A battery gives off explosive hydrogen gas. A spark near a battery can trigger an explosion.

1. Before jump-starting, check that the fluid level is normal and that the battery is not frozen. A maintenance-free battery should not be jump-started if the indicator is red.
2. Only the last connection will cause a spark that could ignite the hydrogen gas. The last connection should, therefore, be a negative (ground) connection at a location remote from the battery to the truck frame. **See Figure 16–3.**

Figure 16–3 Prevent a spark near the battery when jump-starting a vehicle by making the last connection to the engine block. *(Delmar, Cengage Learning)*

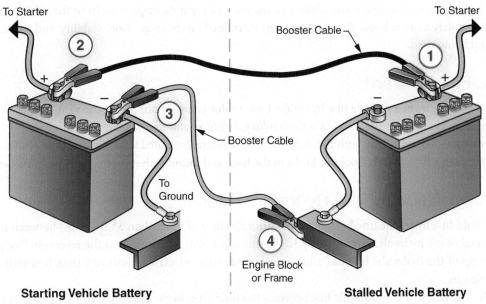

Studs are often installed on large trucks to facilitate jump-starting. Look under the hood or near the driver's side step for a remote positive jump-starting stud, and look for a negative grounding stud on the truck frame.

Turning on the headlights or some other load on the truck with the good battery will reduce the risk of a damaging voltage surge to electronic equipment.

Ways to Reduce High-Risk Driving Hazards

Driving a truck is part of a power line technician's job. The following tips may reduce the risk of typical utility truck accidents.

- Never use the hand valve to apply brakes only to the trailer. Locked brakes on a trailer can cause the trailer to slide sideways/jackknife into other traffic.

- Do not jam on the brakes in an emergency. When the wheels lock up, the truck will go into an uncontrolled skid and steering will be impossible. For a truck without antilock brakes, use "stab" braking in an emergency stop. Emergency braking for a truck with antilock brakes requires steady and increasing pressure on the brake pedal. There may be variations for the most effective braking for trucks with antilock braking; check the owner's manual.

- Watch for restricted roadway height or width. Many derrick diggers and aerial devices are close to the maximum dimensions of 13 ft, 6 in. (4.15 m) in height and 8 ft, 6 in. (2.6 m) in width. A jib left out at an angle or a protruding outrigger can cause a vehicle to be over the height or width limits.

- Be in the correct lower gear before starting down a steep hill. Overreliance on brakes to slow a truck will cause the brakes to overheat and become ineffective. A rule of thumb for going down a steep hill is: "Before starting down the grade, put the truck in the same gear going down the grade as would be needed to go up the hill."

- When extra space is needed to make a turn with a truck and trailer, take the extra space from the street you are entering, not from the street you are leaving.

- To save confusion when backing a trailer, leave your hands on the bottom of the steering wheel. The trailer will go in the direction of your hands. Use an observer when backing. If there is no observer, walk around the vehicle before backing up.

Monitoring a Hydraulic System

Inspecting Hydraulic Equipment

The inspection of any hydraulic equipment should be treated as a separate inspection from the vehicle it is mounted on and recorded separately in a log book. An operator must be trained on the inspection, stability, and capacity of each type of hydraulic unit being used.

Hydraulic Fluid Hazard

Hydraulic fluid coming from a pinhole in a hydraulic hose under pressure can pierce the skin. A leak from a pinhole is not normally visible. Fluid under the skin is a serious injury/infection and needs medical treatment.

Never investigate hydraulic leaks with the system under pressure. Insulated hose without steel reinforcement is especially susceptible to this type of leak. Look for kinks in the hose and for oil gathering in places near the hose.

A Conductive Vacuum in a Hydraulic Hose

The weight of fluid in a hydraulic line having a separation distance of more than 35 ft (11 m) between the oil reservoir and the upper end of the hydraulic line will drop in the line somewhat and drain into the reservoir. The partial vacuum formed at the top of the hydraulic line has a lower resistance to an electrical flashover than hydraulic fluid or air at atmospheric pressure.

Check valves installed in the hydraulic line prevents the fluid level in the line from dropping and creating an electrically conductive vacuum. Vent valves are installed, as a backup, to allow air to enter the hydraulic line and to keep it at atmospheric pressure. The filter in the valve keeps dirt out. These valves need cleaning, and are critical for barehand work where the insulation of the boom and hydraulics provide primary protection.

Troubleshooting a Hydraulic System

Only insulated hydraulic fluids are used in insulated aerial devices. The fluid must be kept clean to prevent an electrical failure.

The touch, sound, look, and smell of hydraulic oil can give clues about impending problems that need correction.

Touch

1. *Caution:* If there is any suspicion of a hydraulic fluid leak and its location is unknown, *do not* try to find the leak by touching any hydraulic lines while the system is under pressure.
2. If the hydraulic pump or a hose is too hot to touch, oxidation in the fluid will cause sludge to form and create more heat and an eventual breakdown. The maximum temperature of hydraulic oil is 135°F (57°C).
3. If a high-frequency vibration can be felt when touching steel fittings, it is a sign of a damaging condition that needs correction.

Sound

1. A loud shotlike sound (water hammer) is caused by a sudden stoppage of the fluid in the system. A resulting pressure surge can be as high as four times the normal pressure. Feathering the controls (easing the valve open slowly) or some corrective plumbing may be needed to prevent this potentially damaging fault. Valves controlled by radio or by fiber optics are designed to prevent shock loading.
2. A sound like the hydraulic pump is pumping marbles indicates that the pump is trying to pump more fluid than the system can supply to it. This causes a partial vacuum in the fluid at the pump (cavitation).
3. A high-pitched whine indicates back pressure at the pump. This often happens when the fluid is not properly warmed up and too thick to pump at the normal rated speed.

Look and Smell

1. Check the fluid level in the hydraulic oil reservoir.
2. Check the appearance of the oil. A milky appearance means the oil is saturated with air or water.
3. If any of the cylinders have jerky or erratic movement, air may be in the system or a cylinder rod may be bent.
4. A burnt fluid smell indicates that the system is being subjected to high temperatures caused by air bubbles in the oil cavitating at the pump.

Causes of Hydraulic Pump Failure

A partial vacuum in the fluid at the pump is the most common cause of cavitation and hydraulic pump failure. Cavitation occurs in the following situations.

- When the pump is over revved or brought up to speed before the fluid has warmed enough to flow adequately
- When the fluid level is low in the reservoir
- When the fluid filter or a suction line is restricted
- When the shut-off valve fails to open after repairs
- When travelling on the highway with the PTO engaged

Proper Warm-Up for Hydraulic Systems

When the temperature is below freezing, the hydraulic pump can be damaged by running it at the rated rpm while the oil is cold.

A typical warm-up would be to set the engine speed at 600 rpm (1,000 rpm for diesel engines), to engage the hydraulic pump, to let it run for about 10 minutes, and then to increase the engine speed gradually until it reaches the specified rpm. *Caution:* The rpms may increase to damaging levels as the unit warms up.

High-Pressure Hoses

Compression tools, wrenches, hot-line tools, and arborist tools being operated from the tool outlet at the bucket level must use hydraulic hoses rated for 10,000 psi and must be nonconducting (generally, orange).

A metal-reinforced hydraulic hose, used by mistake, has caused an electric short that resulted in a ruptured hose and a hydraulic fluid fire in the bucket.

Checking Holding Valves

A holding valve will keep a boom up in the air even if a hydraulic hose fails. The following describes how to do a drift check on the holding valves.

1. Extend the outriggers.
2. Put the boom into position. **See Figure 16–4a** and **b**.

Figure 16–4a A digger derrick boom is elevated for a drift test so that the lift cylinder is under pressure. *(Delmar, Cengage Learning)*

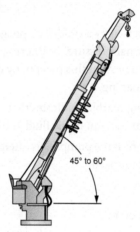

Figure 16–4b An aerial device boom is placed in various positions to put different lift cylinders under pressure. *(Delmar, Cengage Learning)*

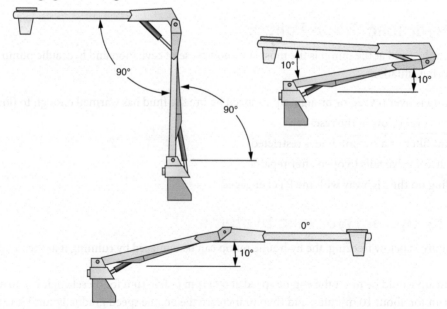

3. Shut off the vehicle.

4. Activate the valves to lower outriggers and booms in the various positions, as shown. Hold the positions for 15 seconds.

Any movement downward indicates that the holding valve needs maintenance or that air is in the system. A formal drift test can be done in the same way, except that the boom is loaded with its rated load and held for a given period of time.

Emergency Lowering of an Aerial-Device Boom

The lower controls are used to bring down the boom in an emergency. The lower controls must be well marked to lessen confusion during an emergency. The lower controls override the upper controls.

It is a given that the crew member on the ground has been trained and has practiced bringing down the boom using the lower controls, removing a casualty from the bucket, and performing necessary CPR and first aid.

Lowering a Boom without Hydraulic Power

When there is a loss of hydraulic power and the boom has to be lowered by manipulating the holding valves, any workers in a bucket should first lower themselves from the bucket to the ground using an approved and practiced method and equipment. It is a given that anyone working from a bucket has been trained and has practiced escaping from the bucket while aloft.

Generally, no-power lowering is best left to a mechanic or specialist because it is very easy to lose control of the process. Many units have an auxiliary battery-operated pump to lower the boom in case of a loss of hydraulic power; this is often called a secondary storage system. Some units have hydraulic couplers that allow the hydraulic system of another truck (preferably a truck with clean insulating fluid to prevent contamination) to connect and operate the disabled system.

Stabilizing a Boom-Equipped Vehicle

Procedure for Stabilizing a Boom-Equipped Vehicle

To reduce the risk of a tip over, set up the truck on a firm, level foundation before using the boom.

Stabilizing a Boom-Equipped Vehicle

Step	Action	Details
1.	Ensure that the mechanical and hydraulic systems of the unit have been maintained.	Ensure that normal maintenance has been carried out. 1. Learn the location and check critical welds for surface cracks and lines of rust. 2. Check pin retainers at pivot points. After any repair or hydraulic oil leak, put the boom through its full cycle to ensure that all hoses are full of oil and that there will be no sudden collapse of the boom because of an air pocket. On a scheduled basis, do a drift test on the holding valves. Keep a log book on the hydraulic unit.
2.	Apply the parking brake and chock blocks.	Hydraulic-braked trucks with boom equipment have both a mechanical parking brake and a hydraulic locking brake (accumulock). Both must be on. *Caution:* Hydraulic parking brakes can bleed off. Should this happen when the job is finished and the outriggers are raised, the truck will be free to roll down a slope unless chock blocks and the mechanical parking brake are applied. If they are properly adjusted, the parking brakes on air-braked trucks are very reliable. With the vehicle on an incline, extending the outriggers reduces the holding power of the tires. Use wheel chocks to help prevent the vehicle from sliding downhill. Actions of the boom sometimes cause the truck to shift on the outrigger planks, especially if the wheels are off the ground. Stability can be increased by anchoring the truck to another truck and/or installing chock blocks, front and back.

(Continued)

Step	Action	Details
3.	Use outrigger pads and other means to provide a solid surface for outriggers.	Loss of stability can occur because the outriggers penetrate into the earth, asphalt, or concrete or because the vehicle slips off the outrigger pads. Each outrigger should be on pads to offer a greater surface-for-weight distribution. **See Figure 16–5**.

Figure 16–5 Pads prevent outriggers from penetrating the ground surface. *(Delmar, Cengage Learning)*

4.	Extend outriggers (stabilizers).	The outriggers should lift all the weight normally supported by the springs. Some telescoping outriggers will have a mark indicating the minimum extension needed to provide stability.
		Units with only two outriggers (trucks with the turret mounted just behind the cab) will sometimes have the front wheels lifted from the ground. Planking or support under the front wheels is needed to prevent a sudden shock-loading drop when the boom is rotated toward the front of the truck.
		Keep the outrigger in sight, or clear the area while lowering the outrigger.
5.	Level the vehicle to less than 5° on a slope.	On sloping ground, extend the low-side outrigger first, then extend the other outrigger(s) until the unit is level.
		It may be necessary to build up the low side with extra timbers/cribbing, or to dig out the high side, or to use the pads specifically made for slopes. If a telescoping-type outrigger cannot be extended far enough, it creates a stability hazard. A radial-type outrigger has an advantage over the telescoping-type outrigger for this problem. **See Figure 16–6**.

Figure 16–6 Curb can prevent the outrigger from extending far enough to provide good stability. *(Delmar, Cengage Learning)*

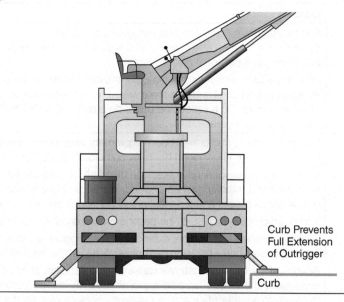

Curb Prevents
Full Extension
of Outrigger

Curb

6.	Trucks without outriggers need to be level and have rated tire pressure.	The truck must be set up as level as possible. A boom over the side can put a truck without outriggers on a 3° tilt. If the truck is set up initially on a 5° slope, the unit could be operating on an 8° tilt.
		The low side can be built up by driving up on pads or similar material. **See Figure 16–7.**
		It is critical to the stability of the truck to maintain proper tire pressure and to learn the inspection points for the stability features of the vehicle.

Figure 16–7 Build up the low side of a truck to make it level. *(Delmar, Cengage Learning)*

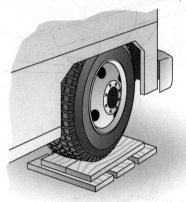

Stability on a Slope

Boom-type equipment is designed to work at a slope of 5° or less. The danger of rotation gear failure or tipping over increases with the slope.

Working over the uphill side or the end of a truck improves stability, but rotating a load uphill or downhill will over-stress the rotation gears. If work is planned for the uphill side of the unit, plan all lifting so that very little load is on the rotation gears.

The stability of some corner-mount digger derricks can be derated by as much as 40% over the downhill side, even if the unit is on a slope of less than 5°. Training and introduction of any unit to a new operator would include any unique stability features.

Electrical Protection for Working with Noninsulated Booms

Types of Contact Hazards with Noninsulated Booms

A vehicle with a noninsulated boom can become an electrical hazard on a job site to people on the ground.

When a Noninsulated Boom Is Hazardous

If	Then
A person on the ground is *in contact* with a utility vehicle, winch line, or power-installed anchor while the noninsulated portion of a boom contacts an energized circuit.	The person is a parallel path to ground and will take *some* share of the current flowing to ground.
A person on the ground is *near* a utility vehicle while the noninsulated portion of a boom accidentally contacts an energized circuit.	The person is exposed to ground gradients at each location where current is entering the earth. **See Figure 16–8.**

(Continued)

If	Then
In a high-induction area, a noninsulated boom is used to lift or handle an *isolated and grounded* conductor.	Unless the boom is bonded to the grounded conductors, the boom can be at a different potential than the conductors being handled.

Figure 16–8 Stay clear of a truck when the boom is in motion because there can be an unexpected electrical contact. *(Delmar, Cengage Learning)*

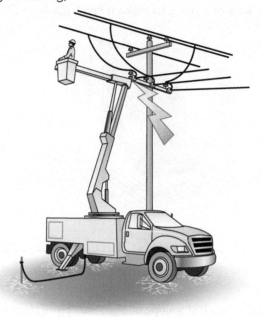

Protection from Shock around Noninsulated Booms

The only real protections available when a boom makes contact with an energized conductor are as follows:

1. *Staying on the vehicle or on a ground-gradient mat* bonded to the vehicle will keep a person at the same potential as the vehicle. As long as a person avoids contact with anything not connected to the vehicle, no current will flow through that person to another object. **See Figure 16–9.**

Figure 16–9 (a) An operator is in an equipotential position when staying on the vehicle. (b) A ground-gradient mat bonded to the truck will put an operator in an equipotential zone. *(a, Delmar, Cengage Learning; b, © Salisbury by Honeywell)*

(a)

(b)

2. *Keeping a safe distance away from a vehicle* will keep a person from making inadvertent contact and will keep a person away from any high ground gradients. A wireless remote-control operating system and/or truck barricading promotes the ability to stay away from the vehicle.

3. *Electrical shock-resistant footwear* is a last-resort safety barrier. The increased resistance offered to a worker will provide some additional protection from being in a parallel path to ground and from ground gradients.

Barricading a Vehicle

The best protection when working near a vehicle with a noninsulated boom (if not on the vehicle) is staying away from the vehicle, the trailer, and any attachments. Barricading promotes keeping a distance from the vehicle.

If the barricades are put in place before they are needed, workers tend to routinely violate the barricades as they prepare a transformer or frame a pole. The operator of a boom should ensure that everyone stays away from the vehicle when the boom is to be moved near an energized circuit.

Reducing the Risk of Accidental Contact with a Conductor

Take steps to reduce the risk of a boom contact with an energized conductor.

1. Use a signal person, other than the equipment operator, to watch the approach distance to exposed lines and equipment and to give timely warnings before the minimum approach distance is reached.

2. A noninsulated boom, with an operator qualified to operate a boom in the vicinity of energized conductors should not get closer than the allowed minimum approach distance unless the truck is grounded, a signal person is available, and protective cover-up is installed.

Communicating with the Derrick Operator

Use hand signals or a two-way radio to communicate with a derrick or crane operator.

- A signal person must be qualified by having completed training and passing an examination.
- Have only one signal person.

- Be sure that the signal person and the operator are using the same signals. **See Figure 16–10.**
- The signal person watches the load, and the crane operator watches the signal person.
- Make sure the load does not pass above workers.
- Watch for the minimum approach distance to the power line.
- Watch for anyone who is not a safe distance from the vehicle or load.

Figure 16–10 These are standard derrick operation hand signals used in the industry. *(Delmar, Cengage Learning)*

Stop Emergency Stop

Swing Boom Up Boom Down

Hoist Up Hoist Down Extend Boom Retract Boom

Stop—With arm extended horizontally to the side, palm down, arm is swung back and forth.

Emergency Stop—With both arms extended horizontally to the side, palms down, arms are swung back and forth.

Swing—With arm extended horizontally, index finger points in direction that boom is to swing.

Boom Up—With arm extended horizontally to the side, thumb points up with other fingers closed.

Boom Down—With arm extended horizontally to the side, thumb points down with other fingers closed.

Hoist Up—With upper arm extended to the side, forearm and index finger pointing straight up, hand and finger make small circles.

Hoist Down—With upper arm extended to the side, forearm and index finger pointing straight down, hand and finger make small circles.

Extend Boom—With hands to the front at waist level, thumbs point outward with other fingers closed.

Retract Boom—With hands to the front at waist level, thumbs point inward with other fingers closed.

Grounding the Vehicle

Grounding the truck promotes a fast trip-out of the circuit if the truck should become energized. A fast trip-out reduces risk by lowering the exposure time to the hazard; it does not eliminate a potentially fatal shock hazard for anyone touching the vehicle.

Note: Anyone touching a vehicle that has become part of the circuit will be a parallel path to ground and will, therefore, have some share of current going through the body. Even when the vehicle is grounded to an excellent ground, like a neutral, the amount of current going through the body could still be fatal.

Following are typical ground electrodes, listed in priority.

1. Permanent ground network such as a station ground, a neutral, or a tower ground

2. Existing ground rod or an anchor rod in earth (*Note:* A truck ground attached to the guy steel could cause the guy to burn off during a fault.)

3. Temporary driven ground rod (*Note:* The pole ground/down-ground on a pole is often too small to carry the available fault current on the circuit.)

On lower-voltage wye distribution systems, the vehicle must be grounded to a neutral to trip the circuit. The resistance of an existing ground rod or anchor rod will not normally be low enough to generate enough fault current to trip out the circuit. A ground rod may suffice on a circuit protected by a circuit breaker or electronic recloser with a ground trip relay.

On a delta circuit, the vehicle should be grounded to a ground rod. Many delta circuits will be protected by a ground fault relay, which will trip out a circuit when there is a line-to-ground fault.

On higher-voltage distribution and subtransmission circuits, the voltage will be high enough to overcome the resistance of the earth around a well-driven ground rod and will cause the circuit to trip out.

Boom Contact with an Isolated and Grounded Conductor

A circuit is not "dead" when it has been isolated and grounded. Current is often flowing in the grounds; a voltage difference often exists between the circuit and a remote ground.

When a noninsulated boom is used to handle conductors on a job where a circuit is isolated and grounded, the boom should be bonded to the protective line grounds.

On a transmission right-of-way that has high induction from energized neighboring circuits, or in the rare case of accidental re-energization, bonding ensures that there will be no potential difference between the vehicle, the boom or its wire rope winch, and conductors.

Electrical Protection for Working with Insulated Booms

Minimum Approach Distance with Insulated Booms

When a qualified operator is in the bucket of an electrically tested bucket truck, that bucket and boom should maintain the same minimum approach distance as a qualified worker must maintain when working on a hot line. Maintaining this distance will reduce the risk of inadvertent contact with other phases or objects at different potentials while working with rubber gloves or barehanded. After protective cover-up is installed, a visible air space should be kept between the insulated portion of the boom and covered conductors.

An unqualified operator in a bucket should maintain a distance of 10 ft (3 m) from a distribution circuit and an additional 4 in. (10 cm) for every 10 kV over that, which works out to 14 ft (4.3 m) for 169 kV, 16 ft (4.9 m) for 230 kV, and 25 ft (7.6 m) for 500 kV.

Maintaining the Insulation Value of the Boom, Bucket, and Jib

To maintain the insulation value of a boom requires continuing care and maintenance.

- Keep the insulated section of the boom, bucket, and jib clean and dry. Maintain a waxed or silicone surface to repel water. Use specified cleaners only; abrasive cleaners can leave tiny scratches, and solvents can soften the surface coatings. A dry boom with a chalky or non-water-repelling surface may pass the dry dielectric test and still fail electrically after a brief shower.

- Keep the boom interior clean and dry. Wash it out with a low-pressure washer (clean water only), and dry it by leaving the boom in a vertical position or by pouring isopropanol down the boom as a drying agent. High-pressure water can cause water to diffuse through the fiberglass, which will require a very long time to dry out.

- Ensure that insulated booms, buckets, liners, and jibs get their scheduled electrical retests. A boom dielectrically tested while it is still wet can cause permanent damage.
- Boom and bucket covers prevent road salt and road wash from contaminating the insulated portions of a unit. They can also prolong the life of a boom by protecting it from the ultraviolet rays of the sun, causing the boom to look fuzzy when the fibers become exposed.
- Cover the jib or store it in a canvas bag or secured on padded holders in a truck bin.
- Boom leakage over 1 mA (1,000 microamperes) can leave track burns, typically inside the boom. If—when checking out the boom for barehand work or carrying out a formal electric test—the reading on the meter shows a return current approaching 1,000 microamperes, stop, clean, or dry the boom.
- Strap down the boom during travel. A boom subjected to vibration and shock loading will shorten the life of fiber-glass, and the boom will be damaged where it sits on the boom rest.

Monitoring Boom Contamination

Barehand work and high-voltage rubber-glove work typically require that the aerial device have a boom-contamination monitoring circuit that will measure and monitor the actual leakage current of the insulated boom.

The initial certifying electrical proof test should be used as a reference point for future contamination monitoring. Keeping a log of every test will show trends that may develop.

The monitoring circuit of the aerial device is checked before doing the boom-contamination test. The circuit can be tested for an open or short circuit using the "test" position on the same meter used to measure the leakage current. **See Figure 16–11.**

Figure 16–11 A boom-contamination monitoring circuit measures the leakage between the upper electrode and the collector band. *(Delmar, Cengage Learning)*

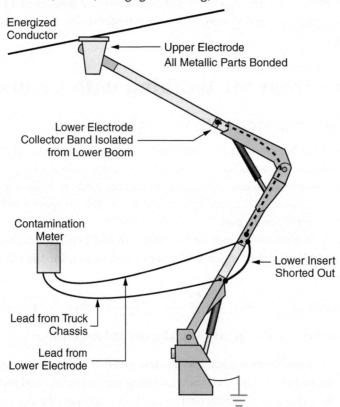

A current test is made daily before starting work and before working on a higher voltage. A metal part of the boom or bucket grid is put in contact with the energized circuit to be worked on at intervals of 1 minute.

The microampere meter shows the amount of current between the upper electrode, which consists of all the metal components bonded together, and the lower electrode, which is the collector band mounted inside and outside of the boom, isolated from the metal lower boom. The leakage current should not exceed 1 microampere per kilovolt of the phase-to-phase voltage of the circuit. Many units will be well below this standard. Note the typical action required for poor contamination meter readings.

Contamination Meter Readings

If	Then
The contamination meter reading is *high* (over 90) or there is a *sudden increase* in the contamination meter reading.	1. Try retesting with the lower-boom position farther away from the conductor (lessen induction). 2. If not successful, clean, dry, and retest the boom.
The contamination meter reading is *fluctuating* up and down.	1. Moisture is probably present. Wipe down the boom and remove the moisture inside of the boom by pouring isopropanol into it. 2. If not successful, there may be a fault in the boom insulation. Remove the truck from service and get it checked by specialists.
The contamination meter reading is showing a *gradual increase* from day to day.	1. There may be dirt contamination on the inside and outside surfaces of the boom. Clean the boom inside and out, let the boom dry out, and then retest. 2. The hydraulic oil may have become contaminated. Shut down the unit for about 15 minutes, drain the water from the reservoir, and then retest. If not successful, change the hydraulic oil and retest. 3. If not successful, there may be a fault in the boom insulation. Remove the truck from service and get it checked by specialists.

Caution: Intentionally letting a boom with high leakage current stay in contact with the circuit as a means to dry the boom will leave carbon tracking inside the boom, which will eventually lead to permanent damage and an electrical failure.

Grounding Vehicles with Insulated Booms

There have been instances of the noninsulated lower boom of an aerial device contacting an energized underbuild or a lateral tap. These vehicles must be grounded to protect workers on the ground. **See Figure 16–12.**

Figure 16–12 An insulated bucket truck with an uninsulated lower boom must be grounded to protect workers on the ground. *(Delmar, Cengage Learning)*

The need to ground an aerial device with an insulated lower boom insert is based on individual company preference. *Cautions:* (1) Some lower-boom inserts may have had their insulation short circuited to use the contamination monitoring system. (2) If a unit with an insulated lower-boom insert (which has not been shorted out) is used for work on a transmission line, a high voltage can be induced on the section of boom above the insert.

Operating a Digger Derrick

Lifting with a Digger Derrick

Reduce the risk of damaging a digger derrick boom by using the lifting charts and boom features available.

1. The lifting capacity chart for each type of digger derrick must be available to the vehicle operator. A lifting chart should be visible from the boom operator's position.
2. There is decreased lifting capacity as the boom is extended away from the turret location. **See Figure 16–13.**

All Booms Retracted	Boom Angle	80°	75°	60°	45°	30°	15°	0°
	Lb. Load	9,500	7,900	5,500	4,300	3,700	3,350	2,950
Intermediate Boom Extended	Elevation	80°	75°	60°	45°	30°	15°	0°
	Lb. Load	7,000	5,900	3,600	2,800	2,500	2,100	1,850
Third Section Capacity	Elevation	80°	75°	60°	45°	30°	15°	0°
	Lb. Load	4,000	3,400	2,000	1,450	1,200	1,100	1,000

Figure 16–13 A typical digger derrick lifting-capacity chart shows the weight that can be lifted is dependent on the length and angle of the boom. *(Delmar, Cengage Learning)*

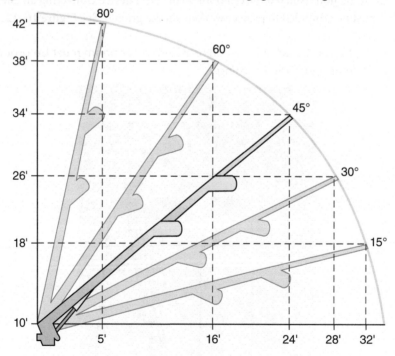

3. The winch and stinger can lift loads in a position that will overload the boom.
4. The lifting capacity changes with the direction of the boom in relation to the truck, outriggers, and slope of the ground. A typical chart will show a 60° rating for a slide slope and 75° rating for an uphill slope when lifting over the rear. **See Figure 16–14.**

Figure 16–14 Note how the lifting capacity is derated on this rear-mount four-outrigger unit on a slope. *(Delmar, Cengage Learning)*

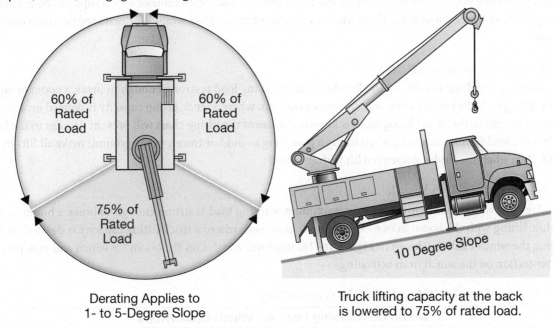

60% of Rated Load

60% of Rated Load

75% of Rated Load

Derating Applies to
1- to 5-Degree Slope

10 Degree Slope

Truck lifting capacity at the back
is lowered to 75% of rated load.

Rotation Gear Failure Hazard

A rotation gear failure allows the boom and load to rotate freely into objects and people. The rotation gears are often the weakest link when pulling an object from a ditch, operating a heavy load with the truck set up on a slope, or installing a power anchor. **See Figure 16–15**.

- Do not pull a load, but rotate the boom toward the load to make the lift.
- Set up so that a power-installed screw anchor can be installed with the need for very little rotation. If rotation is necessary, make sure the boom is moved to follow the anchor as it is installed.
- Do not try to loosen a pole setting for removal by rotating the boom side to side.
- Derate the lifting capacity of the boom when set up on a slope.

Figure 16–15 Do not pull a load by rotating the boom sideways. *(Delmar, Cengage Learning)*

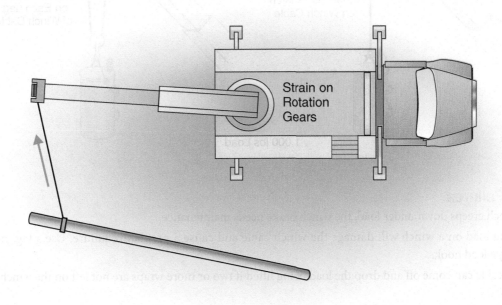

Strain on Rotation Gears

Lifting with a Digger Derrick Winch

The lifting capacity of a boom tip or turret winch and the strength of the winch cable or winch rope can be greater or less than the capability of the boom or load. There shall never be less than two full wraps of winch rope or winch cable on the winch drum.

When the Winch Is Too Strong

A 1/2 in. wire rope winch cable with a 4,200 lbs maximum working load is strong enough to break a boom or tip over a truck while lifting with the boom in one of many weak positions, where a truck lifting capacity is derated on a slope.

When the weight of the object being lifted is known, the use of the lifting chart will prevent damage to the boom or truck. When the load is unknown (e.g., when lifting a pole lying in mud or frozen to the ground) make all lifts by booming up first, then when the load is suspended lift with the winch.

When the Winch Is Too Weak

A 1/2 in. wire rope winch cable with a 4,200 lbs maximum working load is strong enough to break a boom or tip over a truck while lifting with the boom in one of many weak positions, where a truck lifting capacity is derated on a slope. Two-parting the winch cable, as shown in **Figure 16–16,** improves control on the boom tip winch and may prevent the overload protection on the winch from activating.

Typical Digger Derrick Winch Cable Capacities

	Maximum Working Load for Winch Cable/Rope
0.5 in. double-braided polyester rope	2,000 lbs/910 kg
0.75 in. double-braided polyester rope	3,500 lbs/1,590 kg
1 in. double-braided polyester rope	6,000 lbs/2,700 kg
1.25 in. double-braided polyester rope	10,000 lbs/4,500 kg
0.5 in. wire rope	4,200 lbs/1,900 kg

Figure 16–16 Lifting with a boom-tip winch. Note the decreased tension on the cable when using a two-part pull. *(Delmar, Cengage Learning)*

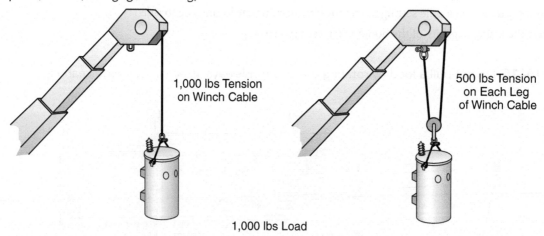

1,000 lbs Tension on Winch Cable

500 lbs Tension on Each Leg of Winch Cable

1,000 lbs Load

Winching Failures

- If the winch creeps down under load, the winch brake needs maintenance.
- A rotating load on a winch will damage the winch cable and cause a premature failure. Use a tag line and/or a swivel-type load hook.
- A winch cable can come off and drop the load being lifted if two or more wraps are not left on the winch drum.

Checking a Unit's Overload Protection

If the unit is equipped with overload protection, certain boom functions will not operate when the unit is overloaded. To check the overload protection, raise the boom to its maximum height until the system bypasses and try to extend the stinger. The stinger will not extend if the overload protection is working. Lowering the boom to the bottom will reset the overload protection.

Pulling a Pole

To remove a pole out of the ground, use the hydraulic pole jack (butt puller). If necessary, use the auger to loosen a pole setting, then lift with the boom only. Do not use the winch, stinger, or boom rotation to loosen the pole in the ground. Also, do not use the auger as a stiff leg.

Operating the Auger

When the auger is unstowed or stowed, there is always a danger of the windup cable or rope breaking during the operation. There is a history of severe injuries when people have been struck by the free-falling auger. Everyone must stand clear during the unstowing and stowing operation. **See Figure 16–17.**

Figure 16–17 There is a risk of the auger windup rope breaking, so always stand clear when it is being lowered or raised. (*Delmar, Cengage Learning*)

When digging a hole, keeping some downward pressure on the auger is normal especially in hard digging. An operator must stay alert that too much downward pressure in certain soil conditions can cause the auger to corkscrew into the ground and overload the boom. To dig a hole straight up and down, there needs to be ample room to extend or retract the stinger/boom extension. **See Figure 16–18.**

Installing a Screw Anchor

When installing a screw anchor, the anchor will draw the boom into the direction the anchor is going into the ground. This has been the cause of boom and rotation gear failures. Operate the boom so that it follows the anchor as it is screwed into the ground. Reduce the complexity of following the anchor by setting up the unit so that it is not necessary to rotate the boom, as well as by booming down. **See Figure 16–19.**

Figure 16–18 When auguring a pole hole with a digger derrick, the boom extension needs to be adjusted to keep the auger straight. *(Delmar, Cengage Learning)*

Figure 16–19 When installing a screw anchor, an operator will need to rotate the boom and move the boom down to follow the anchor into the ground. *(Delmar, Cengage Learning)*

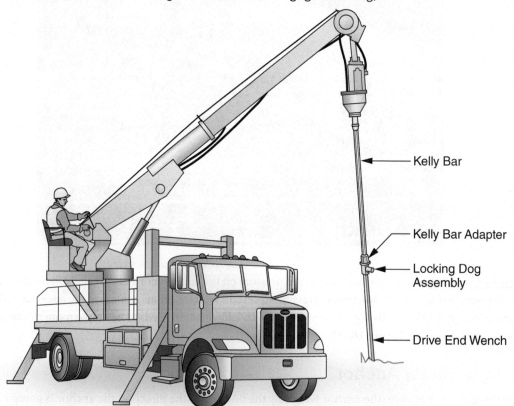

Kelly Bar

Kelly Bar Adapter

Locking Dog
Assembly

Drive End Wench

A good installation requires proper alignment and down pressure. The right amount of down pressure is needed to keep the anchor continuously advancing. Too little down pressure may result in "churning" the soil and result in a loss of the ability for the anchor to hold the expected load. Too much down pressure may bend or even break an anchor helix at torque loads far below the rating. The down pressure on the boom should lift the outrigger on the boom side of the truck. Maintaining down pressure so that the outrigger stays about 2 in. (5 cm) off the outrigger plank will keep a steady down pressure on the installation.

Begin anchor in near vertical position. When anchor has a good start, adjust the boom to the correct anchor angle.

The torque needed to install a screw anchor varies with soil conditions, so it can be too easy to apply more torque than the anchor can withstand. A unit with a "torque limiter" control can be set to limit the torque generated by the auger motor. Screw anchors have a torque rating stamped on them. The risk of boom failure is reduced when the overload protection is working properly.

Operating an Aerial Device

Aerial-Device Boom Capacity

The lifting capacity of a boom is dependent on many variables, including stability, boom angle, and horizontal load.

Stability
The stability of the truck setup affects the ability to lift the total weight shown on the lifting charts. Lifting charts assume a level, stable setup. Some vehicles will have a derating factor for vehicles not parked on the level.

Boom Angles
Vital information about lifting capacity must be taken into account when operating an aerial device. Always use a lifting chart or table for the specific boom being used. **See Figure 16–20.**

Figure 16–20 A sample lifting-capacity chart shows the large variance in lifting capacity and different angles. *(Delmar, Cengage Learning)*

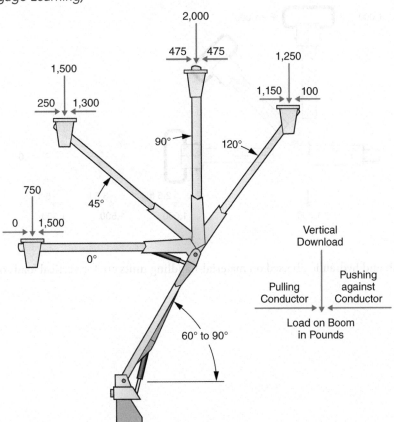

Caution: Determine whether the capacity chart you are using refers to "bare shaft capacity." If so, the total weight of the buckets, jib, and workers must be subtracted from the capacities shown on the chart.

Horizontal Loads

An aerial device is often used to lift conductors into and out of corner structures. Many manufacturers will state that no horizontal load is to be placed on their unit. Horizontal loading-capacity charts are not available for many units.

Moving conductors into a corner is a high-risk operation because the magnitude of the load is usually unknown and the load increases quickly as the conductor is moved into the bisect.

The horizontal loads are sample numbers only. *Note:* Some boom configurations have virtually *no* horizontal load capacity. There is virtually no capacity to pull a conductor into a corner.

Jib Capacity

The loading of a jib is also dependent on the angle and length of extension. While the boom may be at a good angle to make a maximum lift, the jib angle and extension may very well be the weakest link. **See Figure 16–21.**

Figure 16–21 In addition to the capability of a boom at various angles, the jib-capacity chart must also be consulted to make a safe lift. *(Delmar, Cengage Learning)*

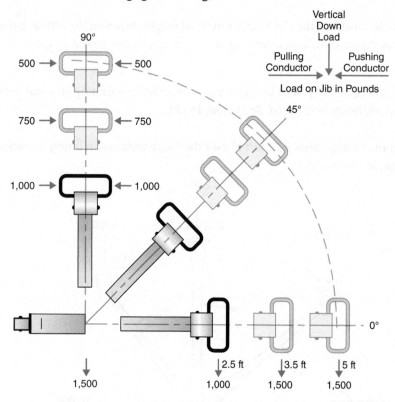

Caution: The winch and hydraulic jib used on material-handling units are for vertical loads only.

Working with a Material-Handling Aerial Device

The risks involved with operating a material-handling aerial device are reduced when complying to the following procedural steps.

1. Keep the load low, and raise or lower it only after reaching the position where the load is to be installed or removed. **See Figure 16–22**.

Figure 16–22 The jib and winch of a material-handling aerial device lifts a transformer from the truck deck. *(Delmar, Cengage Learning)*

2. Remember that the following three lifting variables are known only after reaching the position where the load is to be installed or removed. **See Figure 16–23**.
   - Upper-boom angle (Check the angle indicator on the boom.)
   - Lower-boom angle (Check the angle indicator on the boom.)
   - Load-line radius (How far is the jib extended out from its mounting?)
3. Check the capacity chart for the net load the unit is able to lift in the position. The chart should be easily visible to the operator. **See Figure 16–24**.
4. Side loading is hazardous. Make vertical lifts only.
5. Maintain your minimum approach distance to energized conductors when the winch is in contact with the ground or the pole, for example.
6. A rope winch line eventually becomes conductive because of moisture and contamination.
7. Use a dynamometer to determine the weight of an unknown load.

Figure 16–23 The boom angles must be known in order to calculate the lifting capacity of a material-handling aerial device. *(Delmar, Cengage Learning)*

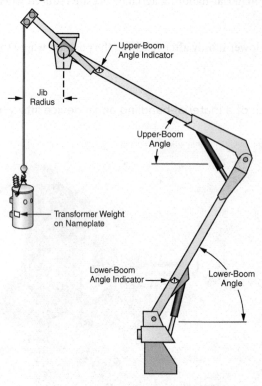

Figure 16–24 Sample Jib Boom Net Load

					Jib Boom Net Load in Pounds						
					Upper-Boom Angles						**Jib**
		0°	**30°**	**45°**	**60°**	**90°**	**120°**	**135°**	**150°**	**180°**	**R. Ft.**
Lower-Boom Angles	81°–100°	15	225	550	1,275	3,000	1,275	550	225	15	1
		0	35	280	800	1,000	800	280	35	0	3
		0	0	80	470	600	470	80	0	0	5
	61°–80°	15	225	550	1,275	3,000	1,275	550	225	15	1
		0	35	280	800	1,000	800	280	35	0	3
		0	0	80	470	600	470	80	0	0	5
	46°–60°	15	225	550	1,275	2,630	815	475	225	15	1
		0	35	280	800	1,000	690	280	35	0	3
		0	0	80	470	600	470	80	0	0	5
	30°–45°	15	225	550	1,275	1,870	575	290	135	15	1
		0	35	280	800	1,000	475	220	35	0	3
		0	0	80	470	600	385	80	0	0	5

8. Watch for "winch pileup and backlash." When the rope falls off the buildup on one edge of a winch, a shock load can be severe enough to cause a rope to fail, a transformer to unload, or an outrigger to sink into the ground and a truck to tip over.

9. The capacity of a winch is specified at its first layer of rope. A buildup of rope will derate the winch capacity significantly.

Review Questions

1. Who is responsible for doing the pretrip air-brake inspection: the fleet mechanic or the driver?

2. When jump-starting a vehicle, why should the last connection not be at the battery?

3. Why is a small hydraulic oil leak a hazard?

4. When should outrigger pads be used?

5. How does working on more than a 5° slope affect an aerial device?

6. There are two fundamental ways a person can be protected working around a vehicle when a boom makes contact with an energized conductor. What are they?

7. What is the purpose of grounding a truck?

8. Why does a truck ground not prevent a person from getting an electric shock when touching a truck in contact with a circuit?

9. How can a boom-tip winch break the boom of a digger derrick?

10. Name three ways an operator can reduce the risk involved with operating a material-handling aerial device.

Chapter 17

Rigging in Power Line Work

Introduction

Rigging in Line Work

One of the more important skills when doing transmission line work is the use of proper rigging. Dead-end tensions are very high, and the loads being lifted are quite heavy. A rigging failure while dead-ending aloft would have grave consequences. Much of the heavy rigging done on transmission lines is engineered.

Although a lot of rigging on distribution line work has been done by instinct, good rigging skills are becoming essential as distribution conductors are getting bigger and heavier.

Working Load Limits

The term working load limit (WLL) is used to identify the capacity of individual rigging components. The working load limit is the maximum load that can be applied to a rigging component and may never be exceeded.

Typically, a WLL is assigned to individual components by the manufacturer. The WLL rating is based on the breaking strength of the equipment divided by a safety factor. For example, if the breaking strength of a rope is 5,000 lbs, and the assigned WLL is 1,000 lbs, then the safety factor is 5.

A safety or design factor refers to the theoretical reserve capability of a component calculated by dividing the breaking strength by the WLL. A safety factor is applied to rigging hardware because, along with wear and tear, whenever a load is picked up, stopped, or moved, increased force results due to dynamic loading. For example, a safety factor of 10 (or 15) is applied to the WLL of a component because it will be used to suspend someone at a specific height.

The WLL for a sling, rope, or winch cable may have a 5-to-1 safety factor, but a transformer gin, a scaffold, stability of an aerial device, or a live-line tool may have different safety factors. There is a higher safety factor assigned to rigging used to suspend people.

The following table lists the sample safety (design) factors assigned to some rigging components by some manufacturers.

Rigging Component	Safety Factor
Nylon rope	9
Polyester rope	9
Polypropylene rope	6
Alloy-steel chain	4
Chain fittings	4
Wire rope	5
Synthetic web sling	5
Wire rope fittings	5

The WLL for individual pieces of equipment may be considered adequate for a load to be lifted but it does not necessarily mean that the configuration of the rigging setup has a safe working load. An additional safety factor may need to be added.

Rigging Equipment

The WLL of each piece of rigging hardware must be labeled or identified by size (diameter) and checked against WLL tables. When using rigging equipment such as rope blocks, hand line, fiber rope, wire rope, nylon slings, blocks, transformer davits (gins), snatch blocks, anchor pulling eyes, conductor grips, and chain hoists, the WLL must be known. Power line technicians should not depend on the WLL shown in tables in this book without checking against the equipment they use.

Using Ropes and Rigging Hardware

Fiber Rope

Fiber ropes refer to ropes that are made from natural or synthetic fibers versus wire rope. Sailors use the term *line* instead of rope.

Types of Construction

Two general categories of rope construction are used in power line work—twisted and braided. Twisted rope is formed by coiling three strands together in the same direction and is the type of rope that has been used for the longest time. Three-strand rope has some natural torque that tends to unravel and kink the rope if one end is free. For example, when new and used for a hand line, it must be run through a hand line pulley quite a few times before the hand line stops twisting.

Three general categories of braided construction exist: solid braid, double braid, and hollow braid. **See Figure 17–1.**

A solid braid is firm and stands up well to chafing of stringing blocks. A single, solid-braid rope cannot be spliced, but when used as part of a double-braid rope it can be spliced.

Figure 17–1 The type of rope construction needs to be recognized in order to choose the correct rope and WLL. *(Delmar, Cengage Learning)*

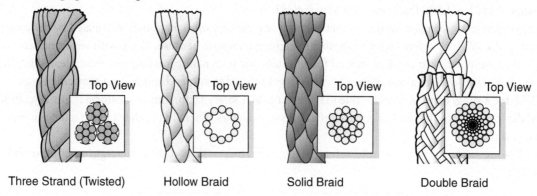

Three Strand (Twisted)   Hollow Braid   Solid Braid   Double Braid

Double braid, also known as braid-on-braid, is two braided ropes combined into one rope. A braided rope core is covered with a braided rope sheath to produce a strong and spliceable rope that is abrasion resistant. This rope has less stretch than most ropes and is often used as a pull rope for stringing. The most common type of rope for a digger derrick winch or a material-handling aerial device winch is a double-braided rope. A common error is often made when the end of a derrick winch rope is used to form a sling to hang a transformer, such that the sling angle and choker overloads the rope.

A hollow braid rope is a braided core with a hollow center that is very quick and easy to splice. It is used in smaller ropes, up to about ½ in. (13 mm) diameter. It is very flexible but can flatten and is not always easy to grip well when used as a hand line.

Types of Fibers

Rope with natural fibers—Manila, sisal, and hemp—are probably not used in the line trade anymore because they tend to be heavy, soak up water, mildew, and decay. Synthetic ropes are lighter and easier to keep dry and clean and are, therefore, better for use around energized lines. Sunlight will deteriorate synthetic rope over time, but by then the rope is probably due for replacement because it has been used and abused. Stringing ropes stored on reels should be covered to protect from sunlight and to keep dry.

There are so many kinds of synthetic ropes that it is hard to distinguish the different types, although there are some important differences between nylon, polypropylene, dacron, and polyethylene. The differences are found mostly in the strength characteristics of stranded, braided, and double-braided rope. For critical work involving a capstan hoist, live-line work, and pulling conductor near energized circuits, it is important to recognize the type of rope being used and its suitability for the application.

Nylon rope has a tensile strength that is nearly three times that of Manila rope, and it also has a lot of stretch. If something breaks or lets go when a rope is under tension, the snapback—especially if the rope is nylon—can cause serious injury. Stay out of the direct line of the rope, or even 45° away from a pull, and also stay out of a bight at a snatch block. The high degree of stretch and the slightly heavier weight make nylon less attractive for use as a long hand line on transmission line work.

Polyester rope is almost as strong as nylon and is more resistant to abrasion, but it stretches less and cannot absorb shock loads as well. Both polyester and nylon rope are recognized by their very fine hairlike fibers.

Polypropylene can take more shock load and has less stretch, but it is weaker than nylon and polyester. It is the only rope that floats. It is more susceptible to high temperatures and should not be used with a friction hitch—such as a taut line—and should not be used on a capstan hoist. It is used as hot-line rope because water on the surface can be removed by shaking and wiping with a cloth. When used as a hot-line rope, it must be used exclusively for that purpose and stored in clean, dry containers with moisture absorbers.

Polyethylene can be as much as three times as strong as nylon and very flexible. Polyethylene and polypropylene rope can be recognized by their bristlelike fibers.

Polydacron rope has good dielectric properties and heat resistance. It can be used on a capstan hoist and with sliding friction hitches, such as a taut-line hitch. It is a favored rope for hand lines.

Ropes used for stringing come in many forms, though some are called simply *poly ropes*. The ropes come on reels that should identify the working load limit.

Rule of Thumb for Fiber Rope Strength

For accurate ratings, consult tables for the WLL of the fiber rope you will be using. If the tables only show breaking strength, that number should be divided by 5 for the WLL, or 10 if it is to support people.

When accuracy is not critical, the following rule of thumb will be a good indicator of the WLL.

Rule of Thumb for WLL

1. Change the rope diameter to eighths.

2. Square the numerator.

3. Multiply by:

   - 40 for three-strand polypropylene rope
   - 50 for hollow-braid polypropylene
   - 50 polydacron rope
   - 70 for three-strand nylon rope
   - 90 for double-braid nylon rope

For example, to find the WLL of ½ in. polypropylene rope, perform the following calculation.

1. Translate ½ in. diameter to eighths of an inch, which is 4/8.

2. Square the numerator 4 = 16.

3. Multiply 16 by the rule-of-thumb factor of 40.

4. WLL is, therefore, 40 × 16 = 640 lbs.

When deciding which rope to use to let down a shield/static wire dead-end, for example, the tables that give the WLL for the rope provide just the start. When the rope contains an eye splice or a knot, derating factors also have to be considered.

Common Knots Used in Line Work

The following knots are some of the more common ones used in line work. Others will be learned informally on the job. This book does not try to describe how to tie the knots. There is really no substitute for learning these knots from an instructor.

The bowline is used to make a fixed-size loop in a rope and is the king of knots for line work. It is the most used knot and can be easily undone if not overstressed. A little ditty that may make it easier to learn or teach is this: "The rabbit comes out of the hole around the tree and then back down his hole." **See Figure 17–2.**

The only way a bowline can become undone is if the loose end of the bowline drops back into the hole. Leave the loose end sticking out at least 12 times the diameter of the rope. A bowline cannot be untied while it is under tension. In certain applications, such as tying a bowline in both ends of the rope that later is tightened, it cannot be released and it can be very inconvenient, as well as embarrassing.

The water bowline is a bowline with an extra loop that is especially useful with natural-fiber ropes such as Manila that could swell up and be difficult to untie. Although a bowline is a good knot to untie, when a big rope is being used on a heavy load, a water bowline may sometimes be much easier to untie. The front loop takes much of the strain. **See Figure 17–3.**

A running bowline is a knot that can be used to make a lasso or noose, but because it stays in position it is also useful for jobs such as tying a rope to a pole when rope guying. A running bowline can be tied loose to the bottom of a pole, slid up the pole with a switch stick, and—when in the desired position—tightened by giving it a pull. **See Figure 17–4.**

Figure 17–2 A bowline is the king of knots. It is easy to untie but cannot be untied while under tension. *(Delmar, Cengage Learning)*

Figure 17–3 A water bowline can be untied even after being subjected to very high tension. *(Delmar, Cengage Learning)*

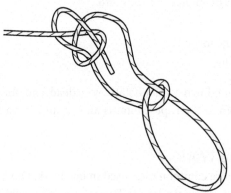

Figure 17–4 A running bowline is used for jobs such as rope guying a pole. *(Delmar, Cengage Learning)*

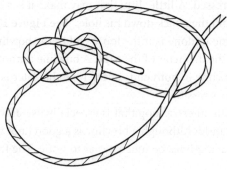

A bowline on a bight is a useful knot when a good, secure knot is needed somewhere between the ends of a rope (on the bight). It also can be used as a suspension seat and has been used by arborists for years before tree saddles were manufactured. This is one of the more difficult knots to learn. **See Figure 17–5**.

Figure 17–5 A bowline on a bight can be tied in the middle of a rope. *(Delmar, Cengage Learning)*

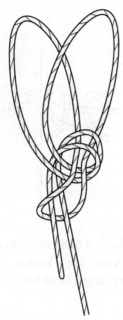

A bowline bend consists of two bowlines tied back to back. It is a very secure way of tying together any two ropes, especially if there is a big difference in the diameters of the ropes. While other knots and bends are available to tie two ropes together, the bowline bend is the most secure and the easiest to untie. **See Figure 17–6.**

Figure 17–6 A bowline bend is a very secure way to tie together two rope ends. *(Delmar, Cengage Learning)*

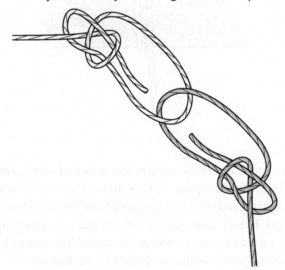

Using two half hitches to tie a rope to an object (also called a snubbing hitch) is one of the more common and easy knots to learn. A heavy load can come apart with only one half hitch. ("One will hold a block, two will hold a person, three will hold the world.") If the rope is hitched to something with a small diameter, go around the object twice and use two half hitches. Go around a cable or conductor six or seven times, then tie two half hitches, and you will have a substitute for a conductor grip. A typical application for two half hitches is to temporarily hold tension on a conductor at a pole. When using two or three wraps around the pole and two half hitches, the hitch can be released when the wire is under tension. **See Figure 17–7.**

Figure 17–7 Two half hitches can be used when the lineman intends to untie the rope while under tension. *(Delmar, Cengage Learning)*

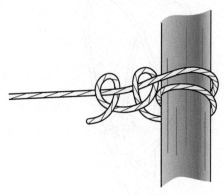

A clove hitch can be tied very quickly and slipped over a bull pin without bending over. Unless a half hitch is added to the tail, the clove hitch should not be used when only one side is loaded. The clove hitch can jam under heavy tension, making it difficult to untie. Two half hitches will do anything a clove hitch will do and can always be used as an alternative. **See Figure 17–8.**

Figure 17–8 Tying a clove hitch is a quick way to attach a rope to a bull pin. *(Delmar, Cengage Learning)*

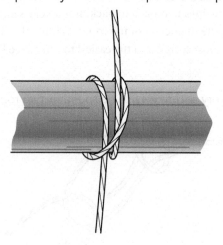

The square knot (reef knot) is easy to tie and will not jam unless loaded very heavily. In applications—such as hanging a stringing block or snatch block after wrapping the rope around an object, such as a pole five or six times—the square knot is used to tie together the two ends of a rope. A square knot should not be used to tie together two ropes that will be under high tension; it will fail. In such cases, use the bowline bend. A square knot is easy to tie incorrectly into a granny knot, where the ends of the rope come out in opposite directions. The granny knot looks something like a square knot but will fail and must never be used intentionally. **See Figure 17–9a, b,** and **c.**

A surgeon's knot is a square-knot variation that has an extra turn at the start to help prevent the knot from loosening before the second turn is made. It is used when a little tension or weight is on the rope. **See Figure 17–10.**

A sheet bend is used instead of a square knot to tie together two ropes, typically when the two ropes are different sizes. It is tied incorrectly (like the granny knot, also called a left-hand sheet bend) when the two ends are not on the same side of the knot. The bowline bend is a good alternative when tension is going to be applied to the ropes. **See Figure 17–11.**

A taut-line hitch is a friction hitch used to tie on to another rope to hold it from moving or running. It has been used by arborists for years as a hitch to hold them in a working position in a tree and to make possible a controlled descent. A little pressure on the hitch will cause it to slide. It can be tied to a conductor or cable to act as a grip, but a high tension can damage an underground cable. The more turns around the wire, the better the grip. **See Figure 17–12.**

Figure 17–9 (a) A square knot is commonly used to hang something on a pole. (b) An application of a square knot is to hang a snatch block on a pole in such a way that it will not slide down. (c) Note how a square knot or two half hitches finishes the lashing or snubbing together of two poles. *(Delmar, Cengage Learning)*

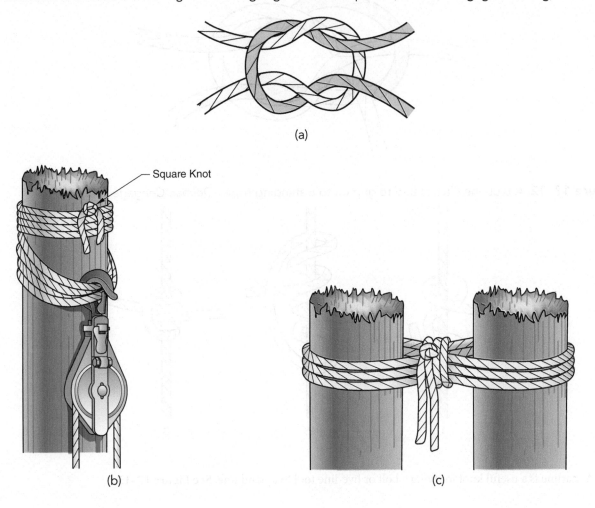

(a)

Square Knot

(b)   (c)

Figure 17–10 A surgeon's knot is used instead of a square knot when something needs to be bound extra tight. *(Delmar, Cengage Learning)*

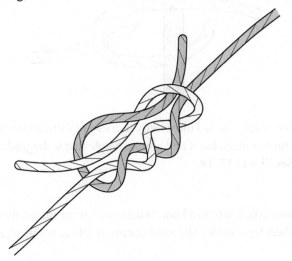

Figure 17–11 A sheet bend is typically used to tie together two ropes that have different diameters. *(Delmar, Cengage Learning)*

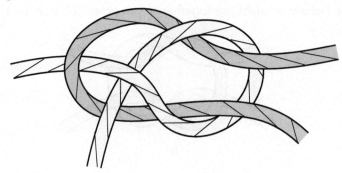

Figure 17–12 A taut-line hitch is tied to grip on to a standing rope. *(Delmar, Cengage Learning)*

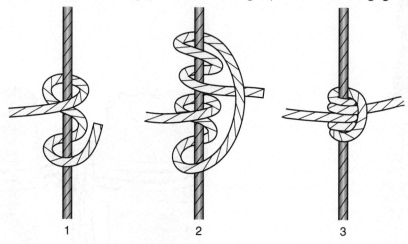

A marline is a useful knot for tying a bolt or live-line tool to a hand line. **See Figure 17–13**.

Figure 17–13 A marline hitch is a quick hitch used to tie objects to a hand line. *(Delmar, Cengage Learning)*

A timber hitch is easy to tie, but when it is tied incorrectly it does not form a loop and it goes back on itself. A running bowline is an alternative to a timber hitch, but if tied around a pole that is dragged through the mud or snow, a timber hitch is much easier to untie. **See Figure 17–14**.

Splicing Fiber Rope

A splice is more permanent and more efficient than a knot. Splices in rope are uncommon because nylon web slings have taken over many of the reasons splices were made. The most common splices are the eye splice, back splice, short splice, and long splice.

Figure 17–14 A timber hitch is used to tie around a pole especially if the pole is to be dragged along the ground. *(Delmar, Cengage Learning)*

Splicing a three-strand rope requires more instruction and skill than splicing a braided rope. For a three-strand rope, inserting the first three strands in the right spot determines if the splice will look good when it is done but does not necessarily affect strength. Three complete tucks are standard for natural-fiber ropes, but four is the recommended number for the more slippery synthetic ropes. **See Figure 17–15.**

Figure 17–15 Three-strand fiber rope can be spliced as shown. The eye splice allows the rope to retain much more of its strength than when using a knot. *(Delmar, Cengage Learning)*

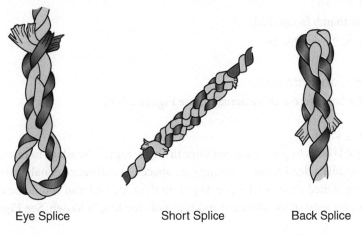

Eye Splice   Short Splice   Back Splice

Splicing a braided rope can actually be learned from the book that comes with the splicing kit. Follow the step-by-step instructions and use a very sharp knife.

Note that Figure 17–15 shows finished splices using three-strand ropes. The long splice is not illustrated because if made properly it looks exactly like the rope. A long splice is like a short splice that is longer and tapered near the ends. Made properly, it will pass through a sheave or block. A short splice is a more permanent and more efficient method of tying together two ropes. The back splice or crown splice is more useful with natural-fiber rope because it cannot be taped and the end fibers cannot be melted as is commonly done with synthetic-fiber rope. The eye splice is probably the only splice still seen in line work.

Derating Factors for the WWL of Fiber Ropes When Using Knots and Splices

- 10% for an eye splice
- 45% for a bowline knot
- 35% for a running bowline knot
- 60% for a square (reef) knot
- 50% for a round turn and two half hitches

Using Wire Rope

It is just as important to recognize the type of wire rope as it is to recognize the type of fiber rope. More than a hundred types of wire rope are manufactured. Wire rope comes with different cores such as Manila, polypropylene, and wire. It comes with different types of steel such as galvanized, stainless, and ungalvanized. It comes with different lays such as regular lay, lang lay, and cable laid. Because of this extensive variation, it is improbable that any power line technician can identify a wire-rope type without a label. When used in a sling configuration, the WLL of the sling must be identified with a label attached to the sling. If the wire rope is a winch, the type and size of winch must be identified and WLL tables consulted.

Rule of Thumb for Wire Rope Strength

Tables must be consulted if you want to obtain an accurate rating of the WLL of the wire rope at hand. When accuracy is not critical, the following rule of thumb will be a good indicator of the WLL.

Rule of Thumb for WLL

1. Change the rope diameter to eighths.
2. Square the numerator.
3. Multiply by 250 for regular-laid wire rope (6 × 19 and 6 × 25) and 150 for cable-laid wire rope (3 × 3 × 19).

For example, to find the WLL of ⅝ in. regular-laid wire rope, perform the following calculation.

1. ⅝ diameter, already has a denominator of 8.
2. Square the numerator 5 = 25
3. Multiply by the rule-of-thumb factor: 250.
4. WLL is, therefore, 250 × 25 = 6,250 lbs.

Types of Wire-Rope Terminations

Wire-rope terminations display different characteristics. **See Figure 17–16.**

Making a Temporary Eye

There is a correct way to use U bolts to put a temporary eye in a wire rope. One adage to ease remembering when installing the U bolts is: "Never saddle a dead horse." The clips are spaced at a distance equal to six times the diameter of the wire. The clips should be tightened again after strain is put on the rope and also should be checked regularly. The clip farthest away from the eye gets most of the vibration and is usually the first to loosen. **See Figure 17–17.**

Specific Hazards When Working with Wire Rope

1. Provide padding, such as wood blocking, when a sling is used around sharp edges.
2. Sharp bends in a wire rope sling will derate the strength of the sling. As a rule of thumb, the diameter (D) of the rope bend in relation to the diameter (d) of the rope itself is as follows:
   - D/d = 10. No reduction in strength: for example, a 1 in. cable bent around a 10 in. object.
   - D/d = 2. The sling is derated to 65% of its original strength: for example, a 1 in. cable bent around a 5 in. object.
   - D/d = 1. The sling is derated to 50% of its original strength: for example, a 1 in. cable bent around a 1 in. object.
3. Wire-rope strength is reduced 20% for a U bolt–clipped eye.
4. Wire-rope strength is reduced 10% for a Flemish eye.
5. Never substitute guy wire for wire rope. Guy wire used as a winch will twist to release a wire-form grip (preform), and it will break if used through a snatch block.
6. Wire rope used for stringing is made of harder steel and must not be made into slings. This steel is not meant for the sharp bends typical for slings.
7. Conductor grips (Chicago style) should not be used on wire rope.
8. Always stay out of the bight of a wire rope under tension as it is changing direction through a block.

Figure 17–16 The efficiency of a wire-rope termination is different for each type. *(Delmar, Cengage Learning)*

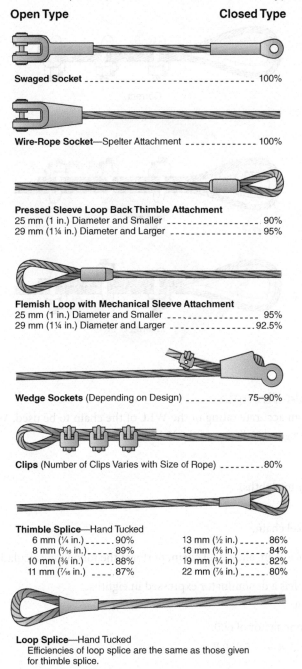

Open Type	Closed Type

Swaged Socket - 100%

Wire-Rope Socket—Spelter Attachment - - - - - - - - - - - - 100%

Pressed Sleeve Loop Back Thimble Attachment
25 mm (1 in.) Diameter and Smaller - - - - - - - - - - - - - - - - 90%
29 mm (1¼ in.) Diameter and Larger - - - - - - - - - - - - - - - 95%

Flemish Loop with Mechanical Sleeve Attachment
25 mm (1 in.) Diameter and Smaller - - - - - - - - - - - - - 95%
29 mm (1¼ in.) Diameter and Larger - - - - - - - - - - - - - 92.5%

Wedge Sockets (Depending on Design) - - - - - - - - - - - 75–90%

Clips (Number of Clips Varies with Size of Rope) - - - - - - - - .80%

Thimble Splice—Hand Tucked
6 mm (¼ in.) - - - - - - .90%   13 mm (½ in.) - - - - - - .86%
8 mm (⁵⁄₁₆ in.) - - - - - 89%   16 mm (⅝ in.) - - - - - - .84%
10 mm (⅜ in.) - - - - - .88%   19 mm (¾ in.) - - - - - - .82%
11 mm (⁷⁄₁₆ in.) - - - - .87%   22 mm (⅞ in.) - - - - - - .80%

Loop Splice—Hand Tucked
Efficiencies of loop splice are the same as those given
for thimble splice.

Using Chains

A common perception in the field is that a chain is the strongest piece of rigging for towing a truck out of mud or lifting something very heavy. Regardless, some serious injuries have occurred when a chain has broken and snapped back. The safety rule of staying out of the direct line of the pull does not work well when the windshield and the driver of the stuck vehicle are in direct line.

When used in a sling configuration, the WLL of the sling must be identified with a label attached to the sling. If a length of chain is used, the type of steel and size must be identified and WLL tables must be consulted.

Figure 17–17 When U bolts are used to install a temporary eye in a wire rope they must be installed properly. *(Delmar, Cengage Learning)*

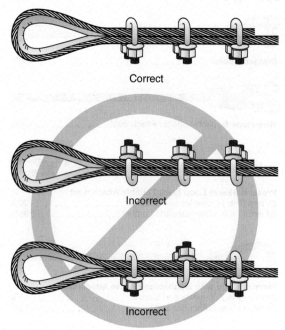

Correct

Incorrect

Incorrect

Rule of Thumb for Chain Strength

Tables should be consulted for an accurate rating of the WLL of the chain to be used. When accuracy is not critical, the following rule of thumb will be a good indicator of the WLL.

Rule of Thumb for WLL

1. Convert the chain diameter to eighths.
2. Square the numerator.
3. Multiply by 600 for alloy steel chain.

For example, to find the WLL of a ⅜ in. alloy steel chain, perform the following calculation.

1. The ⅜ in. diameter already has a denominator expressed in eighths.
2. Square the numerator 3 = 9.
3. Multiply by the rule-of-thumb factor of 600.
4. The WLL is, therefore, is 600 × 9 = 5,400 lbs.

Specific Hazards When Working with Chain

1. Chain slings used for hoisting must not be used for any other purpose. For example, a chain used to bind a load can be subject to severe shock loading.
2. Padding must be used when using chain slings on steel towers.
3. A chain should be used only if the load is known. Avoid using a chain for towing a stuck vehicle. In case of a chain failure, the chain ends can recoil with lethal force.
4. Use the proper end fittings.
5. Never use a knot or any other similar configurations in a chain. **See Figure 17–18.**

Figure 17–18 There are many ways a chain can be used incorrectly. *(Delmar, Cengage Learning)*

Using a Ratchet Chain Hoist

Manually operated lever hoists in line work are commonly ¾, ½, 3, and 6 ton, depending on the manufacturer. Like all hoisting devices, these chain hoists require scheduled maintenance and testing by a competent mechanic. A malfunctioning chain hoist can lead to a rigging failure, as well as a temper tantrum, and they malfunction because they are abused by using them to bind poles on a trailer, by tying the chain into a half hitch behind the head of an anchor rod, by overloading, by using a cheater to lengthen the handle, and by leaving the hoist under tension for prolonged periods. **See Figure 17–19.**

Figure 17–19 If the lower hook of a chain hoist is spread open, then it has been overloaded and needs maintenance. *(Delmar, Cengage Learning)*

Chain hoists are constructed with the lower hook being the weakest part. The lower hook will start to spread under overload before the chain hoist is overloaded. A chain hoist with a bent hook or binding clutch should be removed from the field and repaired.

Using a Web Hoist

A web hoist is lighter in weight than a chain hoist and has become the hoist of choice, as well as the hoist that is most abused. Web hoists were originally intended and reserved for work on or near energized circuits. **See Figure 17–20.**

Figure 17–20 A web hoist is not a live-line tool. An insulated link is required between an energized conductor and a pole or crossarm. A nonconductive handle cannot be used as a live-line tool unless it undergoes regularly scheduled electrical testing. *(Delmar, Cengage Learning)*

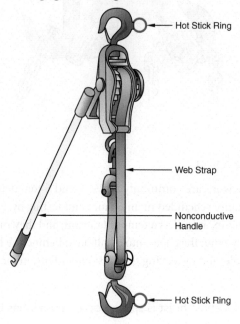

- Hot Stick Ring
- Web Strap
- Nonconductive Handle
- Hot Stick Ring

The nylon material is vulnerable to contamination and must not be connected directly between an energized conductor and a structure without an insulated link. If a web hoist is being used for general duty such as pulling a guy, it should be identified so that it will not be used on an energized circuit.

Working with Blocks

Snatch blocks are used to change the direction of a winch cable, rope, or conductor with no gain in mechanical advantage. Anyone standing in the bight of a winch cable or rope under tension can be severely injured if the snatch block or its anchoring point should fail. It is necessary to know the bisect tension that a snatch block will be holding, the WLL of the block, and the WLL of the anchor point.

The bisect tension on a block at its anchor point depends on tension on the winch or rope and the angle at which the winch travels through the block. When calculating or using a rule of thumb to determine a bisect tension on a block, ensure that the correct angle is measured. **See Figure 17–21.**

Figure 17–21 There are two typical ways that an angle at a block can be measured. Choose the correct angle when calculating bisect tension. *(Delmar, Cengage Learning)*

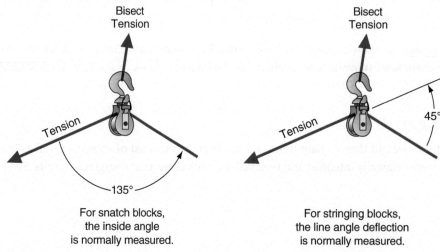

Bisect Tension — Tension — 135° — For snatch blocks, the inside angle is normally measured.

Bisect Tension — Tension — 45° — For stringing blocks, the line angle deflection is normally measured.

A Rule of Thumb for Bisect Tension

The rule of thumb for determining the bisect tension on a snatch block and anchor point by wire rope is to use a deflection angle. **See Figure 17–22.**

Figure 17–22 Load on a snatch block is dependent on the deflection angle. *(Delmar, Cengage Learning)*

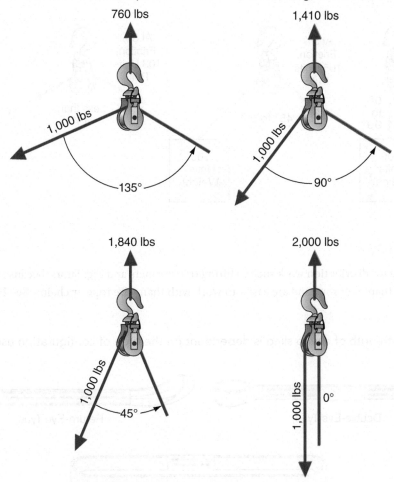

The Strain on a Block and Anchor Point

0.76 × rope tension for 135°

1.0 × rope tension for 120°

1.41 × rope tension for 90°

1.73 × rope tension for 60°

1.84 × rope tension for 45°

2.0 × rope tension for 0°

Matching Rope Blocks and Rope Size

An improperly sized block can derate the WLL of a wire or fiber rope. A rule of thumb for determining the proper size of a block is that the snatch block should have 1 in. of shell diameter for every ⅛ in. of fiber-rope diameter and 1 in. of shell diameter for every 1/16 in. of wire-rope diameter.

Dealing with Friction

There is a loss due to friction any time a rope, wire, chain, and such goes through a block. This friction adds to the force needed on a fall line, a winch, and a conductor puller. When more than one block is used, such as the sheaves in a set of tackle blocks or the stringing blocks over many spans, the friction loss can become significant.

In addition to the physical condition of a block, the greater the length of rope or winch in contact with a block, the greater the friction loss will be. **See Figure 17–23.**

Figure 17–23 Note the typical values that can be used when calculating friction on a block. *(Delmar, Cengage Learning)*

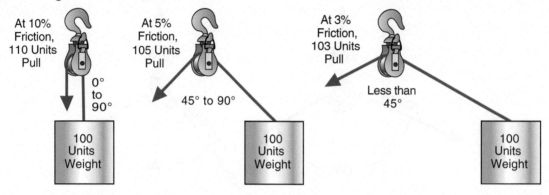

Using Web Slings

Web slings are preferred for distribution work such as lifting transformers and regulators, because they are more resistant to cutting and abrasion than fiber rope and are easier to work with than wire rope or chains. **See Figure 17–24.**

Figure 17–24 The strength of a web sling is dependent on the type of configuration used. *(Delmar, Cengage Learning)*

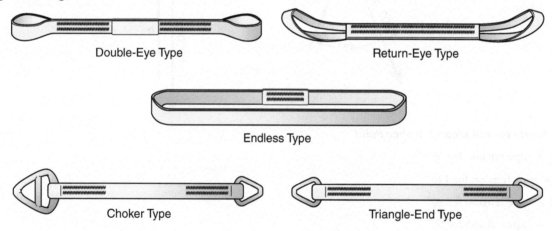

The sling size and rated load are identified on a label, which is usually a sewn-on leather tag. The WLL is based on the configuration that is used. Slings should be discarded if any of the following is true.

1. The label showing the load rating is not visible or is missing.
2. There are holes, tears, cuts, or snags; the wear indicator (red strand) is visible; or there is excessive abrasion.
3. Knots are present in any part of the sling.
4. Excessive color has been lost, which indicates ultraviolet light damage.

Using an Anchor Pulling Eye

When pulling a down guy, the limiting factor could be the anchor pulling eye being used. Some types of anchor pulling eyes have safe working loads of 3,000 lbs (1,400 kg). **See Figure 17–25.**

Figure 17–25 Two types of anchor pulling eyes are shown. When pulling a loaded down guy, the anchor pulling eye may be the weakest link. *(Delmar, Cengage Learning)*

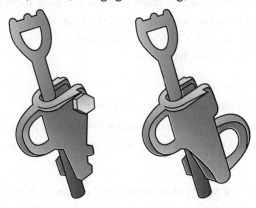

It is very easy to overload the pulling eye, grip, or chain hoist when pulling on a down guy that is in service. Use a derrick boom to help hold or adjust the pole rake.

Working with a Collapsible Bull Wheel

The WLL for a typical collapsible bull wheel (butterfly) is 4,000 lbs (1,800 kg). When used on the extension shaft of a typical digger-derrick boom-tip winch, the WLL is reduced to 800 lbs (360 kg). **See Figure 17–26.**

Figure 17–26 A collapsible bull wheel is typically used to roll up short lengths of wire. *(Delmar, Cengage Learning)*

When taking up a rope under tension, the compressive force of a stretched rope can collapse the bull wheel. The number of turns around the bull wheel should be limited to four. Workers should keep a safe distance and not stand in direct line with a bull wheel.

Lifting a Load

General Lifting Precautions

Sudden starts or stops place much heavier loads on rigging. A load could be increased by 2 to 50 times its actual weight. When lifting a load, the lift should be started very slowly until the sling becomes taut. Then lifting should continue slowly until the load is suspended. The load must be prevented from rotating. Use a tag line to prevent any rotation and to reduce swinging.

Using a Hand Line

A hand line is a hoisting device that is also used by many utilities as a pole-top/tower-top rescue device. For use as a rescue device, it should be made with hardware and rope capable of withstanding a person's weight plus a safety factor. Because of its rugged use, it should have at least a 10-to-1 safety factor.

Following are some precautions to take with hand lines.

- Keep the hand line away from traffic.
- Do not tie the hand-line to a truck unless the worker aloft has the keys.
- While climbing, attach a hand line to a worker's belt so that it breaks away if it gets snagged by passing vehicles or other equipment. Belt hooks that open and release their load when under too much weight are available. **See Figure 17–27.**

Figure 17–27 A collapsible hook is a safety device that will release a hand line attached to a body belt or bucket in case the hand line should be snagged by a passing vehicle. (*Delmar, Cengage Learning*)

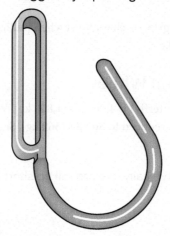

Note that a hand-line pulley and anchor is supporting double the weight of that being lifted. **See Figure 17–28.**

Figure 17–28 The hand line anchor and pulley in this example must be able to withstand double the weight of the load being lifted. (*Delmar, Cengage Learning*)

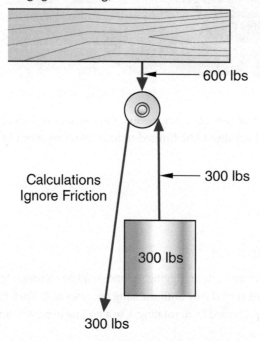

600 lbs

Calculations Ignore Friction

300 lbs

300 lbs

300 lbs

The same principle applies when using a rope and transformer gin to hang a transformer. Power line technicians should check the WLL of the type of transformer gin they are using. The load on the gin is double the weight of the transformer, ignoring that the rope through the block at the top of the pole is at a 0° angle and that therefore the friction adds 10% to the load. The tension on the snatch block and gin will be twice the weight of the transformer. *Note:* Gins are designed for vertical loads only, so any side pulls and tagging out of a load will derate the WLL of the gin. **See Figure 17–29.**

Figure 17–29 A davit with a WLL of 2,000 lbs (900 kg) can lift a maximum load of 900 lbs (400 kg) when using a single sheave in the davit and adding about 10% friction. *(Delmar, Cengage Learning)*

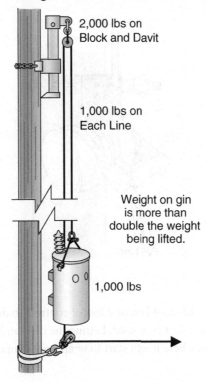

2,000 lbs on
Block and Davit

1,000 lbs on
Each Line

Weight on gin
is more than
double the weight
being lifted.

1,000 lbs

Accidents have happened when a line crew is not alert to the doubling of the weight at the anchor point. Note how the dead-end assembly at a tower failed and the conductor dropped on an energized underbuild a few spans away. **See Figure 17–30.**

Figure 17–30 This dead-end failed when it was subjected to almost double its normal tension. *(Delmar, Cengage Learning)*

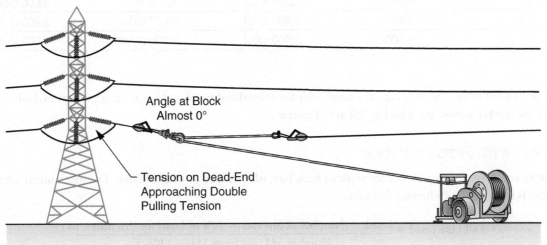

Angle at Block
Almost 0°

Tension on Dead-End
Approaching Double
Pulling Tension

Hoisting with a Capstan Hoist

To control the weight to be lifted by a capstan hoist, the number of wraps around the capstan hoist should be kept so that about 20 to 40 lbs (9 to 18 kg) of pull are on the fall line. More turns will mean a loss of control because releasing tension on the rope coming from the hoist may not stop the load. **See Figure 17–31.**

Figure 17–31 The number of turns on a capstan hoist is dependent on the weight to be lifted. *(Delmar, Cengage Learning)*

Load Line

Fall Line

Trying to add or remove turns during a lift could cause a loss of control. To determine the number of wraps needed for the load to be lifted, use data specific to the hoist being used. Letting the capstan hoist turn without advancing the rope can cause the rope to melt or weld to the drum, where it will start to wrap up the rope like a winch. **See Figure 17–32.**

Figure 17–32 Rope Turns on the Drum for Weight to Be Lifted

Rope Turns on Drum	Ratio of Load Line to Fall Line	Weight on Load Line to Be Lifted, Pounds (kg) Pull on Fall Line		
		20 lbs (9 kg)	30 lbs (14 kg)	40 lbs (18 kg)
3	1:20	400 (180)	600 (270)	800 (360)
3.5	1:30	600 (270)	900 (400)	1,200 (540)
4	1:50	1,000 (450)	1,500 (680)	2,000 (900)
4.5	1:70	1,400 (630)	2,000 (900)	2,800 (1,270)
5	1:100	2,000 (900)	3,000 (1,360)	4,000 (1,800)

Any hoist used to raise workers up to a conductor for specialized barehand work must be kept exclusively for that work, and the worker should use a backup fall arrest system.

Hoisting with Rope Blocks

Rope blocks are often the only choice for work in back lots, islands, and at wide ditches. The mechanical advantage of rope blocks is shown in the following formula.

$$\text{Fall Line Load} = \frac{\text{Load} + (\text{Number of Sheaves} \times 10\% \text{ of Load for Friction Loss})}{\text{Number of Lines from Moving Block}}$$

For example, the 1,000 lbs transformer being raised with a set of three sheave blocks in Figure 17–35 has six lines moving from the moving (bottom) block.

$$\text{Load on the Fall Line} = \frac{1,000 + (6 \times 0.1 \times 1,000)}{6} = 267 \text{ lbs}$$

The friction through the blocks will vary with maintenance and type of bearings. Calculating the mechanical advantage of rope blocks will show a slightly different answer because of the estimated friction. **See Figure 17–33a and b.**

Figure 17–33 (a) Notice that the weight on the gin is the weight of the transformer plus the pull on the fall line. (b) Note how three ropes share the load to lift a 1,000 lbs weight with only 500 lbs on the fall line. Also note that the anchor point is subjected to 1,500 lbs instead of the 2,000 lbs it would have been subjected to without the extra line and block. *(Delmar, Cengage Learning)*

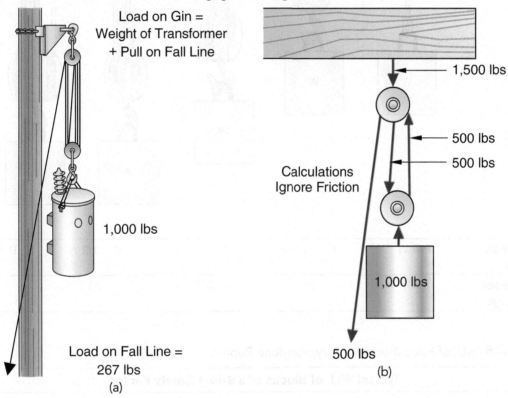

The mechanical advantage changes with each addition of a line between the blocks and the direction of the fall line. **See Figure 17–34.**

Matching Rope Blocks and Rope Size

The shell of the block should be eight times the diameter of the rope. For example, a 0.5 in. (1.3 cm) rope requires 8 × 0.5 in. = 4 in. (10 cm) blocks.

WLL of Rope Blocks

The WLL of a set of blocks will depend on the WLL of the blocks and the size and type of rope. **See Figure 17–35.**

Figure 17–34 The mechanical advantage of the two sheave blocks on the right is four times better than the single sheave pull on the left. *(Delmar, Cengage Learning)*

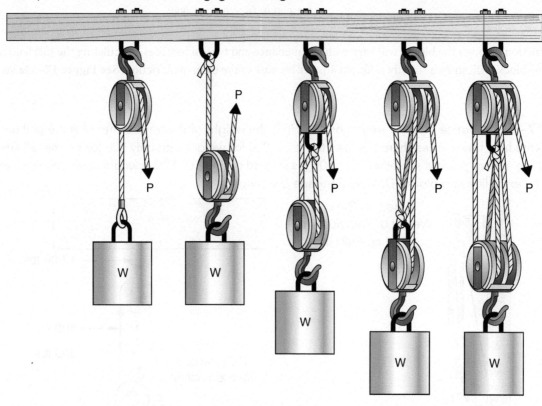

Pull on Fall Line	P = W	P = 1/2 W	P = 1/2 W	P = 1/3 W	P = 1/4 W
Mechanical Advantage	1:1	2:1	2:1	3:1	4:1

Figure 17–35 WLL of Rope Blocks for Polypropylene Rope

Typical WLL of Blocks at a 9-to-1 Safety Factor		
Polypropylene Rope	**Set of 2-Sheave Blocks**	**Set of 3-Sheave Blocks**
1/2 in. (1.3 cm)	1,500 lbs (700 kg)	2,000 lbs (900 kg)
3/4 in. (1.9 cm)	2,400 lbs (1,000 kg)	3,000 lbs (1,400 kg)

Connecting to a Load with a Sling

The strength of a sling depends on the material strength and the manner in which it is hitched to the load. A single-leg vertical sling with proper end fittings is rated at the strength of the rope and the type of end fittings. A tag line is needed when lifting with a single vertical sling because of the tendency for the load to rotate and swing. Allowing the load to rotate can untwist a sling and weaken any hand-tucked eyes. **See Figure 17–36.**

Figure 17–36 Different sling configurations are needed depending on what is being lifted. The WLL of each sling configuration also is different. *(Delmar, Cengage Learning)*

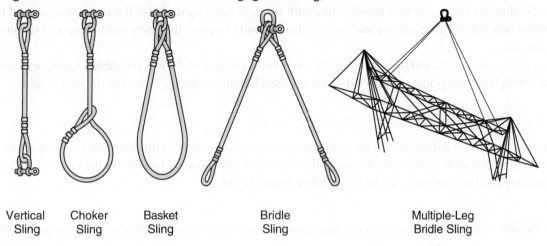

| Vertical Sling | Choker Sling | Basket Sling | Bridle Sling | Multiple-Leg Bridle Sling |

A single-wrap choker hitch is the weakest form of a hitch. When it is not under a load, the hitch can open and release its load. Forcing the choker end down places a sharp bend in the sling and reduces the sling angle. **See Figure 17–37.**

Figure 17–37 Using a choker hitch seems simple but there are many ways to apply it incorrectly. *(Delmar, Cengage Learning)*

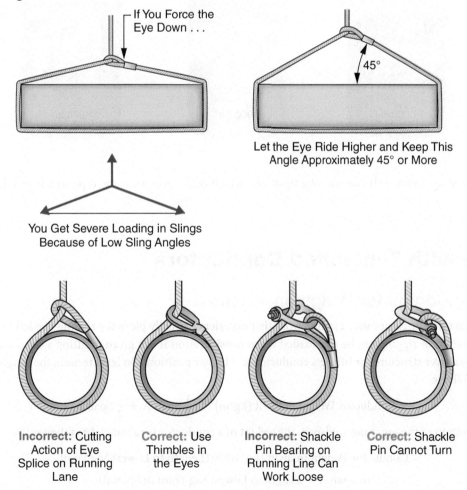

If You Force the Eye Down . . .

Let the Eye Ride Higher and Keep This Angle Approximately 45° or More

45°

You Get Severe Loading in Slings Because of Low Sling Angles

Incorrect: Cutting Action of Eye Splice on Running Lane

Correct: Use Thimbles in the Eyes

Incorrect: Shackle Pin Bearing on Running Line Can Work Loose

Correct: Shackle Pin Cannot Turn

A basket hitch should only be used on straight lifts. Moving the load within the basket hitch can damage and weaken the sling.

A bridle sling is intended to be a two-leg sling with two legs carrying the load. If a single sling is used through a clevis and the two ends are attached to a load, an unbalanced load can cause the heavy end to drop and the sling to slide through the clevis.

When lifting objects with multiple-leg slings (three or four legs), two of the legs should be capable of supporting the total load. Multiple-leg slings must be selected to suit the most heavily loaded leg instead of the total weight.

Effect of Sling Angle on Lifting Capacity

A key factor in determining sling stress is the resultant sling angle when making a lift. Note how the sling angles affect the stress on the sling. For most distribution line work, if a sling is chosen with a WLL of double the weight to be lifted, it will not be unwieldy and the sling angle can be ignored. **See Figure 17–38**.

Figure 17–38 The stress on a sling is highly dependent on the sling angle. *(Delmar, Cengage Learning)*

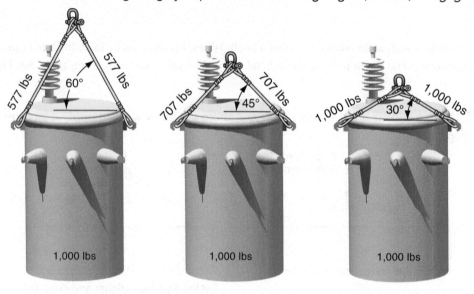

Always use a sling. Using only the end of a tip-boom winch could overstress the rope winch and definitely shorten its life.

Working with Tensioned Conductors

Calculating Conductor Weight in a Span

Conductors are so often lifted or moved in line work that experience usually picks the rigging needed to lift a conductor. However, an aerial lift or rigging can be overloaded when consideration is not given to lifting a conductor from a structure on a hill (breakover structure) or lifting a conductor to a higher position. On level terrain, the weight of a conductor is calculated as follows:

$$\text{Conductor Weight in lbs/ft (kg/m)} \times (\tfrac{1}{2} \text{ Span A} + \tfrac{1}{2} \text{ Span B})$$

When the bottom of the sag is not midspan, the weight of a conductor is calculated as follows:

$$\text{Conductor Weight in lbs/ft (kg/m)} \times (\text{Length to Lowest Sag Point}$$
$$\text{in Span A} + \text{Length to Lowest Sag Point in Span B})$$

A conductor's maximum lift height occurs when the lowest point of sag reaches an adjacent structure. At this point, the conductor weight has doubled. When planning to lift a conductor with an aerial device or live-line tools, multiply the weight to be lifted by a factor of 2. Similarly, as a conductor is lifted to a higher location, the weight increases as the lowest sag point moves farther away. **See Figure 17–39**.

Figure 17–39 The weight of a conductor on a hill can be considerably higher than on the level. *(Delmar, Cengage Learning)*

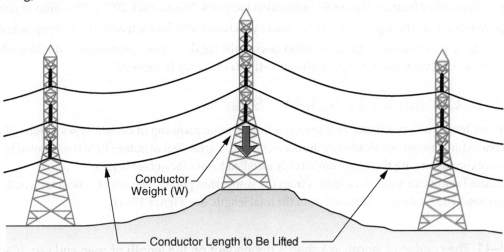

Conductor Weight (W)

Conductor Length to Be Lifted

Calculating Conductor Tension

The tension on a conductor is generally the heaviest weight dealt with by power line technicians. **See Figure 17–40**.

Figure 17–40 The conductor tension is dependent on the weight of the conductor, span length, and sag. *(Delmar, Cengage Learning)*

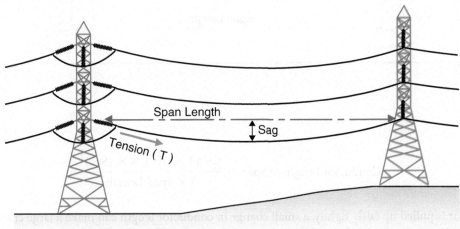

Span Length

Sag

Tension (T)

$$\text{Conductor Tension} = \frac{\text{Conductor Weight in lbs/ft (kg/m)} \times \text{Span Length}^2 \text{ in./ft (m)}}{8 \times \text{Sag in./ft (m)}}$$

Rules of Thumb for Conductor Tension

1. On distribution lines:
   - Line tension will be less than 500 lbs (225 kg) for 3/0 or smaller and less than 750 lbs (340 kg) for 3/0 and larger aluminum conductor.
   - Line tension will be less than 750 lbs (340 kg) for 3/0 or smaller and less than 1,000 lbs (450 kg) for 3/0 and larger copper conductor.
2. At temperatures below freezing, the conductor tension increases 20% for each 20°F (10°C) drop in temperature.
3. At temperatures above freezing, the conductor tension decreases 10% for each 20°F (10°C) drop in temperature.
4. When pulling up on a conductor, the tension increases as the sag decreases. Line tension is doubled when one-half of the sag is removed. Line tension is tripled when two-thirds of the sag is removed.

Calculating Conductor Length in a Span

Knowing the arc length of a conductor in a span can improve the planning of certain types of jobs. If a job requires stringing a conductor between two dead-ends during an outage, a great deal of outage time is prevented by laying out the proper length of conductor with the dead-ends already installed before the outage begins.

The calculated conductor length in a span is from the suspension points. The actual conductor length must have the insulators and dead-end hardware subtracted from the total length. **See Figure 17–41.**

Figure 17–41 The conductor length in a span is dependent on the length of span and sag. (*Delmar, Cengage Learning*)

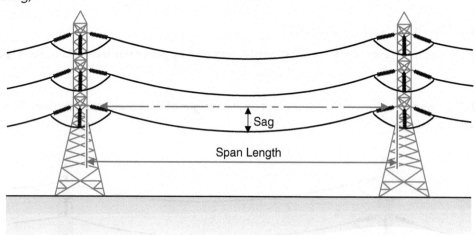

$$\text{Conductor Length in Span} = \frac{\text{Span Length} + 8 \times (\text{Sag})^2}{3 \times \text{Span Length}}$$

When a conductor is pulled up fairly tightly, a small change in conductor length can make a large change in sag and an even larger change in tension. Rigging equipment can be overloaded when pulling up a conductor tighter than planned.

Calculating Tension on a Loaded Down Guy

The tension on a down guy is calculated using the formula,

$$\text{Guy Tension} = \text{Line Tension (T)} \times \frac{\text{Length of the Guy (L)}}{\text{Distance to the Anchor (D)}}$$

For example, calculate the guy tension when the height of the pole is 50 units, the distance to the anchor is 30 units, and the conductor tension is 1,000 units. **See Figure 17–42.**

Figure 17–42 Guy tension will increase quickly when the anchor distance is decreased. *(Delmar, Cengage Learning)*

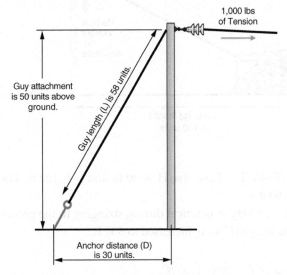

To calculate the guy length, use Pythagoras; calculate the length of the guy when the distance to the anchor is 30 units and the guy is 50 units high, as follows:

$$\text{Length of the Guy} = \sqrt{30^2 + 50^2} = \text{approximately 58 units}$$

Using the Guy Tension Formula

$$\text{Guy Tension} = 1{,}000 \text{ units} \times (58 \div 30) = 1{,}933 \text{ units}$$

The tension on a down guy multiplies fairly quickly as the anchor distance decreases and the conductor tension increases. For example, if the distance to the anchor is reduced to 20 units, notice how quickly the tension on the guy increases. Using Pythagoras, calculate the length of the guy when the distance to the anchor is 20 units and the guy is 50 units high, as follows:

$$\text{Length of the Guy} = \sqrt{20^2 + 50^2} = \text{approximately 54 units}$$

$$\text{Guy Tension} = 1{,}000 \times (54 \div 20) = 2{,}700 \text{ lbs}$$

Vertical Load Applied by a Down-Pull

The vertical load applied to a crossarm, snatch block, or traveler by a down-pull can be excessive when stringing conductor or temporarily dead-ending conductor to the ground. A short lead will exert a high vertical weight, which could overload a structure or its components. **See Figure 17–43.**

$$\text{Vertical Weight on a Structure} = \frac{T \times H}{L}$$

Where:

    T = line tension

    H = height of the attachment above ground

    L = distance from the structure to the anchor

Figure 17–43 The vertical load on a structure is an important consideration when setting up to string conductor. *(Delmar, Cengage Learning)*

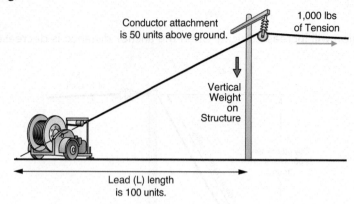

In the example shown in Figure 17–46, T = 1,000 lbs, H = 50 ft, and L = 100 ft. Therefore, the vertical weight on the crossarm = (1,000 × 50)/100 = 500 lbs.

It is best to keep the length (L) as long as practical during stringing in the preceding example because the vertical weight would be only 100 lbs if the length (L) were increased to 500 ft.

Measuring a Line Angle in the Field

A line angle must be known to confirm the proper framing for the structure, and be capable of calculating the bisect tension involved in handling a conductor. To measure an actual line angle in the field, measure out 57 ft in the two directions. The length of line X represents the line angle. This method is reasonably accurate up to 45°. **See Figure 17–44.**

Figure 17–44 A line angle can be measured in the field without a compass. *(Delmar, Cengage Learning)*

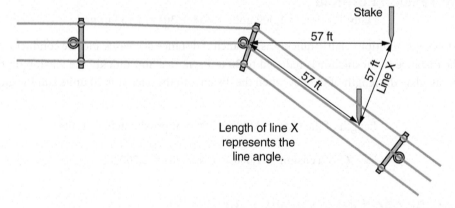

For example, if line X is 30 ft, the line angle is 30°. (For metrics, measure 6 meters in each direction. The line angle is the length of line X in meters × 10. For example, if line X is 3 meters, the line angle is 10 × 3 = 30°.)

Calculating the Bisect Tension of a Conductor at a Corner

Bisect tensions should be known when a conductor is lifted or relocated on a corner structure. The approximate bisect tensions of a conductor can be calculated. A more accurate calculation of the bisect tension is needed as the capacity of the rigging components approaches the bisect tension of the conductor. The actual bisect tension can be calculated as well. **See Figure 17–45.**

Figure 17–45 The bisect tension on conductor is dependent on the line angle and conductor tension. *(Delmar, Cengage Learning)*

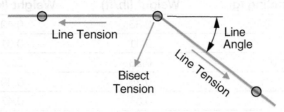

$$\text{Bisect Tension} = \frac{\text{Line Angle}}{60} \times \text{Line Tension}$$

For example, the line tension for a 336 kcmil AL conductor is 600 lbs (272 kg). The line angle is measured as 43°. What is the bisect tension of one conductor on the corner structure?

$$\text{Bisect Tension} = \frac{43}{60} \times 600 = 430 \text{ lbs (195 kg)}$$

From the formula, it can be seen that the bisect tension of a conductor with a line angle of 60° would be equal to the line tension. Prorating this fact allows an approximation of bisect tensions when the line tension and the line angle are known.

There are different factors used to determine an approximate bisect tension. **See Figure 17–46**. *Note:* The bisect tensions on the snatch blocks may appear different; however, the angle of the rope is measured inside the block while line angles tend to be measured as shown in Figure 17–44.

Figure 17–46 Factors to Calculate Bisect Tension

Line Angle (degrees)	Bisect Factor	× Line Tension lb (kg)	= Bisect Tension lb (kg)
900	1.5	1,000	1,500
750	1.25	1,000	1,250
600	1.0	1,000	1,000
450	0.75	1,000	750
300	0.5	1,000	500
150	0.25	1,000	250

Note how the factors are used to calculate the bisect tensions of various angles given a 1,000 lbs line tension. Using this table, an actual line tension can be inserted in the proper row to calculate a bisect tension.

Moving a Conductor into a Corner

The sag and tension of a conductor can change very quickly when moving a conductor into a corner. The tension and sag increase can cause a midspan contact with energized overbuilt conductors, a broken jib, or bucket truck instability. Calculating the bisect tension ahead of time is complex. Keeping a close watch on the sag as the conductor is moved in and out of a corner is the most practical way to prevent excess tensions.

Typical Conductor Data

Data can be used to calculate weight and tension for some conductor sizes. If an estimate is required and the conductor size is not in one of the tables, use the data for a larger conductor. Otherwise, for accurate calculations, consult conductor data tables. **See Figure 17–47.**

Figure 17–47 Conductor Data

Conductor Type	Stranding (g)	Weight (lb/ft)	Weight (kg/m)	Diameter (in./mm)
1/0 copper	7	0.33	0.49	0.368/9.35
2/0 copper	7	0.41	0.61	0.414/10.52
3/0 ACSR*	6/1	0.23	0.34	0.502/12.75
3/0 AACSR**	6/1	0.26	0.39	0.535/13.60
336.4 kcmil AAC***	19	0.31	0.47	0.665/16.89
336.4 kcmil ACSR	26/7	0.46	0.63	0.720/18.3
477.0 kcmil ACSR	26/7	0.66	0.89	0.858/21.8
556.5 kcmil AAC	19	0.52	0.77	0.856/21.74
795.0 kcmil ACSR	26/7	1.1	1.6	1.108/28.14
1,192.5 kcmil ACSR	45/7	1.34	2.0	1.302/33.07
Alumoweld	7 # 8	0.32	0.48	0.375/9.78
Alumoweld	19 # 8	0.88	1.31	0.642/16.3
3/8 160-Grade Steel	7	0.27		0.36/

*ACSR—Aluminum conductor steel reinforced
**AACSR—Aluminum alloy conductor steel reinforced
***AAC—All aluminum conductor

A conductor grip can be overloaded; check a WLL chart to ensure the correct grip is being used. **See Figure 17–48.**

Figure 17–48 Typical WLL for Jaw Grips and Woven Grips for Conductor

Conductor Type		Jaw Grip Manuf. #	WLL (lb/kg)	Woven Grip (in.)	WLL (lb/kg)
1/0 copper	7	1656–30	2,000/900	0.19–0.37	1,300/590
2/0 copper	7	1656–30	2,000/900	0.38–0.62	2,800/1,270
3/0 ACSR*	6/1	1656–30	2,000/900	0.38–0.62	2,800/1,270
3/0 AACSR**	6/1	1656–40	4,000/1,800	0.38–0.62	2,800/1,270
336.4 kcmil AAC***	19	1656–40	4,000/1,800	0.63–0.87	4,000/1,820
336.4 kcmil ACSR	26/7	1656–40	4,000/1,800	0.63–0.87	4,000/1,820
477.0 kcmil ACSR	26/7	1628–30	7,000/3,170	0.63–0.87	4,000/1,820
556.5 kcmil AAC	19	1628–30	7,000/3,170	0.88–1.12	6,120/2,780
795.0 kcmil ACSR	26/7	1628–30	11,000/5,000	0.88–1.12	6,120/2,780
1,192.5 kcmil ACSR	45/7	1628–40	13,000/5,900	1.13–1.37	9,360/4,250
Alumoweld	7 # 8	1656–30 1628–16	2,600/1,200 4,000/1,800	0.38–0.62	2,800/1,270
Alumoweld	19 # 8	1656–40 1628–16	6,000/2,799 11,000/5,000	0.63–0.87	4,000/1,820
3/8 160-Grade Steel	7	1628–5F	6,000/2,500	0.19–0.37	1,300/590

*ACSR—Aluminum conductor steel reinforced
**AACSR—Aluminum alloy conductor steel reinforced
***AAC—All aluminum conductor
Note: Jaw grip data refers to Klein Grips™. Woven grip data refers to Kellem Dua-Pull™.

Compression Sleeves

Typical data stamped on a compression sleeve include the manufacturer's name; conductor size, type, and stranding; and the proper die to be used with the sleeve. There is no standard designation for die sizes. Each manufacturer stamps its own die size on the sleeve.

Conductor Sagging Facts

A ruling span represents the behavior of all the spans in the line section, and using it determines which sag data to use. Sagging to the proper ruling-span data keeps the tension of the conductor in each span relatively equal, regardless of span-length temperature or loading.

$$\text{Ruling Span} = \text{Average Span} + \frac{2}{3}\text{ of Maximum Span} - \text{Average Span}$$

- *Stringing sag* refers to the sag applied by a line crew after stringing the conductor.
- *Final sag* refers to the sag after the conductor has settled and stretched.
- *Maximum sag* refers to the sag under ice loading, wind, and temperature. A conductor under heavy electrical load is typically considered to be at maximum sag when the conductor is at 212°F (100°C).

Aluminum strand conductor may have a die grease coating that can deposit on grip jaws. Conductor and grip jaws should be wiped clean of all grease before use.

Check all parts for distortion or misalignment. Grips should operate smoothly. Spring-loaded "Chicago" grips will lock open when the loop handle is folded over onto itself, and close automatically when the loop handle is returned to the normal position. **See Figure 17–49.**

Figure 17–49 Choose the correct Chicago conductor grip. Grips are designed for different WLL, conductor sizes, and conductor material (e.g., aluminum, steel). *(Delmar, Cengage Learning)*

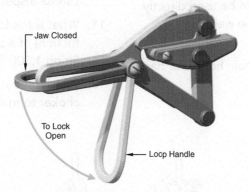

The manufacturer's load rating shall not be exceeded for stringing lines, pulling lines, sock connections, and all load-bearing hardware and accessories.

Conductor grips shall not be used on wire rope unless designed for this application.

Grips are to be used for temporary installation, not for permanent anchorage.

Review Questions

1. What does the term working load limit (WLL) mean when referring to a rigging component?

2. What is the WLL of a component that has a breaking strength of 1,000 lbs and a safety factor of 5?

3. Which knot is considered the king of knots and the one used most frequently in line work?

4. Which of these knots, hitches, or bends can be untied while under tension?
   ( ) bowline
   ( ) clove hitch
   ( ) two half hitches
   ( ) square knot

5. How much is the strength of a rope derated for an eye splice?

6. Which method is correct when installing U bolts to form a temporary eye in wire rope?

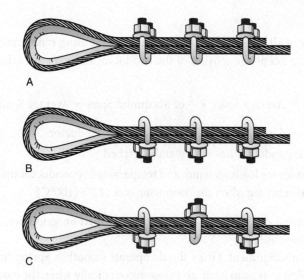

7. Is it an acceptable practice to use a Chicago-type grip to hold tension temporarily on a wire rope?

8. *True or False:* A web hoist can be used directly on an energized line when the webbing has been cleaned.

9. What is an indicator that a chain hoist is or has been overloaded?

10. If the label that identifies a nylon web sling size and rated load is missing, is it still acceptable to use the sling after a careful inspection?

11. What is the load on the fall line (ignoring friction) of a set of three-sheave blocks when lifting 1,500 lbs (kg)?

12. Which methods are correct when using a choker to make a lift?

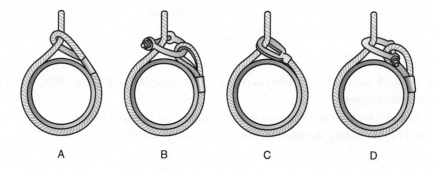

A   B   C   D

13. If a 100 lbs (50 kg) weight is being pulled up on a hand line, what is the weight supported by the hand line pulley and anchor point?

14. What is the tension in a 1,000 ft (300 m) span, with a 795 ACSR (26/7 stranding) that weighs 1.1 lbs/ft (1.6 kg/m) and sagged at 15 ft (5 m)?

15. What is the tension on a down guy where the anchor is 50 ft (15 m) from the pole, the

weight to be held is 2,000 lbs (1,000 kg) at 70 ft (20 m) above the ground?

16. The line tension for a 336 kcmil AL conductor is 800 lbs (400 kg). The line angle is measured as 50°. What is the bisect tension on the corner structure with three 336 kcmil AL conductors?

Chapter 18

Working It Hot

Objectives

After completing this chapter, you should be able to:

1. Examine the task to be carried out to evaluate the option of working hot or working with the circuit isolated and grounded.

2. Take steps to identify and protect yourself from a second point of contact.

3. Perform hot-line work while maintaining the minimum approach distance.

4. Explain the purpose of blocking the automatic reclose on the source circuit breaker while doing hot work.

5. Reduce the risk of a voltage surge when working hot on a transmission line.

6. Explain the hazards involved when using temporary insulated jumpers.

7. Make connections on hot secondary wires.

8. Explain the steps required to prepare for rubber glove work.

9. Explain the hazards and identify the protective barriers to protect from those hazards when doing rubber glove work.

10. Calculate the working load limit on hot line tools.

11. Recognize hazards that may be encountered using hot-line tools.

12. Explain the hazards and the protective barriers needed to perform the various methods of barehand work.

Safety Strategy for Hot-Line Work

Defining Hot-Line Work

Installing a pole, hanging a transformer, stringing conductor, switching, operating a hot-line clamp, and fuse replacement are examples of work in the *vicinity* of hot circuits, but this work is not normally considered "hot-line work."

Hot-line work generally involves work where a conductor more than 750 V is unclamped (untied), moved, cut, or spliced or where solid connections are made or removed.

There are many well-written procedures for doing hot work, especially on transmission lines. On distribution lines, it is much more difficult to write a procedure for each job because the procedure would be different, depending on access with a bucket truck, the use of rubber gloves or hot sticks, or a combination of both, as well as a multitude of different framings on the pole. Because written procedures do not cover much of the details needed for work done on energized distribution lines, preparing a job safety analysis for any jobs other than routine would demonstrate due diligence.

This chapter provides information and discusses some of the preparations needed for live-line work. It is not a replacement for live-line work procedures.

Choosing the Hot-Line Work Option

Do not approach a circuit closer than the minimum approach distance unless you are insulated, the facility is insulated, or the facility has an isolation certification in place based on a formal lockout/tagout procedure.

There is an increased risk when a complex job is carried out "hot." A lot of jobs are carried out with the line in service without even considering alternatives. The option of doing a job, with the circuit isolated and grounded, should at least be considered. However, some jobs are better done "hot" when balanced against the complexity of switching, grounding, and notifying customers. Customer impact and the job complexity should be decision factors when planning a job.

If any of the following are true, arrange for an outage.

1. Customer impact

   - There are no critical customers.
   - There will be no unnecessary economic hardship on commercial customers (hospitals, gasoline stations, restaurants, manufacturing plants, dairy farms).
   - The duration of the outage will not adversely affect the customer (manufacturing plant, shopping mall, ventilation in chicken and pig farms, furnaces or air conditioning on extremely cold or hot days, sump pumps when groundwater is high).

2. Complexity of work

   - There is no approved work method.
   - The weather or lighting conditions are not favorable for hot-line work.
   - The complexity of doing the job hot does not match the needs of the customer.

Protecting from a Second Point of Contact

A second point of contact is where current would leave a person's body if contact were made with a hot circuit. There can be no current flow (electrical burns) if contact is made with a hot circuit and no other part of a person's body is in contact with another object at a different potential.

At distribution voltages, it is always possible to cover, remove, or maintain a minimum approach distance from a second point of contact. This is a fundamental rule for working on or near hot distribution circuits. An insulated boom and bucket provide protection from earth as a second point of contact, but unless the structure, down guys, and other conductors are covered, the protection from a second point of contact is inadequate. **See Figure 18–1.**

When working hot line from transmission line structures, the removal of the second point of contact is not always an option. However, the greater approach distances available reduce the risk of inadvertent contact.

Barehand work is an ultimate example of depending on the second point of contact being removed.

Figure 18–1 A lineman has complete protection from a second point of contact by working from an insulated platform and using the proper amount of protective cover-up. *(iStock.com/Kozmoat98)*

Minimum Approach Distance (MAD) Doing Hot-Line Work

When using hot-line tools, minimum approach distances apply to the length of clear insulating section of the tool. The flashover voltage for a live-line tool is the same as it is for air. A fiberglass live-line tool may be better insulation than air, but the distance needed on a tool is an air gap between the hands on a live-line tool and the energized conductor. **See Figure 18–2.**

Figure 18–2 Typical Minimum Approach Distances for Working with Live-Line Tools

Maximum Phase-to-Phase Voltage (Max. Phase-to-Ground Voltage)	Minimum Approach Distance Phase-to-Ground Exposure	Minimum Approach Distance Phase-to-Phase Exposure
0.05 to 1 kV	Avoid Contact	Avoid Contact
Up to 15 kV (8.7 KV)	2′-1″ (64 cm)	2′-2″ (66 cm)
Up to 36 kV (20.8 kV)	2′-4″ (72 cm)	2′-7″ (77 cm)
Up to 46 kV (26.6 kV)	2′-7″ (77 cm)	2′-10″ (85 cm)
Up to 121 kV	3′-2″ (95 cm)	4′-3″ (1.29 m)
Up to 145 kV	3′-7″ (1.09 m)	4′-11″ (1.5 m)
Up to 169 kV	4′ (1.22 m)	5′-8″ (1.71 m)
Up to 362 kV	8′-6″ (2.59 m)	12′-6″ (3.8 m)
Up to 550 kV	11′-3″ (3.42 m)	18′-1″ (5.5 m)
Up to 800 kV	14′-11″ (4.53 m)	26′ (7.91 m)

The minimum approach distances are often specified by government safety regulations.

Rubber-glove work and barehand work should not be considered exceptions to the minimum approach distances. When in contact with an energized conductor using rubber gloves or barehand techniques, the minimum approach distance still applies, but to the distance between a worker and any second point of contact. When in contact with a hot conductor, a second point of contact with cover-up installed on it may be approached more closely than the distances specified in the minimum approach table. Typically, a utility/employer allows accidental or brush contact only, or requires that an "air gap" be maintained with a covered-up conductor. No cover-up is available for transmission-line work.

A hand stop on a hot-line tool will reduce the risk of inadvertent encroachment (choking the hot stick) into the minimum approach distance.

For some very special transmission-line circumstances, the minimum approach distance for hot-line tool work is reduced when a portable protective gap (PPG) is placed at an adjacent structure.

Some live-line tool configurations, especially for transmission line work, will have a metal fitting such as a splice or a tool-end fitting within the insulated portion of the tool. This metal will introduce electrical stresses that will require extending the minimum approach distance.

Blocking Automatic Reclosing of Protective Switchgear

When hot-line work is to be done on a circuit, the automatic reclosing feature of the source circuit breaker or recloser should be blocked from service and tagged.

- The nonreclose feature does not prevent an accident from occurring and does not guarantee that the circuit will trip out during an accident. When set in a nonreclose position, the breaker or recloser does not operate faster than normal. The nonreclose feature does ensure that a circuit will not be reenergized once it does trip out.

- When the circuit is not automatically reenergized, the exposure time to the electrical hazard is reduced, which will allow a safer rescue effort if needed.

- Damage or injury may be limited by a reduced exposure to the high fault current that can occur during a fault.

- The risk of a restrike, which causes a switching surge (voltage surge), is eliminated if the circuit breaker is put into a nonreclose position.

The procedure to block the automatic reclosing of a breaker involves contacting the controlling operator, establishing continuous communication, and placing tags. On distribution lines, the reclosers are often set in a nonreclose position and tagged by the crew doing the work.

Protecting from Voltage Surge on Transmission Lines

In addition to lightning, a switching surge can cause an overvoltage on a transmission line.

Options to Reduce the Magnitude of a Voltage Surge

To encourage a surge to flash over on a structure where work is not being carried out, consider the following options.

- Surge gap distance at the line terminals can be reduced temporarily for the duration of the work.
- Surge arrestors attached at the line terminals will reduce the magnitude of a voltage surge.
- A temporary portable protective gap (PPG) can be installed at an adjacent structure.
- Some insulators can be shorted out at an adjacent tower.

Reduce the Probability of a Voltage Surge

To reduce the likelihood of a voltage surge, consider the following options.

- Block the automatic reclose feature on all circuit breakers connected to the circuit. This will prevent switching surges after the line trips out.

- Ensure that no relay testing is being done on the circuit while hot-line work is in progress.
- Do not work when lightning is in the vicinity. The controlling station staff for transmission lines should have access to an information system that notifies them of impending electrical storms.

Testing Insulators for Breakdown

When hot-line work (especially barehand) is planned with or near existing suspension-type insulators, each insulator in the string should be tested. **See Figure 18–3.**

Figure 18–3 A hot-line insulator tester will confirm that an insulator string does not have defective units. *(© Hubbell Power Systems, Inc.)*

Another type of insulator tester slides up and down on a hot stick suspended parallel next to an insulator string. This tester takes advantage of the fact that the electrical field surrounding a defective or shorted insulator decreases. It works on polymeric and porcelain/glass insulator strings.

Phasing sticks can be used to test insulators on circuits up to 50 kV. **See Figure 18–4.**

Figure 18–4 When testing insulators with phasing sticks, test between the structure and positions such as 1, 2, and 3 shown. *(Delmar, Cengage Learning)*

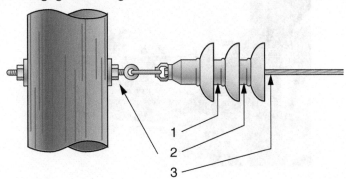

If phasing sticks are used, it will be necessary to establish typical readings for the voltage system and the type of insulators being tested. For example, the following are typical readings for a 27.6 kV system with three porcelain insulators.

1.5 to 2 kV between ground and first insulator

3 to 4 kV between ground and second insulator

14 to 16 kV between ground and third insulator

Use of Temporary Jumpers

When work requires the replacement of any circuit component, such as taps, loops, or sleeves, the component must be bypassed with a temporary **jumper** (red heads, mechanical jumpers, macs). Jumpers used on distribution lines are insulated. **See Figure 18–5.** Unless the jumper is specifically designed to be installed with rubber gloves, a jumper should always be installed with hot-line tools. **See Figure 18–6.**

Before opening or cutting a conductor, use an ammeter to check that the jumper is carrying at least one third of the load current. A larger jumper or cleaning the terminals and conductor may be necessary to achieve the one third load-carrying capability.

There are jumpers with terminals designed to pick up load, typically up to 250 A. They are not designed to drop load. There is hardware available such as a temporary cutout tool, where the jumper is connected to the device and used to drop load. **See Figure 18–7.**

Figure 18–5 Note the thicker insulation in the higher voltage rated jumpers. *(Courtesy of Salisbury by Honeywell)*

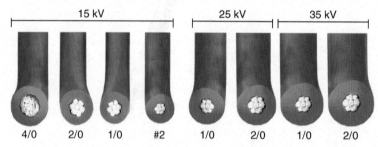

Figure 18–6 A lineman installs a jumper designed for installation with rubber gloves. *(© Salisbury by Honeywell)*

Figure 18–7 A temporary cutout tool can be used to drop load before removing a jumper. *(Reprinted with permission of Hubbell Power Systems, Inc., Centralia, MO USA)*

Hazards Specific to Using Temporary Jumpers

Hazards	Barriers
The jumper insulation may fail.	Ensure that an insulated jumper has been tested as scheduled. Do a visual inspection for damaged rubber. Maintain at least a minimum approach distance from the terminations.
The current-carrying capacity of the jumper is not adequate. An arc is generated when the main conductor is cut.	Inspect the cable near the terminations for broken strands. An insulated or bare jumper should be carrying at least 30% of the load current before cutting a conductor or connection. The jumper conductor size should be equal to the size of the circuit conductor. Following are typical sizes. #2 to carry 200 A #1/0 to carry 260 A #2/0 to carry 300 A #3/0 to carry 350 A #4/0 to carry 400 A
A worker gets between a jumper and a main conductor.	Always make and break connections with a hot-line tool. Rubber jumpers with insulated clamps can be installed and removed when wearing rubber gloves. If wearing rubber gloves, you are wholly dependent on their integrity. In case of an arc being generated, you will be within the flash.
There may be a loss of control while installing a jumper.	Once one end of a jumper is installed, the other end is hot. Use two people with two clamp sticks, or use a jumper with a parking stud so that the other end of the jumper is secured to the same terminal.

Using Hot-Line Rope

Reduce the risks involved with using hot-line rope by using only rope identified as hot-line rope, electrically tested for its full length and retested on a scheduled basis. A soiled, damp rope will become conductive very quickly.

- Store and transport the rope in a special container with a drying agent, such as silica gel, to absorb any moisture.
- Test the rope in the field with a portable hot-stick tester.
- Handle the rope with clean and dry work gloves.
- Clean any snatch blocks or hand-line block that the rope will run through.
- Use a tarpaulin where the rope may touch the ground.

Hardware for Safer Live-Line Work

Hardware that reduces risk when working on energized conductor includes using dead-end clamp (shoe) instead of formed wire dead-ends, insulator clamps instead of tie wire, and saddle clamps designed for installation with live-line tools instead of U bolt–style clamps. **See Figure 18–8.**

Figure 18–8 Insulator clamps and saddle clamps are designed for easier access with live-line tools. *(Delmar, Cengage Learning)*

Working on a Hot Secondary

Specific Hazards for Working a Hot Secondary

Working on hot secondary conductors is not normally considered hot-line work. It does, however, require similar work procedures to working higher voltages hot.

Hazards for Working a Hot Secondary Voltage

Specific Hazards	Barriers
A worker puts self in series with the circuit.	Rubber gloves will give a person a second chance if an open point is accidently bridged.
A worker makes contact while touching a second point of contact.	Rubber gloves will give a person a second chance when another part of the body is in contact with a neutral or other phase.
An uncontrolled wire makes contact with the primary or other phases of the secondary.	When the distance from the primary to the service dead-end is limited, cover up the primary conductors.
Clearance between secondary bars on a pad-mount or underground transformer is close. An accidental short close to the transformer would be very explosive.	Cover up the terminals not being worked on. Work from a rubber mat. In addition to regular PPE, ensure that rubber gloves and eye protection are worn. A ground-gradient mat under the rubber mat will provide protection from a potential rise because of a system fault.

Explosive Flash Hazard Near a Transformer

A step-down transformer can generate a very high current on the secondary side if the secondary wires are accidently shorted. If the secondary wires are shorted, the transformer can briefly carry a load more than 10 times its rating and supply a high current to the faulted location.

The larger the transformer, the greater the capability is to generate a large fault current. Depending on the impedance of the transformer, a short at the secondary terminals of 100 kVA can be as high as 18,000 A. The level of fault current at a work location will depend on the distance from the transformer and the conductor size. **See Figure 18–9**.

If an outage is not practical, use more cover-up when working secondary near the transformer. Wear eye protection when working on a hot secondary because an eye can be permanently damaged by a large flash.

Figure 18–9 Note that the level of fault current available on a secondary conductor drops quickly with distance away from the transformer. *(Delmar, Cengage Learning)*

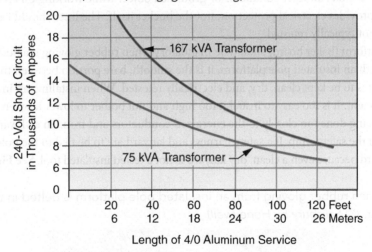

Rubber-Glove Work

Introduction to Rubber-Glove Work

The most common live-line work method on distribution voltages is hot-line rubber-glove work. The voltage range for this work is from 5 to 34.5 kV. Some utilities/employers do not allow rubber-glove work above 5 kV, while others work with rubber gloves all the way up to and including 34.5 kV.

The key to safe rubber-glove work is to isolate the worker from all second points of contact. Rubber gloves should not be the sole barrier between a worker and a potential fatal shock but are considered as the last line of protection.

When covering up live front underground equipment, work from an insulated rubber blanket to reduce the risk of being in contact with a second point of contact. **See Figure 18–10.**

Figure 18–10 Rubber blankets have covered all exposed energized parts on this live front pad-mount equipment. (© *John Bellows*)

Insulated Platforms for Rubber-Glove Work

Working with rubber gloves from an insulated platform is similar to wearing personal protective equipment; it may provide backup protection when things go wrong.

The most common insulated platform is an aerial device bucket. The bucket and boom must be kept clean and dry. While there is a reliance on the insulated boom, the bucket is only considered insulated when an insulated liner is used. A bucket can accidentally contact another conductor or grounded object while working on an energized conductor. It is the removable liner that provides electrical protection, not the bucket itself. The liner should be retested when the boom is retested; the bucket is not typically retested.

An insulated pole platform (baker board) should be mandatory when rubber gloving if a bucket truck is unavailable or cannot get to the pole. With an insulated pole platform, it is the smooth, bare portion between the pole and where a power line technician stands that is to be kept clean, dry, and electrically retested. When installing an insulated platform, it is critical to install it for your height. It is too easy to install it too high and not bother to lower it when it is found to be too high. Meanwhile, the work is being done too closely to the energized conductors and from an awkward position. The tripod or safety rail is used to fasten the safety strap. If a safety harness and lanyard are to be tied to the pole, then there should be an insulated link in the lanyard because even a clean, dry lanyard is not a tested insulated tool. **See Figure 18–11.**

Figure 18–11 A lineman rubber gloving from an insulated pole platform is belted in to an insulated temporary anchor point. (© *Salisbury by Honeywell*)

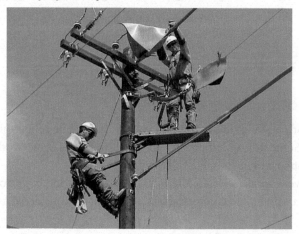

About Rubber Gloves

Rubber gloves are often used as personal protective equipment in case something goes wrong, such as when switching or just working in the vicinity of an energized circuit. In the situations described in this chapter, rubber gloves are used as live-line tools to perform live-line work. Rubber sleeves are used with rubber gloves to provide protection in cases of overreaching or another incident. Instead of sleeves, some rubber gloves are available that extend to the armpits.

Rubber gloves come with different voltage ratings, and it is critical, of course, that the voltage rating (class) be higher than the voltage to be worked on. **See Figure 18–12.**

Figure 18–12 Rubber glove class and voltage ratings, and minimum distance between the top of the protectors and rubber gloves. *(Courtesy of Salisbury by Honeywell)*

Class Color	Proof Test Voltage AC / DC	Max. Use Voltage* AC / DC	Rubber Molded Products Label
00 Beige	2,500 / 10,000	500 / 750	
0 Red	5,000 / 20,000	1,000 / 1,500	
1 White	10,000 / 40,000	7,500 / 11,250	
2 Yellow	20,000 / 50,000	17,000 / 25,500	
3 Green	30,000 / 60,000	26,500 / 39,750	
4 Orange	40,000 / 70,000	36,000 / 54,000	

Glove Class	Min. Distance between Protectors and Rubber Gloves in.	mm
00, 0	1/2	13
1	1	25
2	2	51
3	3	76
4	4	102

Rubber gloves are delicate and must be visually inspected before each use. A portable glove inflator allows a much more thorough check for tears, rips, and punctures. Without an inflator, roll up the cuff tightly, trapping the air, then apply pressure to inflate the glove and look for damage. Most gloves come in two layers of contrasting colors, which allows any cut through one layer to be easily seen. Listening for escaping air is probably not as effective as a good visual inspection. Rubber-glove inspection involves a careful eye for the following:

- Cracking and cutting damage, often from storing rubber gloves folded or pinched too tight. The rubber in a tight fold is said to be stretched equivalent to stretching the rubber to twice its length.
- Checking, often caused by the ultraviolet light from sunshine or fluorescent lighting.
- Embedded wood or metal splinters from climbing or embedded small wire, especially after working with extraflex copper.
- Unusual swelling of the rubber, often caused by exposure to petroleum products.

Between electrical tests, rubber gloves should be washed with clean water only. Add a little corn starch (petroleum products in baby powder can damage the rubber) to make the gloves easier to put on. When not in use, store the gloves in the rubber-glove bag with the finger tips pointed up. *Do not* store the gloves in the truck bin with tool belts and pole climbers.

Rubber gloves come in different sizes. The best way to determine proper size is to try them on. Measure the circumference around the palm, then add 1 in. (25 mm). **See Figure 18–13.**

Figure 18–13 To measure for a rubber-glove size, measure around the palm then add 1 in. *(Delmar, Cengage Learning)*

Measure the Circumference
around the Palm

Leather protectors, matched in size to the rubber gloves, are mandatory to prevent damage to the much more vulnerable rubber. Leather protectors must be inspected for splinters or other embedded objects. Ensure that the rubber glove extends past the leather protector glove. Maintain the minimum required distance of rubber between protector glove and the rolled-up top of the rubber glove.

About Cover-Up Equipment

There are two types of cover-up available: rigid (plastic) and rubber. Rubber cover-up ratings parallel the same ratings as rubber gloves. For example, the use of a class 2 rubber blanket, cover, coupler, line hose, or rubber hood is approved for work on phase-to-phase up to 17,000 V.

Plastic guard equipment for electrical insulation is rated differently and provides protection in a different manner. The voltage rating of rigid plastic cover-up is not based on the thickness of the material, as with rubber, but more on the air space between the conductor and the cover. The terms plastic guard, fiber, and rigid cover-up are used interchangeably, but technically the equipment acts as a guard to keep a certain air space between the worker and the energized line. This air space is less than the minimum approach distance because the human factor is removed from the minimum approach distance formula. The highest-rated plastic guards are rated higher than the highest-rated rubber cover-up. Some plastic guards are designed to couple to similarly rated rubber cover-up. Plastic guards are more difficult to store properly in a truck and are often found abused during safety audits. The voltage or class ratings of plastic guards, unfortunately, are not the same as for rubber. **See Figure 18–14.**

The phase-to-phase rating for plastic guards applies only when both phases are covered. The advantage of plastic guards is the ease of installation from the end of a stick. **See Figure 18–15.**

Figure 18–14 Plastic Guard Equipment Ratings

Class	Phase-to-Phase kV	Phase-to-Ground kV
2	14.6	8.4
3	26.4	15.3
4	36.6	21.1
5	48.3	27.0
6	72.5	41.8

Figure 18–15 Plastic guards are available for many applications, and manufacturers have been willing to design new ones when line hardware changes. (© Salisbury by Honeywell)

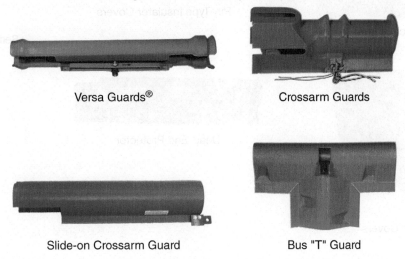

Versa Guards®

Crossarm Guards

Slide-on Crossarm Guard

Bus "T" Guard

Electrical Retesting of Rubber Goods

Electrical testing requirements are designed to meet various standards, and they also are mandated by law. Rubber goods should be retested according to the following schedule, at a minimum. These timelines are mandated by the U.S. Occupational Safety and Health Act (OSHA) as follows:

Rubber gloves: 6 months

Rubber sleeves: 12 months

Rubber-insulating blankets: 12 months

These are minimum requirements, and it is quite common to retest rubber gloves as often as every 2 or 3 months.

Other rubber goods, such as rubber hose, rubber-insulator covers, and rubber-insulated jumpers require tests upon any indication that the insulating value is suspect. Many utilities have scheduled electrical retests for these rubber goods, whereas others have taken the position that a thorough visual inspection will pick up any defect that will cause it to fail electrically. **See Figure 18–16**.

It is important to keep rubber goods clean so that they are easier to inspect, and to reduce the risk of electrical tracking along the surface. Cleaning requires hot water and the use of an approved chemical, followed by a thorough rinsing with clean water.

Figure 18–16 Rubber equipment is designed to fit many types of line hardware. Rubber blankets can always be used when there is no specific rubber equipment available. (© Salisbury by Honeywell)

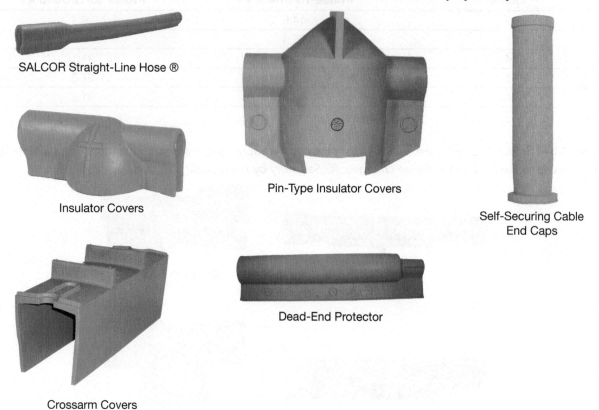

SALCOR Straight-Line Hose ®

Insulator Covers

Pin-Type Insulator Covers

Self-Securing Cable End Caps

Dead-End Protector

Crossarm Covers

Preparation for Rubber-Glove Work

Step	Action	Details
1.	Inspect rubber gloves. **See Figure 18–17**.	• Ensure that the correct class of rubber glove (sleeve) is to be used. • Check to ensure that the rubber gloves (sleeves) are not due for an electrical retest. • Carry out an inspection and air test on the rubber gloves. • Do not wear jewelry, especially a signet-type ring with a raised surface.

Figure 18–17 A more thorough visual inspection can be made after a rubber glove is inflated. (© Salisbury by Honeywell)

2.	Inspect the protective cover-up to be used.	A physical inspection of rubber or plastic cover-up is very effective. In almost all cases, cover-up that has failed an electrical test also has had an obvious physical defect. Use only clean cover-up.
		Roll rubber hose inside and outside to look for cuts, rope burns, and corona damage. Roll rubber blankets from corner to corner, inside and outside, while looking for damage.
		Inspect plastic cover-up, such as conductor guards, insulator guards, crossarm guards, and pole guards for cracks and cuts. The ends and lips of the covers are particularly susceptible to cracks during cold weather.
3.	Inspect insulated jumpers.	In addition to inspecting the insulation on a jumper cable, check for broken strands near the terminals. The insulation can be peeled back somewhat at each terminal for a visual inspection, or you can use the type of continuity tester used to test portable protective grounds. Ensure that the insulation of the jumper is good for the voltage being worked on. Store in appropriate bags when not in use.
4.	Use ground-to-ground, extended reach, or the cradle-to-cradle rubber-glove rule, as mandated by the utility/employer.	*Caution:* A conclusion in many electrical accident investigations is: "The victim was not wearing rubber gloves."
5.	Plan protection against any second point of contact.	Rubber-glove work should be planned and carried out so that a worker is never totally dependent on the integrity of rubber gloves. Having this second line of defense is essential when working on a circuit with rubber gloves. • Work from an insulated aerial device, or if that is not possible. • Work from pole-mounted insulated platforms, but ensure that the top and bottom bare sections are clean, dry, and on an electrical retest schedule. • Stay at a minimum approach distance from all second points of contact or cover-up or remove the neutral, secondary wires, guys, or other phases. • If two people are working at the same structure, they must not work on different phases at the same time because one can become a second contact for the other.

Specific Rubber-Gloving Hazards

Rubber-Gloving Hazards	Barriers
Overreaching exposes an uncovered part of the body to an electrical contact.	1. Work from a position where a slip will not cause an uncovered part of the body to make contact—for example, from a position below the conductor. 2. Wear rubber sleeves to reduce the risk of contact due to overreaching.
Untying and tying-in conductors can be a hazard because of the following: 1. The length of the tie wire 2. Possible puncture of the rubber glove by the sharp end of the tie wire	Roll or bunch up the tie wire in the palm of your rubber glove when untying.
Installing and removing preformed dead-ends or armor wrap can be hazardous, because the preforms for larger conductors are long enough to contact other phases or objects.	For large conductor, use a dead-end clamp (shoe). If it is necessary to remove a long preform, make frequent cuts.
Temporarily removing rubber gloves for fine work or to cool off increases the probability of an inadvertent contact.	A mental lapse can occur anytime, including when rubber gloves have been removed. Tell someone working with you of your intention.
Working on more than one phase at a time is a hazard, because the other phase is a potential second point of contact.	Two people working on the same structure must not work on different phases at the same time. One can become a second point of contact for the other.

(Continued)

Rubber-Gloving Hazards	Barriers
Working on adjacent structures at the same time can create a hazard.	When conductors are moved on adjacent structures at the same time, there must be communication between the two poles.
Small conductor can break while working on it.	Conductors, such as #6 copper, #4 ACSR, and #8A copperweld or smaller, are susceptible to breaking. Consider an outage.
Long metallic tools can bridge or short the rubber-glove cuff.	Long-handled tools, such as a bolt cutter, press, hoist, or hammer, should have nonmetallic handles.
Improvising the temporary placement of a conductor while replacing a crossarm or other support can lead to poor clearance between conductors and to workers.	Use tools such as an auxiliary mast to support conductors while replacing a crossarm or other support items.

Hot-Line Tool Work

Introduction to Hot-Line Tool Work

There are procedures to do almost everything from the end of a stick. Switches can be cut into a line, solid connections can be made, armor wrap can be removed and installed, conductors on transmission lines can be unclamped and transferred into stringing blocks. Many people think that "hot sticking" is the safest method of live-line work, and in some places this is backed up by regulations.

Many power line technicians consider working with live-line tools their favorite work. It takes more skill and practice to become proficient at using live-line tools than it does to untie conductors, install a wedge connector, or dead-end a conductor into a dead-end clamp while working barehand or wearing rubber gloves.

Minimum Approach Distance When Working with Hot-Line Tools

Hot-line tools must be long enough to provide a minimum insulated working distance between any part of a worker and the energized conductor or part being worked on.

Ensure that the clear insulation of conductor support tools, such as link sticks, strain carriers, and insulator cradles, are at least as long as the insulator string or the minimum distance necessary.

Preparation for Hot-Line Tool Work

Step	Action	Details
1.	Identify the weights and tensions to be lifted or pulled.	1. The weights and tensions of conductors to be lifted or pulled can be calculated. Lifting conductor at a hilltop or at a corner can increase the weight and tension on the sticks very quickly. 2. Check the working-load limit of the tool configuration to be used. Check the working-load limit of the components to be used, such as the saddles, tongs, link sticks, and snubbing bands. 3. Check that the work method will allow compliance with the minimum approach distance. 4. Check the conductor attachments at adjacent structures.
2.	Inspect hot-line tools.	1. Check that the tools have had their scheduled electrical tests. 2. Use hot-line tools that are kept clean and dry and in a protective container when not in use. 3. A physical inspection of hot-line tools is very effective. In many cases, a stick that fails an electrical test also has an obvious visual physical defect. Physical damage to the tool, such as surface scratches or a deformed or popped rivet, can allow the moisture to enter. 4. Hollow tools are more susceptible to contamination and may need more frequent electrical testing.

3.	Wipe or clean hot-line tools.	1. Clean and dry your tools with an agent specifically made for this purpose—that is, an agent that evaporates quickly and leaves a treatment on the surface that repels and beads water. 2. Damp and dirty work gloves are the most probable source of surface contamination, especially if both ends of a stick have a tool on it and the worker is using both ends.
4.	Check saddles, lever lifts, and chain tighteners.	1. Use saddles that are stored with the holding clamp in the closed position when not in use. 2. Check for distortions, loose bolts, and loose rivets. 3. When the wire-tong saddles are installed on the hot-line tool, an additional 180° turn on the wing nut may be applied after hand tightening. 4. Use rope blocks to make lifts with the lifting tong. As a second barrier, if rope blocks are removed, tie a rope sling between the saddle and the butt ring. 5. After a lever lift is flipped into position, tie it up to prevent it from dropping during the movement of the conductor.
5.	Check ropes and blocks.	1. Use ropes and rope blocks that are kept exclusively for hot-line tool work.
6.	Protect against any second points of contact.	1. Work from an insulated, pole-mounted platform or an insulated aerial bucket. 2. Cover up or remove the neutral, secondary wires, guys, and other phases. 3. Do not work on more than one phase at a time.

Typical Working-Load Limits on Hot-Line Tools

Check specifications from manufacturers and the utility/employer for actual working-load limits when using hot-line tools.

Tool	Type or Size	Working-Load Limit
A wire tong (pole) is designed for compression loading. All side pulls must be balanced with an opposite side pull. Typically, for a vertical lift the minimum distance between saddles is 5 ft (1.5 m) and the maximum unsupported length above the top saddle is 6 ft (1.8 m).	2.5 in. (6.4 cm) 3 in. (7.6 cm)	The limiting factor for the working-load limit for a lifting pole is the saddles holding it and any horizontal strains being applied to it.
Pole saddles must hold the weight of the lifting tong, the conductor weight, and the pull exerted by the fall line on a set of blocks. For a heavy conductor, hang the set of blocks in a separate snubbing band or sling instead of the saddle.	Saddle without extension Saddle with extension	800 lbs (360 kg) 600 lbs (270 kg)
Lever lifts are used for the heaviest loads to be lifted.	Single-lever lift Double-lever lift	1,000 lbs (450 kg) 1,500 lbs (680 kg)
The allowable tension on a link stick depends on the diameter.	1.25 in. (3 cm) 1.5 in. (4 cm)	3,500 lbs (1,580 kg) 6,500 lbs (2,950 kg)
The working-load limits for suspension sticks and strain carrier poles used to lift or tension conductors on transmission lines vary and should be positively identified and checked out before using.	Typical heavy-duty suspension link stick Typical strain carrier pole	6,500 lbs (3,000 kg) 7,500 lbs (3,400 kg)
Roller link stick	1.25 in. (3 cm)	1,000 lbs (450 kg)
The allowable tension on a snubbing band is shown per ring and the total for the complete unit.	—	500 lbs (230 kg) per ring, the total load not to exceed 1,000 lbs (454 kg)

Calculating the Load on Hot-Line Tools

Most hot-line tool work on transmission lines uses the tools in a straight tension mode, and the maximum safe working load can be determined from the ratings of the individual tools. When live-line tools are used in a configuration, with tools in compression and tension, some measuring and calculations are needed to determine the maximum safe working load. **See Figure 18–18.**

Figure 18–18 To calculate the WLL for this tool configuration, the weight of conductor and length of A, B, and C are needed. *(Delmar, Cengage Learning)*

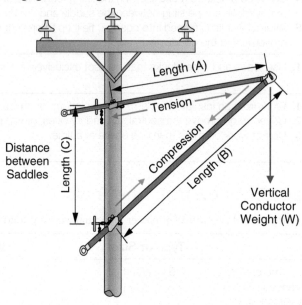

The compression load on lifting tong B =

$$\frac{\text{Length of B} \times \text{Conductor Weight}}{\text{Length of C}} \quad \text{or} \quad \frac{B \times W}{C}$$

The tension on the holding tong A =

$$\frac{\text{Length of A} \times \text{Conductor Weight}}{\text{Length of C}} \quad \text{or} \quad \frac{A \times W}{C}$$

Where:

A = length of holding tong A between the conductor and the saddle

B = length of lifting tong B between the conductor and the saddle

C = distance between the saddles on the pole

W = weight of the conductor in pounds or kilograms

There are other live-line tool configurations, such as lifting a conductor straight up using a single tong or a three-phase lift set (kite). Use the employer or manufacturer's work procedures to ensure that the tools are not overloaded. Remember that the maximum height a conductor can be lifted occurs when the lowest point of sag reaches an adjacent structure. At this point the conductor weight has doubled. When planning to lift a conductor with an aerial device or with hot-line tools, multiply the weight to be lifted by a factor of 2. When the conductor to be lifted is at a corner structure, the counteracting force of a link stick down to a temporary rope guy holding the bisect tension of the conductor must also be added to the weight to be lifted.

Example of Hot-Line Tool Calculations

The example calculations shown will be for two different tool configurations. **See Figure 18–19**.

If the conductor weight is given at 200 lbs, calculate the compression load on the lifting tong and the tension on the holding tong for the initial position and the final position for poles A and B. Note that the distance C can be critical to the tool loading. **See Figure 18–20**.

Figure 18–19 The WLL on live-line tools will change each time a lifting tong or holding tong is moved. *(Delmar, Cengage Learning)*

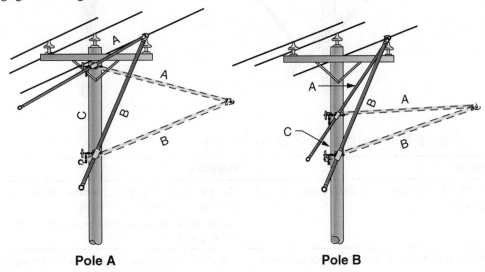

Pole A   Pole B

Figure 18–20 Calculating the WLL on Live-Line Tools

	Pole A		Pole B	
	Initial Position	**Final Position**	**Initial Position**	**Final Position**
Load on Lifting Tong $= \dfrac{B \times W}{C}$	$\dfrac{10 \times 200}{7} = 286$ lbs	$\dfrac{12 \times 200}{7} = 343$ lbs	$\dfrac{10 \times 200}{3} = 667$ lbs	$\dfrac{12 \times 200}{3} = 343$ lbs
Load on Holding Tong $= \dfrac{A \times W}{C}$	$\dfrac{6 \times 200}{7} = 170$ lbs	$\dfrac{10 \times 200}{7} = 286$ lbs	$\dfrac{8 \times 200}{3} = 533$ lbs	$\dfrac{10 \times 200}{3} = 667$ lbs

Conductor Support Equipment

When working with live-line tools or rubber gloves, working from a pole or from a bucket, there is a frequent need to move a conductor to a temporary secure position. There is a lot of hardware available designed for this purpose. **See Figure 18–21**.

Figure 18–21 This equipment is used to temporarily support energized conductors when doing live-line work. (Delmar, Cengage Learning)

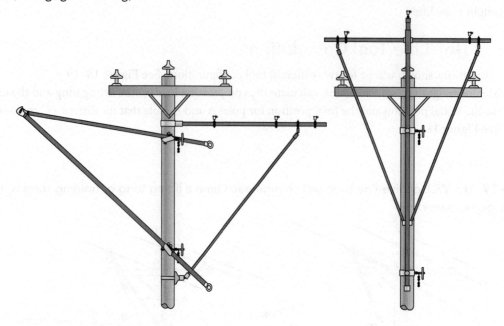

Specific Hazards Involving Hot-Line Tools

Specific Hazards	Barriers
Encroaching on the minimum approach distance on the tool can cause a flashover.	Hand stops mounted just beyond the minimum limit of approach can serve as a barrier. Some utilities require rubber gloves in addition to the hot-line tool.
Using a contaminated damp work glove on a tool can leave a conductive film.	The most likely cause of a tool failure is surface contamination. Use clean, dry work gloves.
Tools may slide through saddles after weight is applied.	Use rope blocks to make the lift. Leave the rope blocks in place to hold the weight or tie a sling between the saddle and the butt ring as a backup to saddles.
Shorting out a tie wire or preform to the structure can cause a flashover and energize the structure.	When untying insulators, cut the tie wire short enough so that it cannot reach any part of the structure. Consider alternative hardware for dead-ending or clamping in.
Hot-line tools left on the circuit can cause tracking and a carbon trail on the tool.	Hot-line tools, especially components such as extension arms, can be left on a circuit for up to 3 weeks in a nonpolluted area. Look for visible signs of tracking before handling them.
Misuse of tools cause a loss of control.	Never use a lifting tong in a tension mode as a link stick. Double-check the attachment of the saddles and snubbing bands before securing a hot conductor to it. There is no backup when a snubbing band lets go.
At transmission line voltages, there is some suspicion that high winds can cause a decrease in air pressure on one side of the tool and the low pressure on that side of the stick causes a flashover.	As a prudent measure, avoid hot-line tool work on transmission lines during high-wind conditions or increase the maximum approach distance on the tools.

Changing Dead-End Insulators

Hot sticks can be used to change out dead-end insulators on a transmission line. This kind of hot stick work requires a detailed written work procedure. **See Figure 18–22.**

Figure 18–22 Live-line work at a transmission line dead-end requires calculating the WLL of the tools. *(© Hubbell Power Systems, Inc.)*

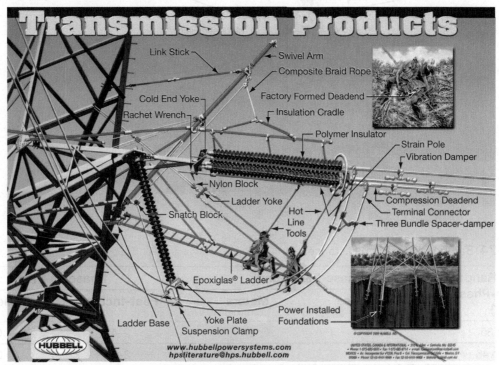

Barehand Work

Introduction to Barehand Work

Like a bird on a wire, a person can make contact with an energized conductor when insulated from all other conductive objects. Also, birds are not seen landing on higher-voltage conductors because the electric field around the conductor will be uncomfortable. Above 50 kV, the electric field for a human, can become uncomfortable and a worker needs to be shielded.

A Faraday cage is used to shield people working barehand on high voltages. Michael Faraday introduced this concept in 1836, stating that an electric field will not penetrate into a conducting sphere; therefore, no current flow will occur inside a sphere. Bonding all the conductive components together, the metal bucket grid, conducting suit, hood, and boots serve as a shield. **See Figure 18–23.**

As shown, the electric field stays outside of the shielded area of the buckets and conductive suits. The electric field is strongest next to the shield and diminishes as it approaches ground. Everything inside that shield is at the same voltage. If working from a ladder or suspended from a rope or link stick, the conductive suit worn by the worker is the Faraday cage.

When bonded on, minimum approach distances are kept from everything not already bonded to the conductor. **See Figure 18–24.**

Figure 18–23 The Faraday cage effect will prevent current flow inside the conducting sphere. *(Delmar, Cengage Learning)*

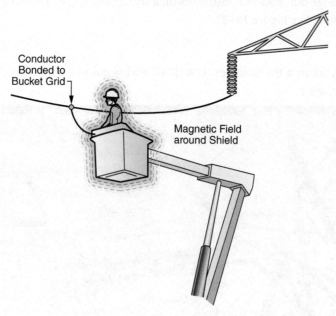

Figure 18–24 Barehand Minimum Distances

Voltage Range Phase-to-Phase (kV)	Phase-to-Ground		Phase-to-Phase	
	Feet–Inches	**Meters**	**Feet–Inches**	**Meters**
30 to 35	2–4	0.7	2–4	0.7
35.1 to 46	2–6	0.8	2–6	0.8
46.1 to 72.5	3–0	0.9	3–0	0.9
72.6 to 121	3–4	1.0	4–6	1.4
138 to 145	3–6	1.1	5–0	1.5
161 to 169	3–8	1.2	5–6	1.7
230 to 242	5–0	1.5	8–4	2.5
345 to 362	7–0	2.1	13–4	4.1
500 to 552	11–0	3.4	20–0	6.1

Barehand work is generally limited to a small number of specialized power line technicians. It is especially useful on transmission line work where the use of very long hot-line tools becomes unwieldy and there is plenty of clearance to other phases and structures.

Barehand work is done from an insulated aerial device, from an insulated ladder/ platform, from a helicopter, suspended from a link stick, and suspended on a hot-line rope. A few utilities allow barehand work on distribution voltages as an alternative to rubber-glove work or hot-line tool work. This chapter provides only an overview of barehand work.

Electrical Integrity of Barehand Equipment

All equipment used for barehand work, such as a bucket truck boom, a live-line rope, or a live-line ladder, are parallel paths to ground and all will have some leakage current (except when using a helicopter). This leakage current on the equipment must be kept below a certain threshold before using it for barehand work. This is true for all live-line work, but with barehand work, leakage current is measured and monitored more frequently because there is no second barrier if there is a failure.

Using an Aerial Device

An aerial device used for barehand work will have a boom contamination meter circuit built where a meter can provide continuous monitoring for leakage. The empty buckets are raised so that some part of the metal grid makes contact with the line. The boom is left in contact for a minimum of 3 minutes. The leakage current is monitored and may not exceed 1 microampere per phase-to-ground kilovolt. For example, a 230 kV line is about 133 kV to ground. The maximum leakage current allowed is 132 microamperes, but, depending on the type of meter, the alarm on the monitoring circuit may sound at 70 microamperes. The meter readings should be recorded in a log because one of the best measures is to see if there is a trend developing. **See Figure 18–25**.

Figure 18–25 A barehand crew is installing in-line disconnect switches in a 46 kV line. (*Delmar, Cengage Learning*)

For work on a transmission line, a corona ring at the upper end of the insulated section of the boom is needed to prevent a concentration of electrical stress similar to the role of the insulation shield of an underground cable.

The inside of the buckets are lined with metal, and all metal in the upper boom area is bonded together, along with the bonding wand and clamp. The first contact on a transmission line is when a power line technician attaches the wand to the conductor. On a 230 kV line, the charging current to energize the bucket grid creates about a 1.5 ft (0.5 m) arc. After bonding on, a more secure clamp is put on the conductor by hand. A spring-loaded breakaway clamp is used in case the workers aloft become incapacitated and the boom has to be lowered with the lower controls.

Using a Ladder or Being Suspended from a Rope or Link Stick

An electrical test can be conducted on live-line ladders, rope, and link sticks immediately before the work by using a portable hot-stick tester. **See Figures 18–26** and **18–27**. The conductive suit and all accessories serve as the shield for the electric field.

Figure 18–26 A hot-stick tool tester can be used in the field immediately before starting work to verify the condition of live-line tools and equipment. (© Hubbell Power Systems, Inc.)

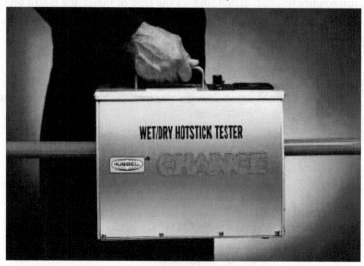

Figure 18–27 Barehand work from a hook ladder is made safer when the ladder is tested with a hot-stick tester. (© Western Area Power Administration)

Barehand Work from a Helicopter

A helicopter is ideal for mid span work on a transmission line. **See Figure 18–28.** The barehand work method is virtually the same as working from other platforms. A power line technician would need to bond on even if working on a de-energized circuit to equalize the voltage between the line and the helicopter.

Figure 18–28 Note the electrical bond between the conductor and the helicopter.

Conductive Clothing and Footwear

The conductive clothing serves as a shield around the worker, keeping the electric field effect outside of the shielding. Coveralls, work gloves, socks (not always worn), and a parka-style hood are made with conducting material. The hood is worn over the hard hat and extends forward somewhat to help shield the face from annoying sparking along the skin. Coveralls have a tail that is used to bond to the metal grid in the bucket. Conducting sole boots are worn, and conducting straps bonded to the boots are bonded to the suit with leg clips.

After the suit is put on, an ohmmeter is used to check the continuity of the suit between the tail and the leg cuffs, the gloves, the hood, and the boot soles. All the measurements should read less than 100,000 ohms.

Hazards Involving Barehand Work

Specific Hazards	Barriers
Electrical failure of equipment.	• Watch for hot-line rope that has become somewhat contaminated with use, especially when humidity is high. • Ensure that there is a check valve and a clean atmospheric vent valve in the hydraulic lines that extend over 35 ft (10.7 m) high to prevent a conductive partial vacuum from forming. The hydraulic motor should not be shut down during barehand work. • When metal is in series with rope or link sticks (a floating electrode), a longer minimum approach is required.
Flashover of insulated tools.	• Wet, dirty work gloves can contaminate the surface of a hot-line tool; so, use clean, dry work gloves. • Avoid barehand work during high winds. There is some suspicion that the slight reduction in air pressure on the down wind side of a boom or tool can reduce the insulating value. • Reduce exposure to a voltage surge. Do not work barehand during electrical storms. • Reduce exposure to a voltage surge due to switching, and have the reclose function of the source breaker blocked.
Flashover of an aerial device boom while doing the contamination tests.	There are conditions, such as high humidity, when the meter readings are not acceptable. If the readings are going down while the boom is in contact, it is probably leakage current that is drying out the boom. Do not use leakage current to dry the boom. This leakage current can leave conductive carbon tracking and lead to a boom flashover.

(Continued)

Specific Hazards	Barriers
Encroaching on the minimum approach distance.	Use a dedicated observer to watch for encroachment on the minimum approach distance. The most likely encroachment has been with the heel (elbow) of an aerial device.
	Measure a hot-line ladder or platform to ensure it will provide a sufficient length of clear insulation for the voltage to be worked on.
Placing bonding clamps on each side of a hot joint or connector. (The typically small bonding wire may carry current and burn off.)	Install a jumper across a hot joint or connector, using hot-line tools, before moving in to bond on.
Forgetting to install a jumper or forgetting to take an ammeter reading on a jumper before cutting.	Forgetting to install a jumper can happen during a repetitious job. Considering the consequences, a crew should use a written form similar to the "switching order."
	Use an ammeter to ensure that the jumper is carrying at least 30% of the load before cutting or opening a conductor barehand.
Shorting out a portion of the insulator string with the buckets or body.	When using slings and hoists near suspension insulators, it is possible to short out some insulators in a string. If some of the insulators are already defective, the risk is compounded. Test the insulators.
	If more than 20% of the insulators have no insulation value, surge-limiting devices can be installed prior to barehand replacements, or insulators can be replaced by other than barehand procedures.
	For work at the energized end of an acceptable insulator string, a bucket or worker should be positioned so that no more than 10% of the insulator string is shorted out by the buckets, worker's body, or tools.
Conductive objects in between a hot circuit and another phase or structure that forms a floating electrode (which can reduce the insulating value across an air gap).	A person in a conductive suit, a metal grid in the buckets, or two link sticks tied together form floating electrodes. A person in a conductive suit working between a tower and conductor on an insulated ladder will be a floating electrode.
	These objects can give off an electrical discharge that reduces the air-gap distance needed to prevent a flashover. Increase the maximum approach distance when a floating electrode is introduced.
Contacting an object before bonding to it.	It is probably not necessary to point this out to an experienced barehand power line technician, but there is a need to bond or bridge to any metallic objects to be touched or brought into the work zone.

Review Questions

1. Why should doing a job "hot" not be the first option?

2. Why is the second point of contact considered a hazard when working hot?

3. What length of clear insulating section of a live-line tool should be maintained when working on a 115 kV circuit?

4. What is the purpose of blocking the automatic reclosing feature of a source circuit breaker or recloser?

5. How can the probability of a voltage surge be reduced while working with live-line tools on a transmission line?

6. An insulated or bare jumper should be carrying at percent of the load current before cutting a conductor.

7. Why is there a higher risk when working on energized secondary next to a transformer rather than a span away?

8. Name three kinds of defects to look for when doing a visual inspection of rubber gloves.

9. Why should rubber glove protectors not be used as work gloves even after it is no longer used as a rubber glove protector?

10. How can the second point of contact be eliminated when doing rubber-glove work?

11. What extra steps can be taken to prevent a wire (lifting) tong from sliding through a saddle?

12. Is it acceptable to use a 2.5 in. (6.4 cm) wire (lifting) tong as a link stick to hold conductor tension on a corner?

13. What is the load (tension) on holding tong A, given that the conductor weight is 500 lbs, length A is 10 ft, length C is 10 ft?

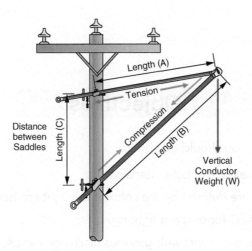

14. *True or False*: The comfort of a power line technician carrying out barehand work depends on an electric field not penetrating into a conducting sphere; therefore, there will be no current flow inside a sphere.

15. *True or False*: Rubber gloves should not be the sole protection between a power line technician and a potential fatal shock.

16. When using plastic guard equipment what kind of insulation does a power line technician depend on for electrical protection?

17. What precautions should be taken to reduce the risk of over reaching when working with rubber gloves?

18. What methods are used to inspect insulated jumpers?

Chapter 19

Outdoor Lighting Systems

Objectives

After completing this chapter, you should be able to:

1. Identify "lumen" as a measure of light intensity.
2. Explain how LED lights are defined by the color of the light emitted.
3. List the advantages of LED for outdoor lighting.
4. Troubleshoot an outage on circuits with gaseous discharge lamps.
5. Perform a test on a ballast.
6. Explain that installing traffic lights must strictly follow the engineering specifications.
7. Explain the difference between DC circuits and series AC circuits when applied to airport runways.
8. Troubleshoot the various controls used on street-lighting systems.
9. Take steps to protect yourself from high-voltage hazards when working with series lighting circuits.

Types of Outdoor Lighting

Applications of Outdoor Lighting

Outdoor lighting can be street lighting, roadway lighting, sports field lighting, airport runway lighting, and traffic lights. While electrical utilities tend to work only on street lights, other lighting systems are discussed in this chapter because power line technicians can be called upon to work on all kinds of outdoor lights. Similar skills and knowledge are required to install and maintain this wide variety of outdoor lighting systems. All types require a power supply, a ballast or LED driver, luminaire, controls, and facilities or structure to install the luminaires.

Street Lights

Street lights were one of the earliest loads for the original central-station type of electric system. The illumination of streets improved traffic safety, pedestrian safety, and security of people and property. A municipality normally owns and is responsible for the street lights within its jurisdiction. Often municipalities have the local utility, or a contractor look after the design, purchase, installation, and maintenance of all their street lights. Street-lighting systems are not typically metered. The rates are fixed (flat rate), and the rate depends on many variables, such as ownership of the pole, type of pole, whether served overhead or underground, type of luminaire, size of lamp, and type of maintenance contract. Because many street lights are on poles that also carry utility lines or fed from utility vaults that also carry high-voltage cable, a contractor must be qualified to work within 10 ft (3 m) of energized primary.

Street lights and street-light circuits are installed in the secondary position on a distribution pole, generally mandated to be at least 40 in. (1 m) above any communications or television cable. A street-light luminaire may be grounded to a neutral, a pole down ground, or both.

Roadway Lighting

Roadway lighting improves visibility at night, especially during poor weather, and reduces night accidents.

Lighting along highways and freeway intersections is owned by the road authorities, and electrical utilities are not normally involved with installation or maintenance of these lights. The utility provides a power supply to a nearby location, and the road authority or contractor installs and maintains the lights.

Sports Field and Area Lighting

Sports field and area lighting promotes business and community activities after daylight hours. Area lighting, especially for parking lots, reduces crime, and adds to the general safety of the public. The electric flood lighting of sports fields is common for all outdoor field sports, for example, racing, football, and tennis. This lighting receives its power from the metered service where contractors install and maintain the lights.

Airport Runway Lights

Lighting is critical at an airfield. An airfield has runway edge lighting, taxiway lighting, approach lighting, obstruction lighting, and beacon lighting circuits.

Traffic Lights

The first four-way electric traffic light was installed in Detroit in 1920. Traffic-signal lighting very quickly became essential at intersections, as can be confirmed when the power to the traffic lights goes off. A police officer cannot do the job as efficiently as traffic lights. The installation of traffic lights requires technical knowledge involving underground construction, hoisting and rigging, traffic signal cabinet installation and hook-up, vehicle detection methods, fiber optics, and the programming of different types of control equipment.

Specifications, Regulations, and Engineering

Designs of both lighting hardware and layout for each of the many outdoor lighting systems are specialized and complex. Skilled workers who install the lights must install them exactly as specified, because the specifications must meet regulatory standards, as well as standard engineering practices.

Complicating factors that require design specialists are as follows:

1. Luminaires are designed to produce a certain pattern of light. The reflector and lens have designs ranging from lighting a narrow laneway to a large intersection on a freeway and have five types that meet standards by the Illuminating Engineering Society of North America (IESNA) and the American National Standards Institute (ANSI). Type 1 is for narrow walkways or bike paths. Type II is for wider walkways, entrance roadways, bike paths, and other long and narrow lighting applications. Type III is ideal for roadway and general parking. Type IV is suited for wall mounting applications and for illuminating the perimeter of parking areas. Type V is for general parking and area lighting applications. Replacement of a luminaire must match the existing lighting. **See Figure 19–1.**

Figure 19–1 Lighting is designed to emit a certain pattern of light suited to its intended application. *(Delmar, Cengage Learning)*

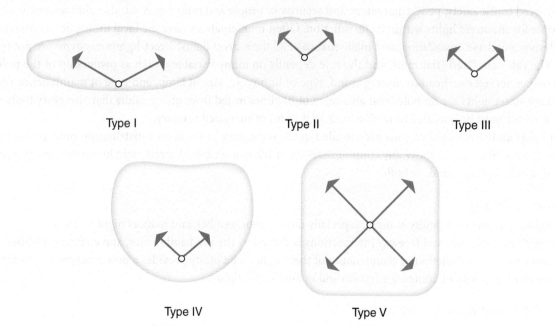

Type I

Type II

Type III

Type IV

Type V

2. Light pollution ("artificial sky glow") is a concern of a public that wants to preserve the natural night environment by preventing outdoor lights from reflecting up into the sky, causing sky glow. In addition to being a waste of energy, it also affects wildlife such as nighttime migrating birds, insects, and sea turtle hatchlings that become disoriented by artificial lighting. This type of pollution has an easy fix, using cutoff lighting. Municipalities will specify the degree that lights are to be shielded (cutoff) so that the light emitted by the fixture is projected below a horizontal plane. Three types of optical systems provide different degrees of control. These include noncutoff, semicutoff, and full cutoff. The full-cutoff light is more severe in directing light to the ground in a fairly tight pattern. Higher mounting and more full-cutoff luminaires are needed to get the same lighting result as possible with semicutoff or noncutoff fixtures. **See Figure 19–2.**

Figure 19–2 Some luminaires are designed to limit the light shining into the sky. *(Delmar, Cengage Learning)*

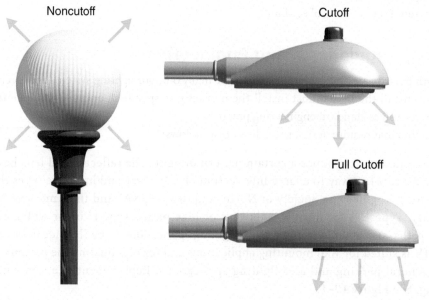

Noncutoff

Cutoff

Full Cutoff

3. There are mounting height restrictions in some municipalities because they are considered out of scale with their environment. The counterargument is that a lighting design in a community with a height restriction results in more poles needed to provide the desired light and higher energy costs. The higher mounting height also reduces glare, as can be seen when high poles are used on sports fields and at freeway intersections.

4. Lighting intensity and uniformity for the type or classification of area being illuminated must meet the standards from organizations such as ANSI and the IESNA. The light location, overhang, pattern, mounting height, type of luminaire, and lamp all contribute to meeting the required standard.

Lumens: A Measure of Light Intensity

The unit used to measure light is the lumen. A lumen is a measure of brightness; more lumens, brighter light. The definition of a lumen involves candelas, foot candles, and luminous flux, which may only be important to the designer of street lighting/parking lot lighting. A street light in a subdivision may be about 5,000 lumens. The lights at an important intersection could be about 10,000 lumens and the tall masts along highways would be much higher. The efficiency of a light is measured in lumens per watt (LPW).

LED (Light-Emitting Diode) Lights

An LED is a semiconductor that emits a light when its two terminals has a voltage applied across them. Multiple LEDs are put together as a package to makes up a light. An example is shown in **Figure 19–3**.

Figure 19–3 An LED lamp made up of multiple light-emitting diodes. (© Van Soelen)

The label on an LED lamp package will have the lumen rating, as shown in **Figure 19–4**. The lumen rating for this bulb is 655 lumens which is about the same number of lumens an old 60 watt incandescent lamp emitted. The energy used by the LED light to emit 655 lumens is shown on the label as 9 watts.

Figure 19–4 A label on an LED light package. *(Source: Cree Lighting)*

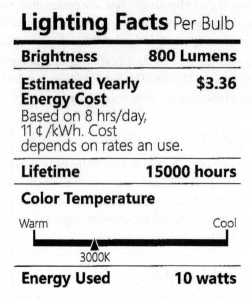

Lighting Facts Per Bulb

Brightness	**800 Lumens**
Estimated Yearly Energy Cost Based on 8 hrs/day, 11 ¢/kWh. Cost depends on rates an use.	**$3.36**
Lifetime	**15000 hours**

Color Temperature

Warm   Cool

3000K

Energy Used	**10 watts**

LED Lights Have a Defined Color

Note that the label in **Figure 19–4** shows the "appearance" of the lamp on a scale from warm to cool. The temperature for this light is shown as 2,700 K (kelvin). Trivia: this temperature rating is related to the temperature of a star. A star that is 2,700° kelvin (2,427° Celsius or 4,400° Fahrenheit) would be a similar color to the LED light described in the label. In other words, the K scale of an LED light refers to its color not its temperature. **Figure 19–5** shows the relationship between color and Kelvin temperature.

Figure 19–5 The LED kelvin color scale.

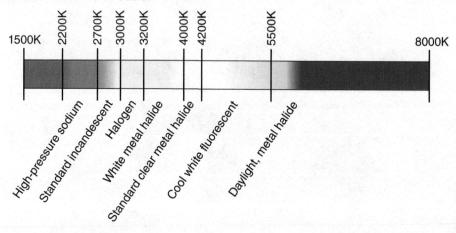

Opposite to our instinct, the hotter the temperature the cooler the light color. High temperature /cool color is not considered good for outdoor lighting. Outdoor lighting is warmer, in the range of 3,000 K. **Figure 19–6** shows the suggested color for various uses.

Any replacement street-light lamps will need to match the existing lamp color or the light will appear different to other lights on a street.

Figure 19–6 LED Light Colors on a K Scale

Color Temperature (Kelvin)	**2,700 K**	**3,000 K**	**5,000 K**
Light Appearance	Warm white	Warm white	Cool daylight
Ambience	Cozy, inviting	Warm welcoming	Crisp, invigorating
Best for	Living rooms, kitchens, bedrooms	Bathrooms, entryways, outdoor	Basements, garages
	Table/floor lamps, pendants, chandeliers	Vanites, overhead lighting	Task lighting, security lighting

Advantages of LED Outdoor Lights

LED lights are very efficient. The lumens per watt (LPW) of an LED light varies with color and design but will range between 80 to 120 lumens per watt (LPW) compared to a halogen light with a range of 10 to 15 LPW.

LED lights need less maintenance. They withstand vibration, cold and heat very well. There is little need for a bulb replacement program; the service life for an LED light is a very high 20,000 to 35,000 hours.

LED lights do not need a ballast, but they do have an LED driver. LED lights need a low voltage (12–24 V) direct current supply. An LED driver transforms and inverts the normal supply voltage of 120 to 277 to the required low-voltage DC power supply. A LED bulb that replaces an incandescent or halogen light in a standard base will have an LED driver incorporated into the bulb. Large fixtures such as stadium lights with an array of LEDs may have a separate driver that can fail prematurely.

Gaseous-Discharge Lamps

Gaseous-discharge lamps were the lamps of choice for outdoor lighting until LED lights took over. Light is produced when electric current passes through a gas (**Figure 19–7**). When a permanent arc is established in a gas-filled lamp, the light produced is referred to as an electric discharge or gaseous discharge. Types of gaseous-discharge lamps include fluorescent, compact fluorescent, low-pressure sodium, mercury-vapor, metal-halide, and high-pressure sodium-vapor lamps and high-intensity discharge.

Except for some specialized purposes, these lights are being phased out and replaced with LED lights. Trouble shooting a gaseous-discharge lamp usually involves changing a bulb or ballast.

The Ballast in a Gaseous Discharge Lamp

This section discusses the ballast in case a power line technician encounters the rare situation in which one needs replacement. Just as a ballast is needed in a ship to stabilize it from the waves on the ocean, a ballast is needed to stabilize the voltage and current in a gas-discharge lamp. In a gaseous-discharge lamp, there is no fixed resistance and the lamp needs a variable voltage supply. It needs a very high voltage to strike an arc between the two electrodes within the vacuum tube. After the arc is struck, the resistance decreases and the normal 120 V, 240 V, 277 V, or 480 V rating will maintain the arc. Because an arc is like a short circuit, the ballast limits the current and prevents the current from continuing to increase and eventually burn out the lamp. The ballast automatically adjusts to supply the proper voltage over the lamp's life. As the lamp gets older, the ballast will supply the needed higher voltage to maintain an arc.

Figure 19–7 A ballast raises the voltage from 200 V to 350 V, to establish an arc in a typical mercury-vapor, gaseous discharge lamp. *(Delmar, Cengage Learning)*

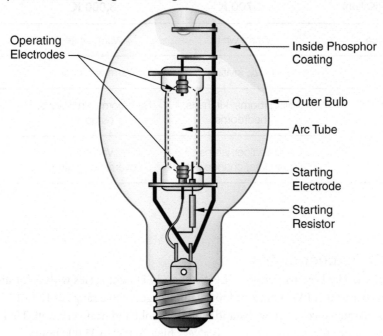

Operating Electrodes

Inside Phosphor Coating

Outer Bulb

Arc Tube

Starting Electrode

Starting Resistor

A ballast is a large inductor or reactor. It takes the place of a resistor and limits current flow while it lowers the voltage as the lamp tube begins to conduct. A ballast consists of a wire coil, an iron core, and sometimes a capacitor. A high-pressure sodium lamp may show a voltage rating of 120 V or 240 V, but it cannot operate without a ballast. A ballast is needed to provide a high voltage to start the lamp. A ballast is also needed to act as a current-limiting device because once this type of lamp is ignited, the arc within the lamp creates a short circuit.

Procedure for Testing a Ballast

Step	Test	Details
1.	Test the insulation of the ballast for short circuit.	To test for a short-to-ground, connect together all the primary and secondary leads and, using a 500 V insulation tester (megger or ohmmeter), test between the leads and ballast case or ground. The insulation reading should be better than 0.5 megohm.
2.	Test for proper output voltage using an open-circuit voltage test.	To check for proper voltage output, disconnect the secondary leads or remove the lamp to provide an open circuit. With the primary connected, the secondary voltage should be as stated on the ballast nameplate. Gaseous discharge lamps need a high voltage to start the initial electric arc. A ballast will step up a 120 V supply voltage to provide a 750 to 4,000 V output. For example, a high-pressure sodium lamp needs a high starting voltage of about 4,000 V. After the lamp starts, the operating voltage of the lamp will vary depending on the type of lamp and wattage rating. Wait until the lamp is at normal brilliance, or approximately 30 seconds, before doing a voltage check on a gaseous-discharge luminaire. The voltage should settle to its normal operating voltage when the light is at normal brilliance.

3.	Determine whether the ballast can limit the current as it is designed to do using a short-circuit test.	Remove the lamp and connect together the secondary leads to form a dead short. Apply a normal supply voltage to the primary and take a current reading on the shorted secondary leads. The purpose of a ballast is to limit the current, so there should be very little current between the secondary leads. Typically, the current for a mercury lamp should be 2 to 2.7 A for a 175 W lamp and 4.2 to 5.3 A for a 400 W lamp.
4.	Check the primary coil for shorted turns or damage with a no-load current test.	Remove the lamp or open the secondary leads so that no load is on the secondary of the ballast. Apply a normal supply voltage to the primary and take a current reading on the primary leads. A reading of more than 50% of the rated primary current indicates a damaged ballast.

Outdoor Lighting Infrastructure Systems

Outdoor Area Lighting

Outdoor area lighting refers to lighting parking lots, highways, and sports fields. While a huge variety of different luminaires, poles, and equipment are available, the trend is to use a high mast for area lighting of parking lots and multilane highway, such as seen in **Figure 19–8**. A typical 150 ft (46 m) light pole can be topped by a 14 ft circumference fixture (halo). Working on these masts requires knowledge of regulations for working at heights as seen in **Figure 19–9**.

Masts for lighting sports such as car racing and football must be very high to reduce nighttime shadows for TV broadcasting. Often, there is a fixed ladder to provide access for maintenance. **See Figure 19–10**.

Figure 19–8 Freeway lighting mounted on very tall structures reduces glare and the number of structures. *(istock.com/Reinhard Krull)*

Figure 19–9 Working on the luminaire from a crane basket. *(moxumbic/Shutterstock.com)*

Figure 19–10 A typical sports field lighting pole is designed with a fixed ladder for climbing and platforms to allow maintenance. *(iStock.com/Reich360)*

Traffic-Light Control Systems

Installing traffic signal lights involves structures such as those in **Figure 19–11**, signal heads, a controller, and vehicle detectors. From installation on a decorative-structure mast arm to suspension on a strand between two poles, the purpose is to install a signal head in a precise location, at the correct level, and aimed in the correct direction.

Specifications must meet standards such as those of the International Municipal Signal Association (IMSA) and Manual on Uniform Traffic Control Devices (MUTCD).

Figure 19–11 This traffic light signal structure requires an engineered foundation and rigging skills. *(iStock.com/Vitpho)*

Figure 19–12 A bucket truck provides the best flexibility and safety for work on a signal head of a traffic signal pole. *(iStock.com/Kozmoat98)*

Power cables that feed a traffic-signal controller will be energized from any nearby overhead or underground secondary. They can be direct burial, in poly-pipe conduit, and overhead.

Cables used to energize and control the signal head and pedestrian head (also called the "ped head") are referred to as IMSA cables and come with various numbers of color-coded wires. **See Figure 19–12.** The following is a standard color-coding scheme.

Red = Northbound

Green = Southbound

Blue = Westbound

Orange = Eastbound

Yellow = Pedestrian movement

White = Left-turn movement

Brown = Spare

The timing and cycle settings in a controller will be predetermined by traffic engineering and are based on traffic studies that will include coordination with other traffic signal lights, traffic volume, traffic speed, and priority traffic. Ideally, traffic-signal coordination means that the signals turn green for approaching traffic.

The traffic-signal controller is the signal's brain and is programmable to take in all the fixed information and variable information, such as a car entering the intersection and activating the vehicle detector signal or by an emergency vehicle triggering a priority.

A vehicle detector signals the controller that a vehicle is present at an intersection. Induction loops that cut into the pavement in each traffic lane have been the most common type. Where video detection is used, a video camera sends its image into a controller that then processes the information. There are also wireless magnetic detectors and self-powered vehicle detectors (SPVDs) that are buried below the surface. They use batteries that need replacement every 4 to 5 years.

Airport Runway Lighting

Installing airport runway lights includes laying long runs underground conduit and cable to feed runway edge lights, beacons, and in-pavement center-line lights. Installing in-pavement taxiway lights requires very precise subgrade elevations.

An AC series circuit or a DC circuit is ideal for lighting an airport runway because of the many lights and long-distance feeds needed to get to all the lights. The utility supplies power to one location, and airport authorities or contractors look after the installation and maintenance of the vast lighting network. **See Figure 19–13.**

Figure 19–13 Series or DC circuit lighting is the only practical method available to feed the long distances involved in airport runway lighting. *(iStock.com/Igmarx)*

With an AC series circuit, a single primary cable fed from a constant current regulator (CCR) can, for example, run out and feed all the lights for 2 miles (3 km) along the edge of one runway and then come back to the source in the same trench or cableway to form a circuit. There would be quite a few individual circuits because each circuit can be controlled separately, providing a lot of switching flexibility. **See Figure 19–14.**

Figure 19–14 Extra loops of series circuit runway lighting provides switching flexibility between the circuits. *(Delmar, Cengage Learning)*

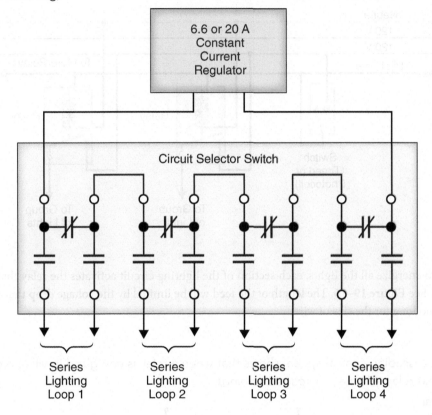

Street Lights

1. Street-light control switches that turn lights on and off can be a time-clock control or manually switched, but a light-dependent resistor (LDR), or photocell, is ideal for initiating the switching for street lights. One control combined with relays can switch on entire streets at once; however, it is also common to have a LDR or photoelectric control in each individual light. A LDR or photoelectric control in each light is easiest to troubleshoot because when this light is not working, the source of the trouble will be at that light or supply.

2. Solar-powered LED lights (automatic street-light systems) are stand-alone without the need for a connection to the grid. Their rechargeable battery supplies power at night. Maintenance includes battery replacement and cleaning the solar panel.

3. Each street light can be fed directly from existing nearby secondary bus or services. Some power-supply systems will have a fuse between the circuit and each light fixture on that circuit. On lights fed from underground, the fusing is done with in-line fuse holders and is accessible at a hand hole in the light pole. Some luminaires will have fuses inside the fixture, but many do not.

4. A pilot-wire street-light system uses a pilot wire (a separate wire strung to all the street lights) to control many lights from one spot so that they all go on and off at the same time, as shown in **Figure 19–15**. The light can be fed from any available and nearby 120/240 V source. The pilot wire is strictly a wire to turn the lights on and off. It is energized from a 120/240 V supply through a light-dependent resistor (LDR) or photocell and will activate a relay when it gets dark. All the lights will turn on or off at the same time.

5. A cascading relay system energizes and controls all the street lights from one transformer and one control. This works well in areas where there are few or no existing transformers or secondary bus and works well on underground supply to lights. A street-light wire is strung with relays installed along its length. To prevent having a very large

Figure 19–15 This pilot-wire control system shows normally open relays that are closed when the pilot wire is energized. *(Delmar, Cengage Learning)*

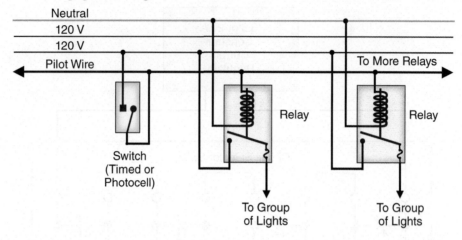

relay (switch) to energize all the lights, each section of the lighting circuit activates the relay that energizes the next lighting circuit. **See Figure 19–16.** The length of the feed will be limited by the voltage drop that will occur when the feed becomes too long for the size of wire.

Figure 19–16 A cascading control system shows that when a relay is energized it will energize a pilot wire, to energize the next relay. *(Delmar, Cengage Learning)*

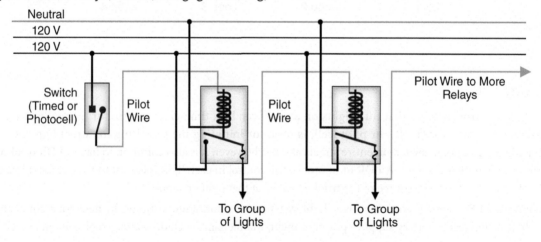

6. At one time, all street lights were installed in a series circuit, but they have generally been replaced. In a series circuit, many lights can be fed with one relatively small and continuous wire at primary voltage. The series circuit wire is fed at a primary voltage and is insulated and placed in a primary position on overhead distribution poles. For an underground installation, the cable would be a primary voltage cable. A high voltage is needed, because in a series circuit the sum of the voltage drop across each individual light is equal to the source voltage. The number of lights in the circuit determines the amount of source voltage needed to feed the lights. It would be impractical to build a transformer with the exact voltage and current output needed for each series lighting circuit. A special constant current transformer, also called a "constant current regulator," is used to raise and lower the voltage automatically, in order to maintain a constant current. A hazard to power line technicians is that the voltage across any break in a series circuit will be equal to the output voltage of the transformer. Maintenance should be done with the circuit isolated and grounded to ensure that a circuit is not opened, cut, or spliced unless a break is "jumpered" to avoid an open-circuit condition.

Maintenance and Troubleshooting Outdoor Lighting Systems

Lighting Maintenance Programs

Troubleshooting "dark" lights one at a time as calls come in is not efficient and is considered poor public relations, because this type of maintenance is always the result of a customer complaint. Some lighting systems, such as those at airports and high-mast highway lighting, have built-in monitoring systems that will notify the maintenance authority of the need for repair. The most common maintenance program adopted by lighting-system owners is the scheduling of lamp replacement and cleaning in geographic blocks.

Regularly scheduled cleaning is necessary because dirt that accumulates on or in luminaires can reduce light output and reduce the life of luminaire components. With the long operating life of LED lights, cleaning of luminaires may be necessary before group lamp replacement is carried out. As lamps age, the number of failures rises, and when a block of lights reaches the average rated life in hours for that type of lamp, group lamp replacement is scheduled or upgraded to LED lighting.

To manage a light maintenance program, the complete lighting system is kept on maps and, in some cases, electrical schematic drawings. The maps show location number, controller locations (relays), conduit locations, junction box locations, transformer locations, type of luminaire light, and type and size of replacement lamp. Structures are numbered on the maps and in the field to aid in locating trouble. A worker can determine the type of light involved for maintenance work before leaving the work center.

Safety and Environmental Hazards Working with Outdoor Lights

Outdoor Lighting Safety Hazards

Working with outdoor lights has its own special hazards. In addition to the usual electrical environment hazard of working in the vicinity of energized circuits, there are also ultraviolet radiation, high-voltage, and chemical hazards.

Ultraviolet-Radiation Hazard

The inner tubes of mercury-vapor and high-pressure sodium-vapor lamps put out an intense ultraviolet radiation. The outer phosphor-coated bulb protects people from exposure to the ultraviolet radiation. If exposed, the eyes will develop the same "sand-in-the-eye" symptoms as those from exposure to an electric arc or welding arc.

Even though the outer bulb is broken, it is possible for the inner bulb to continue to emit an unseen ultraviolet radiation without the outer bulb acting as a filter. The outer bulb can break while replacing a lamp and expose the worker to ultraviolet burns from the inner bulb. Disconnecting the power source will prevent a person from receiving ultraviolet radiation burns.

High-Voltage Hazard Working with Series Lighting Circuits

Do not treat a series lighting circuit as a low-voltage secondary circuit. The source-constant current transformer can supply several thousand volts, depending on the number of lights in the circuit. The voltage across a break in the circuit will be equal to the full-line voltage.

If the series loop is to be opened or repaired, the risk of electrical shock is reduced by opening the source transformer and grounding at the point of work.

Toxic-Chemical Hazard

Mercury, sodium, and phosphor are toxic chemicals. These chemicals can be found inside the outer bulbs of gaseous-discharge lamps and are released when the outer bulbs are broken. Staying outdoors and upwind when working with these lights reduces the risk of exposure. Mercury is hazardous to the human body. Mercury is usually in a gaseous state because it vaporizes at 14°F (10°C). In the gaseous state, it can easily enter the body through the lungs. Sodium is a highly active poisonous chemical that oxidizes very quickly in air. Sodium in contact with water produces a chemical reaction that creates sodium hydroxide (caustic soda), which can damage the lungs when inhaled. Phosphors are poisonous and must not be inhaled.

Outdoor Lighting Environmental Hazards

The capacitors used in older luminaires (before 1979) are impregnated with polychlorinated biphenyls (PCBs). PCBs are not biodegradable and will stay in the environment and end up in the food chain. PCB waste must be disposed of properly according to regulatory requirements.

The mercury and lead found in lamps are toxic substances. Some companies will separate and recycle lamp components such as glass, metal, mercury, and phosphor powder.

Review Questions

1. Using watts and lumens as units, how is one method of the efficiency of a lamp defined?

2. Which of the following lamps are high-intensity discharge lamps?

    ( ) Fluorescent
    ( ) High-pressure sodium
    ( ) Halogen
    ( ) Mercury-vapor
    ( ) Incandescent

3. What is the purpose of an LDR?

4. Why can a break in a series lighting circuit be hazardous to a power line technician?

5. What color of LED lights is preferred for residential street lights?

6. Why are semicutoff or full-cutoff lights specified in some locations?

7. If the outer bulb of a mercury-vapor lamp breaks while replacing the lamp, to what hazard is the worker exposed?

8. What is the function of an LED driver in an LED lamp?

9. What is the primary hazard when working with traffic control lights?

10. What type of electrical supply system is common at airport runways, and why?

Chapter 20

Revenue Metering

Introduction

Every part of an electrical system is metered. Meters are found in transmission and distribution substations to record voltage, current, power, reactive power, and other data needed to operate and monitor the system. Much of the metering is telemetering used to monitor remote generating stations and substations. This chapter, however, deals with the "cash registers" for an electrical utility—namely, revenue metering at the customer.

Determining Cost to the Customer

Three Main Types of Charges to Customers

Typically, a customer can be asked to pay three types of charges:

1. An energy charge, which is the actual kilowatt-hours (kWh) used

2. A capacity (or demand) charge, which can be the kilowatt (kW) demand or the peak kilovolt-ampere (kVA) demand

3. A customer charge that is not a function of either energy used or peak demand but is a charge to cover the cost of the facilities to supply power

Utilities can have dozens of different rates that are variations on these basic charges.

Variables That Affect Cost

Customers supplied by an electrical utility can be residential, industrial, commercial, or another utility. The rates charged to these customers depend on the type of service, type of load, quantity of load, and when the load is used. Some of these variables are measured through metering, and some costs are fixed when the service contract is negotiated.

Fixed Charges

- Minimum charge
- Supply voltage
- Location
- Interruptible power

Metered Charges

- Energy consumption in kilowatt-hours (kWh)
- Peak demand
- Power factor
- Load factor

Minimum Charge for Service

There is a fixed cost to a utility when it supplies power to a customer. A minimum charge relates to the wire and transformer needed to supply the energy a customer *may* require. The minimum charge is to cover the fixed cost of providing voltage regardless of the amount of current used. For a residential customer, the minimum charge saves a utility the extra expense of installing demand meters to track actual demand.

Energy Consumption

Energy consumption is the volts × amperes used by a customer. The voltage supplied is relatively constant; therefore, the amount of current used by the customer is the largest variable measured by a kilowatt-hour revenue meter.

The largest portion of energy consumed by a customer is real or true power. A standard kilowatt-hour meter measures real power, which is equal to volts × amperes × hours × 1,000. Some reactive power is also used by a customer, but it is not normally metered for a residential customer.

Peak Demand

The peak demand is the maximum rate of consumption by a customer during a billing period. The demand for power can be highly intermittent. Meanwhile, the utility has to generate, transmit, and transform the energy to supply the peak load when required.

Most utilities measure the peak demand for larger customers but not for residential customers. The reason most residential rates do not include demand charges is because the metering and tracking of residential demand add to a cost many utilities do not consider worthwhile. Residential consumers have very similar usage patterns, so that a minimum charge can cover the fixed costs without much error.

Commercial and industrial customers have a more varied usage pattern. Their peak demand is measured so that a utility can recover the extra cost of supplying the capability to meet the demand. Measurement of peak demand encourages customers to spread out their need for energy.

1. *Demand* is a measurement of the instantaneous power used by a customer. For example, when ten 100 W lights are on at the same time, they are using 1 kW of demand at that moment.

2. *Peak demand* is the highest amount of instantaneous power used by a customer during a billing period. Peak demand for power is measured as peak kilowatts or peak kilovolt-amperes. For example, if the ten 100 W lights are the largest load drawn during the billing period, the peak demand registered on a demand meter will be 1 kW. The load would have to be on for a given minimum time to register on the demand meter; for example, 15, 30, or 60 minutes are commonly specified demand intervals.

3. *Energy* is the amount of power used over time measured in kilowatt-hours. For example, the energy used by ten 100 W lights after 1 hour will be 1 kWh.

Power Factor

Industrial or commercial customers frequently have large inductive loads that the utility must supply. A low power factor means that the utility must generate, transmit, transform, and distribute extra power to supply the inductive load.

A normal kilowatt-hour meter measures only the resistive load. For customers with a possible low power factor, a meter is installed to measure the kilovolt-ampere peak as well as the kilowatt peak. The ratio of the kilowatt peak to the kilovolt-ampere peak is the power factor.

One method of billing is to charge for the highest of either 100% of the kilowatt peak or 90% of the kilovolt-ampere peak. Usually, a customer with a low power factor takes corrective action to keep the power factor above the level at which there would be extra billing.

Load Factor

An electrical utility needs an equal amount of equipment to supply a customer that consumes a relatively constant supply of power 24 hours a day and a customer that uses a similar amount of power for relatively short periods each day. The customer who uses the power 24 hours a day has a higher load factor and is billed at a lower rate.

The load factor is a percentage indicated by the ratio of the power consumed to the power that could have been consumed if the power had been used continuously.

$$\%\text{Load factor} = \frac{\text{Total consumption (kWh)}}{\text{Peak demand (kW)} \times \text{Hours a month}}$$

Supply Voltage

Residential, small industrial, and commercial customers are supplied with their utilization voltage. Large customers are usually supplied by a line at subtransmission or transmission voltage feeding into a substation owned by the customer. There is less cost to the utility to only supply the circuit feeding the customer without supplying transformation. The customer is, therefore, billed at a lower rate.

Customer Location

It is reasonable for a utility to charge a higher rate to a remote cabin than to a home on a residential street. The cost to get power to the customer is reflected in rates. The rate structure takes into account the customer density at a location. Rural rates are higher than city rates.

Interruptible Power

A large industrial customer can get a better rate if its contract includes an agreement to reduce its demand for power during times when the utility is in short supply. During a utility's peak-load periods or in storm conditions, customers with an interruptible-power contract are asked to cut back their demand for power to previously negotiated levels.

Time of Use

Valley-hour rates or off-peak discounts can be attractive to some industries or residential customers. When load is shifted to off-peak hours, the electric utility benefits, because the utility delays the need to build more generation and energy is being sold at times when generators would otherwise be underutilized. One variable to time-of-use metering is to have a lower-peak-demand penalty if the peak demand is during off-peak hours.

Two Types of Revenue Meter Construction

There are two main types of revenue meter construction: the electromechanical meter and the solid-state electronic/ smart meter. There are locations where the "smart" meter is banned and there are locations where customers can choose not to have a smart meter. However, smart meters are a major contributor to the operation and management of a smart utility grid. The working of electromechanical meter is discussed in some detail. The electronic meter has similarities as to how the KWH is measured.

1. The Electromechanical Meter

An electromechanical meter relies on a disk being turned in the same way that an electric motor turns an armature. The more voltage or current present, the faster the disk rotates. The disk is connected by way of gears to a clock that registers the amount of power used.

The Current Coil in an Electromechanical Meter

The current to be measured flows through a current coil in the meter. The coil is made up of a small number of turns of large wire. In a self-contained meter, all the load current flows through the current coil. For example, a meter commonly referred to as a "100 amp" meter shows a range of 0.75 to 100 A on the nameplate. In an electromechanical meter, a small number of turns in the current coil reduces the amount of inductive reactance created; therefore, the magnetic flux induced on the meter disk is in phase with load current and line voltage. **See Figure 20–1.**

Figure 20–1 A common three-wire, socket base electromechanical residential meter. (© Van Soelen)

The Potential Coil in an Electromechanical Meter

A potential coil inside the meter measures the voltage in a service. The coil is made up of many turns of fine wire. The voltage coil ratings match the various standard voltages to be measured. Typical voltage ratings are 120 V, 240 V, 480 V, and 600 V. These voltages can come directly from the secondary service being measured or from a **potential transformer** that steps down the voltage of a subtransmission or a primary circuit.

In the electromechanical meter, the voltage in the coil induces a magnetic field onto the meter disk. The strength of the magnetic field is dependent on the magnitude of the voltage. The potential coil is an inductive load, and the magnetic field it induces on the meter disk lags the field produced by the current coil by 90°.

Why the Disk Turns in an Electromechanical Meter

The magnetic field from the potential coil induces an electromotive force (emf) on the metal meter disk. The induced emf on the disk has nowhere to go because the disk is not part of a circuit. The emf, therefore, causes current to flow in the form of circular (eddy) currents on the disk.

The north and south poles on the potential coil are 90° out of phase with the north and south poles of the current coil. Interaction of the **eddy currents** being attracted and repelled between the poles of the potential and current coils causes the disk to turn. **See Figure 20–2.**

Figure 20–2 This 200 A, three-wire socket base revenue meter has wireless communication with a utility central control. *(Delmar, Cengage Learning)*

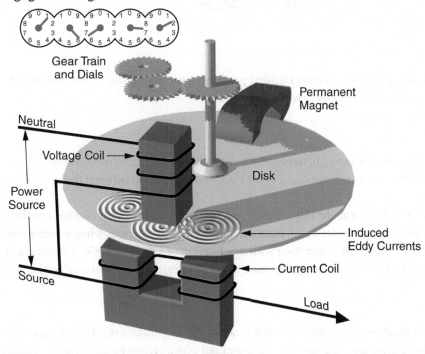

The permanent magnet in the meter provides a magnetic brake on the aluminum disk. Without the damping effect of the "breaking" magnet, the interaction of the electrical fluxes would tend to cause a continuous acceleration of the disk. Although a magnet would not normally affect an aluminum disk, the magnetic field of a permanent magnet will affect the eddy currents that rotate the disk. The retarding torque or drag applied to the disk is proportional to the speed. Moving the magnet outward or inward on the disk, or installing a shunt that bypasses part of the flux of the magnet field, can make adjustments to the braking effect. Changing the braking effect is known as the "full-load" meter adjustment.

Elements of an Electromechanical Meter

One current coil and one potential coil working together are defined as one element. One element will turn one disk in an electromechanical meter. One element can measure the current and voltage of a two-wire service. Three- and four-wire services require more meter elements. The general rule of thumb is that there is a need for one less element than the number of wires in the service to be measured. **See Figure 20–3.**

Figure 20–3 One element of a meter is the combination of a current coil and a potential coil working together. *(Delmar, Cengage Learning)*

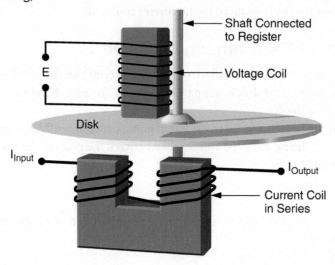

In an electronic meter, each measuring element produces an output pulse. When the energy flowing through the meter increases, the number of pulses increases.

Kh Constant

An electromechanical meter is designed to have a disk rotate within a certain optimum speed to save wear and tear on the meter. A meter is rated for a certain range of expected load, and the gearing between the disk and the register is set up to control the disk speed.

The Kh constant, also known as the test constant, relates the number of disk rotations to the load being registered. The value of the Kh constant is the watt-hours per turn of the disk. For example, when a meter nameplate shows a Kh of 7.2, it means that 7.2 watt-hours have been used with one rotation of a disk, or 1,000 revolutions of the disk = 7.2 kWh. In an electronic meter, the Kh constant is a given value in watt-hours of one pulse issued by the meter.

Kr Constant

While the Kh constant regulates the speed of the disk rotation, the Kr constant regulates the speed of the register. A meter is designed to have the register show the amount of energy used over a certain period of time. A meter is rated for a certain range of expected load, and the register should show the amount of energy used without repeating between-meter-reading intervals.

The Kr constant, also known as the meter multiplier, is a number that represents the ratio between the register gear train and the value registered on the meter dial. A Kr or multiplier of 10 would mean that the amount of energy consumed by the customer is 10 times the amount shown on the register. There is less need for a meter multiplier with electronic metering because with no moving parts, there is no concern about the speed of the register.

A-Base Meter

A revenue meter must be connected into a service so that the current coils are in series and the voltage coil is in parallel. A bottom-connected or a front-connected meter or an A-base meter must have the input and output leads wired into the meter. When connecting an A-base meter under energized conditions, it is critical to connect the current in series. A parallel connection would put a voltage across the low-resistance current coil and result in an explosive dead short. An A-base meter is readily removable. Most A-base meters are used with instrument transformers. **See Figure 20–4.**

Figure 20–4 The source and load hot legs of a 120/240 V service are connected into the bottom of an A-base meter. *(Delmar, Cengage Learning)*

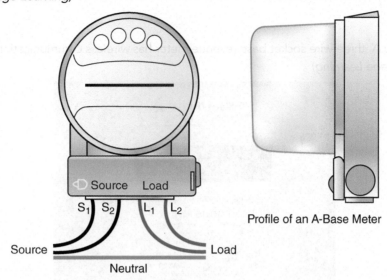

Profile of an A-Base Meter

S-Base Meter

A socket-connected meter (S-base) is plugged into a meter base that has been prewired to match the lugs of the meter to be installed. Most self-contained meters are S-base. **See Figure 20–5.**

The meter is often installed with the top incoming lugs energized. If the meter base is not wired correctly or if there is a short circuit in the customer's wiring, the installation of the meter will energize the short circuit. The integrity of the customer wiring should be tested to avoid having the meter explode during the installation.

Figure 20–5 Connection to a socket-base meter is made when it is plugged into a prewired meter base. *(Delmar, Cengage Learning)*

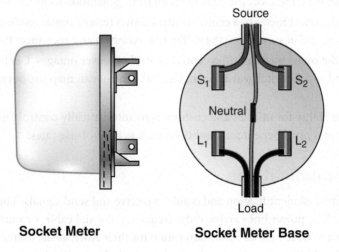

Socket Meter   Socket Meter Base

2. The Electronic/Smart Meter

Figure 20–6 shows a typical 200 A, three-wire socket-based revenue meter with wireless communication to and from the utility central control. In a smart meter, the current and voltage act on electronic metering elements to produce an output pulse with a frequency proportional to the energy flowing through the meter. A pulse initiator produces contact closures (pulses) proportional to the watt-hours being measured. **See Figure 20–6.** The electronic/smart revenue meter is much more than a revenue meter.

Figure 20–6 This 200 A, three-wire socket base revenue meter has wireless communication with a utility central control. *(Delmar, Cengage Learning)*

Depending on the manufacturers and options chosen, smart metering gives a utility the capability to do the following:

- Each smart meter is part of the automated metering infrastructure (AMI) that supplies important data into the utilities smart grid. The smart grid includes the capability of remote switching, outage management and the ability to manage distributed generation from sources such as wind farms. Engineering gets substations and feeder loading without the need to install temporary ammeters.

- The smart meter can be programed to provide consumers with time-of-use rates. Time-of-use rates reduces the peak load during the day which reduces the peak demand from generation to utilization.

- Communication from the smart meter to a central control allows remote meter reading, can measure peak demand, monitor for voltage issues and in some cases the ability to disconnect and reconnect the service.

- Communication from the smart meter notifies central control of power outages. Central control can see how widespread the outage is and which switchgear is open. Central control can map the outage and post it online to keep customers informed.

- Smart meters set up the ability for utilities and customers to automatically control customers' heating and cooling and water heaters, etc. for mutual benefit for conditions such as time-of-use rates.

Smart Meter Communications

Each smart meter has an individual identification and is able to receive and send signals. There are many communication methods such as a telephone line, power line carrier, radio frequency, coaxial cable, or satellite. Many utilities have chosen wireless technologies for a reliable two-way communication for their AMI, as illustrated in **Figure 20–7.**

Collectors/AMI repeaters/AMI routers that receive and send wireless radio signals to and from the meter and the mesh network are installed by power line technicians. Collectors are installed in strategic locations where they can communicate with any meter in its range and be connected to a power supply. Depending on local standards they are installed between the neutral and the primary. The repeater is installed directly onto a primary conductor where the magnetic field induces a power source. **See Figure 20–8.**

Figure 20–7 An advanced metering infrastructure (AMI) mesh network.

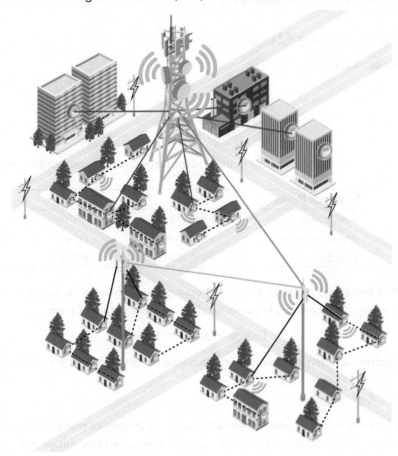

Figure 20–8 (a, b) Collectors connected to a power source provide wireless communication between the customer's meter and the mesh network. (c) Repeaters pick up and send communications from collectors to the mesh network. *(Delmar, Cengage Learning)*

(a)

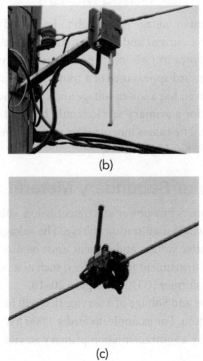

(b)

(c)

Types of Revenue Metering

Three Kinds of Power Are Measured

There are three kinds of electrical power:

1. Real power
2. Apparent power
3. Reactive power

Revenue metering involves the measurement of one or more of the three kinds of power and/or energy used by a customer.

Kinds of Power	Power	Energy
Real power	Peak kW	kWh
Apparent power	Peak kVA	kVAh
Reactive power	Peak kVAR	kVARh

Power is a measurement of energy use at a given instant; for example, real power is the voltage × amperage used by a customer at a given instant. Energy is the amount of power used over a given period of time:

$$\text{Energy} = \text{Power} \times \text{Time}$$

Meters are available to measure a variety of combinations involving the three types of power/energy.

Self-Contained Meters

A self-contained meter is one that can be installed without the use of current or potential transformers. The meter can carry the actual load current through its current coil and can use the actual service voltage in the meter potential coil.

Most customers take power at a secondary voltage. At this voltage, a self-contained meter can measure the actual voltage at the customer entrance. The voltage rating of a self-contained meter is normally 240 V or less, but 600 V meters are available. Self-contained meters are generally 200 A or less, although some higher-rated meters are on the market.

Transformer-Rated Meters

When current or voltage is too high to be carried by a self-contained meter, instrument transformers are used to send a representative current and voltage to the meter. A transformer-rated meter is used to measure a representative current from the current transformer and a representative voltage from the potential transformer.

The outward appearance of a transformer-rated meter is similar to a self-contained meter. The transformer-rated meter, however, has a lower voltage and current rating than the self-contained meter. For example, a transformer-rated meter used for a primary service could be rated at 120 V and less than 10 A. This service does not require a voltage transformer. The values measured by a transformer-rated meter are multiplied by the ratio of the current and potential transformers. **See Figure 20–9**.

Primary or Secondary Metering

Some customers buy power at a transmission, subtransmission, or distribution voltage and use their own transformers to step it down to a utilization voltage. The voltage and the current are too high to run through the meter, and therefore a representative voltage and current must be used. The representative voltage and current that goes through the meter comes from instrument transformers, such as a potential transformer (PT) (also called voltage transformer, or VT) and current transformer (CT). **See Figure 20–10**.

The type and voltage of a service that will be supplied will depend upon the load and the voltage level available in a given location. For example, to feed a 1,000 kW load in an area served by a 12.5/7.2 kV line, a utility may require that it is served by a subtransmission primary service because it would be too much load to add to the 12.5/7.2 kW feeder.

Figure 20–9 Current transformers sends a reduced representative current to this transformer meter. *(Delmar, Cengage Learning).*

Figure 20–10 The PTs and CTs on one circuit of a double-circuit 46 kV line send representative voltage and current to a metering installation. *(Delmar, Cengage Learning)*

(a)

(b)

Meanwhile, if the area was served by a 25/14.4 kV line, the feeder could handle a 1,000 kW load quite easily. Customers such as municipal utilities, trailer parks, or an industry where the customer owns the primary distribution system are also metered as a primary service.

A primary service can be single phase or three phase. It can be overhead or underground. On overhead primary services, the potential transformer(s) and current transformer(s) are typically cluster mounted on a bracket and connected to the primary.

The insulation between the source side and the load side in the primary metering installation may look strange to a power line technician because it appears as though there is not enough insulation for the rated line voltage. However, there is no voltage difference between the line and load side of the current transformer, and therefore more insulation is not needed. Note that the voltage transformers are connected phase to phase. The utility typically supplies, installs, and maintains all primary metering equipment. Instrument transformers are ordered, specifying the CT ratio, VT ratio, and outdoor or indoor types.

Underground primary metering installations can be purchased as a pad-mount cabinet where potential transformer(s) and current transformer(s) are preinstalled in the cabinet. The primary input and output are with elbow connectors similar to a transformer installation, and it will be a dead front installation. The instrument transformers are on the other side of the cabinet with a separate access. The instrument transformer side of the cabinet may have exposed energized primary.

The customer's installation typically includes a fuses and load break disconnect switch, usually a gang-operated disconnect so that the customer can disconnect the service from the system, if desired. The customer is usually responsible for all parts of the electric system beyond the primary metering point. **See Figure 20–11.**

Figure 20–11 A line drawing of a primary metered service shows the switching arrangement used for maintenance. *(Delmar, Cengage Learning)*

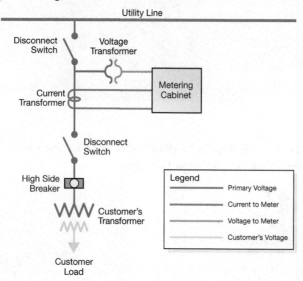

Metering Demand

Demand meters measure the maximum rate at which electricity was used and can be measured as kilowatts (kW), kilovolt-amperes (kVA), or kilovolt-amperes reactive (kVAR). The meter is generally a combination energy and demand meter. **See Figure 20–12.**

Figure 20–12 The pointer in an electromechanical demand meter is turned by heat in a bimetal strip. *(Delmar, Cengage Learning)*

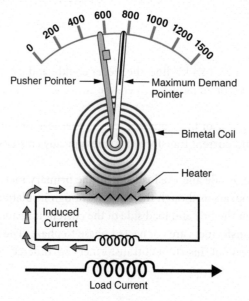

A smart meter can be programmed to record the peak demand. The electromechanical demand meter illustrated shows that a pointer attached to a bimetal strip is heated by a heating element within the meter. A bimetal strip has two metals bonded together. When heated, the metals expand at different rates, which forces the strip to bend. The bimetal strip bends, and it moves a pointer across the meter dial. The pointer attached to the bimetal strip is called a pusher pointer and, in turn, pushes a maximum-demand pointer. The maximum-demand pointer stays in the farthest position it was pushed. The position of the pointer is the maximum demand for power used by the customer between meter readings.

A peak demand that lasts only a short time is not recorded on the demand meter. The heater element has a time-response feature so that the element takes time to reach its maximum heat. Two common elements are 10 and 16 minutes.

Prepayment Metering

Electromechanical prepayment metering is like using a vending machine. A token, a magnetic card, or cash is inserted in the slot of the meter to make payment before the energy is used. A relay will switch off the power when the payment runs out. Some utilities around the world have used these meters more than others, especially for customers with chronic late bill payments and in some cases, for tenants in rental accommodations.

Some smart meters can be switched to a prepayment format to supply the amount of energy equal to the prepayment provided.

Net Metering

Small-scale on-site electrical generation, especially from roof top solar panels is found along many distributed lines, Figure 20–13. Metering the net difference between the power consumed and the power generated is net metering. Electromechanical meters run backwards when the power generated is greater than the power consumed. Bidirectional smart meters can also measure the difference. For very large generation there would be two meters, one measuring consumption and the other measuring the power generated and sold to the utility.

Single-Phase Metering

Single-Phase Meter

Most residential and smaller businesses are supplied with a single-phase service. There are a variety of revenue meters and a lot of metering data that must be recorded by a power line technician before one is installed. Types include self-contained, transformer-type, demand, A-base (bottom connected), S-base, two-wire, three-wire, 100 A, 200 A, and various multiplier single-phase meters.

Two-Wire or Three-Wire Service

On a standard North American single-phase service, the service and the meter are either the two-wire or three-wire type. The connections to an electromechanical meter and a smart meter are the same.

A two-wire service consists of a 120 V leg and a neutral. There is no 240 V supply on a two-wire service. The voltage coil in the meter is energized at 120 V. Only one current coil is needed. **See Figure 20–13**.

A three-wire service is a 120/240 V service. The voltage coil in the meter is connected across 240 V. The load on the two legs of the service will not normally be equal and, therefore, two current coils are needed, one for each leg of the service. **See Figure 20–14**.

Checking Load on a Meter to Allow Safe Removal

An uncontrollable electrical flash can occur when a socket-based meter is removed under load. Some utilities allow a meter to be removed if the load is 10 kW or less. **See Figure 20–15**. Ideally, central control can access a smart meter to check the load on the meter.

Figure 20–13 (a) An illustration of net metering. (b) Typical workings of a single-phase, two-wire, A-base meter. *(Delmar, Cengage Learning)*

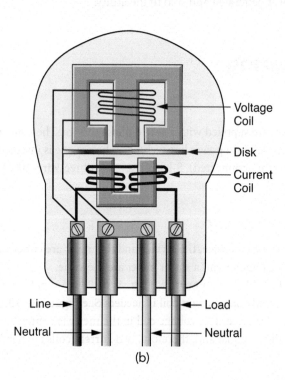

Net metering measures both energy used from the grid and excess energy produced that is sent to the grid

Understanding NET METERING
Solar Photovoltaic Array Example

3 The energy is used in your home, school or business

4 The bidirectional meter indicates energy usage and excess energy produced

1 Solar array converts energy from sunlight into electricity

2 The inverter converts the electricity produced by the solar array from direct current (DC) to alternating current (AC) for use in your home, school or business and measures the energy produced by the solar array

Energy used by your home from the electric grid

Excess energy not used by your home but goes back to the electric grid

Utility pole/ distribution line

The revenue meter only measures the "excess energy" sent to the grid, not the total energy produced by the PV system.

(a)

Voltage Coil

Disk

Current Coil

Line

Load

Neutral

Neutral

(b)

This chart can be used to determine whether a load on an electromechanical meter is more or less than 10 kW. A similar method is not available with an electronic meter. Count the number of disk revolutions in 30 seconds and, with reference to the meter (Kh), a higher number of disk revolutions than shown on the chart indicates the load is more than 10 kW.

For example, the disk of a meter with a Kh of 7.2 will rotate 11 times in 30 seconds to register a load of 10 kW. When the load is more than 10 kW, the customer's main disconnect should be opened before removing the meter.

Figure 20–14 Two-wire and three-wire meters. *(Delmar, Cengage Learning)*

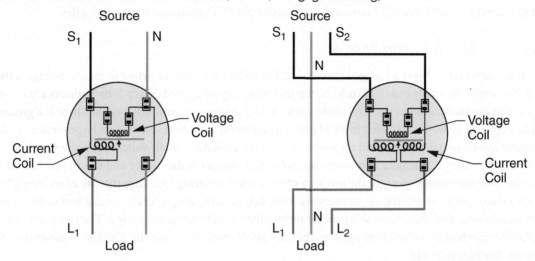

**Two-Wire Self-Contained
Meter**

**Three-Wire Self-Contained
Meter**

Figure 20–15 Determining a 10 kW Load

Kh	Time in Seconds	Disk Revolutions
0.36	30	231
0.6	30	138
0.66	30	126
0.72	30	115
2	30	41
3	30	27
3.33	30	25
3.6	30	23
6	30	14
7.2	30	11

Calculating Load Using the Meter Spin Test

On occasion, it can be beneficial to know the actual instantaneous power demand used by a customer.

The instantaneous kW demand being registered by a self-contained meter can be determined by the following formula:

$$kw = \frac{3.6 \times Kh \times Rev.}{Sec.}$$

Where:

3.6 = the number of seconds per hour ÷ 1,000

Kh = the meter constant found on the meter nameplate

Rev. = the number of disk revolutions (choose 10 for easier calculations)

Sec. = time in seconds for the total number of revolutions counted

The meter multiplier (Kr constant) on a self-contained meter does not affect the spinning of the disk and is, therefore, not applicable to this formula.

To determine the load registered by a transformer-rated meter, multiply the answer in the preceding formula by the product of the current transformer (CT) ratio, voltage transformer (VT) ratio, and meter multiplier.

Tests before Installing a Meter

A socket-connected meter (S-base) is plugged into a meter base which has been prewired to match the lugs of the meter to be installed. The meter is usually installed with the top incoming lugs energized. Many S-base meters are self-contained, which means that the installation or removal of the meter is like closing or opening a switch. There is a greater risk that a fault will be explosive if the service is fed from a large transformer over a short distance on large service conductors.

If the meter installation is protected on the source side by the customer's main switch, open the switch before installing or removing the meter. On installations where the utility transformer is the source and it is impractical or seemingly unnecessary to open the transformer, test the integrity of the customer wiring to reduce the risk of accidentally installing the meter on a short circuit or picking up an excessive load. Before energizing a service with a socket-base meter, test at the meter base to ensure that there is no short circuit or backfeed in the customer's wiring. The insulation tester method and the voltmeter method described here apply to the 120/240 V meter base and the 120/208 V meter base fed from a 208 V network. **See Figure 20–16.**

Figure 20–16 Note the differences in how the two different single-phase, three-wired meter bases are wired and where the test connections are to be made. *(Delmar, Cengage Learning)*

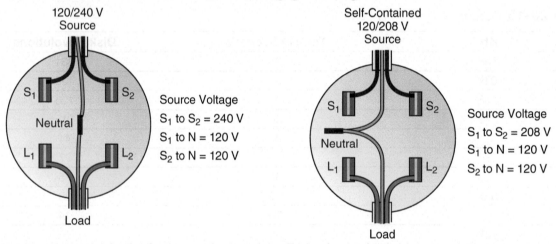

Insulation Tester Method

A 500 V insulation tester, which is like an ohmmeter, can be used at the meter base to test for a short circuit. If the customer's main disconnect is open, there should be an infinity reading between L_1 and L_2, as well as between L_1 and neutral and L_2 and neutral.

Voltmeter Method

1. A voltage check with a zero reading between L_1 and L_2 and the neutral will confirm that the service is not fed back from another source.

2. With the source side energized, and the customer's disconnect *open*, there should be no voltage present between the source terminals, S_1 or S_2, and the load terminals, L_1 and L_2. A voltage reading other than zero indicates there is a connected load or a load wire is shorted to the neutral or ground.

3. To check for a short between the two load wires, install a temporary jumper (some utilities use a special fused jumper) between L_1 and the neutral. The voltage between S_1 and L_2 should read 0 volt. A voltage reading other than zero will mean that there is a connected load or a phase-to-phase short.

Polyphase Metering

Three-Phase Metering

Polyphase metering refers to energy metering on more than a one-phase service. A power line technician normally refers to it as three-phase metering.

Three-phase energy could be measured by metering each phase separately and calculating the resultant total amount of energy used. Three-phase power, however, is normally measured with one unit, which saves space, connections, and calculating the total energy from three separate readings.

The principle of a three-phase meter is similar to a single-phase meter. The meter will have three current coils, one for each phase, connected in series with the load. It will have at least one voltage coil connected between two of the phases.

Three-phase meters, like single-phase meters, can be self-contained, transformer rated, demand, primary or secondary, and A-base or S-base.

Elements in a Three-Phase Meter

One current coil and one potential coil working together to turn one disk are defined as one element in a meter. Two-, three-, and four-wire services require more meter elements. The general rule of thumb is that energy can be metered with one less element than the number of wires in the service.

A four-wire, three-phase service can be measured with a three-element meter, which will have three current coils, three potential coils, and three disks.

Another variation for metering a four-wire, three-phase service is a 2½–element meter. This meter measures the current in all three phases, but the voltage is measured from two phases. The voltage must be reasonably balanced in all three phases for this metering to be accurate.

Element Arrangement

The elements in a meter can be either vertically or horizontally arranged. The arrangement can also be set up to turn one, two, or three disks. **See Figure 20–17.**

Making Metering Connections

The connections to a meter follow certain principles. Variations depend on the manufacturer and regulatory agencies. Drawings should be consulted when making connections to a three-phase meter because a mistake can lead to awkward circumstances for a utility.

- For each phase, the source wires to the current coil and the voltage coil must be the same polarity.
- The load wires from the potential coil must have a polarity opposite the polarity of the current coil. In other words, depending on the type of service and meter, the load side of the potential coil must be connected to another phase or to the neutral.
- Each current coil must be connected in series with the load.
- Each potential coil must be connected in parallel, like a voltmeter.
- The color coding of the wiring must be followed.
  The three phases must be in a proper phase rotation.

Three-Phase Electronic Metering

The smart (electronic or solid-state) revenue meter is available for polyphase services. A multitude of measurements can be programmed into an electronic meter. **See Figure 20–18.**

Figure 20–17 (a) A two-disk vertically arranged four-wire, 1½–element meter measures energy in a three-phase service. (b) A single-disk, three-element horizontal meter can register the energy of a three-phase, four-wire service with a single disk. *(Delmar, Cengage Learning)*

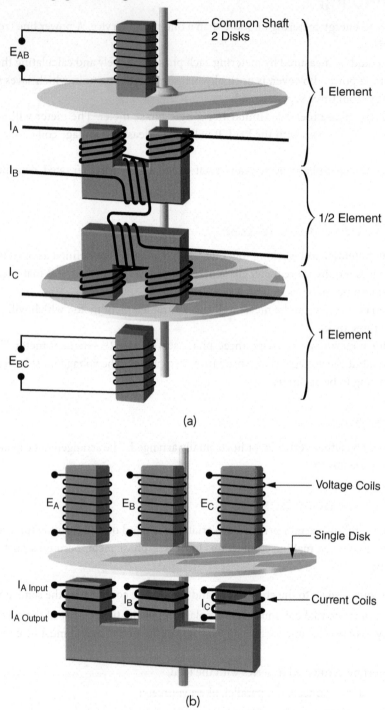

(a)

(b)

Installing a Three-Phase Self-Contained Meter

Installing a self-contained meter is like closing a switch and making sure you are not closing in on a fault. Carry out the insulation tester method or the voltmeter method to ensure that no short circuit is present in the customer's wiring.

The energized installation of a self-contained meter on services more than 240 V should be avoided. If a meter is accidently installed where a customer's wiring is faulted, an explosive fault could occur because of the potentially high fault current available.

Figure 20–18 A three-phase, four-wire, transformer-rated electronic meter needs a CT because it can only measure up to 20 A. *(Delmar, Cengage Learning)*

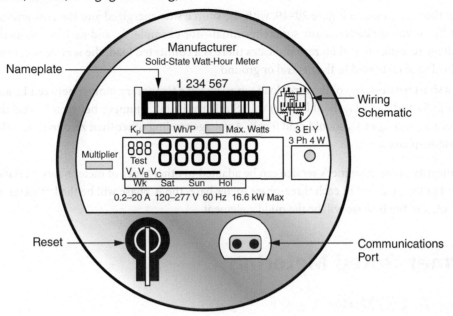

This type of service, especially at higher voltages, often has a main switch on the source side of the meter. The first option should be to isolate the service at the switch instead of energizing the service with the meter.

Insulation Tester Method

To test the meter base shown below, a 500 V insulation tester can be used to test for a short circuit. If the customer's main disconnect is open, there should be an infinity reading between L_A and L_B, L_A and L_C, L_B and L_C, as well as between the neutral and each of L_A, L_B, and L_C. **See Figure 20–19.**

Figure 20–19 The four-wire meter base is for a 120/208 V three-phase service. Current labels show where test connections are located. *(Delmar, Cengage Learning)*

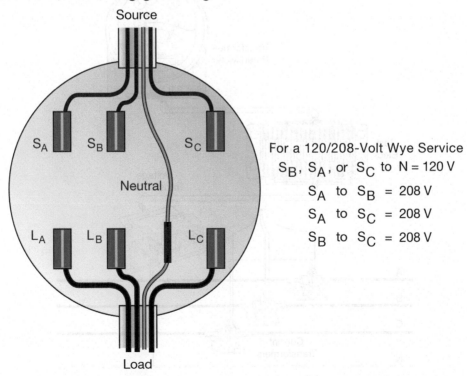

For a 120/208-Volt Wye Service
S_B, S_A, or S_C to N = 120 V
S_A to S_B = 208 V
S_A to S_C = 208 V
S_B to S_C = 208 V

Voltmeter Method

1. A voltage check between L_A, L_B, and L_C and the neutral will confirm that the service is not being fed from another source.

2. When testing the meter base in Figure 20–19, with the source side energized and the customer's disconnect open, there should be no voltage between any source terminal—for example, S_A and each of the load terminals L_A, L_B, and L_C. A voltage reading other than zero indicates there is a connected load, the service is energized from another source, or a load wire is shorted to the neutral or ground.

3. To check for a short between any of the three load wires, install a temporary jumper between L_A and the neutral. The voltage between S_A and L_B or L_C should read 0 volt. Similarly, install a jumper between L_C and the neutral, and the voltage between S_A and L_A or L_B should read 0 volt. A voltage reading other than zero indicates there is a connected load or a phase-to-phase short.

Testing the integrity of the customer's service can be adapted to other types of meter bases. Variations of these meter bases will be found in the field, but in each three-phase meter base, the A phase will be the left element, the C phase will be the right element, and the B phase will be the middle element.

Transformer-Rated Metering

Transformer-Rated Meter Use

A transformer-rated revenue meter is used where the voltage and/or current is too high or impractical for a self-contained meter. A transformer-rated meter measures a current and voltage that have been stepped down by instrument transformers. The secondary of a voltage transformer (VT or PT) and a current transformer (CT) is an accurate representation of the primary voltage and primary current being measured. **See Figure 20–20.** The installation shows a current transformer in series with each phase. It has two voltage transformers that measure the phase-to-neutral voltage on two phases. There is a test block that provides access to do metering tests and to provide fuse protection.

Figure 20–20 A transformer-rated meter installation that has PTs and CTs. *(Delmar, Cengage Learning)*

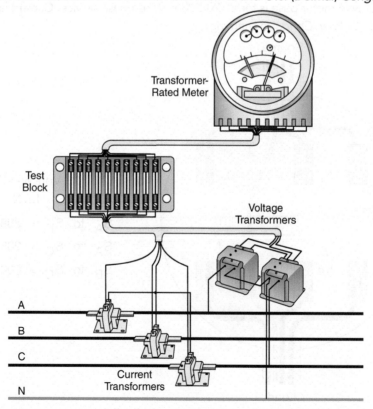

Two interconnected PTs are typical for measuring a three-phase service because any voltage unbalance would be very minor. **See Figure 20–21**.

Figure 20–21 Three-phase metering cabinet shows voltage transformers (360 to 120 V) and current transformers (400 to 5 A) in a metering installation for a 347/600 V service. *(Delmar, Cengage Learning)*

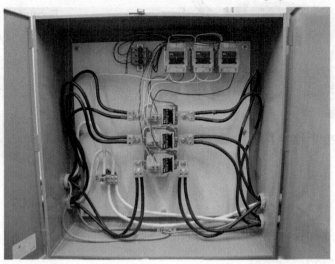

Voltage Transformer

A potential transformer (PT) is used when the energy to be measured is supplied at a high voltage, such as subtransmission, primary, or high secondary voltage. A potential transformer output supplies a voltage at a manageable level to a revenue meter, typically rated at 240 or 120 V.

A multiplier equal to the turns ratio of the transformer calculates the actual voltage of the circuit being measured. For example, a 4,800/120 V PT has a multiplier of 40:1. **See Figure 20–22**.

Figure 20–22 PTs and CTs in a specification drawing could be illustrated as shown here. *(Delmar, Cengage Learning)*

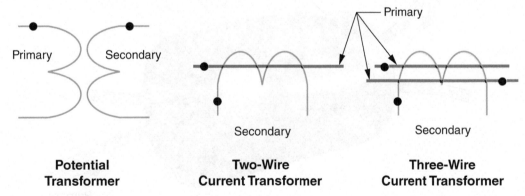

Potential Transformer	**Two-Wire Current Transformer**	**Three-Wire Current Transformer**

Current Transformer

A current transformer reduces the current to a transformer-rated meter, which is typically designed to operate at a maximum of 5 A. A low current allows the use of small wires with minimal losses in the metering circuit. Current transformers for metering are two-wire or three-wire, bar-type, donut-type, or bushing-type designs.

Two-wire current transformers (CTs) have one primary and one secondary winding. The conductor running through the donut-type CT is considered the primary winding. This type of CT is installed in each leg of a 120/240 V service or on each phase of a three-phase service. **See Figure 20–23**.

Figure 20–23 The primary for each CT shown is the one wire running through the center. *(Delmar, Cengage Learning)*

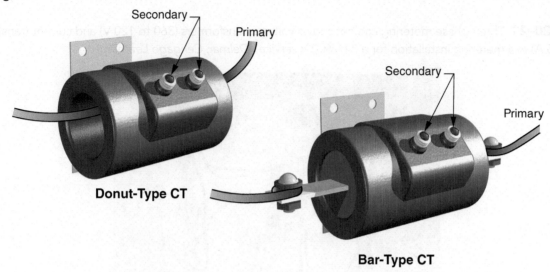

Donut-Type CT

Bar-Type CT

Three-Wire Current Transformer

One CT can be used with a three-wire, 120/240 V service. Both 120 V conductors go through one CT. To have the current in the two opposite-polarity 120 V legs go through the CT in the same direction, the two wires go through the CT from opposite directions. The current from the two legs are added together to double the secondary current. For example, a 200/5 CT becomes a 200/10 CT with this setup.

A three-wire CT used on a 120/240 V service has two primary windings and one secondary winding. **See Figure 20–24.**

Figure 20–24 A three-wire current transformer has two primary wires running through the center and one secondary winding. *(Delmar, Cengage Learning)*

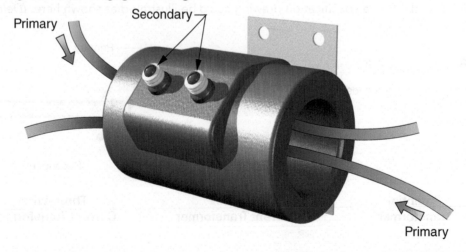

Here, the three-wire CT is actually a two-wire CT with two wires going through it acting as two primary windings. The secondary current from the CT represents the sum of the two 120 V legs of the service.

Another type of service has 120/240 V metered at a customer transformer pole. This central metering installation measures the energy at one location and then allows for services to go from this pole to a house, barn, or other buildings owned by the customer. With this type of service, a customer can install a smaller service entrance panel in each building instead of installing one large-capacity service entrance to feed all the buildings. **See Figure 20–25.**

Figure 20–25 The three-wire CT on a 120/240 V service allows a transformer-rated meter to be installed centrally while services can go to various buildings on this farm. *(Delmar, Cengage Learning)*

Polarity Markings

As with all transformers, the polarity of an instrument transformer must be known and marked on the transformer. For example, when the current enters at the marked primary terminal, the marked secondary terminal is in phase with the primary. The polarity of the primary and secondary terminals of instrument transformers is normally marked with a large dot.

High-Voltage Hazard with Current Transformers

A current transformer is like other transformers:

$$\text{The input } V \times I = \text{The output } V \times I$$

When the current is stepped down, the voltage is stepped up at the same ratio. The voltage would be expected to be high at the terminals, but when there is a load, such as a meter or relay, a voltage drop brings the voltage to a safe level. When the load is removed, the voltage at the secondary terminals and at terminals where relays are installed will be very high. This high voltage can puncture the insulation of the current transformer, and it is a dangerous shock hazard to any person in contact with the secondary circuit.

Therefore, before a load is removed from an energized current transformer, the secondary terminals must be shorted out. Typically, a shorting device at the secondary terminals is part of the current transformer design and can be closed safely while the load is still on the secondary. There are many different types of current transformer designs. If you are uncertain about how to short out the secondary, get help.

The meter base for a socket-mounted, transformer-rated meter often has an automatic shorting device that will close the metering circuit when the meter is removed.

A Multiplier on a Current Transformer

When an instrument transformer is used, the meter is reading and registering a representative voltage and/or current; therefore, the reading must be multiplied by a meter multiplier to obtain the actual status of the energy used.

The **ratio of the current transformer** (CT) and voltage transformer (VT) must also be calculated into the total. The VT multiplier is always equal to the transformer ratio. The CT multiplier depends on how the CT is wired. A bar-type CT multiplier is always equal to the transformer ratio. When one wire goes straight through a donut CT, the multiplier is the same as the transformer ratio. For example, a CT with a ratio of 400:5 (80:1) will have a multiplier of 80.

A CT multiplier can be reduced by looping the primary through the CT more than once. For example, when the primary is looped through the CT twice, the secondary current is doubled. A 400:5 CT will become a 400:10 CT (200:5

or 40:1), which changes the multiplier to 40. The secondary of a 400:5 CT is rated to 5 A. Therefore, a double-looped primary should be used only where the primary is 200 A or less to limit the secondary to 5 A. **See Figure 20–26.**

Figure 20–26 A double-loop primary causes the secondary current to double. *(Delmar, Cengage Learning)*

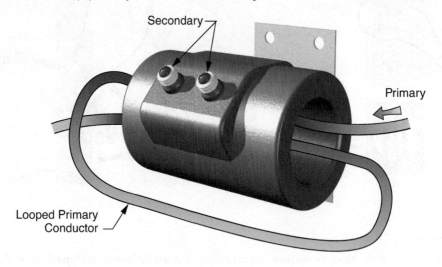

A Meter Socket-Mounted Transfer Switch

A device that is installed behind the meter allows a portable generator to plug into a socket meter base is shown in **Figure 20–27.** When the cord is plugged in, the device automatically prevents feedback into the utility service. The device is installed by utility power line technicians because the meter needs to be pulled, the device installed into the meter base and grounded. The meter is reinstalled into the device.

Figure 20–27 A useful safe device for hooking up a portable generator. *(© Van Soelen)*

Specific Hazards of Installing and Removing Meters

Specific Hazards	Barriers
Striking or pounding on the glass cover can break the glass and cause severe cuts.	Use a meter installer and removal tool on S-base meters to reduce the risk of injury from breaking glass. **See Figure 20–28**.

Figure 20–28 A meter installer and removal tool reduces the hazard of working with socket-based meters on an energized meter base. *(Delmar, Cengage Learning)*

A high-explosive fault current can be available at a meter installation, especially at an installation fed from a large transformer and a short run of large service conductors. Any accidental arc can be explosive.	Using an insulation tester or the voltmeter method, conduct tests to ensure there is no fault in the service. Use a meter installer and removal tool on S-base meters to reduce the risk of injury from explosive faults. Wear eye protection and flame-resistant clothing.
Unknown potential shorting hazards, such as a broken meter lug or dropping an excessive load, can cause an explosive fault while removing a socket-based meter.	Check the load on the meter. If facilities are available, open a source switch, bypass the load, or have the load reduced. Use a meter installer and removal tool on S-base meters to reduce the risk of injury from explosive faults. Wear eye protection and flame-resistant clothing.
Picking up or dropping a load with a 480/277 or 600/348 V, self-contained meter can cause an explosive arc because the available fault current can be very high at these voltages.	Avoid picking up or dropping load with a 480/277 or 600/348 V, self-contained meter. Avoid picking up or dropping load with any bypass facility at these voltages. Use a meter installer and removal tool on a 480/277 or 600/348 V, self-contained meter if it is necessary to install it energized. Wear eye protection and flame-resistant clothing.
There is a high voltage in the meter base when a transformer meter is removed.	Unless there is an automatic shorting device in the meter base, short out the CTs before removing the transformer-rated meter.
A third party makes contact with an unprotected meter base.	If the meter base (socket) is to be left energized, install a meter or cover and seal before leaving the site to protect the public.

Review Questions

1. Energy consumption is the volts × amperes used by a customer. What is the largest variable measured by a kilowatt-hour revenue meter?

2. What is meant by peak demand?

3. Why would a utility charge extra for peak demand?

4. Revenue metering involves the measurement of one or more of the three kinds of power and/or energy used by a customer. What are they?

5. **True or False:** A customer pays a utility more per kWh when power is supplied at a primary voltage.

6. What hazard exists when a socket base meter is plugged into a meter base with the source wires energized?

7. A voltmeter can be used to verify that a socket based meter will not be plugged into a short circuit, true or false?

8. What is the difference between a self contained and a transformer rated meter?

9. Name four advantages an electronic (smart) meter has over an electro mechanical meter.

10. Identify at least five components in this drawing of a typical primary metering service?

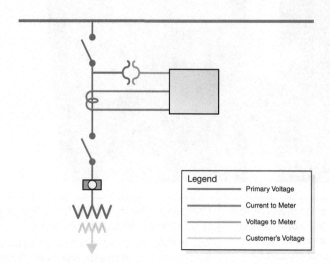

Legend

——————— Primary Voltage

——————— Current to Meter

——————— Voltage to Meter

——————— Customer's Voltage

11. Why is a transformer-rated meter used for some services?

12. How can one CT be used to meter a 120/240 V, three-wire service?

13. Why is there a high-voltage hazard when there is current flowing through the primary of a current transformer while the secondary is open circuited?

Chapter 21

Managing Vegetation in an Electrical Environment

Objectives

After completing this chapter, you should be able to:

1. Describe the reasons why vegetation control in an electrical environment is essential.
2. Identify the elements of a vegetation control program.
3. Implement controls for protection from tree work hazards.
4. Maintain the required safe working distances from energized lines.
5. Explain touch and step potentials as this applies to tree work.
6. Take part in practice exercises in conducting tree top rescue.
7. Describe tree climbing techniques while using a fall arrest system.
8. Identify the procedures used for safe chain saw operation.
9. Identify hazards and controls involved with tree pruning and/or tree felling near energized circuits.

Vegetation Management in Electrical Utilities

Why Electrical Utilities Control Vegetation

A large and critical part of an electric utility maintenance program is vegetation control.

1. Next to lightning, tree contact is the most common cause of outages in an electrical system. Tree contact is the biggest cause of damage to overhead lines after a hurricane, ice storm, or heavy wet snow. Restoration of power because of fallen trees, limbs, and branches is both labor intensive and time-consuming.

2. Brush growth in a right-of-way can cause line outages, especially during peak electrical demand, when the system is most vulnerable. Peak demand and/or warm temperatures can cause a long span on a transmission line to sag an additional 10 to 20 ft. Brush is also a source of fuel for a fire and the ionized air above a fire is conductive and can cause a line outage. Brush growth also restricts access to power line by line and tree crews.

3. The arcing associated with a tree contacting a power line can and does start grass, brush, and forest fires. A fire can also destroy wooden transmission and distribution structures.

4. Fatalities occur every year when members of the public climb into trees, prune trees, or remove trees that are in contact or make contact with power line. During storms, energized conductors that are knocked down by tree damage are an extreme hazard to the public.

5. Vegetation can sustain and spread a fire in or around a substation. Vegetation at a substation can also attract animals, such as squirrels and snakes, that can climb onto the station structure and cause outages.

6. Tall-growing trees can interfere or block microwave-beam paths whose signals are used in monitoring and controlling the status of the system and operating switchgear.

7. To ensure a safe and reliable power supply, regulatory agencies require electric utilities to have a documented vegetation management program that includes the budgeting, scheduling, clearance standards, human resources, and inspections needed to maintain a right-of-way. One of the root causes of the huge Northeast blackout in August 2003 was a very preventable, poor control of vegetation.

8. An electrical utility is responsible for large tracts of land, and regulatory agencies require property owners to control noxious weeds and erosion.

9. Regulatory and good citizenship requirements are enforced to protect the environment, especially as it pertains to water quality and wildlife habitat.

10. The source of much of the face-to-face contact and a large part of an electrical utility public relations program involves vegetation control programs. Customer relations include planning to maintain attractive right-of-ways.

Elements of a Utility Vegetation Management Program

A vegetation management program includes planning, scheduling resources, budgeting for tree pruning on cycle, danger tree removal, brush control, herbicide application, and tree planting.

Managing vegetation on transmission and distribution right-of-ways requires planning, scheduling, clearance standards, human resources, and inspections. An electrical utility would typically have an inventory of the number of right-of-way miles (km) within its jurisdiction broken up into geographic blocks; for example, there would be blocks for tree pruning, and blocks for brush control work and patrol. The number of blocks would correspond with the frequency in a cycle. For example, if there were six pruning blocks, each block would be pruned every 6 years. The workload for each block would be relatively even because the inventory for each block would also take into account factors such as types of trees (fast or slow growing) and density of trees.

Tree work is generated by customer requests and through various patrols and inspections. Light detection and ranging (LIDAR) is an effective method being used to survey right-of-ways. LIDAR is like radar but instead of sending out radio signals and measuring the time it takes to be reflected back, it sends out pulsed laser beams. An aerial survey using LIDAR can map and measure and record the height and encroachment of vegetation. It can see below the brush and tree canopy to identify the height of brush in the right-of-way and identify the location and height of any danger trees. A foot patrol may be needed to verify or gather specific details in the survey.

Quality Pruning

Trees are generally pruned on an established cycle. Pruning must provide for clearance to power line and leave trees in a healthy state. The number of blocks or cycles for pruning will depend on the speed of growth and customer sensitivity. The amount of pruning varies according to the growth rate of the tree species, as well as locations and growth conditions (e.g., soil, water).

Simply cutting off the ends of branches (topping or heading) and leaving stubs used to be a common practice. When a branch was cut off at a crotch, the proper method was to make a flush cut and apply dressing to the wound. However, cut stubs were prone to decay and vigorous growth of water sprouts soon grew at the stub. The sprouts were weak and broke easily during windstorms. Flush cuts and wound dressing have been found to lead to dieback and decay.

The following are the most accepted practices for quality pruning.

- *Directional pruning* (also called natural target pruning, drop crotch pruning, or the Shigo method) has become a preferred standard for pruning near power line. Only the branches that head toward the power line are pruned; those growing down or away from the line are left to grow. Depending on the location of the line and the tree, a directional pruned tree will have a V-shape, an L-shape, or removal of one side. Initially, the tree will have an unbalanced look, but it will grow to correct any lack of balance and will end up healthier than if it had been topped. Directional pruning may not be appropriate for trees that have been topped many times or for trees with a dominating central trunk, such as a conifer. **See Figure 21–1.**

Figure 21–1 Directional pruning involves pruning only the limbs that are growing toward the power line. *(Delmar, Cengage Learning)*

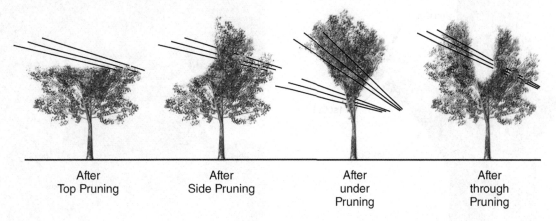

| After Top Pruning | After Side Pruning | After under Pruning | After through Pruning |

- *Avoid topping and leaving no stubs.* If a parent branch (stem or leader) must be shortened, it should be cut back where there is a larger lateral on the tree itself. The parent branch should be cut just above where there is a lateral at least ⅓ diameter of the parent. **See Figure 21–2.**

Figure 21–2 Cut the parent branch back to a lateral that is at least 1/3 the size of the parent being cut. *(Delmar, Cengage Learning)*

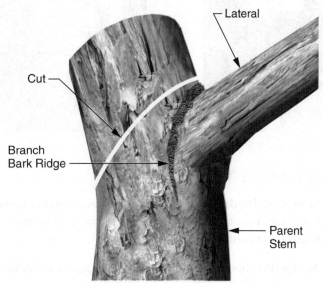

Lateral

Cut

Branch
Bark Ridge

Parent
Stem

- *Make an undercut and a second cut from the top when removing larger branches or limbs.* Make the undercut about 18 in. (0.5 m) from the limb's point of attachment. Make the second cut from the top, a little farther out than the undercut. This method will prevent ripping or tearing of the bark. Small branches lopped off with a pruner would, of course, be one cut.

- Make the final cut just outside the point where the branch meets the trunk, which is the location of the branch collar (branch bark ridge). The branch collar is part of the trunk and should not be removed. This is called the 1-2-3 method of branch removal. **See Figure 21–3.**

Figure 21–3 The 1-2-3 method of branch removal allows a final cut to be made with a reduced risk of damaging the tree. *(Delmar, Cengage Learning)*

Removing or Felling Danger Trees

Any tree removal program involves removing danger trees. A tree removal program cannot be planned and scheduled in the long term because it is dependent on patrols and customer requests for trees to be removed. The term danger tree is used in two different ways in the electrical utility industry.

1. **A Danger to Power line.** Most jurisdictions have regulations that require a utility to remove trees that are a danger to a power line. Any diseased or unstable tree tall enough to fall within or close to striking distance of a power line is considered a danger tree. Felling a danger tree to power line requires an awareness of the electrical hazards and minimum working distances. To reduce the risk of a tree falling into a power line during removal requires excellent tree felling techniques, roping, topping, and/or use of a crane.

2. **A Danger to People.** Trees growing near power line are an electrocution hazard to the public. People pruning their own tree near a power line, children climbing trees, and in at least one case a person picking apples have been electrocuted. Even when the power line is visible, some members of the public do not perceive the hazard, but when tree branches grow and hide the power line, then a trap can be set.

A tree is also defined as a danger tree by government regulations because of climbing or felling hazards. Unhealthy trees, rotten wood, and hangers contribute to hazards that must be identified. Indications that a tree may be suspect include signs of excavation around a tree exposing broken roots, adjacent unhealthy or dead trees, any tree with leaves that have an unusual color, and/or a tree that has been heavily pruned. **See Figure 21–4.**

Training is required to recognize when disease, fungi, or other problems cause a tree to be classified as a danger tree.

Figure 21–4 A power line technician/arborist has to be aware of potential hazards when inspecting a tree. (*Delmar, Cengage Learning*)

Look for Branches intruding from Other Trees

Look for a Heavy Crown

Look for Dead Branches and Hangers

The U.S. Federal Occupational Safety and Health Act (OSHA) regulations have a rule regarding danger trees:

Each danger tree shall be felled, removed, or avoided. Each danger tree, including lodged trees and snags, shall be felled or removed using mechanical or other techniques that minimize employee exposure before work is commenced in the area of the danger tree. A danger tree includes any standing tree that presents a hazard to employees due to conditions such as, but not limited to, deterioration or damage to the tree and direction or lean of the tree. Signs of a danger tree can be:

1. Large, dead branches and hangers in the tree and on the ground.
2. Cracks or splits in the trunk or where branches are attached.
3. Cavities or rotten wood along the trunk or in major branches.

Removing a danger-to-people tree requires techniques that ensure no one is within two tree lengths when it is felled. The techniques used can include attaching a rope or winch from an aerial lift and pulling the tree down or using a crane. Removing a tree that is leaning heavily on a power line (during a power outage) or communications cable is a particularly unpredictable operation, and the tree should be pulled off by mechanical means.

High-Quality Brush Control

Transmission and distribution line owners have a vast amount of property to maintain. In addition to danger trees at the perimeter, controlling brush that has the potential to grow into a power line is a large element of a vegetation program. Brush is woody vegetation and is defined as 4 to 6 in. (10 to 15 cm) in diameter and growing under a power line. Brush that must be removed is any species capable of growing to conductor height. One goal of a vegetation management program is to promote compatible plant species, such as grasses and low-growing shrubs, as a ground cover on the transmission and distribution right-of-ways. Once the right-of-way brush is brought under control, maintenance is minimized and often requires only grass cutting, animal grazing, or farming.

Abiding by environmental regulations for wildlife, erosion, and noxious plants is also part of any brush control program. Selective removal of incompatible brush can be done manually, by mechanical cutting, by grubbing, and/or by spraying herbicide.

Manual Cutting

Manual cutting is effective for spot work on right-of-ways where the brush is fairly well controlled. A powered brush saw is generally used in such circumstances.

Mechanical Cutting

Mechanical cutting with a mower, such as a brushhog or flail mower and other types that are under continuous development, is commonly done where no herbicide is allowed. However, unless followed up with herbicide, cutting brush buys some time but the cut brush of some species comes back with more shoots and faster growth.

Grubbing

Brush can be removed by grubbing, which consists of using a machine such as a bulldozer to dig it out by the roots. To prevent erosion and reduce future maintenance, grubbing is followed up with seeding so that it can be maintained by mowing, grazing, or farming. Dozer blades, such as V-shaped or toothed blades, improve the efficiency of grubbing.

Herbicide

Herbicide spraying is the most effective method for controlling brush.

Proper Use of Herbicides

Different types of herbicides and applications are used to control different vegetation. Herbicide use must be applied only by trained and licensed personnel to ensure that the application meets safety and environmental requirements and is effective. Proper mixtures, control of drift, and accurately calibrated equipment are necessary for effective control and protection of the environment. A herbicide spraying program is usually closely aligned with a public relations program because some members of the public have negative perceptions of herbicides.

Foliage spraying is done with herbicides that kill only broadleaf plants, such as some brush, and thereby promote the growth of grasses. It is essential to get a right-of-way under control so that only compatible species are left to grow. Foliar herbicides may be applied either with ground equipment or aircraft. Spraying from the ground is the only way to spray individual incompatible brush, but aerial spraying is effective for dense brush, especially in rugged terrain. If brush growth is left too long and is 15 to 20 ft (4 to 6 m) tall, it probably must be cut before spraying.

Basal spraying is the application of a herbicide and diesel oil mixture to the lower portion of a trunk to control many species. The entire circumference of a trunk up to 18 in. above ground should be soaked.

Stump treatment (spraying cut stumps, stump injection) at ground level, immediately after cutting, prevents resprouting. Applying herbicide to prevent resprouting can reduce or eliminate future regrowth and, thereby, future cutting.

Soil treatment with a pellet or granular form of herbicide kills many brush species.

Tree growth regulators (TGR) can effectively reduce the need for frequent pruning on fast-growing trees. TGR chemical compounds shorten the amount of limb growth and are applied to the soil around a tree. The chemical stays effective for 3 to 5 years.

Soil sterilization is used to kill all vegetation and is useful in and around a substation.

Planting Trees

Public resistance to tree removal can be somewhat alleviated by having a tree planting program in which compatible trees are planted to replace incompatible trees that were removed. Compatible trees are also planted as screening to reduce objections to substations and transmission lines being visible at road crossings.

A Knowledge of Trees

Arborists doing tree clearing around utility lines can identify and know the features of trees. For example, knowing the growth rate of each species is important so that an arborist will know how far back to prune for a 6-year cycle. A knowledge of dendrology (the study of trees) helps an arborist when talking to customers and discussing the health of the trees being pruned or recommending compatible species that can be planted as well.

The Hazards of Tree Work

Hazards and Controls of Tree Work

Tree Work Hazards	Controls
1. Arborists have a high exposure to musculoskeletal injuries because of climbing, hauling brush, and picking up branches.	Pruning and removing trees are very physical tasks. In addition to keeping fit, every arborist should participate in a back-care program.
2. Arborists have a high exposure to repetitive strain injuries, especially from using pruners and pole saws while leaning out of buckets.	In addition to tool and bucket design, an arborist must be part of an education and exercise program that reduces repetitive strain injury.
3. Arborists have a high exposure to severe cuts when using chain saws, probably the highest-risk power tool ever invented.	Arborists use more personal protective equipment than any other electrical utility worker. In addition to a hard hat and protective footwear, when operating a chain saw an arborist also wears leg, eye, and ear protection, as well as protective gloves.
4. When climbing trees or working from aerial devices, arborists have a high exposure to falls from heights.	Specialized skills, equipment, and training are necessary for climbing trees safely and minimizing the risks of falling. The risk of tree climbing can be reduced by using a fall protection system. One type is a belaying technique that requires very little equipment. This method can be used, for example, when climbing a tree to secure a rope for tree removal.
5. Arborists have a high exposure to electrical hazards. Electrical contact causes about 30% of all fatalities related to tree work. Direct contact with a power line or a tree branch touching a power line increases the risk of receiving a fatal electric burn.	Arborists must be able to recognize the components of the electrical system being worked on and must have extensive training on the proper techniques for pruning and removing trees near energized circuits.

Recognize Your Skill Limits

Pruning and removing trees, especially near energized circuits, is high-risk work. This work should be reserved for trained arborists. Line crews working on no-power calls and in some jurisdictions are trained and expected to do some basic tree pruning and tree removal. It is wise to recognize when a tree job is beyond the skills and equipment available. In such cases, call for a qualified arborist.

Tree Work Near Electrical Circuits

Electrical Awareness

Work near electrical utility circuits is most commonly done by qualified, specialized power line clearance arborists. They are trained in the techniques and knowledge required to prune trees around power line. This training is accomplished through company training programs and or apprenticeship programs.

Occupational safety regulations do not allow an unqualified person to come within 10 ft (3 m) of a distribution electrical circuit and an additional 4 in. (10 cm) for every 10 kV over that, which works out to 14 ft (4.3 m) for 169 kV, 16 ft (4.9 m) for 230 kV, and 25 ft (7.6 m) for 500 kV. Working near electrical circuits is the highest-risk hazard that an arborist must control. Control starts with being able to recognize the voltage level of a line, the safe tool insulation distance, the risk of "tree shock," and step potentials.

An arborist must be able to identify the voltage levels of the power conductors, neutral, open-wire bare secondary, wrapped secondary, tree wire, telephone cable, and streetlight circuits. Some structures will have three or four circuits, each with different voltages. While the size or length of insulation is one indicator, reading the voltage on a transformer nameplate or asking will be more accurate. It is the voltage level that determines the allowable minimum working distance.

An arborist must be able to recognize hazards such as a broken crossarm or insulator, or a fallen conductor. When working under an isolation certification there should be an understanding of the lockout/tagout system and the grounding that provides safe conditions for work.

While direct contact with a conductor obviously can cause electrical burns, an arborist also must be aware of how someone can receive an electrical shock from indirect contact through touch potential, step potential, and tree shock.

Minimum Working Distances for Tree Pruning

The most common and most effective barrier against electrical contact is maintaining a space between the work zone and the electrical conductors. OSHA regulations specify minimum working distances from energized conductors for qualified power line clearance arborists, but these are not necessarily *safe* distances. **See Figure 21–5.**

Figure 21–5 Minimum Working Distances for Tree Pruning

Voltage (Phase to Phase) (in kilovolts)	Minimum Working Distance
2.1–15	2′-1″ (0.6 m)
15.1–35	2′-4″ (0.7 m)
35.1–46	2′-6″ (0.76 m)
46.1–72.5	3′ (0.9 m)
72.6–121	3′-4″ (1 m)
138–145	3′-6″ (1.07 m)
161–169	3′-8″ (1.1 m)
230–242	5′ (1.5 m)
345–362	7′ (2.1 m)
500–552	11′ (3.4 m)
700–765	15′ (4.6 m)

While the length of an insulated and regularly tested pruner is electrically safe for the distance specified, an inadvertent movement or a limb bridging across the pruner can cause electrical contact. Power line technicians should plan to stay farther away than the minimum distance recommended, possibly another 3 ft (1 m). When it is necessary to work *at or close to* the minimum distance, it should be treated like an exception that requires additional planning, such as the following:

1. Stop and plan the approach and movement, preferably keeping the hazard at arm's length.
2. Plan the work so that exposure time will be minimized.
3. Discuss the plan with a person who will act as a dedicated observer for the duration of the work.

A Human Body as an Electrical Path

The human body tolerates electrical current very poorly; 100 mA can be fatal. Considering that a typical household circuit is fused at 15 A (15,000 mA), the risk for a lethal electrical shock exists on all electrical circuits if the voltage is high enough to break down the resistance. A fault on a typical distribution feeder close to a substation can generate 10,000 A (10 million mA).

It takes a certain amount of voltage to break down the initial skin resistance of a human body before a current path is established. Once a current path is established, it is the amount of current and the path the current takes through the body that does the damage. Any voltage over 750 V has very little trouble breaking down skin resistance. A much higher voltage is required to break down the resistance for anyone wearing rubber gloves.

Keep Away from Touch Potential

A touch potential refers to a voltage between a person's hands and feet when the hands are touching an object that has become energized.

A tree in contact with an energized circuit will have current coming down the trunk to ground. The voltage where the tree is in contact will be at full line voltage. Voltage will drop as the current travels down the tree, and the amount of voltage at ground level will depend on many factors, including the type of tree, the season, and the type of earth. A person standing beside the tree and touching it will provide another path for the current to flow to ground. The parallel path provided by the worker will carry some current, and a mere *100 mA can be fatal.* Similarly, standing on the ground and touching a truck that might become energized is a touch potential hazard.

Keep Away from Step Potential

A step potential refers to a voltage between one foot and the other foot when walking on the ground in an area usually safe to walk that has become energized by electrical energy entering the earth.

The hazard is to anyone working at a location where an object in contact with earth contacts an energized conductor. For example, electrical current flows into earth when an energized conductor comes in contact with a tree, a truck boom makes contact, or an energized conductor falls to the ground.

At the point where the current enters earth, the current breaks up and flows in many paths, depending on the makeup and resistance of the earth. The voltage at the current-entry point is higher than the voltage one or two paces away from the entry point. Therefore, a difference of potential is in the earth around the current entry point.

The voltage between the gradient rings lessens as the distance from the contact point is increased. Voltage gradients are also known as ground gradients, potential gradients, and step potentials.

Electrical-resistant work boots are available that—when worn as personal protective equipment—can (depending on negating variables such as the voltage level and if standing in deep mud) provide increased contact resistance with the ground that may make a difference.

Keep Away from Tree Shock

To receive electrical burns when working with trees, it is not always necessary for an arborist to make direct contact with a power line; about half of all electrocution fatalities are the result of indirect contact. Because a tree is not an excellent conductor, there would be a voltage drop as the current flows through the length of the tree where the voltage is highest at the contact or entry point and lowest at the exit point, usually at ground level. A human body in a tree that is in contact with an energized line can provide another path for the current to flow toward the ground. The hands on the tree would be at a different potential than the feet because the voltage drops as it flows through the resistance of the tree. **See Figure 21–6.**

Figure 21–6 There can be a difference of potential between the hands and feet of a person working in a tree that is in contact with an energized line. *(Delmar, Cengage Learning)*

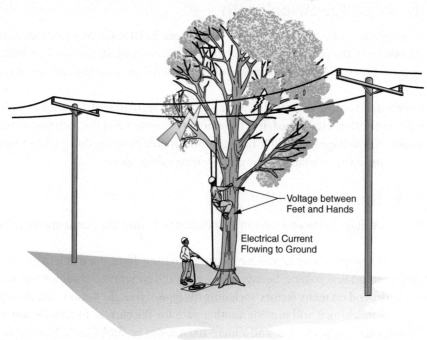

Voltage between
Feet and Hands

Electrical Current
Flowing to Ground

An arborist gets a "tree shock" when the hands are on one voltage gradient and the feet are on another, without being close to the line. In other words, an arborist should not climb a tree that is in contact with an energized circuit. Wearing rubber gloves does not provide adequate protection because many parts of the body are in contact with a tree when climbing.

Keep Away from an Aerial Lift Truck When Boom Is in Motion

When an aerial lift boom is used near energized circuits, the insulated boom and insulated lower boom insert provide good protection for those working near the vehicle on the ground. However, it is not unusual for limbs and branches being dropped by an arborist to short-circuit the lower boom insert. While the probability is low for the lower boom insert to be short-circuited by a branch making contact with the lower boom and an energized circuit, the consequences for anyone touching the truck are high. If it is necessary to contact the truck, the person on the ground should ask the operator to stop all boom movement, check the position of the boom, and then do the task requiring contact with the truck.

While the boom is in the vicinity of energized circuits, the worker on the ground should also keep others away from the truck and any attached equipment, notwithstanding the danger of falling limbs and branches. **See Figure 21–7.**

Conducting a Treetop Rescue

In most jurisdictions, occupational safety regulations require that when working in the vicinity of energized circuits over 750 V, a second person must be present within voice or visual range and trained to perform a rescue. The rescuer must also be trained in first aid and cardiopulmonary resuscitation (CPR).

A rescuer must know and have practiced how to lower a bucket using the lower controls and how to get a victim out of a bucket.

The procedure for rescuing a casualty from a tree without an aerial lift must be practiced at least annually by people who do this kind of work.

Figure 21–7 An aerial lift boom is being used near an energized circuit. The lower boom insulated insert (white) protects people on the ground if any electrical contact occurs above it. *(© Asplundh)*

A Treetop Rescue Procedure

Step	Action	Details
1.	Call loudly to the injured employee.	At the first indication of a problem, call loudly to the worker aloft, "Are you okay?" If the casualty responds but seems stunned or dazed, a rescue is probably needed, but the timing will not be as critical because the victim is breathing. Conscious victims of an electrical contact will often say they can descend on their own. Try to convince such victims to wait for a rescuer to climb up and act as a backup so that the descent will be controlled. *If there is no response, timing is critical.*
2.	Call for help.	At the first indication that a rescue must be performed, call for help using a prearranged call or code words on the company radio or get help from nearby observers. The utility/contractor should have prearranged code words in place that immediately take priority over all other radio traffic. Call 911 and request that EMS head to the location.
3.	Evaluate the situation.	If the casualty is still in contact, call for system control or dispatch to drop the circuit, preferably using prearranged code words to avoid a lot of discussion. If that does not work, the situation is such that the casualty is energized and the tree also may be energized. Climbing the tree presents a risk to the rescuer. Difficult as it may be to accept, from the ground it may be obvious that a rescue is not possible. If the casualty is in a precarious position, within the minimum approach limit, decisions regarding the need for rubber gloves, an insulated pruner or pole saw, or clean, dry rope will be needed to clear the victim. If the casualty is clearly not in contact, proceed with the rescue.
4.	Climb up and attach a lanyard near the victim.	The rescuer climbs the tree with climbing equipment. Fall protection for the rescuer will depend on the tree, the practicality of getting a rope over a crotch quickly, the suitability of using the fall line of the casualty's climbing rope, the availability of work ropes, and the acceptability of a free climb risk. Climb to a position near the casualty and tie in with the lanyard. If the casualty is still in contact or too close to the line to approach safely, use an insulated pruner or pole saw to push or pull the casualty away from the line into a position that will allow the rescuer to approach the casualty.

(Continued)

Step	Action	Details
5.	Assess the casualty's condition.	If the casualty is breathing and/or conscious, the speed of the rescue is less urgent. Reassure the casualty while preparing to lower him or her to the ground. If a casualty insists on descending solo, the rescuer should insist on descending the tree together.
		If the casualty is not breathing, it is urgent to get oxygenated blood to the brain as soon as possible. The best place to do that is on the ground. Before lowering a victim, some utilities/employers specify four quick mouth-to-mouth breaths to fill the lungs, some specify only lowering, and others specify starting CPR aloft.
6.	Rig for lowering the casualty.	Different methods are used. It is important to use and practice the method specified by your utility/employer.
		One method is for the rescuer to crouch in a position above the casualty. The rescuer then descends to the casualty and ties the casualty's saddle to his or her own saddles by either hooking together their carabiners or tying a rope through the two "D" rings of both saddles and connecting them together (using the end of the casualty's fall line). **See Figure 21–8**.

Figure 21–8 The rescuer's friction hitch is used to lower both the rescuer and the casualty. *(Delmar, Cengage Learning)*

Step	Action	Details
7.	Lower the casualty to the ground.	Using the rescuer's friction hitch, the rescuer and the casualty come down together. If more trained workers are present, a worker on the ground could rig up and lower the two people to the ground.
8.	Apply first aid.	When the casualty reaches the ground, apply first aid and, if needed, CPR.

Essential Skills for Tree Work

Climbing a Tree

Climbing and working from a tree are very physical activities, and they are the skills that define an arborist. Tree climbing is also becoming a sport in some areas. While an aerial lift is the preferred method for productive work aloft, climbing is still necessary where even an all-terrain aerial lift cannot access the tree—for example, lines built across residential back lots, wide ditches in wet season, and island work. The climbing discussion, options, and equipment described in the following section all assume that a belaying system will be in place. Belaying is a fall protection system adapted from rock climbing that is very applicable to tree climbing.

Typically, a common climbing rope has been a three-strand ½ in. (12.5 mm) nylon or braided polyester rope with a Dacron exterior sheath, but it could be any other suitable rope that has at least a minimum normal breaking strength of about 5,400 lbs (24 kN) when new, a high heat tolerance, and a maximum working stretch of less than 7%. A knot should be tied to the end of the rope so that when descending, the friction hitch will jam into it, stopping the descent in the rare situation that the climbing rope is too short.

A climbing rope should be a different color or identified by a colored internal strand to keep it distinct from working ropes. Climbing ropes designed for arborists are being manufactured. A climbing rope should not be used as a working rope to raise or lower limbs, but it is often used to safely pull up or lower hand tools or other ropes.

Historically, a bowline on a bight has been used as an arborist's saddle, but much more comfortable nylon saddles that support an arborist are available today. **See Figure 21–9.**

All tree-climbing procedures must be backed up with verification of training—they do not stand alone.

Figure 21–9 (a) A bowline on a bight has been used as a tree saddle for many years. (b) A saddle designed for tree work is much more comfortable and is less prone to an error in its use. *(a, Delmar, Cengage Learning; b, © Buckingham Manufacturing Co Inc)*

Tree-Climbing Procedure

Step	Action	Details
1.	Prepare and discuss a job safety plan.	Typical job safety planning includes identifying such high-risk hazards as electrical contact, danger tree hazards, and falling hazards, then planning to control for these hazards.
2.	Inspect the tree.	1. Identify the location and voltage of any nearby power line and places for any limb contact with the tree. 2. Choose the location for final crotching in. 3. Inspect the tree for dead and broken limbs, splits, and decay to ensure that the tree is not classified as a danger tree. 4. Check for wasp, bee, and hornet nests.
3.	Choose the first and final crotch.	The crotch is the anchor point for safe climbing and for anchoring an arborist in various working positions while descending. Ideally the crotch would be directly above the work area, in a position where a slip would swing the arborist away from any power line. The rope is passed over a branch and around the trunk as high above the ground as possible using branches with a wide crotch. Tight V-shaped crotches should be avoided because they can bind the rope as well as feet and hands. A false crotch can be used in place of a natural crotch. A false crotch (friction saver or cambium saver) is a nylon strap with metal rings at each end, one larger than the other. The climbing rope is threaded through the two rings. A false crotch allows a rope to be placed anywhere on a trunk or branch.

(Continued)

Step	Action	Details
4.	Crotch the rope.	Climbing a tree safely requires getting the climbing rope into the tree over a suitable crotch that will serve as an anchor. A lot of devices have been developed to get the rope into the tree, especially since tree climbing has become a sport. The rope can be installed with a pruner or roping tool. A common option is to install a small rope that will be used later to pull up the climbing rope. Small ropes can be installed with a throwing ball, throw bag, special line gun, special slingshot, bow and arrow, or small coil (monkey fist) of rope made by coiling the end of a rope into a bundle.
		The branches on some conifers are too dense to crotch a rope from the ground. A climber may have to carry the climbing rope up to the point to where it is to be crotched.
5.	Prepare fall protection.	Belaying is a fall protection system adapted from rock climbing that is very applicable to tree climbing. Virtually no extra equipment is needed because the climbing rope is used for fall protection and, later, for work positioning. This method can be used anytime a tree must be climbed no matter which climbing method is used. **See Figure 21–10**.
		After the climbing rope is in the tree, the climber attaches one end of the rope into both D-rings of the saddle and locks the keeper. A carabiner is attached to an anchor point (a tree) with a sling or is attached to both D-rings of a second arborist acting as an anchor. The climber should not be more than 25 lbs (11 kg) heavier than the anchoring arborist.
		The free-hanging part of the climbing rope is hitched to a carabiner with a munter hitch and the carabiner keeper is locked. **See Figure 21–11**.

Figure 21–10 One method of tree-climbing fall protection is the belaying method. *(Delmar, Cengage Learning)*

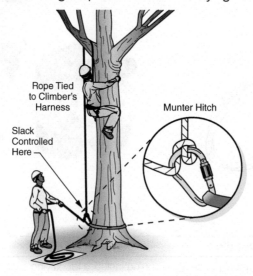

Figure 21–11 A munter hitch will allow the rope to slide through it but can be snubbed with very little pull. *(Delmar, Cengage Learning)*

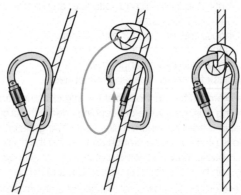

Step	Action	Details
6.	Climb to first crotch.	Climb to the first crotch using any of the various climbing methods available to reach the lower branches. As the climber ascends, regardless of the climbing method, the anchoring arborist pulls the slack through the anchoring carabiner, thereby protecting the climber from falling to the ground.

	(Continued)	The length of the trunk and the equipment available will determine which climbing method would be most suitable. Sectional ladders are often used to get within reach of the lower branches to climb a tree.
		Tree climbers are not used very often anymore because the gaff wounds create entry points for insects and diseases. They are still used on trees that are to be cut down or on some trees that have an outer bark thick enough to withstand the gaff wounds. The gaffs on tree climbers are longer than for pole climbing, but there is always some doubt when climbing on heavy bark about the bark giving way.
		Shinning a short distance to get to the lower branches is not an uncommon practice but should be reserved for small-diameter trees and limited to less than 15 ft (4 m). The climber should be tied in to a belay system every time when climbing, but it is especially necessary when climbing with tree climbers or shinning.
		Body thrusting up a tree using a friction hitch on the climbing rope to maintain each gain in height is a skill that must be learned and practiced. This could almost be called a "self-belay method."
		Some mechanical ascenders are available that are rope-gripping devices that allow a climber to move up a rope. When using limbs to climb, hands and feet should be on different limbs and dead limbs should be broken or cut off on the way up.
7.	Reposition if necessary.	When the first crotch is reached and it is necessary to go higher, the climber puts the lanyard around a main stem and sits back in the arborist saddle. Two hands are available to raise the rope to the next (or final) crotch.
8.	Tie a friction hitch.	When the highest crotch is reached, the climber puts the lanyard around a main stem and then the rope for belaying can be prepared for use as the climbing rope, with a friction hitch attached to allow work while suspended by the rope and seated in the arborist saddle. Depending on the climbing method, the arborist will tie the saddle onto the climbing rope with a friction hitch before leaving the ground or after reaching the intended top position in the tree. Historically, *the taut-line hitch* has been the one that holds an arborist in a working position and allows a controlled descent to other working positions and to the ground. It is a rolling hitch, so it is important to tie a half hitch or a stopper knot in the tail so it will not roll out and come undone. It is easy to slide it down with minimum pressure with one hand to control the speed of descent.
		The Blakes hitch is the newest friction hitch and is becoming more popular. It is a little more complicated to tie and must be tied properly to hold. It can be tied with the free end of a rope and is self-adjusting. Where the tail passes through, the bottom two turns can become a hot spot that will melt if an arborist descends too quickly.
		If transferring from an aerial bucket to a tree, an arborist should tie into a climbing rope with a friction hitch, anchored to a crotch in the tree before removing the lanyard attachment to the boom. **See Figure 21–12**.

Figure 21–12 A taut-line hitch has historically been used by arborists, there is a trend towards the Blakes hitch and mechanical devices. (*Delmar, Cengage Learning*)

9.	Work and descend.	The arborist can descend and stop along the length of the rope and carry out work. The climbing rope, friction hitch, and saddle comprise a work-positioning and fall protection system. If it is necessary to recrotch while in the tree, the arborist uses a lanyard to tie in to the tree while tying into a new crotch.

Operating a Chain Saw

The bottom line for preventing serious accidents with a chain saw is to operate the saw so that no part of the body will make contact with a moving chain. Safety measures designed into the saw, such as a chain brake, chain stop when idling, and safety chains reduce some of the risk when they are in use and maintained. Unlike most other power tools, no guard covers a moving chain. Statistically, experienced operators have more accidents than weekend operators, probably because of high exposure. **See Figure 21–13.**

Figure 21–13 Even though a chain saw is designed with as many safety features as practicable, the chain will be exposed during operation. *(Delmar, Cengage Learning)*

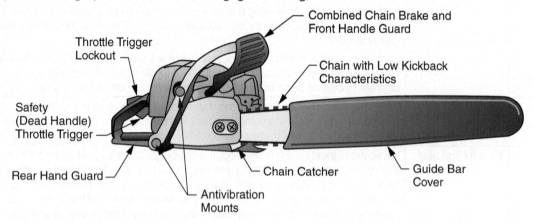

The comprehensive training and regulations regarding chain-saw use reflect the high risk of contacting a moving chain-saw blade. The risk of accidents using a power saw is reduced when the following measures are taken into account.

1. The power saw operator wears head, eye, hearing, hand, foot, and leg protection.
2. The saw is equipped with a chain brake and a protective device that minimizes chain-saw kickback. Kickback occurs when the upper portion of the saw tip comes in contact with another object, causing the saw to jump or kick back toward the operator. **See Figure 21–14.**
3. The saw is well maintained and the chain is sharp. The clutch is adjusted so that it will not engage the chain drive at idling speed and is equipped with a continuous-pressure throttle-control system to stop the chain when pressure on the throttle is released.
4. The saw engine is stopped for all cleaning, refueling, adjustments, and repairs, except for adjustments that cannot be done without the engine running.
5. The saw is started while it is held down firmly on a solid surface, such as the ground or a stump, and the chain brake is engaged.
6. Drop starting of a saw can be done safely only from outside the bucket of an aerial device.
7. When starting or operating a saw, no one is within 6 ft (2 m) of the saw operator. The saw is started at least 10 ft (3 m) away from the fueling area.
8. The saw is not operated with only one hand. An exception to this rule is when using a top handle chain saw designed for one-handed operation. The handle and the trigger are on top of the saw and are used from an aerial device when one hand is needed to direct the limbs in a certain direction. **See Figure 21–15.**
9. Loose material that may catch the saw is removed.
10. A saw weighing more than 20 lbs (9 kg) is never used above the height of an operator's shoulder, in an aerial basket, or in a tree. One reason why safety rules prevent a saw from being used above the shoulders is because a kickback is very hazardous in this position.
11. In an aerial bucket, the saw is stored in a sheath when it is not in use.

Figure 21–14 There are several situations that can cause a hazardous chain-saw kickback. *(Delmar, Cengage Learning)*

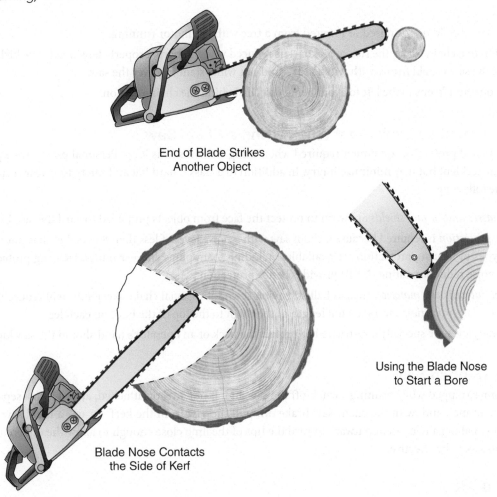

End of Blade Strikes
Another Object

Using the Blade Nose
to Start a Bore

Blade Nose Contacts
the Side of Kerf

Figure 21–15 A top handle chain saw is designed for one-handed operation when working from a bucket. It is properly stored in a sheath. *(Delmar, Cengage Learning)*

12. When working in a tree—except when working from an aerial bucket—a power saw weighing more than 15 lbs (6.8 kg) should be supported by a separate work rope tied in such a way that it will allow the saw to swing clear of the operator.

13. A power saw should not be raised or lowered from a tree with the motor running.

14. In addition to a chain brake, the risk of kickback is reduced when a sharp, properly tensioned "low kickback chain" is used and the saw is held firmly with two hands—and not while standing over the saw.

15. The saw operator keeps to the left to avoid being directly behind the chain rotation.

Personal Protective Equipment When Using a Chain Saw

The list of personal protective equipment required when using a chain saw is long. Personal protective equipment does not prevent an accident but may minimize injury. In addition to a class E hard hat and safety footwear, a chain-saw operator wears the following:

- *Safety glasses and/or face shields* are worn to protect the face from objects propelled toward the face by the saw.
- *Hearing protection* is required because a chain saw with a noise level of less than 85 decibels has not been invented. Many types of hearing protection are available, including foam plugs and ear muffs. Hearing protection with the highest noise reduction rating (NRR) provides better protection.
- *Chaps or cut-resistant material,* such as ballistic nylon sewn into special chain-saw pants, will reduce the risk of cuts to the legs. *Leg protection* covers the full length of the thigh to the top of the boot on each leg.
- *Chain-saw gloves* are specially constructed to protect the back of an operator's hand should the saw kick back.

Freeing a Trapped Saw

If a saw becomes trapped while pruning, switch off the saw and, if not already attached, attach it to a separate tool line. Try lifting the branch and, with the chain-saw brake off, pull the saw from the kerf. Or use a hand saw to release the trapped saw by making a release cut outward toward the tips of the limb close enough to allow the undercut made earlier with the chain saw to be effective.

Pruning a Tree

Pruning a tree to provide adequate clearance to a power line is a task that requires skill and knowledge to do a job well and safely. The directional pruning (Shigo) method has become fairly standard for pruning trees near power line and is defensible when discussing the results with customers.

The principles for safe pruning are similar whether pruning from a bucket or from a tree. Working from a bucket or a tree, an arborist uses an insulated pruner or pole saw (usually hydraulic from a bucket) and stays outside the minimum approach distance.

Pruning Specifics

- A hanger on a conductor can be safely removed from the conductor with an insulated pruner or saw.
- When using a saw cut use the 1-2-3 method of pruning. A pruner cut on limbs that could split or cause the bark to strip should be pruned using the 2-3 method, which is the same as the 1-2-3 method without the undercut. Cut 2 will remove the weight, and cut 3 allows a clean cut in the desired location.
- Many cuts may be needed to shorten branches overhanging power conductors. Each cut should leave the branch short enough that it cannot bridge phase to phase when dropped.

Roping a Limb

When a larger limb is to be removed, the limb should be roped or rigged with rope blocks. A working rope is crotched as high as practical above the limb to be removed. The end of the rope is tied near the tip of the limb. People on the ground will pull up on the limb after the arborist aloft in the tree or bucket cuts the underside about two-thirds through. When the people on the ground have the weight of the limb in hand, the limb is cut off. An optional rope on the butt of the branch can prevent the butt from swinging out toward the power line. The branch is then lowered. A heavy branch may need to be lowered by having the fall line wrapped around an anchor point to control the descent. The rope on the butt can be used as a tag line.

In special circumstances, the limb may need to be roped so that it is pulled sideways instead of up to clear a power line. **See Figure 21–16.**

Figure 21–16 Depending on the situation, rigging skills are needed to properly rope a limb that is being removed. *(Delmar, Cengage Learning)*

The pruner, pole saw, and hydraulic hose (if equipped) should be scheduled regularly for an electric test and also for replacement of any parts that affect the insulated parts. The insulated parts of the pruner and pole saw must be cleaned daily before use. Wiping down the insulated parts with methyl hydrate will ensure that the tools are dry.

Unless the rope in a pole pruner is a live-line rope and tested and treated as such, an insulated link should be installed in the rope. The insulated link will be tested and treated like a live-line tool and should be cleaned daily. **See Figure 21–17.**

Figure 21–17 A pruner with insulated link. *(Delmar, Cengage Learning)*

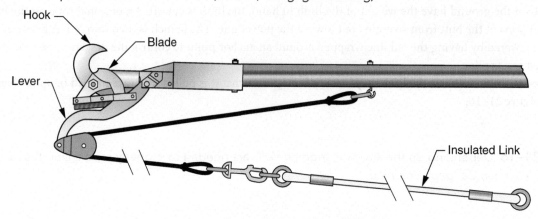

Hanger Hazards

A hanger is a branch that has been pruned and left hanging in the tree. At the end of any pruning job, every tree should be inspected for hangers that may become hazards to the public.

Pruning with an Aerial Saw or Drone

Helicopters trim right of ways using an aerial saw, which can have 10 saw blades hanging vertically from the helicopter. Data from Light Detection and Ranging (LIDAR) and the Global Positioning System (GPS) helps to guide a helicopter pilot when pruning trees with an aerial saw. Drones have also been developed to prune a tree. **See Figure 21–18**. The operator controls the drone while standing safely on the ground.

Figure 21–18 One type of aerial pruning drone.

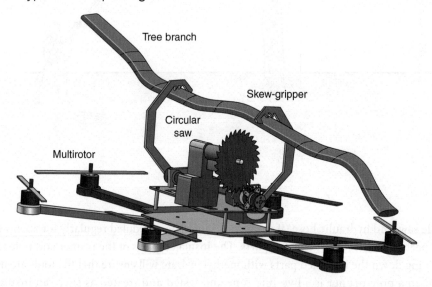

Summary of Electrical Hazards When Pruning Trees

Hazards	Barriers
Working around energized circuits.	An electrical circuit can only be considered de-energized when protective line grounds are installed. Even conductors lying on the ground can be hot or made hot from operating errors, automatic operations, or backfeed from generators. Unless you have the protection afforded by an isolation certification and you can see protective line grounds on the circuit you are working on, rely on tools and procedures that prevent limbs, trees, and workers from making contact.
Inadvertently making contact with a hot conductor.	A distance of 3 ft (1 m) should be added to the distances listed in the minimum working distances tables for routine work. Working with hands or body closer and up to the minimum limit should be considered exceptional circumstances. In such cases, keep exposure time to a minimum and have a second arborist act as a dedicated observer.
Getting a shock while in contact with an insulated conductor.	Learn to recognize the following: 1. **Tree Wire.** Some primary conductor has a polycarbonate cover that protects it from tree abrasion. It is often called "tree conductor." It can withstand a tree contact without immediately shorting out the line. It is not like underground cable with insulation and a grounded outer surface. Arborists and line crews working on this conductor must treat it as though it were bare. 2. **Weatherproof Conductor.** Some conductor is covered, but the covering is just weatherproofing and provides no insulation. It is often found on copper primary and secondary. Unless you have the protection afforded by a lockout/tagout and can see protective line grounds on the circuit you are working on, rely on tools and procedures that prevent limbs, trees, and workers from making contact.
Getting a shock from the tree.	Before going aloft, ensure that no limbs are in contact with the power line. If a limb is making contact, do not climb the tree. A limb or branch can be removed or cut using an insulated pruner or working from an insulated aerial device, while not making bodily contact with any part of the tree. A climber can put weight on a limb and cause it to make electrical contact. Climb on the side of the tree farthest from the conductor and face toward the hazard to keep it visible. Choose a crotch to tie into that would tend to swing you away from the conductor in case of a slip or fall. When climbing a tree that is close to the line, maintain the minimum approach distance for tree work and use a dedicated observer.
A limb making contact with an energized conductor while pruning.	When working from a tree, an arborist can receive a "tree shock" when a limb makes an electrical contact, especially on higher-distribution voltages. Secure the limb with a rope so that it cannot swing into the line. Break back the limb from the conductor with an insulated tool or a rope before cutting with a saw. When working from a bucket, it is standard procedure to use an insulated pole pruner to remove a branch lying on an energized conductor (hanger). This can be done on voltages up to 15 kV. On higher voltages, a limb contacting two conductors might short out the circuit. A short circuit, especially near a substation, will be very explosive. Use eye protection and the full length of the pruner. Ensure that no one stands under the conductor in case it burns off and falls to the ground.
A limb or branch breaking the conductor and an energized wire falling to the ground.	Ensure that no one stands directly under or within 10 ft (3 m) of the conductor within the span being pruned.
Boom insulation bridged out electrically by a falling limb.	The insulated boom of the truck should be treated like personal protective equipment where it is not relied on for electrical protection but may make a difference when an inadvertent contact is made by a branch or limb bridging across.

Typical Tree Felling Procedure

Power line technicians should be aware of a typical tree felling work procedure, but also know that there can be many complications that may require a change to this procedure.

Step	Action	Details
1.	Prepare and discuss a safety job plan.	The plan should identify high-risk hazards and barriers needed to protect from the hazards. High-risk hazards include nearby power line, climbing hazards, hangers, dead wood, a restricted escape route, and a restricted drop area for a tree. If in doubt about the tree condition, use an increment bit to help diagnose the tree.
2.	Make preparations to control the direction of fall.	Acceptable use in a forest may include notching in the desired direction of the fall, varying the thickness of the hinge on one side to control the fall, and/or using a wedge in the backcut; but near power line and buildings a tree should be either winched, roped to a vehicle capable of pulling, or supported by a crane.
		Any tree that cannot be felled directly away from a power line should be taken down piece by piece. Starting at the bottom, limbs are removed and work proceeds toward the crown, as in pruning. If the trunk cannot be felled, it is also removed in pieces.
		Work on problem trees may require skilled arborists and specialized equipment to carry out tasks such as climbing, working aloft, roping, piece-by-piece removal, or using a crane or winch. Problem trees are those that lean the wrong way, are hung up, have heavy tops, have broken or dead limbs (a "widowmaker" or a "chicot"), have a major fork in the trunk, have a falling path that is hard to predict (a "schoolmarm"), and/or have a restricted drop path. Be sure that clearance in the intended direction is adequate for the tree to fall completely to the ground. A lodged tree is very dangerous.
3.	Rope the tree.	When removing a tree near a power line, always rope it.
4.	Prepare a retreat path.	Common causes of accidents include being struck by the butt, rebounding limbs, and broken tops, so it is important to clear a safe work area around the base of the tree.
		Remove limbs, underbrush, and other obstructions. The retreat path should be at approximately a 45° angle to the rear of the notch in case the tree kicks backward off the stump. If the tree is on a slope, the retreat path should be uphill. **See Figure 21–19.**

Figure 21–19 A retreat path must be chosen and planned before starting the cut. *(Delmar, Cengage Learning)*

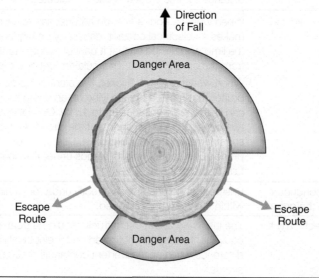

5.	Establish an "exclusion" zone.	An exclusion or danger zone is an area where everyone other than the person designated to carry out a high-risk task is kept at a safe distance. An exclusion zone should be established for everyone other than the arborist making the backcut. Because of the hazard of flying missiles, such as broken branches, a typical exclusion zone should be twice the length of the tree.
		If rigging is used, include the bight of a winch and the area near a snatch block as the exclusion zone.
6.	Cut notch.	A notch (pie cut or bird's mouth) is not normally needed for trees less than 5 to 8 in. (12 to 20 cm) in diameter. When needed, a notch is cut on the fall side of the trunk to a depth of one quarter to one third of the tree diameter. The face opening of a proper notch should be equal to the depth or about 45°. **See Figure 21–20.**
		If the lower cut of the notch is made first, it reduces the risk of a loose wedge of wood from pinching or bending the chain. If the top cut is made first, it is easier to establish the desired hinge width, and an operator can look into the top cut and see when the second cut meets. The upper and lower cut should meet evenly, and a bypass of more than 0.38 in. (1 cm) is considered unacceptable because it weakens the hinge cut.
		An improper notch is one where the notch closes and the tree stops falling.
		1. When the notch closes, it causes the tree to be under strain. Because of the strain, the fibers separate and the tree begins to split. The tree continues to split until it breaks off, leaving a "barber chair."
		2. When the tree stops, the hinge or holding wood must be cut off to drop the tree. Unless the tree is roped, there will be little control of direction of fall.

Figure 21–20 Cutting a proper V notch is a critical step to ensure the tree will fall in the intended direction. *(Delmar, Cengage Learning)*

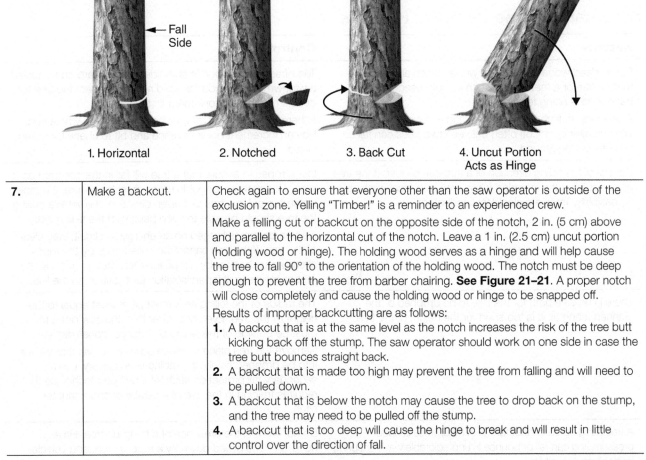

1. Horizontal   2. Notched   3. Back Cut   4. Uncut Portion Acts as Hinge

7.	Make a backcut.	Check again to ensure that everyone other than the saw operator is outside of the exclusion zone. Yelling "Timber!" is a reminder to an experienced crew.
		Make a felling cut or backcut on the opposite side of the notch, 2 in. (5 cm) above and parallel to the horizontal cut of the notch. Leave a 1 in. (2.5 cm) uncut portion (holding wood or hinge). The holding wood serves as a hinge and will help cause the tree to fall 90° to the orientation of the holding wood. The notch must be deep enough to prevent the tree from barber chairing. **See Figure 21–21.** A proper notch will close completely and cause the holding wood or hinge to be snapped off.
		Results of improper backcutting are as follows:
		1. A backcut that is at the same level as the notch increases the risk of the tree butt kicking back off the stump. The saw operator should work on one side in case the tree butt bounces straight back.
		2. A backcut that is made too high may prevent the tree from falling and will need to be pulled down.
		3. A backcut that is below the notch may cause the tree to drop back on the stump, and the tree may need to be pulled off the stump.
		4. A backcut that is too deep will cause the hinge to break and will result in little control over the direction of fall.

(Continued)

Step	Action	Details

Figure 21–21 It is easy to visualize that a barber chair could be hazardous to the chain-saw operator at the base of the tree. *(Delmar, Cengage Learning)*

(Continued)

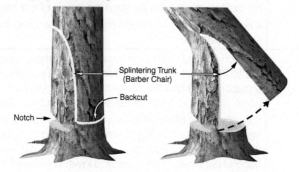

		5. A tree leaning heavily into the direction of the notch may create a barber chair as the backcut is being made. Cutting off the corners on each side of the notch before the backcut is made may help to prevent this from happening.
		Note the value of having a tree roped ahead of time. If a tree must be pulled over, the saw operator must be out of the exclusion zone.
8.	Make your escape.	Saw operators escape or retreat while not turning away and losing sight of the falling tree because misjudged circumstances (wind, heavy crown) can cause a tree to fall in a different direction than planned.
		Give the tree time to settle into its final position before approaching it.

Summary of Tree Removal Hazards

Hazards	Controls
Falling dead wood can strike a worker when an overly mature tree or a mature tree such as oak, beech, or basswood is being felled.	Too often, falling wood is an unidentified hazard and is called a widowmaker (chicot) for good reason. Inspect the tree for dead limbs and remove them first.
A dead top or limb that has fallen and is hung up (a widowmaker or chicot) often between two trees can fall on workers on the ground.	If there is evidence of rotten wood, inspect the whole tree. Use an increment borer or brace and bit to check for rotten wood.
Control of the falling direction of a tree can be lost. If the tree does not fall in the intended felling area, electrical contact is a possibility. Trees are electrical conductors.	Install ropes to ensure that a tree will fall in the intended drop area. The anchors for guide ropes must be secure. If a rope is used to pull the tree in a given direction, the vehicle pulling the tree should be able to keep pace with the falling tree.
	If a tree becomes lodged on an energized circuit, stay clear of the tree. If the tree cannot be pulled clear by the ropes already attached, or by a rope installed with a hot stick and rubber gloves, call for an isolation certification on the line.
There is no safe path for the tree to be dropped or the planned felling area is too short for the height of the tree.	The radius of the felling area must be at least equal to the height of the tree. Everyone other than the saw operator must be outside of the "exclusion/danger zone" radius.
	Remove the tree piece by piece down to a level that allows it to be dropped safely. If climbing is necessary, use a secondary fall protection such as a belaying technique that requires training in the use of a carabiner and a munter hitch.
A tree hung up on conductors or in other trees is under pressure and can fall or bounce in unpredictable ways or spring back when it is cut free.	Do not work in the presence of a hung-up tree. Have hung-up trees pulled down by a rope, winch, and vehicle.

Dead stumps, the wedge cut out from the notch, or other objects in the felling area can become dangerous flying missiles when a felled tree drops on them.	Remove loose objects from the intended drop area.
The saw operator is in danger if the tree butt kicks backward.	A clear escape route must be present at an angle of approximately 45° to the rear of the notch. Never escape directly away from the direction of the fall in case the tree kicks backward off the stump. On a slope, escape in an uphill direction.

The Bore Cut for Felling Trees

The notch and backcut has one big disadvantage. When making the backcut toward the notch, the tree can start to fall at any time; and if it starts to lean too early, the uncut wood between the backcut and the notch tries to hold the tree up, and splits the tree trunk (barber chairs) up the middle, causing hazardous uncontrolled breakage near the saw operator.

The bore cut used by a trained arborist is a safer method, especially for a tree leaning heavily and for trees that are thicker than the length of the chain-saw bar.

The bore cut involves first cutting a notch, one fourth to one third of the tree diameter, in the normal way. The corners (ears) are also cut as small notches. Then using the bore cut method, the chain saw is used to cut into the middle of the tree, leaving uncut a hinge next to the notch and holding wood at the back. It is best to start the bore cut at the back of the tree to reduce the risk of cutting into the area that will serve as a hinge. Start the bore cut by having the chain saw rotated about 45° to reduce the risk of a kickback. The holding wood can withstand a fair amount of tension. With this method, there is time for the chain-saw operator to be more precise in setting up a uniform hinge and holding wood. **See Figure 21–22**.

Figure 21–22 Making a proper bore cut is a safer way to fell a tree only if the arborist is trained in the method. *(Delmar, Cengage Learning)*

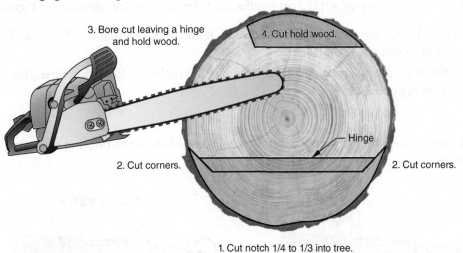

3. Bore cut leaving a hinge and hold wood.

4. Cut hold wood.

2. Cut corners.

Hinge

2. Cut corners.

1. Cut notch 1/4 to 1/3 into tree.

The backcut will release the tree with less potential for surprises. The back (release) cut is made below the bore cut. The lower the backcut is from the higher bore cut, the longer the delay before the tree falls. A lower backcut gives the saw operator more time to escape when the first sounds of wood cracking occur.

Limbing a Tree

When cutting branches from a felled tree, the following steps are essential.

1. The limbs at the base and those on top of the trunk as far up toward the treetop as practical are removed first.
2. The branches resting on the ground are cut next. When cutting these branches, it is likely that the tree will sag or roll. Remove cut branches to keep the area uncluttered so that an escape area is visible.

3. When practical, the chain saw should be operated from the side opposite the limb being cut so that the trunk serves as a barrier between the operator and the saw. **See Figure 21–23.**

4. The saw should be shut down, or the chain break applied, when moving cut limbs or branches out of the way.

Figure 21–23 The tree trunk can act as a barrier while the limbs are cut from a felled tree. *(Delmar, Cengage Learning)*

Bucking a Tree Trunk

When bucking a tree trunk into typical 4 ft (1.2 m) lengths, the following steps are essential. **See Figure 21–24.**

1. Start at the top of the tree and cut sections off the trunk. If practical, raise and chock the trunk to prevent it from rolling. Work on the uphill side of the trunk.

2. If the trunk is flat on the ground, make cuts about three quarters of the way through the tree, then roll it over and cut it through from the opposite side.

Figure 21–24 Avoid touching the ground with the chain-saw blade when bucking a tree trunk. *(Delmar, Cengage Learning)*

2nd Cut 2/3 Dia.

1st Cut 1/3 Dia.

3. To lessen the weight of a tree trunk section, it may need to be cut through in a convenient location—for example, where the trunk is somewhat suspended.

4. Prepare for the likelihood of the trunk rolling and/or a saw kickback by maintaining two hands on the saw and a solid footing.

Review Questions

1. What is directional pruning?

2. Where should the final cut be made when removing a branch?

3. What two types of trees are defined as danger trees?

4. To which five main hazards are arborists exposed?

5. Describe tree shock.

6. How can the munter hitch be used to protect a person climbing a tree from a fall?

7. List the personal protective equipment that must be worn when using a chain saw.

8. How can kickback of a chain saw be prevented?

9. Are pruners and pole saws used around utility lines considered live-line tools requiring the same care as any other live-line tool?

10. How do arborists protect themselves from widowmakers (chicots)?

11. Name five safety features that are part of modern chain saw design.

12. Name two critical methods that are used to control the direction of a tree being felled.

13. Describe an exclusion zone that is established when felling a tree.

14. What is the safest way to remove a tree leaning heavily on conductors or in other trees?

15. What precautions can a saw operator take to reduce the hazard of a tree butt kicking backwards when felling a tree?

Figure A–1 Typical Minimum Approach Distances

Maximum Phase-to-Phase Voltage (Maximum Phase-to-Ground Voltage)	Minimum Approach Distance Phase-to-Ground Exposure	Minimum Approach Distance Phase-to-Phase Exposure
0.05 to 1 kV	Avoid contact	Avoid contact
Up to 15 kV (8.7 kV)	2' 1" (64 cm)	2' 2" (66 cm)
Up to 36 kV (20.8 kV)	2' 4" (72 cm)	2' 7" (77 cm)
Up to 46 kV (26.6 kV)	2' 7" (77 cm)	2' 10" (85 cm)
Up to 72.5 kV	3' (90 cm)	3' 6" (1.05 cm)
Up to 121 kV	3' 2" (95 cm)	4' 3" (1.29 m)
Up to 145 kV	3' 7" (1.09 m)	4' 11" (1.5 m)
Up to 169 kV	4' (1.22 m)	5' 8" (1.71 m)
Up to 242 kV	5' 3" (1.59 m)	7' 6" (2.27 m)
Up to 362 kV	8' 6" (2.59 m)	12' 6" (3.8 m)
Up to 550 kV	11' 3" (3.42 m)	18' 1" (5.5 m)
Up to 800 kV	14' 11" (4.53 m)	26' (7.91 m)

Figure A–2 Proper placement of the pads (electrodes) of an AED to stimulate the heart. *(Courtesy of Coyne First Aid)*

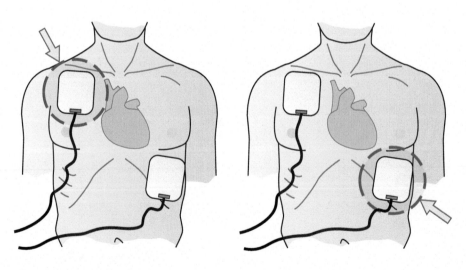

Figure A–3 A current CPR flowchart. *(Courtesy of Coyne First Aid)*

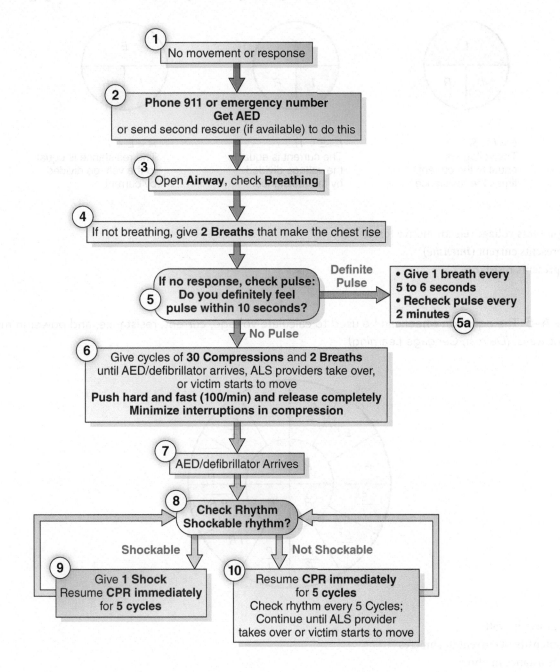

Figure A–4 Ohm's law is the most common electrical formula used by linemen. *(Delmar, Cengage Learning)*

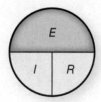
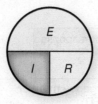
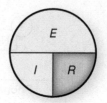

$E = I \times R$
The voltage is equal to the current times the resistance

$I = E \div R$
The current is equal to the voltage divided by the resistance

$R = E \div I$
The resistance is equal to the voltage divided by current

Where:

E represents voltage (**e**lectromotive force)

I represents current (***i**ntensité*)

R represents **r**esistance

Figure A–5 The equation wheel can be used to calculate voltage, current, resistance, and power in many different ways. *(Delmar, Cengage Learning)*

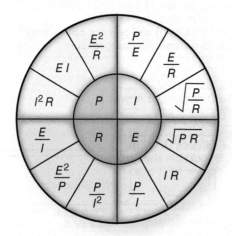

P = power in watts

I = intensity of current in amperes

R = resistance in ohms

E = electromotive force in volts

Figure A–6 Calculations Involving Power

To Find	Direct Current	Single-Phase AC	Three-Phase Wye AC
Kilowatts	$\dfrac{I \times E}{1,000}$	$\dfrac{I \times E \times pf}{1,000}$	$\dfrac{I \times E \times 1.73 \times pf}{1,000}$
Kilovolt-Amperes (kVA)		$\dfrac{I \times E}{1,000}$	$\dfrac{I \times E \times 1.73}{1,000}$
Amperes (when kW are known)	$kW \times \dfrac{1,000}{E}$	$kW \times \dfrac{1,000}{E \times pf}$	$\dfrac{kW \times 1,000}{1.73 \times E \times pf}$
Amperes (when kVA is known)		$\dfrac{kVA \times 1,000}{E}$	$\dfrac{kVA \times 1,000}{1.73 \times E}$

The preceding table gives formulas used to calculate power in three-phase and single-phase circuits. Calculations for field applications use the kilovolt-ampere formulas.

Calculations involving three-phase power use line-to-line (phase-to-phase) voltage.

The square root of 3 is normally used at the value of 1.73.

Calculations with "Handy Numbers"

To be able to do the preceding calculations quickly in the field a "Handy Number" can be used for approximations.

Approximate kVA = Amperes × "Handy Number"

Approximate amperes per phase = kVA ÷ "Handy Number"

Figure A–7 Examples of "Handy Numbers" for Some Voltage Systems

	Three-Phase kV	"Handy Number"	Single-Phase kV	"Handy Number"
kVA = Amps × "Handy Number" or Amps per Phase = kVA/"Handy Number"	230	400		
	115	200		
	69	120		
	46	80		
kVA = Amps × "Handy Number" or Amps per Phase = kVA/"Handy Number"	25	40	14.4	14
	12.5	22	7.2	7
	8.32	14	4.8	5
	0.208	0.36	0.12	0.12

To calculate the "Handy Number" for other voltage systems:

- For three-phase lines, the "Handy Number" = ∅-to-∅ voltage × 1.73 ÷ 1,000
- For single phase, the "Handy Number" = ∅-to-neutral voltage ÷ 1,000

Examples Using "Handy Numbers"

Question: One phase of an 8.3/4.8 kV feeder has 80 amperes more load than the other two phases. To balance the feeder, how many kVA should be transferred to the other two phases?

Answer: Using a "Handy Number," which is 5 for a 4.8-kV single phase line, × 80 amperes = 400 kVA. Therefore, 200 kVA should be transferred to each of the other two phases.

Question: Load on each of the three phases of a 120/208, 50 kVA pad-mount transformer bank are 150 A, 170 A, and 140 A. Is the transformer overloaded?

Answer: The average load on the three phases is 150 + 170 + 140 ÷ 3 = 153 Amperes. Using the "Handy Number," which is 0.36 for 120/208 volt service × 153 = 55 kilovoltamperes. The transformer is only slightly above its rating, but the load should be balanced more between the three phases to prevent one transformer winding from being overloaded.

Figure A–8 Metric Conversions: Miscellaneous

To Convert	Multiply By	To Obtain
Btu/hr	0.2931	watts
Horsepower (hp)	746	watts
Pounds per sq. in.	6.9	kilopascal (kPa)
acre	0.4	hectare

Figure A–9 Geometric Figures

Area of a Circle	Circumference of a Circle	Area of a Triangle	Area of a Sphere	Volume of Sphere
πr^2 ($\pi = 3.14$)	πD or $2\pi r$	Base × 1/2 Height	πD^2 or $4\pi r^2$	$D^3 \times 0.5236$
3.14 × radius squared	3.14 × diameter or 2 × 3.14 × radius	Base of Triangle × 1/2 height	3.14 × diameter squared or 4 × 3.14 × radius squared	Diameter cubed (D × D × D) × 0.5236

$\pi = 3.14$

r = radius

D = diameter

Figure A–10 Weight of Western Cedar Poles in Pounds

Length (ft)	35	40	45	50	55	60	65	70	75	80	85	90	95	100
Class 1	1,050	1,320	1,590	1,790	2,030	2,290	2,820	3,170	3,700	4,400	4,840	5,810	6,750	7,500
Class 2	880	1,150	1,370	1,585	1,760	1,940	2,200	2,640	3,170	3,700	3,960	4,930	5,950	6,550
Class 4	660	790	1,010	1,230	1,410	1,670	1,940	2,290	2,640	3,080	NA	—	—	—
Class 5	570	700	880	1,150	1,410	NA	—	—	—	—	—	—	—	—

Figure A–11 Weight of Southern Pine Poles in Pounds

Length (ft)	35	40	45	50	55	60	65	70	75	80	85	90	95	100
Class 1	1,570	1,880	2,220	2,590	3,000	3,450	4,020	4,620	5,200	6,400	7,200	8,140	6,940	NA
Class 2	1,350	1,620	1,910	2,210	2,570	2,940	3,340	3,780	4,330	5,170	5,750	6,400	6,000	NA
Class 4	1,000	1,220	1,440	1,690	1,930	2,190	2,460	2,730	3,020	3,615	NA	—	—	—
Class 5	860	1,060	1,270	1,500	1,720	1,950	2,240	2,490	NA	—	—	—	—	—

Figure A–12 Weight of Douglas Fir Poles in Pounds

Length (ft)	40	45	50	55	60	65	70	80	90	100	110	120	125
Class H6	—	—	4,370	5,010	5,750	6,490	7,270	8,830	10,580	12,420	14,350	16,330	16,970
Class H5	—	—	3,910	4,550	5,150	5,840	6,530	7,960	9,520	11,180	12,880	14,670	15,640
Class H4		2,990	3,540	4,050	4,650	5,240	5,890	7,180	8,600	10,030	11,640	13,290	14,120
Class H3		2,710	3,170	3,680	4,190	4,740	5,240	6,490	7,730	9,060	10,490	11,960	12,700
Class H2	2,020	2,440	2,850	3,270	3,770	4,230	4,780	5,840	6,950	8,140	9,480	10,760	11,450
Class H1	1,840	2,160	2,570	2,990	3,400	3,820	4,280	5,240	6,300	7,360	8,510	9,710	10,350
Class 1	1,540	1,930	2,220	2,480	2,720	3,060	3,480	4,410	5,140	6,140	7,290	8,440	—
Class 2	1,310	1,610	1,870	2,130	2,430	2,750	3,080	3,960	4,740	5,680	6,660	—	—
Class 3	1,160	1,410	1,620	1,850	2,110	2,430	2,690	3,520	4,170	—	—	—	—
Class 4	1,160	1,410	1,620	1,850	2,110	—	—	—	—	—	—	—	—

Figure A–13 Sample Weight of Concrete Poles

Pole Length (ft)	Pole Class	Weight (lbs)
30	B	1,390
35	A	1,600
35	B	1,750
35	E	2,250
40	E	2,750
45	E	3,230
45	F	3,560
45	H	4,260
50	G	5,000

Figure A–14 Typical Digger Derrick Winch Cable Capacities

Maximum Working Load for Winch Cable/Rope	
0.5-in. double-braided polyester rope	2,000 lbs/910 kg
0.75-in. double-braided polyester rope	3,500 lbs/1,590 kg
1-in. double-braided polyester rope	6,000 lbs/2,700 kg
1.25-in. double-braided polyester rope	10,000 lbs/4,500 kg
0.5-in. wire rope	4,200 lbs/1,900 kg

Figure A–15 Rope Turns on the Drum for Weight to Be Lifted

Rope Turns on Drum	Ratio of Load Line to Fall Line	Weight on Load Line to Be Lifted, Pounds (kg) Pull on Fall Line		
		20 lbs (9 kg)	30 lbs (14 kg)	40 lbs (18 kg)
3	1:20	400 (180)	600 (270)	800 (360)
3.5	1:30	600 (270)	900 (400)	1,200 (540)
4	1:50	1,000 (450)	1,500 (680)	2,000 (900)
4.5	1:70	1,400 (630)	2,000 (900)	2,800 (1,270)
5	1:100	2,000 (900)	3,000 (1,360)	4,000 (1,800)

Figure A–16 WLL of Rope Blocks

Typical WLL of Blocks at a 9-to-1 Safety Factor		
Polypropylene Rope	Set of 2-Sheave Blocks	Set of 3-Sheave Blocks
1/2 in. (1.3 cm)	1,500 lbs (700 kg)	2,000 lbs (900 kg)
3/4 in. (1.9 cm)	2,400 lbs (1,000 kg)	3,000 lbs (1,400 kg)

Figure A–17 Factors to Calculate Bisect Tension

The table shows how the factors are used to calculate the bisect tensions various angles given 1,000 lbs line tension. Using the table, an actual line tension can be inserted in the proper row to calculate a bisect tension.

Line Angle (Degrees)	Bisect Tension (lbs or kg) = Line Tension × Factor		
90	1,500	1,000	1.5
75	1,250	1,000	1.25
60	1,000	1,000	1.0
45	750	1,000	0.75
30	500	1,000	0.5
15	250	1,000	0.25

Figure A–18 Typical Maximum Allowable Pulling Tension for Underground Cable

Cable Size	1 Cables per Duct Pulling		2 Cables per Duct Pulling		3 Cables per Duct Pulling	
	Grip (lbs)	Eye (lbs)	Grip (lbs)	Eye (lbs)	Grip (lbs)	Eye (lbs)
#4	334	334	668	668	668	668
#2	531	531	1,062	1,062	1,062	1,062
1/0	844	844	1,688	1,688	1,688	1,688
2/0	1,000	1,065	2,000	2,130	2,000	2,130
4/0	1,000	1,693	2,000	3,386	2,000	3,386
250	1,000	2,000	2,000	4,000	2,000	4,000
350	1,000	2,800	2,000	5,600	2,000	5,600
500	1,000	4,000	2,000	8,000	2,000	8,000
750	1,000	6,000	2,000	12,000	2,000	12,000
1,000	1,000	8,000	2,000	16,000	2,000	16,000

Formed Wire Distribution Dead-Ends

Manufacturers' tables must be consulted for other types of conductor or jacketed conductor.

Figure A–19 Formed Wire-Distribution Dead Ends

ACSR	All Aluminum	Aluminum Alloy	Compact ACSR	Color Code
#6 6/1	#6 7W	#6 7W	#6 6/1	blue
#4 6/1	#4 7W	#4 7W	#4 6/1	orange
#2 6/1	#2 7W	#2 7W	#2 6/1	red
#1 6/1	#1 7W	#1 7W	#1 6/1	green
1/0 6/1	1/0 7W	1/0 7W	1/0 6/1	yellow
2/0 6/1	2/0 7W	2/0 7W	2/0 6/1	blue
3/0 6/1	3/0 7W	3/0 7W	3/0 6/1	orange
4/0 6/1	4/0 7W	4/0 7W	4/0 6/1	red
266.8, 18/1	266.8 19W	266.8 19W	336.4 18/1	black
336.4, 18/1	336.4 19W	336.4 19W	397.5 18/1	green
397.5, 18/1	450, 19W	397.5, 19W	477, 18/1	orange
477, 36/1	477, 19W		556, 19W	orange
477, 18/1	500, 37W			orange
556.5, 36/1	556.5, 37W	477, 19W	636, 18/1	blue
605, 36/1	636, 37W		795, 19W	blue
636, 18/1	650, 61W			blue
666.6, 36/1	715.5, 37W	636, 37W	874.5, 37W	brown
715.5, 36/1	750, 61W		954, 37W	brown
795, 36/1	795, 61W			brown
874.5, 36/1	874.5, 61W	795, 37W		orange
954, 36/1	954, 61W			orange
1,033.5, 36/1	1,033.5, 61W			orange

Formed Wire-Service Dead Ends 6/1 Bare ACSR

Figure A–20 Formed Wire-Service Dead Ends 6/1 Bare ACSR

Nominal Conductor Size	Conductor Diameter	Color Code
#6	00.169 to 00.198	blue
#5	0.199 to 0.224	white
#4	0.225 to 0.257	orange
#3	0.258 to 0.289	black
#2	0.290 to 0.325	red
#1	0.326 to 0.360	green
1/0	0.361 to 0.400	yellow
2/0	0.401 to 0.450	blue
3/0	0.451 to 0.510	orange
4/0	0.511 to 0.580	red

Kellems Dua Pull Grips

These grips are suitable for ACSR, ACAR, all aluminum and copper conductor, ground wires, messenger strands, and synthetic ropes.

Figure A–21 Kellems Dua Pull Grips

Color	Conductor Diameter Range (in.)	(mm)	Rope Diameter Range (in.)	(mm)	Breaking Strength (approx.) (lbs)	(N)
Black	0.19–0.37	5–9	0.25–0.65	6–17	6,500	28,912
Green	0.38–0.62	10–16	0.50–0.90	13–23	14,000	62,272
Red	0.63–0.87	16–22	0.75–1.10	19–28	20,000	88,960
Blue	0.88–1.12	22–28	1.00–1.50	25–38	30,600	136,109
Yellow	1.13–1.37	29–35	1.25–1.70	32–43	46,800	208,166
Aluminum	1.38–1.90	35–48	1.50–2.10	38–53	66,500	295,792

K-Type Wire-Mesh Grip with Forged Eye

This type of grip is designed for pulling underground cables. The forged eyes are designed to mate with a swivel or shackle.

Figure A–22 K-Type Wire-Mesh Grip with Forged Eye

Description	Cable Diameter Range (in.)	(mm)	Breaking Strength (approx.) (lbs)	(N)
033-01-011	0.50–0.61	13–15	5,600	24,909
033-01-012	0.62–0.74	16–19	6,800	30,246
033-01-024	0.75–0.99	19–25	9,600	42,701
033-01-025	1.00–1.49	25–38	16,400	72,947
033-01-026	1.50–1.99	38–51	16,400	72,947
033-01-027	2.00–2.49	51–63	27,200	120,986
033-01-028	2.50–2.99	64–76	33,000	146,784
033-01-029	3.00–3.49	76–89	41,000	182,368
033-01-031	4.00–4.49	102–114	48,000	213,504
033-01-039	4.50–4.99	114–127	48,000	213,504
033-01-047	5.00–5.99	127–152	48,000	213,504
033-01-045	6.00–6.99	152–178	48,000	213,504

Grips Sizing for Pulling More than One Cable

Figure A–23 Grips Sizing for Pulling More than One Cable

Individual Cable Outside Diameters, in. (mm)					Grip Diameter Range in. (mm)
2	**3**	**4**	**5**	**6 & 7**	
0.30–0.38 (8–10)	0.25–0.31 (6–8)	0.22–0.27 (6–7)	0.19–0.24 (5–6)	0.17–0.22 (4–6)	0.50–0.61 (13–15)
0.38–0.44 (10–11)	0.31–0.36 (8–9)	0.27–0.31 (7–8)	0.24–0.29 (6–7)	0.22–0.26 (6–7)	0.62–0.74 (16–19)
0.44–0.59 (11–15)	0.36–0.49 (9–12)	0.31–0.42 (8–11)	0.29–0.38 (7–10)	0.26–0.34 (7–9)	0.75–0.99 (19–25)
0.59–0.75 (15–19)	0.49–0.63 (12–16)	0.42–0.54 (11–14)	0.38–0.48 (10–12)	0.34–0.43 (9–11)	1.00–1.24 (25–31)
0.75–0.90 (19–23)	0.63–0.76 (16–19)	0.54–0.65 (14–17)	0.48–0.58 (12–15)	0.43–0.52 (11–13)	1.25–1.49 (31–38)
0.90–1.07 (23–27)	0.76–0.89 (19–23)	0.65–0.77 (17–19)	0.58–0.67 (15–17)	0.52–0.60 (13–15)	1.50–1.74 (38–44)
10.7–1.22 (27–31)	0.89–1.02 (23–26)	0.77–0.88 (19–20)	0.67–0.77 (17–19)	0.60–0.69 (15–18)	1.75–1.99 (44–51)
1.22–1.53 (31–39)	1.02–1.28 (26–33)	0.88–1.10 (20–28)	0.77–0.96 (19–24)	0.69–0.86 (18–22)	2.00–2.49 (51–63)
1.53–1.83 (39–46)	1.28–1.53 (33–39)	1.10–1.32 (28–34)	0.96–1.16 (24–29)	0.86–1.03 (22–26)	2.50–2.99 (63–76)
1.83–2.14 (46–54)	1.53–1.79 (39–45)	1.32–1.54 (34–39)	1.16–1.35 (29–34)	1.03–1.20 (26–30)	3.00–3.49 (76–89)
2.14–2.44 (54–62)	1.79–2.05 (45–52)	1.54–1.76 (39–45)	1.35–1.54 (34–39)	1.20–1.37 (30–35)	3.50–3.99 (89–101)
2.44–2.75 (62–70)	2.05–2.30 (52–58)	1.76–1.98 (45–50)	1.54–1.74 (39–44)	1.37–1.55 (35–39)	4.00–4.49 (101–114)
2.75–3.06 (70–78)	2.30–2.56 (58–65)	1.98–2.20 (50–56)	1.74–1.93 (44–49)	1.55–1.72 (39–44)	4.50–4.99 (114–127)

Figure A–24 Use a Ground Cable Size That Will Withstand the Fault Current Available in the Circuit

ASTM Grade	Cable Size	Short Circuit Withstand Rating—Amperes		Continuous Current Rating (kA)
		15 Cycles	30 Cycles	
1	#2 (35 mm²)	14,000	10,000	200
2	1/0 (50 mm²)	21,000	15,000	250
3	2/0 (70 mm²)	27,000	20,000	300
4	3/0 (95 mm²)	34,000	25,000	350
5	4/0 (120 mm²)	43,000	30,000	400
6	250 kcmil (120 mm²) or 2 × 2/0 (70 mm²)	54,000	39,000	450
7	350 kcmil (185 mm²) or 2 × 4/0 (120 mm²)	74,000	54,000	550

Conductor Data

Bare Aluminum Conductor Steel Reinforced (ACSR)

Figure A–25 Bare Aluminum Conductor Steel Reinforced (ACSR)

AWG/kcmil	mm²	Stranding	OD (in.)	Weight (lbs/1,000 ft)	Code Word
4	21.2	6/1	0.250	57.4	Swan
2	33.6	6/1	0.316	91.3	Sparrow
1/0	53.5	6/1	0.398	145.3	Raven
2/0	67.5	6/1	0.447	183.1	Quail
3/0	85	6/1	0.563	230.8	Pigeon
4/0	107	6/1	0.447	291.1	Penguin
266.8	135	18/1	0.609	289.5	Waxwing
266.8	135	26/7	0.642	289.5	Partridge
336.4	170	18/1	0.684	365.3	Merlin
336.4	170	26/7	0.721	462.6	Linnet
477	242	26/7	0.858	656.0	Hawk
477	242	30/7	0.883	747.4	Hen
556.5	282	24/7	0.914	716.8	Parakeet
556.5	282	26/7	0.927	766.0	Dove
636	322	26/7	0.990	875.2	Grosbeak
666.6	338	24/7	1.000	858.9	Flamingo
795	403	26/7	1.108	1,094.0	Drake
954	483	54/7	1.196	1,229.0	Cardinal
1,192.5	604	45/7	1.302	1,344.0	Bunting

Bare All Aluminum Conductor (AAC)

Figure A–26 Bare All Aluminum Conductor (AAC)

Size	Stranding	OD (in.)	Weight (lbs/1,000 ft)	Code Word
1/0	7	0.368	99.1	Poppy
1/0	19	0.371	99.0	Geranium
2/0	7	0.416	124.9	Aster
2/0	19	0.416	125.0	Buttercup
3/0	7	0.464	157.5	Phlox
3/0	19	0.467	158.0	Primrose
4/0	7	0.522	198.7	Oxlip
4/0	19	0.525	198.0	Sunflower
250,000	19	0.574	234.6	Valerian
250,000	37	0.575	235.0	Dandelion
266,800	7	0.586	250.6	Daisy
266,800	19	0.593	250.4	Laurel
336,400	19	0.666	316.0	Tulip
350,000	19	0.679	328.4	Daffodil
350,000	37	0.681	329.0	Gardenia
397,500	19	0.724	373.4	Canna
477,000	19	0.793	447.5	Cosmos
477,000	37	0.795	447.4	Syringa
500,000	19	0.811	469.2	Zinnia
500,000	37	0.813	469.0	Hyacinth
556,500	19	0.856	522.1	Dahlia
556,500	37	0.858	522.0	Mistletoe
600,000	61	0.893	564.0	Lotus
636,000	37	0.918	596.4	Orchid
700,000	61	0.964	656.8	Flag
715,500	37	0.974	672.0	Violet
750,000	61	0.998	704.3	Cattail
795,000	37	1.026	746.4	Arbutus
795,000	61	1.028	746.7	Lilac
954,000	37	1.124	895.8	Magnolia
954,000	61	1.126	896.1	Goldenrod
1,000,000	61	1.152	938.2	Camellia
1,033,500	37	1.170	970.0	Bluebell
1,033,500	61	1.172	970.6	Larkspur

Bare ACSR/AW with Alumaclad Core

ACSR/AW is primarily used as the phase conductor for overhead transmission and distribution. It is preferred over conventional ACSR because of its higher corrosion resistance, lighter weight, longer service life, and reduced power loss.

Figure A–27 Bare ACSR/AW with Alumaclad Core

Size	OD (in.)	Stranding	Weight (lbs/1,000 ft)	Code Word
2	0.316	6/1	86.8	Sparrow/AW
1/0	0.398	6/1	138.2	Raven/AW
2/0	0.447	6/1	174.2	Quail/AW
3/0	0.502	6/1	219.4	Pigeon/AW
4/0	0.563	6/1	276.8	Penguin/AW
266.8	0.609	18/1	283.5	Waxwing/AW
266.8	0.642	26/7	349.6	Partridge/AW
336.4	0.684	18/1	357.7	Merlin/AW
336.4	0.721	26/7	441.1	Linnet/AW
336.4	0.741	30/7	495.1	Oriole/AW
477.0	0.814	18/1	507.2	Pelican/AW
477.0	0.846	24/7	589.4	Flicker/AW
477.0	0.858	26/7	625.4	Hawk/AW
477.0	0.883	30/7	702.0	Hen/AW
556.5	0.879	18/1	591.5	Osprey/AW
556.5	0.914	24/7	687.5	Parakeet/AW
556.5	0.927	26/7	729.1	Dove/AW
556.5	0.953	30/7	819.0	Eagle/AW
636.0	0.940	18/1	676.5	Kingbird/AW
636.0	0.977	24/7	785.6	Rook/AW
636.0	0.990	26/7	833.0	Grosbeak/AW
636.0	1.019	30/19	928.9	Egret/AW
666.6	1.000	24/7	823.7	Flamingo/AW
666.6	1.014	26/7	873.0	Gannet/AW
795.0	1.063	45/7	873.4	Tern/AW
795.0	1.092	24/7	981.8	Cuckoo/AW
795.0	1.093	54/7	982.8	Condor/AW
795.0	1.108	26/7	1,042.0	Drake/AW
795.0	1.140	30/19	1,161.0	Mallard/AW
954.0	1.140	36/1	955.0	Catbird/AW
954.0	1.165	45/7	1,049.0	Rail/AW
954.0	1.196	54/7	1,178.0	Cardinal/AW
1,192.5	1.302	45/7	1,311.0	Bunting/AW
1,272.0	1.345	45/7	1,398.0	Bittern/AW
1,272.0	1.382	54/19	1,570.0	Pheasant/AW

Bare Copper

Figure A–28 Bare Copper

Size	Stranding	OD (in.)	Weight (lbs)/ 1,000 ft
4	Solid	0.204	126.3
2	7 HD	0.292	204.9
1/0	7 MHD	0.368	325.8
2/0	7 HD	0.414	410.8
2/0	19 MHD	0.419	410.9
3/0	7 HD	0.464	518.1
4/0	12 HD	0.552	650.2
4/0	19 HD	0.529	653.3
450	19 HD	0.770	1,389
500	37 MHD	0.813	1,544
1,000	61 MHD	1.152	3,088
1,250	61 MHD	1.288	3,859
1,500	61 HD	1.411	4,631

HD = hard drawn
MHD = medium hard drawn

Bare Alumaclad (Alumoweld) Conductor

These stranded conductors of aluminum-clad steel wires are commonly used for overhead shield wire. Note: The diameter and circular mils (mm^2) of copperweld are the same as for alumoweld with equivalent standing. Copperweld is approximately 25% heavier than alumoweld.

Figure A–29 Bare Alumaclad (Alumoweld) Conductor

Size Designation	# of Strands	AWG of Strands	Diameter (in.)	Diameter (mm)	Lbs/ 1,000 ft	kg/km	(kcmils)	mm^2
37 # 6	37	#6	1.13	28.8	2,222	3,307	971.3	492.2
37 # 8	37	#8	0.899	22.9	1,398	2,080	610.9	309.5
37 # 10	37	#10	0.713	17.9	879.0	1,308	384.2	194.7
19 # 6	19	#6	0.810	20.6	1,134	1,688	498.8	252.7
19 # 8	19	#8	0.642	16.3	713.5	1,062	313.7	158.9
19 # 10	19	#10	0.509	12.9	448.7	666.7	197.3	99.9
7 # 6	7	#6	0.486	12.4	416.3	619.5	183.8	93.1
7 # 8	7	#8	0.385	9.8	261.8	389.6	115.6	58.6
7 # 10	7	#10	0.306	7.8	164.7	245.1	72.68	36.8
7 # 12	7	#12	0.242	6.2	103.6	154.2	45.71	23.2
3 # 6	3	#6	0.349	8.9	178.1	265.0	78.75	39.9
3 # 8	3	#8	0.277	7.0	112.0	166.7	49.53	25.1
3 # 10	3	#10	0.220	5.6	70.43	104.8	31,150	15.8

15 kV AAC Spacer Cable–Tree Wire

Tree cable is an uninsulated conductor. The covering is intended to prevent shorts or flashovers with other objects. Conductors are covered with XLPE or with HMW low-density or high-density polyethylene.

Figure A–30 15 kV AAC Spacer Cable–Tree Wire

Size	Stranding	OD (in.)	Weight (lbs/1,000 ft)		
			XLPE	HMW Poly	HD Poly
2	7	0.583	161.0	144.7	147.3
1/0	19	0.662	214.4	196.5	199.6
2/0	19	0.706	250.5	230.7	234.1
3/0	19	0.756	295.1	272.8	276.5
4/0	19	0.812	349.8	324.5	328.6
250	37	0.858	392.3	369.4	373.7
266.8	19	0.874	416.4	388.2	392.7
336.4	37	0.947	493.5	467.4	472.3
350	37	0.961	510.1	482.9	487.8
400	37	1.006	567.1	538.3	543.6
450	37	1.049	624.1	593.4	598.9
477	37	1.071	654.7	623.0	628.6
500	37	1.089	680.4	648.0	653.8

15 kV ACSR Spacer Cable–Tree Wire

Tree cable is an uninsulated conductor. The covering is intended to prevent shorts or flashovers with other objects. Conductors are covered with XLPE, HMW low-density, or high-density polyethylene.

Figure A–31 15 kV ACSR Spacer Cable–Tree Wire

Size	Stranding	OD (in.)	Weight (lbs/1,000 ft)		
			XLPE	HMW Poly	HD Poly
2	6/1	0.616	200.2	186.0	189.0
1/0	6/1	0.698	276.0	258.9	262.6
2/0	6/1	0.747	327.6	308.7	312.7
3/0	6/1	0.802	391.0	370.0	374.5
4/0	6/1	0.863	468.9	445.6	450.6
266.8	18/1	0.909	472.5	448.7	453.8
266.8	26/7	0.942	552.2	528.0	533.1
336.4	18/1	0.984	568.2	541.6	547.3
336.4	26/7	1.020	668.0	641.2	646.9
477	18/1	1.114	758.0	726.6	733.3
477	26/7	1.158	898.7	867.1	873.9

Guying, Messenger (Strand), and Shield Wire

Zinc-Coated Steel (7 Strand)

Guying, messenger (strand), and shield wire is often galvanized steel (which includes extra-high strength [EHS] and high-strength [HS] Siemens Martin, and utilities grade).

Formed wire dead ends (FWDE) (also called *preforms*) are color coded and labeled. Manufacturers suggest that guy dead ends may be removed and reapplied twice after initial installation to retension guy strands. Should it become necessary to remove a guy dead end after it has been installed for a period of three months or longer, it should be replaced with a new dead end.

Figure A–32 Zinc-Coated Steel (7 Strand)

Size	OD (in.)	Weight (lbs/1,000 ft)	Preform Color Code	Utilities Grade Strength
5/16	0.327	225	Black	6,000
3/8	0.360	273	Orange	11,500
7/16	0.499	388	Green	18,000
1/2	0.517	510	Blue	25,000

Alumaclad Steel "Type M Guy Strand"

Figure A–33 Alumaclad Steel "Type M Guy Strand"

Designation	AWG (equiv.)	OD (in.)	Weight (lbs/1,000 ft)
5/16 in. MG3	3/7 AWG	5/16	142.7
10 MG	7/10 AWG	0.306	165.1
5/16 in. MG	—	5/16	171.6
11.5 MG	—	0.330	192.0
12.5 MG	7/9 AWG	0.343	206.2
3/8 in. MG	—	3/8	228.4
14 MG	—	0.363	232.2
16 MG	7/8 AWG	0.386	260.0
18 MG	—	0.417	306.6
7/16 in. MG	7/7 AWG	7/16	333.6
20 MG	—	0.444	347.5
1/2 in. MG	—	1/2	432.0
25 MG	—	0.519	474.8

Electrical Connectors

Ampacts

83282-1
Pub No.
Rev. 10/20/04

Code letter of number indicates the larger groove in each wedge.

This chart cross-references MAIN LINE conductors (rows) with TAP LINE conductors (columns). The cell values are Ampact catalog numbers. Columns are grouped by cartridge:

WHITE CARTRIDGE #69338-5 — taps: #6 (3/16), #4 (1/4), #2 (9/32), 1/0 (3/8 · 5/16)
BLUE CARTRIDGE #65338-1 — taps: 2/0 (7/16), 3/0 (1/2), 4/0 (9/16), 250 CU (9/16), 250 CU · 266.8 ACSR · AAC (5/8)
YELLOW CARTRIDGE #69338-4 — taps: 336.4 (5/8 · 3/4), 500 CU · 477 ACSR · 556 AAC (7/8)

MAIN LINE (groove)	#6 (3/16)	#4 (1/4)	#2 (9/32)	1/0 (5/16)	2/0 (7/16)	3/0 (1/2)	4/0 (9/16)	250 CU (9/16)	266.8 ACSR/AAC (5/8)	336.4 (5/8)	500 CU/477 ACSR/556 AAC (7/8)
#6 (3/16)	602283-4										
#4 (1/4)	602283-4	602283-3									
#2 (9/32)	602283-2	602283-2	602283-1								
1/0 (5/16)	602283-2	602283-1	602283	600403							
2/0 (7/16)	600446	600447	600403	600448	600411						
3/0 (1/2)	600446	600447	600448	600411	600458	600459					
4/0 (9/16)	600455	600456	600411	600458	600459	600465	600466				
250 CU (9/16)											
266.8 ACSR/AAC (5/8)	602046-1	602046-2	602046-3	602046-4	602046-5	602046-6	602046-7	602046-7	602046-9		
336 (5/8)	6020314	6020313	6020000	6020001	6020002	6020003	6020004	6020006	6020006	6020007	
500 CU/477 ACSR/556 AAC (7/8)	1-602031-0	602031-9	602031-8	1-602031-9	1-602031-8	1-602031-7	1-602031-6	1-602031-5	1-602031-4	1-602031-3	

MULTIPLE CONNECTIONS

NO. WIRES		TYPE	GROOVE SIZE
2	1/4	strd.Alum#2 (S.B.)	2/0
3	1/4	strd.Alum#2 (S.B.)	4/0
2	1/4	strd.Alum#4 or #4 ACSR	1/0
3	1/4	strd.Alum#4 or ACSR	2/0
2	9/32	strd.Copper #2 or Alum. Std. Round	3/0
2	9/32	strd.Copper #2 or Alum. Std. Round	266.8
2	9/32	ACSR #2 Std. Round	4/0
3	9/32	ACSR #2 Std. Round	266.8
2	9/32	strd.Alum#2 (S.B.)	3/0
3	9/32	strd.Alum#2 (S.B.)	266.8
2	5/8	strd.Alum 250 (S.B.)	556.5

Tyco Electronics Energy Division
800 Purfoy Road
Fuquay-Varina, NC 27526-9349
T:800-327-6996
F:800-527-8350
www.tyco.com

Tyco Electronics Canada Ltd.
20 Esna Park Drive
Markham, Ontario L3R 1E1
Phone: (905) 475-6222
Fax: (905) 470-5271
www.tyco.com

All fractions are comparable steel wire sizes.

THIS CHART TO BE USED FOR A.C.S.R., A.A.C. & STRANDED COPPER ONLY.
For solid copper, smooth body and other wire types, see master chart (GP 1931).

Figure A–34

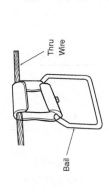

Thru Wire

Bail

AMPACT STIRRUPS

PART NUMBER	CABLE RANGE	BAIL SIZE
602585 (TYPE II)	#6	#2
602585 (TYPE II)	#4, #2	
600464	1/0, 2/0	#2
275436-1	1/0, 2/0	1/0
600468	2/0, 3/0	#2
600469	3/0, 4/0	#2
275435-1		1/0
602173		2/0
600463	266.8	2
602201		1/0
602502	350.0	#2
276476-1		1/0
600474	336.4	1/0
602142		2/0
602136		4/0
602047	477.0	1/0
602143		2/0
602247		4/0
602104	556.5	1/0
602248	636.0 AAC	2/0
602115		4/0
602147	636.0 ACSR	2/0
275074		4/0
602162	795.0	2/0
602163		4/0
602237	1033.5	4/0

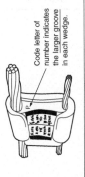

Code letter of number indicates the larger groove in each wedge.

P

MAIN LINE	795.0	795.0 CU	556.5 AAC / 636.0 AAC	477.0 / 500.0 CU	336.0
#6	1-602121-4	1-602121-4	2-602031-2	1-602031-0	602014
#4 AAC/ACSR	1-602121-3	1-602121-3	2-602031-1	602031-9	602013
#2 AAC/ACSR	1-602121-2	1-602121-2	2-602031-0	602031-8	602000
1/0 AAC/ACSR	1-602121-1	1-602121-1	2-602031-8	1-602031-9	602001
2/0 AAC/ACSR	1-602121-0	1-602121-0	2-602031-7	1-602031-8	602002
3/0 AAC/ACSR	602121-9	602121-9	2-602031-6	1-602031-7	602003
4/0	602121-8	602121-8	2-602031-5	1-602031-6	602004
250 / 266.8 CU	602121-7	602121-7	2-602031-4	1-602031-5	602006
336.4 AAC/ACSR	602121-6	602121-6	2-602031-3	1-602031-4	602007
500.0 / 477.0 CU/ACSR	602121-4	1-602121-5	2-602031-2	1-602031-3	
556.5 AAC	602121-3	1-602121-4			
636.0 AAC	602121-2	1-602121-3			
750.0 CU	602121-1	1-602121-2			
795.0	602121				

YELLOW CARTRIDGE #69338-4

THIS CHART TO BE USED FOR A.C.S.R., A.A.C. & STRANDED COPPER ONLY.
For solid copper, smooth body and other wire types, see master chart (GP 1931).

Figure A–35

Insulated Serv-ens (Cooper Power)

Figure A–36 Insulated Serv-ens (Cooper Power)

	End A				End B		
Color	**Solid**	**Strand**	**ACSR**	**Color**	**Solid**	**Strand**	**ACSR**
Green	#6	#8	—	Green	#6	#8	—
Blue	–4	–5 & 6	#6-6/1	Brown	#8	#10	—
Blue	–4	#5 & 6	#6-6/1	Green	#6	#8	—
Blue	#4	#5 & 6	#6-6/1	Blue	#4	#5 & 6	#6-6/1
Orange	#2	#3 & 4	#4-6/1-7/1	Green	#6	#8	—
Orange	#2	#3 & 4	#4-6/1-7/1	Blue	#4	#5 & 6	#6-6/1
Orange	#2	#3 & 4	#4-6/1-7/1	Orange	#2	#3 & 4	#4-6/1-7/1
Red	—	#1 & 2	#2-6/1-7/1	Green	#6	#8	—
Red	—	#1 & 2	#6-6/1-7/1	Blue	#4	#5 & 6	#6-6/1
Red	—	#1 & 2	#2-6/1-7/1	Orange	#2	3 & 4	#4-6/1-7/1
Red	—	#1 & 2	#2-6/1-7/1	Red	—	#1 & 2	#2-6/1-7/1
Yellow	—	1/0	1/0 –6/1	Blue	#4	#5 & 6	#6-6/1
Yellow	—	1/0	1/0 – 6/1	Orange	#2	#3 & 4	#4-6/1-7/1
Yellow	—	1/0	1/0 – 6/1	Red	—	#1 & 2	#2-6/1-7/1
Yellow	—	1/0	1/0 – 6/1	Yellow	—	1/0	1/0 – 6/1

Glossary

AAAC All-aluminum-alloy conductor

AAC All-aluminum conductor

AACSR Aluminum-alloy conductor, steel-reinforced

ACAR Aluminum Conductor, Aluminum Alloy Reinforced

ACCC Aluminum conductor composite core

ACCC/TW Aluminum conductor composite core/Trapezoid wires

ACCR Aluminum composite conductor-reinforced

ACSR Aluminum Conductor Steel Reinforced

ACSR/AW Aluminum Conductor, Aluminum Clad Steel Reinforced

ACSR/TW ACSR conductor made using Trapezoidal Wire construction

ACSS Aluminum Conductor Steel Supported. This is a conductor that is generally used for overhead transmission construction. ACSS is often preferred over ACSR because of its superior sag characteristics.

aerial cable A cabled group of insulated conductors installed on a pole or similar overhead structures. It may be self-supporting or attached to a messenger cable.

air-blast breakers A variety of high-voltage circuit breakers that use a blast of compressed air to break the load and short-circuit current when the contacts are open. Normally, such breakers are only built for transmission class circuit breakers.

alley-arm A side brace mounted to the side of a cross arm that is not evenly balanced.

American Wire Gauge (AWG) A standardized system used in the United States and Canada for categorizing the size of an electrical conductor based on the diameter of the wire.

ampacity The amp-carrying capacity of an electrical conductor. The contraction of "amp" and "capacity" creates an electrical term for how much current a conductor can carry safely.

ampere The unit of electrical current flow, often abbreviated as amp or just A. It is a measure of the movement of electrons as 1 coulomb per second past a fixed point.

anchor A device that supports and holds conductors in place when they end at a pole or structure. The anchor is buried and attached to the pole by a guy wire to counteract the mechanical pull of these conductors.

anode (1) The positive electrode that emits positive ions and attracts negative ions, within a voltaic cell or other such device. (2) The positive pole of a battery or electrolytic cell.

apparent power The amount of power that is delivered to an AC circuit. It is calculated by multiplying line voltage and line current. It is the amount of power apparently consumed but it is not the true power of a circuit with reactive components.

arc A discharge of electricity through air or gas between two conductive electrodes.

armor rod An outer layer of metal applied to a cable for mechanical protection. Armor rods are made of factory formed wire, designed to be applied to a range of conductor sizes.

arrestor Short for Surge Arrestor, a device that limits surge voltage by diverting it, thereby protecting equipment from the effects of overvoltage.

ASC Aluminum-stranded conductor

autotransformer A transformer in which at least two windings have a shared section. They are used to either "buck" or "boost" the incoming line voltage.

AW Aluminum-weld

AWG American Wire Gauge

B

balanced load Refers to an equal loading on each of the three phases of a three-phase system.

bank A group of electrical devices, usually transformers or capacitors, connected in a way to increase capacity or to interconnect as three phase.

basic impulse level (BIL) A reference voltage for the insulation strength of a piece of equipment or the maximum peak voltage it can withstand. It is used to express the ability of electrical equipment such as transformers to withstand certain levels of voltage surges.

bayonet A fusing device frequently used to protect distribution apparatus such as transformers and downstream devices. A Bay-O-Net fuse may include a Partial Range Current Limiting Fuse in series with an under-oil fuse link.

blackout The complete interrupting of load to an electric utility customer or group of customers.

bonding The joining of metallic parts to form an electrically conductive path that will ensure electrical continuity and the capacity to conduct any current to be present in a safe manner.

brownouts Refers to a reduction of voltage on the system. This dims the lights as a means of conserving energy.

bundled Several cables used to form one phase of an overhead circuit.

bus A conductor, which may be a solid bar or pipe, normally made of aluminum or copper, used to interconnect one or more circuits in an electric power system. An example would be the bus used to connect a substation transformer to the outgoing circuits.

bushings An insulator with a conductor in its center, used to connect equipment to a power source.

C

capacitance (1) The ability to hold a charge. (2) The ratio of a charge put on a conductor to the corresponding change in potential. (3) The ratio of the charge on either conductor of a capacitor to the resulting potential difference between the conductors.

cathode (1) The negative electrode that emits electrons or gives off negative ions and toward which positive ions move or collect in a voltaic cell or other such device. (2) The negative pole of a battery or electrolytic cell.

circuit breaker A device that can be used to manually open or close a circuit and to automatically open a circuit at a prechosen level of overcurrent to prevent damage.

circular mils (kcmil) The area of a circle with a diameter of one mil (1/1000 inch), used to describe the cross-sectional area of a conductor. One cmil equals approximately 0.0000008 square inches.

clamp stick A specialized hot stick that can hook to certain types of clamps and devices. It is also called a "Grip All" stick.

clearing time The total time needed for a protective device such as a fuse or circuit breaker to completely open and clear a fault.

conductor shield A semiconducting material, applied over the conductor to provide a smooth and compatible interface between the conductor and insulation.

conduit An opening through which conductors and cables pass and are protected, made of metal or an insulating material, usually circular in cross-section like a pipe. Also referred to as a Duct.

connector (1) A conductive coupling device used to connect conductors together. (2) A device used to terminate a conduit.

continuity test A test performed on a conductor to determine if it is a complete, unbroken circuit through which current flows.

coordination Relating to the protection of the power system, the process of coordinating the fuse, breakers and reclosers of a system so to allow the downstream devices to operate first.

core loss Power loss in a transformer due to excitation of the magnetic circuit (core). No load losses are present at all times when the transformer has voltage applied.

corona loss An electrical discharge at the surface of a conductor accompanied by the ionization of the surrounding atmosphere. It can be accompanied by light and audible noise.

cross-linked polyethylene (XLPE) A common thermoset insulation material for wire and cable. It undergoes a cross-linking chemical reaction during a curing process that causes the compound molecules to bond, forming heavier molecules.

current transformer A transformer used to measure the amount of current flowing in a circuit by sending a lower representative current to a measuring device such as a meter.

current-limiting fuse A fuse designed to reduce damaging extremely high current.

cutoff Luminaire light distribution is classified as cutoff when the candlepower per 1000 lamp lumens does not numerically exceed 25 (2.5%) at an angle of 90 degrees above nadir (horizontal), and 100 (10%) at a vertical angle of 80 degrees above nadir.

D

dead-front Generally refers to equipment that is connected without exposed conductor. Dead-front equipment is normally connected with elbows.

de-energized Free from any electrical connection to a source of potential difference and from electrical charge. A circuit is not truly de-energized until protective grounds have been installed.

delta A three-phase connection where each phase is connected in series with the next, separated by a phase rotation of 120 degrees.

delta-wye Refers to a transformer that is connected Delta on the primary side and Wye on the secondary.

dielectric (1) Any electrical insulator between two conductors. (2) A material used to provide electrical isolation or separation.

direct bury system The cables are buried directly into the ground instead of in ducts.

disconnect switch A switch that is used to disconnect an electrical circuit. It may or may not have the ability to open while the circuit is loaded.

distribution system A term used to describe the final part of an electric power system that distributes the electricity to end users from a bulk power location such as a substation. It includes all lines and equipment beyond the substation fence.

distribution transformer A transformer that reduces voltage from the supply lines to a lower voltage needed for direct connection to operate end users' devices.

distribution voltage A nominal operating voltage of up to 38 kV.

E

eddy current The current that is induced in a transformer core due to the induced voltage in each lamination. It is proportional to the square of the lamination thickness and to the square of the frequency.

elbow A device used to connect a medium voltage cable (4-35 kV nominal) to an electrical component such as a switch or transformer. Its name is derived from "L" or elbow shape. Elbows are available in ratings of 200, 600, and 900 Amperes.

electrical current Electrical current is the flow of electrons and can be measured with an ammeter.

EPR Ethylene Propylene Rubber, a synthetic rubber compound that is used as cable insulation.

extra-high-voltage An electrical system or cable designed to operate at 345 kV (nominal) or higher.

F

fault indicator A device used on a conductor to indicate if current exceeded the indicator's current rating. Fault indicators detect the magnetic field induced by load current and are typically used for troubleshooting on an underground system.

feeders A three-phase distribution line circuit used as a source to other three-phase and single-phase circuits.

ferroresonance In transformers, an unstable, over-voltage condition that can occur when the core is excited through capacitance in series with the inductor. This is especially common in transformers that have very low core losses.

flashover An unintended electrical discharge to ground or another phase. Flashovers can occur between two conductors, across the surface or around insulators to ground or equipment bushings to ground.

frequency The number of complete waveforms completed in 1 second. Frequency is a measure of how often the waveform is completed in relation to time, expressed in cycles per second. The unit of measure is the hertz (Hz), named after Heinrich Hertz.

fuse link A replaceable fuse element used in a Fused Cutout.

G

ground fault An unwanted current between ground and an electrical potential.

H

harmonic A sinusoidal component of the voltage that is a multiple of the fundamental wave frequency.

harmonic distortion The presence of harmonics that change an AC waveform from sinusoidal to complex. They can cause unacceptable disturbance to electronic equipment.

hertz The SI unit of frequency replacing cycles per second.

hipot High voltage applied across a conductor to test the reliability of an insulation system.

I

impedance The vector sum of the oppositions found in some AC circuits. The vectors may include inductive reactance, capacitive reactance, and resistance. The vector addition of these components results in the total opposition to the AC current flow. Impedance is measured in ohms and is represented by the letter Z in formulas.

induced current Current in a conductor resulting from a nearby electromagnetic field.

induced voltage A voltage produced in a circuit from a nearby electric field.

induction The property of an electric circuit displayed when a varying current induces an electromotive force in that circuit or in a neighboring circuit. A circuit has inductance when magnetic induction occurs.

insulation (1) A nonconductive material used to separate conducting materials in a circuit. (2) The nonconductive material used in the manufacture of insulated cables.

isolation link A metal link used in series with a fusing device that will melt during a transformer failure and prevents refusing/re-energization.

isolation To be electrically separate from other structures or materials. A measure of the strength of the dielectric providing the electrical division or separation.

J

jumper An electrical connection between two points; a short length of conductor used to bypass part of a circuit.

K

kilowatt-hour The use of 1000 watts for one hour.

kVA (1) Apparent Power expressed in 1000 Volt-Amps. (2) the Kilovolt Ampere rating indicates the output which a transformer can deliver at rated voltage and frequency without exceeding a specified temperature rise.

kVAR kVAR is the measure of additional reactive current flow which occurs when the voltage and current flow are not perfectly in phase.

L

lags The condition where the current is delayed in time with respect to the voltage in an ac circuit (for example, an inductive load).

leads The condition where the current leads in time with respect to the voltage in an ac circuit (for example, a capacitive load).

live-line tool An insulated stick, usually made of fiberglass, that is used to work energized overhead conductors and operate electrical equipment.

lubricant A chemical compound used to reduce pulling tension by lubricating a cable when pulled into a duct or conduit.

luminaire An electrical device used to create illumination, consisting of a light source with a means of distribution (reflector and/or refractor), lamp positioning (socket), lamp protection (housing), and a provision for power connection.

M

medium-voltage An electrical system or cable designed to operate between 2001–35 kV.

metal-clad (switchgear) An expression used by some manufacturers to describe an assembly of medium voltage switchgear units where the circuit breakers are all enclosed in grounded, sheet-steel enclosures.

N

neutral conductor In polyphase circuits, the conductor used to carry uneven current. In single-phase systems, the conductor used for a return current path.

O

off-load tap changer (no-load tap changer) A tap changer that is not designed for operation while the transformer is supplying load.

Ohm's law $E = IR$; $I = E/R$; $R = E/I$; where E = voltage, I = current flowing in a circuit and R = circuit resistance. Ohm's law is used for calculating voltage drop, fault current, and other characteristics of an electrical circuit.

ohms A unit of electrical resistance, often shown as the Greek letter omega (Ω).

on-load tap changers A tap changer that can be operated while the transformer is supplying load.

overcurrent relay A protection relay designed to prevent overload damage by tripping when the measured current exceeds a set value.

overload An excessive load; the specified maximum magnitude that can be input for a period of time without causing damage.

P

pad-mount A pad-mount transformer is installed above ground instead of in a vault.

pilot line A cord or rope used to pull a heavier rope that will in turn assist in pulling a conductor into place.

polarity (1) The electrical term used to denote the voltage relationship to a reference potential (+). (2) With regard to Transformers, Polarity is the indication of the direction of the current flow through the high-voltage terminals with respect to the direction of current flow through the low-voltage terminals.

potential transformer A transformer used to lower the voltage at a set ratio so that the voltage can be measured by instruments and meters at a safe representative level.

pothead Slang for a device used to transition an overhead conductor to underground. Potheads are normally porcelain and have been largely replaced with nonceramic and synthetic rubber.

power factor A factor applied to the apparent power to yield true power. The power factor decimal is expressed as a percentage. It is also the ratio of watts divided by volt-amperes expressed as a percentage of true watts compared to apparent volt amperes.

power line carrier communication A transmission of information over a power transmission line by using a carrier frequency superimposed on the normal power frequency, the benefit of which is two application for a single system.

protection scheme The protection of one or more elements in a power system through isolation of faulted parts. A protection scheme may be comprised of several protection systems.

pulling rope Heavy line used to pull wire or cable into a conduit or into an overhead configuration.

R

ratio of the current transformer The value of current through a meter can be a representation governed by a given ratio of a current transformer.

reactance The amount of opposition to current flow resultant from a magnetic field in an inductor or an electrostatic field in a capacitor.

real power The average value of the instantaneous product of volts and amps over a fixed period of time in an AC circuit.

recloser An automatic switch used to open and then quickly reclose a power switch which has been opened by an overcurrent on a distribution voltage (medium voltage) line.

regulator A device that is used to control the voltage of a circuit by raising and lowering voltage to maintain a constant level.

riser pole A pole used to transition from overhead and underground cables.

S

sag The amount of vertical displacement of an overhead conductor between support points. Sag is a consideration when designing a pole or tower line and will be a determining factor in the overall height of the structure. The sag in conductor lines varies with temperature.

SCADA Supervisory Control and Data Acquisition.

schematic A diagram which uses graphic symbols to indicate the electrical connections and functions of a circuit.

sectionalizer A self-contained circuit-opening device that automatically opens the main electrical circuit after sensing and responding to a preset number of successive main current impulses; thereby allowing the rest of a circuit to remain energized after a section has been isolated.

semiconductor The semi-conducting material extruded over the insulation on medium voltage insulated cables. It referred to as Semi Con.

sheath The outermost layer of a cable providing overall protection. Also referred to as jacket.

skin effect The tendency of an ac current to distribute itself so that the outer portion of a conductor carries more of the current as the frequency of the current increases.

stringing block A sheave used to support a cable that is being installed while allowing movement. These are normally used overhead but there are also specialized designs used at the entrance to a conduit system.

subtransmission line A line that carries high voltages from a higher transmission line system, reduced for more convenient transmission to nearby load centers, and delivers power to distribution substations or the largest industrial plants.

Sulfur-Hexafluoride (SF_6) A very dense, inert, nonconducting gas used inside high-voltage equipment to insulate conducting components from surfaces at ground potential. It also is used as an interrupting medium in high-voltage circuit breakers.

surge (lightning) arrestor A device that protects power lines and equipment against high-voltage lighting surges and switching surges. Connected from line to ground potential, the device has a very high resistance to current flow at normal voltages.

switchgear A general term covering the combination of switching and interrupting devices with associated control, metering, protective, and regulating devices.

switching surge A high-voltage spike that occurs when current flowing in a highly inductive circuit, or a long transmission line, is suddenly interrupted.

T

tap changer A device usually fitted to the primary winding of a transformer, to alter the turns ratio of the transformer by small discrete amounts over a defined range, thereby regulating the output voltage for required levels.

termination (1) The act of preparing the connection or transition of an insulating cable. (2) The device that transitions an underground cable to an overhead cable or wire.

three-phase Three phase refers to one circuit consisting of three conductors where the current and voltage in each conductor (phase) is 120° out of phase with each other phase.

thumper A high-voltage device used to locate an underground cable fault. The device applies a high voltage to the faulted cable with a resulting discharge to ground at the location of the fault.

transformer An electro-magnetic device used to change the voltage in an alternating current electrical circuit.

transient overvoltage A transient (sometimes called impulse) is an extremely fast interference (millionths of a second to a few milliseconds) indicated by a sharp change in voltage.

treeing Water treeing is a form of damage to cable insulation when microchannels develop, often appearing as a treelike structure in the insulation, due to a complex interaction of water, electrical stress, impurities, and imperfections.

turns ratio The ratio of primary volts divided by secondary volts.

U

unbalanced load A three phase system where the individual phases are loaded to unacceptable unequal values adversely affecting the circuit protection.

underground residential distribution (URD) Refers to the system of electric utility equipment that is installed below grade.

V

vacuum circuit breakers Circuit breakers with minimal arcing, normally applied at medium voltages, that use vacuum interrupters to extinguish the electrical arc and shut-off current.

vacuum container A sealed "bottle" containing contacts of a switch inside a very high vacuum. When the contacts are parted in the vacuum, as there is no gas in the bottle for ionization, so the current flow is quickly stopped.

vault system Equipment such as transformers and switchgear installed in vaults underground or at customer's premises.

voltage drop The reduction of voltage in a circuit when current flows.

voltage regulation Maintenance of a near constant voltage in a system, compensating for transformer and/or line voltage deviation caused by varying load current. The voltage change is affected by the magnitude and the power factor of the load current.

volts The electromotive force that pushes electrons through the conductors, wires, or components of a circuit. It is similar to the pressure exerted on a system of fluid using pipes. The higher the pressure, the more flow. Specifically, the volt is the amount of work done per coulomb of charge (volts = joules per coulomb) and is represented by the letter V. In calculations, voltage is represented by the letter E. Remember that voltage is the force required in creating flow but that volts do not flow through the circuit.

W

wye A three-phase, four-wire electrical configuration where each of the individual phases is connected to a common point, the "center" of the Y. This common connection point normally is connected to an electrical ground.

X

XLPE Cross-Linked Polyethylene. A thermo set plastic compound that is used for insulation of wire and cable.

Z

Z Impedance is the total resistance to current in an AC circuit.

Index

A

AAAC (all-aluminum-alloy conductor), 258
AAC (all-aluminum conductor), 258
AACSR (aluminum-alloy conductor, steel-reinforced), 259
A-base meters, 590–591
A bus, 94
ACAR (aluminum conductor, aluminum-alloy-reinforced), 259
accident investigations, 3
AC circuits
  amperage lagging voltage, 115
  capacitance, 115–116
  capacitive reactance, 117, 118, 119
  capacitors, 116–117
  frequency on reactance, 119
  impedance in, 60, 113
  induction in, 114
  inductive reactance, 114
  loads fed, 113
  reactance in, 113
  resistance in, 113
  voltage, current leads, 117–118
ACCR (aluminum composite conductor-reinforced), 259–260
ACCR/TW (aluminum composite conductor-reinforced, trapezoid-shaped wire), 260
AC current
  active power, 119
  apparent power, 119–120
  characteristics of, 109–113
  overview of, 108
  power factors, 120–121
  power triangles, 120
  reactive power, 120
  values of, 111–112
ACSR self-dampening conductor, 261
ACSR/TW (aluminum conductor, steel reinforced, trapezoid-shaped wires), 259
active power, 119
AC voltage rating, conductors, 253
additive transformer polarity, 404–406

ADSS (all-dielectric self-supporting) fiber optic cable, 292
aeolian vibration, 256–257
aerial cable, secondary bus, 214, 264–265
aerial devices
  aerial baskets, 144
  barehand work, 565–567
  digger derricks, 500–505
  emergency boom lowering, 491
  hydraulic systems, 488–491
  insulated booms, electrical protection, 497–500
  noninsulated booms, electrical protection, 493–497
  operating, 505–508
  operator communication, hand signals, 495–497
  stabilizing vehicles, 491–493
  traffic hazards and, 19
  tree work, 620
  vehicle grounding and bonding, 154–158, 496–497
  working aloft from, 34
aerial spacer cable, 262
air, ionization of, 300
air-blast circuit breakers, 101, 300, 313
air-brake inspection, trucks, 487
air core dry-type reactors (ACRs), 96
air-insulated substations, 92
airport runway lights, 571, 580–581
alarm systems, 249
all-aluminum-alloy conductor (AAAC), 258
all-aluminum conductor (AAC), 258
all-dielectric self-supporting (ADSS) fiber optic cable, 292
alley-arm braces, 200
allowable tension (AT) strength, guy wires, 190
alloy-steel chain, safety factor, 511. See also rigging
alternating current (AC). See AC current
aluminum
  bare conductors, types of, 257–262
  as electrical conductor, 79
  galvanic series, 372
  overhead conductors, advantages of, 257

aluminum-alloy conductor, steel-reinforced (AACSR), 259
aluminum-clad conductor, 262
aluminum composite concentric-lay stranded (ACSS), 260
aluminum composite conductor-reinforced (ACCR), 259–260
aluminum composite conductor-reinforced, trapezoid-shaped wire (ACCR/TW), 260
aluminum conductor, aluminum-alloy-reinforced (ACAR), 259
aluminum conductor, steel-reinforced (ACSR), 258
aluminum conductor, steel reinforced, trapezoid-shaped wires (ACSR/TW), 259
aluminum oxide, formation of, 265
aluminum tie wires, 208
aluminum-weld (AW) conductor, 262
American National Standards Institute (ANSI), 52, 571, 573
American Wire Gauge (AWG), conductor size, 254–255
ammeters, 54–55, 467–468
amorphous steel, 408
ampacity
  conductors, 252
  underground cables, 280
  underground transmission lines, 246
amperage lagging voltage, 115
amperes
  defined, 51, 54
  field calculations, 134
  impedance and, 113
  power, 63
  values during boom contact, 156–157
  when kVA are known, 134
  when kW are known, 134
analyses, safety design, 4, 5
anchor pulling eyes, 526–527
anchors, 170
  choosing/installing, 195–196
  poles, 190–195
anchor structures, 218
angle-iron braces, 200
angular displacement, transformer banks, 428–431

antennas, 162
apparent power, 119–121, 594–597
arc flash exposure, 12
arcing, 101, 104, 299–301, 321–324.
          See also switchgear
arc-rated (AR) clothing, 12
arc reflection method (ARM), 288
area lighting, 571
ARM (arc reflection method), 288
armature magnets, 110
arm guys, 190
armless framing, 200
armor rods, 209
asbestos, chemical hazards and, 13
ASCR (aluminum conductor, steel-
          reinforced), 258
atoms, 50
audits and assessments, safety, 4, 5
augered foundations, 218, 219
augers, operating, 503
automated external defibrillators
          (AED), 40
automatic reclosers
     blocking, 546
     types, 316
automatic splices, overhead
          conductors, 270–271
autotransformers, 88, 440, 456–463
auxiliary equipment, underground
          transmission lines, 249
average value, AC and voltage, 112
AWG (American Wire Gauge),
          conductor size, 254–255

B

backfeed, transformers, 446–447
backfills
     pole foundations, 183
     underground transmission line
          construction, 246
back guys, 190
balanced three-phase wye circuit,
          153
ball-and-socket studs, 379
ballasts, outdoor lighting, 576–577
banana and birdcaged splice, overhead
          conductors, 270
bandwidth, voltage regulator, 460
banger, 288
bare conductors, types of, 257–262
barehand procedures, 164, 544–546,
          563–568
barriers, traffic hazards and, 19

basic impulse level (BIL), insulation
          coordination, 364
Basic Insulation Level (BIL) rating,
          insulators, 202
basic life support (BLS), 40
basket hitch, 534
basket slings, 534
batteries, 102, 487–488
battery energy storage systems (BESS),
          77
bayonet fuse, 332
B bus, 94
Belleville washers, 266
belt hooks, 528
bending radius, underground cable, 238
BIL (basic impulse level), insulation
          coordination, 364
bill of materials, 196–197
biological hazards, 29
biomasses, 76
bisect tension, 524
Blackout of 2003, 339–340
blackouts, 81
bleeding wounds, 40
blocks, rigging, 523, 525–526, 530–532
bloodborne pathogens, 40
bodily fluids, 40
body belts, sizing of, 38–39
body temperature, RF energy, 162
bonding
     bonding principle, 375, 381–386
     down guys, 194
     equipotential grounding/bonding
          procedure, 388–390
     underground cable, 398–399
     vehicles, 154–158
     workers on the ground, 393–394
boom contamination meter, 144,
          498–499
boom equipment
     aerial devices, operating, 505–508
     communication, hand signals,
          495–497
     digger derricks, 500–505
     insulated booms, electrical
          protection, 497–500
     noninsulated booms, electrical
          protection, 493–497
     vehicle grounding and bonding,
          149–150, 154–158, 497
     vehicles, stabilizing, 491–493
bowline knots, 513–515, 519
bracket grounds, 387, 390–391, 393
braided rope, 511–513

braid-on-braid ropes, 512–513
brake inspection, trucks, 487
brass, 372
breakers, 57, 102
bridle slings, 534
broken neutrals, 153
brownouts, 81
brush control, 616
bucket rescues, 42–43
buckets. *See* aerial devices
bundled conductors, 263–264
burns, arc-rated clothing and, 12
bus, 94
     secondary
          overhead conductors, 264–265
          stringing, 214–215
          underground distribution systems,
               237
          urban grids, 444–445
     substation, 89
     switchyards and, 80
bushings, 101–102, 130
Bus T guard, 554–555
butterfly (collapsible bull wheel), 527
bypass, protective switchgear, 326–327

C

cabinets, substation, 88
cable radar, 288
cables
     aerial spacers, 262
     cable feeding sheave, 239
     cable guides, 238–239
     color-coded marking system, 237
     fiber optics
          electrical tracking, 293
          hardware, 294
          overview of, 291
          reasons for use, 290–291
          splicing, 294–295
          types of, 292–293
          working with, 293
     grounding cable, 377–380
     tree wire, 262
     underground cable
          cable design, 275–276
          cable faults, locating, 286–288
          continuity, 282
          erratic secondary cable fault, 289
          faulted section, locating, 285–286
          insulating gas SF_6, 282
          insulation megohmmeter, using,
               286

cables (*continued*)
    locating and tracing, 283–285
    pulling and laying, 237–241
    testing, 289
    trenches, identifying cable, 282–283
    types of, 273–275
    underground transmission lines
      ampacity, 280
      cable failure, causes of, 277
      cable shielding, 276–277
      classification, 245
      current-carrying capacity, 246
      current in cable sheath, 280–281
      pulling cables, 248
      splicing and terminations, 277–280
      voltage in cable sheath, 281
cable-separable connectors (elbows), 329–331
cadmium, 372
Cadweld, 271
calculations
    field, 134
    handy numbers, 135
    power, 134
cancer risk, electromagnetic fields, 158
candela, 573
cant hooks, 151
capacitance, 115–116, 358–359
capacitive discharge set, 288
capacitive induction, 145–146
capacitive loads, 60, 113
capacitive reactance
    in AC circuits, 113
    applications of, 118
    ferroresonance, 476–477
    overview, 117
    substations and, 96, 97
capacitive reactive power, 120
capacitor banks, 97
capacitors
    construction of, 464
    distribution feeder voltage, power quality and, 456
    distribution substation voltage, power quality, 456
    equipment on poles, 198
    overview, 116–117
    power quality and, 463–467
    reactance in AC circuits, 113
    reactive power, 120
    substations, 87, 97
    underground distribution systems, 241

capacity charges, 586. *See also* revenue metering
capital project planning, 167–168, 170–171
capstan hoist, 512–513, 530
cardiopulmonary resuscitation (CPR), 40
carts, transmission line, 226
cascading, 218, 581–582
cathodic protection, 373
cellular antennas, 162, 173
cement growth, installing insulators, 201
ceramic insulators, 201, 203, 221–223
chain, rigging uses, 521–522
chain hoist, ratchet, 523, 524
chain saw operation, tree work, 5, 626–628
chainsaw-resistant footwear, 12
chain slings, 52
chemical effect, electrical energy, 67
chemical hazards, 12–13
Chicago grips, 541
chocker type web sling, 526
choker slings, 534
chromated copper arsenate (CCA), 176
chromated copper arsenate/oil emulsion (CCA-ET), 176
chromated copper arsenate/polymer additive (CCA-PA), 176
circuit breakers, 299. *See also* circuit breakers; circuit protection
    distribution circuit protection, 340–345
    loop primary, 83–84
    metal-clad, 314–315
    operating, 313–314
    substation power flow, 100
    substations and, 87
    switchgear, 93
    transmission system protection, 338–340, 339
circuit identification, switchgear, 307–308
circuit protection
    corrosion, 371–373
    distribution protection
      fuses, 351–353
      high-voltage conversion, 355
      hydraulic recloser frame size, 347–348
      loop protection, 353
      network protection, 354
      overcurrent trouble calls, 355

      overview, 340–345
      planning for, 345–347
      reclosers, 348–351
      sectionalizers, 353
      trip coil size, 348
    grounding, ground rods, 369–371
    ground *vs.* neutral connections, 367–368
    international grounding standard, 367
    neutral, 368–369
    overcurrent protection, 369
    overview of, 366–367
    selection of, 369
    surge protection, 369
    overvoltage
      causes of, 355–356
      effects of, 359–360
      equipment voltage-time curves, 364
      insulation coordination, 364
      lightning, 355–357
      live-line work, shorting insulators, 364–365
      problem locations, 365–366
      shield wire, 364
      surge arrestors, 360–362
      switching surges, 358–359
    purpose of, 337
    transmission system, 338–340
circuits. *See also* circuit breakers; circuit protection
    AC circuits
      amperage lagging voltage, 115
      capacitance, 115–116
      capacitive reactance, 117, 118, 119
      capacitors, 116–117
      frequency on reactance, 119
      impedance in, 60, 113
      induction in, 114
      inductive reactance, 114
      loads fed, 113
      reactance in, 113
      resistance in, 113
      voltage, current leads, 117–118
    defined, 55
    delta circuits
      converting to wye circuits, 133
      grounding, 376–377
      single-phase transformers, 417
      vehicle grounding, 156, 496–497
    distribution, capacity rating, 253
    in phase, 113
    switchyards and, 80

transmission, capacity rating, 252–253

voltage conversion, preparing for, 438–440

wye circuits
converting from delta circuits, 133
grounding, 376–377
leakage current, 144
overview, 129
parallel contact and, 141
single-phase transformers, 417
vehicle grounding, 496–497

circular mil, defined, 254–255

civil work, underground distribution lines, 236–237

civil work, underground transmission lines, 246–247

clamp-on ammeter coils, 55

clamp-on tester, ground rod, 370–371

clamps
cable-separable connectors (elbows), 329–331
grounding clamps, 377–380
hot-line, operating, 313

clay, resistivity, 370

clearance, isolation certification, 16

clearance, line design, 171

clearance from the ground, sagging conductors, 205

climbing
buckets, 34
climber gaffs, 37–38
climber guards, 37
climbers, sizing and maintaining, 37–38
fall protection systems, 30–34
hook ladders, 33–34
ladders, 34–35
poles, 36–40
techniques, 36
transmission structures, 32–33
trees, 622–625

clip-in conductors, 226–227

clips, wire rope termination, 520

clothing, safety and, 146, 567

clove hitch knot, 516

coax, control cables, 99

cogeneration, 71–72

coils, reactors and, 96

cold, working in, 24–25

cold flow, 265

cold load pickups, 57

collapsible bull wheel, 527

collector substation, 91

color-coded marking systems
rubber gloves, 8
traffic-light control systems, 579
underground utilities, 174, 237

combined cycle systems, gas turbines, 71

commissioning, 228–229

common mode, power quality, 450

communication
color-coded marking system, 174
distribution substations, 82
hand signals, 495–497
health and safety promotion, 5
managing emergencies and, 40
between stations, 80

communication cables
color-coded marking system, 237
as control cables, 99
safety, 29
wave traps, 98

community antennae television (CATV), joint use, 173

compact fluorescent lamps, 575

compensation settings, voltage regulator, 460

compensator starter, 440

completely self-protecting (CSP) transformers, 413

composite backfills, pole foundations, 183

composite poles, 179. See also overhead powerline construction

compressed air, 76

compression, guy strain insulators, 194

compression sleeves, 540

concrete pads, 105

concrete poles, 152, 178–179, 181. See also overhead powerline construction

conductance, 60

conducting-sole boots, 146

conductive harmonic interference, 473

conductors
ampacity, 252
boom contact with, 157–158
bundled conductors, 263–264
and cable work, 251
capacity rating, 252–253
conductor creep, 205
conductor grips, wire rope and, 520
conductor loss, 408
corona loss, 255
dead-ending transmission lines, 227

distribution circuit, capacity rating, 253

distribution feeder voltage, 455–456

distribution substation voltage, power quality, 455–456

electric current in, 252

electric field effect on, 146

magnetic field induction, 147

overhead
aluminum, advantages of, 257
bare conductors, types of, 257–262
energized circuit connections, 266–268
exothermic connections, 271
implosive sleeves, 271
secondary bus and service drops, 264–265
splices, 225, 265, 268–271
splices and connections, tools for, 271–273
tap connections, 266
types of, 257

paralleling, 113

power quality and, 453

resistance, 60

selection of, 252

sizes, 254–255

skin effect, 256

small conductors, 268

string, 223–225

tensioned, working with
bisect tension, corners, 538–539
compression sleeves, 540
conductor length in span, calculating, 536
conductor weight in span, 534–535
corner, moving to, 539
line angle, measuring, 538
tension on loaded down guy, calculating, 536–537
vertical load, down-pull, 537–538

tree wire and aerial spacer cable, 262

underground transmission cables, 245–246

vibration and galloping, 209, 256–257

voltage drop, 253–254

voltage drop, three-phase system, 254

voltage rating, 253

conduit, underground distribution lines, 236

confined spaces, 5, 19–23
connections
  exothermic, 271
  overhead conductors, 265, 268–271
  single-phase transformers, 415–418
  tap connections, 266
  transformers, differing polarities, 406
constant current regulator (CCR), 580
constant-current transformers, 440
construction. See overhead powerline construction; underground powerline construction
construction drawings, staking a project, 172–173
continuity testing, transformers, 406
contractor, safety policies, 4, 5
control cables, 99
convection thunderstorms, 356
converters, DC transmission line, 122
converter stations, 88, 90
copper, galvanic series, 372
copper conductors, 257
copper loss, 408
copper oxide, formation of, 265
copper tie wires, 208
copper-to-aluminum connections, 372–373
core loss, 408
corner conductors, bisect tension, 538–539
corner stringing blocks, 210
corner structures, 218
corner transmission-lines, guy wires, 191
coronas
  bundled conductors, 263–264
  conductors, 255
  conductor voltage rating, 253
  dissolved gas analysis, 101
  lightning, 358
  radio and television interference, 482–483
corona shields, 227
corrective maintenance (CM), 169
corrosion
  bare conductors, types of, 257–262
  protection from, 371–373
costs, capital project planning, 170–171
counter-electromotive force (cemf), 114, 256
counterpoise, transmission line grounding, 218

coupling capacitive voltage transformer (CCVT), 80
cover-ups, 14, 554–555
CPR, safety training, 5
cradle to cradle rule, rubber gloves, 7
crane operation, hand signals, 495–497
creosote, wood poles, 176
crossarm covers, 555
crossarms, 199–200, 554–555
cross-linked polyethylene (XLPE), 275
crow's foot counterpoise, 218
CSP (completely self-protecting) transformers, 413
culture, line work, 6
current. See also AC current; circuit protection; power quality; safety; transformers
  capacitive reactance and, 117
  in conductors, 252
  delta-connected systems, 130
  delta systems, 131
  electrical power, 63
  electromagnetic induction, protective grounds, 386–387
  EMF measurement, 160–161
  grounding principle, 375–380
  Joule's law and, 63
  magnetic field, 147, 159–160
  on a neutral, 152–154
  Ohm's law, 59–60
  overcurrent, grounding protection, 369
  parallel circuits, 141
  personal protective grounds, 375
  phases 120° apart, 126
  rise and fall of, 110
  series circuit, 138
  three-phase circuits, 125
  transformers, effect of, 407–409
  underground cable sheath, 280–281
  values during boom contact, 156–157
  voltage drop, 53
  in a wye system, 132
current-carrying capacity, underground transmission lines, 246
current coil, revenue meters, 588
current-limiting fuses, 324–326, 332–333, 412
current-limiting reactors (CLR), 97, 115
current transformers
  potential and, 98–99
  protective relaying, 338

revenue meters, 594
substations and, 87
transformer-rated meters, 604–609
customer charge, 586
customer connections, 82, 83
customer electric service, 216–217. See also revenue metering
customer service staff, 86
customer substations, 88
custom power, 451
cutoff lighting, 572
cutout, distribution, 320–322
cutout mounted recloser, 317
cutout tests, gaffs, 38
cycles per second, 109

D

damping reactors, 97
DC current
  applications for, 123
  characteristics of, 122–123
  Hertz, 159
  polarity and, 109
  power formulas, 134
  stray-current corrosion, 373
  transmission
    disadvantages of, 124
    monopolar and bipolar, 122–123
DC voltage rating, conductors, 253
dead-end protector, 555
dead-ends
  anchoring, 195
  automatic splices, 270–271
  clamps, hot-line safety, 550
  conductors, procedure for, 227
  crossarms, 199
  fiber optic cable, 294
  framing, 196
  guys, 190, 191, 194
  implosive sleeves, 271
  insulation materials, 201, 203
  pole selection, 176, 180
  sagging conductors, 206
  secondary bus, stringing, 214
  service to customer, stringing, 217
  structures, line design, 218
  tension stringing procedure, 214
dead-front arrestors, 362
dead-front transformers, grounding, 396–398
dead-man anchors, 195
de-energize tags, 16
defensive driving, safety training, 5

delta circuits
  converting to wye circuits, 133
  grounding, 376–377
  single-phase transformers, 417
  vehicle grounding, 156, 496–497
delta-connected three-phase systems, 128, 130–131
delta connections, transformers
  angular displacement, wye-delta and delta-wye, 428–430
  delta-delta banks, 424
  delta-wye banks, 425
  open-wye/open delta, open-delta/open-delta, 425–427
  primary delta connections, 422
  three-phase secondary voltage arrangements, 431–436
  wye-delta banks, 424–425
delta secondary service, 433, 434–435, 437
delta systems, 129
delta towers, 218
delta/wye transformer banks, 131
demand charges, 586. *See also* revenue metering
demand meters, 596–597
derricks, 186, 495–497, 500–505.
  See also aerial devices
design engineers and technicians, 86
design standards, sag charts, 205
diagnostic tests, substation maintenance, 101
dielectric hydraulic hoses, 227–228
dielectric tests, 101
differential mode, power quality, 450
digital cellular phone service, 162
DIP switch, 349
direct burial-type fiber optic cable, 292
direct-bury systems, 85, 234–235
directional pruning, trees, 613
disaster exercises, utilities, 175
discharge resistors, 466–467
disconnect switches, 299
  load break disconnects, 312
  nonload break, 310–311
  operating at ground level, 302–303
  substation, 88, 101
dissolved gas analysis (DGA), 101
distributed energy resources (DER), 77–78
distribution cables, 275
distribution class arrestor, 362
distribution static compensator (DSTATCOM), 451

distribution systems, 81
  bracket grounding procedure, 390–391
  cables, types of, 273–275
  circuit capacity rating, 253
  circuit protection, 340–345
  cutouts, 320–322
  designs, 83
  distribution substation voltage, 455–456
  equipotential grounding/bonding procedure, 388–390
  European secondary systems, 441–443
  feeders, 89, 115
  power cables, 95
  protection schemes
    automation switchgear, 452
    fuses, 351–353
    high-voltage conversion, 355
    hydraulic recloser frame size, 347–348
    loop protection, 353
    network protection, 354
    overcurrent trouble calls, 355
    planning for, 345–347
    reclosers, 348–351
    sectionalizers, 353
    substations, 82, 88–91, 454–455
    transformers, 82, 83, 402–404
    trip coil size, 348
    tying or clipping-in, 208–209
  underground
    cable, pulling and laying, 237–241
    civil work, trenching, vaults and pads, 236–237
    installing equipment, 241
    maintenance on, 244
    reasons for, 233–234
    secondary, 242–244
    storm hardening, 244
    switchgear, types of, 328–329
    terminations and splices, 242
    tools and hardware, 242
    types of, 234–235
  voltage, 82, 165
  voltage, typical North American, 54
diversity, cold load pickups, 57
documentation, job planning, 13
domino effect, transmission line designing, 218
dose, EMF, 161

double braid ropes, 512–513
double-circuit poles, guy wires, 193
double-eye web sling, 526
Douglas Fir poles, 177
downstream protection schemes, 344
down-thrust loads, 218
drawings, 196–198, 309
driving hazards, 488
drop crotch pruning, trees, 613
drowning, 25–26
dry-well canister fuses, 332–333
DSTATCOM (distribution static compensator), 451
duct, underground cable pulling and laying, 238
duct and vault systems, 85, 234–235
duct rodder, 238
duct rods, safety and, 21
duct-type fiber optic cable, 292
dynamic voltage restorer (DVR), 451
dynamometers, 206

E

earplugs, 12, 13
earth resistance testers, 58
eddy current, 408, 589
effective power, 119
effective value, AC and voltage, 112
efficiency, transformer, 410
elasticity
  bare conductors, types of, 257–262
  sagging conductors, 205
elbow arrestors, 362
elbows, cable-separable connectors, 329–331
elbow terminations, 278–279
electric, color-coded marking system, 174, 237
electrical burns, 63
electrical circuits. *See* circuits
electrical clearance distance, 164
electrical conductive footwear, 12
electrical factor, minimum approach distance, 164
electrical generators, 51
electrical hazard (EH) boots, 11
electrical hazard (EH) overshoe footwear, 11
electrical hazard (EH)–rated footwear, 11
electrical potential, 51–54
electrical power pools, 81
electrical retesting, rubber goods, 555–556

electrical safety. *See* safety

electrical schematic drawings, 309

electrical surges, 87

electrical system emergencies, 48

electric field induction, 145–146

electric heating, Joule's law, 63

electricity, overview of

    currents, 50, 54–57

    distribution, overview of, 81–86

    electrical potential, 51–54

    electrical power, 63–65

    electrical resistance, 57–62

    generation of

        cogeneration, 71–72

        gas turbines, 71–72

        hydroelectric, 68–69

        nuclear, 70–71

        steam, 69–70

        turbines, 67–68

        wind power, 72–74

    from nontraditional sources, 76

    overview of, 67

    solar energy, 74–78

    from stored sources, 76–77

    transmission of, 78–81

electric puller monitors, 240–241

electric reclosers. *See* reclosers

electric shock, first aid for, 42

electrodes, choice of ground, 156

electrolysis, 67

electromagnetic fields (EMF)

    dose, 161–163

    induction, protective grounds and, 386–387

    measurement of, 160–161

    overview, 158

    reactive power, 120

    reducing the strength of, 160

    sources of power frequency, 159–160

    underground cables, 280–281

    underground transmission lines, 244–245

electromagnetic induction, 145, 386–387

electromagnetic microwaves (MW), 162

electromechanical revenue meter, 588–591

electromotive force (emf), 51, 589

electronic meters, 588, 591–593

electronic reclosers, 319–320, 348

electronic sectionalizer, 320

electrons, 50, 54

electrostatic fields, 120

electrostatic induction, 374

ELF (extremely low frequency), 159

emergencies, managing. *See also* safety

    basic life support, 40

    bucket rescues, 42–43

    communicating, 40

    electrical system emergencies, 48

    electric shock first aid, 42

    fires in electrical environments, 46–47

    pole tops, tower tops, or substation structure rescues, 44–46

emergency preparedness plans, 4, 5

emergency standards, utilities, 175

EMF. *See* electromagnetic fields (EMF)

emf. *See* electromotive force (emf)

employee

    safety, 3. *See also* safety

    training, 3, 4

employee assistance programs (EAPs), 5

employer, safety, 2–3

enclosed spaces, 19–23

enclosures, underground distribution systems, 235

endless type web sling, 526

energized lines. *See* hot-line work

energy. *See also* electricity, overview of

    hysteresis loss, 408

    overview of, 67

    transformers, 402

energy charges, 586. *See also* revenue metering

energy consumption, 586

engines, 76

environment, outdoor. *See* outdoor environment

equipment

    inspection and maintenance of, 5

    on poles, 198–199

    substations, voltage control, 454

    voltage-time curves, 364

equipotential grids, constructing, 105–106

equipotential grounding/bonding procedure, 388–390, 391–392

equipotential mats, 149

equipotential zones, 93, 375–376, 381–386

equivalent span, sagging conductors, 205

ergonomic analysis, 5

erratic secondary cable fault, 289

estimating capital projects, 171

European secondary systems, 441–443

evacuation plans, 5

excavations, 23–24, 237

excitation loss, energy, 408

existing lines, major projects, 227–228

exothermic connections, 271

expansion anchors, 195

explosive flash, transformers, 447

expulsion-cutout fuse, 320–323

extended reach rule, rubber gloves, 7

extension ladders, 34

extra-high-voltage (EHV) lines, 78–79, 93

extreme high voltage, 53

extreme low voltage, 53

extremely low frequency (ELF) waves, 159

extruded dielectric, polyethylene (XLPE) pipe cable, 245–247

eye protection, 10–11

eye splice, 519

F

FACTS (flexible AC transmission systems), 451

fall arrest systems, 30–31

falling/dropping hazards, 32–33

fall of potential test, 370–371

fall protection systems, overview, 30–34

Faraday, Michael, 563

Faraday cage, 146, 563

farads, 116, 117

fatigue resistance, bare conductors, 257–262

fault currents. *See also* circuit protection; power quality

    distribution circuit protection, 340–345

    erratic secondary cable fault, 289

    ground fault, 56, 131, 149, 341–342, 378

    overview, 57

    transmission system protection, 338–340

fault indicators, 285–286

fault tamer fuses, 324

fault to ground, step potential, 152

Federal Highway Administration, 18

feeder protection, substation, 344

feeder voltage regulators, 455

fencing

    substation grids, 105

    substation inspection of, 102

ferroresonance, 475–478

ferrule, 377, 379

fiberglass crossarms, 199

fiberglass energized-line tool, 15

fiberglass live-line tool, 164

fiber optic cables

  control cables, 99

  electrical tracking, 293

  hardware, 294

  overview of, 290–291

  reasons for use, 290

  shield wires, 225

  splicing, 294–295

  types of, 292–293

  working with, 293

fiber optic communications, 80, 98

fiber rods, guy strain insulators, 194

fiber ropes

  knots, types of, 513–519

  rope blocks, lifting with, 530–532

  splicing, 518–519

  strength of, 513

  types of, 511–513

filter lenses, 10

final sag, 541

finger lines, stringing blocks, 210

fires

  in electrical environments, 46–47

  extinguishers, classes of, 47

  prevention plans, 5

  response, safety training, 5

  vaults and, 23

first aid

  basic life support, 40

  for electric shock, 42

  for frostbite and hypothermia, 25

  for heating, 162

  response plans, 5

  safety training, 5

fission, nuclear, 71

five-wire service, 433

fixed charges, 586

fixed ladders, 34

flammable gases, enclosed spaces, 19–23

flash, safety, 57

flash-bang method, lightning

    estimation, 357

flash hazard analysis, 12

flashovers, 15, 164, 202, 359–360

flat-strap braces, 200

Flemish loop, wire rope termination,

    521

flexible AC transmission systems

    (FACTS), 451

flexible mandrel, 239

floating D-rings, 38

fluorescent lights, 60, 574–575

FM and television, 159

footings, transmission line

    construction, 218

footwear, 11–12, 146, 567

forestry and tree crews

  chain saw operation, 626–628

  cutting trees, 616

  danger trees, 614–615

  near electrical circuits, 618–622

  grubbing, 616

  hanger hazards, 630

  hazards of, 617

  planting trees, 617

  tree climbing, 622–625, 635–636

  tree felling, 632–634, 635

  tree pruning, 628–631

  tree removal hazards, 634–635

  tree trunk bucking, 636

  treetop rescue procedures, 620–622

  vegetation management, 611–617

forklift operation, safety training, 5

foundation plan, substations, 105

four-wire (lighting) service, 433

fractures, first aid for, 41

framing, poles, 196–198

frequency (Hz)

  alternating currents, 109

  capacitive reactance, 117

  defined, 158

  effect on reactance, 119

  electromagnetic spectrum, 159

  ferroresonance, 475–478

  harmonics, 472–473, 475–478

  inductive reactance, 114

friction

  rigging and, 525–526

  rope blocks, 531–532

  from stringing blocks, 210

friction hitch, 513

frontal thunderstorms, 356

frostbite, first aid for, 25

fuel cells, 76

full-body harness, 34, 38

"full-load" meter adjustment, 589

full-range current-limiting fuse, 326

fused cutouts

  delta-connected systems, 130

  operating, 320–323

  wye-connected systems, 132

fused elbows, 330–331

fuse link, 300

fuses

  cold load pickups, 57

  current-limiting fuses, 324–326

  distribution protection, 347, 351–353

  limiters, 445

  overview, 323–324

  transformers, 412–413

G

galloping

  bare conductors, types of, 257–262

  conductors, 256–257

  sagging conductors, 205

galvanic corrosion, 372

galvanic series, line hardware, 372

galvanized steel, bare conductors,

    257–262

gamma rays, 159

gas chromatography, 101

gas-insulated (SF_6) circuit breakers, 101

gas-insulated (SF_6) equipment, 101

gas-insulated (SF_6) transformers, 95

gas-insulated lines (GILs), 245

gas-insulated substations, 92, 104

gas turbines, electrical generation

    overview, 71–72

gas utilities

  color-coded marking system, 174,

    237

  joint use, 173

  safety, 29

gaseous-discharge lamps, 575–577, 584

gauss (Gs), 160

general-purpose current-limiting fuse,

    326

generators, backfeed grounding, 395

geographic-based maps, 309

geographic information systems (GIS),

    172–173

geomagnetically induced current (GIC),

    338

geothermals, 76

gigawatts (GW), 64

GILs (gas-insulated lines), 245

gins

  hand lines, lifting with, 529

  poles, 186

  rope blocks, lifting with, 530–532

glass insulators

  installing, 201

  overview, 203

  transmission line construction,

    221–223

global information system (GIS) technology, 168

global positioning systems (GPS), 172–173, 630

granite, resistivity, 370

granny knot, 516

graphite, 372

grid-scale battery station, 91

grids, electrical, 81, 83

grillage foundations, 218, 219

ground connections

    disconnect switch operation, 302–303

    down guys, 194

    grounded conductors, boom contact, 157–158

    grounded networks, 93, 102

    ground electrodes, 156, 395

    ground fault protection, 56, 131, 149, 341–342

    ground gradient control, parallel circuits, 144

    grounding principle, 375–380, 396–398

    grounding transformers, 440

    ground rods, 105, 106, 369–371, 376–377

    ground selection, 369

    ground set tester, 379

    ground to ground rule, rubber gloves, 7

ground *vs.* neutral connections, 367–368

    ground wire, defined, 367–368

    installing, 210

    international grounding standard, 367

    magnetic field induction, 147

    neutral, 368–369

    overcurrent protection, 369

    overview of, 366–367

    personal protective grounds

        applying, procedures for, 388–393

        bonding principle, 381–386

        electromagnetic induction, 386–387

        grounding hazards, 393–395

        grounding principle, 375–380

        reasons for, 374–375

        underground cable, 395–399

    power quality, 451

    steel towers, 218

    substation inspection of, 102

    surge protection, 369

    telescopic switch stick, use of, 304

    tingle voltage, 478–482

    transformers, 414

    vehicles, 154–158

ground-gradient effect, lightning, 357

ground-gradient mats, 149, 211, 214

grounding gradients, 149–152, 155, 394–395

guard poles, installing, 223

guys

    guying a pole, 190–195

    guy wire, 520, 526–527

    strain insulators, 194

    tension, calculating, 536–537

H

halogen lamps, 574–575

hand lines, 528

hand-operated press, 272

hand-ratchet crimpers, 271

hand signals, derrick and crane operation, 495–497

handy numbers, 135

hang insulators, 221–223

hard hats, 10

hardware, underground distribution lines, 242

harmonic frequencies, ferroresonance, 475–478

harmonic interference, power quality, 472–473

harness, safety, 41, 552

harness hang syndrome, 41

hazard analysis, 5

hazardous chemicals, 12–13, 584

hazardous materials management, 5

hazard registry, 5

hazards. *See* safety

head guys, 190

head protection, 10

hearing protection, 12

heat

    conductors, 252

    from currents, 63, 252

    electrical energy, 67

    RF energy, 162

    underground cables, 280

    working in, 24–25

heavy-duty class arrestor, 362

heights, working at. *See also* aerial devices; tree work

buckets, 34

fall protection systems, 30–34

hook ladders, 33–34

ladders, 34–35

poles, 36–40

transmission structures, 32–33

helicopters

    inspections of, 168–169

    safety, 26–29

    setting poles, 186

henrys, 114

hepatitis B virus (HBV), 40

herbicides, proper use of, 616

hertz (Hz), 109, 114

high-induction areas, grounding hazards, 394

high-pressure, fluid-filled (HPFF) pipe cable, 245, 273

high-pressure, gas-filled (HPGF) cable, 273

high-pressure, gas-filled (HPGF) pipe cable, 245

high-pressure sodium lamps, 576, 583

high-resistance fault, locating, 288

high-resistance materials, 61

high-voltage bus, 94, 95, 100

high-voltage conversion, circuit protection, 355

high-voltage direct current (HVDC), 451

high-voltage environment, electric field effect, 146

high-voltage lines, conductor voltage rating, 253

high-voltage surges, 202, 355–356, 356–358, 360–362. *See also* lightning; surge arrestors

hipot test, 289–290

hitch, friction, 513

hitch knots, 516, 518

hoists. *See also* rigging

    capstan hoists, 512–513, 530

    chain hoist, 523, 524

    hand lines, 528

    rope blocks, 530–532

    web hoist, 523, 524

"hold off" tags, 16

home wiring, circuit protection example, 341–342

hook ladders, 33–34

hooks, belt, 528

horsepower (HP), 64

hot-line clamps, 313, 326–327

hot-line work
  automatic reclosing, blocking, 546
  barehand work, procedures and
      hazards, 563–568
  choice of, 544
  defined, 543–544
  electrical retesting, rubber goods,
      555–556
  hot-line rope, using, 549
  hot secondary, 550–551
  insulator testing, 547–548
  minimum approach distance (MAD),
      545–546
  overhead conductor connections,
      266–268
  rubber gloves, hazards, 557–558
  rubber gloves, overview, 551–554
  rubber gloves, preparation for use,
      556–557
  safety hardware, 550
  second point of contact, 544
  setting poles in, 187–190
  temporary jumpers, 548–549
  tool work, procedures and hazards,
      558–563
  voltage surge protection,
      transmission lines, 546–547
hot-stick tool tester, 566
HPFF (high-pressure, fluid-filled) pipe
      cable, 245
HPGF (high-pressure, gas-filled) cable,
      273
HPGF (high-pressure, gas-filled) pipe
      cable, 245
human body
  electric field effect, 146
  as electrical path, 619
  parallel circuits, 144–145
  series circuits, 140
  values during boom contact,
      156–157
human factor, minimum approach
      distance, 165
human immunodeficiency virus (HIV), 40
HVDC (high-voltage DC), 451
hydraulic press, 272
hydraulic reclosers, 317–319, 347–348.
      See also reclosers
hydraulic sectionalizer, 320
hydraulic systems, utility vehicles,
      488–491
hydraulic tampers, pole foundations,
      183
hydroelectric generation, 68–69

hydroelectric plants, 64
hypothermia, 25
hypress, 273
hysteresis loss, energy, 408

I

ice
  galloping, conductors, 256–257
  guy wires, 190, 191
  pole strength, 176
  safety, 25–26
  sagging conductors, 205
  structure foundations and, 218
identification, pole, 181–182
Illuminating Engineering Society of
      North America (IESNA), 571,
      573
impedance
  AC circuits, 113
  fault current, 344
  ground return, 369
  overview, 51, 60
  secondary fault current,
      transformers, 408
  skin effect, conductors, 256
  transformers, 409–410
implosive sleeves, 271
impulse generators, 288
incandescent lamps, 573–575
incident investigations, 4, 5, 6
independent work, 2
induction, 114, 374
inductive harmonic interference, 473
inductive loads, 113
inductive reactance, 60, 96, 97, 113,
      114, 476
inductive reactive power, 120
inductors, 98
industrial substations, 88
infrared visible light, 159
in phase, 113
inrush currents, 57
in-span disconnect, 310–311
inspections
  line, 168
  maintenance and, 4, 5
  rubber gloves, 9–10
  substation, 101
instantaneous value, AC and voltage,
      111
instrument transformers, 440
insulated booms, leakage current, 144.
      See also aerial devices

insulation, 199
  cable shielding, 276–277
  conductor voltage ratings, 253
  covers, 555
  dead-end, changing, 563
  guy strain, 194
  hot-line work, testing for
      breakdown, 547–548
  installing, 200–204
  insulated compression sleeves, 217
  insulated pole platform, 552
  live-line work, shorting insulators,
      364–365
  overview, 60–61
  overvoltage protection, 364
  radio and television interference,
      482–483
  resistance testers, 58
  substations, 92, 102, 103
  testers, 58
  transformers, 95, 406, 410
  transmission line, 78, 221–223
  transmission line construction,
      218–221
  underground cables, 245, 273–275,
      282
insulation megohmmeter, 286
insulinks, 217
interfacial tension tests, 101
intermediate class arrestor, 362
International Municipal Signal
      Association (IMSA), 578–579
International Radiation Protection
      Organization, 162
interrupters, zero-awaiting/forcing, 300
interruptible power, 587
investigations, incidents, 3, 4, 5
ionization of air, 300
ionizing radiation, 159
ions, 50
I2R loss, 408
iron, 372
iron loss, 408
irrigation, color-coded marking system,
      174, 237
isolation certification, 16
IT grounding/earthing system, 367

J

jib capacity, 506–508
job briefing, 5–6, 13–14
job planning, 13–14
job-related injury/illness, 3

job safety planning, 4. *See also* safety
job site analysis (JSA), 4
joint health and safety committees, 4, 5
joint use, with other utilities, 173–174
Joule's law, 63
jumper grounds, 377–380
jumpers, series connections, 139
jumpers, temporary, 548–549
jump-starting utility trucks, 487–488
justifying capital projects, 170–171

K

keillng hitch, 46
Kh constant, revenue meters, 590
kilovolt-amperes (kVA), 64, 120.
        See also revenue metering
kilovolt-amperes reactive (kVAR),
        116, 120. *See also* revenue
        metering
kilovolts (kV), 79
kilovolts per meter (kV/m), 160
kilowatt-hours (kWh), 51, 65. *See also*
        revenue metering
kilowatts (kW), 51, 64, 119, 134
K-link fuses, 323–324
knots, types of, 513–519
Kr constant, revenue meters, 590

L

labor laws, substation entry, 100
ladders
    barehand work, hot-line, 566
    extension ladders, 34
    fixed ladders, 34
    hazards and controls, 35
    hook ladder, 33–34
laminated wood poles, 176
lateral taps, 84, 395
layouts, staking a project, 172
lead, 13, 275, 372
leakage current, 144, 202, 289
leather gloves, 8, 10
LEDs (light-emitting diodes), 573–575
left-hand sheet bend, 516
length, cables. *See* cables; sag
        conductors
length, poles, 176, 179
LIDAR (light detection and ranging),
        612, 630
lifting loads. *See also* rigging
    capstan hoist, 512–513, 530
    chain hoist, 523, 524

general precautions, 527
    hand lines, 527–528
    rope blocks, 530–532
    slings, 532–534
    web hoist, 523, 524
light, electrical energy, 67
light detection and ranging (LIDAR),
        612, 630
light-duty class arrestor, 362
light-emitting diodes (LED), 573–575
lighting, outdoor
    airport runways, 580–581
    area lighting, 577–582
    ballasts, 576–577
    cascading control system, 581–582
    gaseous-discharge lamps, 575–577
    halogen lamps, 574–575
    hazards of, 583–584
    incandescent lamps, 573–575
    light-emitting diodes, 573–575
    light pollution, 572
    lumens, 573
    luminaires, 571
    maintenance and troubleshooting,
        583
    pilot-wire control systems, 581–582
    series systems, 582
    specifications, regulations and
        engineering, 571–573
    structures for, 577–582
    traffic-light control systems, 578–580
    types of, 570–577
lighting (four-wire) service, 433
lighting transformer, 427–428
lightning
    circuits, effect on, 358
    direct stroke, effects of, 356–357
    distance from, estimating, 357
    ground-gradient effect, 357
    live-line work, 364–365
    overcurrent, causes of, 343–344
    problem locations, 365–366
    safety tips, 357–358
    substations, 87
    thunderstorms, types of, 356
    tracking storms, 358
light pollution, 572
limbing a tree, 635–636
limestone, resistivity, 370
limiters, 445
line angle, measuring, 538
line crews, distribution system
        management, 86
line drop compensation, 460

line equipment, preventative
        maintenance on, 169
line inspections, types of, 168–169
line loss, voltage drop, 53
lines. *See also* cables; distribution
        systems; transmission systems
    designing, 171–172, 217–218
    staking out, 172–173
line-to-ground clearance, transmission
        lines, 175
line-to-ground faults, 93
line traps, 98
line work
    overview, 1–2, 6
    risk in, 6–7
linkit press, 271
link sticks, 566
lip-roller block, 239
liquid filled fuses, 324
live-front transformer, grounding, 398
live-lines. *See* hot-line work
loads, electric
    calculating, meter spin test, 599–600
    connecting in parallel, 141–145
    connecting in series, 138–141
    on the guy, 191–193
    hot-line tools, calculating, 559–561
    load break disconnects, 312
    load break elbows, 329–331
    load bust tool, 322
    load currents, 56, 407–409
    load factor, revenue metering, 587
    transformers, 411
loads, lifting. *See also* rigging
    chains, 521–522
    collapsible bull wheel, 527
    derating factors, 519
    general precautions, 527
    hand lines, 528
    rope blocks, 530–532
    slings, 532–534
    web slings, 526
    wire rope, 520
    working load limits, overview,
        510–511
load trees, concrete poles, 179
loam, resistivity, 370
locking snap hooks, 39
lockout/tagout, 3, 4
lockout/tagout procedure, 16–17
lock to lock rule, rubber gloves, 7
long ties, 208
loop primary, 83–84
loop protection, 353

loop splice, wire rope termination, 521
low-pressure oil-filled (LPOF) cables, 273
low-pressure sodium lamps, 575
low-resistance faults, underground cable, 287–288
LTO, isolation certification, 16
lumens, outdoor lighting, 573
luminaires, outdoor lighting, 570–572
luminous effect, electrical energy, 67

M

magnesium, 372
magnet, breaking, 589
magnetic fields. See also transformers
    currents and, 111
    electrical energy, 67
    EMF, measurement of, 160–161
    exposure levels and risks, 161–163
    induction, 114, 145, 147–148, 387
    overview, 158
    power, generation of, 109–110
    reducing the strength of EMF, 160
    sources of power frequency, 159–160
    three-phase currents, 126
main bus, 94
maintenance
    line inspections, 168
    planning for, 167–168
    types of, 169–170
management
    commitment, 4
    and supervisors, 4
management culture, safety, 4–5
mandrel, flexible, 239
Manual on Uniform Traffic Control Devices (MUTCD), 578
maps, 309–310
mark up, isolation certification, 16
marline knot, 518
marsh, resistivity, 370
maximum permissible exposure (MPE), 162
maximum sag, 541
mechanical protective footwear, 11
medical and exposure records, 3
medium-voltage bus, 94, 95, 100
meetings, safety, 4, 5, 6
megavolt-amperes (MVA), 64
megawatts (MW), 64
meggers, 58
megohmmeter, cable fault, 286
mercury, hazards of, 584

mercury-vapor lamps, 575
messenger cable, 214, 256–257
metal-clad circuit breakers, 314–315
metal-halide lamps, 575
metallic armor, 277
metallic glass, 408
metal-oxide varistors (MOV), 362
meter bases, 217
metered charges, 586
metering. See also revenue metering
    meter equipment, 100
    meter multiplier, 590
    meter readings, substation inspection, 102
    substation, 81, 87, 88
methyl hydrate (methanol), 13
mho, conductance, 60
microteslas, 159, 160–161
microwave
    communications, 80
    control cables, 99
    dishes, 162
milligauss (mG), 159
milliwatts per squared centimetre (mV/cm^2), 162
minimum approach distance
    as a barrier, 163–165
    as electrical energy control, 14–16
    hot-line tools, 558
    hot-line work, 545–546
    typical, 16
minimum charge, 586
minimum safe distance, calculating, 162
minimum working distance, tree pruning, 618–619
mobile substations, 88, 89–90
motors, 60, 119
multicircuit configurations, 160
multicircuit poles, guying corners, 193
multimeters, 51, 54, 55, 58
multiple-leg bridle slings, 532–534
multiple-let slings, 534
multiplier, current transformer, 607–608
multipoint junctions, 332–334
multi-sheave stringing blocks, 210, 223

N

National Manual on Uniform Traffic Control Devices (MUTCD), 18
natural target pruning, trees, 613
net metering, 597

network protection, 354
network protectors, urban grids, 444–445
network transformers, urban grids, 444–445
neutral connections
    neutralizing transformers, 440
    neutral potential, 368–369
    overview, 368–369
    transformers, 414
    working with, 152–154
    wye and delta grounding, 132–133, 376–377
new employees, orientation of, 5
noise, bundled conductors, 264
noisy ground, power quality, 450
no-load loss, 408
no-load tap changers, 402–404
nominal voltage (V), 53
nonexpulsion (NX) current-limiting fuses, 324
nonexpulsion (NX) switches, 331
nonload break disconnect, 310–311
normal-duty class arrestor, 362
normal high voltage, 53
normal low voltage, 53
normal mode, power quality, 450
nuclear generations, 70–71
NX switches, 331
nylon rope
    characteristics of, 511–513
    safety factor, 511
nylon slings, 226

O

occupational health, 4, 5
Occupational Safety and Health Administration (OSHA), 3
off-circuit tap changers, 95
off-load tap changers, 402–404
ohmmeters, 58
ohms
    capacitive reactance, 117
    defined, 51, 57
    effect of frequency on reactance, 119
    inductive reactance, 114
    power, 64
Ohm's law
    AC circuits and, 113
    overview, 59–60
    power formulas, 63–64
    working with neutrals, 153

oil circuit breakers, 93, 101, 313–314

oil-filled cables, 273

oil lines, color-coded marking system, 237

oil-pumping stations, 249

One Call Center, 29

one-line distribution drawings, 309

on-load tap changers, 95

open circuit

    detecting, 288

    distribution circuit protection, 340–345

    transmission circuit protection, 338–340

open neutrals, 153

open-wire bus, 214, 264–265

operating handles, disconnect switch, 302–303

operations and maintenance (O&M), 167

operations drawings, 102

OPGW (optical ground wires), fiber optic cables, 292

optical devices, 98

optical fibers. See fiber optic cables

optical ground wires (OPGW), 225, 292

orientation, new employee, 5

outdoor environment, 2

outdoor lighting

    airport runways, 580–581

    area lighting, 577–582

    ballasts, 576–577

    cascading control system, 581–582

    gaseous-discharge lamps, 575–577

    halogen lamps, 574–575

    hazards of, 583–584

    incandescent lamps, 573–575

    light-emitting diodes, 573–575

    lumens, 573

    luminaires, 571

    maintenance and troubleshooting, 583

    pilot-wire control systems, 581–582

    series systems, 582

    specifications, regulations and engineering, 571–573

    structures for, 577–582

    traffic-light control systems, 578–580

    types of, 570–577

out of phase, 112

outriggers, 155, 491–493

overcurrent. See also overvoltage surge; power quality

    causes of, 343–344

    grounding protection, 369

    transformers, fuse protection, 412–413

    trouble calls, 355

overhead conductors

    aluminum, advantages of, 257

    bare conductors, types of, 257–262

    bundled conductors, 263–264

    connections, 265–266, 271–273

    energized circuit connections, 266–268

    exothermic connections, 271

    implosive sleeves, 271

    insulation, 61

    secondary bus and service drops, 264–265

    small conductors, 268

    splices, procedures, 265, 268–271

    splices, tools for, 271–273

    tree wire and aerial spacer cable, 262

    types of, 257

overhead powerline construction. See also overhead conductors

    designing a line, 171–172

    joint use with other utilities, 173–174

    line inspections, 168–169

    maintenance, 167–168, 169–170

    planning a new transmission line, 174–175

    pole lines

        anchors, 195–196

        conductors, stringing, 204

        crossarms, installing, 199–200

        equipment on poles, 198–199

        facing wood poles, 190

        framing, 196–198

        guying a pole, 190–195

        insulators, installing, 200–204

        pole foundations, 182–183

        pole selection, 176–182

        roadways, stringing across, 205

        sagging conductors, 205–207

        secondary bus, stringing, 214–215

        service to customers, stringing, 216–217

        setting poles, 185–187

        setting poles in energized lines, 187–190

        slack stringing, 204–205

        storm hardening, design for, 184

        tension stringing, 209–214

        tying/clipping-in on distribution lines, 208–209

        in wildfire vulnerable areas, constructing, 184–185

    project planning, 167–168, 170–171

    staking a project, 172–173

    transmission-line refurbishment, 174

    transmission lines

        assembling/erecting structures, 218–221

        clearing the right-of-way, 218

        commissioning, 228–229

        constructing structure foundations, 218

        dead ending conductors, 227

        deciding the route, 217

        designing the line, 217–218

        hang insulators and stringing blocks, 221–223

        major projects with existing lines, 227–228

        rider poles, installing, 223

        sag/clip-in conductors, 226–227

        string conductors, 223–225

        wind turbines, installing, 229–231

overhead systems. See also overhead conductors

    electrical distribution of, 84–86

    no-power, troubleshooting, 468–469

overheating, connections, 101

overheating, transformers, 411

overload protection, booms, 503

overshoe boots, 11

overvoltage surge. See also lightning; power quality

    causes of, 355–356

    convection and frontal thunderstorm, 356

    distribution circuit protection, 340–345

    effects of, 359–360

    equipment voltage-time curves, 364

    hot-line work, protection from, 546–547

    insulation coordination, 364

    live-line work, shorting insulators, 364–365

    minimum approach distance, 164

    problem locations, 365–366

    shield wire, 364

    surge arrestors, 360–362

    switching surges, 358–359

    transformers, 413–414

oxygen deficiency, enclosed spaces, 19–23
ozone gas (O₃), 255

P

pad foundations, 218, 219, 236–237, 241
pad-mount switchgear, 85
parallel circuit
 characteristics of, 142
 defined, 141
 grounding principle, 375–376
 total resistance, 143
parallel-connected loads, 144–145
parallel-connected three-phase systems, 131–133
parallel connections, transformers, 418
paralleling conductors, 113
parallel shunt, 375
partial discharge, 277
partial-range current-limiting fuse, 325
patent anchors, 195
peak demand, 586–587
peak load currents, 56–57
peak value, AC and voltage, 111
penstocks, 69
pentachlorophenol, wood pole, 176
PE (polyethylene) conduit, 236
permanent faults, 343
permits, enclosed space entry, 21
personal communications services (PCS), 162
personal flotation devices, 25
personal protective equipment
 arc-rated (AR) clothing, 12
 chain saw operation, 628
 chemical hazards, 12–13
 eye protection, 10–11
 footwear, 11–12
 head protection, 10
 hearing protection, 12
 rubber gloves and sleeves, 7–10
personal protective grounds
 applying, procedures for, 388–393
 bonding principle, 381–386
 electromagnetic induction, 386–387
 grounding hazards, 393–395
 grounding principle, 375–380
 reasons for, 374–375
 underground cable, 395–399
phase, 112–113
phase designations, 127
phase rotation, 128

phases 120° apart, 126–127
phase-shifting transformers (PST), 97
phase-to-ground exposure, 16, 163
phase-to-ground fault, 133, 343–344
phase-to-ground resistance, 101
phase-to-ground voltage, 163
phase-to-neutral fault, 343–344
phase-to-neutral systems, transformer connections, 417
phase-to-neutral voltage, 132
phase-to-phase connections, 130, 376–377
phase-to-phase exposure, 16, 163
phase-to-phase resistance, 101
phase-to-phase voltage, 16, 132, 163
phasing, confirmation of, 304
phasing sticks, 52
phasing testers, 52, 304–307
phasing tools, 52
phi (F), 125
phosphor, hazards of, 584
photocells, outdoor lighting, 581–582
photovoltaic cells, 75–76
pier foundations, 218, 219
pigtail ties, 208
pike poles, setting, 185
pilot lines (ropes), 210–211, 224
pilot-wire control systems, 581–582
pin-type insulator covers, 555
pin-type insulators, 202–203
pipeline cathodic protection hazard, 373
planning engineers and technicians, 86, 170
planning maps, 310
plants, capital project planning, 170
plastic cover-up equipment, 554–555
plate anchors, 195
point-of-work grounds, 383–384
polar compounds, 101
pole climbing climbers, 37
pole/down-grounds, 218
pole lines, constructing
 anchors, 195–196
 assembling and erecting structures, 218–221
 conductors, stringing, 204
 crossarms, installing, 199–200
 equipment on poles, 198–199
 facing wood poles, 190
 framing, 196–198
 guying a pole, 190–195
 insulators, installing, 200–204
 pole foundations, 182–183

pole selection, 176–182
 roadways, stringing across, 205
 sagging conductors, 205–207
 secondary bus, stringing, 214–215
 service to customers, stringing, 216–217
 setting poles, 185–187
 setting poles in energized lines, 187–190
 slack stringing, 204–205
 tension stringing, 209–214
 tying/clipping-in on distribution lines, 208–209
poles
 body belt sizing, 38–39
 climbers, sizing and maintaining of, 37–38
 climbing techniques, 36
 falling from, 31
 falling with, 31–32
 fires, 46–47
 foundations, 182–183
 grounding hazards, 395
 pole straps, 39
 wood, grounding hazards, 394
 work positioning, 40
pole setting, energized distribution lines, 149
pole straps, 39
pole tongs, energized distribution lines, 151
pole tops, rescue from, 44–46
policies, safety, 3–5. *See also* safety
polychlorinated biphenyls (PCBs), 12, 583
polydacron rope, 513
polyester rope, 511–513
polyethylene (PE) conduit, 236
polyethylene rope, 513
polymeric insulators, 201–202, 204, 221–223, 275
polyphase metering, 601–604
 insulation tester method, 603
 three-phase metering, 601–604
 voltmeter method, 604
polypropylene rope, 511–513, 532
poly ropes, 513
polyvinyl chloride (PVC), 236
pools, electrical, 81
porcelain insulators, 194, 201–202, 203, 204, 221–223
porcelain pin insulators, 202
portable generator, backfeed grounding, 395

portable protective gap (PPG), 546

post insulators, 204, 221–223

potential coil, revenue meter, 588–589

potential energy, 68

potential gradients. *See* voltage gradients

potential tests, 51

potential transformers, 588
current transformers and, 98–99
overview, 458–459
protective relaying, 338
revenue meters, 594–596
substations and, 87
transformer-rated meters, 604–609

potheads, underground transmission lines, 246

power, overview
calculations, 134
electrical, 63–65
generation of, 109–110
generation of three-phase, 126

power cables, 95, 99

power factors, 120–121, 587

power flow control reactors, 97

power fuses, substation, 88

powerline carrier communication (PLCC), 98

powerlines. *See also* overhead powerline construction; overhead systems; transmission lines; underground powerline construction; underground systems
Hertz, 158
vibration and galloping, conductors, 256–257

power pools, 81

power quality
capacitors, 463–467
conductors, 453
daily change, 453
distribution feeder voltage, 455–456
distribution substation voltage, 454–456
feeder voltage regulators, 456–463
ferroresonance, 475–478
harmonic interference, 472–473
overview of, 450
radio and television interference, 482–483
reactance, 453
seasonal change, 453

tingle voltage, 478–482
transmission lines, voltage control, 454
troubleshooting
ammeter/voltmeter use, 467–468
high-voltage, 471–472
individual customer service, 467
low-voltage, 470–471
no-power, overhead system, 468–469
no-power, underground system, 469–470
voltage
control and regulation, 452
flicker, 473–475
as measure of, 451–452
voltage drop, 452

power transformer, 427

power triangles, 120

PPG (portable protective gap), 546

predictive maintenance, 101, 169

pre-job briefing, 14

prepayment metering, 597

prescription safety glasses, 10

pressed sleeve loop back thimble attachment, 521

pressure leaks, 101

preventative maintenance (PM), 169

primary feeders, distribution systems, 82, 83

primary metering, revenue meters, 594–596

primary networks, 84

primary step-down installations, voltage conversions, 439–440

procedures, safety, 3–5. *See also* safety

proposed excavation, color-coded marking system, 174, 237

protective equipment, 4, 313–314, 377–380

protective grounding hardware, 377–380

protective switchgear, 299. *See also* switchgear

pruning, trees, 612–614, 628–631

public safety, 4, 5, 29–30

puller monitors, electric, 240–241

pulling cable, underground distribution lines, 237–241

pulling ropes, 210, 213, 224, 238

pulling tension, 210, 238

pulling wire, 238

pulse closer fault interrupter, 316

pumped storage, 77

puncture, voltage surge, 359–360

push-pull braces, 190

PVC (polyvinyl chloride), 236

Q

quadrant block, 239

R

radar microwave, 159

radial feeders, 83

radial systems, 83

radio, control cables, 99

radio frequency (RF) waves, 162

radio interference, 482–483

radio noise, bundled conductors, 264

railroad crossings, guard structures, 223

rain, effect on insulators, 61

raising horse, setting poles, 185

ratchet chain hoist, 523

rated breaking strength (RBS), 190

rated tensile strength (RTS), 205, 256–257

ratio of the current transformer, 607

ratio test, transformers, 402

reactance
in AC circuits, 113
frequency on, 119
power quality, 453

reactive power, 120, 594–597

reactors
air core, 96
substations and, 87, 96–97
underground distribution systems, 241

real power, 119, 594–597

reclosers, 299
automatic reclosers, 316, 546
bypassing, 326–327
cutout mounted, 317
downstream reclosers, 350–351
electronic reclosers, 319–320
frame size, 347–348
hydraulic reclosers, 317–319
operating, 314
recloser protection zone, 348
recloser speed and sequence, 349–350
substations, 102
TCC curve, 349–350

recognition, safety performance, 5
recovery voltage, 139
rectifiers, DC transmission line, 122
reef knot, 516, 519
refurbishment, transmission-line, 174
regulation, voltage, 451–452
regulators
    distribution feeder voltage, power
        quality and, 456
    distribution substation voltage,
        power quality, 456
    equipment on poles, 198
    operating, 461
    power quality and, 456–463
    reverse feed, 463
    substations and, 87
    troubleshooting, 462–463
relay equipment, 87, 100, 338–340
reliability-centered maintenance
        (RCM), 101, 169
remote control, substation operation,
        102
remote terminal units (RTUs), 301
reporting incidents and accidents, 6
resagging, existing transition lines,
        227–228
rescue devices, hand lines, 528
resistance
    in AC circuits, 113
    conductors, 253
    electrical, 57–62
    ground rods, 370–371
    heat from, 63
    Joule's law and, 63
    measurement of, 51
    Ohm's law, 59–60
    parallel circuits, 142
    power, 64
    series circuit, 139
    soils, differences in, 369
    transformers, 406
    voltage drop, 53
resistive loads, 113
resonance, ferroresonance, 475–478
restringing, existing transition lines,
        228
retesting, rubber goods, 555–556
return-eye web sling, 526
return flows, delta circuits, 131
return-wave methods, sagging
        conductors, 207
revenue metering
    cost variables, 586–588
    polyphase metering, 601–604

single-phase metering, 597–600
transformer-rated metering, 604–609
types of, 586, 594–597
reverse feed, regulator, 463
RF/MW transmitters, induced current,
        162
rider poles, installing, 223
rigging
    anchor pulling eyes, 526–527
    blocks, 523, 525–526
    chains, using, 521–522
    collapsible bull wheel, 527
    fiber ropes
        knots, types of, 513–519
        splicing, 518–519
        strength of, 513
        types of, 511–513
    friction, 525–526
    lifting loads
        capstan hoist, 530
        general precautions, 527
        hand lines, 528
        rope blocks, 530–532
        slings, 532–534
    ratchet chain hoist, 523
    tensioned conductors
        bisect tension, corners, 538–539
        calculating conductor tension,
            535–536
        compression sleeves, 540
        conductor length in span, 536
        conductor weight in span,
            534–535
        corner, moving to, 539
        line angle, measuring, 538
        sagging facts, 541
        tension on loaded down guy,
            536–537
        vertical load, down-pull, 537–538
    underground cable, pulling, 238–239
    web hoist, 523
    web slings, 526
    wire rope
        hazards of, 520
        temporary eye, creating, 520
        terminations, 520
        working load limits, 520
    working load limits, 510–511
right-of-way, 218
rigid (plastic) cover-up equipment,
        554–555
ring bus, 94
risers, 246, 396
risks, 5, 6–7, 13–40. See also safety

roadways
    guard structures, 223
    lighting, 571
    minimum ground clearance, 217
    stringing across, 205
    traffic protection and, 18–19
rock anchors, 196
rock soil, anchors, 196
rod gap surge protection, 361
rope ladders, 33
ropes. See also rigging
    block size and, 525
    climbing trees, 622–625
    fiber ropes
        splicing, 518–519
        strength of, 513
        types of, 511–513
    hot-line rope, using, 549
    knots, types of, 513–519
    rope blocks, lifting with, 530–532
    wire rope
        hazards of, 520
        temporary eye, creating, 520
        terminations, 520
        working load limit, 520
    working load limits, 510–511, 520
roping a tree limb, 628–629
rotation gear failure, 501
RTUs (remote terminal units),
        switchgear, 301
rubber cover-up equipment, 554–555
rubber gloves
    color code, 8
    high-voltage environment, 146
    leakage current, 144
    minimum approach distance, 164,
        545–546
    overview of use, 551–554
    parallel contact, 141
    as personal protective equipment,
        7–10
    poles in energized distribution lines,
        151
    preparation for use, 556–557
    safety and, 14
    series contact, 141
    types of, 8–9
rubber sleeves, 553–554
rugged work, 2
rules and regulations, 3, 4
ruling span, 205, 541
running boards, 224
running bowline knot, 513, 519
rural substations, 88

S

saddle, tree climbing, 622–625

saddle clamps, 209, 226, 550

safety. See also personal protective
  grounds
  chains, working with, 521–522
  current transformers, 607–608
  driving, utility trucks, 488
  electrical current, 56
  electrical potential, 52
  electrical resistance and, 62
  electrical retesting, rubber goods,
    555–556
  electric field induction, 145–146
  EMF exposure, 161–163
  flash, 57
  ground faults, 149
  grounding hazards, 393–395
  guy wires, cutting of, 194
  hand lines, 528
  hot-line work
    automatic reclosing, blocking,
      546
    barehand work, procedures and
      hazards, 563–568
    choice of, 544
    defined, 543–544
    hot-line rope, using, 549
    hot secondary, 550–551
    insulator testing, 547–548
    minimum approach distance
      (MAD), 545–546
    rubber gloves, hazards, 557–558
    rubber gloves, overview, 551–554
    rubber gloves, preparation for
      use, 556–557
    safety hardware, 550
    second point of contact, 544
    temporary jumpers, 548–549
    tool work, procedures and
      hazards, 558–563
    voltage surge protection,
      transmission lines, 546–547
  loads in parallel, 144–145
  magnetic field induction, 147–148
  managing, 3–5
  minimum approach distance as
    barrier, 163–165
  noninsulated booms, 493–497
  outdoor lighting hazards, 583–584
  poles in energized distribution lines,
    149–151
  responsibilities and expectations, 2–3

revenue meter installation and
  removal, 609

safety committee, 3

safety factor, working load limits,
  510–511, 532

safety meetings, 5

series connections, 140–141, 579

setting poles in energized lines,
  187–190

slack stringing, 204–205

static electricity, 145

step and touch potentials, 152

stringing across roadways, 205

substation entry, 100

substation work, 104

tension stringing, 209–214

transformer hazards, 446–448

transmission line projects, 227–228

tree work
  chain saw operation, 626–628
  common hazards, 617
  cutting trees, 616
  danger trees, 614–615
  near electrical circuits, 618–622
  grubbing, 616
  hanger hazards, 630
  planting trees, 617
  tree climbing, 622–625, 635–636
  tree felling, 632–634, 635
  tree pruning, 628–631
  tree removal hazards, 634–635
  tree trunk bucking, 636
  treetop rescue procedure,
    620–622
  vegetation management, 611–617

vehicle grounding and bonding,
  154–158

vehicles, stabilizing, 491–493

wire rope hazards, 520

working with neutrals, 153–154

safety belts, 31

safety data sheets (SDS), 12

safety glasses, 10

safety harness, 552

safety strap, 552

safe work observations, 4

sag, 171

sag boards, 206

sag conductors
  bare conductors, types of, 257–262
  conductor selection, 252
  conductor weight, calculating,
    534–535
  length in span, calculating, 536

overview, 205–207, 226–227
  sagging facts, 541
  tension, calculating, 535–536
  vibration and galloping, 256–257

SALCOR straight-line hose, 555

sand, resistivity, 370

sandstone, anchors, 196

sandstone, resistivity, 370

sandy soil, anchors, 195

satellite, control cables, 99

S-base meters, 591

SCADA (supervisory control and data
  acquisition) systems, 301

SCFF (self-contained, fluid-filled) cable,
  273

SCFF (self-contained, fluid-filled) pipe
  cable, 245

scheduled maintenance, 169

screw anchors, 195, 503–505

screw-in foundations, 219

seasonal changes, power quality, 453

secondary, hot-line work safety,
  550–551

secondary bus
  overhead conductors, 264–265
  stringing, 214–215
  underground distribution systems,
    237
  urban grids, 444–445

secondary cables, 95, 275

secondary class arrestor, 362

secondary fault current, transformers,
  408

secondary metering, revenue meters,
  594–596

secondary network (banked
  secondary), 445–446

secondary systems, distribution
  systems, 82, 83, 441–445

secondary voltage, three-phase
  arrangements, 431–436

second point of contact, 62, 544

sectionalizers, 320, 353

self-contained, fluid-filled (SCFF) cable,
  245, 273

self-contained meters, 594

self-dampening ACSR conductors, 261

self-inductance, 114

self-securing cable end caps, 555

semiconducting glaze (SCG) insulators,
  202

semiconducting shield, 276–277

series, loads placed in, 139

series capacitors, 465

series circuit
  airport runway lighting, 580–581
  characteristics of, 139
  defined, 138
  hazards of, 583
  outdoor lighting systems, 582
series-connected three-phase systems, 130–131
service drops, overhead conductors, 264–265
setting depth, wood poles, 182
setting poles, 149–151, 185–187.
        See also overhead powerline construction
sewer lines, color-coded marking system, 174, 237
shale, anchors, 196
sheet bend knot, 516
shielding, 160, 276–277, 364
shield wires, 225
Shigo method, tree pruning, 613
short-circuit
  overview, 57
  parallel circuits, 142
  transformers, primary coil, 408
  transmission circuit protection, 338–340
  underground cable, locating, 287–288
shunt capacitors, 465
shunt reactors, 96, 115
side arms, 200
side guys, 190
sidewalk guys, 190
sidewalks, minimum ground clearance, 217
silver, 372
sine waves
  AC power, 110–111
  amperage lagging voltage, 115
  current leads the voltage, 117–118
  phases 120° apart, 126–127
  values of AC and voltage, 111–112
single-line diagrams, 102
single phase, voltage standards, 53
single-phase AC, power formulas, 134
single-phase kV, 135
single-phase metering, 597–600
  insulation tester method, 600
  three-wire service, 597
  two-wire service, 597
  voltmeter method, 600
single-phase service, troubleshooting, 436

single-phase transformer connections, 415–418
single-phase wye circuit, return current, 153
single-point grounding, 388–390
single-pole wishbone, 218–221
single wire earth return (SWER), 133
single-wrap choker hitch, 534
skin effect, conductors, 256
slack stringing, 204–205
slack stringing operation, traffic hazards and, 19
sleeves, 7–10, 14, 271, 540, 553–554
slide-on crossarm guard, 554–555
slings, 511, 526, 532–534. See also rigging
slip-resistant footwear, 12
slope angles, trenches, 23–24
slug anchors, 195
smart metering, 253, 591–593. See also revenue metering
snatch blocks, 239, 524, 525–526, 529
snubbing hitch knot, 515
socket-based meters, 217, 591, 608
sodium, hazards of, 584
soil conditions, 195–196, 369, 394
soils, excavations and trenches, 23–24
solar energy, 72
  into electrical energy, 74–78
solar PV utility scale power plant, 75–76
solar thermal power plants, 74–75
solid braid ropes, 512–513
solid-material-filled power fuses, 323
solid-state breakers, 451
solid-state revenue meters, 591–593
spacer dampers, 226–227, 228, 263
spark gap surge protection, 361
sparkover, 359–360
specialists' culture, 5
specifications
  framing, 196, 198
  sag charts, 205
Spelter attachment, wire rope termination, 521
spills response plans, 5
splices
  cable shielding, 277
  fiber optic cable, 294–295
  overhead conductors, 265, 268–271
  underground cables, 242, 246, 277–280
  vibration-resistant conductors, 261
  wire rope termination, 521

split bolt connectors, 217
sports field lighting, 571, 577–578
square-based latticed towers, 218
square knot, 516, 519
staking a project, 172–173
staking data, guying a pole, 190–195
STATCOM (static synchronous compensator), 122
static electricity, 12, 33, 145
static VAR compensator reactors (SVCs), 96, 122
station class arrestor, 362
station service, 98, 99
station yards, 93
statistical analysis, incident summary reports, 4, 5
stator magnets, 110
Stat-X First Responder, 23
steam, power generation, 68, 69–70
steam lines, color-coded marking system, 237
steel. See also overhead powerline construction
  amorphous steel, 408
  bare conductors, types of, 257–262
  braces, 200
  crossarms, 199
  galvanic series, 372
  guy wires, 190
  poles, 152, 179–180, 182, 218, 395
  steel-lattice structures, transmission lines, 218–221
  towers, grounding, 218
  transformer core configurations, 401
step and touch potentials, 152
step-down transmission substations, 88
step potentials. See also voltage gradients
  defined, 152
  pole setting safety and, 189
  tree work, 619
step-up transmission substations, 88
step-voltage regulator, 456–463
storm guys, 190, 192
storm restoration, 175. See also lightning
strand cable, vibration and, 256–257
strap, safety, 552
strategic planning, 4
stray-current corrosion, 373
stray voltage, 478–482
street lights, 571, 581–582. See also outdoor lighting

streets, minimum ground clearance, 217
strength
  composite poles, 179
  pole, 176
  steel poles, 179–180
  suspension insulators, 203
  wire rope, 520
stringing
  conductors, 204
  across roadways, 205
  sag, 541
  secondary bus, 214–215
  service to customers, 216–217
stringing blocks, 210, 221–226
structure foundations, constructing, 218
structure numbers, 308
strut guys, 190
stubs, enclosed spaces, 19–23
submarine cables, types of, 275
submarine lines, 234
subspan galloping conductors, 256–257
substations
  angular displacement, substation transformers, 430
  collector, 91
  construction of, 92–93
    auxiliary equipment, wiring, 107
    construction drawings, 105
    equipotential grid, constructing, 105–106
    foundation plan, 105
    heavy substation equipment, installing, 106
    locating, 104
    station structures, erecting, 106
  distribution circuit protection, 340–345
  distribution voltage, 454–455
  elements of
    capacitors, 97
    control cables, 99
    high-voltage and medium-voltage bus, 94, 95
    inductors, 98
    meter and relay equipment, 100
    phase-shifting transformers, 97
    potential and current transformers, 98–99
    power cables, 95
    reactors, 96–97
    station service, 98
    station yards, 93

  switchgear, 93–94
  transformers, 95
  voltage regulation equipment, 95
  wave traps, 98
  entering, 100
  feeder protection, 344
  hazards of, 104
  inspecting, 101–102
  maintenance programs, 101
  metering at, 81
  mobile, 88, 89–90
  operating, 102–103
  overview of, 87
  power flow, 100
  preventative maintenance on, 169
  rescue from, 44–46
  storm-hardening, 92–93
  switchgear identification, 307–310
  system capacitors, 466
  transmission system, 79–80
  transmission voltage control, 454
  types of, 88–91
subtractive transformer polarity, 404–406
subtransmission, 82, 95, 454–455
subtransmission line voltages (kV), 54
sulfur hexafluoride (SF$_6$) gas, 92, 93, 282, 313–314
Supervisory Control and Data Acquisition (SCADA), 82, 87, 301
surge arrestors
  delta-connected systems, 130
  overview of, 360–362
  substations, 87, 88, 102
  transformers, 412–414
  wye-connected systems, 132
surgeon's knot, 516
surges. See also power quality; surge arrestors
  ground protection, 369
  high-voltage, 202
  surge generators, 288
surveys, staking a project, 172
survey tape, color-coded marking system, 237
survival suits, 25
suspension insulator strings, swings of, 218
suspension systems, 31
suspension trauma, 41
suspension-type insulators, 203
swaged socket wire rope termination, 521

swampy soil, anchors, 195
switched capacitors, 465
switchgear, 299
  arc hazards, 299–300
  automatic reclosing, blocking, 546
  bypassing protective switchgear, 326–327
  circuit breakers, 313–314
  current-limiting fuses, 324–326
  disconnect switch operation, 302–303
  distribution cutouts, 320–322
  distribution protection, 347
  distribution substations, 82
  equipment on poles, 198
  fuses, overview, 323–324
  isolating switchgear, operation, 310–313
  locating, 307–310
  nomenclature, 309
  nonreclose feature, breakers and reclosers, 327
  operation of, 302
  phasing, confirmation of, 304
  phasing tester, use of, 304–307
  reclosers, 314
  remote control of, 301
  SCADA systems, 301
  sectionalizers, 320
  substation maintenance, 101
  substation power flow, 100
  substations, 88
  substations and, 93–94
  telescopic switch stick operation, 304
  three-pole switches, 301
  transformer fuse protection, 412
  types of, 299
  underground systems
    cable-separable connectors (elbows), 329–331
    dry-well canister fuses, 332–333
    grounding, 396–398
    multipoint junctions, 332–334
    Nx switches, 331
    types of, 234–235
    types of switchgear, 328–329
    under-oil bayonet-style fuses, 332
    undercover-style switchgear, 335
switching
  isolation certification, 17
  overvoltage, 356
  surge, 164, 358–359
  switching orders, 17–18
switching stations. See substations

switchyards, 79, 80
swivel joints, 213
symbols, electrical schematic drawings, 309
synchronized generators, 113
synchronous condenser, 122
synthetic ropes, types of, 511–513
synthetic web sling, safety factor, 511
system controls, substations, 100, 102

T

tailboard conference plans, 15
tailraces, 69
tampers, pole foundations, 183
tap changers, 95, 101
tap connections, 266
taps, transformers, 402–404
tar, wood pole treatment, 176
target method, sagging conductors, 206
taut-line hitch knot, 516, 625
TCC (time current characteristic) curve, 349–350
TDR (time-domain reflectometer), 288
Techweld Cadweld, 271
telecommunications systems, 80, 98, 100, 173
telescopic switch stick operation, 304
television interference, 482–483
temperature
    resistivity and, 369
    sagging conductors, 206–207
    transformer overheating, 411
    working in heat and cold, 24–25
temporary jumpers, hot-line work, 548–549
temporary survey tape, color-coded marking system, 174, 237
tension
    bisect tension, 525, 538–539
    conductor tension, calculating, 535–536
    guy wires, 191, 536–537
    load on hot-line tools, calculating, 560–561
    measuring, 206–207
    stringing, 19, 209–214
    underground cable, pulling and laying, 238
tensioners, 211
terminations
    cable shielding, 277
    implosive sleeves, 271

metering at, 81
midspan service termination, 264–265
overhead conductors, procedure for, 268–271
transformer polarity, 404–406
underground cables, 277–280
underground distribution lines, 242
underground transmission lines, 246, 248–249
teslas (T), 160
test constant, revenue meters, 590
T guards, 554–555
thermal effect, electrical energy, 67
thermal expansion, bare conductors, 257–262
thermal resistivity, underground cables, 280
thermographic (infrared) surveys, 101
Thermoweld, 271
thimble splice, wire rope termination, 521
three-phase, gang-operated switches, 312
three-phase circuits. See also three-phase connections; three-phase power
    characteristics of, 125–129
    delta or series-connected three-phase systems, 130–131
    generation of power, 126
    overview of, 125
    phase designations, 127
    phase rotation, 128
    phases 120° apart, 126–127
    three-phase four-wire, voltage standards, 53
    three-phase gang-operated switch, 310–311
    three-phase lines, 160
    voltage drop, 254
    wye and delta systems, 129
    wye circuits, return currents, 153
    wye-or parallel-connected three-phase systems, 131–133
three-phase connections. See also three-phase circuits; three-phase power
    overview, 128–129
    transformers
        angular displacement, substation transformers, 430
        angular displacement, wye-delta and delta-wye, 428–430

backfeed, 447
banking, 420–422
delta-delta banks, 424
delta-wye banks, 425
open-wye/open delta, open-delta/open-delta, 425–427
overview, 419–420
parallel banks, 428
primary delta connections, 422
primary wye connections, 422–423
wye-delta banks, 424–425
wye or delta connections, 422
wye-wye banks, 423
three-phase power
generation of, 126
overview, 134–135
secondary voltage arrangements, 431–436
substations and, 88
three-phase kV, 135
three-phase metering, 601–604
three-wire voltage standard, 53
voltage arrangements, transformers, 431–436
wye AC power formulas, 134
three-pole switches, operating, 301
three-wire current transformer, 606–607
three-wire 120/240 V secondary service, 153
throw lines, stringing blocks, 210
thumper, 288
thunderstorms, 161, 356, 358, 365. See also lightning
thyristor valves, 122
tidal power, 76
timber hitch knot, 518
time current characteristic (TCC) curve, 349–350
time delay switch, voltage regulator, 460
time-domain reflectometer (TDR), 288
time-of-use metering (TOU), 592
tingle voltage, 478–482
tires, ground gradient hazards, 155
T-link fuses, 323–324
TN grounding/earthing system, 367
tools
    hot-line tools, load calculation, 559–561
    hot-line tools, minimum approach distance, 558
    inspection/maintenance of, 5
    leakage current, 144

tools (*continued*)
   overhead splices and connections, 271–273
   phasing tools, 52
   underground distribution line construction, 242
total resistance, parallel circuits, 143
touch potentials
   defined, 152
   setting poles, energized lines, 187
   tree work, 619
TOU (time-of-use) metering, 592
tower loading diagrams, 218
tower structures, 32–33, 218
tower tops, rescue from, 44–46
traffic, safety, 5, 18–19. *See also* safety
traffic lights, 571, 578–580
trailers, ground gradient hazards, 155
training, safety, 3, 4, 5
transfer bus, 94
transformer fires, 46–47
transformer poles, grounding hazards, 395
transformer-rated meters, 594, 604–609
transformers
   components of, 401
   core configurations, 401
   current, effect on, 407–409
   current-limiting fuses, 324–326
   delta-connected systems, 130
   distribution substations, 82, 456
   efficiency, 410
   energy in, 402
   equipment on poles, 198
   European secondary systems, 441–443
   hazards of, 446–448, 551
   insulation resistance and continuity testing, 406
   losses and impedance, 409–410
   maintenance, 101
   other applications, 440
   parallel-connected loads, 144–145
   polarity, 404–406
   protection of, 410
   protective relaying, 338
   reactive power and, 120
   secondary network (banked secondary), 445–446
   single-phase connections, 415–418
   substations, 83, 87, 88, 95, 100, 101, 456
   taps, 402–404

three-phase, secondary voltage arrangements, 431–436
three-phase connections
   angular displacement, 428–431
   banking, 420–422
   delta-delta banks, 424
   delta-wye banks, 425
   open-wye/open delta, open-delta/open-delta, 425–427
   overview, 419–420
   parallel banks, 428
   primary delta connections, 422
   primary wye connections, 422–423
   wye-delta banks, 424–425
   wye or delta connections, 422
   wye-wye banks, 423
transformer stations. *See* substations
transient faults, 343
transits, 226
transmission line construction. *See also* transmission lines
   assembling/erecting structures, 218–221
   clearing the right-of-way, 218
   commissioning, 228–229
   constructing structure foundations, 218
   dead ending conductors, 227
   deciding the route, 217
   designing the line, 217–218
   hang insulators and stringing blocks, 221–223
   installing rider poles, 223
   installing wind turbines, 229–231
   major projects with existing lines, 227–228
   sag/clip-in conductors, 226–227
   string conductors, 223–225
   underground
      auxiliary equipment, 249
      cable classification, 245
      cable current-carrying capacity, 246
      pulling cable, 248
      reasons for, 244–245
      splices, 246
      terminations, 246, 248–249
      trenching and vaults, 246–247
transmission lines. *See also* transmission line construction
boom contact, 157
bracket grounding procedure, 393
circuit capacity rating, 252–253

equipotential grounding/bonding procedure, 391–392
overhead conductors
   aluminum, advantages of, 257
   bare conductors, types of, 257–262
   bundled conductors, 263–264
   energized circuit connections, 266–268
   exothermic connections, 271
   implosive sleeves, 271
   overview, 257
   secondary bus and service drops, 264–265
   small conductors, 268
   splices, 265, 268–271
   splices and connections, tools for, 271–273
   tap connections, 266
   tree wire and aerial spacer cable, 262
   planning a new, 174–175
   power cables, 95
   refurbishment, 174
   underground
      ampacity, 280
      cable design, 275–276
      cable failure, causes of, 277
      cable faults, locating, 286–288
      cable shielding, 276–277
      cable splicing and terminations, 277–280
      cable types, 273–275
      continuity, 282
      current in cable sheath, 280–281
      erratic secondary cable fault, 289
      faulted section, locating, 285–286
      insulating gas SF_6, 282
      insulation megohmmeter, 286
      locating and tracing, 283–285
      testing, 289
      trenches, identifying cable, 282–283
      voltage in cable sheath, 281
vibration and galloping, conductors, 256–257
voltages (kV), 54, 165
voltage surge protection, hot-line work, 546–547
transmission-line steel poles, 180–181. *See also* overhead powerline construction
transmission systems, 81, 338–340
transmission system substations, 79–80
transpositions, 147–148, 160

travel restraint systems, 31
tree climbing climbers, 37
tree crews (arborists)
    distribution system management, 86
tree-retardant cross-linked
        polyethylene (TR-XLPE), 275
tree saddle, 622–625
tree shock, 619–620
tree wire cable, 262
tree work
    chain saw operation, 626–628
    cutting trees, 616
    danger trees, 614–615
    distribution system management, 86
    near electrical circuits, 618–622
    grubbing, 616
    hanger hazards, 630
    hazards of, 617
    planting trees, 617
    tree climbing, 622–625, 635–636
    tree felling, 632–634, 635
    tree pruning, 628–631
    tree removal hazards, 634–635
    tree trunk bucking, 636
    treetop rescue procedures, 620–622
    vegetation management, 611–617
trench boxes, 23–24
trenches
    enclosed spaces, 19–23
    identifying cable, 282–283
    joint use, 173
    safety training, 5
    underground distribution lines,
        236–237
    underground transmission lines,
        246–247
    working in, 23–24
triangle-end type web sling, 526
trip coils, 348
triple-circuit poles, guy wires, 193
troubleshooting, 436–437
truck gloves, 7
trucks. See vehicles
true power, 119
TR-XLPE (tree-retardant cross-linked
        polyethylene), 275
TT grounding/earthing system, 367
turbines, electrical generation
        overview, 67–68
turns ratio, 401–402
    urban secondary network grid,
        444–445
    voltage conversion, working on,
        438–440

TVI (radio and television interference),
        482–483
twin-pole gulf port structures,
        218–221
twin-pole wood (H-frame) structures,
        218–221
twisted rope, 511–513
twisted (T-2) conductors, 261
two half hitches knot, 516

U

U bolts, 520
ultrasonic tests, 101
ultraviolet radiation, 159, 163, 583
undercover-style switchgear, 335
underground cable. See also
        underground powerline
        construction; underground
        systems
    ampacity, 280
    cable faults, locating, 286–288
    cable shielding, 276–277
    continuity, 282
    current in cable sheath, 280–281
    design of, 275–276
    erratic secondary cable fault, 289
    failure, causes of, 277
    faulted section, locating, 285–286
    insulation megohmmeter, using,
        286
    lightning strikes, 358
    locating and tracing, 283–285
    phasing testers, using, 306
    protective grounding, 395–399
    splicing and terminations, 277–280
    testing, 289
    trenches, identifying cable, 282–283
    types of, 273–275
    voltage in cable sheath, 281
underground powerline construction
    distribution lines
        cable, pulling and laying,
            237–241
        civil work, trenching, vaults and
            pads, 236–237
        installing equipment, 241
        maintenance on, 244
        reasons for, 233–234
        secondary, 242–244
        storm hardening, 244
        terminations and splices, 242
        tools and hardware, 242
        types of, 234–235

transmission lines
    auxiliary equipment, 249
    cable classifications, 245
    cable current-carrying capacity,
        246
    pulling cable, 248
    reasons for, 244–245
    splices, 246
    terminations, 246, 248–249
    trenching and vaults, 246–247
underground systems
    conductors, 61, 252, 253
    distribution switchgear
        cable-separable connectors
            (elbows), 329–331
        dry-well canister fuses, 332–333
        multipoint junctions, 332–334
        types of, 328–329
        undercover-style switchgear, 335
        under-oil bayonet-style fuses, 332
    electrical distribution of, 84–86
    electromagnetic fields, 160
    no-power, troubleshooting, 469–470
    residential distribution (URD), 235
    revenue metering, 594–596
    service wire size, 216
    substation feeders, 88
    transformer fuse protection, 412
    voltage conversions, 538–539
underground taps, 84
underground utility vaults, 19–23
underload tap changers (ULTC), 454
under-oil backup current-limiting
        fuses, 323
under-oil bayonet-style fuses, 332
under-oil expulsion fuses, 323
under-oil submersible partial-range
        current-limiting fuse, 325
uninterruptible power supply (UPS), 77
Universal Precautions, 40
uplift loads, 218
upstream documentation, 13
uranium, 71
urban secondary network grid,
        444–445
U.S. National Electrical Safety Code,
        217
utilities, color-coded marking systems,
        174
utility enclosures, underground
        distribution systems, 235
utility operating diagrams, 54
utility voltages, 53–54
utilization voltages (V), 54

V

VacuFuse interrupter, 317
vacuum circuit breakers, 93, 101,
    313–314
vacuum leaks, 101
valve-type arrestor, 361
vapors, enclosed spaces, 19–23
vaults, 20–21, 241, 246–247
vector diagrams, 120
vectors, transformer banking, 420–422
vegetation management, 611–617
vehicles
    aerial devices, operating, 505–508
    air-brake inspections, 487
    batteries, jump-starting, 487–488
    boom-equipped, stabilizing,
        491–493
    communication, hand signals,
        495–497
    daily inspection, 485–487
    digger derricks, 500–505
    driving hazards, 488
    grounding and bonding, 154–158,
        497, 499–500
    hand signals, 495–497
    hydraulic systems, 488–491
    insulated booms, electrical
        protection, 497–500
    noninsulated booms, electrical
        protection, 493–497
ventilation, enclosed spaces, 20, 22
Versa guards, 554–555
vertical load, down-pull, 537–538
vertical slings, 532–534
VHF (very high frequency), 159
vibration
    bare conductors, types of, 257–262
    conductor, 209, 256–257
    dampers, transmission lines, 226–227
    sagging conductors, 205
    tests for, 101
    vibration-resistant conductors, 261
VLF (very low frequency) AM radio,
    159
voltage. *See also* circuit protection;
        grounding; power quality;
        substations; transformers
    bonding principle, 381–386
    capacitive reactance and, 117
    conductors, voltage rating, 253
    conversion, working on, 438–440
    corona loss, 255
    current leads, 117–118

delta systems, 131
electrical potential, 51–54
electrical power, 63
electric fields, 159
electromagnetic induction,
        protective grounds, 386–387
electromotive force and, 51
flicker, 473–475
high-voltage, troubleshooting,
        471–472
high-voltage conversion, circuit
        protection, 355
hot-line work, surge protection,
        546–547
impedance and, 113
induction in AC circuits and, 114
low-voltage, troubleshooting,
        470–471
minimum working distance, 14–15
on a neutral, 152–154
Ohm's law, 59–60
outdoor lighting hazards, 583
overvoltage, 202, 355–356, 356–358,
        360–362
parallel circuit, 142
persona protective grounds, 375
phases 120° apart, 126
phasing testers, 304–307
power, 64
regulators, 82, 95, 101, 440,
        456–463
RF/MW transmitters, 162
rise and fall of, 110
step and touch potentials, 152
surge protection, transformers,
        413–414
three-phase circuits, 125–126
transformers, effect on current,
        407–409
transmission line, 79
underground cable sheath, 281
values during boom contact,
        156–157
values of, 111–112
voltage drop, 53, 139, 252, 253–254
in a wye system, 132
voltage-changing stations, 79
voltage gradients, 149–152, 155, 212,
        214, 394
voltage survey, transformers, 410
voltage transformers (VT), 51, 98–99,
        604–609
volt-amperes, 120
volt-amperes reactance (VAR), 87

volt-amperes reactive (VAR), 120
voltmeters, 51, 467–468
volts per meter (V/m), 159, 160
V towers, 218

W

washers, tap connections, 266
water
    crossings, underground systems, 245,
        248
    electricity and, 47
    insulators and, 61
    utilities, color-coded marking system,
        174, 237
water bowline knot, 513
watt-hours, revenue meters, 590
watt-seconds, 63
watts (W), 51, 63, 119. *See also*
        revenue metering
wavelengths, 158, 159
wave traps (line traps), 80, 98
weather, working in, 24–25. *See also*
        lightning
weatherproof covering, conductors,
        257
web hoist, 523, 524
web slings, 526
wedge sockets, wire rope termination,
        521
winches. *See also* rigging
    bisect tension, 525
    cable, working load limits, 510–511
    digger derrick, 502
wind
    galloping, conductors, 256–257
    guy wires, 190
    pole foundations, 83
    pole strength, 176
    sagging conductors, 205
    structure foundations and, 218
windmills, 72
wind power, 72–74
wind turbines, installing, 229–231
wireless local area networks (LAN), 80
wire rope, 224, 511, 520–521. *See also*
        rigging
wires. *See also* cables
    customer service wire size, 216
    guy, 190–191, 194
    ties, 208
WLL. *See* working load limits (WLL)
wood, guy strain insulators, 194
wooded crossarms, 199

wooden conductor reels, 212

wood poles. *See also* overhead powerline construction

    characteristics of, 176–177

    facing, 190

    foundations for, 182

    grounding hazards, 394–395

    resistance, 152

    strength of, 176

    transmission line grounding, 218

    voltage gradients, 152

work, rate of, 51

worker culture, safety, 3–4

working aloft. *See also* aerial devices

    buckets, 34

    fall protection systems, 30–34

    hook ladders, 33–34

    ladders, 34–35

    poles, 36–40

    transmission structures, 32–33

working alone, safety, 16

working load limits (WLL)

    chains, 521–522

    collapsible bull wheel, 527

    derating factors, 519

    hot-line tools, 559–563

overview, 510–511

rope blocks, 531–532

web slings, 526

wire rope, 520

work observations, 4

work permits, 16

work positioning systems, 30, 40

work procedures, 4

World Health Organization, 162

written job sequence, 13

wye circuits

    converting from delta circuits, 133

    grounding, 376–377

    leakage current, 144

    overview, 129

    parallel contact and, 141

    single-phase transformers, 417

    vehicle grounding, 496–497

wye-connected three-phase systems, 128, 131–133

wye connections, transformers

    angular displacement, wye-delta and delta-wye, 428–430

    delta-wye banks, 425

    open-wye/open delta, open-delta/ open-delta, 425–427

primary wye connections, 422–423

three-phase secondary voltage arrangements, 431–436

wye-delta banks, 424–425

wye-wye banks, 423

wye secondary service, 432, 433, 434–435, 436, 437

wye/wye transformer banks, 131

X

XLPE (extruded dielectric, polyethylene) pipe cable, 245–247

X-rays, 159

Y

Y towers, 217

Z

zero-awaiting interrupters, 300

zero-forcing interrupters, 300

zinc, 372

zone protection, 339